T0156291

Quantum Noise in Mesoscopic Physics

NATO Science Series

A Series presenting the results of scientific meetings supported under the NATO Science Programme.

The Series is published by IOS Press, Amsterdam, and Kluwer Academic Publishers in conjunction with the NATO Scientific Affairs Division

Sub-Series

I. Life and Behavioural Sciences	IOS Press
II. Mathematics, Physics and Chemistry	Kluwer Academic Publishers
III. Computer and Systems Science	IOS Press
IV. Earth and Environmental Sciences	Kluwer Academic Publishers
V. Science and Technology Policy	IOS Press

The NATO Science Series continues the series of books published formerly as the NATO ASI Series.

The NATO Science Programme offers support for collaboration in civil science between scientists of countries of the Euro-Atlantic Partnership Council. The types of scientific meeting generally supported are "Advanced Study Institutes" and "Advanced Research Workshops", although other types of meeting are supported from time to time. The NATO Science Series collects together the results of these meetings. The meetings are co-organized bij scientists from NATO countries and scientists from NATO's Partner countries – countries of the CIS and Central and Eastern Europe.

Advanced Study Institutes are high-level tutorial courses offering in-depth study of latest advances in a field.
Advanced Research Workshops are expert meetings aimed at critical assessment of a field, and identification of directions for future action.

As a consequence of the restructuring of the NATO Science Programme in 1999, the NATO Science Series has been re-organised and there are currently Five Sub-series as noted above. Please consult the following web sites for information on previous volumes published in the Series, as well as details of earlier Sub-series.

http://www.nato.int/science
http://www.wkap.nl
http://www.iospress.nl
http://www.wtv-books.de/nato-pco.htm

Series II: Mathematics, Physics and Chemistry – Vol. 97

Quantum Noise in Mesoscopic Physics

edited by

Yuli V. Nazarov
Delft University of Technology, The Netherlands

Kluwer Academic Publishers

Dordrecht / Boston / London

Published in cooperation with NATO Scientific Affairs Division

Proceedings of the NATO Advanced Research Workshop on
Quantum Noise in Mesoscopic Physics
Delft, The Netherlands
2–4 June 2002

A C.I.P. Catalogue record for this book is available from the Library of Congress.

ISBN 1-4020-1239-X (HB)
ISBN 1-4020-1240-3 (PB)

Published by Kluwer Academic Publishers,
P.O. Box 17, 3300 AA Dordrecht, The Netherlands.

Sold and distributed in North, Central and South America
by Kluwer Academic Publishers,
101 Philip Drive, Norwell, MA 02061, U.S.A.

In all other countries, sold and distributed
by Kluwer Academic Publishers,
P.O. Box 322, 3300 AH Dordrecht, The Netherlands.

Printed on acid-free paper

Printed in the Netherlands.

Contents

Preface vii

Introduction ix

PART ONE: SHORT NOISE

M. Büttiker/ Reversing the Sign of Current-Current Correlations 3

J. M. van Ruitenbeek/ Shot Noise and Channel Composition of Atomic-sized Contacts 33

A. Martín-Rodero, J. C. Cuevas, A. Levy Yeyati, R. Cron, M. F. Goffman, D. Esteve and C. Urbina/ Quantum Noise and Mutiple Andreev Reflections in Superconducting Contacts 51

B. Reulet, D.E. Prober and W. Belzig/ Shot Noise of Mesoscopic NS Structures : the Role of Andreev Reflection 73

E.V. Bezuglyi, E.N. Bratus', V.S. Shumeiko, and G. Wendin/ Current Noise in Diffusive SNS Junctions in the Incoherent MAR Regime 93

C. Strunk and C. Schönenberger/ Shot noise in diffusive superconductor/ normal metal heterostructures 119

D.C. Glattli, Y. Jin, L.-H Reydellet and P. Roche/ Photo-Assisted Electron-Hole Partition Noise in Quantum Point Contacts 135

E. Sukhorukov, G. Burkard, and D. Loss/ Shot Noise of Cotunneling Current 149

PART TWO: QUANTUM MEASUREMENT AND ENGLANGLEMENT

R.J. Schoelkopf,A.A. Clerk, S. M. Girvin, K.W. Lehnert, and M.H. Devoret/ Qubits as Spectrometers of Quantum Noise 175

A.N. Korotkov/ Noisy Quantum Measurement of Solid-State Qubits: Bayesian Approach 205

D.V. Averin/ Linear Quantum Measurements 229

J. C. Egues, P. Recher, D. S. Saraga, V. N. Golovach, G. Burkard E. V. Sukhorukov, and D. Loss / Shot Noise for Entangled and Spin-Polarized Electrons 241

W. D. Oliver, G. Feve, N. Y. Kim, F, Yamaguchi and Y. Yamamoto/ The Generation and Detection of Single and Entangled Electrons in Mesoscopic 2DEG Systems 275

U. Gavish, Y. Imry, Y. Levinson and B. Yurke/ What Quantity Is Measured in an Excess Noise Experiment? 297

T. Martin, A. Crepieux and N. Chtchelkatchev/ Noise Correlations, Entanglement, and Bell Inequalities 313

G. Johansson,P. Delsing, K. Bladh, D. Gunnarsson, T. Duty, A. Käck, G. Wendin, and A. Aassime/ Noise in the Single Electron Transistor and its Back Action during Measurement 337

A. Shirman and G. Schön/ Dephasing and Renormalization in Quantum Two Level Systems 357

PART THREE: FULL COUNTING STATISTICS

L.S. Levitov/ The Statistical Theory Of Mesoscopic Noise 373

I. Klich/ An Elementary Derivation of Levitov's Formula 397

M. Kindermann and Yu. V. Nazarov/ Full Counting Statistics in electric circuits 403

D. A. Bagrets and Yu. V. Nazarov/ Multiterminal Counting Statistics 429

W. Belzig/ Full Counting Statistics of Superconductor--Normal-Metal Heterostructures 463

D. B. Gutman, Y. Gefen and A.D. Mirlin/ High Cumulants of Current Fluctuations out of Equilibrium 497

Preface

This book is written to conclude the NATO Advanced Research Workshop "Quantum Noise in Mesoscopic Physics" held in Delft, the Netherlands, on June 2-4, 2002. The workshop was co-directed by M. Reznikov of Israel Institute of Technology, and me. The members of the organizing committee were Yaroslav Blanter (Delft), Chirstopher Glattli (Saclay and ENS Paris) and R. Schoelkopf (Yale). The workshop was very successful, and we hope that the reader will be satisfied with the scientific level of the present book. Before addressing scientific issues I find it suitable to address several non-scientific ones.

The workshop was attended by researchers from many countries. Most of them perform their activities in academic institutions, where one usually finds the necessary isolation from the problems and sores of the modern world. However, there was a large group of participants for which such isolation was far from perfect. War, hatred, and violence rage just several miles away of their campuses and laboratories, poisoning everyday life in the land of Israel.

Science and scientists can hardly help to resolve the situation. Albeit there is something we can do. We witness various actions that differ much in means but have a common goal: to undermine scientific and cultural ties between Israel and the rest of the world. Just two examples. Before the workshop, on April 6, 2002, 120 university professors published a letter in *Gardian* calling a moratorium on partnership of Israel in EU projects. After the workshop, on July 31, 2002, a bomb hidden in a handbag exploded in a crowded cafeteria at Hebrew University in Jerusalem. Four foreign exchange students were among the fallen. These actions can and must be confronted. That is why our workshop was in partnership with Israel, and we were proud to profit from a traditionally high level of Israeli research in the field of quantum noise.

We shall remain firm and hopeful, and we overcome. To support this with an example, I recall my first scientific discussion with the Israeli co-director of the present workshop, Michael Reznikov. It took place in 1981 in a Soviet military training facility. At that time, neither the circumstances of our life nor the general political situation inspired us to do research. We were trained to contribute to a large-scale nuclear weapons game, and our future looked dim if not apocalyptic. Yet the Almighty was merciful to all people, and us, so that the evil was defused. One could doubt the efficiency and the moral foundations of the North Atlantic Treaty Organization. However, it was one of the means, which defused the evil. In 1981 we were imaginative and ambitious young people. However, we could not think of co-directing a scientific conference supported by NATO. This fact from the past enormously enhances our appreciation of NATO support.

I would like to conclude by acknowledging other persons and organizations. The workshop would not take place without Yaroslav Blanter, who generously invested his time, energy and enthusiasm into the venture. Christian Glattli and Rob

Schoelkopf contributed much with their advice and organizational efforts. The financial support of Royal Netherlands Academy of Arts and Sciences (KNAW) and Foundation for Fundamental Research on Matter (FOM) is gladly appreciated. Delft University of Technology was so kind as to serve our debts. Yvonne Zwang was always ready to assist us without asking for reimbursement.

Last but not least, I would like to thank all the participants of the workshop and contributors of this volume. Your enthusiastic response exceeded the expectations of the organizers. This is the best award for our activities.

Yuli V. Nazarov,
Delft University of Technology

Introduction

The field of quantum noise in mesoscopic physics has been intensively developing for more than a decade. Remarkably, the developments do not seem to slow down yet. This book presents a collection of mini-reviews where the leading research teams summarize the most recent results and chart new directions. I was pleasantly surprised by enthusiasm of contributors who were willing to invest their time and energy in writing. This shows that the book is timely. Taken together, the reviews give a fairy representative snapshot of the modern state of the field. This state is still not coherent; an attentive reader will notice not only different approaches and conflicting views but contradictory results and interpretations as well. The contradictions should be present in the book since they drive the rapid evolution of the field. A general reader would possibly be more interested in driving forces and high intellectual content than in concrete results.

The book is divided into three parts: shot noise, quantum measurement and entanglement, and full counting statistics. These are the three main research streams. Since many contributions are simultaneously related to two or three streams, their placing may be rather subjective.

The whole field was pioneered in late eighties by Lesovik, Büttiker, Beenakker and Levitov; I am happy that two of the four could contribute to the book. The shot noise part opens with an article of Büttiker that presents a broad study of the sign of noise correlation and its (un)relation to Fermi statistics. Two subsequent contributions illustrate all the power and beauty of Landauer-Büttiker scattering approach. Van Ruitenbeek introduces atomic-size contacts as a brilliant experimental realization of a few mode scatterer. Not only PIN-code of transmission eigenvalues can be measured experimentally, the PIN-code works revealing harmony of noise properties even in such complicated situations as multiple Andreev reflection (A. Martin-Rodero et al.)! The scattering becomes much more complicated in realistic diffusive conductors, this is revealed by Andreev reflection from near superconductors. High-frequency noise measurements (Reulet et al.) pinpoint these coherent effects, quantum circuit theory is required to reach a harmony between experiment and theory and calls on the full counting statistics. The theory of Bezuglyi et al. and the experiment of Strunk and Schönenberger pertain the incoherent multiple Andreev reflection regime that occur in longer and hotter NS structures and considerably enhances the noise. Glattli et al. present their pioneering experimental observation of non-transport photo-assisted partition noise. The contribution of Sukhorukov et al. concerns strongly interacting Coulomb blockade systems and reveals unexpected deviations of cotunneling noise from Poisson statistics.

The words "quantum measurement" and "entanglement" used to sound very abstract just a few years ago. The attempts to realize quantum manipulation at meso- and nanoscale have put these topics in the focus of practical research. It was soon recognized that the progress in quantum manipulation is impossible without deepening our knowledge about quantum noise that hinders the manipulation and affects the measurement. It was also recognized that a noise measurement may provide indispensable information concerning the entanglement, and the result of the manipulation. This initiated a new and most active research direction in the field. Yale collaboration presents a qubit as an ideal spectrum analyzer for quantum noise, their contribution being a fairly complete experimental proposal. Korotkov explicates the continuous measurement of a single qubit. His contribution would heal many from the plague of the field: making (carriers in) physics from semantic traps that arise from (mis)interpretation of quantum mechanics. The contribution of Averin possesses similar healing properties: linear measurement is explained. Egues et al. present a detailed description of the last crusade of Basel group, this is aimed to demonstrate the use of shot noise for detection of spin entanglement and polarization. Oliver at al. review their recent experiments in generating and detection of electron entanglement. Gavish et al. address the measurement of excess noise under conditions of amplification. Martin et al. propose experimental check of Bell inequalities for electrons by means of noise correlation measurement. Johanssson et al. investigate the feasibility of Single Electron Transistor for measuring qubits, this includes careful analysis of SET noise and back-action. Shnirman and Schön review dephasing and renormalization in a qubit placed into a dissipative environment.

I have felt in love with the full counting statistics, and probably would not be objective describing this research stream. My excuse is that I share this love with rapidly increasing number of people, some even seemed too prominent to experience such feelings. So that, full counting statistics is everything for us. It provides ultimate knowledge about charge transfer, it appears to be a solid foundation of the whole quantum transport, it may provide ultimate information concerning muli-particle entanglement ... I wished the first experiment in FCS was presented in the book. Unfortunately, it will appear elsewhere. Levitov, the founder of the FCS, introduces the FCS and reviews his achievements. Klich presents a novel derivation of Levitov's formula, this short contribution being pedagogically indispensable. Kindermann and me begin with a general analysis of quantum measurement aspects of the FCS. This quickly brings us to practical description of the FCS in electric circuits. Bagrets and me address the FCS in multi-terminal circuits treating the limits of non-interacting and Coulomb-blockaded electrons. Belzig discusses FCS for Andreev reflection. Gutman et al. review FCS of non-equilibrium electrons.

I hope that this book can serve as a good introduction to the field, both for a novice and an expert.

PART ONE

Short Noise

REVERSING THE SIGN OF CURRENT-CURRENT CORRELATIONS

MARKUS BÜTTIKER
Département de Physique Théorique, Université de Genève,
CH-1211 Genève 4, Switzerland

1. Introduction

Dynamic fluctuation properties of mesoscopic electrical conductors provide additional information not obtainable through conductance measurement. Indeed, over the last decade, experimental and theoretical investigations of current fluctuations have successfully developed into an important subfield of mesoscopic physics. A detailed report of this development is presented in the review by Blanter and Büttiker [1].

In this work we are concerned with the correlation of current fluctuations which can be measured at different terminals of multiprobe conductors. Of particular interest are situations where, as a function of an externally controlled parameter, the sign of the correlation function can be reversed.

Electrical correlations can be viewed as the Fermionic analog of the Bosonic intensity-intensity correlations measured in optical experiments. In a famous astronomical experiment Hanbury Brown and Twiss demonstrated that intensity-intensity correlations of the light of a star can be used to determine its diameter [2]. In subsequent laboratory experiments of light split by a half-silvered mirror statistical properties of light were further analyzed [3]. Much of modern optics derives its power from the analysis of correlations of entangled optical photon pairs generated by non-linear down conversion [4]. The intensity-intensity correlations of a thermal Bosonic source are positive due to statistical bunching. In contrast, anti-bunching of a Fermionic system leads to negative correlations [5].

Concern with current-current correlations in mesoscopic conductors originated with Refs. [6, 7]. The aim of this work was to investigate the fluctuations and correlations for an arbitrary multiprobe conductor for which the conductance matrix can be expressed with the help of the scattering matrix [8, 9]. Refs. [6, 7] provided an extension of the discussions of shot noise by Khlus [10] and Lesovik [11] which applies to two-terminal conductors. These authors assumed from the outset that the transmission matrix is diagonal and provided expressions for the

Y. V. Nazarov (ed.), Quantum Noise in Mesoscopic Physics, 3–31.

two terminal shot noise in terms of transmission probabilities. It turns out that even for two probe conductors, shot noise can be expressed in terms of transmission probabilities only in a special basis (eigen channels). Such a special basis does not exist for multiprobe conductors and we are necessarily left with expressions for shot noise in terms of quartic products of scattering matrices [7, 12]. There are exceptions to this rule: for instance correlations in three-terminal one-channel conductors can also be expressed in terms of transmission probabilities only [13].

The reason that shot noise, in contrast to conductance, is in general not simply determined by transmission probabilities is the following: if carriers incident from different reservoirs (contacts) or quantum channels can be scattered into the same final reservoir or quantum channel, quantum mechanics demands that we treat these particles as indistinguishable. We are not allowed to be able to distinguish from which initial contact or quantum channel a carrier has arrived. The noise expressions must be invariant under the exchange of the initial channels [14–18]. The occurrence of exchange terms is what permitted Hanbury Brown and Twiss to measure the diameter of the stars: Light emitted by widely separated portions of the star nevertheless exhibits (a second order) interference effect in intensity-intensity correlations [2].

Experiments which investigate current-correlations in mesoscopic conductors have come along only recently. Oliver et al. used a geometry in which a "half-silvered mirror" is implemented with the help of a gate that creates a partially transparent barrier [19]. Henny et al. [20] separated transmission and reflection along edge states of a quantum point contact subject to a high magnetic field. In the zero temperature limit an electron reservoir compactly fills all the states incident on the conductor. A subsequent experiment by Oberholzer at al. [21] uses a configuration with two quantum point contacts, as shown in Fig. 1. This geometry permits to thin out the occupation in the incident electron beam and thus allows to investigate the transition in the correlation as we pass from degenerate Fermi statistics to dilute Maxwell-Boltzmann satistics. Anti-bunching effects vanish in the Maxwell-Boltzmann limit and the current-current correlation tends to zero as the occupation of the incident beam is diminished. The fact that in electrical conductors the incident beam is highly degenerate is what made these Hanbury Brown and Twiss experiments possible. In contrast, an emission of electrons into the vacuum generates an electron beam with only a feeble occupation of electrons [22] and for this reason an experiment in vacuum has in fact just been achieved only very recently [23]. Below we will discuss the experiments in electrical conductors in more detail.

Within the scattering approach, in the white noise limit, it can be demonstrated, that current-current correlations are negative, irrespective of the voltages applied to the conductor, temperature and geometry of the conductor [7, 12]. The wide applicability of this statement might give the impression, that in systems of Fermions current correlations are always negative. However, the proof

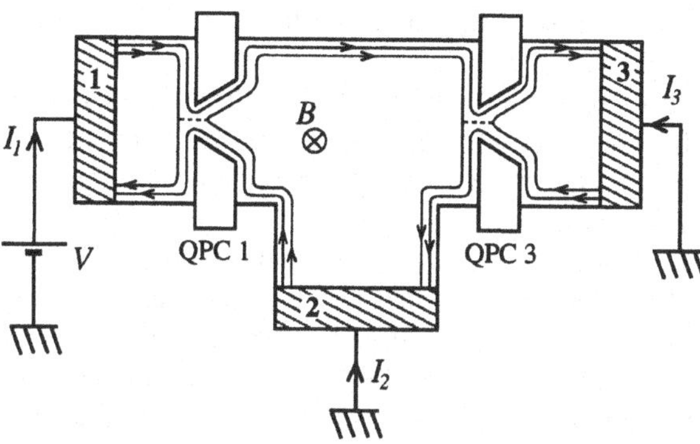

Figure 1. Experimental arrangement of Oberholzer at al. Current is injected at contact 1. One edge channel is perfectly transmitted (and noiseless) the other is partially transmitted at QPC 1 with probability T_1 and partially transmitted at QPC 3 with probability T_3 into contact 3 and reflected with probability $R_3 = 1 - T_3$ into contact 2. Of interest is the correlation of currents measured at contacts 2 and 3.

rests on a number of assumptions: in addition to the white-noise limit (low frequency limit) it is assumed that the terminals are all held at a constant (time-independent) terminal-specific potential. This is possible if the mesoscopic conductor is embedded in a zero-impedance external circuit. No general statement on the sign of correlations exists if the external circuit is characterized by an arbitrary impedance.

In this work we are interested in situations for which the above mentioned proof does not apply. For instance, a voltmeter ideally has infinite impedance, and a conductor in which one of the contacts is connected to a voltmeter presents a simple example in which it is possible to measure *positive* current-current correlations [24]. In steady state transport the potential at a voltage probe floats to achieve zero net current. If the currents fluctuate the potential at the voltage probe must exhibit voltage fluctuations to maintain zero current at every instant. As has been shown by Texier and Büttiker, the fluctuating potential at a voltage probe can lead to a change in sign of a current-current correlation [24].

A voltage probe also relaxes the energy of carriers, it is a source of dissipation [25–28]. Probes which are non-dissipative are of interest as models of dephasors. At low temperatures dephasing is quasi-elastic and it is therefore reasonable to model dephasing in an energy conserving way. This can be achieved by asking that a fictitious voltage probe maintains zero current at every energy [29]. Ref. [17] presents an application of this approach to noise-correlations in chaotic cavities.

It is of interest to investigate current-correlations in the presence of such a dephasing voltage probe and to compare the result with a real dissipative voltage probe. No examples are known in which a dephasing probe leads to positive correlations. However, there exists also no proof that correlations in the presence of dephasing voltage probes are always negative.

The proof that correlations in Fermionic conductors are negative also does not apply in the high-frequency regime. We discuss the frequency-dependence of equilibrium fluctuations in a ballistic wire to demonstrate the ocurrence of positive correlations at large frequencies.

Another form of interactions which can induce positive correlations comes about if a normal conductor is coupled to a superconductor. Experiments have already probed shot noise in hybrid normal-superconducting two-terminal structures [30–34]. In the Bogoliubov de Gennes approach the superconductor creates excitations in the normal conductor which consist of correlated electron-hole pairs. The process which creates the correlation is the Andreev reflection process by which an incident electron (hole) is reflected as a hole (electron). In this picture it is the occurrence of quasi-particles of different charge which makes positive correlations possible [35–37]. The quantum statistics remains Fermi like since the field operator associated with the Bogoliubov de Gennes equations obeys the commutation rules of a Fermi field [38]. Alternatively the superconductor can be viewed as an injector of Cooper pairs [39]. In this picture is the brake-up of Cooper pairs and the (nearly) simultaneous emission of the two electrons through different contacts which makes positive correlations possible. Our discussion centers on the conditions (geometries) which are necessary for the observation of positive correlations in mesoscopic normal conductors with channel mixing. Boerlin et al. [40] have investigated the current-correlations of a normal conductor with a channel mixing central island seprated by tunnel junctions from the contacts and the superconductor. Samuelsson and Büttiker [41] consider a chaotic dot which can have completely transparent contacts or contacts with tunnel junctions. Interestingly while a chaotic cavity with perfectly transmitting normal contacts and an even wider perfect contact to the superconductor exhibits positive correlations, application of a magnetic flux of the order of one flux quantum only is sufficient to destroy the proximity effect and is sufficient in this particular geometry to change the sign of correlations from positive to negative [41]. Equally interesting is the result that a barrier at the interface to the superconductor helps to drive the correlations positive [41].

2. Quantum Statistics and the sign of Current-Current Correlations

In this section we elucidate the connection between statistics and current-current correlations in multiterminal mesoscopic conductors and compare them with intensity-intensity correlations of a multiterminal wave guide connected to black body radiation sources [7, 12]. We start by considering a conductor that is so small and at such a low temperature that transmission of carriers through the conductor can be treated as completely coherent. The conductor is embedded in a zero-impedance external circuit. Each contact, labeled $\alpha = 1, 2, ...$, is characterized by its Fermi distribution function f_α. Scattering of electrons at the conductor is described by a scattering matrix S. The S-matrix relates the incoming amplitudes to the outgoing amplitudes: the element $s_{\alpha\beta,mn}(E)$ gives the amplitude of the current probability in contact α in channel m if a carrier is injected in contact β in channel n with amplitude 1 (see [12] for a more precise definition). The modulus of an S-matrix element is the probability for transmission from one channel to another. We introduce a *total* transmission probability (for $\alpha \neq \beta$)

$$T_{\alpha\beta} = \mathrm{Tr}\left\{s^{\dagger}_{\alpha\beta}(E)s_{\alpha\beta}(E)\right\} . \tag{1}$$

Here the trace is over transverse quantum channels and spin quantum numbers. This permits to write the conductance in the form [8, 12]

$$G_{\alpha\beta} = -\frac{e^2}{h}\int \mathrm{D}E\,(-df/dE)T_{\alpha\beta} . \tag{2}$$

where f is the equilibrium Fermi function. The diagonal elements of the conductance matrix can be expressed with the help of $s_{\alpha\alpha}$. With the help of the *total* reflection probability $R_{\alpha\alpha} = N_\alpha - \mathrm{Tr}\left\{s^{\dagger}_{\alpha\alpha}(E)s_{\alpha\alpha}(E)\right\}$ where N_α is the number of quantum channels in contact α we have $G_{\alpha\alpha} = e^2/h \int \mathrm{D}E\,(-df/dE)[N_\alpha - R_{\alpha\alpha}]$. Alternatively, since $\sum_\beta G_{\alpha\beta} = \sum_\alpha G_{\alpha\beta} = 0$ the diagonal elements can be obtained from the off-diagonal elements. The average currents of the conductor are determined by the transmission probabilities and the Fermi functions of the reservoir

$$I_\alpha = \frac{e}{h}\int dE[(N_\alpha - R_{\alpha\alpha})f_\alpha - \sum_{\alpha\beta} T_{\alpha\beta}(E)f_\beta] . \tag{3}$$

In reality the currents fluctuate. The total current at a contact is thus the sum of an average current and a fluctuating current. We can express the total current in terms of a "Langevin" equation

$$I_\alpha = \frac{e}{h}\int dE[(N_\alpha - R_{\alpha\alpha})f_\alpha - \sum_{\alpha\beta} T_{\alpha\beta}(E)f_\beta] + \delta I_\alpha . \tag{4}$$

We have to find the auto - and cross-correlations of the fluctuating currents δI_α such that at equilibrium we have a Fluctuation-Dissipation theorem and such that in the case of transport the correct non-equilibrium (shot noise) is described by the fluctuating currents. The first part of Eq. (4) represents the average current only in the case that the Fermi distributions are constant in time. This is the case if the conductor is part of a zero-impedance external circuit. If the external circuit has a finite impedance, the voltage at a contact fluctuates and consequently the distribution function of such a contact is also time-dependent. In this section we consider only the case of constant voltages in all the contacts.

We compare the current fluctuations of the electrical conductor with the intensity fluctuations of a (multi-terminal) structure for photons in which each terminal connects to a black body radiation source characterized by a Bose-Einstein distribution function f_α. Like the electrical conductor the wave guide is similarly characterized by scattering matrices $s_{\alpha\beta}(E)$.

The noise spectrum is defined as $P_{\alpha\beta}(\omega)2\pi\delta(\omega+\omega') = \langle \delta\hat{I}_\alpha(\omega)\delta\hat{I}_\beta(\omega') + \delta\hat{I}_\beta(\omega')\delta\hat{I}_\alpha(\omega)\rangle$ with $\delta\hat{I}_\alpha(\omega) = \hat{I}_\alpha(\omega) - \langle\hat{I}_\alpha(\omega)\rangle$, where $\hat{I}_\alpha(\omega)$ is the Fourier transform of the current operator at contact α. The zero frequency limit which will be of interest here is denoted by: $P_{\alpha\beta} \equiv P_{\alpha\beta}(\omega=0)$. The scattering approach leads to the following expression for the noise [6, 7, 12]

$$P_{\alpha\beta} = \frac{2e^2}{h}\int \mathrm{DE} \sum_{\gamma,\lambda} \mathrm{Tr}\left\{A^{\alpha}_{\gamma\lambda}A^{\beta}_{\lambda\gamma}\right\} f_\gamma(1\mp f_\lambda)\,. \tag{5}$$

The matrix $A^{\beta}_{\lambda\gamma}$ is composed of the matrix elements of the current operator in lead β associated with the scattering states describing carriers incident from contact λ and γ and is given by

$$A^{\alpha}_{\gamma\lambda} = \delta_{\alpha\gamma}\delta_{\alpha\lambda} - s^{\dagger}_{\alpha\gamma}(E)s_{\alpha\lambda}(E)\,. \tag{6}$$

In Eq. (5) the upper sign refers to Fermi statistics and the lower sign to Bose statistics.

To clarify the role of statistics it is useful to split the noise spectrum in an equilibrium like part $P^{eq}_{\alpha\beta}$ and a transport part $P^{tr}_{\alpha\beta}$ such that $P_{\alpha\beta} = P^{eq}_{\alpha\beta} + P^{tr}_{\alpha\beta}$. We are interested in the correlations of the currents at two different terminals $\alpha \neq \beta$. The equilibrium part consists of Johnson-Nyquist noise contributions which can be expressed in terms of transmission probabilities only [7, 12]

$$P^{eq}_{\alpha\beta} = -\frac{2e^2}{h}\int \mathrm{DE}\,(T_{\alpha\beta}f_\beta(1\mp f_\beta) + T_{\beta\alpha}f_\alpha(1\mp f_\alpha)\,. \tag{7}$$

Since both for Fermi statistics and Bose statistics $f_\alpha(1\mp f_\alpha) = -kTdf_\alpha/dE$ is positive, the equilibrium fluctuations are *negative* independent of statistics. The transport part of the noise correlation is

$$P^{tr}_{\alpha\beta} = \mp\frac{2e^2}{h}\int \mathrm{DE} \sum_{\gamma,\lambda} \mathrm{Tr}\left\{s^{\dagger}_{\alpha\gamma}s_{\alpha\lambda}s^{\dagger}_{\beta\lambda}s_{\beta\gamma}\right\} f_\gamma f_\lambda\,. \tag{8}$$

To see that this expression is negative for Fermi statistics and positive for Bose statistics one notices that it can be brought onto he form [7, 12]

$$P^{tr}_{\alpha\beta} = \mp\frac{2e^2}{h}\int \mathrm{DE}\ \mathrm{Tr}\left\{[\sum_{\gamma} s_{\beta\gamma}s^{\dagger}_{\alpha\gamma}f_\gamma][\sum_{\lambda} s_{\alpha\lambda}s^{\dagger}_{\beta\lambda}f_\lambda]\right\}\,. \tag{9}$$

The trace now contains the product of two self-adjoint matrices. Thus the transport part of the correlation has a definite sign depending on the statistics.

It follows that current-current correlations in a normal conductor are negative due to the Fermi statistics of carriers whereas for a Bose system we have the possibility of observing positive correlations, as for instance in the optical Hanbury Brown and Twiss experiments [2, 3].

There are several important assumptions which are used to derive this result: It is assumed that the reservoirs are at a well defined chemical potential. For an electrical conductor this assumption holds only if the external circuit has zero impedance. The above considerations are also valid only in the white-noise (or zero-frequency limit). We have furthermore assumed that the conductor supports only one type of charge, electrons or holes, but not both. Below we are interested in examples in which one of these assumptions does not hold and which demonstrate that also in electrical purely normal conductors we can, under certain conditions, have positive correlations.

3. Coherent Current-Current Correlation

We now consider the specific conductor shown in Fig. 1. It is a schematic drawing of the conductor used in the experiment of Oberholzer et al. [21]. The sample is subject to a high magnetic field such that the only states which connect one contact to another one are edge states [42, 43]. We consider first the case when there is *only one edge state* (filling factor $\nu = 1$ away from the quantum point contacts). The edge state is partially transmitted with probability T_1 at the left quantum point contact and is partially transmitted with probability T_3 at the right quantum point contact. The potential $\mu_1 = \mu + eV$ at contact 1 is elevated in comparison with the potentials $\mu_2 = \mu_3 = \mu$ at contact 2 and 3. Thus carriers enter the conductor at contact 1 and leave the conductor through contact 2 and 3. Application of the scattering approach requires also the specification of phases. However, for the example shown here, without closed paths, the result is independent of the phase accumulated during traversal of the sample and the result can be expressed in terms of transmission probabilities only.

At zero temperature we can directly apply Eq. (9) to find the cross-correlation. Taking into account that only the energy interval between μ_1 and μ is of interest we see immediately that $P_{23} = \mp\frac{2e^2}{h}|eV|[s_{31}s^{\dagger}_{21}s_{21}s^{\dagger}_{31}]$ which is equal to $P_{23} = \mp\frac{2e^2}{h}|eV|[s^{\dagger}_{21}s_{21}s^{\dagger}_{31}s_{31}]$. But $s^{\dagger}_{21}s_{21} = T_1R_3$, where $R_3 = 1 - T_3$ and $s^{\dagger}_{31}s_{31} = T_1T_3$ and thus

$$P_{23} = -\frac{2e^2}{h}|eV|T_1^2R_3T_3\,. \tag{10}$$

Transmission through the first quantum point contact thins out the occupation in the transmitted edge state. This edge state has now an effective distribution $f_{eff} = T_1$. The correlation function has thus the form $P_{23} = -\frac{2e^2}{h}|eV|f^2_{eff}R_3T_3$.

For $T_1 = 1$ we have a completely occupied beam of carriers incident on the second quantum point contact and the correlation is maximally negative with $P_{23} = -\frac{2e^2}{h}|eV|R_3T_3$. In this case the correlation is completely determined by current conservation: Denoting the current fluctuations at contact α by δI_α we have $\delta I_1 + \delta I_2 + \delta I_3 = 0$. Consequently since the incident electron stream is noiseless $\delta I_1 = 0$ we have $P_{23} = -P_{22} = -P_{33}$. Therefore if the first quantum point is open the weighted correlation $p_{23} = P_{23}/(P_{22}P_{33})^{1/2} = -1$. The fact that an electron reservoir is noiseless is an important property of a source with Fermi-Dirac statistics [20].

If the transmission through the first quantum point contact is less than one the diminished occupation of the incident carrier beam reduces the correlation. Eventually in the non-degenerate limit f_{eff} becomes negligibly small and the correlation between the transmitted and reflected current tends to zero. This is the limit of Maxwell-Boltzmann statistics.

The experiment by Oberholzer et al. [21] measured the correlation for the entire range of occupation of the incident beam and thus illustrates the full transition from Fermi statistics to Maxwell-Boltzmann statistics. The experiment by Oliver et al. [19] even though it is for a different geometry (and at zero magnetic field) is discussed by the authors in terms of the same formula Eq. (10). The range over which the contact which determines the filling of the incident carrier stream can be varied is, however, more limited than in the experiment by Oberholzer et al..

Before continuing we mention for completeness also the auto-correlations

$$P_{33} = \frac{2e^2}{h}|eV|T_3T_1(1 - T_3T_1)\,, \tag{11}$$

$$P_{22} = \frac{2e^2}{h}|eV|T_1R_3(1 - T_1R_3)\,, \tag{12}$$

For $T_1 = 1$ this is the partition noise of a quantum point contact [44, 45].

We are now interested in the following question: Carriers along the upper edge of the conductor have to traverse a long distance from quantum point contact 1 to quantum point contact 3 (see Fig. 1). How would quasi-elastic scattering (dephasing) or inelastic scattering affect the cross correlation Eq. (10)? For the case treated above where only one edge state or a spin degenerate edge is involved the answer is simple: the cross correlation remains unaffected by either quasi-elastic or inelastic scattering. The question (asked by B. van Wees) becomes interesting if there are two or more edge states involved. It is for this reason that Fig. (1) shows two edge channels.

4. Cross correlation in the presence of quasi-elastic scattering

Incoherence can be introduced into the coherent scattering approach to electrical conduction with the help of fictitious voltage probes. (see Fig. 3). Ideally a voltage

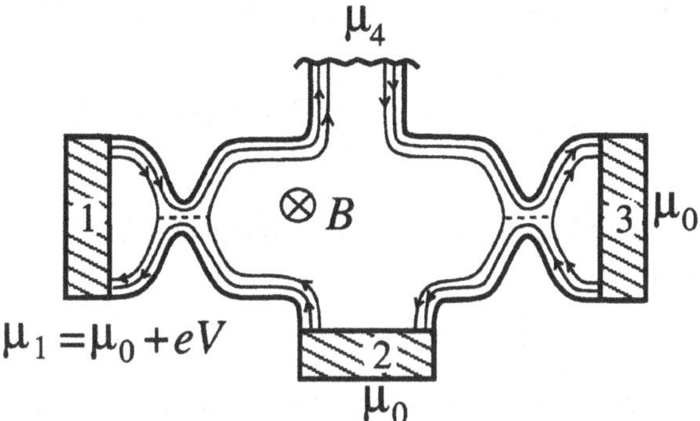

Figure 2. A voltage probe at the upper edge generates inelastic scattering or dephasing depending on whether the total instantaneous current or additionally the current at every energy is set to zero. After Texier and Büttiker [24].

probe maintains zero net current at every instant of time. [A realistic voltmeter will have a finite response time. However since we are concerned with the low-frequency limit this is of no interest here.] A carrier entering a voltage probe will thus be replaced by a carrier entering the conductor from the voltage probe. Outgoing and incoming carriers are unrelated in phase and thus a voltage probe is a source of decoherence. A real voltage probe is dissipative. If we wish to model dephasing which at low temperatures is due to quasi-elastic scattering we have to invent a voltage probe which preserves energy. de Jong and Beenakker [29] proposed that the probe keeps not only the total current zero but that the current in each energy interval is zero at every instant of time. Noise correlations in the presence of a dephasing voltage probe have been investigated by van Langen and the author for multi-terminal chaotic cavities [17].

In the discussion that follows we will assume, as shown in Fig. 1 that the outer edge channel is perfectly transmitted at both quantum point contacts. Only the inner edge channel is as above transmitted with probability T_1 at the first quantum point contact and with probability T_3 at the second quantum point contact. Elastic inter-edge channel scattering is very small as demonstrated in experiments by van Wees et al. [46], Komiyama et al. [47], Alphenaar et al. [48] and Mueller et al. [49] and below we will not address its effect on the cross correlation. For a discussion of elastic interedge scattering in this geometry the reader is referred to the work by Texier and Büttiker [24]. We wish to focus on the effects of quasi-elastic scattering and inelastic scattering. The addition of the outer edge channel has no effect on the noise in a purely quantum coherent conductor. Edge channels with perfect transmission are noiseless [6].

To model quasi-elastic scattering along the upper edge of the conductor we now introduce an additional contact (see Fig. 3). To maintain the current at zero for each energy interval we re-write the Langevin equations Eq. (4) for each energy interval dE,

$$\Delta I_\alpha(E,t) = \frac{e}{h}[(N_\alpha - R_{\alpha\alpha}(E))f_\alpha(E,t) - \sum_{\beta\neq\alpha} T_{\alpha\beta}(E)f_\beta(E,t)] + \delta I_\alpha(E,t)\,. \tag{13}$$

At the voltage probe we have $\Delta I_4(E,t) = 0$ and thus the distribution function of contact 4 is given by

$$f_4(E,t) = \bar{f}_4(E) + \delta f_4(E,t)\,, \tag{14}$$

where the time-independent part of the distribution function is given by

$$\bar{f}_4(E) = \frac{1}{N_\alpha - R_{\alpha\alpha}}[\sum_{\alpha=1}^{\alpha=3} T_{4\alpha} f_\alpha]\,, \tag{15}$$

and the fluctuating part of the distribution function is

$$\Delta f_4(E,t) = \frac{1}{N_\alpha - R_{\alpha\alpha}}\frac{h}{e}\delta I_4(E,t)\,. \tag{16}$$

Here we have taken into account that the distribution functions at contact 1, 2 and 3 are time-independent (equilibrium) Fermi functions. Only the distribution at contact 4 fluctuates. Additionally, its time-averaged part is a non-equilibrium distribution function. For the simple example considered here it is given by

$$\bar{f}_4(E) = \frac{1+T_1}{2}f_1(E) + \frac{R_1}{2}f_2(E)\,. \tag{17}$$

It is a two step distribution function [50, 17] as shown in Fig. 4. Since there is now also a fluctuating part of the distribution function the total fluctuating current at contact α contains according to Eq. (13) also a term $-(e/h)T_{\alpha\beta}(E)\Delta f_4(E,t)$. We take the transmission probabilities to be energy independent. Integration over energy gives thus for the fluctuating current at contact α

$$\Delta I_\alpha = \delta I_\alpha - \frac{G_{\alpha 4}}{G_{44}}\delta I_4\,. \tag{18}$$

As a consequence the correlation between the currents at contacts α and β in the presence of a quasi-elastic voltage probe is $P^{qe}_{\alpha\beta} = \langle \Delta I_\alpha \Delta I_\beta \rangle$ with

$$P^{qe}_{\alpha\beta} = P_{\alpha\beta} - \frac{G_{\alpha 4}}{G_{44}}P_{\beta 4} - \frac{G_{\beta 4}}{G_{44}}P_{\alpha 4} + \frac{G_{\alpha 4}G_{\beta 4}}{G^2_{44}}P_{44}\,. \tag{19}$$

Here $P_{\alpha\beta}$ are the auto-correlations and cross-correlations of the fluctuating currents in the energy resolved Langevin equation Eq. (13). The spectra are evaluated with the help of Eqs. (5) that apply for a completely coherent conductor except that we use the distribution functions f_1, f_2, f_3 and $\bar{f}_4(E)$. In this procedure we neglect thus the fluctuations of the distribution function in the evaluation of the intrinsic noise powers $P_{\alpha\beta}$. [This is appropriate for the second order correlations of interest here, but not for the higher order cumulants [51, 52]].

In contrast to Eq. (9) the current-current correlation Eq. (19) is not necessarily negative. Taking into account that the off-diagonal conductances are negative and that the intrinsic spectra $P_{\alpha\beta}$ and $P_{4\beta}$ are negative for cross-correlations, it is clear that the first three terms in Eq. (19) are negative. The forth term, due to the fluctuating distribution in the dephasing contact, is positive. For all examples known to us, it turns out that for a dephasing voltage probe, the first three terms win and the resulting correlation is negative [17]. For the inelastic (physical) voltage probe this is not the case as we demonstrate below.

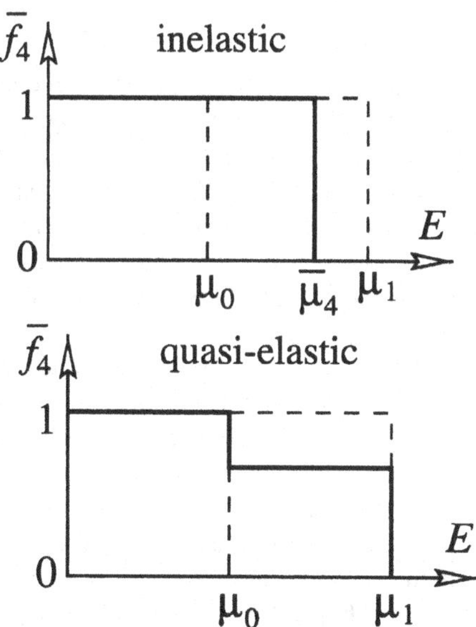

Figure 3. Distribution functions in the voltage probe reservoir. For inelastic scattering this an equilibrium distribution function with a potential μ_4 determined such that the average current vanishes. For a dephasing voltage probe this is a two step function determined such that the current at the probe vanishes at every energy.

We will not discuss the most general result of Ref. [24] here but instead focus on the fact that such a dephasing probe can generate shot noise even in the case where the quantum coherent sample is noiseless.

5. Quasi-elastic partition noise

Consider the conductor for which the quantum point contacts are both closed for the inner edge channel $T_1 = 0$ and $T_3 = 0$. In this case, at zero temperature, the quantum coherent sample is noiseless. Transmission along each edge state is either one or zero. Now consider the conductor with the dephasing probe. Under the biasing condition considered here, the distribution function $\bar{f}_4(E)$ is still a *non-equilibrium* distribution function and given by $\bar{f}_4(E) = \frac{1}{2} f_1(E) + \frac{1}{2} f_2(E)$. The distribution function at the dephasing contact is similar to a distribution at an elevated temperature with $kT = eV/4$ with eV the voltage applied between contact 1 and contacts 2 and 3. We have $\int dE \bar{f}_4(E)(1 - \bar{f}_4(E)) = e|V|/4$. As a consequence the bare spectra $P_{\alpha\beta}$ and $P_{\beta 4}$ are now non-vanishing. Evaluation of the correlation function Eq. (19) gives [24],

$$P_{23}^{qe} = -\frac{e^2}{h}|eV|\frac{1}{4} \,. \tag{20}$$

The electron current incident into the voltage probe from contact 1 is noise-less. Similarly, the hole current that is in the same energy range incident from contact 2 is noiseless. However, the voltage probe has two available out-going channels. The noise generated by the voltage probe is thus a consequence of the partitioning of incoming electrons and holes into the two out-going channels. In contrast, at zero-temperature, the partition noise in a coherent conductor is a purely quantum mechanical effect. Here the partioning invokes *no quantum coherence* and is a classical effect.

We emphasize that a dephasing voltage probe generates zero-temperature, incoherent partition noise whenever it is connected to channels which in a certain energy range are not completely filled. In our example the nonequilibrium filling of the channels incident on the voltage probe arises since the incident channels are occupied by reservoirs at different potentials. For instance a dephasing voltage probe connected to a ballistic wire (with adiabatic contacts) will generate incoherent partition noise if the probabilities for both left and right movers to enter the voltage probe are non-vanishing. If we demand that a dephasing voltage probe sees only left movers [53, 54] (we might add a dephasing voltage probe which sees only right movers) we have a dephasing probe that not only conserves energy but also generates only forward scattering. As long as all incident channels are equally filled such a forward scattering dephasing probe will not generate partition noise.

6. Voltage probe with inelastic scattering

We next compare the results of the energy conserving voltage probe with that of a real (physical) voltage probe. At such a probe only the total current vanishes. The

Langevin equations are

$$I_\alpha(t) = \sum_\beta G_{\alpha\beta} V_\beta(t) + \delta I_\alpha(t) \,. \tag{21}$$

where $G_{\alpha\beta}$ are the elements of the conductance matrix and $V_\beta(t)$ is the voltage at contact β. The voltages at contacts $1, 2$ and 3 are constant in time $eV_1 = \mu_1$ and $eV_2 = eV_3 = \mu_0$. But the voltage at contact 4 is determined by $I_4(t) = 0$ and is given by

$$V_4(t) = \bar{V}_4 + \delta V_4(t) \,. \tag{22}$$

The time-independent voltage is

$$\bar{V}_4 = -G_{44}^{-1} \sum_{\beta \neq 4} G_{4\beta} V_\beta \,, \tag{23}$$

and the fluctuating voltage is

$$V_4(t) = -G_{44}^{-1} \delta I_4(t) \,. \tag{24}$$

The distribution function in contact 4 consists of a time-independent part and a fluctuating part. The time-independent distribution is an *equilibrium* Fermi distribution at the potential $e\bar{V}_4 = \bar{\mu}_4$. For our example we have

$$\bar{V}_4 = \mu_0 + \frac{1}{2}(1 + T_1)e|V| \,. \tag{25}$$

We remark that that this potential is independent of the transmission of quantum point contact 3. The fluctuating currents at the contacts of the sample are

$$\Delta I_\alpha(t) = \delta I_\alpha(t) - \frac{G_{\alpha 4}}{G_{44}} \delta I_4 \,. \tag{26}$$

As a consequence the correlations of the currents are given by an equation which is similar to Eq. (19)

$$P_{\alpha\beta}^{\rm in} = P_{\alpha\beta} - \frac{G_{\alpha 4}}{G_{44}} P_{\beta 4} - \frac{G_{\beta 4}}{G_{44}} P_{\alpha 4} + \frac{G_{\alpha 4} G_{\beta 4}}{G_{44}^2} P_{44}, \tag{27}$$

but with the important difference that the bare noise spectra are evaluated with the equilibrium Fermi functions f_α with $\alpha = 1, 2, 3, 4$.

As in the quasi-elastic case, the first three terms are negative and the forth term is positive due to the auto-correlations of the fluctuating voltage in this contact. In the inelastic case we can not consider the case where both T_1 and T_3 are zero since this implies that $\mu_4 = \mu_1$. Thus T_1 must be non-vanishing. On the other hand we are stil free to choose T_3 to simplify the problem. It is now interesting to consider the case $T_3 = 0$. In this case shot noise is generated at QPC 1 and the voltage probe generates fluctuating populations in the the two out-going edge channels. The outer edge state leads carriers to contact 3 and the inner edge state leads carriers to contact 2. Interestingly, with this choice the first three terms in $P^{\rm in}_{23}$ vanish and the only non-zero term is the forth term arising from the auto-correlations of the voltage fluctuations in contact four. The correlation $P^{\rm in}_{23}$ is thus positive!

In the presence of the voltage probe and for $T_3 = 0$, the correlation at contacts 2 and 3 is [24]

$$P^{\rm in}_{23} = +\frac{e^2}{h}|eV|\frac{1}{2}T_1 R_1 \,. \tag{28}$$

The autocorrelations are $P^{\rm in}_{22} = P^{\rm in}_{33} = P^{\rm in}_{23}$. Current conservation is obeyed since $P^{\rm in}_{12} = P^{\rm in}_{13} = -2P^{\rm in}_{23}$ and $P^{\rm in}_{11} = 4P^{\rm in}_{23}$. Thus the normalized correlation function is $p^{in}_{23} \equiv P^{\rm in}_{23}/(P^{\rm in}_{22}P^{\rm in}_{33})^{1/2} = +1$. Clearly, as a consequence of the fluctuating voltage electrons are injected into the two edge channels in a correlated way.

In the introduction we have remarked that thermal fluctuations are always anti-correlated. Therefore, as we increase the temperature in this conductor but keep the voltage fixed thermal fluctuations should eventually overpower the correlations due to the fluctating potential of the voltage probe. As a consequence, with increasing temperature, the correlation function should change sign. Indeed, a calculation gives [24]

$$P^{\rm in}_{23} = \frac{e^2}{h}\left[- k_B T\,(2 + R_1 + R_1 T_1) + \frac{R_1 T_1}{2} eV \coth\frac{eV}{2k_B T}\right] . \tag{29}$$

If $k_B T = 0$ we recover the positive result $P^{\rm in}_{23} = (e^2/h)|eV|R_1 T_1/2$ for the shot noise, and if $V = 0$ we find $P^{\rm in}_{23} = -(2e^2/h)k_B T(1 + R_1/2)$, which is the result of the fluctuation-dissipation theorem: $P^{\rm in}_{23} = 2k_B T(G^{\rm in}_{23} + G^{\rm in}_{32})$ where [54] $G^{\rm in}_{\alpha\beta} = G_{\alpha\beta} - (G_{\alpha 4}G_{4\beta}/G_{44})$ is the conductance of the three-terminal conductor in the presence of incoherent scattering.

We define T_c, the critical temperature above which the correlations $P^{\rm in}_{23}$ are negative. For small transmission $T_1 \ll 1$ we find: $k_B T_c \simeq |eV|T_1/6$, and for large transmission $R_1 \ll 1$: $k_B T_c \simeq |eV|R_1/4$. The transmission that maximizes the critical temperature is [24] $T_1 = 3 - \sqrt{6} \simeq 0.55$. In this case we have: $k_B T_c^{\rm max} \simeq |eV|(5\sqrt{6} - 12)/(2(6\sqrt{6} - 12)) \simeq |eV|/21.8$. Clearly, it would be intersting to see an experiment which investigates the reverasl of the sign of such a correlation function as a function of temperature. Another possibility is to perform the experiment at a fixed temperature but to make transmission into the voltage probe variable (for instance with the help of a gate). At temperatures so low that intrinsic inelastic scattering can be neglected the theory predicts a negative correlation if the connection is closed. As the contact to the voltage is opend there must exist a critical transmission probability at which the correlation vanishes. Finally for sufficiently large transmission the correlation is positive.

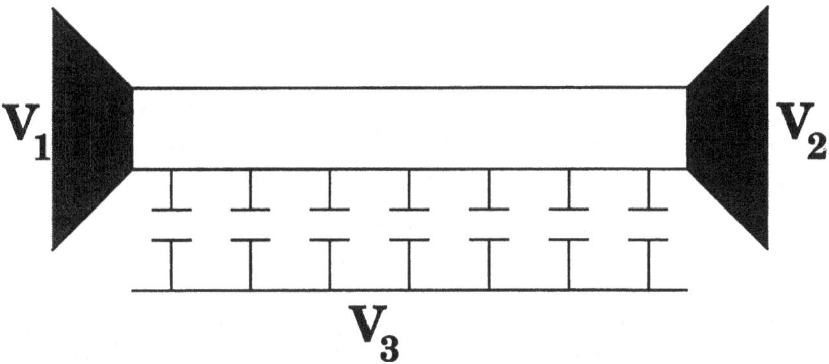

Figure 4. Ballistic one-channel wire coupled to reservoirs and capacitively coupled to a gate. After Blanter, Hekking and Büttiker, [55].

7. Dynamic Reversal of the sign of a Current-Current Correlation

The proof that current-current correlations in normal conductors are negative applies only to the white-noise (low-frequency) limit. At finite frequencies it is possible to have positive current-current correlations even at equilibrium. To illustrate this we consider a one-channel ballistic conductor connected adiabatically to two electron reservoirs and capacitively coupled to a gate with capacitance c per unit length. The gate is connected to ground without additional impedance. The conductance matrix of this wire was calculated by Blanter, Hekking and Büttiker [55] combining scattering theory with dynamic screening to determine the electrostatic potential self-consistently in random phase approximation. The conductance matrix determined in this way agrees with a theory based on a Tomonaga-Luttinger Hamiltonian and bosonization [56]. The wire has a length L and a density of states (per unit length) of $\nu_F = 2/hv_F$ where v_F is the Fermi velocity. The interaction is described by the parameter

$$g^2 = \frac{1}{1 + e^2\nu_F/c} \tag{30}$$

which is 1 in the limit of a very large capacitance c (non-interacting limit) and tends to zero as the capacitance c becomes very small. The parameter g and the density of states determine the static, electro-chemical capacitance [57] of the wire vis-a-vis the gate, $c_\mu = g^2e^2\nu_F$. The dynamic conductance matrix is defined as $G_{\alpha\beta}(\omega) = \delta I_\alpha(\omega)/\delta V_\beta(\omega)$ where $\delta I_\alpha(\omega)$ and $\delta V_\beta(\omega)$ are the Fourier coefficients of the current at contact α and the voltage at contact β. Here α and β label the contacts, the reservoirs 1 and 2 and the gate 3. At equilibrium the dynamic conductance matrix is related to the current-current fluctuations via the

Fluctuation-Dissipation theorem,

$$P_{\alpha\beta}(\omega) = \hbar\omega\ coth\frac{\hbar\omega}{2kT}G'_{\alpha\beta}(\omega) \tag{31}$$

where $G'_{\alpha\beta}(\omega) = (1/2)(G_{\alpha\beta}(\omega)+G^{\star}_{\beta\alpha}(\omega))$ is the real part of the element $G_{\alpha\beta}(\omega)$ of the conductance matrix. Consider now the current-current correlation $P_{12}(\omega)$. In terms of the effective wave-vector $q \equiv \omega g/v_F$ Ref. [55] finds

$$G'_{12}(\omega) = -\frac{e^2}{h}\frac{16g^2cos(qL)}{16g^2cos^2(qL) + 4(1+g^2)^2sin^2(qL)}\,. \tag{32}$$

In the zero-frequency limit we have $G'_{12}(\omega) = -\frac{e^2}{h}$ which is negative as it must be for a conductance determined by a transmission probability. At a critical frequency

$$\omega_c = v_F\pi/Lg\,. \tag{33}$$

determined by $qL = \pi$, the real part of this conductance element becomes positive. Hence for $\omega > \omega_c$ there exist frequency windows for which the equilibrium currents are positively correlated. We note that in the non-interacting limit $g = 1$ this frequency is determined by the time an electron takes to traverse half of the length of the wire, $\tau = L/2v_F$. At this frequency the wire is charged by carriers coming in simultaneously from both reservoirs. Increasing the interaction suppresses charging and thus increases this frequency. On the other hand the frequency is inversely proportional to the length of the wire and the frequency tends to zero as the wire length tends to infinity.

Since much of the discussion based on Luttinger theory and bosonization does not take into account the finite size of the sample, we can expect [58] that such theories would in fact predict positive correlations! Indeed if we consider for a moment a Luttinger liquid coupled at a point $x = 0$ to a tunneling contact an electron inserted into the wire gives with probability $1/2$ rise to a left (right) going plasma excitation with charge e(1-g)/2 and a right (left) going excitation with charge eg/2. This charges lead to positively correlated currents at $x = \pm L/2$ with a noise spectrum proportional to $(1/4)g(1-g)$ (for a detailed discussion see Ref. [59]). Since the transition from a Luttinger liquid to a normal region leads to reflection of plasma excitations [56] we can expect that a full treatment of the contacts would restore the expected negative correlations.

The positive dynamic correlations discussed here are only accessible at high frequencies. Is it possible to observe positively correlated currents at low frequencies? The answer is yes and we will now discuss two geometries.

8. Positive Correlations of Dynamic Screening Currents

Consider the classical electrical circuit shown in Fig. 8 in which a node with potential U is at one branch connected via a resistor with resistance R to terminal 1 and at the other branches via capacitances C_1 and C_2 to terminals 2 and 3. We are interested in the low-frequency behavior and expand the classical ac-conductance matrix $\mathbf{G}$ in powers of the frequency

$$\mathbf{G}(\omega) = -i\omega\mathbf{C} + \omega^2\mathbf{K} + \dots . \tag{34}$$

The first term is purely capacitive with a capacitance matrix $\mathbf{C}$. The second term, which is of interest here, is the lowest order in frequency term which is dissipative. For the classical circuit of Fig. 8 it is given by

$$\mathbf{K} = R \begin{pmatrix} C_\Sigma^2 & -C_1 C_\Sigma & -C_2 C_\Sigma \\ -C_1 C_\Sigma & C_1^2 & C_1 C_2 \\ -C_2 C_\Sigma & C_1 C_2 & C_2^2 \end{pmatrix} . \tag{35}$$

Here $C_\Sigma = C_1 + C_2$. The key point is of course that the off-diagonal elements $K_{23} = K_{32}$ are positive. In view of the fluctuation-dissipation theorem this implies that the correlation of currents at the two capacitive terminals are positively correlated, $P_{23} = 2kT\omega^2 K_{23} = 2kT\omega^2 C_1 C_2 R$. A current fluctation leads to charging of the capacitors C_1 and C_2 which in turn generates simultaneously current flowing through the two terminals 2 and 3 to compensate this charge.

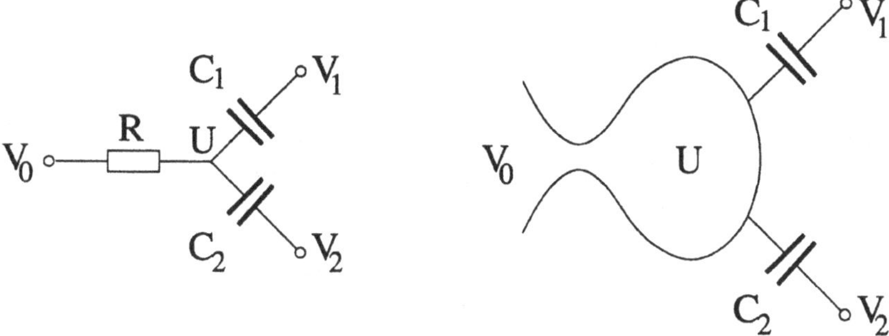

Figure 5. Left: A classical three-terminal circuit: One branch is coupled via a resistor R and two branches are coupled with capacitances C_1 and C_2 to the node. Right: A cavity coupled via a narrow lead to a reservoir at voltage V_0 and coupled capacitively to two gates with geometric capacitances C_1 and C_2. U is the voltage at node (left) and inside the cavity (right).

The classical circuit can, for example, be viewed as a simple model for the ac currents of a mesoscopic (chaotic) cavity coupled capacitively to two gates with geometrical capacitances C_1 and C_2 and connected via a quantum point contact to a particle reservoir. There are mesoscopic corrections to the geometrical capacitances and they are replaced by electrochemical capacitances $C_{\mu,1}$ and $C_{\mu,2}$. Similarly, the classical two terminal resistance R is replaced by a *charge relaxation* resistance [60]. In a theory that determines the internal potential U of the cavity in random phase approximation both the electrochemical capacitances and the charge relaxation resistances can be expressed in terms of elements of the Wigner-Smith, Jauch-Marchand, time delay matrix [61–63],

$$N_{\beta\gamma} = \frac{1}{2\pi i} \sum_{\alpha} s^{\dagger}_{\beta\alpha} \frac{ds_{\gamma\alpha}}{dE}. \tag{36}$$

that characterizes fully the low-frequency charge fluctuations on the cavity [64]. For the mesoscopic cavity the four dynamical transport coefficients of interest are,

$$\begin{aligned} D = e^2 \mathrm{Tr} N, \quad & C_{\mu,1} = \tfrac{C_1 D}{C_\Sigma + D}, \\ C_{\mu,2} = \tfrac{C_2 D}{C_\Sigma + D}, \quad & R_q = \tfrac{h}{2e^2} \frac{(\mathrm{Tr} N^2)}{(\mathrm{Tr} N)^2}. \end{aligned} \tag{37}$$

Replacing C_1, C_2 and R in Eqs. (34) and (35) with $C_{\mu,1}, C_{\mu,2}$ and R_q gives the low frequency response of the mesoscopic cavity. Thus the equilibrium current correlations $P_{12} = P_{21}$ of the mesoscopic cavity are also positively correlated.

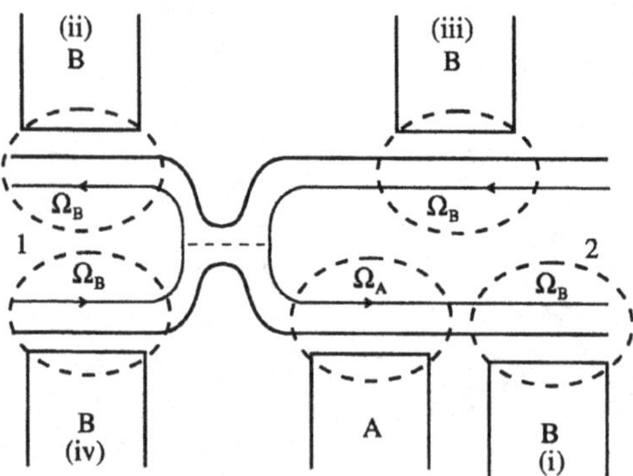

Figure 6. Quantum point contact in the quantum Hall regime: a single edge is partially transmitted and reflected. Charging of the edge state is probed capacitively with a capacitor at A and second capacitor B which can be in positions I-IV. After Martin and Büttiker [65].

Thus far we have considered frequency dependent equilibrium fluctuations. Martin and Büttiker [65] have investigated the correlation of dynamic screening currents in open conductors. The example considered is shown in Fig. 8. A Hall bar in a high magnetic field is connected to two reservoirs 1 and 2. The magnetic field corresponds to a filling factor $\nu = 1$ such that the wire is in the integer quantum Hall regime. Backscattering is generated by a quantum point contact. The lines along the edges of the conductor indicate the edge states. Their chirality is indicated by arrows. Two gates A and B are used to probe charge fluctuations capacitively. Four positions I, II, III, IV are considered for gate B whereas gate A is always held at the same position. Charge fluctuations on an edge state will induce capacitive currents on the gates A and B.

To keep the discussion simple it is assumed that charge pile up occurs only in the proximity of the gates A and B and that the remaining part of the conductor is charge neutral. The regions where charge pile up can occur are indicated by the volumes Ω_A for gate A and Ω_B for gate B. The geometrical capacitance of the gates to the edge states are denoted by C_A and C_B. Again it is possible to express the charge fluctuations with the help of a generalized Wigner-Smith matrix. Whereas in the example of the cavity discussed above, the charge in the entire cavity was of interest, here we are interested only in the charge pile up in the regions Ω_A and Ω_B. We are thus interested in *local* charge fluctuations. As a consequence we now have to consider functional derivatives of the scattering matrix with regard to the local potential [66],

$$\mathcal{N}_{\delta\gamma}^{(\eta)} = \frac{-1}{2\pi} \sum_{\nu} \int_{\Omega_\eta} d^3\mathbf{r} \left[s_{\nu\delta}^{\star}(E, U(\mathbf{r})) \frac{\delta s_{\nu\gamma}(E, U(\mathbf{r}))}{e\delta U(\mathbf{r})} \right] \tag{38}$$

where $\eta = A$ or $\eta = B$ and $\mathbf{r}$ is in the volume Ω_η and $U(\mathbf{r})$ is the electrostatic potential at position $\mathbf{r}$. For example $\mathcal{N}_{12}^{(A)}(\mathbf{r})$ is the electron density, at position $\mathbf{r}$ in volume Ω_A, associated with *two electron current amplitudes* incident from contacts 1 and 2. The explicit relation of the charge operator to local wave functions is given in [66] and a detailed derivation is found in Ref. [67].

The density of states of the edge state in region A and B and the electrochemical capacitances are

$$N_\eta = \sum_{\gamma} \mathcal{N}_{\gamma\gamma}^{\eta}(\mathbf{r}), \tag{39}$$

$$C_{\mu_\eta} = \frac{e^2 N_\eta C_\eta}{C_\eta + e^2 N_\eta}. \tag{40}$$

The current correlation at equilibrium at a temperature kT can be brought into the form

$$S_{I_\alpha I_\beta} = 2\omega^2 C_{\mu_\alpha} C_{\mu_\beta} R_q^{\alpha\beta} kT\,, \tag{41}$$

where

$$R_q^{\alpha\beta} = \frac{h}{2e^2} \frac{\sum_{\gamma\delta} \mathrm{Tr}\left[\mathcal{N}_{\delta\gamma}^{(\alpha)} (\mathcal{N}_{\delta\gamma}^{(\beta)})^{\dagger}\right]}{\mathrm{Tr}\left[\sum_\gamma \mathcal{N}_{\gamma\gamma}^{(\alpha)}\right] \mathrm{Tr}\left[\sum_\gamma \mathcal{N}_{\gamma\gamma}^{(\beta)}\right]}. \tag{42}$$

For $\alpha = \beta = A$ these equations determine the auto-correlation and for $\alpha = A$ and $\beta = B$ the cross-correlation. In the zero temperature limit in the presence of

TABLE I. Sign of equilibrium ($S^q_{I_A I_B}(\omega)$) and non- equilibrium ($S^V_{I_A I_B}(\omega)$) current correlations between gates A and B for the four positions of gate B relative to gate A.

	(i)	(ii)	(iii)	(iv)
$S^q_{I_A I_B}(\omega)$	>0	$=0$	≥ 0	≥ 0
$S^V_{I_A I_B}(\omega)$	≥ 0	≤ 0	$=0$	$=0$

an applied voltage $|eV|$ between contact 1 and 2 we can bring the current-current correlation into the form

$$S_{I_\alpha I_\beta}(\omega) = 2\omega^2 C_{\mu_\alpha} C_{\mu_\beta} R_V^{\alpha\beta} |eV| \,, \tag{43}$$

where

$$R_V^{\alpha\beta} = \frac{h}{2e^2} \frac{\mathrm{Tr}\left[\mathcal{N}_{12}^{(\alpha)} (\mathcal{N}_{12}^{(\beta)})^\dagger\right] + \mathrm{Tr}\left[\mathcal{N}_{21}^{(\alpha)} (\mathcal{N}_{21}^{(\beta)})^\dagger\right]}{\mathrm{Tr}\left[\sum_\gamma \mathcal{N}_{\gamma\gamma}^{(\alpha)}\right] \mathrm{Tr}\left[\sum_\gamma \mathcal{N}_{\gamma\gamma}^{(\beta)}\right]} . \tag{44}$$

The electrochemical capacitances are positive and the sign of the current correlations at the two gates is thus determined by $R_q^{\alpha\beta}$ at equilibrium and by $R_V^{\alpha\beta}$ in the zero temperature limit in the presence of transport.

For the geometry I Ref. [65] finds,

$$R_q^{\alpha\beta} = (h/2e^2) \quad \text{and} \quad R_V^{\alpha\beta} = (h/e^2) TR, \tag{45}$$

independent of the choice of α and β. Here T is the transmission probability through the quantum point contact and R is the reflection probability. Thus at equilibrium the charge relaxation resistance is universal and given by $h/2e^2$. This results from the fact that a charge accumulated on the edge state near gate A and B can leave the sample only through contact 2 where we have an interface resistance $h/2e^2$. In the presence of transport, in the zero temperature limit considered here, the charge fluctuations reflect the shot noise and are proportional to $T(1-T)$. In geometry I we find thus both at equilibrium and in the presence of shot noise a positive correlation.

Consider next geometry II. Here gate A and B tests charge accumulation due to transmitted and reflected particles. These are mutually exclusive events and

Ref. [65] finds,

$$R_q^{AA} = R_q^{BB} = (h/2e^2), \tag{46}$$

$$R_q^{AB} = R_q^{BA} = 0, \tag{47}$$

$$R_V^{AA} = R_V^{BB} = -R_V^{AB} = -R_V^{BA} = (h/e^2)TR. \tag{48}$$

The equilibrium correlations proportional to R_q^{AB} are zero, whereas the non-equilibrium correlations given by Eq. (48) are negative. The results for the different geometries are summarized in Table *I*.

The direct relation between charge fluctuations and the resistances R_q and R_v (see Eqs. (37) and Eqs. (42),(44) makes these quantities useful for many problems. Ref. [67–69] link these quantities to dephasing times in Coulomb coupled open conductors and Ref.[70] demonstrates, that the dephasing time and relaxation time of a closed double quantum dot capacitively coupled to a mesoscopic conductor is governed by these resistances.

9. Cooper pair partition versus pair breaking noise

Hybrid-structures [30–34] consisting of a normal conductor and superconductor provide another system in which interactions play an important role in current-current correlations. At a normal-superconducting interface an electron (hole) is reflected as a hole (electron) if it is incident with an energy below the gap of the superconductor. This process, known as Andreev reflection, *correlates* excitations with different charge. Currents at the normal contacts of such a structure can be written as a sum an electron current (e) and hole current (h). Thus the correlation function $P_{\alpha\beta}$ can be similarly decomposed into four terms,

$$P_{\alpha\beta} = P_{\alpha\beta}^{ee} + P_{\alpha\beta}^{hh} + P_{\alpha\beta}^{eh} + P_{\alpha\beta}^{he} \tag{49}$$

corresponding to correlations of currents of the same type of quasi-particles and correlations between electron and hole currents. It can be shown that P^{ee} and P_{hh} are negative and P^{eh} and P^{he} are positive. The sign of the correlation depends on the strength of the different contributions. Indeed Anantram and Datta [35] showed that for a simple one-channel normal structure in which the normal part and the superconducting part form a loop penetrated by a flux Φ, that the correlation measured at two normal contacts changes sign as a function of flux. Subsequent investigations based on a single channel Y-structure with a wave splitter which depends on a coupling parameter [71] found that the correlation changes sign and becomes positive as the coupling to the superconductor is decreased [37, 72]. Investigation of a highly asymmetric geometry of an NS-structure in which one of the normal contacts is a tunneling tip found similarly restrictive conditions for positive correlations and moreover indicated that with increasing

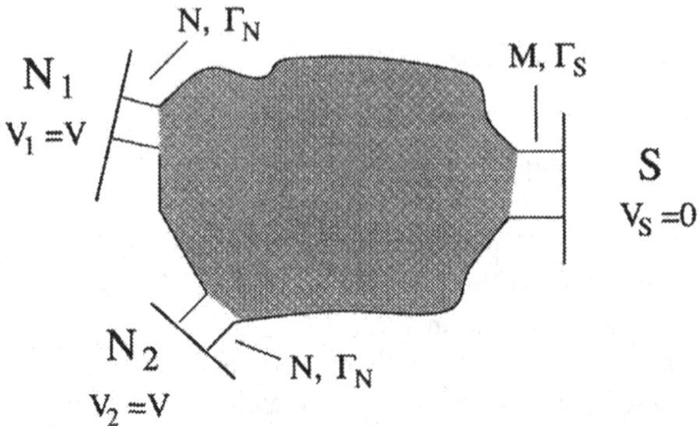

Figure 7. A chaotic cavity is connected to two normal (N_1 and N_2) reservoirs and and to a superconductor S via quantum point contacts. After Samuelsson and Büttiker [41].

channel number of the normal conductor it is less and less likely to observe positive correlations [38].

These results pose the question of whether positive correlations are indeed a feature of few channel ballistic systems only and could in fact not be seen in multi-channel systems which are typically also channel mixing. Indeed Nagaev and the author [73] investigating diffusive normal structures, perfectly coupled to the superconductor, and neglecting the proximity effect, found that correlations are manifestly negative, as in purely normal structures. In view of the current interest in sources of entangled massive particles and the detection of entanglement, understanding the correlations generated in hybrid structures is of particular interest [74–76].

To investigate the sign of current correlations in channel mixing hybrid structures for a wider range of conditions Samuelsson and the author [41] analyzed current correlations in a chaotic cavity using random matrix theory. The system is shown in Fig. 9. A chaotic cavity is coupled via quantum point contacts with $N_1 = N$ and $N_2 = N$ open channels at the normal contacts and with M channels to the superconducting contact.

The result of the random matrix calculation is depicted in Fig. 9. In the absence of the proximity effect (broken line in Fig. 9) the ensemble averaged cross-correlation is negative over the entire range of the ratio $2N/M$. This situation is the analog for the chaotic cavity of the negative correlations found in diffusive conductors by Nagaev and the author [73]. The result is dramatically different if the proximity effect plays a role (solid curve of Fig. (9). Now, at least in the limit where the cavity is much better coupled to the superconductor than to the normal reservoirs, the correlations are positive. Due to the multitude of processes contributing to these results a detailed microscopic explanation is difficult. In Ref. [41] Samuelsson and the author present explanations for the limiting behavior ($N \ll M$ and $M \gg N$).

A simple picture emerges if a barrier of strength Γ_s is inserted into the contact between the cavity and a superconductor (see Fig. 9). The case where there are tunnel barriers in all contacts has been investigated by Boerlin et al. [40]. Here we focus our attention to the case where the contacts to the normal reservoirs are perfect quantum point contacts and only the contact to the superconductor contains a barrier (as shown in the inset of Fig. 9). A simple result is obtained in the limit $2N/M \gg 1$. In this case injected quasi-particles scatter at most once from the superconductor-dot contact and the resulting scattering matrix simplifies considerably. The resulting correlation function is

$$\frac{\langle P_{12} \rangle}{P_0} = \frac{M}{2N} R_{eh}(1 - 2R_{eh}) \tag{50}$$

where $R_{eh} = \Gamma_S^2/(2-\Gamma_S)^2$ is the Andreev reflection probability of quasiparticles incident in the dot-superconductor contact. There is a crossover from negative to positive correlations that takes place already for $R_{eh} = 1/2$, i.e $\Gamma_S = 2(\sqrt{2}-1) \approx 0.83$, in agreement with the full numerics in Fig. 9. The fact that a "bad" contact reducing the Andreev reflection is favorable in generating positive correlations seems at first counter intuitive. Below we give a simple discussion to explain this result.

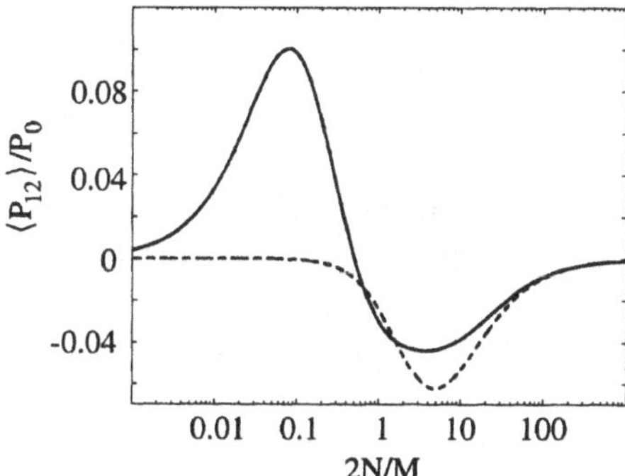

Figure 8. Current-current correlation of the chaotic dot junction coupled to the superconductor with M open channels and with normal contacts of N open channels as a function of $2N/M$. In the presence of the proximity effect (solid line) the correlation is positive for $2N < M$. Application of a magnetic flux of the order of a flux quantum suppresses the proximity effect and leads to negative correlations (broken line) independent of the ratio of N and M. After P. Samuelsson and M. Büttiker [41].

Eq.(50) is the cross-correlation averaged over an ensemble of cavities. Since the proximity effect plays no role, it must be possible, to derive this result from a purely semiclassical discussion. This statement holds of course not only for the particular geometry of interest here but of all the results obtained in the absence of the proximity effect. A semiclassical theory for chaotic-dot superconductor systems is presented in Ref. [77] not only for the current-current correlations but also for the higher cumulants. Below we focus on the simple result described by Eq. (50).

In the presence of the tunnel barrier at the superconductor-dot contact we can view the superconductor as an injector of Cooper pairs [39]. This picture differs from the Andreev-Bogoliubov-de Gennes picture of correlated electron-hole processes. The (mathematical) transformation between these two pictures is of interest and will be discussed elsewhere. The argument presented below expands a suggestion by Schomerus [78]. We divide time into intervals such that the n-th time slot might contain a Cooper pair $\sigma_n = 1$ which has successfully penetrated through the barrier and entered the cavity or the n-th time slot is empty, $\sigma_n = 0$, if the Cooper pair has been reflected. Clearly, we have

$$< \sigma_n > = < \sigma_n^2 > = R_{eh} \tag{51}$$

where R_{eh} is the Andreev reflection probability. Once the Cooper pair has entered the cavity two processes are possible: either the entire Cooper pair is transmitted into one of the normal contacts giving raise to *Cooper pair partition noise* or the Cooper pair is split up and one electron leaves through contact 1 and the other electron leaves through contact 2. We refer to the contribution to the correlation function by this second process as *pair breaking noise*. To proceed we assume that each electron has a probability T_1 to enter contact 1 and probability $T_2 = 1-T_1$ to enter contact 2. Thus for a symmetric junction $T_1 = T_2 = 1/2$ an incident Cooper pair contributes with probability $1/2$ to the partition noise and with probability $1/2$ to the pair breaking noise.

We now want to write the correlation function in a way that permits us to separate these processes. The charge Q_1 transferred into contact 1 over large number of time slots is

$$Q_1 = \sum_n \sigma_n (p_n + q_n) \tag{52}$$

and the charge Q_2 transferred into contact 2 is

$$Q_2 = \sum_n \sigma_n (1 - p_n + 1 - q_n). \tag{53}$$

Here q_n and p_n denote the two particles comprising the cooper pair. For pair partition we have $p_n = q_n = 1$ or $p_n = q_n = 0$ and for pair breaking we have $p_n = 1, q_n = 0$ or $p_n = 0, q_n = 1$.

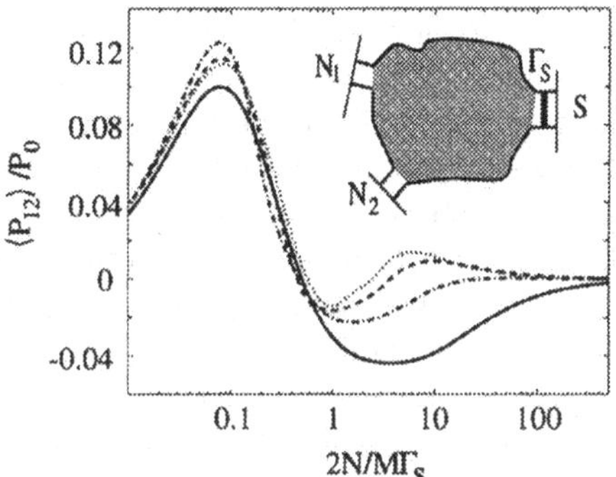

Figure 9. Current-current correlation of a chaotic cavity connected to the superconductor via a contact with conductance proportional to $\Gamma_s M$ and to two normal reservoirs with N open channels as a function of $2N/M\Gamma_s$. The contact transparencies are $\Gamma_s = 1$ (solid line) 0.8 (dashed dotted) 0.6 (dashed line). For large $2N/M\Gamma_s$ the correlation crosses over from negative to positive as the transparency Γ_s is reduced. After P. Samuelsson and M. Büttiker [41].

Next we consider the fluctuations of the transferred charge $\Delta Q_i = Q_i - < Q_i >$. The average transmitted charge is $< Q_i > = R_{eh}$. For the correlation we find,

$$< \Delta Q_1 \Delta Q_2 > = < \Delta Q_1 \Delta Q_2 >_p + < \Delta Q_1 \Delta Q_2 >_e \qquad (54)$$

where the index p denotes the contribution of the pairs which are transmitted in their entierty into lead 1 or 2 and e is the average only over the pairs which are broken up and an electron is emitted into each contact. For pair transmission we can distinguish events which emit a pair through the upper lead 1. In this case $\Delta Q_1 = 2\sum_n \sigma_n - R_{eh}$ and $\Delta Q_2 = -R_{eh}$ One quarter of all events are of this type. Similarly we can treat the case of pairs emitted through lead 2. Taking into account that $< \sum_n \sigma_n > = R_{eh}$ we find a pair partition noise

$$< \Delta Q_1 \Delta Q_2 >_p = -(1/2) R_{eh}^2 \qquad (55)$$

Like the partition noise of single electrons in a normal conductor it is negative.

Consider next the pair breaking events. For these events $\Delta Q_1 = \sum_n \sigma_n - R_{eh}$ and $\Delta Q_2 = \sum_m \sigma_m - R_{eh}$. Onehalf of the time-slots with a Cooper pair are of this type. The correlation contains four terms, $< \sum_n \sum_m \sigma_n \sigma_m > = R_{eh}$, $-R_{eh} < \sum_n \sigma_n > = -R_{eh} R_{eh}$, $-R_{eh} < \sum_n \sigma_n > = -R_{eh} R_{eh}$, and a term $+R_{eh} R_{eh}$. Thus simultaneous emission of electrons gives rise to a noise

$$< \Delta Q_1 \Delta Q_2 >_e = (1/2) R_{eh} (1 - R_{eh}). \qquad (56)$$

This contribution to the cross-correlation is positive.

Notice that the pair partition noise is negative and quadratic in R_{eh}. The pair breaking process gives a contribution which is linear in R_{eh} for small R_{eh} and thus wins in this limit. To achieve positive correlations it is thus favorably to have a small Andreev reflection probability. Only in this limit can the pair breaking processes overcome the negative partition noise of Cooper pairs and give rise to positive correlations. The full counting statistics is discussed in Ref. [77].

In hybrid superconducting normal structures there are thus several possibilities for a sign reversal of current-current fluctuations. For a cavity that is well coupled to a superconductor (see Fig. (9) application of a magnetic flux reverses the sign from positive to negative. As a function of temperature or applied voltage we can have a sign reversal both for the cavity that is well coupled to the superconductor (see Fig. 9) as well as for the cavity that connects to the superconductor via a tunnel contact (see Fig. (9). For temperatures and voltages large compared to the superconducting gap the structure considered here behaves like a normal structure and exhibits negative correlations.

10. Summary

For non-interacting particles injected from thermal sources there is a simple connection between the sign of correlations and statistics. In contrast to photons, electrons are interacting entities, and we can expect the simple connection between statistics and the sign of current-current correlations to be broken, if interactions play a crucial role.

The standard situation consists of a normal conductor embedded in a zero-frequency external impedance circuit such that the voltages at the contacts can be considered to be constant in time. Under this condition the low frequency current-current cross-correlations measured at reservoirs are negative independent of the geometry, number of contacts and the bias applied to the conductor (as long as we do not depart to far from equilibrium). The negative correlations are a consequence of Fermi-statistics and the unitarity of the scattering matrix. Under these conditions the fluctuations in the potential play no role. We have shown that already the voltage fluctuations at a real voltage contact are sufficient to change the sign of correlations in certain special situations. Carriers injected by the voltage probe are correlated by the fluctuations of the potential of the voltage probe and can in the situation considered overcome the anti-bunching generated by Fermi statistics. We have also pointed out that displacement currents (screening currents) are positively correlated even at small frequencies. The electron-hole correlations generated in a normal conductor by a superconductor can similarly generate positive correlations in situations in which the pair partition noise is overcome by the pair breaking noise.

The fact that interactions can have a dramatic effect on current-current correlations (change even their sign) clearly makes them a promising subject of further theoretical and experimental investigations.

Acknowledgements

The work presented here is to a large extent based on collaborations with Christophe Texier, Andrew Martin and Peter Samuelsson. I thank S. Pilgram, K. E. Nagaev and E. V. Sukhorukov for valuable comments on the manuscript. The work is supported by the Swiss National Science Foundation, the Swiss program for Materials with Novel Properties and the European Network of Phase Coherent Dynamics of Hybrid Nanostructures.

References

1. Ya. M. Blanter and M. Büttiker, Physics Reports, **336**, 1-166 (2000).
2. R. Hanbury Brown and R.Q. Twiss, Nature **177**, 27 (1956);
3. R. Hanbury Brown and R. Q. Twiss, Nature **178**, 1447 (1956).
4. A. M. Steinberg, P. G. Kwiat, and R. Y. Chiao, Phys. Rev. Lett. **71**, 708 (1993).
5. E. M. Prucell, Nature **178**, 1449 (1956).
6. M. Büttiker, Phys. Rev. Lett. **65**, 2901 (1990).
7. M. Büttiker, Physica B**175**, 199 (1991).
8. M. Büttiker, Phys. Rev. Lett. **57**, 1761 (1986).
9. M. Büttiker, IBM J. Res. Developm. **32**, 317 (1988).
10. V. A. Khlus, Zh. Éksp. Teor. Fiz. **93** (1987) 2179 [Sov. Phys. JETP **66** (1987) 1243].
11. G. B. Lesovik, Pis'ma Zh. Éksp. Teor. Fiz. **49** (1989) 513 [JETP Lett. **49** (1989) 592].
12. M. Büttiker, Phys. Rev. B**46**, 12485 (1992).
13. R. Landauer and Th. Martin, Physica B **175** (1991) 167; T. Martin and R. Landauer, Phys. Rev. B **45**(4), 1742 (1992).
14. M. Büttiker, Phys. Rev. Lett. **68**, 843, (1992).
15. T. Gramespacher and M. Büttiker, Phys. Rev. Lett. **81**, 2763 (1998).
16. Ya. M. Blanter and M. Büttiker, Phys. Rev. B**55**, 2127 (1997).
17. S. A. van Langen and M. Büttiker, Phys. Rev. B **56**, 1680 (1997).
18. E. V. Sukhorukov and D. Loss, Phys. Rev. B**59**, 13054 (1999).
19. W. D. Oliver, J. Kim, R. C. Liu, and Y. Yamamoto, Science **284**, 299 (1999).
20. M. Henny, S. Oberholzer, C. Strunk, T. Heinzel, K. Ensslin, M. Holland, and C. Schönenberger, Science **284**, 296 (1999).
21. S. Oberholzer, M. Henny, C. Strunk, C. Schonenberger, T. Heinzel, K. Ensslin, and M. Holland, Physica E **6**, 314 (2000).
22. T. Kodama, N. Osakabe, J. Endo, and A. Tonomura, K. Ohbayashi, T. Urakami, S. Ohsuka, H. Tsuchiya, Y. Tsuchiya and Y. Uchikawa Phys. Rev. A **57**, 2781 (1998).
23. H. Kiesel, A. Renz and F. Hasselbach, Nature **418**, 392 (2002).
24. C. Texier and M. Büttiker, Phys. Rev. B **62**, 7454 (2000).
25. M. Büttiker, Phys. Rev. B **32**, 1846 (1985).
26. M. Büttiker, Phys. Rev. B **33**, 3020 (1986).
27. C. W. J. Beenakker and M. Büttiker, Phys. Rev. B**46**, 1889 (1992).
28. R. C. Liu and Y. Yamamoto, Phys. Rev. B**50**, 17411 (1994).

29. M. J. M. de Jong and C. W. J. Beenakker, Physica A **230**, 219 (1996).
30. X. Jehl, M. Sanquer, R. Calemczuk and D. Mailly, Nature (London) **405** 50 (2000).
31. A.A. Kozhevnikov, R.J. Shoelkopf, and D.E. Prober, Phys. Rev. Lett. **84** 3398 (2000).
32. X. Jehl and M. Sanquer, Phys. Rev. B **63**, 052511 (2001).
33. B. Reulet, A. A. Kozhevnikov, D. E. Prober, W. Belzig, Yu. V. Nazarov, "Phase Sensitive Shot Noise in an Andreev Interferometer", cond-mat/0208089
34. F. Lefloch, C. Hoffmann, M. Sanquer and D. Quirion, "Doubled Full Shot Noise in Quantum Coherent Superconductor - Semiconductor Junctions", cond-mat/0208126
35. M. P. Anantram and S. Datta, Phys. Rev. B **53**, 16390 (1996).
36. Th. Martin, Phys. Lett. A**220**, 137 (1966).
37. J. Torrs and Th. Martin, Eur. Phys. J. B **12**, 319 (1999).
38. T. Gramespacher and M. Büttiker, Phys. Rev. B **61**, 8125 (2000).
39. P. Recher, E. V. Sukhorukov, and D. Loss, Phys. Rev. B **63**, 165314 (2001); C. Bena, S. Vishveshwara, L. Balents, and M. P. A. Fisher, Phys. Rev. Lett. **89**, 037901 (2002).
40. J. Börlin , W. Belzig, and C. Bruder , Phys. Rev. Lett. **88**, 197001 (2002).
41. P. Samuelsson and M. Büttiker, Phys. Rev. Lett. **89**, 046601 (2002).
42. B. I. Halperin, Phys. Rev. B **25**, 2185 (1982).
43. M. Büttiker, Phys. Rev. B **38**, 9375 (1988).
44. M. I. Reznikov, M. Heiblum, H. Shtrikman, and D. Mahalu, Phys. Rev. Lett. **75** 3340 (1995).
45. A. Kumar, L. Saminadayar, D. C. Glattli, Y. Jin, and B. Etienne, Phys. Rev. Lett. **76**, 2778 (1996).
46. B. J. van Wees, E. M. M. Willems, L. P. Kouwenhoven, C. J. P. M. Harmans, J. G. Williamson, C. T. Foxon, and J. Harris, Phys. Rev. B **39**, 8066 (1989).
47. S. Komiyama, H. Hirai, S. Sasa, and T. Fujii, Solid State Commun. **73**, 91 (1990).
48. B. W. Alphenaar, P. L. McEuen, R. G. Wheeler, and R. N. Sacks, Phys. Rev. Lett. **64**, 677 (1990).
49. G. Müller, D. Weiss, A. V. Khaetskii, K. von Klitzing, S. Koch, H. Nickel, W. Schlapp, and R. Lösch, Phys. Rev. B **45**, 3932 (1992).
50. K. E. Nagaev, Phys. Lett. A **169**, 103 (1992).
51. K. E. Nagaev, "Boltzmann - Langevin approach to higher-order current correlations in diffusive metal contacts", cond-mat/0203503
52. K. E. Nagaev, P. Samuelsson and S. Pilgram, "Cascade approach to current fluctuations in a chaotic cavity", (unpublished). cond-mat/0208147
53. M. Büttiker, "Noise in Mesoscopic Conductors and Capacitors", Proceedings of the 13th International Conference on Noise in Physical Systems and 1/f-Fluctuations, eds. V. Bareikis and R. Katilius, (Word Scientific, Singapore, 1995). p. 35 - 40.
54. M. Büttiker, in "Resonant Tunneling in Semiconductors: Physics and Applications", edited by L. L. Chang, E. E. Mendez, and C. Tejedor, (Plenum Press, New York, 1991). p. 213-227.
55. Ya. M. Blanter, F.W.J. Hekking, and M. Büttiker, Phys. Rev. Lett. **81**, 1925 (1998).
56. I. Safi, Ann. Phys. (Paris) **22**, 463 (1997).
57. M. Büttiker, J. Phys. Condensed Matter **5**, 9361 (1993).
58. M. Büttiker (unpublished).
59. A. Crepieux, R. Guyon, P. Devillard and T. Martin, (unpublished). cond-mat/0209291
60. M. Büttiker, H. Thomas, and A. Prêtre, Phys. Lett. **A180**, 364 (1993).
61. E. P. Wigner, Phys. Rev. **98**, 145 (1955); F. Smith, Phys. Rev. **118**, 349 (1960).
62. J. M. Jauch and J. P. Marchand, Helv. Physica Acta **40**, 217 (1967).
63. Since we are interested in the charge density within a specific region the derivative should be taken with respect to the local (electrostatic) potential, (see Eq. (38)) and not energy. The difference is important in situations where a WKB-approximation does not apply. See also M. Büttiker, Phys. Rev. B**27**, 6178 (1983).

64. M. H. Pedersen, S. A. van Langen, M. Büttiker, Phys. Rev. **B 57**, 1838 (1998).
65. A. M. Martin and M. Büttiker, Phys. Rev. Lett. **84**, 3386 (2000).
66. M. Büttiker, J. Math. Phys. **37**, 4793 (1996).
67. M. Büttiker, in "Quantum Mesoscopic Phenomena and Mesoscopic Devices", edited by I. O. Kulik and R. Ellialtioglu, (Kluwer, Academic Publishers, Dordrecht, 2000). Vol. 559, p. 211. cond-mat/9911188
68. M. Büttiker and A. M. Martin, Phys. Rev. **B61**, 2737 (2000).
69. G. Seelig and M. Büttiker, Phys. Rev. B **64**, 245313 (2001).
70. S. Pilgram and M. Büttiker, (unpublished). cond-mat/0203340.
71. M. Büttiker, Y. Imry and M. Ya. Azbel, Phys. Rev. A **30**, 1982 (1984).
72. G. B. Lesovik , T. Martin, and G. Blatter , Eur. Phys. J. B **24**, 287 (2001).
73. K. Nagaev and M. Büttiker, Phys. Rev. B **63** 081301, (2001).
74. G. Burkard, D. Loss, and E. V. Sukhorukov, Phys. Rev. B **61**, 16303 (2000).
75. F. Taddei, and R. Fazio, Phys. Rev. B 65, 134522 (2002).
76. P. Recher and D. Loss, "Superconductor coupled to two Luttinger liquids as an entangler for electron spins", cond-mat/0204501
77. P. Samuelsson and M. Büttiker, "Semiclassical theory of current correlations in chaotic dot-superconductor systems", (unpublished). cond-mat/0207585
78. Private communication by H. Schomerus

SHOT NOISE AND CHANNEL COMPOSITION OF ATOMIC-SIZED CONTACTS

JAN M. VAN RUITENBEEK
Kamerlingh Onnes Laboratorium, Universiteit Leiden,
Postbus 9504, 2300 RA Leiden, The Netherlands
ruitenbe@phys.leidenuniv.nl

Abstract. Experiments on shot noise in the current through atomic-sized metallic contacts reveal information on the composition of the contact in terms of transmission eigenmodes. Results for the monovalent metal gold and the *sp* metal aluminum are presented and discussed in relation to the expected number of conductance channels and the information obtained from the analysis of the superconducting subgap structure for such contacts.

1. Introduction

Using remarkably simple experimental techniques it is possible to gently break a metallic contact and thus form conducting nanowires. During the last stages of the pulling a neck-shaped wire connects the two electrodes, the diameter of which is reduced to single atom upon further stretching. Although the atomic structure of the contacts can be quite complicated, as soon as the weakest point is reduced to just a single atom the complexity is removed. The properties of the contacts are then dominantly determined by the nature of this atom. This has allowed for quantitative comparison of theory and experiment for many properties, and atomic contacts have proven to form a rich test-bed for concepts from mesoscopic physics.

We will discuss the quantum properties of the conductance of such contacts and address the question of how many channels contribute to the conductance through an atom. Exploiting measurements of multiple Andreev reflection and shot noise, among other methods, allows to investigate this problem experimentally.

Y.V. Nazarov (ed.), Quantum Noise in Mesoscopic Physics, 33–50.

2. The concept of eigenchannels

For a ballistic conductor at low temperatures, the conductance can be expressed as [17]

$$G = \frac{2e^2}{h} \mathrm{Tr}(\hat{t}^\dagger \hat{t}) \tag{1}$$

This is known as the Landauer formula and it shows that the linear conductance can be evaluated from the coefficients t_{nm}, which give the outgoing amplitude for mode m in lead 2 at unity amplitude of the incoming mode n in lead 1. Although $\hat{t}$ is not in general a square matrix (the number of modes in each lead need not to be identical) the matrix $\hat{t}^\dagger \hat{t}$ is always a $N_1 \times N_1$ square matrix (assuming $N_1 < N_2$). Current conservation certainly requires that $T_{12} = T_{21} = \mathrm{Tr}\left((\hat{t}')^\dagger \hat{t}'\right)$. This property is a simple consequence of time reversal symmetry of the Schrödinger equation which ensures that $t_{nm} = (t'_{mn})^*$.

Being the trace of an Hermitian matrix, T_{12} has certain invariance properties. For instance, there exists a unitary transformation $\hat{U}$ such that $\hat{U}^{-1}\hat{t}^\dagger \hat{t}\hat{U}$ adopts a diagonal form. Due to hermiticity of $\hat{t}^\dagger \hat{t}$ its eigenvalues τ_i, with $i = 1, ..., N_1$, should be real. Moreover, due to the unitarity of the scattering matrix one has $\hat{t}^\dagger \hat{t} + \hat{r}^\dagger \hat{r} = \hat{I}$ and then both $\hat{t}^\dagger \hat{t}$ and $\hat{r}^\dagger \hat{r}$ should become diagonal under the same transformation $\hat{U}$. As also both $\hat{t}^\dagger \hat{t}$ and $\hat{r}^\dagger \hat{r}$ are positive definite it is then easy to show that $0 \le \tau_i \le 1$ for all i.

The eigenvectors of $\hat{t}^\dagger \hat{t}$ and $\hat{r}^\dagger \hat{r}$ are called *eigenchannels*. They correspond to a particular linear combination of the incoming modes which remains invariant upon reflection on the sample. In the basis of eigenchannels the transport problem becomes a simple superposition of independent single mode problems without any coupling, and the conductance can be written as

$$G = \frac{2e^2}{h} \sum_i \tau_i. \tag{2}$$

At this point the definition of eigenchannels may seem somewhat arbitrary and dependent on the number of channels of the perfect leads attached to the sample. For instance, the dimension of the transmission matrix $\hat{t}^\dagger \hat{t}$ can be arbitrarily large depending on the number of modes introduced to represent the leads, which suggests that the number of eigenchannels is not a well defined quantity for a given sample.

In order to convince ourselves that this is not the case, let us consider a situation where the sample is a narrow cylindrical constriction between two wide cylindrical leads as shown in Fig. 1. Let us call N_c the number of propagating modes at the Fermi energy on the constriction. Clearly one has $N_c \ll N_1, N_2$. This geometry can be analyzed as two 'wide-narrow' interfaces connected in series. In such an interface the number of conduction channels with non-vanishing

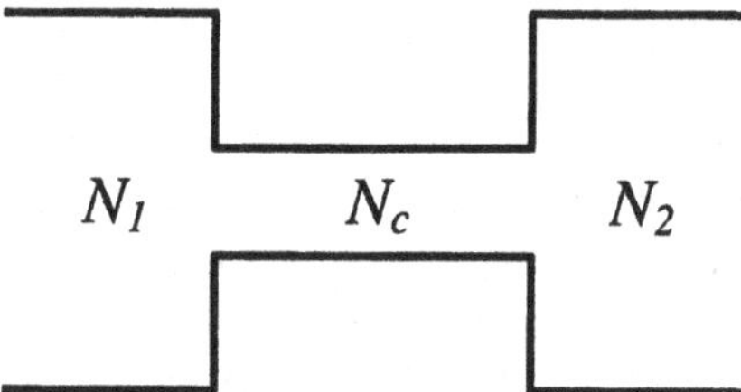

Figure 1. Two perfect cylindrical leads connecting to a sample in the form of a narrow cylindrical conductor.

transmission is controlled by the number of propagating modes in the narrowest cross-section. This property is a simple consequence of current conservation along each conduction channel. Mathematically, one can show that the non-vanishing eigenvalues of $\hat{t}^\dagger\hat{t}$ (a $N_1 \times N_1$ matrix) should be the same as those of $\hat{t}\hat{t}^\dagger$ (a $N_c \times N_c$ matrix). Therefore, there should be $N_1 - N_c$ channels with zero transmission. By applying the same reasoning to the second "narrow-wide" interface we conclude that only a small fraction of the incoming channels can have a non-zero transmission. The number of relevant eigenchannels is thus determined by the narrowest cross-section of the constriction.

For a constriction of only one atom in cross section one can estimate the number of conductance channels as $N_c \simeq (k_F a/2)^2$, which is between 1 and 3 for most metals. We shall see that the actual number of channels is determined by the valence orbital structure of the atoms.

3. Valence-Orbital-Based Description of the Conductance Modes

Cuevas, Levy Yeyati and Martín-Rodero [9] investigated the number of conductance channels for single-atom contacts using a tight binding calculation, for a geometry of two atomic pyramids touching at the apex through a single atom. They argue that it is very important to make the tight binding calculation self-consistent, by which they mean that local charge neutrality is maintained at each atomic site, by iteration and adjustment of the site energy for each individual atom in the model configuration. They find that the conductance channels can be described in terms of the atomic valence orbitals. Aluminum has a configuration [Ne]$3s^2 3p^1$, and a total of four orbitals would be available for current transport: one s orbital and three p orbitals, p_x, p_y and p_z. They identify in their calculation three contributions to the conductance, one which originates from a combination of s and p_z orbitals (where the z coordinate is taken in the current direction), and two smaller identical contributions labelled p_x and p_y. The degeneracy of these two channels is due to the symmetry of the problem, and can be lifted by changing the local environment for the central atom. The fourth possible channel, an antisymmetric combination of s and p_z, is found to have a negligible transmission probability. Their calculation confirms the experimental observation by Scheer *et al.* [29] that three channels contribute to the conductance for a single aluminum atom, as will be briefly discussed below. It was also found that the total conductance for the three channels is of order 1 G_0. The results are very robust against changes in the atomic configuration; only the transmission of each of the modes varies somewhat between different choices for the atomic geometry.

The analysis was extended to other metals [9, 28], by which it was shown that the number of conductance channels for an atom of a given metallic element depends on the number of valence orbitals. It is 1 for the monovalent metals Au, Na, K ,..., the number is 3 for the *sp* metals such as Al and Pb, and it is 5 or 6 for the transition metals with partially filled *d* shells.

4. Experimental Techniques

The experimental tools for fabricating atomic-scale contacts are mostly based on a piezoelectric actuator for the adjustment of the contact size between two metal electrodes. Standard Scanning Tunnelling Microscopes (STM) are often used for this purpose [1, 11, 24, 25]. The tip of the STM is driven into the surface and the conductance is recorded while gradually breaking the contact by retracting the tip.

A practical tool for the purpose of studying metallic quantum point contacts is the Mechanically Controllable Break Junction (MCBJ) technique [23]. By breaking a macroscopic metal wire at low tempertures, two clean fracture surfaces are exposed, which remain clean due to the cryo-pumping action of the low-temperature vacuum can. This method circumvents the problem of surface contamination of tip and sample in STM experiments, where a UHV chamber with surface preparation and analysis facilities are required to obtain similar conditions. The fracture surfaces can be brought back into contact by relaxing the force on the elastic substrate, while a piezoelectric element is used for fine control. The roughness of the fracture surfaces results in a first contact at one point. In addition to a clean surface, a second advantage of the method is the stability of the two electrodes with respect to each other.

5. Conductance of Atomic-Scale Contacts

Figure 5 shows some examples of the conductance measured during breaking of a gold contact at low temperatures, using an MCBJ device. The conductance decreases by sudden jumps, separated by plateaus, which have a negative slope, the higher the conductance the steeper. Some of the plateaus are remarkably close to multiples of the conductance quantum, G_0; in particular the last plateau before loosing contact is nearly flat and very close to 1 G_0. Closer inspection, however, shows that many plateaus cannot be identified with integer multiples of the

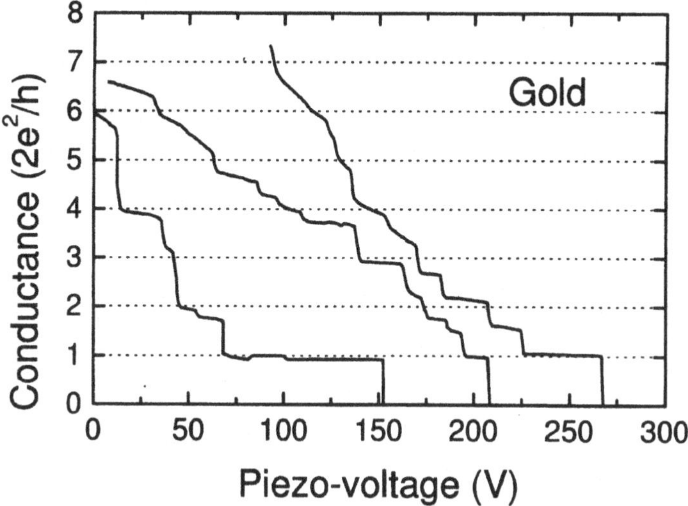

Figure 2. Three typical recordings of the conductance G measured in atomic-sized contacts for gold at helium temperatures, using the MCBJ technique. The electrodes are pulled apart by increasing the piezo-voltage. The corresponding displacement is about 0.1 nm per 25 V. After each recording the electrodes are pushed firmly together, and each trace has new structure. (After J.M. Krans [15])

quantum unit, and the structure of the steps is different for each new recording. Also, the height of the steps is of the order of the quantum unit, but can vary by more than a factor of 2, where both smaller and larger steps are found. Drawing a figure such as Fig. 5, with grid lines at multiples of G_0, guides the eye to the coincidences and may convey that the origin of the steps is in quantization of the conductance. However, in evaluating the graphs, one should be aware that a plateau cannot be farther away than one half from an integer value, and that a more objective analysis is required. For *sp* metals and transition metals with partially filled *d* orbitals we still find steps in the conductance, as illustrated for aluminum in Fig. 5, but the structure of the steps is more complicated and the plateaus are often inclining strongly upward.

For all metals studied the transition between the plateaus is very sudden and sharp. The jumps find their origin in sudden rearrangements of the atomic structure of the contact. Upon stretching of the contact, the stress accumulates elastic energy in the atomic bonds over the length of a plateau. This energy is suddenly released in a transition to a new atomic configuration, which will typically have a smaller contact size. Such atomic-scale mechanical processes were first described by Landman *et al.* [18] and by Sutton and Pethica [32]. The first direct proof for atomic rearrangements at conductance steps was provided in an experiment

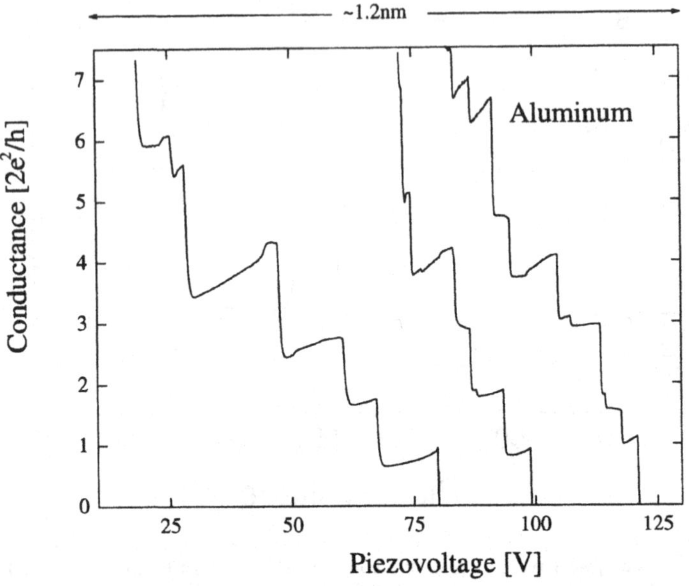

Figure 3. Three examples of conductance curves for aluminum contacts measured at 4.2 K as a function of the piezovoltage. In most cases the last 'plateau', before contact is lost, has an anomalous positive slope. An approximate scale for the range of the displacement is indicated at the top of the graph. From [14].

by Rubio *et al.* [27], where the conductance for atomic-sized gold contacts was measured simultaneously with the force on the contacts. The stress accumulation on the plateaus and the coincidence of the stress relief events with the jumps in the conductance can be clearly distinguished in those measurements. From this and related experiments we can conclude that the last plateau before contact is lost is generally formed by a single atom. We will now exploit these simple techniques to produce and study single-atom contacts.

6. Subgap structure in superconducting contacts

The most powerful method to analyze the eigenchannel composition of the conductance through an atom was introduced by Scheer *et al.* [29] and relies on the leads forming the contact being in the superconducting state. The principle can be illustrated by considering first a contact having a single mode with low transmission probability, $\tau \ll 1$. For $\tau \ll 1$ we have essentially a tunnel junction, and the current–voltage characteristic for a superconducting tunnel junction is known to directly reflect the gap, Δ, in the density of states for the superconductor. No current flows until the applied voltage exceeds $2\Delta/e$ (where the factor 2 results from the fact that we have identical superconductors on both sides of the junction), after which the current jumps approximately to the normal-state resistance line. For $eV > 2\Delta$ single quasi-particles can be transferred from the occupied states at $E_F - \Delta$ on the low voltage side of the junction to empty states at $E_F + \Delta$ at the other side. For $eV < 2\Delta$ this process is blocked, since there are no states available in the gap.

However, when we consider higher order tunnel processes a small current can still be obtained. A process that is allowed for $2\Delta > eV > \Delta$ consists of the simultaneous tunnelling of *two* quasiparticles from the low bias side to form a *Cooper pair* on the other side of the junction. The onset of this process causes a step in the current at half the gap value, $V = 2\Delta/2e$. The height of the current step is smaller than the step at $2\Delta/e$ by a factor τ, since the probability for two particles to tunnel is τ^2. In general, one can construct similar processes of order n, involving the simultaneous transfer of n particles, which give rise to a current onset at $eV = 2\Delta/n$ with a step height proportional to τ^n. This mechanism is known as multiple particle tunnelling and was first described by Schrieffer and Wilkins [30]. It is now understood that this is the weak coupling limit of a more general mechanism which is referred to as multiple Andreev reflection [13, 3, 4, 10, 6]. The theory has only been tested recently, since it requires the fabrication of a tunnel junction having a single tunnelling mode with a well-defined tunnelling probability τ. For atomic-sized niobium tunnel junctions the theory was shown to give a very good agreement [36, 21], describing up to three current steps, including the curvature and the slopes, while the only adjustable parameter is the tunnel probability, which follows directly from the normal state resistance.

Since the theory has now been developed to all orders in τ, Scheer *et al.* [29] realized that this mechanism offers the possibility of extracting the transmission probabilities for contacts with a finite number of channels contributing to the current, and is ideally suited to analyzing atomic-sized contacts. Roughly speaking, the current steps at $eV = 2\Delta/n$ are proportional to $\sum \tau_m^n$, with m the channel index, and when we can resolve sufficient details in the current–voltage characteristics, we can fit many independent sums of powers of τ_m's. When the τ_m's are not small compared to 1, all processes to all orders need to be included for a description of the experimental curves. In practice, the full expression for the current–voltage characteristic for a single channel is numerically evaluated from theory [4, 10, 6] for a given transmission probability τ_m (Fig. 6, inset), and a number of such curves are added independently, where the τ_m's are used as fitting parameters. Scheer *et al.* tested their approach first for aluminum contacts. As shown in Fig. 6, all current–voltage curves for small contacts can be very well described by the theory. However, the most important finding was that at the last 'plateau' in the conductance, just before the breaking of the contact, typically three channels with different τ's are required for a good description, while the

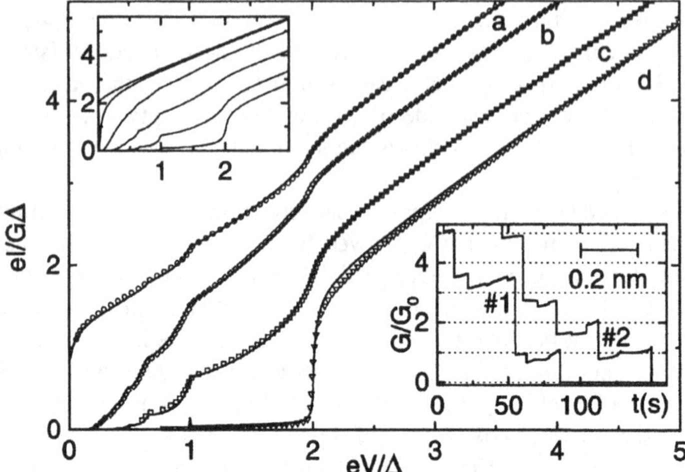

Figure 4. Current–voltage characteristics for four atom-size contacts of aluminum using a lithographically fabricated mechanically controllable break junction at 30 mK (*symbols*). The *right inset* shows the typical variation of the conductance, or total transmission $g = \sum \tau_n = G/G_0$, as a function of the displacement of the electrodes, while pulling. The bar indicates the approximate length scale. The data in the main panel have been recorded by stopping the elongation at the last stages of the contact (**a–c**) or just after the jump to the tunnelling regime (**d**) and then measuring the current while slowly sweeping the bias voltage. The current and voltage are plotted in reduced units, $eI/G\Delta$ and eV/Δ, where G is the normal state conductance for each contact and Δ is the measured superconducting gap, $\Delta/e = (182.5 \pm 2.0)\mu$V. The *left inset* shows the current–voltage characteristics obtained from first-principles theory for a single channel junction [4, 10, 6] with different values for the transmission probability τ (from bottom to top: τ=0.1, 0.4, 0.7, 0.9, 0.99, 1). The units along the axes are the same as for the main panel. The *full curves* in the main panel have been obtained by adding several theoretical curves and optimizing the set of τ values. The curves are obtained with: (**a**) three channels, τ_1=0.997, τ_2=0.46, τ_3=0.29 with a total transmission g =1.747, (**b**) two channels, τ_1=0.74, τ_2=0.11, with a total transmission g =0.85, (**c**) three channels, τ_1=0.46, τ_2=0.35, τ_3=0.07 with a total transmission g =0.88. (**d**) In the tunnelling range a single channel is sufficient, here $g = \tau_1$=0.025. From Scheer *et al.* [29].

total conductance for such contacts is of order 1 G_0 and would in principle require only a single conductance channel.

In the experiments it is found that the number of channels obtained for single-atom contacts is 1 for Au, 3 for Al and Pb, and 5 for Nb. Note that gold is not a superconductor, and a special device was fabricated which allowed the use of proximity induced superconductivity [28]. The device is a nanofabricated version of a break junction, having a thick superconducting aluminum layer forming a bridge with a gap of about 100 nm. This small gap was closed by a thin gold film in intimate contact with the aluminum. Superconducting properties were thereby induced in the gold film, and by breaking the gold film and adjusting an atomic-sized contact, the same subgap analysis could be performed.

The number of channels and the total conductance at the last plateau before breaking found in the experiment agree very well with the theory, for which the results can be summarized in the following way. Single atom contacts for monovalent metals have a single valence orbital available for current transport, giving rise to a single channel with a transmission probability close to unity. The total conductance for such contacts is thus expected to be close to 1 G_0. For s-p metals, including Al and Pb, three channels give a noticeable contribution to the current. Niobium is a d-metal with a configuration [Kr]$4d^4 5s^1$ having 6 valence orbitals: 1 s and 5 d. The theory again predicts one combination with a negligible contribution and that the remaining five channels should add up to a total conductance of about 2.8 G_0, again in good agreement with the experiment.

7. Shot noise in single-atom contacts

In single-channel quantum point contacts characterized by a transmission probability τ, shot noise is expected to be suppressed by a factor proportional to $\tau(1-\tau)$. This quantum suppression was first observed in point contact devices in a 2-dimensional electron gas [26, 16]. An expression for the shot noise power for an arbitrary set of transmission probabilities, at temperature T and bias voltage V is given by [12]

$$S_I(T,V) = \frac{2e^2}{h}\left[2k_\mathrm{B}T\sum_n \tau_n^2 + eV\coth\left(\frac{eV}{2k_\mathrm{B}T}\right)\sum_n \tau_n\left(1-\tau_n\right)\right]. \quad (3)$$

Since this general expression for shot noise in a multi-channel contact depends on the sum over the second power of the transmission coefficients, this quantity is independent of the conductance, $G = G_0\sum \tau_n$, and simultaneous measurement of these two quantities should give information about the channel distribution.

For the detection of the small noise signals the experimental setup schematically drawn in Fig.7 is often used. In order to suppress the noise of the pre-amplifiers one employs two sets of pre-amplifiers in parallel and feeds the signals into a spectrum analyzer that calculates the cross-correlation power spectrum.

7.1. GOLD CONTACTS: SATURATION OF CHANNEL TRANSMISSION

First we discuss the results for the monovalent metal gold, for which a single atom contact is expected to transmit a single conductance mode. In Fig. 7.1 the experimental results for a number of conductance values are shown (filled circles), where we plot the measured shot noise relative to the classical shot noise value $2eI$,

$$s_I = \frac{S_I}{2eI} = \frac{\sum_{n=1}^{N}\tau_n\left(1-\tau_n\right)}{g}. \quad (4)$$

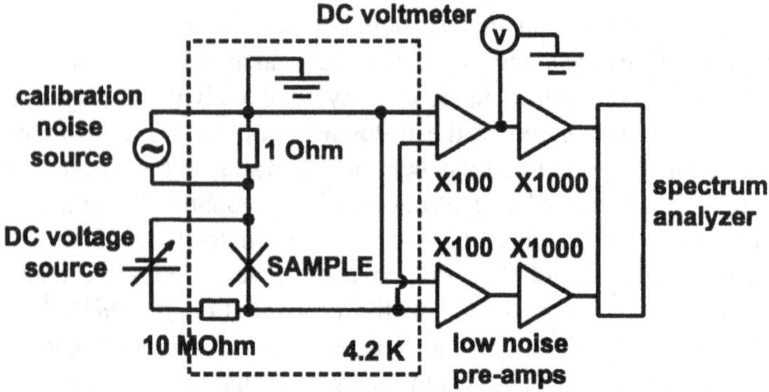

Figure 5. Wiring diagram of the experimental setup used for shot noise measurements on atomic-sized point contacts in Ref. [34]. The area enclosed by the dashed lines indicates the part at 4.2 K. In order to determine the electronic transfer function of the system a white noise signal can be injected from a calibration source through a 1 Ω series resistance.

Here

$$g = \sum_{n=1}^{N} \tau_n = G/G_0,$$

which is the conductance in units of the conductance quantum, is plotted along the horizontal axis. All data are strongly suppressed compared to the full shot noise value, with minima close to 1 and 2 times the conductance quantum.

For given conductance g and number of channels N the minimum shot noise level is obtained for all τ_n equal to either 0 or 1, except for a single one, say τ_j, that takes the remaining fraction, $s_I^{\min} = \tau_j(1-\tau_j)/g$. For future use it also useful to point out that the maximum shot signal arises when we take all τ_n equal, giving $s_I^{\max} = 1 - g/N$.

We compare the data in Fig. 7.1 to a model that assumes a certain evolution of the values of τ_n as a function of the total conductance. In the simplest case, the conductance is due to only fully transmitted modes ($\tau_n = 1$) plus a single partially transmitted mode, which gives the minimum shot noise level $s_I^{\min}$ (full curve). The model illustrated in the inset gives a measure for the deviation from this ideal case in terms of the contribution x of other partially open channels; the corresponding behavior of the shot noise as a function of conductance is shown as the dashed curves in Fig. 7.1. This model has no physical basis but merely serves to illustrate the extent to which additional, partially open channels are required to describe the measured shot noise. For a more physical model fitting the data of Fig. 7.1 see [7].

We see that for $G < G_0$ the data are very close to the $x = 0\%$ curve, while for $G_0 < G < 2\,G_0$ the data are closer to the $x = 10\%$ curve. For $G > 2\,G_0$ the contribution of other partially open channels continues to grow.

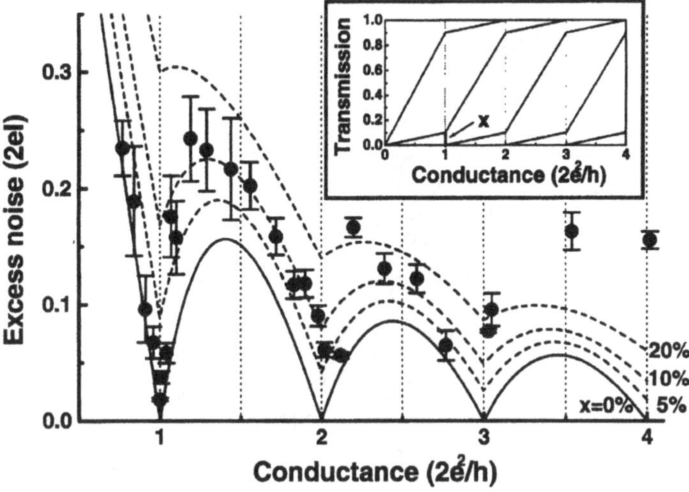

Figure 6. Measured shot noise values for gold (filled circles) with a bias current of 0.9 μA. Comparison is made with the noise power calculated assuming the channels open one-by-one (full curve). This sets the minimum shot noise level at each conductance value. The amount of admixture of additional channels can be estimated using an assumed evolution of channel transmission as illustrated in the inset (dashed curves). The model gives a measure for the deviation from the ideal case of channels opening one by one, by means of a fixed contribution $(1 - \tau_{n-1}) + \tau_{n+1} = x$ of the two neighboring modes. As an illustration the case of a $x = 10\%$ contribution from neighboring modes is shown. From [35].

For all points measured on the last conductance plateau before the transition to tunnelling, which is expected to consist of a single atom, the results are well-described by a single conductance channel, in agreement with the fact that gold has only a single valence orbital. For larger contacts there is a distinct tendency to open the channels on after the other, which has been referred to as the saturation of channel transmission [20].

7.2. ALUMINUM CONTACTS

When the experiment is repeated for aluminum contacts, a different behavior is observed. For contacts between 0.8 G_0 and 2.5 G_0 the shot noise values vary from 0.3 to 0.6 $(2eI)$, which is much higher than for gold (see Fig. 7.2). A systematic dependence of the shot noise power on the conductance seems to be absent. From the two measured parameters, the conductance, G, and the shot noise, S_I, one cannot determine the full set of transmission probabilities. However, the shot noise values found for aluminum, especially the ones at conductance values close to G_0, agree with Eq. (3) only if we assume that more than one mode is transmitted. The maximum shot noise $s_I^{max} = 1 - g/N$ that can be generated by two, three or

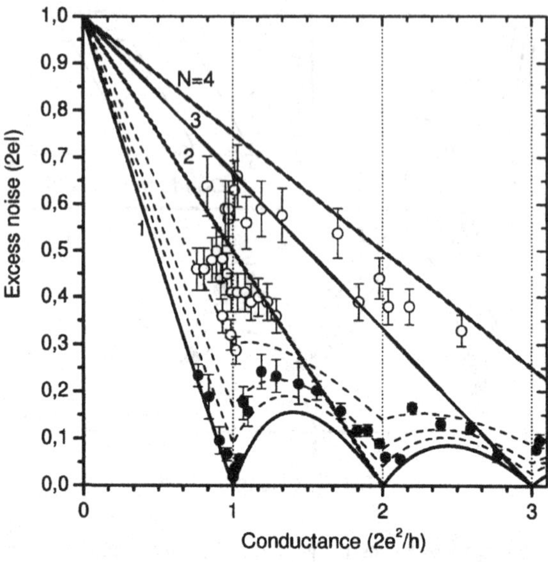

Figure 7. Measured shot noise values for aluminum (open circles), compared to the data from Fig. 7.1 (full circles), recorded at 4.2 K with a bias current of 0.9 μA. Comparison is made with the maximum shot noise that can be produced by N modes (gray curves), as explained in the text. The minimum shot noise is given by the full black curve. In the limit of zero conductance the theoretical curves all converge to full shot noise, as expected for the tunnelling regime. From [33].

four modes as a function of conductance is plotted as the gray curves in Fig. 7.2; the minimum shot noise in all cases is given by the full black curve. Hence, for a contact with shot noise higher than indicated by the gray N-mode line at least $N+1$ modes are contributing to the conductance. From this simple analysis we can see that for a considerable number of contacts with a conductance close to $1\,G_0$, the number of contributing modes is at least three. This is consistent with the number of modes expected based on the number of valence orbitals, and with results of the subgap structure analysis. Note that the points below the line labelled $N=2$ should not be interpreted as corresponding to two channels: the noise level observed requires at least two channels, but there may be three, or more.

A more detailed observation of the absence of a response in the shot noise to the ratio of the conductance and the quantum conductance unit, is obtained by following the evolution of the shot noise while stretching a single atom contact. For gold, the conductance of a single atom contact is nearly independent of the strain imposed on the contact. This is seen when measuring the conductance as a function of contact elongation, which decreases step-wise and drops to about $1\,G_0$ at the last plateau before jumping to tunnelling, and this last plateau has a nearly constant conductance, independent of contact elongation, over the length of the

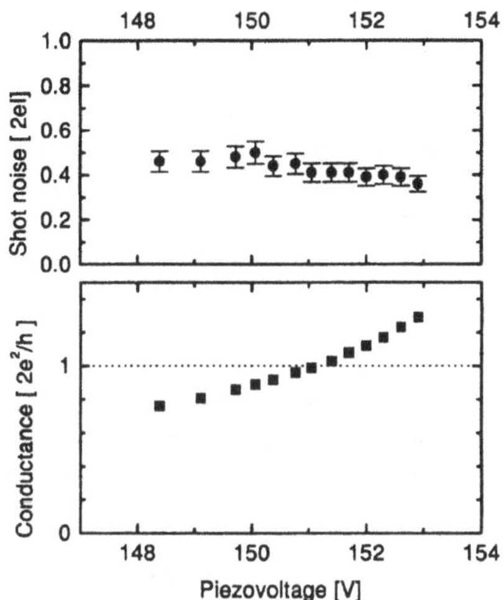

Figure 8. The conductance for an aluminum contact on the last plateau before the jump to tunnelling, which corresponds to a single atom contact (bottom panel). In contrast to gold, which shows a nearly constant conductance on the last plateau, the conductance increases on further stretching of the contact. Simultaneous measurements of the shot noise at each point (top panel) shows a nearly constant and fairly high level of shot noise. The measurements were taken at 4.2 K with a bias current of up to 0.9 μA. From [33].

plateau. In contrast, aluminum shows an anomalous *increase* in the conductance over the last 'plateau' [14], which starts at a conductance below 1 G_0, and gradually rises above it. Fig. 7.2 shows an example of such conductance behavior, where the shot noise was measured simultaneously at each point. The noise signal is high, and it shows no anomalous response at the crossing of the unit conductance value. This is consistent with a conductance being the result of several partially transmitted channels.

7.3. LARGER CONTACTS

The measurements on gold contacts, discussed above are performed on the smallest contacts only. We already noted that the contribution of partially open channels grows with increasing conductance. This raises the question, what happens at higher conductance values. For long conductors a crossover is expected to the diffusive regime, where the shot noise suppression is 1/3 [5], or to the interacting hot-electron regime, where the shot noise grows to $\sqrt{3}/4\,(2eI)$ [31]. In the limit of a macroscopic conductor, shot noise is absent because of inelastic scattering [5].

Experiments for gold contacts larger than the ones discussed above are shown in Fig. 7.3. Nearly all shot noise data for contacts above $G = 3\,G_0$ are in the range

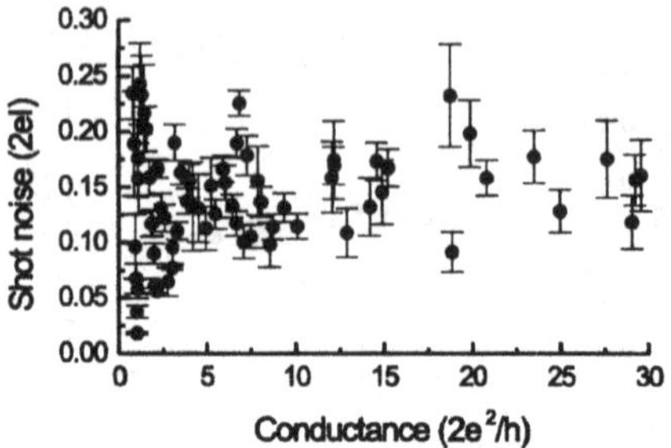

Figure 9. Measured shot noise values for larger gold contacts at 4.2 K. The gold data from Fig. 7.1, with conductance below 3 G_0, are also shown. Note that above this value, nearly all data are in the range from 0.10 – 0.20 ($2eI$), without an apparent systematic dependence on the conductance.

from 0.10 – 0.20 ($2eI$), without a systematic dependence on the conductance. This means that the number of partially transmitted modes continues to grow with contact size. This can be seen by realizing that according to Eq. 4 the single mode contribution to the shot noise decreases with increasing conductance. We are not aware of any universal predictions for the shot noise in large ballistic contacts, but the experiment suggests a nearly constant value of $s_I = 0.15 \pm 0.05$ applies for gold contacts. More experiments are needed to verify whether this depends on the type of metal being studied.

8. Shot noise and the 'mesoscopic PIN code'

More recently, shot noise measurements by Cron *et al.* [8] have provided a very stringent experimental test of the multichannel character of the electrical conduction in Al. In these experiments the set of transmissions τ_n are first determined independently by the technique of fitting the subgap structure in the superconducting state, discussed in Sect. 6. In words of the authors of Ref. [8], these coefficients constitute the 'mesoscopic PIN code' of a given contact. The knowledge of this code allows a direct quantitative comparison of the experimental results on the shot noise with the theoretically expected signal for the given set of transmission values, and is a critical test for the validity of the analysis. The experiments were done using Al nanofabricated break junctions, which exhibit a large mechanical stability. The superconducting IV curves for the smallest contacts were measured well below T_c, which is about 1 K and then a magnetic field of 50 mT was applied in order to switch into the normal state. The measured voltage dependence of S_I is shown in Fig. 8a for a typical contact in the normal state at three different temperatures, together with the predicted behavior based on the the mesoscopic pin code τ_n that was measured independently. The noise measured at the lowest temperature for four contacts having different sets of transmission coefficients is

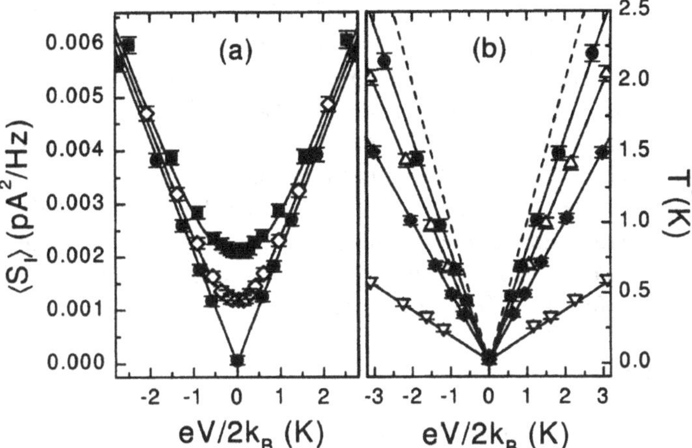

Figure 10. (a) Symbols: measured average current noise power density $\langle S_I \rangle$ and noise temperature T^*, defined as $T^* = S_I/4k_BG$, as a function of reduced voltage, for a contact in the normal state at three different temperatures (from bottom to top: 20, 428, 765 mK). The solid lines are the predictions of Eq. (3) for the set of transmissions $\{0.21,0.20,0.20\}$ measured independently from the IV in the superconducting state. (b) Symbols: measured effective noise temperature T^* versus reduced voltage for four different contacts in the normal state at $T = 20$ mK. The solid lines are predictions of Eq. (3) for the corresponding set of transmissions (from top to bottom: $\{0.21,0.20,0.20\}$, $\{0.40,0.27,0.03\}$, $\{0.68,0.25,0.22\}$, $\{0.996,0.26\}$. The dashed line is the Poisson limit. From [8].

shown in Fig. 8b, together with the predictions of the theory. This excellent agreement between theory and experiments provides an unambiguous demonstration of the presence of several conduction channels in the smallest Al contacts and serves as a test of the accuracy that can be obtain in the determination of the τ's from the subgap structure in the superconducting IV curve. The experimental results for shot noise in the superconducting state will be discussed in a separate chapter by Levy Yeyati *et al.*

9. Conclusion

The results obtained from superconducting subgap structure [28], shot noise measurements [34], conductance fluctuations [20] and thermopower [22] are in agreement and show that the conductance for single-atom contacts of monovalent metals (gold) is carried by a single mode. For increasing contact size it is found that the transmission for the first mode goes to unity, before the second mode opens, then the second mode tends to go fully open, before the third opens, and so on. This property of atomic-sized contacts has been described as 'saturation of the channel transmission' and holds to very good approximation (a few percent) up to $G = 1G_0$, but deviations increase to 20% admixture of the next channels at $G = 3G_0$. The fact that the shot noise for aluminum near $G = 1G_0$ cannot be described by transport through a single conductance channel, in contrast to gold, is fully consistent with recent model calculations for single atom junctions [9, 19, 37]. A saturation of channel transmission is not observed for Al, and is expected to be absent in all metals other than simple s-metals. By combining a determination of the 'mesoscopic PIN code' from the analysis of the supgap structure and measurement of the shot noise power on the same single-atom contact [8] it has been possible to demonstrate quantitative agreement, which strongly reinforces our confidence in both procedures.

Acknowledgments

This brief review is partly based a larger review that I recently finished together with Nicolas Agraït and Alfredo Levy Yeyati [2], and I thank them for their permission to use part of the material.

References

1. Agraït, N., J. Rodrigo, and S. Vieira: 1993, 'Conductance steps and quantization in atomic-size contacts'. *Phys. Rev. B* **47**, 12345–12348.
2. Agraït, N., A. L. Yeyati, and J. van Ruitenbeek: 2002, 'Quantum properties of atomic-sized conductors'. *Phys. Rep.* submitted; preprint http://xxx.lanl.gov/abs/cond-mat/0208239.
3. Arnold, G.: 1987, 'Superconducting tunneling without the tunneling Hamiltonian. II. Subgap harmonic structure'. *J. Low Temp. Phys.* **68**, 1–27.
4. Averin, D. and A. Bardas: 1995, 'ac Josephson effect in a single quantum channel'. *Phys. Rev. Lett.* **75**, 1831–1834.
5. Beenakker, C. and M. Büttiker: 1992, 'Suppression of shot noise in metallic diffusive conductors'. *Phys. Rev. B* **46**, 1889–1892.
6. Bratus', E., V. Shumeiko, E. Bezuglyi, and G. Wendin: 1997, 'dc-current transport and ac Josephson effect in quantum junctions at low voltage'. *Phys. Rev. B* **55**, 12666–12677.
7. Bürki, J. and C. Stafford: 1999, 'Comment on "Quantum suppression of shot noise in atom-size metallic contacts"'. *Phys. Rev. Lett.* **83**, 3342.
8. Cron, R., M. Goffman, D. Esteve, and C. Urbina: 2001, 'Multiple-charge-quanta shot noise in superconducting atomic contacts'. *Phys. Rev. Lett.* **86**, 4104–4107.
9. Cuevas, J., A. Levy Yeyati, and A. Martín-Rodero: 1998, 'Microscopic origin of conducting channels in metallic atomic-size contacts'. *Phys. Rev. Lett.* **80**, 1066–1069.
10. Cuevas, J., A. Martín-Rodero, and A. Levy Yeyati: 1996, 'Hamiltonian approach to the transport properties of superconducting quantum point contacts'. *Phys. Rev. B* **54**, 7366–7379.
11. Gimzewski, J. and R. Möller: 1987, 'Transition from the tunneling regime to point contact studied using scanning tunneling microscopy'. *Physica B* **36**, 1284–1287.
12. Khlus, V.: 1989, 'Excess quantum noise in 2D ballistic point contacts'. *Sov. Phys. JETP* **66**, 592. [Zh. Eksp. Teor. Fiz.**93** (1987) 2179].
13. Klapwijk, T., G. Blonder, and M. Tinkham: 1982, 'Explanation of subharmonic energy gap structure in superconducting contacts'. *Physica B* **109 & 110**, 1657–1664.
14. Krans, J., C. Muller, I. Yanson, T. Govaert, R. Hesper, and J. van Ruitenbeek: 1993, 'One-atom point contacts'. *Phys. Rev. B* **48**, 14721–14274.

15. Krans, J. M.: 1996, 'Size effects in atomic-scale point contacts'. Ph.D. thesis, Universiteit Leiden, The Netherlands.
16. Kumar, A., L. Saminadayar, D. Glattli, Y. Jin, and B. Etienne: 1996, 'Experimental test of the quantum shot noise reduction theory'. *Phys. Rev. Lett.* **76**, 2778–2781.
17. Landauer, R.: 1970, 'Electrical resistance of disordered one-dimensional lattices'. *Phil. Mag.* **21**, 863–867.
18. Landman, U., W. Luedtke, N. Burnham, and R. Colton: 1990, 'Atomistic mechanisms and dynamics of adhesion, nanoindentation, and fracture'. *Science* **248**, 454–461.
19. Lang, N.: 1995, 'Resistance of atomic wires'. *Phys. Rev. B* **52**, 5335–5342.
20. Ludoph, B., M. Devoret, D. Esteve, C. Urbina, and J. van Ruitenbeek: 1999, 'Evidence for saturation of channel transmission from conductance fluctuations in atomic-size point contacts'. *Phys. Rev. Lett.* **82**, 1530–1533.
21. Ludoph, B., N. van der Post, E. Bratus', E. Bezuglyi, V. Shumeiko, G. Wendin, and J. van Ruitenbeek: 2000, 'Multiple Andreev reflection in single atom niobium junctions'. *Phys. Rev. B* **61**, 8561–8569.
22. Ludoph, B. and J. van Ruitenbeek: 1999, 'Thermopower of atomic-size metallic contacts'. *Phys. Rev. B* **59**, 12290–12293.
23. Muller, C., J. van Ruitenbeek, and L. de Jongh: 1992, 'Conductance and supercurrent discontinuities in atomic-scale metallic constrictions of variable width'. *Phys. Rev. Lett.* **69**, 140–143.
24. Olesen, L., E. Lægsgaard, I. Stensgaard, F. Besenbacher, J. Schiøtz, P. Stoltze, K. Jacobsen, and J. Nørskov: 1994, 'Quantised Conductance in an Atom-Sized Point Contact'. *Phys. Rev. Lett.* **72**, 2251–2254.
25. Pascual, J., J. Méndez, J. Gómez-Herrero, A. Baró, N. García, and V. T. Binh: 1993, 'Quantum contact in gold nanostructures by scanning tunneling microscopy'. *Phys. Rev. Lett.* **71**, 1852–1855.
26. Reznikov, M., M. Heiblum, H. Shtrikman, and D. Mahalu: 1995, 'Temporal correlation of electrons: suppression of shot noise in a ballistic quantum point contact'. *Phys. Rev. Lett.* **75**, 3340–3343.
27. Rubio, G., N. Agraït, and S. Vieira: 1996, 'Atomic-sized metallic contacts: Mechanical properties and electronic transport'. *Phys. Rev. Lett.* **76**, 2302–2305.
28. Scheer, E., N. Agraït, J. Cuevas, A. Levy Yeyati, B. Ludoph, A. Martín-Rodero, G. Rubio Bollinger, J. van Ruitenbeek, and C. Urbina: 1998, 'The signature of chemical valence in the electrical conduction through a single-atom contact'. *Nature* **394**, 154–157.
29. Scheer, E., P. Joyez, D. Esteve, C. Urbina, and M. Devoret: 1997, 'Conduction channel transmissions of atomic-size aluminum contacts'. *Phys. Rev. Lett.* **78**, 3535–3538.
30. Schrieffer, J. and J. Wilkins: 1963, 'Two-particle tunneling processes between superconductors'. *Phys. Rev. Lett.* **10**, 17–20.
31. Steinbach, A., J. Martinis, and M. Devoret: 1996, 'Observation of Hot-Electron Shot Noise in a Metallic Resistor'. *Phys. Rev. Lett.* **76**, 3806–3809.
32. Sutton, A. and J. Pethica: 1990, 'Inelastic flow processes in nanometre volumes of solids'. *J. Phys.: Condens. Matter* **2**, 5317–5326.
33. van den Brom, H.: 2000, 'Fluctuation phenomena in atomic-size contacts'. Ph.D. thesis, Universiteit Leiden, The Netherlands.
34. van den Brom, H. and J. van Ruitenbeek: 1999, 'Quantum suppression of shot noise in atom-size metallic contacts'. *Phys. Rev. Lett.* **82**, 1526–1529.
35. van den Brom, H. and J. van Ruitenbeek: 2000, 'Shot noise suppression in metallic quantum point contacts'. In: D. Reguera, G. Platero, L. Bonilla, and J. Rubí (eds.): *Statistical and Dynamical Aspects of Mesoscopic Systems*. Berlin Heidelberg, pp. 114–122.
36. van der Post, N., E. Peters, I. Yanson, and J. van Ruitenbeek: 1994, 'Subgap Structure as

Function of the Barrier in Atom-Size Superconducting Tunnel Junctions'. *Phys. Rev. Lett.* **73**, 2611–2613.

37. Wan, C., J.-L. Mozos, G. Taraschi, J. Wang, and H. Guo: 1997, 'Quantum transport through atomic wires'. *Appl. Phys. Lett.* **71**, 419–421.

QUANTUM NOISE AND MUTIPLE ANDREEV REFLECTIONS IN SUPERCONDUCTING CONTACTS

A. MARTÍN-RODERO, J. C. CUEVAS* AND A. LEVY YEYATI
Departamento de Física Teórica de la Materia Condensada
Universidad Autónoma de Madrid, E-28049 Madrid, Spain

R. CRON, M.F. GOFFMAN, D. ESTEVE AND C. URBINA
Service de Physique de l'Etat Condensé, Commissariat à
u'Energie Atomique, Saclay, F-91191 Gif-sur-Yvette Cedex, France

Abstract. The mechanism of multiple Andreev reflections (MAR) leads to a rather complex behavior of the noise spectral density in superconducting quantum point contacts as function of the relevant parameters. In this contribution we analyze recent theoretical and experimental efforts which have permitted to clarify this issue to a great extent. The theoretical description of noise in the coherent MAR regime will be summarized, discussing its main predictions for equilibrium and non-equilibrium current fluctuations. We then analyze noise measurements in well characterized superconducting atomic contacts. These systems allow for a direct test of the theoretical predictions without fitting parameters. In particular, the increase of the effective charge corresponding to the openning of higher order Andreev channels has been verified.

1. Introduction

Non-equilibrium current fluctuations provide a powerful probe of the transport mechanisms in mesoscopic structures. Indeed, in contrast to the universal equilibrium *thermal noise*, even the low-frequency power spectrum of these nonequilibrium fluctuations, or "shot-noise", contains a wealth of information on the interactions and quantum correlations between electrons [1]. When the current I is made up from independent shots, the low frequency spectrum acquires the well-known Poissonian form $S = 2qI$, where q is the "effective charge" transferred at each shot. This result was derived by Schottky as early as in 1918 for a vacuum diode [2]. In the case of normal, i.e. nonsuperconducting, metallic reservoirs, the charge of the shots is simply the electron charge e. Interactions and correlations lead to large deviations from this value. One of the most striking examples is the fractional charge of quasiparticles in the fractional quantum Hall regime, as it was evidenced through noise measurements [3]. The mechanism giving rise

Y.V. Nazarov (ed.), Quantum Noise in Mesoscopic Physics, 51–71.

to superconductivity is another source of correlations among electrons. How big are the shots when superconducting electrodes are involved? And more generally, what can we learn from the analysis of noise in superconducting nanostructures? These are the central questions that we shall address in this contribution.

Noise phenomena when superconductors are involved is of particular interest due to the peculiar nature of the charge transfer mechanisms arising from the presence of a Cooper pair condensate. Thus, for instance, the current between a superconducting reservoir and a normal one connected by a short normal wire proceeds through the process of Andreev reflection in which charge is transferred in shots of $2e$, thus resulting in a doubling of the noise with respect to the normal case [4]. When two superconducting electrodes connected through structures such as tunnel junctions or short weak links are voltage biased on an energy scale eV smaller than the superconducting gap Δ, the current proceeds through multiple Andreev reflections (MAR) [5]. Fig. 1 illustrates the lowest order processes contributing to quasiparticle transport in a superconducting junction. In a n-order MAR process, which has a threshold voltage of $eV = 2\Delta/n$, an electron (or hole) is created on the right (left) electrode after $n-1$ Andreev reflections, the total charge transmitted in the whole process being ne. For a given voltage many such processes can contribute to the current, but roughly speaking, "giant" shots, with an effective charge $q \sim e(1 + 2\Delta/eV)$ are expected at subgap energies. Of course, the exact value of q, like all other transport properties of a coherent nanostructure, depends on its "mesoscopic pin code", i.e. the set of transmission coefficients τ_i characterizing its conduction channels.

Although the MAR mechanism in superconducting junctions was first introduced nearly two decades ago [5], a complete understanding of transport in the coherent MAR regime has only recently been achieved due to the combination of theoretical and experimental developments. On the theoretical side, fully quantum mechanical calculations have allowed to obtain detailed quantitative predictions for different transport properties [6–9, 11, 10, 12, 13].

At the same time, the development of atomic size contacts by means of break junction and STM techniques [14–21], has opened the possibility of a direct comparison between theory and experiments [17–21]. These systems are caracterised by a few conduction channels whose transmission coefficients can be determined experimentally with great accuracy [17–19].

In the present paper we shall review recent advances in the understanding of noise phenomena in nano-scale superconducting devices. In the first section we summarize the theoretical description of noise in the coherent MAR regime. An approach based on non-equilibrium Green functions techniques will be discussed with some detail. We then present the main theoretical predictions for equilibrium and non-equlibrium current fluctuations. The experimental studies of noise in SNS nanostructures are reviewed in Sect. III. We shall mainly concentrate in the discussion of noise measurements using well characterized superconducting atomic contacts. Finally, in Sect. IV we present our concluding remarks.

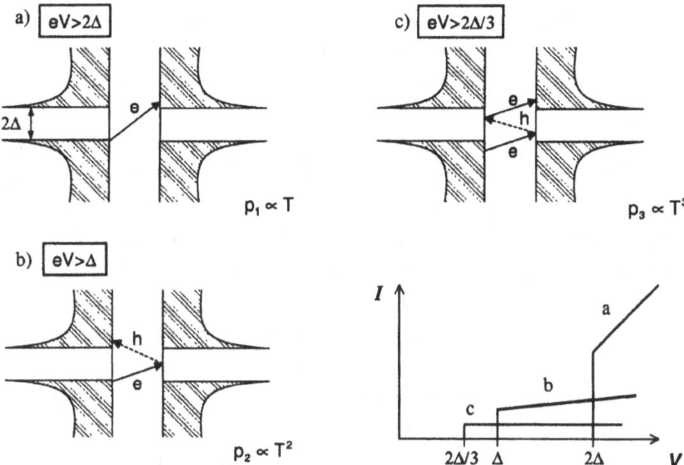

Figure 1. Schematic representation of the MAR processes. We have represented the density of states of both electrodes. The transmission probability is is denoted by T in the figure. In the left upper panel we describe the single-quasiparticle process in which an electron tunnels through the system overcoming the superconducting gap due to a voltage $eV \geq 2\Delta$. The lower left panel shows an Andreev reflection in which an electron is reflected as a hole, which can reach an empty state in the left electrode if the voltage is $eV \geq \Delta$. During this process, which for small transmission has a probability T^2, two electron charges are transferred as a Cooper pair from left to right. The upper right pannel shows an Andreev reflection of order 3 in which an Andreev reflected hole is still Andreev reflected as an electron, which finally reaches an empty state in the right electrode. In this process, whose threshold voltage is $eV = 2\Delta/3$, three charges are transferred with a probablity T^3. The contribution of these processes to the dc current is schematically depicted in the lower right pannel.

2. Theoretical description of noise in the coherent MAR regime

The MAR mechanism was introduced by Klapwijk *et al.* [5] to explain the subgap structure observed in different types of superconducting junctions. That first approach was based on semi-classical arguments which neglected quantum interference between different processes. An earlier microscopic theory due to Schrieffer and Wilkins [22] attempted to introduce the contribution of multiparticle processes like the ones depicted in Fig. 1 by means of lowest order perturbation theory in the tunnel Hamiltonian coupling two superconducting leads. In spite of being a fully quantum mechanical approach, this Multiparticle Tunneling Theory (MPT), is plagued with divergencies which can only be avoided by carrying out the calculation up to infinite order. This goal was achieved more recenlty by the so called Hamiltonian approach [9], based on a Green functions formalism, which demonstrated the essential equivalence of the MPT and MAR arguments.

Recent calculations in the coherent MAR regime have been based on two different approaches: the Hamiltonian approach mentioned above and the scattering approach developed in Refs. [7] and [8]. Although both approaches yield equivalent results in the limit when the energy dependence of the normal transmission can be neglected, each one has its own advantages. Thus, although the scattering approach might appear as conceptualy simpler, the Hamiltonian approach is more rigorous as it does not rely on ad-hoc assumptions on boundary conditions. In addition the use of Green function techniques allows to deal, for instance, with electron correlation effects within this approach. In what follows we briefly sketch the basic ingredients in the Hamiltonian approach for the description of a voltage biased superconducting contact. We shall consider the case of a superconducting quantum point-contact, i.e. a short ($L \ll \xi_0$) mesoscopic constriction between two superconducting electrodes with a constant applied bias voltage V. For the range $eV \sim \Delta$ one can neglect the energy dependence of the transmission coefficients and all transport properties can be expressed as a superposition of independent channel contributions. Thus, we will concentrate in analyzing a single channel model which can be described by the following Hamiltonian [9]

$$\hat{H}(t) = \hat{H}_L + \hat{H}_R + \sum_{\sigma} \left(v e^{i\phi(t)/2} c^{\dagger}_{L\sigma} c_{R\sigma} + \ h.c. \right), \tag{1}$$

where $\hat{H}_{L,R}$ are the BCS Hamiltonians for the left and right uncoupled electrodes, $\phi(t) = \phi_0 + 2eVt/\hbar$ is the time-dependent superconducting phase difference, which after a gauge transformation enters as a phase factor in the hopping terms describing electron transfer between both electrodes. To obtain the transport properties we use a perturbative Green functions approach including processes up to infinite order in the hopping parameter v (all MAR processes of arbitrary order are in this way naturally included). In the normal state, the one-channel contact is characterized by its transmission coefficient τ, which as a function of v adopts the form $\tau = 4(v/W)^2/(1+(v/W)^2)^2$, where $W = 1/\pi\rho_F$, ρ_F being the electrodes density of states at the Femi energy [9]. Within this model the current operator is given by

$$\begin{aligned} \hat{I}(t) \ &= \ \frac{ie}{\hbar} \sum_{\sigma} \left[v e^{i\phi(t)/2} c^{\dagger}_{L\sigma}(t) c_{R\sigma}(t) \right. \\ &\qquad \left. - v^* e^{-i\phi(t)/2} c^{\dagger}_{R\sigma}(t) c_{L\sigma}(t) \right] . \end{aligned} \tag{2}$$

The current noise spectral density is defined as

$$\begin{aligned} S(\omega, t) \ &= \ \hbar \int dt' \ e^{i\omega t'} \langle \delta\hat{I}(t+t')\delta\hat{I}(t) + \delta\hat{I}(t)\delta\hat{I}(t+t') \rangle \\ &\equiv \ \hbar \int dt' \ e^{i\omega t'} \ K(t, t'), \end{aligned} \tag{3}$$

where $\delta\hat{I}(t) = \hat{I}(t) - \langle \hat{I}(t) \rangle$ is the time-dependent fluctuations in the current.

The relevant quantities to be determined can be expressed in terms of non-equilibrium Keldysh Green functions [23] in a Nambu representation $\hat{G}^{+-}_{ij}(t,t')$ and $\hat{G}^{-+}_{ij}(t,t')$ where $i,j \equiv L,R$ defined as

$$\hat{G}^{+-}_{i,j}(t,t') = i \begin{pmatrix} \langle c^{\dagger}_{j\uparrow}(t')c_{i\uparrow}(t)\rangle & \langle c_{j\downarrow}(t')c_{i\uparrow}(t)\rangle \\ \langle c^{\dagger}_{j\uparrow}(t')c^{\dagger}_{i\downarrow}(t)\rangle & \langle c_{j\downarrow}(t')c^{\dagger}_{i\downarrow}(t)\rangle \end{pmatrix}, \tag{4}$$

and obey the relation $\hat{G}^{-+}_{i,j}(t,t') = -\hat{\sigma}_x \left[\hat{G}^{+-}_{j,i}(t',t)\right]^T \hat{\sigma}_x$, $\hat{\sigma}_x$ being the Pauli matrix.

Then, the mean current and the kernel $K(t,t')$ in the noise spectral density are given by

$$\begin{aligned} < \hat{I}(t) > &= \frac{e}{\hbar} \mathrm{Tr}\left[\hat{\sigma}_z \left(\hat{v}(t)\hat{G}^{+-}_{RL}(t,t) - \hat{v}^{\dagger}(t)\hat{G}^{+-}_{LR}(t,t)\right)\right] \\ K(t,t') &= \frac{2e^2}{\hbar^2}\Big\{\mathrm{Tr}\Big[\hat{\sigma}_z\hat{v}^{\dagger}(t)\hat{G}^{+,-}_{LL}(t,t')\hat{\sigma}_z\hat{v}(t')\hat{G}^{-,+}_{RR}(t',t) - \\ &\quad \hat{\sigma}_z\hat{v}^{\dagger}(t)\hat{G}^{+,-}_{LR}(t,t')\hat{\sigma}_z\hat{v}^{\dagger}(t')\hat{G}^{-,+}_{LR}(t',t)\Big] + (t \to t')\Big\}, \end{aligned} \tag{5}$$

where $\hat{\sigma}_z$ is the Pauli matrix, Tr denotes the trace in the Nambu space and $\hat{v}$ is the hopping in this representation

$$\hat{v}(t) = \begin{pmatrix} v e^{i\phi(t)/2} & 0 \\ 0 & -v^{*}e^{-i\phi(t)/2} \end{pmatrix}. \tag{6}$$

The expression of the Kernel in terms of one-particle Green functions (Eq. 5) has been obtained using Wick's theorem. This is equivalent to neglect correlations beyond the BCS mean-field theory. As mentioned above, in order to determine the Green functions we follow a perturbative scheme and treat the coupling term in Hamiltonian (1) as a perturbation. The unperturbed Green functions, $\hat{g}$, correspond to the uncoupled electrodes in equilibrium. Thus, the retarded and advanced components adopt the BCS form: $\hat{g}^{r,a}(\epsilon) = g^{r,a}(\epsilon)\hat{1} + f^{r,a}(\epsilon)\hat{\sigma}_x$, where $g^{r,a}(\epsilon) = -(\epsilon^{r,a}/\Delta)f(\epsilon) = -\epsilon^{r,a}/W\sqrt{\Delta^2 - (\epsilon^{r,a})^2}$, where $\epsilon^{r,a} = \epsilon \pm i\eta$, and η is a small energy relaxation rate that takes into account the damping of quasiparticle states due to inelastic processes inside the electrodes. This parameter can be estimated from the electron-phonon interaction to be a small fraction of Δ [24]. To determine the Keldysh Green functions appearing in the current and noise expression we first use their relation with the advanced and retarded functions $G^{a,r}$,

$$\hat{G}^{+,-}(t,t') = \left(\hat{1} + \hat{G}^r \otimes \hat{\Sigma}^r\right) \otimes \hat{g}^{+,-} \otimes \left(\hat{1} + \hat{\Sigma}^a \otimes \hat{G}^a\right), \tag{7}$$

where the $\otimes$ product stands for an integration over the common time variable. The self-energy in this problem is simply given by $\hat{\Sigma}^{r,a}_{LL} = \hat{\Sigma}^{r,a}_{RR} = 0$ and $\hat{\Sigma}^{r,a}_{LR} =$

$(\hat{\Sigma}^{r,a}_{RL})^\dagger = \hat{v}(t)$. The unperturbed Green function $\hat{g}^{+,-}$ are given by $\hat{g}^{+,-}(\epsilon) = [\hat{g}^a(\epsilon) - \hat{g}^r(\epsilon)] f(\epsilon)$, where $f(\epsilon)$ is the Fermi function. Finally, the functions $G^{r,a}$ satisfy the Dyson equations

$$\hat{G}^{r,a}(t,t') = \hat{g}^{r,a} + \hat{g}^{r,a} \otimes \hat{\Sigma}^{r,a} \otimes \hat{G}^{r,a}. \tag{8}$$

In order to solve the above integral equations it is convenient to work in the energy space. Thus, we Fourier transform the Green functions with respect to the temporal arguments

$$\hat{G}(t,t') = \frac{1}{2\pi} \int d\epsilon \int d\epsilon' \; \mathrm{e}^{-i\epsilon t} \mathrm{e}^{i\epsilon' t'} \hat{G}(\epsilon,\epsilon'). \tag{9}$$

Due to the special time dependence of the coupling elements (see Eq. 6), every Green function admits a Fourier expansion of the form

$$\hat{G}(t,t') = \sum_n \mathrm{e}^{in\phi(t')/2} \int \frac{d\epsilon}{2\pi} \mathrm{e}^{-i\epsilon(t-t')} \hat{G}(\epsilon, \epsilon + neV), \tag{10}$$

which, in other words, means that $\hat{G}(\epsilon,\epsilon') = \sum_n \hat{G}(\epsilon,\epsilon+neV)\delta(\epsilon-\epsilon'+neV)$. Thus, the calculation of the different transport properties is reduced to the determination of the Fourier components $\hat{G}^{r,a}_{nm}(\epsilon) \equiv \hat{G}^{r,a}(\epsilon+neV, \epsilon+meV)$. Eq. 10 indicates that the different transport properties of this system should contain terms oscillating with all the harmonic of the Josephson frequency. In particular, $S(\omega,t) = \sum_n S_n(\omega) \exp[in\phi(t)]$. At finite bias voltage, we shall concentrate ourselves in the dc part of the noise, i.e. S_0, which for simplicity will be denoted as $S(\omega)$. This noise component can be expressed in terms of the Fourier components of the Green functions as follows

$$\begin{aligned} S(\omega) \; = \; & \frac{2e^2}{h} \sum_{\pm\omega,n} \int d\epsilon \; \mathrm{Tr} \left\{ \hat{\sigma}_z \hat{D}^{+,-}_{RL,0n}(\epsilon \pm \omega) \hat{\sigma}_z \hat{D}^{-,+}_{LR,n0}(\epsilon) - \right. \\ & \left. \hat{\sigma}_z \hat{D}^{+,-}_{RR,0n}(\epsilon \pm \omega) \hat{\sigma}_z \hat{D}^{-,+}_{RR,n0}(\epsilon) \right\}, \end{aligned} \tag{11}$$

where we have defined $\hat{D}(t,t') \equiv \hat{v}(t)\hat{G}(t,t')$ to simplify the notation. The last step in the calculation is the computation of the Fourier components $\hat{G}^{r,a}_{nm}$.. They can be determined by Fourier transforming Eq. 8 and using the relation of Eq. 10. Thus for instance, it can be shown that the components $\hat{G}^{r,a}_{RR,nm} \equiv \hat{G}_{nm}$ fulfill the following algebraic linear equation (assuming that the contact is symmetric)

$$\hat{G}_{nm} = \hat{g}_{nm}\delta_{n,m} + \hat{\mathcal{E}}_{nn}\hat{G}_{nm} + \hat{\mathcal{V}}_{n,n-2}\hat{G}_{n-2,m} + \hat{\mathcal{V}}_{n,n+2}\hat{G}_{n+2,m}, \tag{12}$$

where the matrix coefficients $\hat{\mathcal{E}}_{nn}$ and $\hat{\mathcal{V}}_{n,m}$ can be expressed in terms of the Green's functions of the uncoupled electrodes, as

$$\hat{\mathcal{E}}_n \; = \; v^2 \begin{pmatrix} g_n g_{n-1} & f_n g_{n+1} \\ f_n g_{n-1} & g_n g_{n+1} \end{pmatrix}$$

$$\hat{\mathcal{V}}_{n,n+2} = -v^2 f_{n+1} \begin{pmatrix} f_n & 0 \\ g_n & 0 \end{pmatrix}$$
$$\hat{\mathcal{V}}_{n,n-2} = -v^2 f_{n-1} \begin{pmatrix} 0 & g_n \\ 0 & f_n \end{pmatrix}, \quad (13)$$

In these equations the notation $g_n = g_i(\epsilon + neV)$ is used. Notice that this set of linear equations is analogous to those describing a tight-binding chain with nearest-neighbor hopping parameters $\hat{\mathcal{V}}_{n,n+2}$ and $\hat{\mathcal{V}}_{n,n-2}$. A solution can then be obtained by standard recursive techniques (see Ref. [9] for details), which permits to obtain analytical results in some limits, and in any case an efficient numerical evaluation of the Fourier components. In the case of zero bias, all the harmonics of noise give a finite contribution. In this limit the calculation can be greatly simplify noticing that the problem becomes stationary (the time-dependent Green functions only depend on the time difference) and using the equilibrium relations

$$\hat{G}^{+,-}(\epsilon) = \left[\hat{G}^a(\epsilon) - \hat{G}^r(\epsilon)\right] f(\epsilon), \quad (14)$$
$$\hat{G}^{-,+}(\epsilon) = \left[\hat{G}^a(\epsilon) - \hat{G}^r(\epsilon)\right] [f(\epsilon) - 1]. \quad (15)$$

In this limit one can obtain analytical results, as we shall detail in the next section.

3. Theoretical results

3.1. THERMAL NOISE

In the limit of vanishing bias voltage the current is due to Cooper pair tunneling (non-dissipative Josephson current). One would naively expect that thermal noise at temperatures $k_B T \ll \Delta$ would be negligible. This is certainly the case for a tunnel junction with a exponentially small barrier transparency. However, in a superconducting point contact the situation can be radically different due to the presence of the Andreev states inside the gap. Fluctuations in the population of these states can lead to a huge increase of the noise in certain conditions. The noise spectral density in this regime was calculated in Ref. [11] using the formalism discussed in the previous section. The spectrum at subgap frequencies can be understood in terms of a simple two-level model describing the Andreev states. Let us recall that for a single channel contact of transmission τ there are two bound states at energies $\pm\epsilon_S$, where $\epsilon_S = \Delta\sqrt{1 - \tau \sin^2 \phi/2}$. The zero temperature supercurrent is just given by $I(\phi) = -(2e/\hbar)\partial_\phi \epsilon_S$. In such a two-level system at finite temperature the upper level can be thermally populated giving rise to a reverse in the sign of the supercurrent. It is important to notice that in a real system the Andreev bound states are affected by a long but finite life-time fixed by the typical inelastic tunneling rate of the system $\eta \ll \Delta$ [24]. The current fluctuations

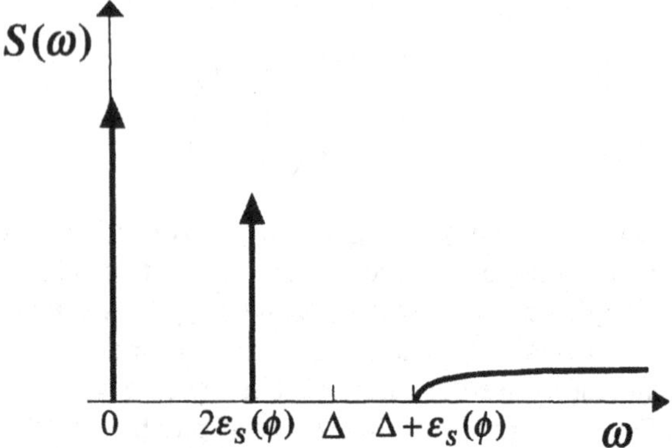

Figure 2. Schematic representation of the thermal noise spectrum. Notice the discrete character of the spectrum for $\omega < \epsilon_S + \Delta$. Only two sharp resonances (width $\sim \eta$) at $\omega = 0$ and $\omega = 2\epsilon_S$ appear due to the contribution of the subgap states.

thus correspond to a type of telegraph noise in which the system switches between positive and negative current with a characteristic time given by $\hbar/\eta$ [10].

As the gap between the Andreev states decreases with increasing contact transmission one could expect a large increase of the noise. In fact, the results of Ref. [11] show that the noise exhibits a huge increase when $\Delta\sqrt{1-\tau} \ll k_B T$. The exact result for the zero-frequency noise is found to be given by

$$S(0,\phi) = \frac{2e^2}{h}\frac{\pi}{\eta}\frac{\Delta^4\tau^2\sin^2(\phi)}{\epsilon_S^2} f(\epsilon_S)\left[1 - f(\epsilon_S)\right]. \tag{16}$$

¿From this expression one can verify that the thermal noise approaches its maximum value, $e^2\Delta^2/\hbar\eta$, when $\phi \to \pi$ and $\tau \to 1$, for any finite temperature. This result is strictly valid in the limit of small inelastic tunneling rate, $\eta \ll \tau\Delta$, and in the low transmission regime differ strongly from the noise spectrum that is obtained for tunnel junctions using standard tunnel theory [25], which yields $S(0) \sim \tau(1+\cos\phi)\ln(\Delta/\eta)$. As shown in [26], the reason for this discrepancy is that the limits $\tau \to 0$ and $\eta \to 0$ actually do not commute: when $\eta \ll \tau\Delta$ the main contribution to the noise comes from the MAR processes building up the Andreev states and should be taken into account up to infinite order; while for $\eta \gg \tau\Delta$ higher order MAR processes become heavily damped and the lowest order perturbation theory in the tunnel Hamiltonian gives the correct answer.

It is interesting to point out that the contact linear conductance, $G(\phi)$, can be obtain from Eq. (16) through the fluctuation dissipation theorem, which states that $S(0,\phi) = 4k_BTG(\phi)$. The expression of the linear conductance can also be obtained by a direct calculation of the current in the limit $V \to 0$ [26]. The full noise spectrum in the zero bias limit exhibits also an additional peak at $\omega = 2\epsilon_s$

associated with excitations from the lower to upper Andreev state. The weight of this peak is found to be given by [11]

$$S(2\epsilon_S) = \frac{2e^2}{h}\frac{\pi}{\eta}\frac{\Delta^4\tau^2(1-\tau)\sin^4(\phi)}{\epsilon_S^2}\left[f^2(\epsilon_S) + f^2(-\epsilon_S)\right] \tag{17}$$

It is worth noticing that in the zero temperature limit this is the only remaining subgap feature in the noise spectrum. This expression clearly shows that $S(2\epsilon_S)$ is proportional to the square of the zero temperature supercurrent with a Fano reduction factor $(1-\tau)$.

3.2. SHOT NOISE

When a finite bias voltage is applied, the current is due to quasiparticle tunneling mediated by MAR processes. In this section we shall concentrate in the analysis of the shot noise regime ($eV \gg k_BT$). As discussed in the introduction, shot noise in the Poissonian limit provides a measure of the charge of the quasiparticles being transmitted. The question then arises on whether this relation still holds for superconducting junctions in the coherent MAR regime. As we discuss below, the situation is far more complex in this case. Only in the low transmission regime one can clearly identify the charge which is being transmitted for a given voltage bias. In general, it is not possible to obtain a compact expression for the noise spectrum in the non-equilibrium situation. The zero-frequency noise for arbitrary transmission and voltage has been calculated in Refs. [12, 13]. The numerical results are summarized in Fig. 3, where we also show the dc current for comparison. As can be observed, the most prominent features in the shot noise are: (i) the presence of a strongly pronounced subgap ($V \leq 2\Delta$) structure, which remains up to transmissions close to one (in the dc current this structure is only pronounced for low transmissions). In the low transparency limit the shot noise subgap structure consists of a series of steps at voltages $eV_n = 2\Delta/n$ (n integer) as in the case of the dc current. (ii) The shot noise can be much large than the Poisson noise ($S_{Poisson} = 2eI$), as can be seen in Fig. 4. (iii) For higher transmissions there is a steep increase in the noise at low voltages. (iv) For perfect transmission the shot noise is greatly reduced. (v) In the large voltage limit ($eV \gg \Delta$) there is an excess noise with respect to the normal case.

The shot noise can be analyzed with more detail in the two opposite limits: $\tau \rightarrow 0$ (tunnel) and $\tau \rightarrow 1$ (ballistic regime). Let us start by considering the low transmission regime ($\tau \ll 1$). In this limit the electronic transport can be analyzed as a multiple sequential tunneling process in which the dc current can be written as the addition of the tunneling rates corresponding to different MAR processes: $I_0(V) = \sum_n ne\Gamma_n(V)$, with $\Gamma_n = (2/h)\int d\epsilon R_n(\epsilon)$, where the probability of an

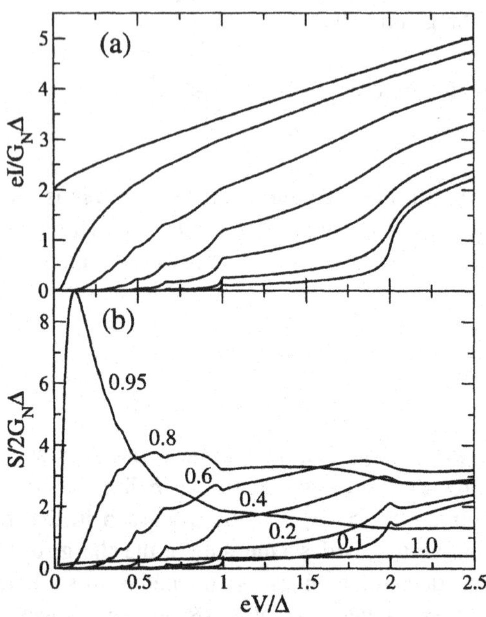

Figure 3. (a) Current-voltage and (b) noise-voltage characteristics for different transmissions at zero temperature. The values of the transmission are the same in both panels. $G_N = (2e^2/h)\tau$ is the normal conductance.

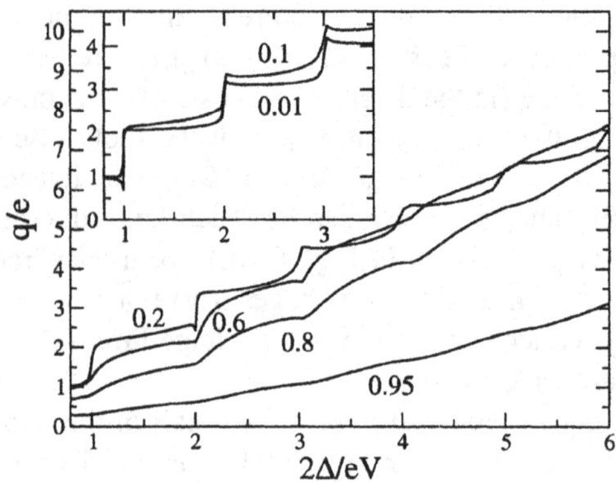

Figure 4. Effective charge $q = S/(2I)$ as a function of the reduced inverse voltage for different transmissions.

nth-order Andreev process R_n is given by [9]

$$R_n(\epsilon) = \frac{\pi^2 \tau^n}{4^{n-1}} \left[\prod_{k=1}^{n-1} |p(\epsilon - keV)|^2 \right] \rho(\epsilon - neV)\rho(\epsilon) \; ; \quad \text{where } \epsilon \in [\Delta, neV - \Delta], \tag{18}$$

where $\rho(\epsilon) = |\epsilon|/\sqrt{\epsilon^2 - \Delta^2}$ is the dimensionless BCS density of states and $p(\epsilon) = \Delta/\sqrt{\Delta^2 - \epsilon^2}$ is the Cooper pair creation amplitude. This expression for R_n clearly displays the different ingredients in a MAR processes, i.e. it is proportional to the initial and final density of states, to the probabitlity of creating $n-1$ Cooper pairs and to the probability of a quasiparticle crossing n times the interface (τ^n). This current expression already suggests that the transmitted charge associated with each MAR process is well defined in this limit. In contrast to a normal channel, characterized by a binomial distribution (the electron is either emitted or reflected), in the present case the quasiparticles can tunnel through many different channels (corresponding to the different MAR processes) giving rise to a multinomial distribution. Consequently, the shot noise can be written as

$$S(0,V) = \frac{4e^2}{h} \int d\epsilon \left[\sum_{n=1}^{\infty} n^2 R_n - \left(\sum_{n=1}^{\infty} n R_n \right)^2 \right] . \tag{19}$$

The MAR probability, R_n, is proportional to τ^n and is finite only for $eV > 2\Delta/n$. Therefore, when $\tau \to 0$ in the voltage interval $[2\Delta/en, 2\Delta/e(n-1)]$ the main contribution to the current comes from a n-order MAR, and the current distribution becomes Poissonian with a different charge depending on the voltage range. In fact, the effective charge defined as $q = S(0,V)/2I$ exhibits in this limit a staircase behavior given by $q(V)/e = \text{Int}[1 + 2\Delta/eV]$. In Fig. 4 we show the effective charge calculated for different transmissions. As can be observed $q(V)$ increases for decreasing bias as $1/V$, its shape becoming progressively steplike for decreasing transmission.

In the ballistic regime the calculation of the amplitudes for the MAR processes is simplified due to the absence of interference effects due to backscattering. Averin and Imam [10] derived the following expression for the spectral density in this limit

$$S(\omega, V) = \frac{4e^2}{h} \sum_{\pm\omega} \int d\epsilon \, F(\epsilon) \left[1 - F(\epsilon \pm \hbar\omega)\right] \times \left[1 + 2 \, \text{Re} \sum_{k=1}^{\infty} \prod_{l=1}^{k} a(\epsilon + leV) a^*(\epsilon + leV \pm \hbar\omega) \right] . \tag{20}$$

Here $a(\epsilon) = (\epsilon + i\sqrt{\Delta^2 - \epsilon^2})/\Delta$ is the amplitude of Andreev reflection from the superconductors, and F is the nonequilibrium distribution of quasiparticles in the point contact,

$$F(\epsilon) = f(\epsilon) + \sum_{n=0}^{\infty} \prod_{m=0}^{n} |a(\epsilon - meV)|^2 \times [f(\epsilon - (n+1)eV) - f(\epsilon - neV)] . \quad (21)$$

At zero frequency and zero temperature, this expression gives rise to the rather featureless curve as a function of bias shown in Fig. 3 for $\tau = 1$. The great reduction of the noise is due to the fact the probabilities of the different MARs are equal to 1 inside the gap. However, the noise does not vanish completely, like in the normal case, because the MAR probabilities are less than 1 outside the gap. As mentioned above, an interesting feature of the noise is the huge enhancement close to perfect transparency at small voltages (see curve $\tau = 0.95$ in Fig. 3). Following the analysis of Naveh and Averin [13], the behavior of the noise in this limit can be understood in terms of Landau-Zener transitions between the two ballistic subgap states. Within this picture the noise is originated by the stochastic quantum-mechanical nature of the transitions between the two states. As explained in the previous section, these two states carry the currents $\pm(e\Delta/\hbar)\sin\phi/2$, where ϕ is the Josephson phase difference across the junction, with $\phi = 2eV/\hbar$. In each period of the Josephson oscillations the junction either stays on one of the Andreev levels and carries the current of the same sign during the whole oscillation period, or either makes a transition between the two states at $\phi = \pi$ so that the current changes sign for the second half of the period. The first case occurs with probability $p = \exp\{-\pi R\Delta/eV\}$, where $R = 1 - \tau$, and then the charge

$$Q_0 = \frac{e\Delta}{\hbar} \int_0^{T_0} dt \sin\phi/2 = \frac{2\Delta}{V},$$

where $T_0 = \pi\hbar/eV$ is the period of the Josephson oscillations, is transferred through the junction. In the second case the probability is $1 - p$ and no net charge is transferred. Therefore statistics of charge Q transferred through the junction during the large time interval $T \gg T_0$ is characterized by a binomial distribution with probability p. The noise is then given by

$$S(0,V) = \frac{2(\langle Q^2\rangle - \langle Q\rangle^2)}{T} = \frac{8e\Delta^2}{\pi\hbar V} p(1-p) . \quad (22)$$

The noise reaches a maximum at small voltages $eV \simeq \pi R\Delta$ and its peak value increases with decreasing R as $1/R$. One can check that Eq. (22) describes accurately the low-voltage behavior of the curves with small R in Fig. 3. It is also interesting to analyze the large voltage limit, in which the zero-frequency noise behaves as $S(eV \gg \Delta = (4e^2/h)\tau(1-\tau)V + S_{exc}$, i.e. the shot-noise of a normal contact with transmission τ plus a voltage-independent "excess noise" S_{exc}. The excess noise as a function of transmission is shown in Fig. 5. S_{exc} has the same

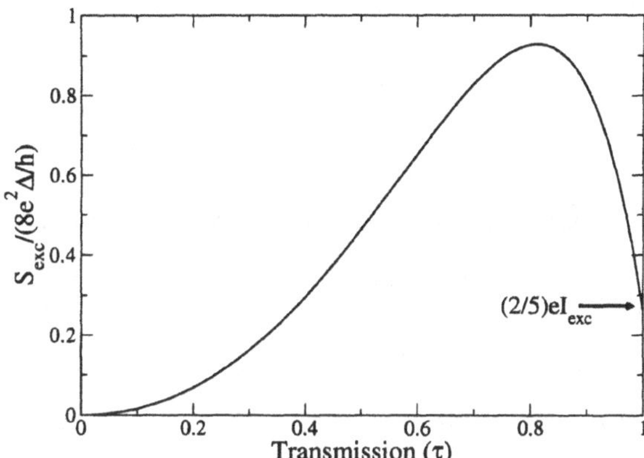

Figure 5. Excess noise as a function of the transmission at zero temperature.

physical origin as the excess current (I_{exc}), which arises from the contribution of the first order Andreev process. We obtain that at zero temperature S_{exc} is twice the excess noise of a N-S contact [27] with the same transmission. In particular, this relation yields $S_{exc} = 2/5eI_{exc}$ for the perfect ballistic case [28].

3.3. NON-EQUILIBRIUM NOISE AT FINITE FREQUENCIES

Our knowledge about shot noise in the coherent MAR regime is mainly restricted to the zero frequency limit. However, a rich frequency-dependency is to be expected according to the strong non-linear behavior of the IV characteristics. This dependency should contain very valuable information on the dynamics of quasiparticles in the MAR regime. Additonal interest on the full noise spectrum arises from its connection with the Coulomb blockade phenomena as recently pointed out in Ref. [29]. In this work it was shown that the deviation from the Ohmic behavior of normal a coherent conductor due to its electromagnetic environment can be expressed in terms of the conductor noise spectrum $S(\omega, V)$. Thus, the dynamical Coulomb blockade in a circuit containing a quantum point contact should vanish in the same way as shot noise when the contact transmissions approach unity. It is to be expected that a similar relation should hold in the superconducting case. In order to get an idea about the behavior of the full noise spectrum we have extended the calculations of the previous section to the finite-frequency domain. In Fig. 6 we show some numerical results for the finite-frequency shot noise for different transmissions. The most important feature is the splitting of the subharmonic gap structure, which now takes place at voltages $eV = (2\Delta \pm \hbar\omega)/n$, with n integer. This is specially clear at low transparencies and at frequencies $\hbar\omega < \Delta$ (see Fig. 6(a)). This result can be understood with the sequential analysis described in the previous section. As in the case of the zero-frequency noise,

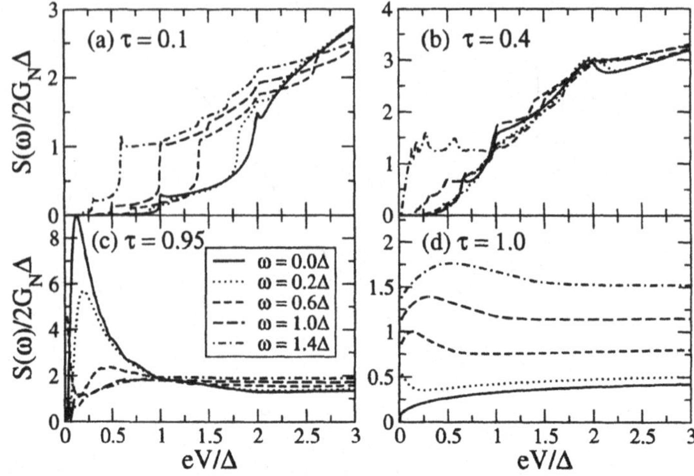

Figure 6. Finite frequency shot noise as a function of the voltage for different transmission and frequencies at zero temperature.

the subharmonic gap structure is progressively washed out as the transparency is increased. Close to perfect transparency the main feature is the suppression of the peak at low voltages. When $\tau = 1$ the noise spectrum is almost featureless and exhibits a linear increase with frequency. The consequences of these findings on the dynamical Coulomb blockade of multiple Andreev reflections is currently under investigation.

4. Experimental studies of noise in SNS nanostructures

Motivated by some of the previously discussed theoretical predictions, shot noise has been studied experimentally in different types of SNS nanostructures. In 1997, Dieleman et al. observed what seemed to be a divergence of the effective charge at low voltages in NbN/MgO/NbN tunnel junctions [30]. The measured junctions consisted probably of parallel SNS point contacts due to the presence of small deffects acting as "pin-holes" in the tunnel barrier. This interpretation was confirmed by the observation of a finite subgap current exhibiting the typical structure at $eV = 2\Delta/n$. The mean transmission of the pin-holes was estimated to be $\tau \simeq 0.17$. In spite of the rather large error bars in the noise determination it was possible to observe a clear increase of the effective charge with decreasing voltage. Dieleman et al. developed a qualitative explanation of their data within the framework of the semiclassical theory of MAR of Refs. [5]. The increase of the effective charge at low voltages has also been observed in diffusive SNS junctions by Hoss et al. [31]. They used high transparency Nb/Au/Nb, Al/Au/Al and Al/Cu/Al junctions prepared by lithographic techniques. Although being diffusive, the normal region is these junctions was smaller than the phase coherence lenght, which allowed to observe the coherent MAR regime. On the other hand, the junctions presented a very small critical current, which permitted to reach the low voltage regime. The excess noise in these experiments exhibited a pronounced peak at very low voltages (of the order of a few μV) corresponding to an effective charge increasing much faster than the predicted $1/V$ behavior. It should be pointed out that there is at present no clear theory for diffusive SNS junctions in the coherent MAR regime.

4.1. MEASUREMENTS OF SHOT NOISE IN WELL CHARACTERIZED ATOMIC CONTACTS

As we emphasized in the introduction, atomic-size contacts provide an almost ideal "test-bed" for many of the predictions of the theory. Since all their characterstic dimensions are of the order of the Fermi wavelenght, atomic contacts are perfect quantum conductors, even at room temperature, and accomodate only a small number of conduction channels. For one-atom contacts the number of conduction channels is directly related to the number of valence orbitals of the central atom [18]. For example gold one-atom contacts contain only one channel, while aluminum and lead have three, and niobium five. For such a small number of channels it is possible to determine with good accuracy the mesoscopic code [17] from the precise measurement of the current-voltage characteristic in the superconducting state. The discovery that their mesoscopic "PIN-code", i.e. the set of transmission eigenvalues $\{\tau_n\}$, could be accurately decoded, paved a way to a new generation of quantum transport experiments in which the measured quantities could be compared to the theoretical predictions without adjustable parameters.

It is worth mentionning that van den Brom and van Ruitenbeek [32], reversing this point of view, have performed shot noise measurements in atomic-size contacts in the normal state, in order to get information about the number of conduction channels and their transmission probabilities(see article by JVR in this book). For 27 different gold contacts they measured a spectral density well below the poissonian value, indicating that current is mostly carried by well transmitting channels. The values of the conductance and the shot noise density are related respectively to the first and second moment of the transmission probability distribution. Because from two parameters the code can be disentangled only if the contact contains no more than two conduction channels, their results were quantitative only for total conductances below two conductance quanta. For a single gold atom contact the conductance is about one and their shot noise measurements established that the contribution of partially transmitted conduction channels is only a few percent.

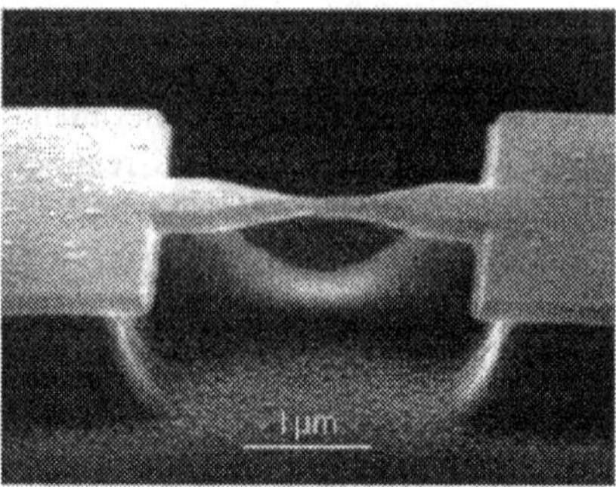

Figure 7. Micrograph of a nanofabricated break junction used in the noise measurements of Ref. [21].

In a recent work, we have used nanofabricated Al break junctions for analyzing shot noise in atomic-size contacts. A micrograph of a nanofabricated junction is shown in Fig. 7. It consists of a metallic bridge clamped to an elastic substrate and suspended over a few micrometers between two anchors. The bridge presents in its center a constriction with a diameter of approximately 100 nm. In order to obtain an atomic-size contact, the substrate is first bent till the bridge breaks at the constriction. The two resulting electrodes are then slowly brought back into contact. The high mechanical reduction ratio of the bending bench allows to control the number of atoms forming the contact one by one; in this way, single atom contacts can be produced in a controlled fashion. Nanofabricated atomic-size contacts are extremely stable and can be maintained for days. Due to its high mechanical stability, this technique [33] is particularly suitable for shot-noise measurements at very low bias currents.

The set-up used to measure shot noise is depicted in Fig 8. It consists basically of one coaxial line, used to bias the on-chip grounded atomic contact, and of two twisted-pair lines used to obtain two independent measurements of the voltage across with two sets of low-noise amplifiers. With this set-up current fluctuations are thus not directly measured, but instead inferred from the fluctuations of the voltage across the contact. The current and voltage fluctuations spectral densities S_I and S_V respectively, are related, at a given voltage V through $S_V(V) = R_D^2 S_I(V)$, where $R_D(V) = \partial V/\partial I(V)$ is the differential resistance. In the normal state, this differential resistance is essentially constant in the voltage range in which the experiments are carried out, and equals R_N, the normal resistance of the contact. In the superconducting state, the differential resistance can be highly non-linear and is determined using a lock-in amplifier for each point at which the noise is measured.

All noise sources along the measurement lines, like the Johnson-Nyquist thermal noise of the resistors or the current and voltage noise of the amplifiers input

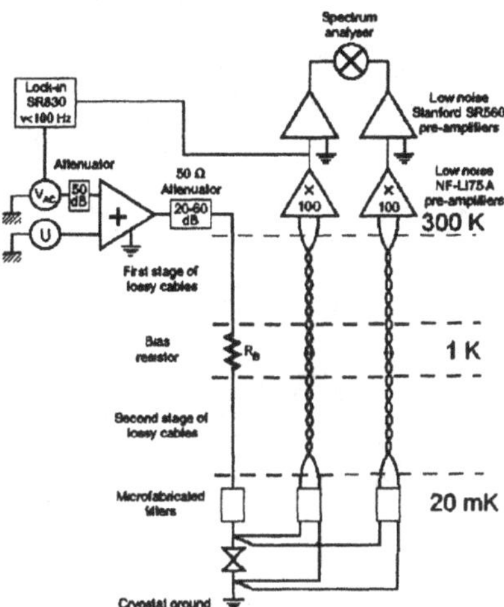

Figure 8. Schematic representation of the measurement set-up consisting of a coaxial line to bias the atomic-size contact (two triangle symbol) and of two twisted-pair lines used to measure twice the voltage across it. The spectrum analyzer calculates the cross-correlation of these two signals.

stages, induce fluctuations that poison the shot noise signal. Because of that, the measurement lines and the bias line were carefully designed and built so as to limit and keep under control this additional noise. As mentionned before, the voltage across the atomic-size contact is measured twice and the real part of the cross-correlation spectrum of the two amplified signals is calculated in real time by a spectrum analyzer. This cross-correlation technique allows one to get rid of the voltage noise coming from the preamplifiers and the measurement lines that poison the white noise signal. Typically, the spectra were measured over 800 points in a frequency window $[360, 3560 Hz]$ and averaged 1000 times in 4 min (a detailed discussion of the setup calibration can be found in [34]).

Once the PIN-code of a given contact has been determined, a first check of consistency is obtained from the measurement of the current fluctuations in the normal state. The normal state is recovered without changing the temperature by applying a small magnetic field (of the order of 50 mT) which does not affect the transmissions. The measured low frequency spectral density as a function of

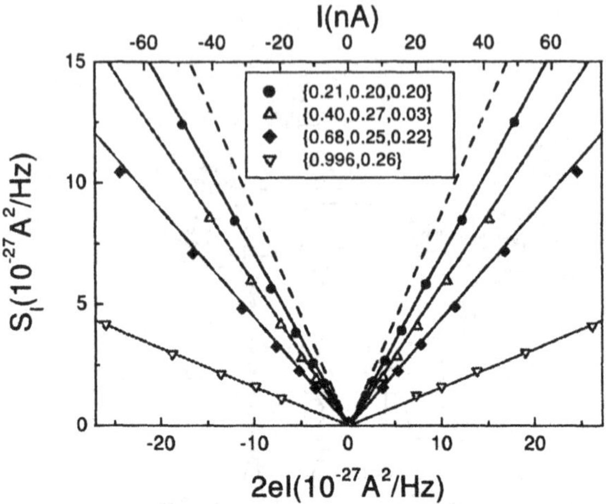

Figure 9. Symbols: measured low frequency spectral density of aluminum atomic-size contacts versus poissonian spectral density $2eI$. Solid lines are prediction of for the corresponding mesoscopic codes. The dashed line is the poissonian limit.

the average current is shown in Fig. 9 for different contacts. These results can be compared with the predictions of the theory for noise in normal contacts [27]

$$S_I(V,T,\{\tau_i\}) = 2eV \coth(\frac{eV}{2k_BT})G_0 \sum_i \tau_i(1-\tau_i) + 4k_BTG_0 \sum_i \tau_i^2 \, . \tag{23}$$

The full lines in Fig. 9 corresponds to Eq. (23) using the set of PIN-codes extracted from the analysis of the superconducting IV curves. As can be observed, there is a remarkable agreement between theoretical and experimental results. It should be emphasized that there are no fitting parameters in this comparison as the transmission coefficients have been determined independently. This good agreement provides additional proof of the accuracy that can be obtained in the determination of the contact PIN-code.

The following step in the experiments was the measurement of the noise in the superconducting state. Fig. 10 shows the measured spectral densities as a function of bias voltage for three different contacts. In contrast to the behavior in the normal case, the measured S_I is markedly nonlinear and for high enough voltages it is above the value determined in the normal state. For comparison, the theoretical predictions for the corresponding PIN-codes are shown as full lines.

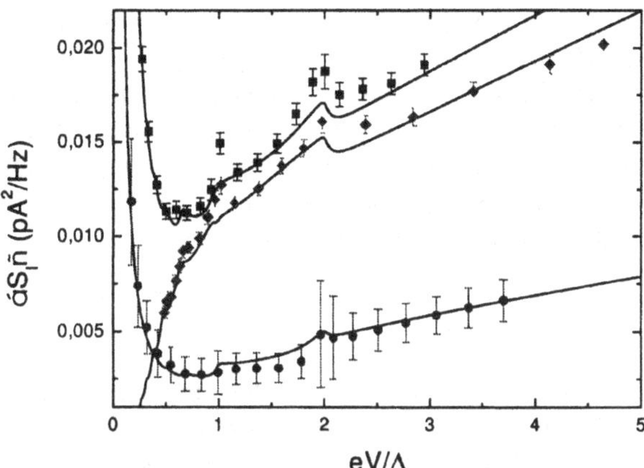

Figure 10. Dots: Measured current fluctuation spectral density as a function of reduced voltage of three atomic-size contacts. Mesoscopic PIN codes: $\{0.98, 0.55, 0.24, 0.22\}$ (squares), $\{0.68, 0.25, 0.22\}$ (diamonds), $\{0.996, 0.26\}$ (circles). Full curves: theoretical predictions of the MAR theory using the mesoscopic code.

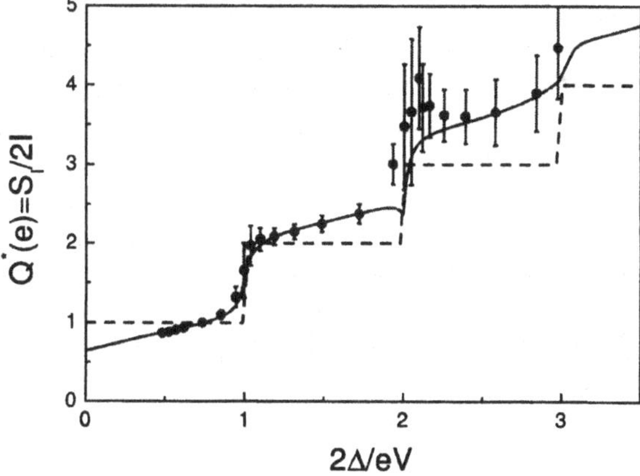

Figure 11. Effective size of the shot-noise "pellets", in units of e, as a function of the inverse reduced voltage for a contact in the superconducting state. Dashed line : MAR theory prediction in the tunnel limit. As the voltage increases, MAR processes of lower order set-in one by one leading to this perfect staircase pattern. Dots : Data for an aluminum atomic contact with mesoscopic PIN code $\{0.40, 0.27, 0.03\}$. Full line : MAR theory prediction for this code.

As can be observed the agreement between theory and experiment is quantitative both regarding the subgap structure and the excess noise at large bias.

The data can be presented in a more intuitive way, by plotting the effective charge $q = S_I/2I$ as a function of the inverse voltage. Fig. 11 shows the data for a contact having an intermediate transmission. Let us recall that only in the tunnel limit the theory predicts that q should increase in a perfect staircase pattern with decreasing bias. Our setup sensitivity was not enough to measure the noise in this limit of very small bias currents. However, the emergence of a staircase pattern can already be recognized in the data shown in Fig 11.

5. Conclusions

We have analyzed several aspects of noise phenomena in superconducting quantum point contacts. The more remarkable features both in the equilibrium and non-equilibrium current fluctuations appear as a consequence of the underlying MAR mechanism for electronic transport. At zero bias voltage huge supercurrent fluctuations are predicted due to transitions between the subgap Andreev states. At finite bias the noise spectrum exhibits a very pronounced subgap structure which corresponds to the onset of higer order MAR processes when the voltage is reduced. The effective charge generally increase as $1/V$ but is only quantized in the tunnel limit when the interference between different MAR processes can be neglected. All these properties exhibit a highly non-trivial dependence on the contact transparency. On the experimental side, noise measurements in well characterized superconducting atomic contacts have provided an unambiguous test of the theoretical predictions. The possibility of obtaining the corresponding set of transmissions (the PIN-code) from the analysis of the subgap structure in the superconducting IV curves has allowed a direct comparison with theory without fitting parameters. The quantitative agreement found for the noise completes a comprehensive series of tests which includes the measurement of the supercurrent [20] and the current-voltage characteristics [17]. Finally, the study of non-equilibrium noise in the finite-frequency domain appears as a very promising avenue of reaserch.

* Present address: Institut für Theoretische Festkörperphysik, Universität Karlsruhe, 76128 Karlsruhe, Germany

References

1. Ya.M. Blanter and M. Büttiker, Phys. Rep. **336**, 1 (2000).
2. W. Schottky, Ann. Phys. (Leipzig) **57**, 541 (1918).
3. L. Saminadayar, D.C. Glattli, Y. Lin and B. Etienne, Phys. Rev. Lett. **79**, 2526 (1997); R. dePicciotto, M. Reznikov, M. Heiblum, V. Umansky, G. Bunin and D. Mahalu, Nature **389**, 162 (1997); M. Reznikov *et al.*, Nature **399**, 238 (1999).
4. X. Jehl *et al.*, Nature **405**, 50 (2000).

5. T.M. Klapwijk, G.E. Blonder and M. Tinkham, Physica B **109&110**, 1657 (1982); M. Octavio, G.E. Blonder, M. Tinkham, and T.M. Klapwijk, Phys. Rev. B **27**, 6739 (1983).
6. G.B. Arnold, J. Low Temp. Phys. **68**, 1 (1987).
7. E.N. Bratus, V.S. Shumeiko, and G. Wendin, Phys. Rev. Lett. **74**, 2110 (1995).
8. D. Averin and A. Bardas, Phys. Rev. Lett. **75**, 1831 (1995).
9. J.C. Cuevas, A. Martín-Rodero and A. Levy Yeyati, Phys. Rev. B **54**, 7366 (1996).
10. D.V. Averin and H.T. Imam, Phys. Rev. Lett. **76**, 3814, (1996).
11. A. Martín-Rodero, A. Levy Yeyati and F.J. García-Vidal, Phys. Rev. B, **53**, 8891 (1996).
12. J.C. Cuevas, A. Martín-Rodero and A. Levy Yeyati, Phys. Rev. Lett. **82**, 4086 (1999).
13. Y. Naveh and D.V. Averin, Phys. Rev. Lett. **82**, 4090 (1999).
14. C.J. Muller, J.M. van Ruitenbeek and L.J. de Jongh, Physica C **191**, 485 (1992).
15. N. van der Post, E.T. Peters, I.K. Yanson, and J.M. van Ruitenbeek, Phys. Rev. Lett. **73**, 2611 (1994).
16. M.C. Koops, G.V. van Duyneveldt, and R. de Bruyn Ouboter, Phys. Rev. Lett. **77**, 2542 (1996).
17. E. Scheer, P. Joyez, D. Esteve, C. Urbina and M.H. Devoret, Phys. Rev. Lett. **78**, 3535 (1997).
18. E. Scheer, N. Agraït, J.C. Cuevas, A. Levy Yeyati, B. Ludoph, A. Martín-Rodero, G. Rubio, J.M. van Ruitenbeek and C. Urbina, Nature **394**, 154 (1998).
19. B. Ludoph *et al.*, Phys. Rev. B **61**, 8561 (2000).
20. M.F. Goffman *et al.*, Phys. Rev. Lett. **85**, 170 (2000).
21. R. Cron *et al.*, Phys. Rev. Lett. **86**, 4104 (2001).
22. J.R. Schrieffer and J.W. Wilkins, Phys. Rev. Lett. **10**, 17 (1963).
23. L.V. Keldysh, Sov. Phys. JETP **20**, 1018 (1965).
24. S.B. Kaplan *et al.*, Phys. Rev. B **14**, 4854 (1976).
25. D. Rogovin and D.J. Scalapino, Annals of Physics **86**, 1 (1974).
26. A. Martín-Rodero, A. Levy Yeyati, and J.C. Cuevas, Physica B **218**, 126 (1996); A. Levy Yeyati, A. Martín-Rodero, and J.C. Cuevas, J. Phys.: Condens. Matter **8**, 449 (1996).
27. V.A. Khlus, Sov. Phys. JETP **66**, 1243 (1987).
28. J.P. Hessling, V.S. Shumeiko, Yu. M. Galperin and G. Wendin, Europhys. Lett. **34**, 49 (1996).
29. A. Levy Yeyati, A. Martín-Rodero, D. Esteve, and C. Urbina, Phys. Rev. Lett. **87**, 046802 (2001).
30. P. Dieleman *et al.*, Phys. Rev. Lett. **79**, 3486 (1997).
31. T. Hoss *et al.*, Phys. Rev. B **62**, 4079 (2000).
32. H.E. van den Brom and J.M. van Ruitenbeek, Phys. Rev. Lett. 82, 1526 (1999).
33. J.M. van Ruitenbeek, A. Alvarez, I. Piñeyro, C. Grahmann, P. Joyez, M.H. Devoret, D. Esteve, and C. Urbina, Rev. Sci. Instrum. 67, 108 (1996).
34. Ronald Cron, PhD dissertation, Université de Paris, that can be downloaded from http://www-drecam.cea.fr/drecam/spec/Pres/Quantro/Qsite/index.htm

SHOT NOISE OF MESOSCOPIC NS STRUCTURES : THE ROLE OF ANDREEV REFLECTION

B. REULET[1,2] and D.E. PROBER[1]
[1] *Departments of Applied Physics and Physics, Yale University New Haven CT 06520-8284, USA*

[2] *Laboratoire de Physique des Solides, associé au CNRS bâtiment 510, Université Paris-Sud 91405 ORSAY Cedex, France*

W. BELZIG[3]
[3] *Department of Physics and Astronomy, University of Basel Klingelbergstr. 82, 4056 Basel, Switzerland*

1. Introduction

When a mesoscopic wire made of normal metal N is in contact with a superconducting reservoir S, Andreev reflection (AR) occurs [1]. This affects the electronic properties of the wire [2]. In this article we address both experimentally and theoretically the following question: how does Andreev reflection manifest itself in shot noise measurements, and what physics can we deduce from such measurements ? Our discussion will rely on high frequency measurements performed on various NS structures, that can be found in refs. [3–5].

The shot noise is a direct consequence of the granularity of the electric charge. Even though an electric current exists only if there are free charged carriers, the value of this charge does not directly affect how much current flows through a sample when biased at a finite voltage. One has to investigate the fluctuations of the current to determine the charge of the carriers. As a consequence the measurement of the shot noise offers direct access to the elementary excitations of any system, through the determination of their effective charge. An effective charge different from $1e$ directly reflects how, due to their interactions, the electrons are

Y.V. Nazarov (ed.), Quantum Noise in Mesoscopic Physics, 73–91.

correlated, as in a superconductor, a 2D electron gas in the fractionnal quantum Hall regime or a 1D Luttinger liquid.

In an NS system, electrons can enter the superconductor only in pairs. Thus, the elementary charge participating in the electric current is no longer e, but $2e$. Thus, a naive expectation is that the shot noise should be doubled in the presence of an NS interface, as compared to the case of a normal metal. In the following we explore this expectation, to see how Andreev reflection affects current noise. In particular we show how a deviation of the effective charge from $2e$ reflects the existence of correlations among *pairs* of electrons.

In equilibrium the current noise power S_I is given by the Johnson-Nyquist formula, which relates the current fluctuations to the conductance of the sample G: $S_I(V = 0, T) = 4k_BTG$, where T is the electron temperature. Whether the fluctuating current is made of single electrons or pairs affects equilibrium noise only through the conductance, which may depend on the state (N or S) of the metal. This picture is valid at low frequency ($\hbar\omega \ll k_BT$). It is the 'classical' regime. At finite frequency ω a quantum mechanical treatment of noise is necessary. The general expression for the equilibrium current noise is given in terms of the reduced frequency $w = \hbar\omega/(2k_BT)$ by [6]:

$$S_I(V = 0, T, \omega) = 4k_BTG(T, \omega)g(w) \tag{1}$$

where the function $g(x)$ is defined as: $g(x) = x \coth x$. As in the classical regime, all the physics is contained in the conductance G, but here G is the real part of the complex and frequency dependent admittance of the sample. The $g(w)$ term accounts for statistical distribution of an excitation of energy $\hbar\omega$ at thermal equilibrium. Through $g(w)$ the finite frequency adds an energy scale at which the classical-to-quantum noise crossover takes place: $\hbar\omega = 2k_BT$, or $w = 1$.

Since we are interested here in shot noise, let us discuss first the case of the normal tunnel junction. The shot noise of the tunnel junction at low frequency is given by [7]:

$$S_I^{tunnel}(V, \omega = 0, T) = 4k_BTGg(v) \tag{2}$$

with the reduced voltage $v = qV/(2k_BT)$. Here $g(v)$ interpolates between Johnson noise ($g(0) = 1$) and shot noise ($g(v \gg 1) = v$). Hence the charge q appears at two levels: it can be measured through the equilibrium-to-shot noise crossover occuring at $qV = 2k_BT$, or through the magnitude of the noise at high voltage, such that $S_I^{tunnel} = 2qI$. At finite frequency, the noise emitted by a tunnel junction is:

$$S_I^{tunnel}(V, \omega, T) = 2k_BTG(g(v + w) + g(v - w)) \tag{3}$$

Thus, a finite frequency investigation offers another way to measure q, through the classical-to-quantum noise crossover occuring at $qV = \hbar\omega$.

In the case of a diffusive mesoscopic wire, the shot noise is reduced by the disorder, but the principles above are still valid. Thus, AR should show up in the *shape* of the noise spectrum as a doubling of the classical-to-quantum noise crossover frequency, occuring at $\hbar\omega = 2eV$. To perform such a measurement as a function of frequency, one needs to have a precise knowledge of $G(\omega)$ and of the frequency response of the experimental setup at high frequencies. (For example, $T = 100$mK corresponds to $V = 8.6\mu$V and $\omega/2\pi = 2.1$GHz.) As a consequence, the direct measurement of the frequency dependence of the noise spectrum at fixed voltage has never been accomplished. The noise measured in a narrow frequency band is expected to show the same crossover, but as a function of the applied voltage. This is a much easier (but still difficult) experiment, which we shall report in section 3. The measurement has been performed so far on a normal metal wire, but could also be carried out on an NS sample. In Section 4 we report measurements of photon assisted noise in an NS wire, which provide an alternative to the measurement of the crossover frequency. In this experiment a high frequency excitation is applied to the sample. The shot noise develops features as a function of voltage V each time qV is a multiple of the energy of the incident photons $\hbar\omega$. For the NS wire, $q = 2e$.

The discussion above treats the consequences of the AR due to the energy qV, as compared to k_BT or $\hbar\omega$. It investigates the effect of AR on the *distribution statistics* of the electronic excitations (which involve pairs of electrons) rather than the *effective charge* they carry. Specifically , the phenomena that have been measured and discussed above are related to steps in the distribution function, as will be discussed in section 6. A better measurement of the effective charge is in the fully developed shot noise regime ($eV \gg k_BT$). Here the noise is determined by the fact that electrons are paired and also by the interferences and the correlations that can exist among pairs. In section 5 we report measurements of phase dependent shot noise in an Andreev interferometer, which point out such a sensitivity of the effective charge to pair correlations. Section 6 is devoted to the theoretical investigation of the effective charge deduced from the shot noise. Section 2 contains information about sample preparation and experimental setup, common to the experiments reported in subsequent sections.

2. Experimental considerations

The experiments we report in the next sections use samples prepared with similar methods, and measured with similar detection schemes. Each sample consists of a metallic wire or loop between two metallic reservoirs, either normal or superconducting. The measurements are performed through contacts to the two reservoirs. This allows dc characterization of the sample (two-contact differential resistance $R_{diff} = dV/dI$, measured at $\sim$ 200Hz) as well as high frequency measurements. Even for the measurement of the effective charge of the Andreev interferometer, which does not intrinsically call for the use of rf techniques, high frequency measurements have been chosen for their extremely high sensitivity (the signal-to-noise ratio is proportionnal to the square root of the bandwidth times the integration time of each measurement).

The measurements were performed in a dilution refrigerator at a mixing chamber temperature $T \sim 50$ mK. At low temperature the electron energy relaxation is dominated by electron-electron interactions [8] and the associated inelastic length L_{ee} is larger than L, so the transport in the device is elastic. We have not conducted weak localization measurements on these samples. From these, the phase coherence length L_φ could have been extracted (L_{ee} and L_φ coincide if the dominant phase relaxation mechanism is electron-electron interaction, which is likely in our samples, otherwise $L_{ee} > L_\varphi$) [9]. Nevertheless, we observe significant harmonic content of the R vs. flux curve of the interferometer (data not shown). The nth harmonics decays as $\exp(-nL/L^*(T))$, where L is the distance between the two reservoirs. The empirical characteristic length L^* includes phase beaking mechanisms (L_φ) as well as thermal averaging (usually described by the thermal length $L_T = (\hbar D/k_B T)^{1/2}$). We obtain $L^*(T = 50\text{mK}) = 800\text{nm}$ of the order of L_T, which implies that $L_\varphi \gg L_T$, i.e., L_φ is much larger than the sample size $L = 540\text{nm}$. Other measurements on a longer interferometer ($L \sim 1\mu\text{m}$) also gave evidence that $L_{ee} > L$ in that device. The wires described in sections 3 and 4 are shorter and thus are also likely in the regime $L < L_\varphi$.

2.1. SAMPLE PREPARATION

The samples studied have been patterned by e-beam lithography. All are made of thin (10nm) evaporated gold wires between thick metallic reservoirs. The N wire and N reservoirs are deposited using a double angle evaporation technique [10] in a single vacuum pump down. Sputtered Nb is used for the thick (80 nm) S reservoirs. The transparency of the NS interface has been achieved by ion beam cleaning before Nb deposition. We estimate that the interface resistance is less than $1/10$ of the wire resistance. Theoretical calculations which consider this extra resistance show that its effect is negligible. The Au wires have a temperature independent sheet resistance of the order of $\sim 10\ \Omega$ per square. The Au reservoirs are 70 nm thick and have a sheet resistance of less than $\sim 0.5\ \Omega$ per square.

2.2. EXPERIMENTAL SETUP FOR HIGH FREQUENCY NOISE MEASUREMENTS

The experimental setup we used to perform the noise measurements on the Andreev interferometer is depicted in fig. 2.2. The current fluctuations S_I in the sample are measured in a frequency band Δf from 1.25 to 1.75GHz using an impedance matched cryogenic HEMT amplifier. The noise emitted by the sample passes through a cold circulator, employed to isolate the sample from amplifier emissions. It is then amplified by the cryogenic amplifier and rectified at room temperature after further amplification. The detected power is thus given

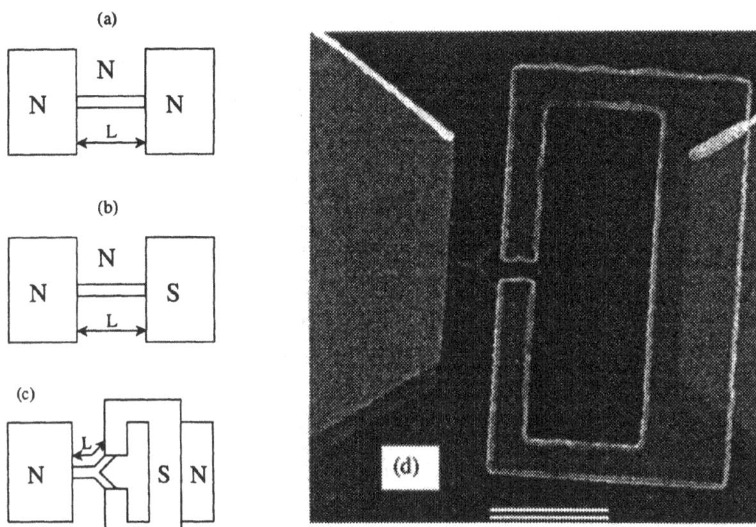

Figure 1. Schematics of the different samples that have been measured. (a) N wire between N reservoirs (L = 200nm, D = 40 cm^2/s). (b) N wire between N and S reservoirs (L = 280nm, D = 30cm^2/s). (c) Andreev interferometer (L = 540nm, D = 33cm^2/s). (d) SEM picture of the Andreev interferometer; N contact on right not shown. The scale bar corresponds to 1μm.

by $P_{det} = G\Delta f(k_B T_{out} + k_B T_A)$ where G is the gain of the amplification chain, $T_A \sim 6.5$K is the noise temperature of the amplifier, and T_{out} is the effective temperature corresponding to the noise power coming from the sample ($T_{out} = 0.04 - 0.6$K for $V = 0 - 150\ \mu$V for the case of an NS interface). We determine $G\Delta f$ and T_A by measuring the sample's Johnson noise vs. temperature at $V = 0$ and its shot noise at $eV \gg (k_B T, E_C)$. ($E_c = \hbar D/L^2$ is the Thouless energy of a diffusive wire of length L; it is the energy corresponding to the inverse of the diffusion time along the wire). We modulate the current through the sample to suppress the contribution of T_A. We measure dP_{det}/dI. This gives dT_{out}/dI. T_{out} is related to the noise emitted by the sample through:

$$T_{out} = (1 - |\Gamma|^2)T_N + |\Gamma|^2 T_{in} \tag{4}$$

and S_I is given by $S_I = 4k_B T_N \mathrm{Re} Z_{diff}^{-1}$. Here T_N is the sample's noise temperature, Z_{diff} is the complex differential impedance of the sample at the measurement frequency, Γ is the amplitude reflection coefficient of the sample and T_{in} the external noise incoming to the sample. In eq. (4), the first term on the right represents the noise emitted by the sample which is coupled to the amplifier. The second term represents the external noise the sample reflects.

In order to determine S_I at finite frequency, it is necessary to know both Z_{diff} and Γ at the measurement frequency. Z_{diff} is deduced from the measurement of

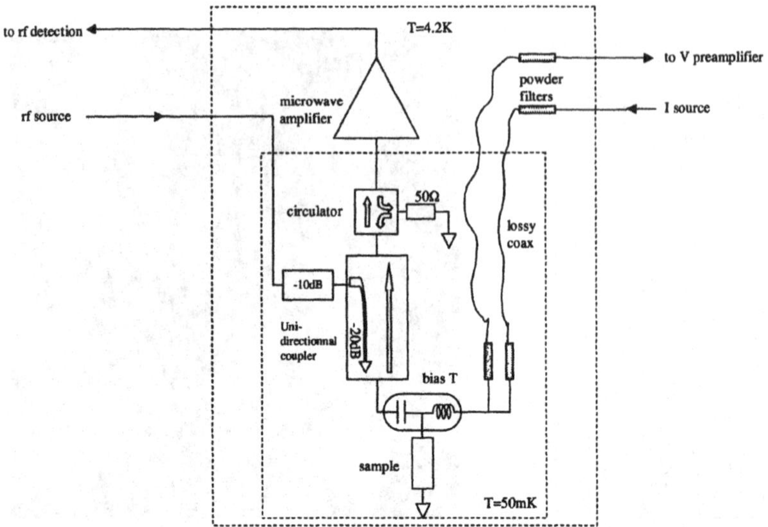

Figure 2. Experimental setup used for high frequency measurements. The inner dotted line correspond to the mixing chamber at $T = 50$mK, the outer one to the He bath or the 4K stage of the dilution refrigerator. The position of the coupler and the circulator are those used for the measurement performed on the Andreev interferometer only.

Γ through:

$$\Gamma(\omega) = \frac{Z_{diff}(\omega) - Z_0(\omega)}{Z_{diff}(\omega) + Z_0(\omega)} \tag{5}$$

where Z_0 the impedance of the measurement apparatus. Z_0 is ideally real and equal to 50Ω. In practice it has an imaginary part and is frequency dependent (due to finite return loss of the amplifier or isolator, parasitic capacitance in parallel with the sample, inductance of the wire bond, etc.). Thus a careful calibration is necessary to have a reliable measurement of Z_{diff}[11]. However, since our samples have a resistance close to 50Ω, the amplitude of $|\Gamma|^2$ is of the order of a few percent, and can be neglected in the first term of eq. (4). This is not always the case for the second term. In the measurement performed on the normal wire (section 3), no circulator was used. $T_{in} \sim 30K \gg T_N$ for this broadband (20 GHz) amplifier. The impedance of the sample (a very short gold wire) is voltage and frequency independent, so that the noise reflected by the sample adds up to the total as a voltage independent (but frequency dependent) constant. (T_{in} depends on frequency because the amplifier emission does). For the NS wire (section 4), a circulator placed in liquid helium has been used. The circulator attenuates by 20dB the noise emitted by the narrowband amplifier towards the sample ($T_{emit} \sim$ 2K). In that case T_{in} is equal to the temperature $T =$ 4K of the 50Ω termination of the circulator. $T_{in} = (T_{emit}/100) + 4K > T_{emit}$, but T_{emit} drifts, which adds drift to the measurement. For this experiment one was interested only in the voltage dependence of the features of S_I, the corrections due to T_{in} were not significant.

For the precise measurement of the effective charge in the Andreev interferometer, the magnitude of S_I is of interest; it is thus crucial for T_{in} to be minimized and stable. For this experiment we therefore placed the circulator at $T = 50$mK. We also measured the variations of $|\Gamma|^2$ by sending white noise to the sample through the unidirectionnal coupler (see fig. 2.2) and detecting the change in the noise power. This determines the relative variation of the amplitude of the reflection coefficient (as a function of voltage or magnetic flux) over the bandwidth used for the noise measurement. We draw the following conclusions: i) the variations of $|\Gamma|^2$ are small enough to be neglected, allowing us to take $\Gamma = 0$ in the data analysis. This is confirmed by the fact that at $V = 0$, where $T_N = T$, R_{diff} and Γ^2 are flux dependent; yet T_{out} at $V = 0$, does not depend on the flux (see eq.(4)); ii) the impedance of the sample at the measurement frequency is different from its dc value. This conclusion is seen from the following : if $Z_{diff}(\omega)$ were equal to its dc value $R_{diff} = dV/dI$, then whatever $Z_0(\omega)$ is (i.e. whatever the imperfections of the experiment are), $|\Gamma|^2$ plotted as a function of R_{diff} should collapse into a single curve for all the values of flux and voltage. As shown on fig. 2.2, this doesn't occur. This means that the impedance of the sample is not equal to $R_{diff} = dV/dI$ measured at low frequency. It has an flux- or voltage-dependent imaginary part, or its real part is not simply proportionnal to R_{diff}. This observation deserves more study, through measurements of the amplitude and phase of the reflection coefficient as a function of flux, voltage and frequency. However, this effect is small, and for our present study of noise, we simply use R_{diff} to determine S_I from T_N. This is also justified by the fact that in our short phase-coherent samples transport is elastic. Thus, the ac conductance is given by the dc $I(V)$ characteristics shifted by $\pm\hbar\omega/e \approx \pm 6\mu$V [7]. Since the characteristic scale for changes of dV/dI is $\sim 30\mu$V, finite frequency corrections to R_{diff} should be small.

In the experiment described in section 3, the noise needs to be measured over a broad frequency range. Thus, a broadband ($1 - 20$ GHz) cryogenic amplifier has been used, even though it has a higher noise temperature ($T_A \sim 100$K) than a narrow band amplifier. Also a circulator cannot be used. The noise at different frequencies is obtained by measuring the low frequency noise power after heterodyne mixing (at room temperature) the amplified signal from the sample against a variable frequency oscillator.

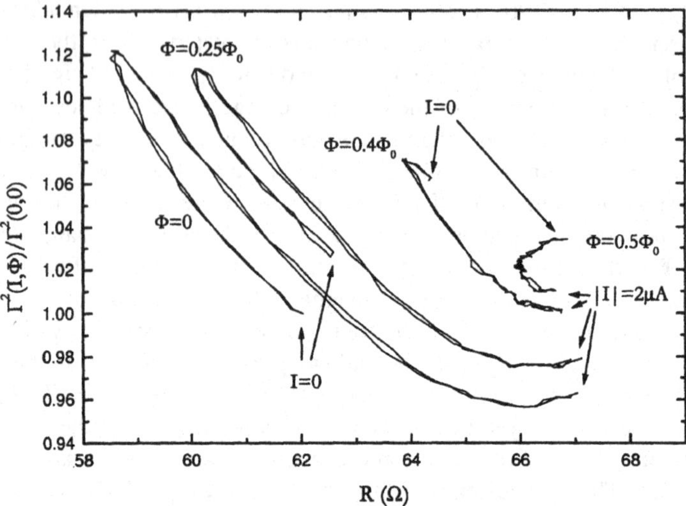

Figure 3. Measurement of the variations of the power reflection coefficient $|\Gamma|^2$ vs. dc resistance of the Andreev interferometer. Each curve corresponds to a fixed magnetic flux and a varying current, between -2μA and 2μA. The reflection coefficient has been arbitrarily rescaled to its value at $V = 0, \Phi = 0$. Conclusions are given in the text.

3. Measurement of the frequency dependence of the shot noise in a diffusive N wire

In this section we report measurements of the frequency dependence of the out-of-equilibrium ($V \neq 0$) noise in a normal metal wire between two N reservoirs (see fig. 1(a)) [3]. In such a system the current noise at finite frequency is given by [12]:

$$S_I(\omega, T, V) = 4k_BTG\left(\frac{\eta}{2}\left(g(v+w) + g(v-w)\right) + (1-\eta)g(w)\right) \quad (6)$$

The $\eta = 1/3$ factor corresponds to the shot noise reduction due to disorder (Fano factor)[12]. The shot noise corresponds to the v terms whereas equilibrium noise is given by $v = 0$. As shown by eq. (6), the total noise is *not* given by the sum of equilibrium noise (classical Johnson or quantum) and shot noise, even at zero frequency ($w = 0$). The unusual nature of this superposition can be emphasized by examining the fluctuations predicted by eq. (6) as a function of voltage for different frequencies (see fig. 4 left). At zero frequency (full line), there is a transition from Johnson noise to the linearly rising shot noise at $eV \sim k_BT \sim 2$ μeV. At 20 GHz (dotted line), the fluctuations are dominated by quantum noise and do not increase from their value at equilibrium, until the voltage $V_c = \hbar\omega/e \sim 80\mu$V is

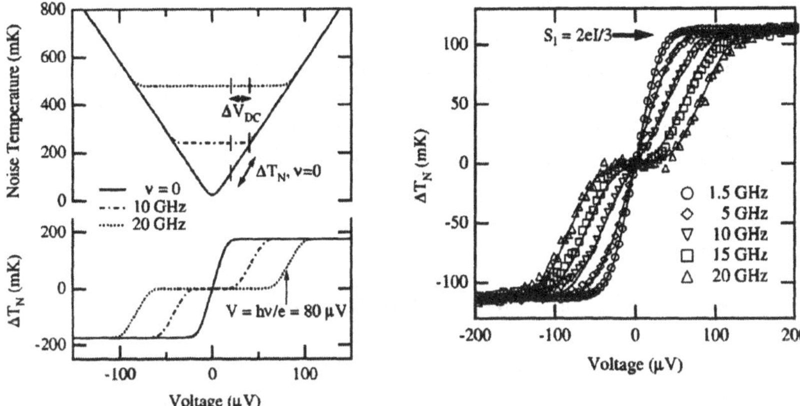

Figure 4. Left: predicted dc bias voltage dependence of noise (top) for three frequencies at a bath temperature of 25 mK. The current spectral density predicted by eq. (6) has been converted to an equivalent noise temperature T_N through the relation $T_N = S_I/(4k_BG)$. Note that the noise is independent of bias voltage for $e|V| < \hbar\omega$. The bias voltage modulation technique we employed is shown schematically and the expected differential noise ΔT_N for a 30μV p.p. modulation is also displayed (bottom). Right: measured differential noise for frequencies of $\nu = \omega/2\pi$ =1.5, 5, 10, 15, and 20 GHz, with mixing chamber temperature of 40 mK. Solid lines show the predictions of eq. (6) for an electron temperature of 100 mK, and accounting for the voltage modulation of 30 μV p.p.

exceeded, even though the condition $eV > k_BT$ is fulfilled and the low-frequency fluctuations (solid line) are increasing rapidly. Only above V_c does the dc voltage provide enough energy to increase the emitted photon noise.

The differential noise theoretically expected for the diffusive conductor, under the example conditions of $T = 25$ mK, and a small ($\Delta V = 30\mu$V p.p.) square-wave voltage modulation, are shown in the bottom left of fig. 4. The square-wave modulation contributes significantly to the width of the rises. The noise measurements are reported as a variation, ΔT_N, of the sample's noise temperature due to the voltage modulation ΔV. The measured values of ΔT_N for frequencies of 1.5, 5, 10, 15, and 20 GHz, taken at a mixing chamber temperature of $T = 40$ mK, are shown in fig. 4 right. While we see that ΔT_N for the low-frequency noise (circles) changes rapidly with voltage, approaching its linear asymptote at voltages only a few times k_BT/e, the curves become successively broader for increasing frequency. The noise for the highest frequencies has a clearly different shape, displaying the expected plateau around $V = 0$. Also shown in fig. 4 right (full lines) are theoretical curves based on eq. (6), accounting for the finite voltage difference used, and for an electron temperature of 100 mK. The asymptotic value of ΔT_N has been arbitrarily scaled (since the frequency dependent system gain is not known to better than about 30%) to be 112 mK for each frequency, corresponding to the expected reduction factor of $\eta = 1/3$ (see eq.(6)). Note that such a reduction of the shot noise could also be attributed to the heating of the electrons (hot electron regime)[13]. See ref.[3] for a detailed discussion of this possibility.

Though it has not been measured, in an NS geometry the same qualitative behaviour is expected, with the scale eV replaced by $2eV$. For the NS system there might be corrections at a frequency of the order of the Thouless energy, in the same way there is a signature at $\sim 4E_c$ in the voltage dependence of the noise measured at low frequency (see section 5).

4. Observation of photon assited noise in an NS wire

In this section we report measurements of low frequency noise (measured in the bandwidth $1.25 - 1.75$GHz) emitted by an NS sample which experiences both a dc and an ac bias [4]. The sample is a thin Au wire between N and S reservoirs, as depicted in fig. 1(b). In the presence of an ac excitation at frequency ω, the noise emitted by a diffusive wire is:

$$S_I(V,T) = 4k_BTG\left((1-\eta) + 2\eta \sum_{n=-\infty}^{+\infty} J_n^2(\alpha)g(v+nw)\right) \tag{7}$$

where $\alpha = 2eV_{ac}/(\hbar\omega)$ is a dimensionless parameter measuring the amplitude of the ac excitation voltage (note that ω denotes here the frequency of the ac bias; the the frequency at which the noise is measured is considered to be dc). This formula is valid at low ($E \ll E_c$) and high ($E \gg E_c$) energy, where G is voltage- and frequency-independent. In between, the energy dependence of the Andreev process may give rise to corrections, as is the case for the conductance and for the effective charge (see section 6).

In the absence of ac excitation, the measured differential noise dT_N/dV vs. bias voltage for the N-S device is, within 5%, twice as big as that measured when the device is driven normal by a magnetic field of 5 T [14]. This is a measure of the doubling of the effective charge, which will be discussed more in section 5. We now turn to the noise measured in the presence of an ac excitation. If the transport is still elastic in the presence of ac excitation, the shot noise is expected to develop features at bias voltages such that $qV = \pm n\hbar\omega$. The location of these features should be independent of ac power. In contrast, if the transport is inelastic, no photon-assisted features should occur. The derivative of the noise vs. bias voltage was measured with ac excitation at $\omega/2\pi = 34$ GHz, at different levels of ac power. Figures 5(a) and (b) show the predicted and observed derivative of the noise temperature vs. dc bias voltage for several levels of ac power. To see the features more clearly, we plot in fig. 5(c) the second derivative d^2T_N/dV^2 obtained by numerical differentiation of the experimental data. With no ac excitation, d^2T_N/dV^2 has a peak at $V = 0$. With ac excitation, the sidebands of this peak are

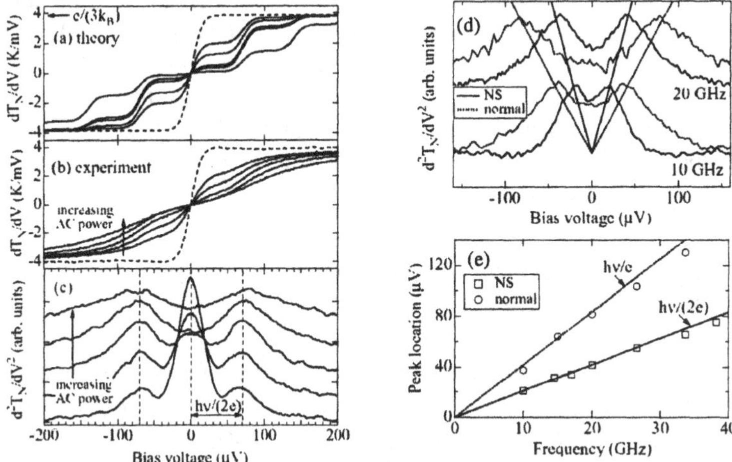

Figure 5. Predicted and observed shot noise of an N-S device vs. bias voltage without ac bias and at different powers of ac excitation at 34 GHz: (a) theory for dT_N/dV at $T = 100$ mK with no ac (dashed line) and with ac excitation at $\alpha = 1.1, 1.4, 1.7, 2.2, 2.8$ (solid lines); (b) experimentally measured dT_N/dV with no ac bias (dashed line) and with ac excitation powers differing by 2 dB and corresponding to the above values of α (solid lines); (c) d^2T_N/dV^2 obtained by numerical differentiation of data in (b). (d) d^2T_N/dV^2 vs. bias voltage at $B = 0$ (solid lines) and at $B = 5$ T (dotted lines) with ac excitation at $\hbar\omega/2\pi = 10$ and 20 GHz. The curves are offset vertically by an amount proportional to frequency. The solid straight lines mark the expected peak locations for the N-S case (at $B = 0$): $V_{peak} = \hbar\omega/(2e)$; the dotted straight lines mark the expected peak locations for a normal device (at $B = 5$ T): $V_{peak} = \hbar\omega/e$; (e) peak location vs. frequency for $B = 0$ and $B = 5$ T; the solid and the dotted straight lines are $V_{peak} = \hbar\omega/(2e)$ and $V_{peak} = \hbar\omega/e$, respectively.

clearly evident at $V = \pm n\hbar\omega/(2e)$. The sideband locations are power independent, which further argues that the structure is due to a photon-assisted process. The magnitude of d^2T_N/dV^2 at $V = 0$ displays oscillatory (roughly $\sim J^2(\alpha)$) behavior vs. ac excitation amplitude (not shown), which is another hallmark of a photon-assisted process. We note that photon assisted processes are seen clearly in SIS tunnel junctions[15]. The features there are in the quasiparticle current, so the charge involved in that case is $1e$. They are centered at the gap voltage $V = 2\Delta/e$, since at low temperature pair breaking must occur for a quasiparticle to tunnel.

The most convincing evidence of the photon-assisted nature of the observed effects is the dependence of the voltage location of the sideband peak on the frequency of the ac excitation. Measurements of the shot noise were made at several different frequencies of ac bias, both in zero magnetic field and at $B = 5$ T, for which the Nb reservoir is driven normal. Figure 5(d) shows the second derivative of the shot noise power vs. bias voltage for $\hbar\omega/2\pi = 10$ and 20 GHz at $B = 0$

(solid lines) and for the same device at $B = 5$ T (dotted lines), where the sample is driven normal. The solid and dotted straight lines are the expected peak positions for the N-S and normal cases, respectively. The peak locations clearly follow the theoretical predictions $V = \hbar\omega/q$ with $q = 2e$ in the case of the N-S device and $q = e$ in the case of the device driven normal. Figure 5(e) shows the peak locations for a number of different ac excitation frequencies at $B = 0$ and $B = 5$ T. The solid and dotted lines are theoretical predictions with no adjustable parameters.

5. Measurement of the phase dependent effective charge in an Andreev interferometer

In this section we show precise measurements of the effective charge q_{eff}, and how deviations from $q_{eff} = 2e$ are phase sensitive. We relate these deviations to correlations of the charge transfer during Andreev reflection [5]. After an Andreev process, the reflected hole carries information about the phase of the superconducting order parameter of the S reservoir at the N-S interface. When two S reservoirs are connected to the same phase-coherent normal region, a phase gradient develops along the normal metal, resulting in phase-dependent properties. In an Andreev interferometer - a device containing a mesoscopic multiterminal normal region with a (macroscopic) superconducting loop, all the electronic properties are periodic with the magnetic flux Φ enclosed by the loop, with a period of the flux quantum, $\Phi_0 = h/(2e)$. The sample used for this experiment is depicted in fig. 1(c) and (d).

From the noise measurements we deduce the effective charge, q_{eff}=$(3/2)(dS_I/dI)$; see fig. 6(a). By considering dS_I/dI rather than dS_I/dV we eliminate the trivial effect of a non-linear $I(V)$ characteristic. At finite energy ($E > k_BT$) the effective charge reflects the charge transferred but also includes the effects of correlations in the transfer process. The voltage dependence of dS_I/dI yields information on energy-dependent correlations between charge transfers. Figure 6(b) gives the theory results based on full counting statistics. The inset shows the theory for $\Phi = 0$ and $\Phi = \Phi_0/2$ at $T = 0$. The effective charge is seen in the theory to be independent of the phase difference at bias voltages larger than ~ 100 μV, with significant phase modulation of q_{eff} in the bias voltage range $10 - 80$ μV. The maximum magnitude of the observed dip of q_{eff} vs. voltage is $\sim 10\%$, and occurs for $\Phi \sim \Phi_0/4$. There is no dip for $\Phi = \Phi_0/2$. For $T = 0$, q_{eff} returns to $2e$ as $V \rightarrow 0$ (see inset of fig.6). At finite temperature, q_{eff} goes to zero for $eV \ll k_BT$. This is because S_I reduces to Johnson noise at $V = 0$. Thus, the decrease of q_{eff} at finite temperature and at very low voltages is not related to Andreev physics. In contrast, the dip near $4E_c \sim 30$ μeV (with $E_c = \hbar D/L^2$) is due to the energy dependence of the Andreev processes.

The experimental results are in fairly good agreement with the theoretical predictions. As expected, there is no phase modulation of q_{eff} at large energies

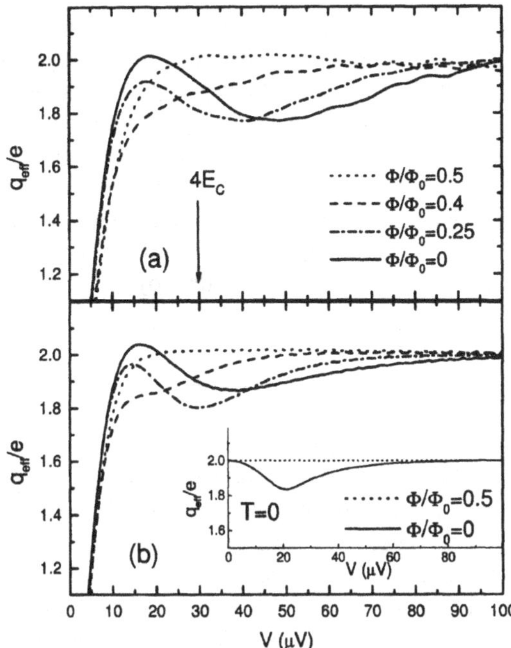

Figure 6. (a) Experimentally measured effective charge q_{eff} for several values of magnetic flux. (b) Theoretical predictions for $E_C = 7.5\mu$eV and $T = 43$mK. The dip in q_{eff} is predicted to occur at $\sim 4E_c$. The inset shows the theory for $\Phi = 0$ and $\Phi = \Phi_0/2$ at $T = 0$. Note that our definition $E_c = \hbar D/L^2$ uses L for the full length of the normal region, and thus differs from the definition in Ref. [5].

$eV \gg E_c$, and here $q_{eff} = 2e$. At $E \sim 4E_c$, the effective charge is smaller for integer flux than for half-integer flux. The non-trivial energy- and flux dependence predicted (crossings of the different curves) is seen in the experiment, though the agreement is not perfect. The magnitude of the dip of q_{eff} in the data is also close to the theoretical prediction.

To understand the origin of the dip of S_I seen for $\Phi = 0$, we have also solved a generalized Boltzmann-Langevin (BL) equation. In such an approach correlations due to the superconductor enter only through the energy- and space-dependent conductivity, which gives $I(V, \Phi)$. Thus, the BL result is not complete, and we will compare its predictions to that of the full-counting-statistics theory, to help understand those predictions. At $T = 0$, the BL result for all flux values is simply $S_I^{BL} = (2/3)2eI(V, \Phi)$, i.e., $q_{eff} = 2e$ at all energies. This implies that the deviation of the effective charge from $2e$, measured and predicted by the full theory, must be due to fluctuation processes which are not related to single-particle scattering, on which the BL approach is based. We believe that the higher-order process which is responsible for the dip of S_I is a two-pair correlation process. At high energies ($E \gg E_c$) the electron-hole pair states have a length $\sim (\hbar D/E)^{1/2}$, shorter than L. This results in uncorrelated entry of pairs into the normal region. For $E \sim E_C$ the pair size is larger, and the spatial overlap prevents fully random entry, suppressing S_I. Suppressed shot noise is a signature of anti-correlated charge entry [12]. At yet lower energies (at $T = 0$) the effective charge is predicted to return to 2e; we do not yet have a physical interpretation of this. In any case, for the case of $\Phi = \Phi_0/2$, the dip of q_{eff} is fully suppressed, according to the theory. This means that the phase gradients destroy the pair correlation effect responsible for the dip.

6. Theoretical approach to the effective charge

We now turn to our theoretical approach to current noise in mesoscopic proximity effect structures. Our goal in this section is to explain the predictions and the meaning of the effective charge. We shall see that the 'dip' in the effective charge seen for the interferometer is also seen in wires, and arises from similar pair correlation effects. A characteristic feature of the NS structures is that the phase coherent propagation of Andreev pairs in the normal metal is influenced by the proximity effect. One consequence is the so-called reentrant behaviour of the conductance of a normal diffusive wire in good contact to a superconducting terminal [2]. The conductance is enhanced at energies of the order of $\sim 5E_c$. At higher and lower energies the conductance approaches its normal state value [16]. In a fork geometry, as discussed in the previous section, the proximity effect can be tuned by a phase difference between the two superconducting terminals.

We are interested in the bias-voltage dependence of the current noise in a diffusive NS structure. There is, however, a simple energy dependence resulting from the proximity-induced energy-dependent conductivity. In contrast, the correlations of interest are not due to this energy-dependence of the conductivity. They can be distinguished in the following way. We note that for the average transport properties, i.e. the differential conductance, the kinetic equation takes the very simple form [16–19]

$$\nabla\sigma(\mathbf{x}, E)\nabla f_T(\mathbf{x}, E) = 0\,. \tag{8}$$

The energy- and space-dependent conductivity $\sigma(\mathbf{x}, E)$ includes the proximity effect, and $f_T(\mathbf{x}, E) = 1 - f(\mathbf{x}, E) - f(\mathbf{x}, -E)$ is the symmetrized distribution function. Due to the induced superconducting correlations, σ is enhanced above its normal state value σ_N and its energy- and space dependence is obtained from the spectral part of the Usadel equation [16, 20]. For the geometry in fig. 7 the spectral conductance is $G^{-1}(E) = \int dx/\sigma(x, E)$. Note that $G(E) \neq 0$ even for

$E \ll \Delta$ as a consequence of the proximity effect. The current for a given bias voltage V and temperature T is then given by

$$I(V,T) = \frac{1}{2e}\int_{-\infty}^{+\infty} G(E) f_T^N(E,V,T) dE \qquad (9)$$

where $f_T^N(E,V,T)$ is the symmetrized distribution function in the normal metal reservoir and we have accounted for the boundary condition $f_T^S(E \ll \Delta, V, T) = 0$ at the superconducting terminal.

The form of the kinetic equation (8) suggests that electrons and holes (i.e. positive and negative energy quasiparticles) obey *independent* diffusion equations, which are only coupled through the boundary condition $f_T^S = 0$ at the superconducting terminal. Thus, we may try to apply the semiclassical Boltzmann-Langevin (BL) approach [21, 22]. The only modification is that we have to account for an energy- and space-dependent conductivity. For that purpose it is convenient to introduce the characteristic potential, defined as the solution of the equation :

$$\nabla\sigma(x,E)\nabla\nu(x,E) = 0 \qquad (10)$$

with the boundary condition that $\nu = 1$ at the normal terminal and $\nu = 0$ at the superconducting terminal. The solution for our quasi-one dimensional geometry is :

$$\nu(x,E) = G(E)\int_x^L \frac{dx}{\sigma(E,x)} \qquad (11)$$

The current noise can then be expressed in the familiar form :

$$S_I^{BL}(V) = 4\int dE \int dx\sigma(x,E)\left(\nabla\nu(x,E)\right)^2 f(x,E)(1 - f(x,E)) \qquad (12)$$

where the distribution function is given by $f(x,E) = \nu(x,E) f^N(E,V,T)$. As a result we find for the noise at zero temperature (i.e., $f_T^N(E,V,T = 0)$ = -sign(eV) for $|E| < |eV|$ and zero otherwise) :

$$S_I^{BL}(V) = \frac{4e}{3} I(V) \qquad (13)$$

where the current is given by $I(V) = (1/2e)\int_{-eV}^{eV} dE G(E)$. Thus, we find that the current noise depends in a non linear fashion on the voltage. The nonlinearity is given by the $I(V)$ characteristic. However, this dependence is in some sense trivial, since the only way the electron-hole coherence enters is through the energy dependent conductivity.

We note that the doubling of the shot noise in comparison to the normal case results from the *energy integration* from $-eV$ to $+eV$, instead of the interval between 0 (i.e., the Fermi energy) and eV, as it would be in the normal case. On the other hand, the average 'noisiness' (coming from the spatial integral in

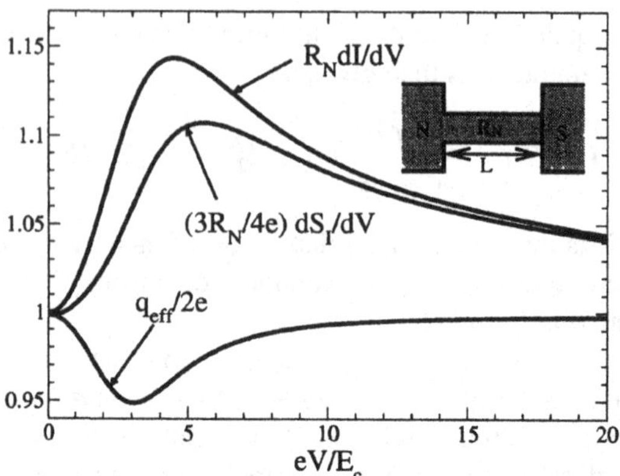

Figure 7. Transport characteristics of a proximity wire. The inset shows the layout. The relevant energy scale is the Thouless energy $E_c = \hbar D/L^2$. In the main plot the differential conductance, the differential noise $dS_I(V)/dV$, and the effective charge as defined in (14) at zero temperature are shown.

eq. (12) alone) is the *same* in the normal and the superconducting cases [22]. Thus, the doubling of the noise needs not be interpreted as a direct consequence of the doubled charge transfer involved in an Andreev reflection process. It reflects the particle-hole symmetry in the superconducting terminal. Nevertheless we adopt below the notion of an effective charge, since it is a convenient measure of the deviation from the independent electron fluctuations.

These arguments can also be used to explain the experiments on photon-assisted noise described previously. In the presence of an ac voltage of frequency ω the electron distribution in the normal terminal acquires side-bands, i.e., additional steps at energies $\pm n\hbar\omega$. The noise of the diffusive wire depends essentially on a superposition of left and right distribution functions differing by the voltage eV in the normal state and by $2eV$ in the NS case. It is clear that the noise properties change as a function of voltage, when sideband features match the other steps in the distribution function. As a consequence, photon-assisted steps occur when the voltage matches $n\hbar\omega/e$ in the normal case and $n\hbar\omega/2e$ in the superconducting case. This is what is observed in the experiments (see section 4).

A correct calculation of the noise requires that we go beyond the independent electron- and hole-fluctuations in the Boltzmann-Langevin approach. This can be accessed by the extended Green's function approach [23]. To describe the fluctuations, we utilize the results of the full counting statistics and define an effective charge

$$q_{eff}(V,T) = \frac{3}{2}\frac{\partial S_I(V,T)}{\partial I(V,T)}, \tag{14}$$

which takes the value $2e$ for the Boltzmann-Langevin result (see eq. (13)). It follows that the energy dependence of the effective charge gives information about the correlated electron-hole fluctuation processes.

In order to illustrate these considerations, we show in fig. 7 results for the conductance dI/dV, the differential noise $dS_I(V)/dV$, and the effective charge at zero temperature for a one-dimensional diffusive wire between a normal and a superconducting reservoir. Note: q_{eff} is proportional to $dS_I/dI = (dS_I/dV)$ $(dI/dV)^{-1}$. These results were obtained by a numerical solution of the quantum-kinetic equation. The differential conductance shows the well-known reentrance (peak) behaviour [16]. At low and high energy the conductance approaches the normal state value G_N. At intermediate energies of the order of the Thouless energy the conductance is enhanced by $\sim$ 15% above G_N. A similar result is found for the interferometer when $\Phi = 0$ [5]. We observe that the differential noise dS/dV has a roughly similar energy dependence[23], although it is quantitatively different. The deviation of the voltage-dependent effective charge from $2e$ demonstrates the energy dependence of the higher order correlations, which are not contained in the independent electron-hole picture, the BL approach. Around $E \sim 4E_c$ the effective charge is suppressed below $2e$, showing that the higher order correlations result in a reduced noise in comparison to the BL case of uncorrelated electron and hole fluctuations, for which $q_{eff} = 2e$. As $V \rightarrow 0$, q_{eff} approaches $2e$ again.

The physics of this effective charge is clarified if we consider the full counting statistics [23–26], instead of the current noise only. In full counting statistics, we obtains the *distribution of transferred charges*, which clearly contains direct information on both the statistics and the nature of the charge carriers. There it follows that *all* charge transfers at subgap-energies occur in units of $2e$ [27]. However, this does not necessarily result in a doubled effective charge using our definition of q_{eff}. The effective charge also includes the effect of correlations between the different charge transfers. Our work in section 5 shows that these correlations are phase-dependent.Nevertheless, in the limit of *uncorrelated* transfer of Andreev pairs, the effective charge at $E \gg k_B T$ is simply $2e$.

7. Conclusion

In this paper we have discussed how Andreev Reflection affects shot noise of mesoscopic NS structures. High frequency measurements provide a very powerfull tool since they are very sensitive, as is essential to the measurement of the small voltage- and flux dependence of the effective charge, and these measurements access an energy domain in which interesting phenomena occur, when $\hbar\omega > eV, k_B T$. These techniques are very promising for investigation of even more subtle quantities like cross-correlations in the noise or higher moments of the current fluctuations [28]. The measurements have revealed the existence of correlations in pair charge transfers which are not accessible through conductance measurements. The full counting statistics method gives access to the full distribution of the charge transfers, and is essential for understanding the physics of such NS structures. This method sheds light on the correlations revealed by the shot noise, and allows as well the investigation of cross-correlations and higher moments.

Acknowledgements

The authors thank R.J. Schoelkopf, A.A. Kozhevnikov, P.J. Burke and M.J. Rooks for collaboration on some of the experiments reported, and also M. Devoret and I. Siddiqi and Yu V. Nazarov for useful discussions. This work was supported by NSF DMR grant 0072022. The work of W.B. was supported by the Swiss NSF and the NCCR Nanoscience.

References

1. A.F. Andreev, Sov. Phys. JETP **19**, 1228 (1964).
2. For a review, see D. Esteve *et al.* in *Mesoscopic Electron Transport*, edited by L.L. Sohn, L.P. Kouwenhoven and G. Schön (Kluwer, Dordrecht, 1997); B. Pannetier and H. Courtois, J. Low Temp. Phys. **118**, 599 (2000).
3. R.J. Schoelkopf *et al.*, Phys. Rev. Lett. **78**, 3370 (1997).
4. A.A. Kozhevnikov, R.J. Schoelkopf and D.E. Prober, Phys. Rev. Lett. **84**, 3398 (2000).
5. B. Reulet *et al.*, cond-mat/0208089.
6. For a review, see Kogan, S.M. (1996) *Electronic Noise and Fluctuations in Solids*. Cambridge: Cambridge University Press
7. D. Rogovin and D.J. Scalapino, Ann. Phys. **86**, 1 (1974).
8. B.L. Altshuler, A.G. Aronov and D.E. Khmelnitsky, J. Phys. C **15**, 7367 (1982).
9. G. Bergman, Phys. Rep. **107**, 1 (1984).
10. T.A. Fulton and G.J. Dolan, Phys. Rev. Lett. **59**, 109 (1987).
11. J.B. Pieper, J.C. Price and J.M. Martinis, Phys. Rev. **B45**, 3857 (1992). J.B. Pieper and J.C. Price, Phys. Rev. Lett. **72**, 3586 (1994).
12. Y.M. Blanter and M. Büttiker, Phys. Rep. **336**, 1 (2000).
13. A.H. Steinbach, J.M. Martinis and M.H. Devoret, Phys. Rev. Lett. **76**, 3806 (1996).
14. A.A. Kozhevnikov *et al.*, J. Low Temp. Phys. **118**, 671 (2000).
15. J.R. Tucker and M.J. Feldman, Rev. Mod. Phys. **57**, 1055 (1985).
16. Yuli V. Nazarov and T. H. Stoof, Phys. Rev. Lett. **76**, 823 (1996).
17. A. F. Volkov, A. V. Zaitsev and T. M. Klapwijk, Physica C **59**, 21 (1993).
18. W. Belzig *et al.*, Superlattices Microst. **25**, 1251 (1999).
19. Yu. V. Nazarov, Superlattices Microst. **25**, 1221 (1999).
20. G. Eilenberger, Z. Phys. **214**, 195 (1968); A. I. Larkin and Yu. N. Ovchinnikov, Sov. Phys. JETP **26**, 1200 (1968); K. D. Usadel, Phys. Rev. Lett. **25**, 507 (1970).
21. K. E. Nagaev, Phys. Lett. A **169**, 103 (1992).
22. K. E. Nagaev and M. Büttiker, Phys. Rev. B **63**, 081301 (2001).
23. W. Belzig and Yu. V. Nazarov, Phys. Rev. Lett. **87**, 067006 (2001).

24. W. Belzig and Yu. V. Nazarov, Phys. Rev. Lett. **87**, 197006 (2001).
25. L. S. Levitov, H. W. Lee, and G. B. Lesovik, J. Math. Phys. **37**, 4845 (1996).
26. Yu. V. Nazarov, Ann. Phys. (Leipzig) **8**, SI-193 (1999).
27. B. A. Muzykantskii and D. E. Khmelnitzkii, Phys. Rev. B **50**, 3982 (1994).
28. B. Reulet, J. Senzier and D.E. Prober, unpublished.

CURRENT NOISE IN DIFFUSIVE SNS JUNCTIONS IN THE INCOHERENT MAR REGIME

E.V. BEZUGLYI and E.N. BRATUS'
B. Verkin Institute for Low Temperature Physics and Engineering, 61103 Kharkov, Ukraine

V.S. SHUMEIKO and G. WENDIN
Chalmers University of Technology, S-41296 Gothenburg, Sweden

1. Introduction

1.1. MULTIPLE ANDREEV REFLECTIONS

During last decade considerable progress has been made in the investigation and understanding the mechanisms of current transport in mesoscopic superconducting junctions. The term mesoscopic here refers to the junctions where bulk superconducting electrodes in equilibrium (reservoirs) are connected by small non-superconducting region with the size smaller than any inelastic mean free path. Such junctions include metallic atomic-size contacts, tunnel junctions, diffusive (metallic) and ballistic (2D electron gas) SNS junctions. A common feature of all these structures concerns the fact that the quasiparticles injected in the junction at zero temperature cannot escape into the reservoir unless the applied voltage is larger than the superconducting energy gap in the reservoir, $eV > 2\Delta$. In 1963, Schrieffer and Wilkins suggested that the necessary energy for the quasiparticle transmission at subgap voltage, $eV < 2\Delta$ can be provided by transferring Cooper pairs between the reservoirs [1].

The microscopic mechanism for such multiparticle transport, multiple Andreev reflections (MAR), was suggested in 1982 by Klapwijk, Blonder, and Tinkham [2, 3]. According to the MAR scenario formulated in terms of the scattering theory, injected quasiparticles repeatedly undergo Andreev reflections from the superconducting reservoirs, gaining energy eV during each traversal of the junction, which allows them to eventually escape from the junction, see Fig. 1. As the

Y.V. Nazarov (ed.), *Quantum Noise in Mesoscopic Physics*, 93–118.

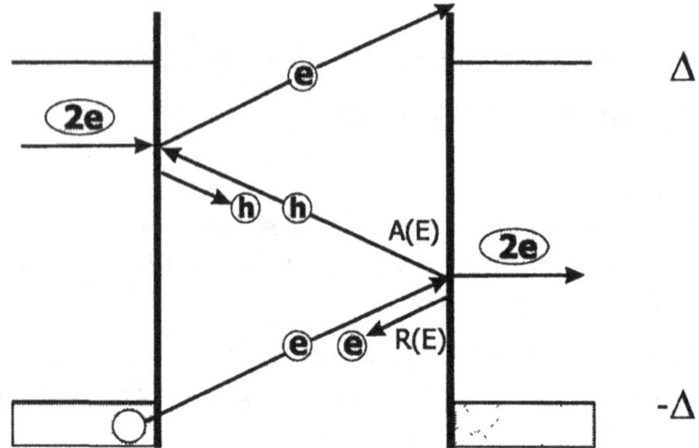

Figure 1. Multiple Andreev reflections in superconducting junction. Quasiparticle energy increases by eV after every traversal of the junction, generating spectral flow across the energy gap, $(-\Delta, \Delta)$. At the interface, quasiparticle undergoes Andreev reflection with probability $A(E)$, and normal reflection with probability $R(E)$.

result, a spectral flow across the energy gap is generated, which creates strongly non-equilibrium quasiparticle distribution within the contact area. This mechanism explains the nature of the dissipative current in voltage biased junctions. Also it allows one to anticipate considerable enhancement of the current shot noise. Indeed, transmission of one quasiparticle across the energy gap at applied voltage V requires $N = \text{Int}(2\Delta/eV)$ Andreev reflections [Int(x) denoting the integer part of x], which transfer the total charge, $q^{\mathit{eff}} = (N+1)e$, between the electrodes. This enhancement of the transmitted charge gives rise to the enhancement of the current shot noise compared to the case of normal junction, according to the Schottky formula, $S = 2q^{\mathit{eff}} I$, where S is the spectral density of the noise at zero frequency.

The mechanism of MAR is general for all kinds of superconducting junctions. However, the MAR transport regime appears differently in junctions having small and large lengths. In short junctions where the distance between the superconducting electrodes is small compared to the superconducting coherence length, $\xi_0 = \hbar v_F/\Delta$, such as point contacts and tunnel junctions, consequent Andreev reflections are fully coherent. Essential feature of this coherent MAR regime is the ac Josephson effect, and also highly non-linear I-V characteristics with the subharmonic gap structure (SGS) – sequence of current structures at voltages $eV = 2\Delta/n$ (n is an integer) [4].

The same features, the ac Josephson effect and SGS, appear also in ballistic SNS junctions and short diffusive SNS junctions [5, 6]. In diffusive SNS junctions, the electron-hole coherence in the normal metal persists over a distance of the coherence length $\xi_E = \sqrt{\hbar\mathcal{D}/2E}$ from the superconductor ($\mathcal{D}$ is the diffusion constant). The overlap of coherent proximity regions induced by both SN inter-

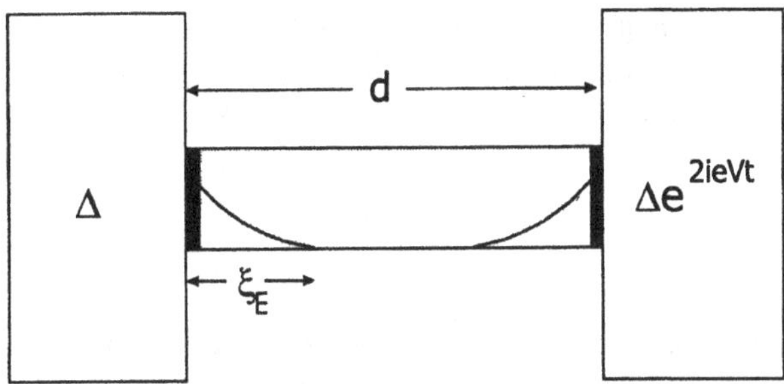

Figure 2. Proximity effect in long diffusive SNS junctions. Shaded region indicates diffusive metal connecting superconducting reservoirs marked with Δ; bold lines indicate interface resistances. Superconducting correlations exponentially decay over length ξ_E, which is small compared to the junction length d

faces creates an energy gap in the electron spectrum of the normal metal. In short junctions with a small length d compared to the coherence length, $d \ll \xi_\Delta$, and with a large proximity gap of the order of the energy gap Δ in the superconducting electrodes, the phase coherence covers the entire normal region.

Rather different, incoherent MAR regime occurs in long diffusive SNS junctions with a small proximity gap of the order of Thouless energy $E_{Th} = \hbar\mathcal{D}/d^2 \ll \Delta$ [7]. If E_{Th} is also small on a scale of applied voltage, $E_{Th} \ll eV$, then the coherence length ξ_E is much smaller than the junction length at all relevant energies $E \sim min(eV, \Delta)$. In this case, the proximity regions near the SN interfaces become virtually decoupled, as shown in Fig. 2, and the Josephson current oscillations are strongly suppressed. At the same time, the quasiparticle distribution inside the energy gap is strongly non-equilibrium as soon as an inelastic mean free path exceeds the junction length, $l_\varepsilon \gg d$, because the subgap electrons must undergo many (incoherent) Andreev reflections before they enter the reservoir.

It is interesting that in the junctions with transparent interfaces, the complex transport mechanism of incoherent MAR is not clearly revealed by I-V characteristics of the junction, which are close to the Ohm's law. Thus the excess shot noise becomes the central characteristic of interest in this regime. Quasi-linear behavior of I-V characteristics in the incoherent MAR regime can be understood from the following argument. The MAR transport in real space through the junction with normal resistance R_N is associated with a spectral current flow along the energy axis through a series of $N + 1$ resistors with the total effective resistance $R = (N + 1)R_N$. Since only electrons incoming within the energy layer eV below the gap edge, $-\Delta$, participate in the MAR transport, the spectral current I_p is given by the equation $I_p = V/R$. However, each pair of consecutive Andreev reflections transfers the charge $2e$ through the junction, and the real current I is therefore $N+1$ times larger than the spectral current: $I = (N+1)I_p = V/R_N$. [1] It is clear from this argument, that particular nature of the normal resistance (e.g. tunnel resistance instead of diffusive metallic resistance) plays no essential role in the behavior of the current.

1.2. CURRENT SHOT NOISE

Shot noise in mesoscopic conductors has been extensively studied during recent years (for a review see [8]). In ballistic normal electron systems with tunnel barriers, the shot noise is produced by electron tunnelling through the barriers, and it is described by the Schottky formula $S = 2eI$ [9]. In diffusive metallic wires the shot noise is due to elastic electron scattering by the impurities, and in this case, an additional factor $1/3$ appears in the Schottky formula, $S = (2/3)eI$ [10–12]. This Fano factor is modified in long wires whose length exceeds the inelastic scattering length due to the effect of electron-electron and electron-phonon relaxation [10, 13, 14].

In normal-superconducting (NS) systems, the current shot noise produced by the impurity scattering is enhanced at subgap voltages, $eV < \Delta$, by the factor of two compared to the normal conductors [15, 16]. This is because the subgap current transport involves one Andreev reflection, which results in the transfer through the junction of an elementary charge $2e$, instead of e. More elaborated analysis shows further enhancement of the shot noise at small voltage due to the contribution of the proximity region near the NS interface [17].

In superconducting junctions, the shot noise power is tremendously enhanced at small voltage by the factor $q^{eff}/e \sim 2\Delta/eV \gg 1$, because of the MAR. For the coherent MAR regime this effect has been well theoretically established [18–20] and experimentally investigated [21–23]. For the incoherent MAR regime, similar effect of the shot noise enhancement has been theoretically predicted in Refs. [25–27]. The experimental observation of multiply enhanced shot noise in long SNS junctions was reported in Refs. [23, 24, 28]. Although there is the same mechanism of the noise enhancement in both cases, there is remarkable difference between the behavior of the noise power at small voltage in the ballistic and diffusive junctions. In the ballistic junctions, the current exponentially decreases with voltage [4] while the effective charge grows inversely proportionally to the voltage, hence the shot noise power also exponentially decreases. On the other hand, in the diffusive junctions, the current decay follows a power law. In particular, in the incoherent MAR regime, the current approximately follows the Ohm's law, and therefore the noise power approaches a constant value at zero voltage, $S \sim \Delta/R_N$. This effect, which results from the enhanced effective charge, can

[1] In this argument we neglected the small effect of proximity corrections, which are responsible, together with the interface resistance, for the SGS in the incoherent MAR regime [7].

be also explained as the result of strong electron non-equilibrium distribution developed by the MAR process with an effective noise temperature $T_0 \sim \Delta$.

The finite noise level at zero voltage is a very interesting property of the incoherent MAR regime, which can be employed for the investigation of the effect of inelastic relaxation on MAR. The MAR transport regime assumes the time spent by a quasiparticle within the junction area, τ_{dwell}, to be small compared to the inelastic relaxation time τ_ε. However, at $eV \to 0$, the dwelling time infinitely increases because of MAR, and inelastic relaxation unavoidably starts to play a role. The inelastic scattering suppresses the spectral flow upwards in energy generated by MAR, and thus the MAR regime is destroyed. As the result, the normal region of the junction becomes an equilibrium reservoir, and the SNS junction turns into two NS junctions connected in series. This is manifested by the decrease of the noise level at small voltage, the cross over being controlled by the ratio $\tau_\varepsilon/\tau_{dwell}$. Precisely this behavior of the shot noise has been observed in the experiment [24, 28]. It is important that because of the absence of the Josephson effect in the incoherent MAR regime, a small-voltage region of the dissipative current branch, $E_{Th} \ll eV \ll \Delta$, can be experimentally accessed without switching to the Josephson branch.

In this paper, we outline a theory of the current shot noise in the incoherent MAR regime. We will discuss the "collisionless" limit of perfect MAR as well as the effect of inelastic relaxation. While theoretical analysis of the current shot noise in the coherent MAR regime can be done on the basis of the scattering theory [18, 19, 29], the present case requires different approach, which operates with electron and hole diffusion flows rather than with ballistic quasiparticle trajectories, and considers the Andreev reflections as the relationships between these diffusive flows rather than the scattering events. Such formalism has been developed in Ref. [7], and it is outlined in the next section.

2. Circuit theory of incoherent MAR

2.1. MICROSCOPIC BACKGROUND

The system under consideration consists of a normal channel ($0<x<d$) confined between two voltage biased superconducting electrodes, with the elastic mean free path ℓ much shorter than any characteristic size of the problem. In this limit, the microscopic analysis of current transport can be performed in the framework of the diffusive equations of nonequilibrium superconductivity [30] for the 4×4 matrix Keldysh-Green function $\check{G}(t_1 t_2, x)$,

$$[\check{H}, \check{G}] = i\hbar\mathcal{D}\partial_x \check{J}, \quad \check{J} = \check{G}\partial_x\check{G}, \quad \check{G}^2 = \check{1}, \tag{1}$$

$$\check{H} = \check{1}[i\hbar\sigma_z\partial_t - e\phi(t) + \hat{\Delta}(t)], \quad \hat{\Delta} = \Delta e^{i\sigma_z\chi} i\sigma_y, \tag{2}$$

where Δ, χ are the modulus and the phase of the order parameter, and ϕ is the electric potential. The Pauli matrices σ_i operate in the Nambu space of 2×2

matrices denoted by "hats", the products of two-time functions are interpreted as their time convolutions. The electric current I per unit area is expressed through the Keldysh component $\hat{J}^K$ of the matrix current $\check{J}$,

$$I(t) = (\pi\hbar\sigma_N/4e)\,\mathrm{Tr}\,\sigma_z\hat{J}^K(tt,x), \tag{3}$$

where σ_N is the conductivity of the normal metal.

At the SN interface, the matrix $\check{G}$ satisfies the boundary condition [31]

$$\left(\sigma_N\check{J}\right)_{\pm 0} = (2R_{SN})^{-1}\left[\check{G}_{-0}, \check{G}_{+0}\right], \tag{4}$$

where the indices ± 0 denote the right and left sides of the interface and R_{SN} is the interface resistance per unit area in the normal state. Within the model of infinitely narrow potential of the interface barrier, $U(x) = H\delta(x)$, the interface resistance is related to the barrier strength $Z = H(\hbar v_F)^{-1}$ as $R_{SN} = 2\ell Z^2/3\sigma_N$ [32]. It has been shown in Ref. [33] that Eq. (4) is valid either for a completely transparent interface ($R_{SN} \to 0$, $\check{G}_{+0} = \check{G}_{-0}$) or for an opaque barrier whose resistance is much larger than the resistance $R(\ell) = \ell/\sigma_N$ of a metal layer with the thickness formally equal to ℓ.

According to the definition of the matrix $\check{G}$,

$$\check{G} = \begin{pmatrix} \hat{g}^R & \hat{G}^K \\ 0 & \hat{g}^A \end{pmatrix}, \quad \hat{G}^K = \hat{g}^R\hat{f} - \hat{f}\hat{g}^A, \tag{5}$$

Eqs. (1) and (4) represent a compact form of separate equations for the retarded and advanced Green's functions $\hat{g}^{R,A}$ and the distribution function $\hat{f} = f_+ + \sigma_z f_-$. Their time evolution is imposed by the Josephson relation $\chi(t) = 2eVt$ for the phase of the order parameter in the right electrode (we assume $\chi = 0$ in the left terminal). This implies that the function $\check{G}(t_1t_2, x)$ consists of a set of harmonics $\check{G}(E_n, E_m, x)$, $E_n = E + neV$, which interfere in time and produce the ac Josephson current. However, when the junction length d is much larger than the coherence length ξ_E at all relevant energies, we may consider coherent quasiparticle states separately at both sides of the junction, neglecting their mutual interference and the ac Josephson effect. Thus, the Green's function in the vicinity of left SN interface can be approximated by the solution $\hat{g} = \sigma_z\,\mathrm{ch}\,\theta + i\sigma_y\,\mathrm{sh}\,\theta$ of the static Usadel equations for a semi-infinite SN structure [34], with the spectral angle $\theta(E, x)$ satisfying the equation

$$\mathrm{th}[\theta(E,x)/4] = \mathrm{th}[\theta_N(E)/4]\exp(-x/\xi_E\sqrt{i}), \tag{6}$$

with the boundary condition

$$W\sqrt{i\Delta/E}\,\mathrm{sh}(\theta_N - \theta_S) + 2\,\mathrm{sh}(\theta_N/2) = 0. \tag{7}$$

The indices S, N in these equations refer to the superconducting and the normal side of the interface, respectively.

The dimensionless parameter W in Eq. (7),

$$W = \frac{R(\xi_\Delta)}{R_{SN}} = \frac{\xi_\Delta}{dr}, \quad r = \frac{R_{SN}}{R_N}, \tag{8}$$

where $R_N = R(d) = d/\sigma_N$ is the resistance of the normal channel per unit area, has the meaning of an effective barrier transmissivity for the spectral functions [35]. Note that even at large barrier strength $Z \gg 1$ ensuring the validity of the boundary conditions Eq. (4) [33], the effective transmissivity $W \sim (\xi_\Delta/\ell)Z^{-2}$ of the barrier in a "dirty" system, $\ell \ll \xi_\Delta$, could be large. In this case, the spectral functions are virtually insensitive to the presence of a barrier and, therefore, the boundary conditions Eqs. (4) can be applied to an arbitrary interface if we approximately consider highly transmissive interfaces with $W > \xi_\Delta/\ell \gg 1$ as completely transparent, $W = \infty$.

The distribution functions $f_\pm(E,x)$ are to be considered as global quantities within the whole normal channel determined by the kinetic equations

$$\partial_x[D_\pm(E,x)\partial_x f_\pm(E,x)] = 0, \tag{9}$$

with dimensionless diffusion coefficients

$$\begin{aligned} D_+ &= (1/4)\,\mathrm{Tr}\left(1 - \hat{g}^R\hat{g}^A\right) = \cos^2 \mathrm{Im}\,\theta, \\ D_- &= (1/4)\,\mathrm{Tr}\left(1 - \sigma_z\hat{g}^R\sigma_z\hat{g}^A\right) = \mathrm{ch}^2\,\mathrm{Re}\,\theta. \end{aligned} \tag{10}$$

Assuming the normal conductance of electrodes to be much larger than the junction conductance, we consider them as equilibrium reservoirs with unperturbed spectrum, $\theta_S = \mathrm{Arctanh}(\Delta/E)$, and equilibrium quasiparticle distribution, $\hat{f}_S(E) = f_0(E) \equiv \mathrm{th}(E/2T)$. Within this approximation, the boundary conditions for the distribution functions at $x = 0$ read

$$\sigma_N D_+ \partial_x f_+(E,0) = G_+(E)[f_+(E,0) - f_0(E)], \tag{11}$$

$$\sigma_N D_- \partial_x f_-(E,0) = G_-(E) f_-(E,0), \tag{12}$$

where

$$G_\pm(E) = R_{SN}^{-1}\left(N_S N_N \mp M_S^\pm M_N^\pm\right), \tag{13}$$

$$N(E) = \mathrm{Re}(\mathrm{ch}\,\theta), \; M^+(E) + iM^-(E) = \mathrm{sh}\,\theta. \tag{14}$$

At large energies, $|E| \gg \Delta$, when the normalized density of states $N(E)$ approaches unity and the condensate spectral functions $M^\pm(E)$ turn to zero at both sides of the interface, the conductances $G_\pm(E)$ coincide with the normal barrier conductance; within the subgap region $|E| < \Delta$, $G_+(E) = 0$.

Similar considerations are valid for the right NS interface, if we eliminate the time dependence of the order parameter in Eq. (1), along with the potential of right electrode, by means of a gauge transformation [36]

$$\check{G}(t_1t_2,x) = \exp(i\sigma_z eVt_1)\tilde{\check{G}}(t_1t_2,x)\exp(-i\sigma_z eVt_2). \tag{15}$$

As a result, we arrive at the same static equations and boundary conditions, Eqs. (6)-(14), with $x \to d - x$, for the gauge-transformed functions $\tilde{\hat{g}}(E,x)$ and $\tilde{\hat{f}}(E,x)$. Thus, to obtain a complete solution, e.g. for the distribution function f_-, which determines the dissipative current,

$$I = \frac{\sigma_N}{2e}\int_{-\infty}^{\infty} dE\, D_- \partial_x f_-, \tag{16}$$

one must solve the boundary problem for $\hat{f}(E,x)$ at the left SN interface, and a similar boundary problem for $\tilde{\hat{f}}(E,x)$ at the right interface, and then match the distribution function asymptotics deep inside the normal region by making use of the relationship following from Eqs. (5), (15),

$$\hat{f}(E,x) = \tilde{\hat{f}}(E + \sigma_z eV, x). \tag{17}$$

2.2. CIRCUIT REPRESENTATION OF BOUNDARY CONDITIONS

In order for this kinetic scheme to conform to the conventional physical interpretation of Andreev reflection in terms of electrons and holes, we introduce the following parametrization of the matrix distribution function,

$$\hat{f}(E,x) = 1 - 2\begin{pmatrix} n^e(E,x) & 0 \\ 0 & n^h(E,x) \end{pmatrix}, \tag{18}$$

where n^e and n^h will be considered as the electron and hole population numbers. Deep inside the normal metal region, they acquire rigorous meaning of distribution functions of electrons and holes, and approach the Fermi distribution in equilibrium. In this representation, Eqs. (9) take the form

$$D_\pm(E,x)\partial_x n_\pm(E,x) = \text{const} \equiv -I_\pm(E)/\sigma_N, \tag{19}$$

where $n_\pm = n^e \pm n^h$, and they may be interpreted as conservation equations for the (specifically normalized) net probability current I_+ of electrons and holes, and for the electron-hole imbalance current I_-. Furthermore, the probability currents of electrons and holes, defined as $I^{e,h} = (1/2)(I_+ \pm I_-)$, separately obey the conservation equations. The probability currents $I^{e,h}$ are naturally related to the electron and hole diffusion flows, $I^{e,h} = -\sigma_N \partial_x n^{e,h}$, at large distances $x \gg \xi_E$

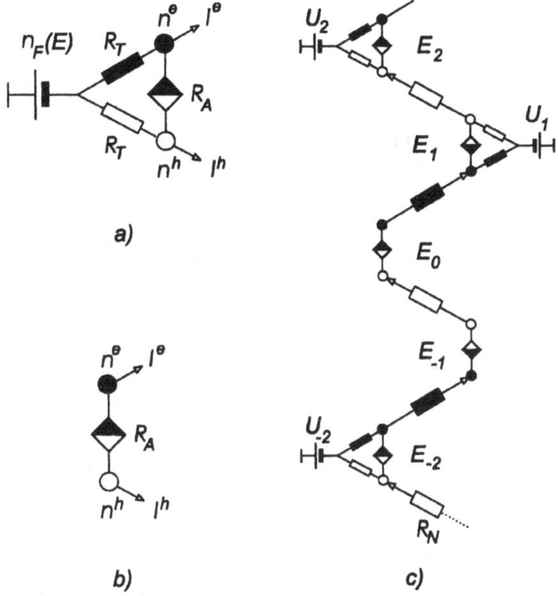

Figure 3. Elementary equivalent circuits representing boundary conditions in Eq. (21) for the electron and hole population numbers $n^{e,h}(E,0)$ and probability currents $I^{e,h}(E)$, at energies outside the gap, $|E| > \Delta$ (a), and within the subgap region, $|E| < \Delta$ (b); equivalent network in (x, E)-space for incoherent MAR in SNS junction (c). Filled and empty symbols stand for electron- and hole- related elements, respectively; half-filled squares denote Andreev resistors; $U_n = n_F(E_n)$.

from the SN boundary. Within the proximity region, $x < \xi_E$, each current $I^{e,h}$ generally consists of a combination of both the electron and hole diffusion flows, which reflects coherent mixing of normal electron and hole states in this region,

$$I^{e,h} = -(\sigma_N/2)\left[(D_+ \pm D_-)\partial_x n^e + (D_+ \mp D_-)\partial_x n^h\right]. \tag{20}$$

In terms of electrons and holes, the boundary conditions in Eqs. (11), (12) read

$$I^{e,h} = G_T\left(n_F - n^{e,h}\right) \mp G_A\left(n^e - n^h\right), \tag{21}$$

where

$$G_T = G_+,\ G_A = (G_- - G_+)/2. \tag{22}$$

Each of the equations (21) may be clearly interpreted as a Kirchhoff rule for the electron or hole probability current flowing through the effective circuit (tripole) shown in Fig. 3(a). Within such interpretation, the nonequilibrium populations of electrons and holes $n^{e,h}$ at the interface correspond to "potentials" of nodes attached to the "voltage source" – the Fermi distribution $n_F(E)$ in the superconducting reservoir – by "tunnel resistors" $R_T(E) = G_T^{-1}(E)$. The "Andreev resistor" $R_A(E) = G_A^{-1}(E)$ between the nodes provides electron-hole conversion (Andreev reflection) at the SN interface.

The circuit representation of the diffusive SN interface is analogous to the scattering description of ballistic SN interfaces: the tunnel and Andreev resistances play the same role as the normal and Andreev reflection coefficients in the ballistic case [32]. For instance, at $|E| > \Delta$ [Fig. 3(a)], the probability current I^e is contributed by equilibrium electrons incoming from the superconductor through the tunnel resistor R_T, and also by the current flowing through the Andreev resistor R_A as the result of hole-electron conversion. Within the subgap region, $|E| < \Delta$ [Fig. 3(b)], the quasiparticles cannot penetrate into the superconductor, $R_T = \infty$, and the voltage source is disconnected, which results in detailed balance between the electron and hole probability currents, $I^e = -I^h$ (complete reflection). For the perfect interface, R_A turns to zero, and the electron and hole population numbers become equal, $n^e = n^h$ (complete Andreev reflection). The nonzero value of the Andreev resistance for $R_{SN} \neq 0$ accounts for suppression of Andreev reflection due to the normal reflection by the interface.

Detailed information about the boundary resistances can be obtained from their asymptotic expressions in Ref. [7]. In particular, $R_{\pm}(E)$ turns to zero at the gap edges due to the singularity in the density of states which enhances the tunnelling probability. The resistance $R_-(E)$ approaches the normal value R_{SN} at $E \to 0$ due to the enhancement of the Andreev reflection at small energies, which results from multiple coherent backscattering of quasiparticles by the impurities within the proximity region. [2]

The proximity effect can be incorporated into the circuit scheme by the following way. We note that the diffusion coefficients $D_{\pm}$ in Eq. (10) turn to unity far from the SN boundary, and therefore the population numbers $n^{e,h}$ become linear functions of x,

$$n^{e,h}(E,x) \approx \overline{n}^{e,h}(E,0) - R_N I^{e,h}(E)x/d. \tag{23}$$

This equation defines the renormalized population numbers $\overline{n}^{e,h}(E,0)$ at the NS interface, which differ from $n^{e,h}(E,0)$ due to the proximity effect, as shown in Fig. 4. These quantities have the meaning of the true electron/hole populations which would appear at the NS interface if the proximity effect had been switched off. It is possible to formulate the boundary conditions in Eq. (21) in terms of these population numbers by including the proximity effect into renormalization of the tunnel and Andreev resistances. To this end, we will associate the node potentials with renormalized boundary values $\overline{n}^{e,h}(E,0) = (1/2)[\overline{n}_+(E,0) \pm \overline{n}_-(E,0)]$ of the population numbers, where $\overline{n}_{\pm}(E,0)$ are found from the exact solutions of

[2] This property is the reason for the re-entrant behavior of the conductance of high-resistive SIN systems [37, 38] at low voltages. In the MAR regime, one cannot expect any reentrance since quasiparticles at all subgap energies participate in the charge transport even at small applied voltage.

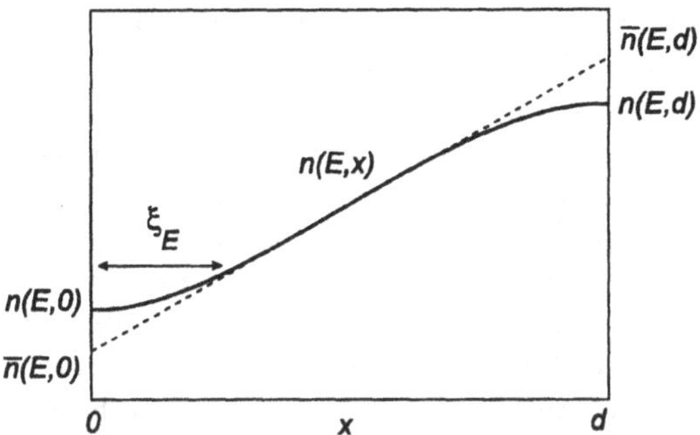

Figure 4. Qualitative behavior of population numbers within the normal channel (solid curve). The edge distortions of the linear x-dependence of population numbers, Eq. (23), occur within the proximity regions. The difference between the boundary population numbers $n(E,0)$, $n(E,d)$ and their effective values $\overline{n}(E,0)$, $\overline{n}(E,d)$ for true normal electrons and holes is included in the renormalization of the boundary resistances, Eq. (26).

Eqs. (19),

$$\overline{n}_{\pm}(E,0) = n_{\pm}(E,0) - m_{\pm}(E)I_{\pm}(E). \tag{24}$$

Here $m_{\pm}(E)$ are the proximity corrections to the normal metal resistance at given energy for the probability and imbalance currents, respectively,

$$m_{\pm}(E) = \pm R_N \int_0^{\infty} \frac{dx}{d} \left| D_{\pm}^{-1}(E,x) - 1 \right|. \tag{25}$$

It follows from Eq. (24) that the same Kirchhoff rules as in Eqs. (21), (22) hold for $\overline{n}^{e,h}(E,0)$ and $I^{e,h}(E)$, if the bare resistances $R_{\pm}$ are substituted by the renormalized ones,

$$R_{\pm}(E) \to \overline{R}_{\pm}(E) = R_{\pm}(E) + m_{\pm}(E). \tag{26}$$

In certain cases, there is an essential difference between the bare and renormalized resistances, which leads to qualitatively different properties of the SN interface for normal electrons and holes compared to the properties of the bare boundary. Let us first discuss a perfect SN interface with $R_{SN} \to 0$. Within the subgap region $|E| < \Delta$, the bare tunnel resistance R_T is infinite whereas the bare Andreev resistance R_A turns to zero; this corresponds to complete Andreev reflection, as already explained. However, the Andreev resistance for normal electrons and holes, $\overline{R}_A(E) = 2m_-(E)$, is finite and negative, [3] which leads to enhancement of the normal metal conductivity within the proximity region [38, 39]. At $|E| > \Delta$, the bare tunnel resistance R_T is zero, while the renormalized tunnel

[3] In the terms of the circuit theory, this means that the "voltage drop" between the electron and hole nodes is directed against the probability current flowing through the Andreev resistor.

resistance $\overline{R}_T(E) = m_+(E)$ is finite (though rapidly decreasing at large energies). This leads to suppression of the probability currents of normal electrons and holes within the proximity region, which is to be attributed to the appearance of Andreev reflection. Such a suppression is a global property of the proximity region in the presence of sharp spatial variation of the order parameter, and it is similar to the over-the-barrier Andreev reflection in the ballistic systems. In the presence of normal scattering at the SN interface, the overall picture depends on the interplay between the bare interface resistances $R_\pm$ and the proximity corrections $m_\pm$; for example, the renormalized tunnel resistance $\overline{R}_T(E)$ diverges at $|E| \to \Delta$, along with the proximity correction $m_+(E)$, in contrast to the bare tunnel resistance $R_T(E)$. This indicates complete Andreev reflection at the gap edge independently of the transparency of the barrier, which is similar to the situation in the ballistic systems where the probability of Andreev reflection at $|E| = \Delta$ is always equal to unity.

2.3. MAR NETWORKS

To complete the definition of an equivalent MAR network, we have to construct a similar tripole for the right NS interface and to connect boundary values of population numbers (node potentials) using the matching condition in Eq. (17) expressed in terms of electrons and holes,

$$n^{e,h}(E,x) = \widetilde{n}^{e,h}(E \pm eV, x). \tag{27}$$

Since the gauge-transformed distribution functions $\widetilde{f}_\pm$ obey the same equations Eq. (9)-(14), the results of the previous Section can be applied to the functions $\widetilde{n}^{e,h}(E)$ and $-\widetilde{I}^{e,h}(E)$ (the minus sign implies that $\widetilde{I}$ is associated with the current incoming to the right-boundary tripole). In particular, the asymptotics of the gauge-transformed population numbers far from the right interface are given by the equation

$$\widetilde{n}^{e,h}(E,x) \approx \overline{\widetilde{n}}^{e,h}(E,d) + R_N \widetilde{I}^{e,h}(E)\,(1 - x/d)\,. \tag{28}$$

After matching the asymptotics in Eqs. (23) and (28) by means of Eq. (27), we find the following relations,

$$I^{e,h}(E) = \widetilde{I}^{e,h}(E \pm eV), \tag{29}$$

$$\overline{n}^{e,h}(E,0) - \overline{\widetilde{n}}^{e,h}(E \pm eV, d) = R_N I^{e,h}(E). \tag{30}$$

¿From the viewpoint of the circuit theory, Eq. (30) may be interpreted as Ohm's law for the resistors R_N which connect energy-shifted boundary tripoles, separately for the electrons and holes, as shown in Fig. 3(c).

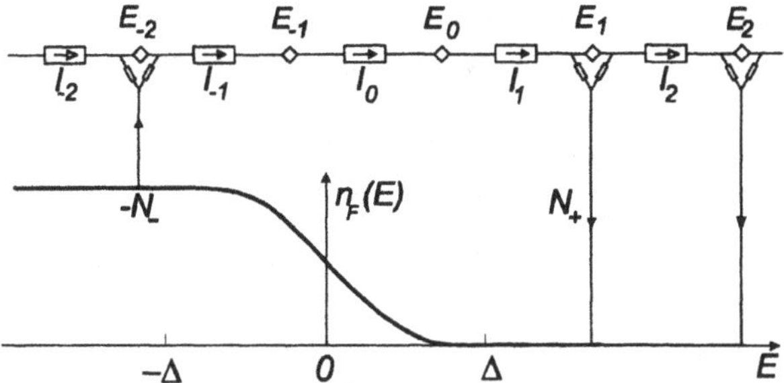

Figure 5. MAR network of Fig. 3(c) in energy space. The nodes outside the gap are connected with the distributed voltage source $n_F(E)$ (bold curve); the subgap nodes are disconnected from the voltage source.

The final step which essentially simplifies the analysis of the MAR network, is based on the following observation. The spectral probability currents $I^{e,h}$ yield opposite contributions to the electric current in Eq. (16),

$$I = \frac{1}{2e} \int_{-\infty}^{\infty} dE \left[I^e(E) - I^h(E) \right], \tag{31}$$

due to the opposite charge of electrons and holes. At the same time, these currents, referred to the energy axis, transfer the charge in the same direction, viz., from bottom to top of Fig. 3(c), according to our choice of positive eV. Thus, by introducing the notation $I_n(E)$ for an electric current entering the node n, as shown by arrows in Fig. 3(c),

$$I_n(E) = \begin{cases} I^e(E_{n-1}), & n = 2k+1, \\ -I^h(E_n), & n = 2k, \end{cases}, \quad E_n = E + neV, \tag{32}$$

we arrive at an "electrical engineering" problem of current distribution in an equivalent network in energy space plotted in Fig. 5, where the difference between electrons and holes becomes unessential. The bold curve in Fig. 5 represents a distributed voltage source – the Fermi distribution $n_F(E)$ connected periodically with the network nodes. Within the gap, $|E_n| < \Delta$, the nodes are disconnected from the Fermi reservoir and therefore all partial currents associated with the subgap nodes are equal.

Since all resistances and potentials of this network depend on $E_n = E + neV$, the partial currents obey the relationship $I_n(E) = I_k[E + (n-k)eV]$ which allows us to express the physical electric current, Eq. (31), through the sum of all partial currents I_n flowing through the normal resistors R_N, integrated over an elementary energy interval $0 < E < eV$,

$$I = \frac{1}{2e} \int_{-\infty}^{\infty} dE \left[I_1(E) + I_0(E) \right] = \frac{1}{e} \int_0^{eV} dE\, J(E), \quad J(E) = \sum_{n=-\infty}^{+\infty} I_n(E). \tag{33}$$

The spectral density $J(E)$ is periodic in E with the period eV and symmetric in E, $J(-E) = J(E)$, which follows from the symmetry of all resistances with respect to E.

As soon as the partial currents are found, the population numbers can be recovered by virtue of Eqs. (19), (21), (23), and (32),

$$n^{e,h}(E,x) = \overline{n}^{e,h}(E,0) \mp R_N I_{1,0}(E)x/d, \tag{34}$$

$$\overline{n}^{e,h}(E,0) = n_F - (1/2)\left[\overline{R}_+(I_1 - I_0) \pm \overline{R}_-(I_1 + I_0)\right] \tag{35}$$

at $|E| > \Delta$. Within the subgap region, Eq. (35) is inapplicable due to the indeterminacy of product $\overline{R}_+(I_1 - I_0)$. In this case, one may consider the subgap part of the network as a voltage divider between the nodes nearest to the gap edges, having the numbers $-N_-$, N_+, respectively, where

$$N_\pm = \mathrm{Int}[(\Delta \mp E)/eV] + 1. \tag{36}$$

Then the boundary populations at $|E| < \Delta$ become

$$\overline{n}^{e,h}(E,0) = n^{L,R}(E_{\pm N_\pm}) \pm I_0 \left[N_\pm R_N + \sum_{k=1}^{N_\pm - 1} R_A(E_{\pm k})\right], \tag{37}$$

where R, L indicate the right (left) node of the tripole, irrespectively of whether it relates to the left (even n) or right (odd n) interface. The physical meaning of $n^{R,L}(E_n)$, however, depends on the parity of n,

$$n^{R,L}(E_n) = \begin{cases} \overline{n}^{e,h}(E_n, 0), & n = 2k, \\ \overline{\overline{n}}^{h,e}(E_n, d), & n = 2k+1. \end{cases} \tag{38}$$

The values $n^{R,L}$ in Eq. (37) can be found from Eq. (35) which is generalized for any tripole of the network in Fig. 5 outside the gap as

$$n^{R,L}(E_n) = n_F(E_n) - \frac{1}{2}\left[\overline{R}_+(E_n)(I_{n+1} - I_n) \pm \overline{R}_-(E_n)(I_{n+1} + I_n)\right]. \tag{39}$$

As follows from Eqs. (35), (37), the energy distribution of quasiparticles has a step-like form (Fig. 6), which is qualitatively similar to, but quantitatively different from that found in OTBK theory [3]. The number of steps increases at low voltage, and the shape of the distribution function becomes resemblant to a "hot electron" distribution with the effective temperature of the order of Δ. This distribution is modulated due to the discrete nature of the heating mechanism of MAR, which transfers the energy from an external voltage source to the quasiparticles by energy quanta eV.

The circuit formalism can be simply generalized to the case of different transparencies of NS interfaces, as well as to different values of Δ in the electrodes. In

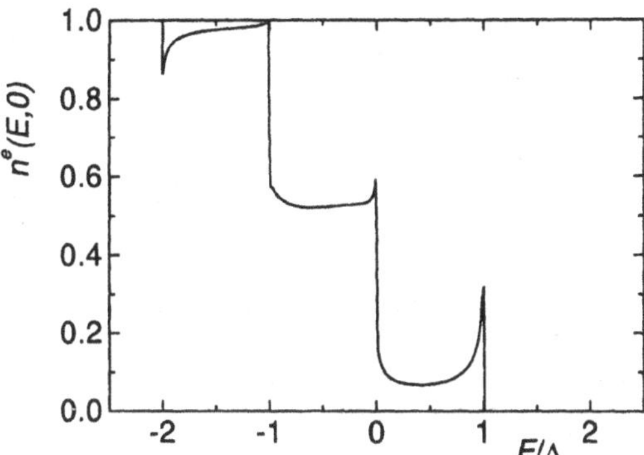

Figure 6. Energy dependence of the electron population number $n^e(E, 0)$ at the left interface of the SNS junction with $R_{SN} = R_N$ and $d = 5\xi_\Delta$, at $V = \Delta/e$ and $T = 0$.

this case, the network resistances become dependent not only on E_n but also on the parity of n. As a result, the periodicity of the current spectral density doubles: $J(E) = J(E+2eV)$, and, therefore, $J(E)$ is to be integrated in Eq. (33) over the period $2eV$, with an additional factor $1/2$.

3. Current shot noise in the MAR regime

3.1. NOISE OF NORMAL CONDUCTOR

The main source of the shot noise in long diffusive SNS junctions with transparent interfaces ($R_{NS} \ll R_N$) is the impurity scattering of electrons in the normal region of the junction. In this section we apply the circuit theory to calculate the noise of the normal conductor in the junction [26]. The effect of proximity regions near NS interfaces can be neglected for $\xi_\Delta \ll d$ since their length is small compared to the length of the diffusive conductor. The noise of the diffusive normal region can be calculated within a Langevin approach [40]. Following Ref. [41], in which the Langevin equation was applied to the current fluctuations in a diffusive NS junction, we derive an expression for the current noise spectral density in SNS junctions at zero frequency in terms of the nonequilibrium population numbers $n^{e,h}(E, x)$ of electrons and holes within the normal metal, $0 < x < d$,

$$S = \frac{2}{R_N} \int_0^d \frac{dx}{d} \int_{-\infty}^{\infty} dE \left[n^e (1 - n^e) + n^h \left(1 - n^h\right)\right]. \tag{40}$$

The electric current through the junction, Eq. (16), is given by

$$I = \frac{d}{2eR_N} \int_{-\infty}^{\infty} dE \, \partial_x \left(n^e - n^h\right). \tag{41}$$

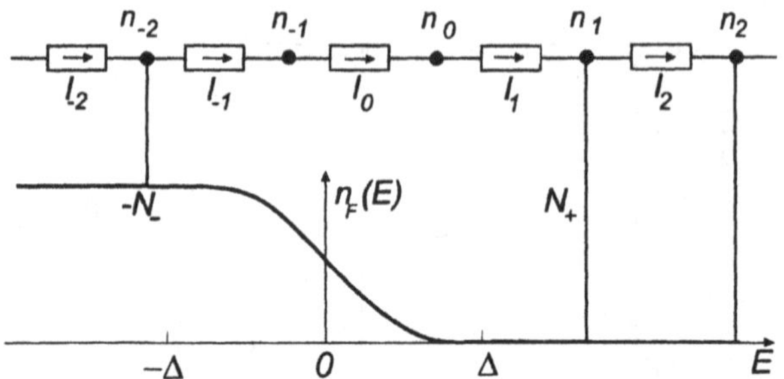

Figure 7. Equivalent MAR network in energy space in the limit of negligibly small normal reflection and proximity regions at the NS interfaces.

The MAR network for the present case, in which the contributions of the proximity effect and normal scattering at the interfaces have been neglected, is shown in Fig. 7. Within such model i) the renormalized resistances coincide with the bare ones, ii) the Andreev reflection inside the gap is complete: $R_A = 0$, $G_T = 0$ at $|E| < \Delta$, and iii) the over-the-barrier Andreev reflection, as well as the normal reflection, are excluded: $G_A = 0$, $R_T = 0$ at $|E| > \Delta$. As the result, the network is essentially simplified and represents a series of Drude resistances connected periodically, at the energies $E_k = E + keV$, with the distributed "voltage source" $n_F(E)$. The "potentials" n_k of the network nodes with even numbers k represent equal electron and hole populations $n_0^{e,h}(E_k)$ at the left NS interface, whereas the potentials of the odd nodes describe equal boundary populations $n_d^{e,h}(E_k \mp eV)$ at the right interface. The "currents" I_k entering k-th node are related to the probability currents $n'(E_k) = \partial n(E_k)/\partial x$ as $I_k(E) = -\sigma_N n^{e\prime}(E_{k-1})$ (odd k) and $I_k(E) = \sigma_N n^{h\prime}(E_k)$ (even k), and represent partial electric currents transferred by the electrons and holes across the junction, obeying Ohm's law in energy space, $I_k = (n_{k-1} - n_k)/R_N$. Within the gap, $|E_k| < \Delta$, i.e., at $-N_- < k < N_+$, the nodes are disconnected from the reservoir due to complete Andreev reflection and therefore all currents flowing through the subgap nodes are equal.

The analytical equations corresponding to the MAR network are as follows: At $|E| > \Delta$, the boundary populations $n_{0,d}(E) = n(E,x)|_{x=0,d}$ are local-equilibrium Fermi functions, $n_0^{e,h}(E) = n_F(E)$, $n_d^{e,h}(E) = n_F(E \pm eV)$ (we use the potential of the left electrode as the energy reference level). At subgap energies, $|E| < \Delta$, the boundary conditions are modified in accordance with the mechanics of complete Andreev reflection which equalizes the electron and hole population numbers at a given electrochemical potential and blocks the net probability current through the NS interface [7],

$$n_0^e(E) = n_0^h(E),\; n_d^e(E - eV) = n_d^h(E + eV),$$
$$n_0^{e\prime}(E) + n_0^{h\prime}(E) = 0,\; n_d^{e\prime}(E - eV) + n_d^{h\prime}(E + eV) = 0, \qquad (42)$$

where $n'_{0,d}$ are the boundary values of the electron and hole probability flows $\partial n/\partial x$. The recurrences for boundary populations and diffusive flows within the subgap region read

$$n_0^{e,h}(E - eV) - n_0^{e,h}(E + eV) = \mp 2dn^{e,h\prime}(E \mp eV),$$
$$n_0^{e,h\prime}(E - eV) = n_0^{e,h\prime}(E + eV). \tag{43}$$

Due to periodicity of the network, the partial currents obey the relationship $I_k(E) = I_m(E_{k-m})$, and the boundary population n_0 is related to the node potentials n_k as $n_0(E_k) = n_k(E)$. This allows us to reduce the integration over energy in Eqs. (40) and (41) to an elementary interval $0 < E < eV$,

$$I = \frac{1}{e}\int_0^{eV} dE\, J(E), \quad J(E) = \sum_{k=-\infty}^{\infty} I_k, \tag{44}$$

$$S = \frac{2}{R_N}\int_0^{eV} dE \sum_{k=-\infty}^{\infty}\left[2n_k(1-n_k) + \frac{1}{3}(R_N I_k)^2\right]. \tag{45}$$

The "potentials" of the nodes outside the gap, $|E_k| > \Delta$, are equal to local-equilibrium values of the Fermi function, $n_k(E) = n_F(E_k)$ at $k \geq N_+$, $k \leq -N_-$. The partial currents flowing between these nodes,

$$I_k = \left[n_F(E_{k-1}) - n_F(E_k)\right]/R_N, \quad k > N_+, \ k \leq -N_-, \tag{46}$$

are associated with thermally excited quasiparticles. The subgap currents may be calculated by Ohm's law for the series of $N_+ + N_-$ subgap resistors,

$$I_k = \frac{n_- - n_+}{(N_+ + N_-)R_N}, \quad -N_- < k \leq N_+, \tag{47}$$

where $n_\pm(E) = n_F(E_{\pm N_\pm})$. From Eqs. (46) and (47) we obtain the current spectral density in Eq. (44) as $J(E) = 1/R_N$, which results in Ohm's law, $V = IR_N$, for the electric current through the junction. This conclusion is related to our disregarding the proximity effect and the normal scattering at the interface. Actually, both of these factors lead to the appearance of SGS and excess or deficit currents in the I-V characteristic, with the magnitude increasing along with the interface barrier strength and the ratio ξ_0/d [7].

The subgap populations can be found as the potentials of the nodes of the subgap "voltage divider",

$$n_k = n_- - (n_- - n_+)\frac{N_- + k}{N_+ + N_-}. \tag{48}$$

By making use of Eqs. (45)-(48), the net current noise can be expressed through the sum of the thermal noise of quasiparticles outside the gap and the subgap

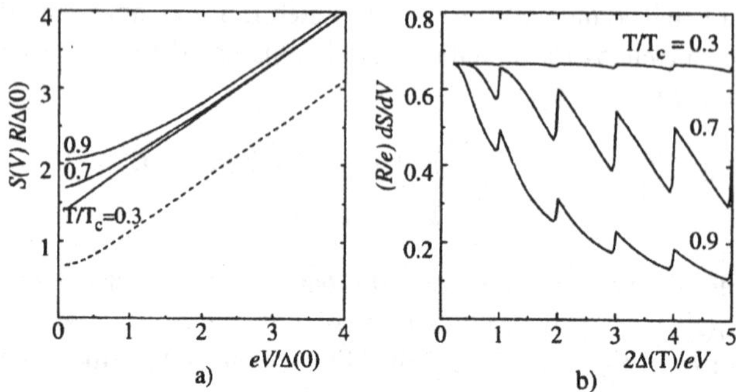

Figure 8. Spectral density S of current noise vs voltage (a) and its derivative dS/dV vs inverse voltage (b) at different temperatures. Dashed line shows the result for normal metal junction [10] at $T = 0.3T_c$.

noise, $S = S_> + S_\Delta$, where

$$S_> = \frac{4T}{3R_N}\left\{2\left[n_F(\Delta)+n_F(\Delta+eV)\right]+\left[\frac{eV}{T}+\ln\frac{n_F(\Delta+eV)}{n_F(\Delta)}\right]\coth\frac{eV}{2T}\right\}, \quad (49)$$

and

$$S_\Delta = \frac{2}{3R_N}\int_0^{eV} dE\,(N_+ + N_-)\left[f_{+-} + f_{-+} + 2(f_{++} + f_{--})\right],$$
$$f_{\alpha\beta} = n_\alpha(1 - n_\beta). \quad (50)$$

At low temperatures, $T \ll \Delta$, the thermal noise $S_>$ vanishes, and the total noise coincides with the subgap shot noise, which takes the form

$$S = \frac{2}{3R_N}\int_0^{eV} dE\,(N_+ + N_-) = \frac{2}{3R_N}(eV + 2\Delta), \quad (51)$$

of $1/3$-suppressed Poisson noise $S = (2/3)q^{\mathit{eff}} I$ for the effective charge $q^{\mathit{eff}} = e(1+2\Delta/eV)$ [26]. At $V \to 0$, the shot noise turns to a constant value $4\Delta/3R$. At finite voltages, this quantity plays the role of the "excess" noise, i.e. the voltage-independent addition to the shot noise of a normal metal at low temperatures [see Fig. 8(a)]. Unlike short junctions, where the excess noise is proportional to the excess current [29], in our system the excess current is small and has nothing to do with large excess noise.

Results of numerical calculation of the noise at finite temperature are shown in Fig. 8. While the temperature increases, the noise approaches its value for normal metal structures [10], with additional Johnson-Nyquist noise coming from thermal excitations. In this case, the voltage-independent part of current noise may be qualitatively approximated by the Nyquist formula $S(T) = 4T^*/R$ with the effective temperature $T^* = T + \Delta(T)/3$. The most remarkable phenomenon at nonzero temperature is the appearance of steps in the voltage dependence of the derivative dS/dV at the gap subharmonics $eV = 2\Delta/n$ [Fig. 8(b)], which reflect discrete transitions between the quasiparticle trajectories with different numbers of Andreev reflections. The magnitude of SGS decreases both at $T \to 0$ and $T \to T_c$, which resembles the behavior of SGS in the I-V characteristic of long ballistic SNS junction with perfect interfaces within the OTBK model [2, 3]. A small "residual" SGS in current noise, similar to the one in the I-V characteristic [7], should occur at $T \to 0$ due to normal scattering at the interface or due to proximity effect [see comments to Eq. (47)].

3.2. NOISE OF TUNNEL BARRIER

It is instructive to compare the shot noise of a distributed source considered in the previous section with the one produced by a localized scatterer, e.g. opaque tunnel barrier with the resistance $R \gg R_N$ inserted in the normal region [25], see Fig. 9. In this case, the potential drops at the tunnel barrier, and the population number is almost constant within the conducting region, while undergoes discontinuity at the barrier. However, the recurrences in Eqs. (42), (43) are not sensitive to the details of spatial distribution of the population numbers, and therefore the result of the previous section, Eq. (48), must be also valid for the present case. General equation for the noise in superconducting tunnel junctions has been derived in Ref. [42]. In our case of long SNINS junction, this equation takes the form,

$$S = \int_{-\infty}^{+\infty} \frac{dE}{R} \left[f(E) + f(E+eV) - f(E)f(E+eV) \right], \tag{52}$$

where $f = n^e + n^h$. Taking into account the distribution function in Eq. (48), the noise power (52) at zero temperature becomes

$$S = \frac{2}{R} \int_{-\Delta-eV}^{-\Delta} \frac{dE}{3} \left[N_+ + N_- + \frac{2}{N_+ + N_-} \right]. \tag{53}$$

At voltages $eV > 2\Delta$ this formula gives conventional Poissonian noise $S = 2eI$. At subgap voltages, the noise power undergoes enhancement: it shows a piecewise linear voltage dependence, $dS/dV = (2e/3R)[1 + 4/(\mathrm{Int}\,(2\Delta/eV) + 2)]$, with kinks at the subharmonics of the superconducting gap, $eV_n = 2\Delta/n$ (see Fig. 10). At zero voltage, the noise power approaches the constant value $S(0) = 4\Delta/3R$, coinciding with the noise power of the diffusive normal region without tunnel barrier. However, in contrast to the latter case, the voltage dependence of the noise here exhibits SGS already at zero temperature, which consists of a step-wise increase of the effective charge $q^{\mathit{eff}}(V) = S(V)/2I$ with

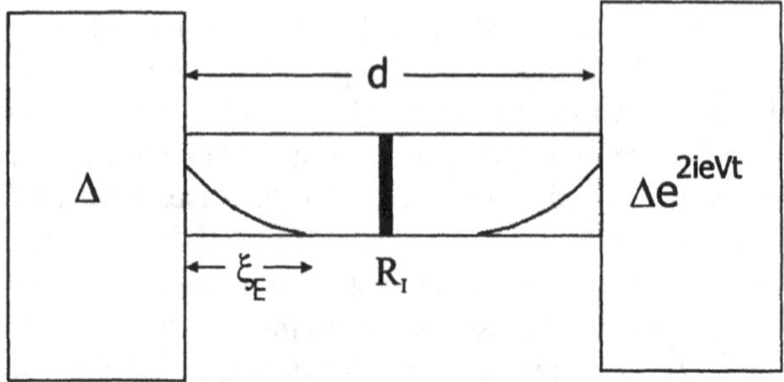

Figure 9. Diffusive mesoscopic SNS junction with a tunnel barrier (bold line).

decreasing voltage,

$$\frac{q^{eff}(n)}{e} = \frac{1}{3}\left(n + 1 + \frac{2}{n+1}\right) = 1, \frac{11}{9}, \frac{22}{15}, \ldots \tag{54}$$

At $eV \to 0$ the effective charge increases as $q^{eff}(V)/e \approx (1/3)(1 + 2\Delta/eV)$.

Calculation presented in this section shows that the $1/3$-factor in the expression for the noise has no direct relation to the Fano factor for diffusive normal conductors but is a general property of the incoherent MAR regime. Appearance of this factor results from the infinitely increasing chain of the normal resistances in the MAR network at small voltage, in close analogy with the case of multibarrier tunnel structures considered in Ref. [9].

4. Effect of inelastic relaxation on MAR

During the previous discussion, an inelastic electron relaxation within the normal region of the contact has been entirely ignored. This is legitimate when inelastic mean free path exceeds the junction length, $l_\varepsilon = \sqrt{\mathcal{D}\tau_\varepsilon} \gg d$, or, equivalently, when the inelastic relaxation time exceeds the diffusion time, $\tau_\varepsilon \gg \hbar/E_{Th}$. [4] In the opposite limiting case, $\tau_\varepsilon \ll \hbar/E_{Th}$, the energy relaxation completely destroys the MAR regime because the quasiparticle distribution within the normal region becomes equilibrium, and the SNS junction becomes equivalent to the two NS junctions connected in series through the equilibrium normal reservoir. Thus the appearance of the MAR regime is controlled by parameter $W_\varepsilon = E_{Th}\tau_\varepsilon/\hbar$. The relaxation time in these estimates must be considered for the energy of order Δ, $\tau_\varepsilon(\Delta)$, since non-equilibrium electron population under the MAR regime develops within the whole subgap energy interval. However, to build up such a nonequilibrium population at small voltage, a long time, $\tau_{dwell} = (\hbar/E_{Th})(2\Delta/eV)^2$, is actually needed, which exceeds the diffusion time $\hbar/E_{Th}$ by squared number of

[4] For the tunnel barrier, this estimate changes to $\tau_\varepsilon \gg (R/R_N)(\hbar/E_{Th})$ [25].

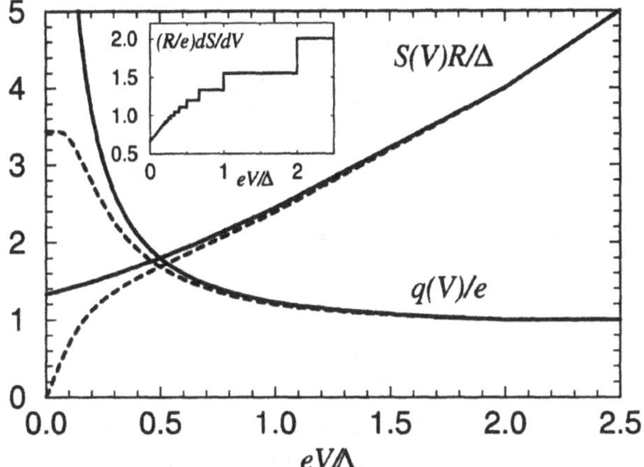

Figure 10. Spectral density $S(V)$ of current shot noise and effective transferred charge $q(V)$ as functions of applied voltage V. In the absence of inelastic collisions (solid lines), the shot noise power approaches the finite value $S(0) = 4\Delta/3R$ at $eV \to 0$, and the effective charge increases as $q(V) = (e/3)(1+2\Delta/eV)$. The effect of inelastic scattering is represented by dashed lines for the nonequilibrium parameter $W_\varepsilon = 5$. The dependence $S(V)$ contains kinks at the gap subharmonics, $eV = 2\Delta/n$, as shown in the inset.

Andreev reflections (see below Eq. (57) and further text). Thus, the condition for the MAR regime, $\tau_\varepsilon > \tau_{dwell}$, is more restrictive than $W_\varepsilon > 1$, and in the limit of zero voltage the MAR regime is always suppressed. Hence the effect of the shot noise enhancement disappears at small voltage, and the noise approaches the thermal noise level, the noise temperature being equal to the physical temperature T if the inelastic scattering is dominated by electron-phonon interaction (assuming that the phonons are in equilibrium with the electron reservoir). If the electron-electron scattering dominates, the noise temperature may exceed the temperature T of the electron reservoir if this temperature is small, $T \ll \Delta$ (hot electron regime). The reason is that at low temperature, the subgap electrons are well decoupled from the reservoir (electrons outside the gap) due to weak energy flow through the gap edges. In the cross over region of a small applied voltage, the non-equilibrium electron population appears as the result of the two competing mechanisms: the spectral flow upward in energy space driven by MAR and spectral counter flow due to inelastic relaxation. In this section we briefly consider the behavior of the shot noise in this situation.

To include the inelastic relaxation into the consideration we add the inelastic collision term I_ε to the diffusion equation (9),

$$\mathcal{D}\frac{\partial^2 n}{\partial x^2} = I_\varepsilon(n). \tag{55}$$

At small voltage, $eV \ll \Delta$, the spatial variation of the population number is small, and the function $n(x)$ in the collision term can be replaced by the boundary values, $n_0^{e,h}(E) \approx n_d^{e,h}(E) \equiv n(E)$. Then including the collision term into the recurrences, Eq. (43), and combining them with equations (42), we arrive at the generalized recurrence,

$$\mathcal{D}\left[n(E+2eV)+n(E-2eV)-2n(E)\right]=4I_\varepsilon[n(E)]. \tag{56}$$

Within the same approximation, this recurrence is to be considered as differential relation, which results in the diffusion equation for $n(E)$,

$$D_E \frac{\partial^2 n}{\partial E^2} = I_\varepsilon(n), \tag{57}$$

where $D_E = (eV)^2 E_{Th}/\hbar$ is the diffusion coefficient in energy space. [5]

To demonstrate the effect of electron-phonon scattering and suppression of the current shot noise, it is sufficient to assume the relaxation time approximation in the collision term, $I_\varepsilon(n) = (1/\tau_\varepsilon)[n - n_F(T)]$. The results of numerical calculations within this approximation of the shot noise at zero reservoir temperature are presented in Fig. 10 by dashed curves. The rapid decrease of $S(V)$ at low voltage is described by the following analytical approximation,

$$S(V) = S(0)\frac{3}{\alpha}\left(\frac{\operatorname{th}\alpha}{2+\frac{\alpha-\operatorname{sh}\alpha}{\operatorname{sh}^2\alpha}}\right), \quad \alpha = \frac{\Delta}{eV\sqrt{W_\varepsilon}}, \tag{58}$$

and it occurs when the length of the MAR path in energy space interrupted by inelastic scattering, $eV\sqrt{W_\varepsilon}$, becomes smaller than 2Δ.

4.1. HOT ELECTRON REGIME

In the case of dominant electron-electron scattering, equation (57) describes the crossover from the "collisionless" MAR regime to the hot electron regime as function of the parameter $D_E\tau_{ee}(\Delta)/\Delta^2$. In the hot electron limit, $\Delta^2 \gg D_E\tau_{ee}$, the collision integral dominates in Eq. (57), and therefore the approximate solution of the diffusion equation is the Fermi function with a certain effective temperature $T_0 \ll \Delta$. The value of T_0 can be found from Eq. (57) integrated over energy within the interval $(-\Delta, \Delta)$ with the weight E, taking into account the boundary conditions $n(\pm\Delta) = n_F(\pm\Delta/T)$ and the conservation of energy by the collision integral,

$$D_E\left[1-2\left(e^{-\Delta/T}+\frac{\Delta}{T_0}e^{-\Delta/T_0}\right)\right]+2\int_\Delta^\infty dE\,E\,I_\varepsilon = 0. \tag{59}$$

[5] The finite resistance R_{NS} of the NS interfaces, which partially blocks quasiparticle diffusion, can be taken into consideration by renormalization of the diffusion coefficient, $D_E \to D_E[1 + (d/\xi_0)(R_{NS}/R)^2]^{-1}$ (see Ref. [7]).

At zero temperature of the reservoir and $T_0 \ll \Delta$, Eq. (59) takes the form

$$D_E \tau_{ee} \Delta^2 = \tag{60}$$

$$2 \int_{\Delta}^{\infty} dE\, E \int_{E-2\Delta}^{\infty} dE' \int_{E-E'-\Delta}^{\infty} d\omega\, n_F(E - E' - \omega) n_F(E') n_F(\omega),$$

and can be reduced to an asymptotic equation for T_0,

$$(eV)^2 W_\varepsilon \exp(\Delta/T_0) = T_0 \Delta (1 + T_0/\Delta), \tag{61}$$

which shows that the effective temperature of the subgap electrons decreases logarithmically with decreasing voltage. The noise of the hot subgap electrons is given by the Nyquist formula with temperature T_0,

$$S(V) = (4T_0/R)\left[1 - 2\exp(-\Delta/T_0)\right], \tag{62}$$

where the last term is due to the finite energy interval available for the hot electrons, $|E| < \Delta$. Equations (61), (62) give a reasonably good approximation to the result of the numerical solution of Eq. (57) [27].

5. Summary

We have presented a theory for the current shot noise in long diffusive SNS structures with low-resistive interfaces at arbitrary temperatures. In such structures, the noise is mostly generated by normal electron scattering in the N-region. Whereas the I-V characteristics are approximately described by Ohm's law, the current noise reveals all characteristic features of the MAR regime: "giant" enhancement at low voltages, pronounced SGS, and excess noise at large voltages. The most spectacular feature of the noise in the incoherent MAR regime is a universal finite noise level at zero voltage and at zero temperature, $S = 4\Delta/3R$. This effect can be understood as the result of the enhancement of the effective charge of the carriers, $q^{eff} = 2\Delta/V$, or, alternatively, as the effect of strongly non-equilibrium quasiparticle population in the energy gap region with the effective temperature $T_0 = \Delta/3$. Appearance of the excess noise is controlled by a large value of the parameter $\tau_\varepsilon/\tau_{dwell} \gg 1$, where τ_ε is the inelastic relaxation time, and τ_{dwell} is the time spent by quasiparticle in the contact. At small applied voltage this condition is always violated, and the excess noise disappears. Under the condition of dominant electron-electron scattering, the junction undergoes crossover to the hot electron regime, with the effective temperature of the subgap electrons decreasing logarithmically with the voltage. Calculation of the noise power has been done on the basis of circuit theory of the incoherent MAR [7], which may be considered as an extension of Nazarov's circuit theory [43] to a system of voltage biased superconducting terminals connected by normal wires in the absence of supercurrent. The theory is valid as soon as the applied voltage is much larger than the Thouless energy: in this case, the overlap of the proximity regions near the NS interfaces is negligibly small, and the ac Josephson effect is suppressed.

References

1. Schrieffer, J.R., and Wilkins, J.W. (1963) Two-particle tunneling processes between superconductors, *Phys. Rev. Lett.*, **Vol. no. 10**, pp. 17-19.
2. Klapwijk, T.M., Blonder, G.E., and Tinkham, M. (1982) Explanation of subharmonic energy gap structure in superconducting contacts, *Physica B+C*, **Vol. no. 109-110**, pp. 1657-1664.
3. Octavio, M., Tinkham, M., Blonder, G.E., and Klapwijk, T.M. (1988) Subharmonic energy-gap structure in superconducting constrictions, *Phys. Rev. B*, **Vol. no. 27**, pp. 6739-6746; Flensberg, K., Bindslev Hansen, J., and Octavio, M. (1988) Subharmonic energy-gap structure in superconducting weak links, *ibid.*, **Vol. no. 38**, pp. 8707-8711.
4. Bratus', E.N., Shumeiko, V.S., and Wendin, G. (1995) Theory of subharmonic gap structure in superconducting mesoscopic tunnel contacts, *Phys. Rev. Lett.*, **Vol. no. 74**, pp. 2110-2113; Averin, D., and Bardas, A. (1995) ac Josephson effect in a single quantum channel, *ibid.*, **Vol. no. 75**, pp. 1831-1834; Cuevas, J.C., Martín-Rodero, A., and Levy Yeyati, A. (1996) Hamiltonian approach to the transport properties of superconducting quantum point contacts, *Phys. Rev. B* , **Vol. no. 54**, pp. 7366-7379.
5. Bardas, A., and Averin, D.V. (1997) Electron transport in mesoscopic disordered superconductor-normal-metal-superconductor junctions, *Phys. Rev. B*, **Vol. no. 56**, pp. 8518-8521; Zaitsev, A.V., and Averin, D.V. (1998) Theory of ac Josephson effect in superconducting constrictions, *Phys. Rev. Lett.*, **Vol. no. 80**, pp. 3602-3605.
6. Zaitsev, A.V., (1990) Properties of "dirty" S-S'-N and S-S'-S structures with potential barriers at the metal boundaries, *JETP Lett.*, **Vol. no. 51**, pp. 41-46.
7. Bezuglyi, E.V., Bratus', E.N., Shumeiko, V.S., Wendin, G., and Takayanagi, H. (2000) Circuit theory of multiple Andreev reflections in diffusive SNS junctions: The incoherent case, *Phys. Rev. B*, **Vol. no. 62**, pp. 14439-14451.
8. Blanter, Ya.M., and Büttiker, M. (2000) Shot noise in mesoscopic conductors, *Physics Reports*, **Vol. no. 336**, pp. 1-166.
9. de Jong, M.J.M., and Beenakker, C.W.J. (1996) Semiclassical theory of shot noise in mesoscopic conductors, *Physica A*, **Vol. no. 230**, pp. 219-249.
10. Nagaev, K.E. (1992) On the shot noise in dirty metal contacts, *Phys. Lett. A*, **Vol. no. 169**, pp. 103-107.
11. Beenakker, C.W.J., and Büttiker, M. (1992) Suppression of shot noise in metallic diffusive conductors, *Phys. Rev. B* **Vol. no. 46**, pp. 1889-1892; González, T., González, C., Mateos, J., Pardo, D., Reggiani, L., Bulashenko, O.M., and Rubí, J.M. (1998) Universality of the 1/3 shot noise suppression factor in nondegenerate diffusive conductors, *Phys. Rev. Lett.*, **Vol. no. 80**, pp. 2901-2904; Sukhorukov, E.V., and Loss, D. (1998) Universality of shot noise in multiterminal diffusive conductors, *ibid.*, **Vol. no. 80**, pp. 4959-4962.
12. Schoelkopf, R.J., Burke, P.J., Kozhevnikov, A.A., Prober, D.E., and Rooks, M. J. (1997) Frequency dependence of shot noise in a diffusive mesoscopic conductor, *Phys. Rev. Lett.*, **Vol. no. 78**, pp. 3370-3373; Henny, M., Oberholzer, S., Strunk, C., and Schönenberger, C. (1999) 1/3-shot-noise suppression in diffusive nanowires, *Phys. Rev. B*, **Vol. no. 59**, pp. 2871-2880.
13. Nagaev, K.E. (1995) Influence of electron-electron scattering on shot noise in diffusive contacts, *Phys. Rev. B*, **Vol. no. 52**, pp. 4740-4743; (1998) Frequency-dependent shot noise as a probe of electron-electron interaction in mesoscopic diffusive contacts, *ibid.*, **Vol. no. 58**, pp. 7512-7515; Naveh, Y., Averin, D.V., and Likharev, K.K. (1998) Shot noise in diffusive conductors: A quantitative analysis of electron-phonon interaction effects, *ibid.*, **Vol. no. 58**,

pp. 15371-15374; Naveh, Y., Averin, D.V., and Likharev, K.K. (1999) Noise properties and ac conductance of mesoscopic diffusive conductors with screening, *ibid.*, **Vol. no. 59**, pp. 2848-2860.

14. Liefrink, F., Dijkhuis, J.I., de Jong, M.J.M., Molenkamp, L.W., and van Houten, H. (1994) Experimental study of reduced shot noise in a diffusive mesoscopic conductor, *Phys. Rev. B*, **Vol. no 49**, pp. 14066-14069; Steinbach, A.H., Martinis, J.M., and Devoret, M.H. (1996) Observation of hot-electron shot noise in a metallic resistor, *Phys. Rev. Lett.*, **Vol. no. 76**, pp. 3806-3809.
15. Khlus, V.A. (1987) Current and voltage fluctuations in microjunctions between normal metals and superconductors, *Sov. Phys. JETP*, **Vol. no. 66**, pp. 1243-1249; Muzykantskii, B.A., and Khmelnitskii, D.E. (1994) On quantum shot noise, *Physica B*, **Vol. no. 203**, pp. 233-239; de Jong, M.J.M., and Beenakker, C.W.J. (1994) Doubled shot noise in disordered normal-metal-superconductor junctions, *Phys. Rev. B*, **Vol. no. 49**, pp. 16070-16073.
16. Jehl, X., Payet-Burin, P., Baraduc, C., Calemczuk, R., and Sanquer, M. (1999) Andreev reflection enhanced shot noise in mesoscopic SNS junctions, *Phys. Rev. Lett.*, **Vol. no. 83**, pp. 1660-1663; Kozhevnikov, A.A., Schoelkopf, R.J., and Prober, D.E. (2000) Observation of photon-assisted noise in a diffusive normal metal-superconductor junction, *ibid.*, **Vol. no. 84**, pp. 3398-3401.
17. Belzig, W., and Nazarov, Yu.V. (2001) Full counting statistics in diffusive normal-superconducting structures, *Phys. Rev. Lett.*, **Vol. no. 87**, 197006.
18. Hessling, J.P. (1996) Fluctuations in mesoscopic constrictions, *PhD Thesis*, Göteborg.
19. Naveh, Y., and Averin, D.V. (1999) Nonequilibrium current noise in mesoscopic disordered superconductor-normal-metal-superconductor junctions, *Phys. Rev. Lett.*, **Vol. no. 82**, pp. 4090-4093.
20. Cuevas, J.C., Martín-Rodero, A., and Levy Yeyati, A. (1999) Shot noise and coherent multiple charge transfer in superconducting quantum point contacts, *Phys. Rev. Lett.*, **Vol. no. 82**, pp. 4086-4089.
21. Dieleman, P., Bukkems, H.G., Klapwijk, T.M., Schicke, M., and Gundlach, K.H. (1997) Observation of Andreev reflection enhanced shot noise, *Phys. Rev. Lett.*, **Vol. no. 79**, pp. 3486-3489.
22. Cron R., Goffman M.F., Esteve D., and Urbina C. (2001) Multiple-charge-quanta shot noise in superconducting atomic contacts, *Phys. Rev. Lett.*, **Vol. no. 86**, pp. 4104-4107.
23. Hoss, T., Strunk, C., Nussbaumer, T., Huber, R., Staufer, U., and Schönenberger, C. (2000) Multiple Andreev reflection and giant excess noise in diffusive superconductor/normal-metal/superconductor junctions, *Phys. Rev. B*, **Vol. no. 62**, pp. 4079-4085.
24. Strunk, C. and Schnenberger C. (2002) Shot noise in diffusive superconductor/normal metal heterostructures, *This volume*.
25. Bezuglyi, E.V., Bratus', E.N., Shumeiko, V.S., and Wendin, G. (1999) Multiple Andreev reflections and enhanced shot noise in diffusive superconducting-normal-superconducting junctions, *Phys. Rev. Lett.*, **Vol. no. 83**, pp 2050-2053.
26. Bezuglyi, E.V., Bratus', E.N., Shumeiko, V.S., and Wendin, G. (2001) Current noise in long diffusive SNS junctions in the incoherent multiple Andreev reflections regime, *Phys. Rev. B*, **Vol. no. 63**, 100501.
27. Nagaev, K.E. (2001) Frequency-dependent shot noise in long disordered superconductor–normal-metal–superconductor contacts, *Phys. Rev. Lett.*, **Vol. no. 86**, pp. 3112-3115.
28. Hoffmann C., Lefloch F., and Sanquer M., (2002) Inelastic relaxation and noise temperature in S/N/S junctions, *Preprint*.
29. Hessling, J.P., Shumeiko, V.S., Galperin, Yu.M., and Wendin, G. (1996) Current noise in biased superconducting weak links, *Europhys. Lett.*, **Vol. no. 34**, pp. 49-54.
30. Larkin, A.I., and Ovchinnikov, Yu.N. (1975) Nonlinear conductance of superconductors in

the mixed state, *Sov. Phys. JETP*, **Vol. no. 41**, pp. 960-965; (1977) Nonlinear effects during motion of vortices in superconductors, *ibid.*, **Vol. no. 46**, pp. 155-162.

31. Kupriyanov, M.Yu., and Lukichev, V.F. (1988) Influence of boundary transparency on the critical current of "dirty" SS'S structures, *Sov. Phys. JETP*, **Vol. no. 67**, pp. 1163-1168.
32. Blonder, G.E., Tinkham, M., and Klapwijk, T.M. (1982) Transition from metallic to tunneling regimes in superconducting microconstrictions: Excess current, charge imbalance, and supercurrent conversion, *Phys. Rev. B*, **Vol. no. 25**, pp. 4515-4532.
33. Lambert, C.J., Raimondi, R., Sweeney, V., and Volkov, A.F. (1997) Boundary conditions for quasiclassical equations in the theory of superconductivity, *Phys. Rev. B*, **Vol. no 55**, pp. 6015-6021.
34. Zaikin, A.D., and Zharkov, G.F. (1981) Theory of wide dirty SNS junctions, *Sov. J. Low Temp. Phys.*, **Vol. no. 7**, pp. 375-379.
35. Bezuglyi, E.V., Bratus', E.N., and Galaiko, V.P. (1999) On the theory of Josephson effect in a diffusive tunnel junction, *Low Temp. Phys.*, **Vol. no. 25**, pp 167-174.
36. Artemenko, S.N., Volkov, A.F., and Zaitsev, A.V. (1979) Theory of nonstationary Josephson effect in short superconducting contacts, *Sov. Phys. JETP*, **Vol. no. 49**, pp. 924-931.
37. Volkov, A.F., and Klapwijk, T.M. (1992) Microscopic theory of superconducting contacts with insulating barriers, *Phys. Lett. A*, **Vol. no. 168**, pp. 217-224.
38. Volkov, A.F., Zaitsev, A.V., and Klapwijk, T.M. (1993) Proximity effect under nonequilibrium conditions in double-barrier superconducting junctions, *Physica C*, **Vol. no. 210**, pp. 21-34.
39. Nazarov, Yu.V., and Stoof, T.H. (1996) Diffusive conductors as Andreev interferometers, *Phys. Rev. Lett.*, **Vol. no. 76**, pp. 823-826; Stoof, T.H., and Nazarov, Yu.V. (1996) Kinetic-equation approach to diffusive superconducting hybrid devices, *Phys. Rev. B*, **Vol. no. 53**, pp. 14496-14505; Volkov, A.F., Allsopp, N., and Lambert, C.J. (1996) Crossover from mesoscopic to classical proximity effects induced by particle-hole symmetry breaking in Andreev interferometers, *J. Phys.: Cond. Matter*, **Vol. no. 8**, pp. L45-L50.
40. Kogan, Sh.M., and Shul'man, A.Ya. (1969) Theory of fluctuations in a nonequilibrium electron gas, *Sov. Phys. JETP*, **Vol. no. 29**, pp. 467-474; Nagaev, K.E. (1998) Long-range Coulomb interaction and frequency dependence of shot noise in mesoscopic diffusive contacts, *Phys. Rev. B*, **Vol. no. 57**, pp. 4628-4634.
41. Nagaev, K.E., and Büttiker, M. (2001) Semiclassical theory of the shot noise in disordered SN contacts, *Phys. Rev. B*, **Vol. no 63**, 081301.
42. Larkin, A.I., and Ovchinnikov, Yu.N. (1984) Current damping in superconducting junctions with nonequilibrium electron distribution functions, *Sov. Phys. JETP*, **Vol. no. 60**, pp. 1060-1067.
43. Nazarov, Yu.V. (1994) Circuit theory of Andreev conductance, *Phys. Rev. Lett.*, **Vol. no. 73**, pp. 1420-1423; (1999) Novel circuit theory of Andreev reflection, *Superlattices Microstruct.*, **Vol. no. 25**, pp. 1221-1231.

SHOT NOISE IN DIFFUSIVE SUPERCONDUCTOR/ NORMAL METAL HETEROSTRUCTURES

CHRISTOPH STRUNK
Institute of Experimental and Applied Physics,
University of Regensburg
Universitätsstr. 31, D-93040 Regensburg, Germany

CHRISTIAN SCHONENBERGER
Institute of Physics,
University of Basel
Klingelbergstr. 82, CH-4056 Basel, Switzerland

1. Introduction

According to the scattering theory of quantum transport, the electrons propagate through mesoscopic conductors between large charge reservoirs similar to photons in wave guides. The conductance G is described in terms of transmission modes or transport channels with transmission coefficients $\mathcal{T}_n$. The stochastic transmission of electrical charge through the scattering region causes fluctuations of the current, i.e. current noise. This is called shot noise similar to the noise in vacuum tubes.

The archetypical model system of mesoscopic physics is the quantum point contact. It consists of a constriction in a metallic conductor, which is usually defined by split gates. The number of transport channels and their transmission coefficients can be tuned by the gate voltage V_G. The resulting step-like conductance characteristics $G(V_G)$ reflects the successive opening of the conductance channels. For a single channel the spectral density $S_I(\omega)$ of the classical shot noise is given by $S_I(0) = 2eI \propto \mathcal{T}$ in the limit of small frequency ω. This result holds for small transmission probabilities $\mathcal{T} \ll 1$ where the statistics of current fluctuations is poissonian. As $\mathcal{T}$ increases the statistics crosses over to a binomial distribution which leads to a suppression of the shot noise $S_I(0) = 2eI(1 - \mathcal{T}) \propto \mathcal{T}(1 - \mathcal{T})$ [1–4]. In the limit $\mathcal{T} \to 1$ and temperature $T \to 0$ the noise vanishes as a con-

Y.V. Nazarov (ed.), Quantum Noise in Mesoscopic Physics, 119–133.

sequence of the Pauli principle. At finite temperatures not only the transmission is stochastic, but also the occupation number of the single particle states in the reservoirs. In equilibrium this leads to the thermal (Johnson-Nyquist) noise with $S_I(0) = 4k_BTG$, where G is the conductance of the sample.

As opposed to ballistic point contacts, the scattering region of a *diffusive* conductor contains usually many transport channels whose transmission coefficients obey a *universal* probability distribution which favors values of $\mathcal{T}_n$ close to zero and close to unity [5]. On average this leads to a universal suppression factor of $1/3$ with respect to the Poisson limit: $S_I(0) = 1/3 \cdot 2eI$ for diffusive wires [6]. This result is universal in the sense, that it does depend on neither the shape nor the material of the conductor; even spatial variations of the conductivity are allowed [7–9]. The noise suppression in the diffusive limit has been predicted quite long ago, but was only recently verified experimentally [10, 11].

When the electrons can be treated quasiclassically there is an alternative, very different approach to the noise, which is based on the quasiclassical kinetic theory [12, 13]. In this approach the noise is expressed in terms of the quasiclassical distribution function $f(E)$ of the electrons. The binary alternative is the occupation or emptyness of the single particle states rather than the transmission or reflection of electrons. Since $f(E)$ varies between 0 and 1, we have again a binomial distribution. Accordingly, the noise is given by the variance of the occupation number $f(1-f)$ averaged over energy and space, i.e.:

$$S_I(0) = \frac{4G}{\Omega} \int d^3r\, dE\, f(E,x)\,(1 - f(E,x)) \ , \tag{1}$$

where Ω is the sample volume.

In this paper we will consider the non-equilibrium noise of narrow wires with length L between large reservoirs, as sketched in Fig. 1a. The main advantage of the kinetic theory is, that it allows the calculation of $f(E)$ via the Boltzmann equation. Taking into account the inelastic scattering it reads in one dimension

$$D\frac{d^2}{dx^2} f(E,x) + \mathcal{I}(E,x) = 0 \ . \tag{2}$$

Here the elastic scattering has been absorbed in the diffusion constant D, while the inelastic scattering is contained in the collision integral $\mathcal{I}$. The latter depends on $f(E,x)$ and an energy dependent scattering kernel which is specific for the nature of the scattering processes [14]. In the absence of inelastic scattering, $f(E,x)$ varies linearly in space, while for fixed position $f(E,x)$ is a weighted average of the distribution functions in the reservoirs. For a narrow wire at low temperatures $f(E)$ has a two step shape, reflecting the superposition of the two fermi functions in the reservoirs at each end of the wire. This two-step shape is sketched in Fig. 1b and has recently been observed experimentally [15]. If $f(E,x)$ is inserted into Eq. 1, the 1/3-suppression of the shot noise is recovered. The sharpness of the

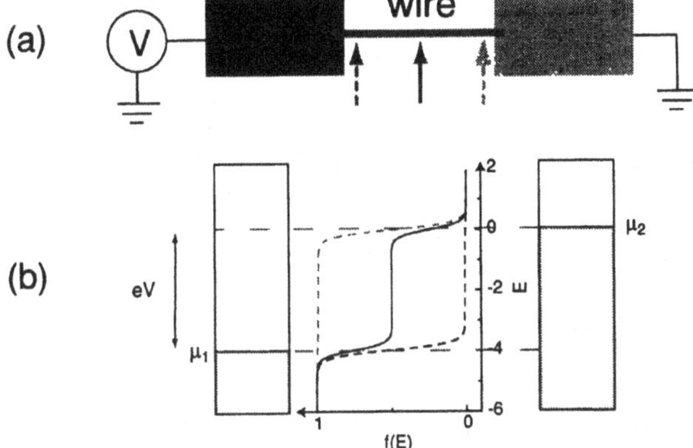

Figure 1. (a) Sketch of a typical sample consisting of a narrow wire made from a diffusive normal metal and two thick and large reservoirs. (b) Schematic of the quasiparticle distribution function near the reservoirs (dashed lines) and in the center of the wire (solid line). In the center a typical two step shape emerges, which results from the superposition of quasiparticles coming from the left and right reservoir.

steps is determined by the temperature of the reservoirs. In the presence of inelastic scattering the steps are further smeared until local thermodynamic equilibrium is obtained, i.e. $f(E,x) = f_0(T(x), \mu(x))$, where f_0 is the fermi function. In this limit, the noise assumes another universal value: $S_I(0) = \sqrt{3}/4 \cdot 2eI$ [12, 16, 17]. We note, that in the absence of inelastic scattering the results of the quasiclassical approach and of the scattering approach based on random matrix theories are identical. This holds not only for diffusive wires, but also for chaotic cavities, which are characterized by the universal suppression factor 1/4 [18–20].

2. Andreev- and multiple Andreev reflection

Several new features are expected, when one or both of the normal conducting (N) reservoirs are replaced by a superconductor (S). Electrons with energies smaller than the gap energy Δ cannot enter the superconductor because there are no single particle states available. Instead, charge is transferred by the process of Andreev reflection (AR) [21]. An electron with energy E (with respect to the chemical potential of the superconductor) is retroreflected as a hole with energy $-E$ at the NS interface and a Cooper pair enters the superconductor. Since two particles are involved, Andreev reflection is a second order process with a probability $\propto \mathcal{T}^2$. This implies that the Andreev reflection can only be observed if the NS-interface is sufficiently transparent. Interestingly, the charge transferred by a single Andreev reflection process is $2e$ and the conductance of an ideally transmitted channel doubles [22]. The same doubling is expected for the shot noise of a SNN-structure,

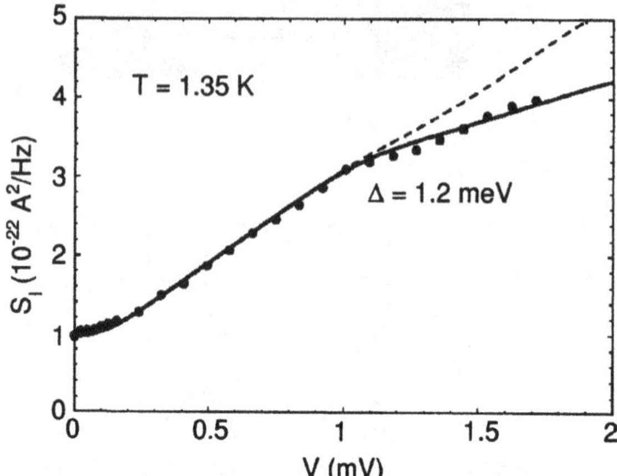

Figure 2. Shot-noise measurements (dots) in an NNS (Cu-Nb) junction compared to the predictions (solid line) from the semiclassical theory [26] with the superconducting gap Δ as only parameter (reproduced from [28]). For $eV < \Delta$ the predicted doubled shot noise is confirmed experimentally. The dashed line simulates a doubled shot noise above Δ to quantitatively emphasize the difference with the normal case. For $eV > \Delta$ an excellent quantitative agreement with the theory is found.

where one of the reservoirs is replaced by a superconductor [23]. This has been confirmed experimentally by Jehl *et al.* [24] and by Kozhevnikov *et al.* [25].

As in the NNN-case, the shot noise can be expressed in terms of the quasiparticle distribution function [26]. The key point is that a quasiparticle diffusing at an energy $|E| < \Delta$ in the vicinity of the NS-interface hits the interface many times, implying many conversions from electron to hole and vice versa. Because the diffusion times as an electron and as a hole are random, but *equal on average*, one obtains $f(E) = 1/2$ near the NS-interface, provided that the number of electrons impinging from the normal reservoir is equal above and below the chemical potential of the superconductor [27]. The noise calculated with this distribution function is $S_I = 2/3 \cdot 2eI$ in agreement with the naive expectation based on charge doubling. Figure 2 shows the noise of a diffusive NNS-junction as measured by Jehl *et al.* [28]. The experiment clearly shows the expected shot noise doubling for $eV < \Delta$. The kink at $eV = \Delta$ is caused by the onset of quasiparticle transmission into the reservoirs for $eV > \Delta$. In absence of inelastic scattering, the semiclassical theory agrees again with the quantum theory [23].

If both reservoirs are superconducting, quasiparticles are confined between the two NS-interfaces within an energy range 2Δ. In a naive classical picture the quasiparticles are diffusing back and forth between the two superconductors, taking Cooper pairs from one side and delivering them on the other. This peculiar 'shuttle' process is called *multiple* Andreev reflection (MAR). A more rigorous

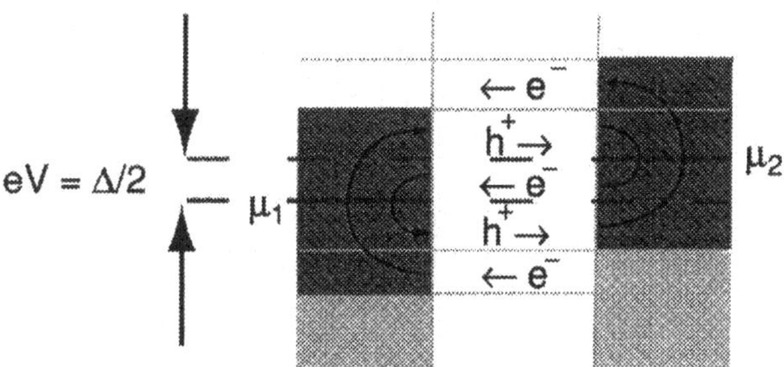

Figure 3. Schematic of the multiple Andreev reflection process for $eV = \Delta/2$. The dark shaded regions correspond to the energy gap. The light shaded regions correspond to occupied quasiparticle states in the superconductors. The electrochemical potential of the two superconductors are labelled μ_1 and μ_2. Electron-like quasiparticles are injected from the right. Each Andreev reflection at the left interface converts electron-like quasiparticles with energy $-E$ with respect to μ_1 into holes with energy E and vice versa. At the right interface the Andreev reflection takes place with respect to μ_2. A total charge of $5e$ per quasiparticle is transferred by this particular MAR cycle.

quantum mechanical calculation reveals the formation of pairs of Andreev bound states at zero applied voltage [29]. If a phase bias is applied between the superconductors the Andreev bound states carry a net supercurrent at temperatures $k_BT \lesssim E_c$. Here $E_c = \hbar/\tau_D = \hbar D/L^2$ is the Thouless energy which is determined by the diffusion time τ_D through the sample. In this paper we mainly want to restrict our discussion to the simplest case $k_BT \gg E_c$, where the supercurrent is averaged out.

At finite voltages, the MAR manifests itself in a different way. The quasiparticles are no longer confined indefinitely, but can escape after a finite number of Andreev reflections. This is explained in Fig. 3. If the applied voltage eV is an integer fraction of 2Δ, i.e. $n = 2\Delta/eV$, the energy range in which Andreev reflections occur splits up into $n+1$ branches. For each energy and position along the wire we have both electron- and hole-like quasiparticles arriving from the right, respectively, the left superconductor.

It is essential that the charge transfer q^* per injected quasiparticle increases with the number of Andreev reflections, i.e., $q^* = e(1+n)$. Hence, a step-like increase of the current at integer fractions of 2Δ is expected, when the bias voltage is reduced. This corresponds to peaks in the differential conductance [30] at voltages $V = 2\Delta/ne$, known as the *subharmonic gap structure*. At low voltage many Andreev reflections are needed, before a quasiparticle can escape into one of the superconducting banks. Subharmonic peaks up to many orders have been observed [31] in ballistic SNS contacts based on InAs quantum wells. The above argument does not depend on whether the quasiparticle motion is ballistic or diffusive. Hence, the subharmonic gap structures are also expected to occur in diffusive systems and have indeed been found experimentally [32, 33]. However, the average diffusion times $\tau_D = L^2/D$ are much longer than the ballistic

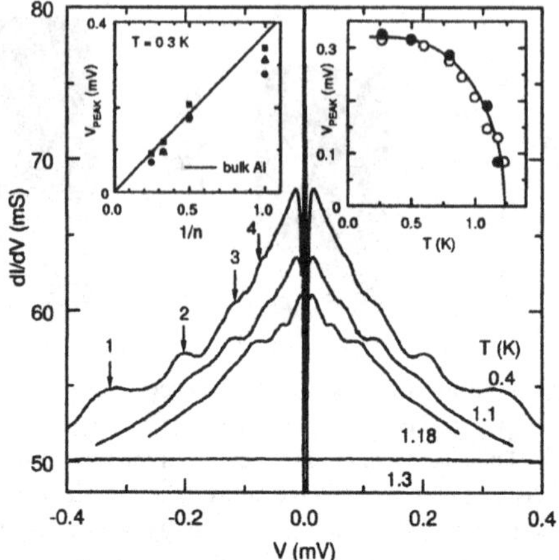

Figure 4. Differential conductance per wire dI/dV vs. voltage V of a chain of 16 Cu wires between Al reservoirs for several temperatures. The Cu wires are 0.9 μm long, 160 nm wide and 18 nm thick and have a diffusion constant $D \simeq 72$ cm^2/s. The thickness of the Al reservoirs is 150 nm. The arrows indicate subharmonic gap structures corresponding approximately to integer fractions of 2Δ. The structure at zero bias signals the presence of a supercurrent at the lowest temperatures. Left inset: Position of the conductance peaks vs. $1/n$ for three different samples. The solid line indicates the scaling for the gap of bulk Al. Right inset: Position of the 2Δ conductance peak vs. temperature for two samples with different normal state conductance (• : 50 mS, ○ : 29.3 mS). The solid line is a BCS fit for 2Δ =325 μeV and T_c = 1.23 K.

traversal times $\tau_{ballistic} = L/v_F$. Therefore the observation of subharmonic gap structures in diffusive samples is more easily hindered by inelastic scattering. An example for the subharmonic gap structures is shown in Fig. 4, where peaks up to fourth order can be seen. The peak positions scale well with $\Delta(T)$ and $1/n$. Sometimes the peaks are slightly split (i.e. for $n = 3$), a feature not understood yet. Details of the sample design and the measurement can be found elsewhere [33].

3. Effect of multiple Andreev reflection on the distribution function of diffusive SNS samples

The confinement of quasiparticles by MAR is expected to have a drastic effect on the distribution function. We start the discussion at the high voltage limit and first neglect inelastic scattering. Figure 5 shows a schematic of the distribution function in this limit. If the junction is biased at a high voltage, e.g., $eV = 4\Delta$, there are two energy regimes where (single) Andreev reflection can occur, i.e., the gap regions of the two superconductors. Since $eV > 2\Delta$ there is an intermediate regime of regular diffusion. As in the NS-case discussed above, the distribution function near the interfaces is reduced to $f(E) = 1/2$, if E is in one of the gap regions.

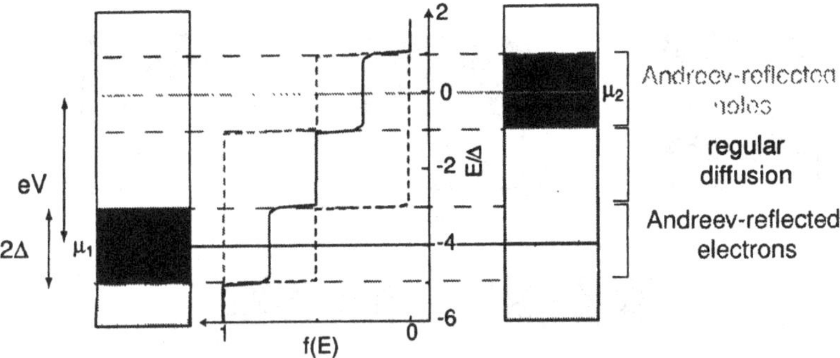

Figure 5. Schematic of the quasiparticle distribution function in an SNS sample at high bias voltage $eV = 4\Delta$. The Andreev reflection in the gap regions of the superconductors near the reservoirs (dashed lines) leads to the generation of holes at the left interface and to the generation of electrons at the right interface. As discussed in the text, $f(E) = 1/2$ in the gap regions. By the superposition of the two boundary functions a four-step shape of the distibution function emerges in the center of the wire (solid line).

The superposition of these two distributions with equal weight (corresponding to the middle of the wire) results in four steps in $f(E)$ as opposed to the two-step distribution function obtained in the case of normal reservoirs. This peculiar shape of $f(E)$ has been observed directly in a very recent experiment by Pierre *et al.* [34]. Figure 6 illustrates the evolution of $f(E)$ as a function of voltage. As the bias voltage is reduced the intermediate zone of regular diffusion shrinks and vanishes at $eV = 2\Delta$. For lower voltages the two gap regions overlap and multiple Andreev reflections occur, resulting in even more steps [35]. As the voltage is reduced further, the steps become finer and finer and their number increases, as seen in the last panel of Fig. 6. In the limit of $V \to 0$, $f(E)$ becomes a straight line between 1 at $E = -\Delta$ and 0 at $E = +\Delta$.

In order to calculate the noise, $f(E)$ has to be averaged according to Eq. 1. Then all step-like features disappear and the noise becomes a straight line [35, 36]:

$$S_I(0) = \frac{2G}{3}\,(eV + 2\Delta)\ . \tag{3}$$

This is just the result for a diffusive wire between normal reservoirs, but offset by $4G\Delta/3$. The finite offset of the noise at $V \to 0$ reflects the finite broadening of $f(E)$ at $V \to 0$. To compare this result with the Poisson noise it is possible to introduce a voltage dependent Fano-factor defined by

$$S_I(0) = F \cdot 2eI\ . \tag{4}$$

Comparison with Eq. 3 results in

$$F(V) = \frac{1}{3}\left(1 + \frac{2\Delta}{eV}\right)\ . \tag{5}$$

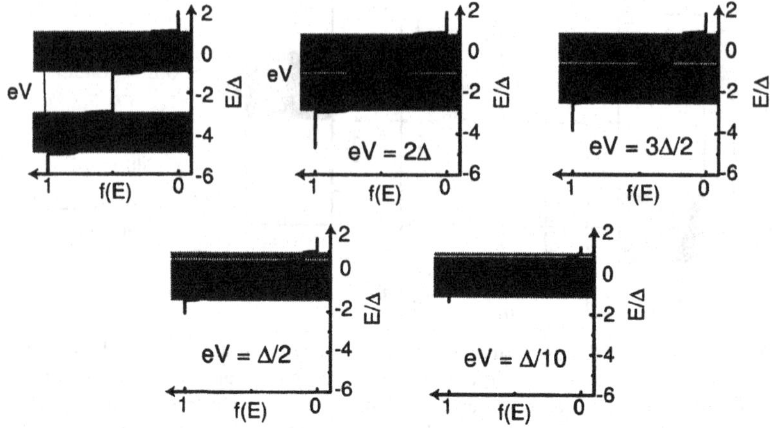

Figure 6. Evolution of the distribution function with decreasing voltage. If the gap regions overlap for $eV < 2\Delta$, multiple Andreev reflections occur. More and more steps appear at low voltages, resulting eventually in a minimal width of the distribution function of 2Δ.

This precisely coincides with the naive expectation $S_I(0) = 1/3 \cdot 2q^* I$ with $q^* = e(1+n)$ and $n = 2\Delta/eV$ being the number of Andreev reflections as defined above.

4. Effect of inelastic scattering

The observation that $S_I(0) \to 4G\Delta/3$ in the limit $V \to 0$ (see Eq. 3) at first sight appears to be in contradiction to the fluctuation dissipation theorem, which states $S_I(0) = 4k_B GT$, i.e., $S_I = 0$ at $T = 0$. This contradiction is resolved in presence of inelastic scattering, when energy is removed from the quasiparticles confined in the wire. We have to distinguish the effects of inelastic electron-electron (el-el) scattering and electron-phonon (el-ph) scattering. The el-el scattering leads to a smearing of the steps in $f(E)$ towards local thermal equilibrium, including the occupation of states with energies $|E| > \Delta$. In this way a part of the dissipated power can be removed into the reservoirs. The el-ph scattering additionally dumps energy into the bath and thus leads also to a *narrowing* of the distribution function. The width of the distribution function is governed by the balance between the energy input via the current on one side and the energy drain into the contacts and the phonon bath on the other side. Hence, the scattering rates have an important influence on the shape of the distribution function.

In order to calculate the effect of inelastic scattering, the full Boltzmann equation (Eq. 2) has to be solved. This can be done numerically [15]. The scattering integral in Eq. 2 contains a kernel $K(\epsilon)$ depending on the energy exchange ϵ in

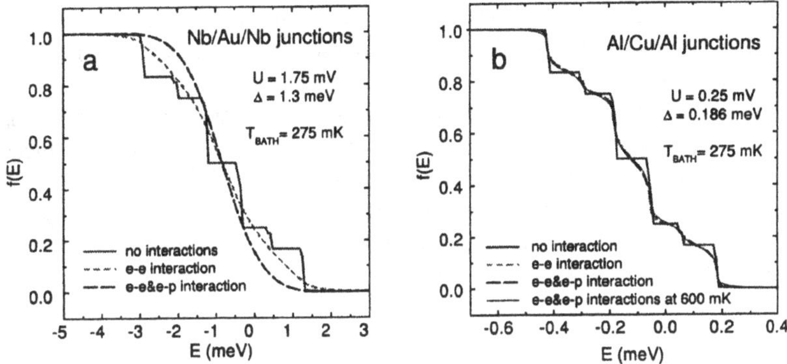

Figure 7. Effect of inelastic scattering on the distribution function of (a) a Nb/Au/Nb contact and (b) a Al/Cu/Al-contact. The ratio $eV/\Delta = 1.35$ is the same in both panels. The figure illustrates the drastic difference in the efficiency of the inelastic scattering on the thermalization and the cooling of the quasiparticles depending on the absolute value of Δ.

the collision [37]. The dependence is usually algebraic, i.e. $K(\epsilon) = \kappa\,\epsilon^{\alpha}$, where κ is a constant determined by the scattering matrix element and the exponent α is characteristic for the scattering mechanism. For the electron-electron scattering one obtains $\alpha = -3/2$ in the case of disorder enhanced el-el interaction [38] and $\alpha = -2$ in the case of impurity-spin mediated el-el interaction [39]. In contrast, for the el-ph scattering $\alpha = 2$ is found [40], when the disorder is ignored. The negative sign of α reflects the importance of quasi-elastic scattering processes characteristic for the el-el interactions [38], while $\alpha = 2$ for the el-ph scattering reflects the energy dependence of the phonon density of states (DOS) in the Debye model. In any case, the collision integral is greatly enhanced with increasing applied voltage, i.e. increasing width of the distribution function, because the phase space available for the scattering is greatly extended.

In Fig. 4 we show $f(E)$ calculated for two examples using Nb (Al) as the superconductor and Au (Cu) as the normal metal. To see the effect of different strengths of inelastic scattering we kept the ratio $eV/\Delta = 1.35$ constant in both cases. Because of the much larger gap of Nb ($\Delta_{Nb} \simeq 1.3$ meV) compared to Al ($\Delta_{Al} \simeq 0.186$ meV), the case of Nb involves much higher energies than the case of Al. Hence, the scattering is expected to be much more effective in Nb-based junctions as compared to Al-based junctions. Figure 4a shows $f(E)$ of the Nb based junctions for the three cases of absence of inelastic scattering, inclusion of el-el scattering, and inclusion of both el-el and el-ph scattering. We have used $\alpha_{el-el} = -2$ and values of κ_{el-el} and κ_{el-ph} inferred from weak localization measurements [41]. It is seen that the el-el scattering completely smears the steps expected from independent electron model, while the width of the distribution function is not changed. The inclusion of el-ph scattering leads in addition to a pronounced narrowing of the distribution function, i.e. a cooling of the electron system by the bath.

On the other hand, for the Al-based junctions the effect of el-el scattering is much less pronounced as illustrated by Fig. 4b. The steps are somewhat smeared, but still clearly visible. More important, there is no visible effect of the el-ph scattering. The reason for the apparent irrelevance of the el-ph scattering is the rapid decrease of the phonon DOS at low energies. These observations hold nearly unchanged up to $T = 600$ mK.

5. Noise measurements on SNS junctions

We now turn to our main experimental results, i.e., the effect of quasiparticle confinement on the shot noise of diffusive SNS junctions. In the case of the Nb-based junctions (see Fig. 8a) we observe a very steep increase (filled circles) of the noise with increasing voltage, which levels off above $V \gtrsim 1$ mV. As a reference, we have also measured the noise in a magnetic field of 6 T, where the superconductivity of the Nb is completely suppressed. In the latter case (open squares) we see a linear increase with the 1/3 suppression factor, in agreement with the expectation for diffusive wires (dashed line). Above $eV = 2\Delta_{Nb}$ the two curves should come very close together because the Andreev reflection rapidly dies out above the gap. This is indeed observed. The solid line is the result of the simulation. The agreement is surprisingly good, regarding the fact that no adjustable parameters have been used. The theoretical curve is nearly independent of κ_{el-el}, but mainly determined by Δ_{Nb} and κ_{el-ph}. Therefore, our measurement demonstrates in particular the cooling of the confined quasiparticles by the el-ph scattering. Similar results have also been obtained by Hoffmann *et al.* [42].

When looking at the Al-based junctions (see Fig. 8b) everything is shifted to much smaller energies. The filled dots are the noise measurements in the superconducting state, showing a roughly linear increase of the noise with the voltage and with an apparently finite intercept on the vertical axis. The open squares represent again the reference measurement in the normal state. In this case the simulation (solid line) does neither depend on κ_{el-el} nor on κ_{el-ph}, but directly reflects Eq. 3, which contains only on Δ_{Al}. This shows that in the investigated voltage range neither the el-el nor the el-ph scattering have an effect on the noise. On the other hand, the simulation and the experimental data agree well only at higher voltages $eV \gtrsim \Delta$. We believe that the apparent suppression of the measured noise with respect to the simple model of incoherent MAR is caused by the proximity effect, which we have neglected so far. This issue will discussed further in the next section.

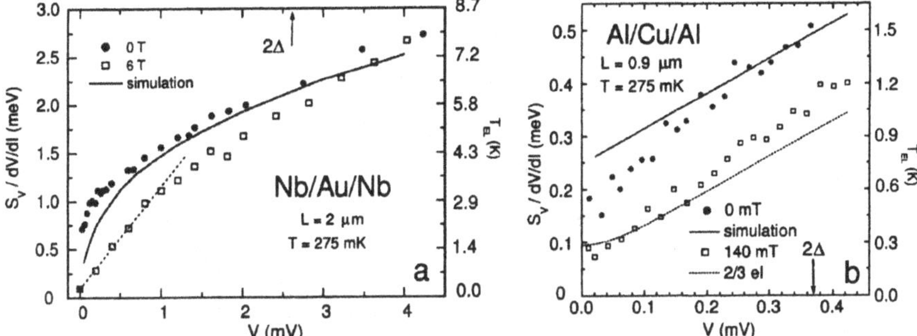

Figure 8. (a) Scaled excess noise $S_V/(dV/dI)$ as a function of voltage for a series of 9 Nb/Au/Nb junctions of 2 μm length, 200 nm width and 15 nm thickness for superconducting (•) and normal (□) Nb reservoirs. The arrow indicates $V = 2\Delta/e = 2.6$ mV. The solid line represents a numerical solution of Eq. 2 assuming for the interaction parameters $\kappa_{el-el} = 4$ ns^{-1}, $\kappa_{el-ph} = 13$ ns^{-1}meV^{-3} and a diffusion time $\tau_D = 0.56$ ns. The dotted line shows the usual 1/3-shot noise suppression for diffusive wires between normal reservoirs. (b) Scaled excess noise $S_V/(dV/dI)$ as a function of voltage for the same device as in Fig. 4 with superconducting (•) and normal (□) Al reservoirs. The arrow indicates $V = 2\Delta/e = 0.37$ mV. The solid line represents a numerical solution of Eq. 2 assuming for the interaction parameters $\kappa_{el-el} = 1$ ns^{-1}, $\kappa_{el-ph} = 5$ ns^{-1}meV^{-3} and a diffusion time $\tau_D = 0.11$ ns. The dotted line shows the usual 1/3-shot noise suppression for diffusive wires between normal reservoirs.

6. Multiple charges?

So far we have completely neglected the proximity effect, i.e. the phase coherent propagation of quasiparticles and the Josephson coupling between the superconducting reservoir. At small voltages and our lowest measurement temperature, however, this is not justified, as indicated by the supercurrent apparent in the divergence of the differential conductance near zero bias voltage (see Fig. 4). At $T \simeq 400$ mK the Cooper pairs penetrate into the normal metal on the length scale $L_T = \sqrt{\hbar D/2\pi k_B T} \simeq 300$ nm. It is plausible that the penetration of the superconducting condensate into the normal wire gives rise to a supercurrent component in the total current which reduces the current noise.

In order to check this hypothesis, we have measured the low voltage noise also at elevated temperatures, where the supercurrent is exponentially suppressed. In Fig. 6, we have plotted the Fano factor, respectively the effective charge $q^* = 3e\,F(V)$ as a function of $1/V$ for different temperatures. The voltage range has been restricted to $V \gtrsim 2$ μV, where the change of the differential conductance dI/dV are kept smaller than 2 in order to avoid errors when converting the measured voltage noise into current noise. Despite the significant scatter of the data, we find a roughly linear increase of q^* with $1/V$ up to very large values of $q^* \approx 50$ and more. The solid line is the prediction of the semiclassical theory [35, 36], i.e. Eq. 5. The agreement between the semiclassical theory and the data means that

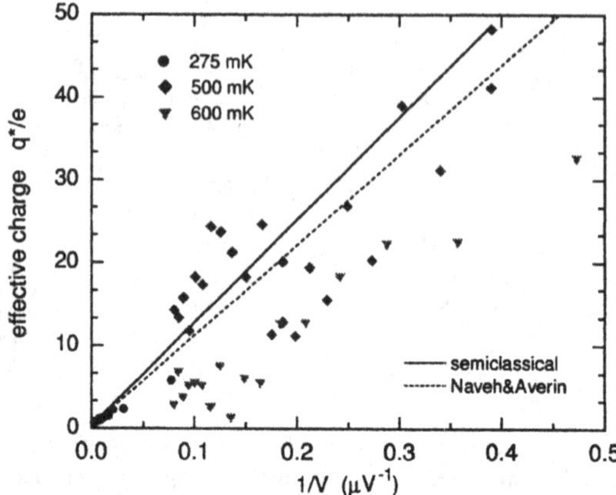

Figure 9. Effective multiple charge $q^* = S_I/2I$ as a function of $1/V$. Despite the considerable scatter of the data, a linear increase of q^* can be seen. The solid line indicates the theoretical estimate for q^* for the quasiclassical theory. The dashed line is the prediction of Naveh and Averin [44].

even at very low voltages no significant suppression of the shot noise by inelastic processes occurs. The dashed line is the prediction of a fully quantum mechanical calculation by Naveh and Averin [44], which results in a slightly different Fano factor $F = (1 - 1/\sqrt{2})\ 2\Delta/eV \simeq 0.29 \cdot 2\Delta/eV$. This seems to be another example for the surprisingly good agreement between quasiclassical and quantum mechanical approaches, although one should keep in mind that the latter calculation is valid in the limit $E_c \gg \Delta$, while our samples are in the opposite limit. Further experiments are needed to explore the full temperature and magnetic field dependence of these effects.

How literally should we take the simple explanation of the enhanced low voltage noise in terms of multiple charges produced by the MAR? Each injected quasiparticle indeed generates a whole avalanche of $2\Delta/eV$ Andreev reflections. Although this process cannot be viewed as the transmission of a coherent charge package, it appears at a single shot in the frequency range where our experiment is operative. We measure at frequencies around 100 kHz, i.e., on the timescale of several microseconds. The diffusion times across our samples are typically fractions of nanoseconds, implying that we cannot detect the correlation times induced by avalanches of 50 MARs and more.

7. Further developments

In this paper we have discussed the effects induced by *incoherent* multiple Andreev reflection. However, there are numerous additional effects expected to arise from *coherent* MAR, i.e., the proximity effect, which we have completely neglected so far. In a NNS-geometry with one superconducting and one normal reservoir it is known that there are phase coherent contributions to the conductance leading to resistance oscillations in Andreev interferometers [45] and the re-entrance of the proximity-enhanced conductance at low temperatures $k_B T \lesssim E_C$ [46]. The analog of these phenomena in the noise are discussed by Reulet *et al.* in these proceedings. A phase sensitive contribution to the Fano factor is developed near $eV \simeq E_c$, where F falls below its semiclassical value of $F = 2$ [47].

In SNS-geometries with short wires, the Josephson effects are expected to lead to considerable complications. In fact in the vicinity of the transition to the zero resistance state our samples have exhibited a dramatic supercurrent noise peak [33]. Previous work on shunted Josephson junctions [48] suggests that this may be quantum noise mixed down to lower frequencies. Further work is required to settle this issue and to investigate possible deviations from the RSJ model.

Another very interesting direction of research is the study of extremely small, i.e. *atomic* point contacts as reported by Cuevas *et al.* in these proceedings. Using a mechanical break junction technique, metallic constrictions are prepared which contain only a few conduction channels [49]. The full quantum theory of multiple Andreev reflection is required to describe dc transport and shot noise in atomic dimensions [44, 50, 51]. The quantitative agreement between theory and experiment is excellent [52].

8. Conclusions

We have shown, that the simple effect of quasiparticle confinement by Andreev scattering in diffusive SN- and SNS-junctions has drastic consequences for the quasiparticle distribution function. The changes of the distribution function are reflected in the shot noise. When the effect of the inelastic scattering is weak, the noise can mostly be described in terms of an effective charge q^*, introduced by the Andreev reflection. In the case of SNS-junctions, the width of the distribution function at low voltages is finite and of the order of 2Δ, provided that the electron-phonon scattering is weak. On the other hand, when the electron-phonon scattering becomes important, the quasiparticles are thermalized and cooled down. Short Al- and Nb-based junctions are examples for these limiting cases.

Acknowledgements

We thank T. Hoss, T. Nussbaumer, U. Staufer and R. Huber for their contributions to the measurements. In particular, we acknowledge F. Pierre and H. Pothier, who provided their simulation program for calculating the distribution functions. We also benefitted from inspiring discussions with C. Bruder, X. Jehl, H. Pothier, M. Sanquer, V. Shumeiko, C. Urbina and B. van Wees. This work was supported by the Swiss National Science Foundation.

References

1. G. B. Lesovik, JETP lett. **49**, 592 (1989).
2. M. Büttiker, Phys. Rev. Lett. **65**, 2901 (1990).
3. M. Reznikov, M. Heiblum, H. Shtrikman, and D. Mahalu, Phys. Rev. Lett. **75**, 3340 (1995).
4. A. Kumar, L. Saminadayar, D. C. Glattli, Y. Jin, and B. Etienne, Phys. Rev. Lett. **76**, 2778 (1996).
5. O. N. Dorokhov, Solid state Comm. **51** 381 (1984).
6. C. W. J. Beenakker and M. Büttiker, Phys. Rev. B **46**, 1889 (1992).
7. Yu. V. Nazarov, Phys. Rev. Lett. **73**, 134 (1994).
8. E. V. Sukhorukov and D. Loss, Phys. Rev. Lett. **80**, 4959 (1998).
9. E. V. Sukhorukov and D. Loss, Phys. Rev. B **59**, 13054 (1999).
10. R. J. Schoelkopf, P. J. Burke, A. A. Kozhevnikov, D. E. Prober, and M. J. Rooks, Phys. Rev. Lett. **78**, 3370 (1997).
11. M. Henny, S. Oberholzer, C. Strunk, and C. Schönenberger, Phys. Rev. B **59**, 2871 (1999); M. Henny, H. Birk, R. Huber, C. Strunk, A. Bachtold, M. Krüger, and C. Schönenberger, Appl. Phys. Lett. **71**, 773 (1997).
12. K. E. Nagaev, Phys. Lett. A **169**, 103, (1992).
13. K. E. Nagaev, Phys. Rev. B **52**, 4740 (1995).
14. Here we consider the *angle averaged* distribution function depending only on energy. This is reasonable, because the diffusive motion makes the angle dependent distribution function nearly isotropic.
15. H. Pothier, S. Gueron, N. O. Birge, D. Esteve, and M. Devoret, Phys. Rev. Lett. **79**, 3490 (1997).
16. V. I. Kozub and A. M. Rudin, Phys. Rev. B **52**, 7853 (1995).
17. A. Steinbach, J. M. Martinis, and M. H. Devoret, Phys. Rev. Lett. **76**, 3806 (1996).
18. R. A. Jalabert, J.-L. Pichard, and C. W. J. Beenakker, Europhys. Lett. **27**, 255 (1994).
19. Ya. M. Banter and E. V. Sukhurukov, Phys. Rev. Lett. **84**, 1280 (2000).
20. S. Oberholzer, E. V. Sukhorukov, C. Strunk, and C. Schönenberger, Phys. Rev. Lett. **86**, 2114 (2001).
21. A. A. Andreev, JETP **19**, 1228 (1964).
22. G. E. Blonder, M. Tinkham, and T. M. Klapwijk, Phys. Rev. B **25**, 4515 (1982).
23. M. J. M. de Jong and C. W. J. Beenakker, Phys. Rev. B **49**, 16070 (1994).
24. X. Jehl, M. Sanquer, R. Calemczuk, and D. Mailly, Nature **405**, 50 (2000).
25. A. A. Kozhevnikov, R. J. Schoelkopf, D. E. Prober, Phys. Rev. Lett., **84**, 3398 (2000).
26. K. E. Nagaev and M. Büttiker, Phys. Rev. B **63**, 081301(R) (2001).
27. This seems trivial in case of one normal and one superconducting reservoir, but is not the case in presence of two superconductors discussed below.
28. X. Jehl and M. Sanquer, Phys. Rev. B **63**, 052511 (2001).
29. I. O. Kulik, Sov. Phys. JETP, **30**, 944 (1973); [Zh. Eksp. Teor. Fiz. **57**, 1745 (1969)].
30. T. M. Klapwijk, G. E. Blonder, and M. Tinkham, Physica **109&110B**, 1657 (1982); W. M. van Huffelen *et al.*, Phys. Rev. B **47**, 5170 (1993); A. W. Kleinsasser *et al.*, Phys. Rev. Lett. **72**, 1738 (1994).
31. A. Chrestin, T. Matsuyama, and U. Merkt, Phys. Rev. B 55, 84578465 (1997).
32. J. Kutchinsky, R. Taboryski, T. Clausen, C. B. Sørensen, A. Kristensen, P. E. Lindelof, J. Bindslev Hansen, C. Schelde Jacobsen, and J. L. Skov, Phys. Rev. Lett. **78**, 931 (1997).
33. T. Hoss, C. Strunk, T. Nussbaumer, R. Huber, U. Staufer, and C. Schönenberger, Phys. Rev. B, 4079 (2000).
34. F. Pierre, A. Anthore, H. Pothier, C. Urbina, and D. Esteve, Phys. Rev. Lett. **86**, 1078 (2001).
35. K. E. Nagaev, Phys. Rev. Lett. **86**, 3112 (2001).

36. E. V. Bezuglyi, E. N. Bratus, V. S. Shumeiko, and G. Wendin, Phys. Rev. B **63**, 100501(R) (2001).
37. F. Pierre, H. Pothier, D. Esteve, and M. Devoret, J. Low. Temp.Phys. **118**, 437 (2000).
38. For a review, see B. L. Altshuler and A. G. Aronov in *Electron-Electron Interactions in disordered systems*, Edts. A. l. Efros and M. Pollak, Elsevier Science Publishers B. V. (1985).
39. A. Kaminski and L. I. Glazman, Phys. Rev. Lett. **86**, 2400 (2001).
40. J. M. Ziman, *Principles of the theory of solids*, Cambridge University Press, Cambrigde (1979).
41. A. B. Gougam, F. Pierre, H. Pothier, D. Esteve, and N. O. Birge, J. Low. Temp.Phys. **118**, 447 (2000).
42. C. Hoffmann, F. Lefloch, and M. Sanquer, cond-mat/0209310.
43. H. Courtois *et al.*, Phys. Rev. Lett. **76**, 130 (1996).
44. Y. Naveh and D. V. Averin, Phys. Rev. Lett. **82**, 4090 (1999).
45. A. Dimoulas, J. P. Heida, B. J. v. Wees, T. M. Klapwijk, W. v. d. Graaf, and G. Borghs, Phys. Rev. Lett. **74**, 602 (1995).
46. P. Charlat, H. Courtois, Ph. Gandit, D. Mailly, A. F. Volkov, and B. Pannetier, Phys. Rev. Lett. **77**, 4950 (1996).
47. B. Reulet, A. A. Kozhevnikov, D. E. Prober, W. Belzig, Yu. V. Nazarov, cond-mat/0208089.
48. C. M. Falco, W. H. Parker, S. E. Trullinger, and P. K. Hansma, Phys. Rev. B **10**, 1865 (1974); R. H. Koch, D. J. Van Harlingen, and J. Clarke, Phys. Rev. B **26**, 74 (1982).
49. E. Scheer, P. Joyez, D. Esteve, C. Urbina, and M.H. Devoret, Phys. Rev. Lett. **78**, 3535 (1997).
50. D. Averin and H. Imam, Phys. Rev. Lett. **76**, 3814 (1996).
51. J. C. Cuevas, A. Martín-Rodero, and A. Levy-Yeyati, Phys. Rev. Lett. **82**, 4086 (1999).
52. R. Cron, M. F. Goffman, D. Esteve, and C. Urbina, Phys. Rev. Lett. **86**, 4104 (2001).

PHOTO-ASSISTED ELECTRON-HOLE PARTITION NOISE IN QUANTUM POINT CONTACTS

D.C. GLATTLI[1,3], Y. JIN[2] L.-H REYDELLET[1]AND P. ROCHE[1]
[1]*Service de Physique de l'Etat Condensé,*
CEA Saclay, F-91191 Gif-sur-Yvette, France

[2]*Laboratoire de Photonique et Nanostructures,*
CNRS, Route de Nozay, F-91460 Marcoussis, France.

[3]*Laboratoire de Physique de la Matière Condensée,*
24 rue Lhomond, F-75231 Paris 05, France.

Abstract. The quantum partition noise of quasiparticles can be measured when no d.c. current flows trough a quantum conductor. This new approach, which looks paradoxical as usually shot noise is associated with a current, has been applied to a ballistic conductor, a quantum point contact realized in a 2D electron system. Irradiation by radio-frequency photons has been used to provide a non-transport determination of the Fano factor characterizing the partitioning of the photo-created quasi-particles. The article discuss the fundamental physical mechanisms as well as recent experimental results.

1. Introduction: transport and non-transport partition noise:

During the last few years very reliable measures of the Fano factor characterizing the partitioning of electrons scattered by a quantum conductor have been obtained using Shot Noise [1, 2]. In all experiments which have been done, a net d.c. current flows through the conductor and the resulting shot noise proportional to the current is measured. The ratio of the noise over the current gives a direct measure of the Fano factor. Here, we address the possibility to measure the Fano factor *without* d.c. current flowing through the conductor. We discuss recent experiments [3] which show that extracting the Fano factor from a *non-transport* shot noise experiment is indeed possible by irradiating a contact by a radio-frequency voltage.

Y. V. Nazarov (ed.), Quantum Noise in Mesoscopic Physics, 135–148.

Electrical shot noise as a long history [4]. Shot noise refers to the current noise associated with a non-equilibrium conditions and is to be distinguished from the Johnson-Nyquist noise or thermal noise observed at equilibrium. More precisely shot noise is the part of non-equilibrium current noise resulting from electrons elastically scattered by the conductor and *partitioned* between source and drain. In nearly all shot noise experiments done so far, non-equilibrium was produced by applying a constant electro-chemical potential difference between the electron reservoirs (or contacts) resulting in a net d.c. current I. We will call this noise *tranport* shot noise as opposed to the *non-transport* shot noise considered below. Transport shot noise is of practical importance. In most electronic devices considered by engineers, the random transfer of electrons between contacts obeys a poissonian statistical law and the low frequency current noise in a bandwitdth Δf is $\overline{\Delta I^2} = 2eI.\Delta f$, the Schottky formula [4]. At a more fundamental level, the recent understanding of electrical conduction at the quantum scale has shown that, in good conductors, the statistical law of electron partitioning is in general sub-poissonian because electrons have a low probability to be backscattered [1, 2, 5]. As, in addition the contacts are fundamentally noiseless sources of electrons, shot noise identifies to *partition noise*. The *partition noise* characterizes the scattering properties of the conductor and its measure relative to the poissonian noise is quantified by the Fano factor F, in analogy with quantum optics for the noise of photon beams.

Non-equilibrium situations generating a current noise can also be obtained when heating or irradiating one of the two reservoirs while no bias voltage is applied between the contacts. In this *non-transport* noise regime, the questions to which we would like to answer are : -1) is the current noise really shot noise (does it contains partitioning and not simply reservoir noise)? -2) is it possible to simply extract from noise measurements the Fano factor F characterizing the partitioning of electrons? Here, there is no possibility to compare the noise to the Shottky formula as a current is not necessarily produced by the non-equilibrium situation. Even if a thermo- or photo-current is generated, there is no general relation between the d.c. current and the current noise contrary to ordinary transport shot noise. A linear variation between current and noise is even not expected. Indeed, the current results from thermo- or photo-assisted processes and, by definition, occurs only when the transmissions of the electronic modes are energy dependent. To first order, the current is proportional to the derivative of the transmission with respect to energy, while the noise is a function of the transmissions and contains no first order derivative. In order to better understand the noise mechanism it is therefore easier to consider the case of negligible energy dependence of the transmission such that no d.c. current occurs.

In the next sections we will answer questions 1) and 2) and present a recent experimental answer. While heating generates a mixed situation of reservoir noise and partition noise, we will see that photon irradiation is a source of pure partition noise (at zero temperature). Also we will focus on photo-assisted noise. We will experimentally show that it provides a *direct measure* of the Fano factor. The contribution is organized as follows. In section II we discuss the ordinary transport shot noise and briefly review some of the most important results obtained so far. This gives the basic notations useful for section III where non-transport shot noise, mostly photo-assisted, is discussed. Section IV presents the mixed situation where both transport and non-transport shot noise compete. Finally in section V we discuss recent experimental results on non-transport and transport photo-assisted shot noise obtained using a Quantum Point Contact.

2. Transport partition noise:

We consider for simplicity the case of a two terminal single mode conductor at zero temperature. Applying an electrochemical potential difference eV between left and right reservoirs gives rise to a d.c. current I. According to our recent understanding of quantum transport, electrons in the energy range eV above the Fermi energy of the right reservoir are regularly emitted from the left reservoir at a frequency eV/h, as a result of the Fermi statistics [6]. The resulting incoming current $I_0 = e(eV/h)$ splits into transmitted and reflected current I and $I_0 - I$ respectively. If D is the transmission of the conductor, the net current is $I = DI_0$ and the conductance is $G = De^2/h$. As the regular injection of electron is noiseless, the only source of noise is the partition noise generated by electrons scattered by the conductor and being either transmitted or reflected. The resulting fluctuations for an frequency bandwidth Δf are $\overline{\Delta I^2} = 2eI_0D(1-D)\Delta f$, where the term $D(1-D)$ is the variance of the binomial statistics of the partitioning. For weak transmission, when the binomial statistics identifies to Poisson's statistics, the Schottky's formula $2eI\Delta f$ is recovered. For finite transmission, we can define the Fano factor $F = \overline{\Delta I^2}/2eI\Delta f = 1 - D$ which measures the departure from Poisson's statistics.

For a general conductor elastically scattering electrons and characterized by multiple electronic modes with transmission D_n, the contribution of each mode adds temporally incoherently and the total noise is $\overline{\Delta I^2} = 2eI_0\Delta f \sum_n D_n(1 - D_n)$. The resulting Fano factor is [5]

$$F = \frac{\sum_n D_n(1 - D_n)}{\sum_n D_n}. \tag{1}$$

It is always lower than unity : subpoissonian noise is a hallmark of quantum conductors. Because the reservoir are noiseless the quantum partition limit is reached.

The fundamental quantum reduction of noise due to the Fermi statistics has been observed in a variety of quantum conductors [1, 2]. The transmission dependence of the Fano factor has been experimentally successfully tested using

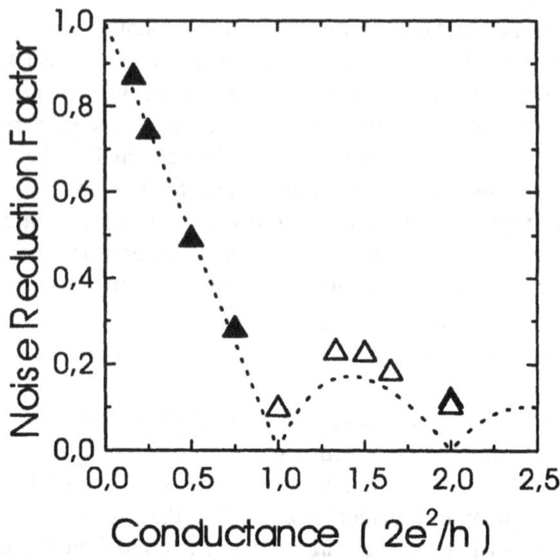

Figure 1. A transport noise measurement of the Fano factor F. A ballistic quantum point contact allows to vary from 0 to 1 the transmission of the electronic modes one by one. The figure shows F versus the total transmission (conductance in units of the zero field quantum of conductance $2e^2/h$)for the first two modes. The dashed line is comparison with the theory (adapted from Ref.[8]).

Quantum Point Contacts (QPC) realized in 2D electron systems [7, 8]. The opening of the point contact is controlled by a gate voltage. Sweeping its voltage allows to select one by one the number of transmitted modes and to accurately control their transmission D_n [9]. Measuring simultaneously the conductance $G = 2\frac{e^2}{h}\sum_n D_n$, the current and the noise allows to test Eq.(1) without adjustable parameters. Figure 1 shows a typical determination of F for the first two modes, in perfect agreement with (1). For the first mode $F = 1 - D_1$ and for the second mode $F = D_2(1-D_2)/(1+D_2)$, ... etc, assuming the $n+1^{th}$ mode starts to have non negligible transmission when the nth mode is fully transmitted, a reasonable assumption in good quantum point contacts. Remarkably for total transmission $\sum_n D_n = 1$ and 2, the noise suppression observed is higher than 90%. In the following we will show that *non-transport* photo-assisted shot noise in a similar quantum point contact gives an experimental determination of the Fano factor with accuracy even better than that of figure 1 [3].

Measurements of the Fano factor versus transmission in recent transport shot noise experiments in atomic Point Contact give results similar to those obtained using QPC [10]. However here the transmission cannot be easily changed as in 2D electron systems. Transport shot noise has been also studied in various metallic conductors and is now well documented and understood. In diffusive metallic systems the ensemble average value of the Fano factor is $\langle F \rangle = 1/3$ [11, 12] and has been measured by [13–15]. A similar value $\langle F \rangle = 1/4$ has been predicted [16] for the current noise of an open quantum dot (or chaotic cavity) and measured by [17]. Generalization of the shot noise to superconducting-normal mesoscopic conductors has been also done and the corresponding Fano factor measured (for a review see: [1, 2]).

3. Non-transport shot noise :

While shot noise is often associated with a d.c. current, we show here that transport is not necessary to observe partition shot noise and to extract the Fano factor. Of course non-equilibrium and scattering is necessary for the observation of the partition noise. Historically, non-transport shot noise is implicitly considered in the theoretical works on photo-assisted noise[18, 19] and thermo-assisted noise [20]. Noise in the mixed regime of transport and photo-assisted noise has been experimentally observed in Ref [15, 21]. Non-transport photo-assisted noise will be experimentally demonstrated in this paper.

Non-equilibrium can be produced when no d.c. voltage bias is applied to the conductor but one of the two contacts, say the left, is heated or irradiated by photons. If in addition the energy dependence of the transmissions is negligible, *no d.c. current* flows through the conductor. In the case of heating, the total noise results from the thermal noise of the heated reservoir (reservoir noise) and from the partition noise. The partition noise comes from electrons and holes activated respectively above and below the Fermi energy of the right reservoir and emitted by the left reservoir. Because the right reservoir emits no electrons and no holes at the corresponding energies, the partitioning is not inhibited by the Pauli principle and a partition noise proportional to $\sum_n D_n(1 - D_n)$ occurs.

Here, we focus on photo-assisted noise, rather than thermo-assisted noise, as, ideally, no reservoir noise is produced and the total noise comes from partitioning only. To understand the mechanism we will consider a two terminal single mode conductor with transmission D as above. An electron emitted from the left reservoir with an energy $\epsilon \leq h\nu$ below the Fermi energy can be either pumped to an energy $h\nu - \varepsilon$ above the Fermi energy with probability $\mathcal{P}_1$ or unpumped with probability $\mathcal{P}_0$. Unpumped electrons cannot generate current fluctuations as the right reservoir also emits electrons at the same energy. The Pauli principle imposes that both right and left outgoing states be filled with one electron. This leads to no current and hence no fluctuation. However the photo-pumped incoming electrons and holes do generate noise: the right reservoir does not emit electrons nor holes at energies $h\nu - \epsilon$ and $-\epsilon$ respectively, so that partition noise is not inhibited. The different processes are schematically shown in Figure 2. The electron and hole incoming currents are $I_0^{(e)} = \mathcal{P}_1 h\nu\, e/h$ and $I_0^{(h)} = -I_0^{(e)}$, are sources of independent temporal current fluctuations $\overline{\Delta I^{(e)2}} = 2eI_0^{(e)} D(1 - D)\Delta f$ and $\overline{\Delta I^{(h)2}} = \overline{\Delta I^{(e)2}}$ respectively. They add incoherently to give the total shot noise:

$$\overline{\Delta I^2} = 4h\nu \frac{e}{h} D(1 - D)\mathcal{P}_1 \Delta f. \tag{2}$$

Formula (2) is the single mode, zero temperature, zero voltage bias limit derived in [18, 19]. The generalization to multiple modes and multiple photon absorption processes is straightforward and gives:

$$\overline{\Delta I^2} = 4\frac{e}{h} \left(\sum_{n=1}^{\infty} D_n(1 - D_n) \right) \left(\sum_{l=1}^{\infty} (lh\nu)\mathcal{P}_l \right) \Delta f, \tag{3}$$

where $\mathcal{P}_l$ is the probability to absorb l photons.

Observation of this non-transport photoassisted noise has awaited until very recently [3] and a presentation of these experimental results will be done in the second part of this contribution. Before we will discuss the mixed situation were both transport and non-transport noise coexist.

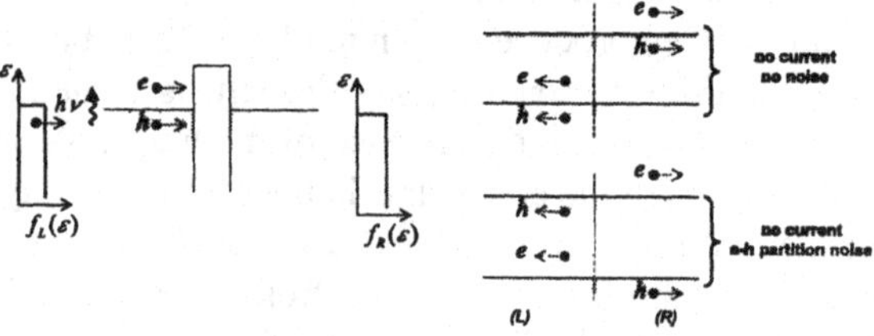

Figure 2. Schematic view of single photon absorption noise mechanisms entering in the non-transport shot noise. Left: an electron emitted by the left reservoir according to the Fermi distribution f_L absorbs a photon. The electron is promoted to an energy above the Fermi energy while a hole is left below the Fermi energy. Right: possible scattering events. Only the last two generate partition noise.

4. Photo-assisted noise with transport:

In [18, 19] the double non equilibrium situation was considered where both irradiation and a voltage bias are applied to the sample. The competition between transport and non-transport partition noise gives rise to a series of singularities of the noise derivative each time the voltage bias crosses a multiple of the value $h\nu/e$. The singularities have been observed in the second derivative of the noise using a diffusive metallic contact by [15, 21].

To understand the mechanism we will again consider a single mode conductor at zero temperature and only single photon absorption. The left reservoir is irradiated by the radiofrequency and biased at an electrochemical potential eV ($eV > 0$) above the right reservoir. As shown in Figure 3, for small bias $eV < h\nu$, we can separate the electrons emitted by the left reservoir in the energy range eV into two classes: 1) pumped electrons in the energy interval $[-h\nu, -eV]$ below the Fermi energy of the left reservoir (i.e. below the Fermi energy of the right reservoir); 2) pumped and unpumped electrons in the energy range $[-eV, 0]$ below the Fermi energy of the left reservoir (i.e. above the Fermi energy of the right reservoir). The first class of electrons do not contribute to the net current but generate non-transport shot noise as described previously. The second class contributes to ordinary transport shot noise independently of photon irradiation. For $eV > h\nu$, only the second class of electrons remains : the non-transport electron-hole partition noise is *suppressed* by ordinary transport noise. The resulting shot noise is thus:

$$\begin{aligned}\overline{\Delta I^2} &= 4(h\nu - |eV|)\frac{e}{h}D(1-D)\mathcal{P}_1\Delta f + |eV|\,\frac{e}{h}D(1-D)\Delta f \text{ for } |eV| \leq h\nu \\ &= |eV|\,\frac{e}{h}D(1-D)\Delta f \text{ for } |eV| > h\nu\end{aligned}$$

It is easy to generalize this zero temperature formula to multimodes and multi-photon absorption. A general formula including finite temperature can be found in [18, 19]. It predicts thermally rounded singularities at $eV = lh\nu$ in the derivative of the noise which signals the suppression of the partition noise for electron-hole pairs having absorbed l quanta to the benefit of simple electron transport shot noise.

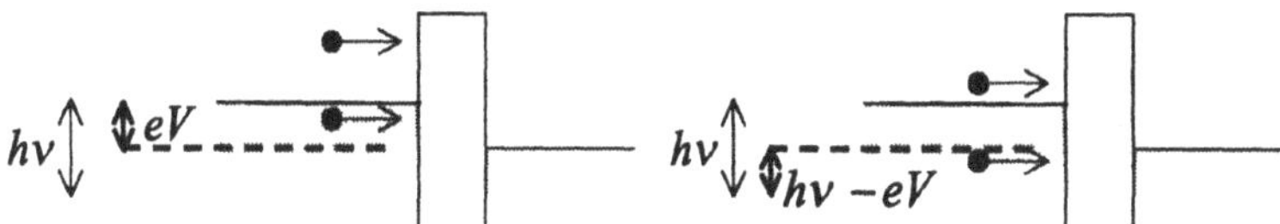

Figure 3. Processes involved in the mixed regime of non-transport photo-assisted noise and transport noise. For the later case shot noise is not an observable sensitive to photo-absorption, also the contribution to the noise is identical to that of ordinary transport noise.

Before presenting the recent measurements on non transport photo-assisted shot noise, it is important to conclude this presentation of the physics by considering the mechanism of photo absorption. Ref.[18] considered that the left lead connecting the scattering region of the conductor to the reservoirs meanders as an almost close loop. A radio-frequency a.c. flux $\Phi(t) = \Phi_{ac}\cos(2\pi\nu t)$

pierces the loop such that the phase $\phi(t)$ of an electron incoming from the left is modulated as $\phi(t) = 2\pi(\Phi_{ac}/\Phi_0)\cos(2\pi\nu t)$ where $\Phi_0 = h/e$, the flux quantum. The time dependence of the phase immediately projects a wavefunction of energy ε over wavefunctions of energies $\varepsilon \pm lh\nu$ with the spectral weight $P_l = J_l^2(2\pi\Phi_{ac}/\Phi_0)$, where J_l is the lth order Bessel function. Experimentally, it is easier to applied an a.c. electrochemical potential $eV(t) = eV_{ac}\cos(2\pi\nu t)$ to, say, the left contact[15, 21, 19, 3]. Note that the a.c. potential is the mean field potential experienced by electrons which includes screening. Indeed plasma frequencies are usually much faster that the frequencies considered here such that screening is instantaneous. In QPCs the screening length is short enough to consider the potential approximately flat between the short length scattering region and the wide region separating the scatter from the effective left reservoir (all the potential drop is at the QPC). As a consequence the effect of the potential is simply to add a time dependent phase to electrons incoming from the left $\phi(t) = (eV_{ac}/h\nu)\cos(2\pi\nu t)$. The l photo-absorption probability is $P_l = J_l^2(eV_{ac}/h\nu)$. Note that both approaches are equivalent if one identifies V_{ac} to $2\pi\nu\Phi_{ac}$.

Finally, irradiation by an a.c. potential rises another question. Consider for simplicity the case of zero bias and adding a low frequency a.c. potential V_{ac} $\cos(2\pi\nu t)$ to the left contact of the previous single mode conductor. This gives rise to a low frequency a.c. current at the same frequency. While there is no average current ($\overline{I} = 0$), there is an average shot noise ($\langle S_I\rangle \propto \overline{|I|} \neq 0$). Is this trivial shot noise distinguishable from the non-transport photo-assisted noise considered above? Quantitatively, the time average noise :

$$\left\langle \Delta I^2 \right\rangle = \frac{2}{\pi}.eV_{ac}\frac{e}{h}D(1-D)\Delta f$$

is to be compared with the photo-assisted shot noise

$$\overline{\Delta I^2} = 4\frac{e}{h}\left(\sum_{l=1}^{\infty}(lh\nu)J_l^2(eV_{ac}/h\nu)\right)D(1-D)\Delta f.$$

Using the properties of Bessel function, it is easy to show that in the limit of $\alpha = eV_{ac}/h\nu \gg 1$ both expressions are equal but when $\alpha \leq 5$ a sensible departure between the two numerical values is observable.

5. The observation of non-transport photo-assisted shot noise in a ballistic conductor:

Transport photo-assisted noise as predicted in Refs.[18, 19] has been reported in Refs.[15, 21] using a diffusive sample. While the authors definitely have shown the existence of photo-assisted processes, the technique however allowed only to measure the derivative of the noise with bias and was not able to measure the full shot noise in the zero current regime. Also, because the system being diffusive, the total transmission resulted from a statistical distribution of transmission and it was not possible to vary the transmission for an accurate test of the theory. In the next sections we present recent measurements obtained by measuring non-transport and transport photo-assisted noise in a ballistic system. The device is a quantum point contact realized in a 2D electron gas (2DEG). It allows to vary the transmission accurately and so the Fano factor in order to test the predictions. In addition, the noise detection set up measures the *total* noise. The experiment shows that electrons pumped to higher energy by photo-absorption do generate shot noise even when no bias voltage is applied between reservoirs (and hence no current flows through the conductor). The non-transport noise measurements allow to determine the Fano factor of the noise. In the doubly non-equilibrium regime, where both rf and finite bias voltage are applied, the $eV = h\nu$ noise singularity is also observed, although rounded, providing further evidence that photo-pumping is the basic underlying mechanism.

The QPC is realized using a 2D electron gas in GaAs/AlGaAs with $8\,10^5$ cm^2/Vs mobility and $4.8\,10^{11} cm^{-2}$ density. Special etching of the mesa prior to evaporation of the QPC metallic gates provides significant depletion at the QPC with zero gate voltage. At low temperature, well defined conductance plateaus for gate voltages ranging from $-55\,mV$ to $30\,mV$ allow accurately tuning of the transmission probability of the first two modes. The noise measurements were performed using a cross correlation technique [22] in the 2.6 to $4.2\,kHz$ range. The current noise power S_I is calculated from the voltage noise power S_V measured across the sample : $S_I = G^2 S_V$ where G is the differential conductance recorded simultaneously. A $5.2 \times 10^{-28} A^2/Hz$ background current noise results from the amplifier current noise and the room temperature $100M\Omega$ current source. The sensitivity of our experimental setup is checked within 2% both by measuring the quantum reduction of shot noise [7, 8] at transmission $1/2$ and also by measuring the thermal noise for temperature varying from $200\,mK$ to $600\,mK$. The base electronic temperature is $94 \pm 5\,mK$ for a $28\,mK$ refrigerator temperature. The difference arises from the low loss coaxial cable carrying the rf which brings a wide bandwidth of high temperature black body radiation to the sample [22].

The first step of the experiment is to determine the frequencies giving the highest coupling between the radiofrequency and the sample. This is achieved by measuring a weak photocurrent which never exceed $0.2\,nA$ in the explored rf power range (the equivalent open circuit voltage is always lower than k_BT such that its effect on noise can be neglected in the experiments described below). This study shows that the coupling is sharply enhanced at two frequencies 17.32 and $8.73\,GHz$ which therefore will be used for photon irradiation.

It is interesting to compare the rf period with the transit time of electrons between reservoirs. The distance between ohmic contacts being $30\,\mu m$ and the elastic mean free path $9\,\mu m$ we estimate the transit time to be $0.4\,ns$. This is

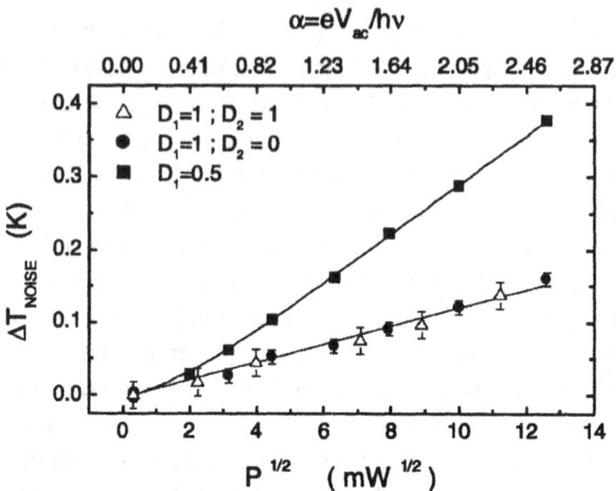

Figure 4. Excess noise temperature as a function of the rf power P on the top of the fridge at 17.32 GHz. From the noise increase at transmission 1, we deduced the electronic temperature increase due to dissipation. The solid line for $D = 0.5$ is a fit using Eq.(4) when taking into account the temperature increase. It gives the proportionality between α and $P^{1/2}$

shorter than the coherence time and much longer than the rf period such that application of photo-assisted model is legitimate.

We now discuss the results of the observation of photo-assisted electron and hole partition noise with *no applied bias voltage* obtained in Ref.[3]. In the limit where $h\nu \gg k_B T$, the expected noise [18, 19, 15] re-written in terms of equivalent noise temperature $T_N = S_I/4Gk_B$ is:

$$T_N = T\left(J_0^2(\alpha) + \frac{\sum_n D_n^2}{\sum_n D_n}(1 - J_0^2(\alpha))\right) + \sum_{l=1}^{+\infty} \frac{lh\nu}{k_B} J_l^2(\alpha) \frac{\sum_n D_n(1 - D_n)}{\sum_n D_n} \quad (4)$$

Here D_n is the transmission probability of the nth mode, $\alpha = eV_{ac}/h\nu$, J_l the integer Bessel function of order l and V_{ac} the rf voltage amplitude. The first term represents the thermal noise of unpumped and pumped electrons. The second term (which interests us here) is the partition noise of photo-created electrons and holes scattered by the QPC as discussed in the beginning ($\mathcal{P}_1 = J_1^2$ here). When the modes are either fully transmitted or reflected ($D_n = 1$ or 0), the noise is Johnson-Nyquist noise : $T_N = T$ and does not depend on rf power. However, in a real experiments heating of the reservoir by the rf power can not be excluded [15]. In Ref.[3] the measurements have been first performed on the first conductance plateau ($D_1 = G/G_0 = 1$) and on the second plateau ($D_1 = D_2 = 1$, $G/G_0 = 2$,

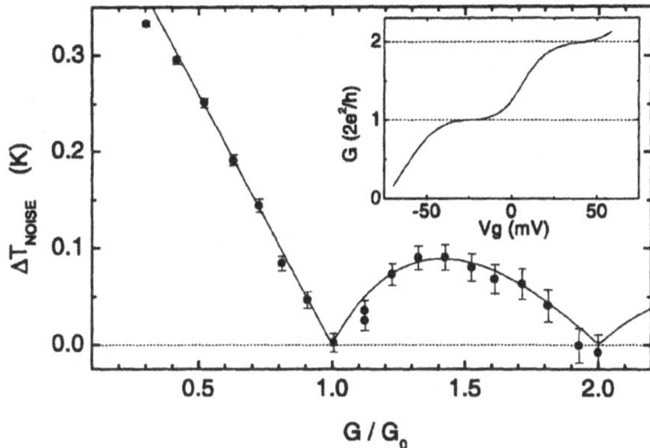

Figure 5. Noise temperature increase as a function of the transmission G/G_0 when applying a $17.32\,GHz$ ac excitation with $\alpha = 2.3$. The effect due to heating deduced from Fig.(4) has been removed. The solid line is the quantum suppression of the noise $\sum_n D_n(1 - D_n)/\sum_n D_n$. Inset: conductance versus gate voltage.

where $G_0 = 2e^2/h$). An increase of the noise is indeed observed when increasing the rf power as shown in fig.(4). Starting from a base electron temperature of $94\,mK$, the noise temperature increases and reaches $150\,mK$ for the highest power used in the experiments. The increase is the same on the first and second plateau. This indicates that heating occurs at the contact and is not related to the physics of scattering at the QPC. It is likely that heating of the contact results from rf absorption in the lossy coaxial lines.

The small heating being now characterized and quantitatively measured, the next step is to look at the partition noise regime expected for partial mode transmission. Fig.(4) shows a much larger increase of the noise temperature for $D_1 = 1/2$ than for $D_1 = 1$ and $D_1 = D_2 = 1$. This difference can safely be attributed to partition noise of photo-pumped electrons rather than to thermally assisted shot noise. Indeed, it is straightforward to give a quantitative estimation of the latter process. For an average temperature increase of the left and right reservoirs $\Delta T = (\Delta T_{left} + \Delta T_{right})/2$, simple calculation shows that the noise temperature increase should never exceed $\Delta T_N^{th} \leq \Delta T + (1 - D_1)[2\ln 2 - 1]\Delta T$. According to the study at transmission 1 and 2, ΔT is at most $150\,mK$, which gives $\Delta T_N^{th} \leq 179\,mK$ at half transmission, i.e. only $29\,mK$ above the noise temperature increase observed on the plateaus. The much larger noise observed at $D_1 = 1/2$ strongly suggests the presence of photo-assisted process. Taking the heating into account Eq.(4) fits the experimental results extremely well. From this we deduce $\alpha = eV_{ac}/h\nu$ to be a function of the square root of the applied rf

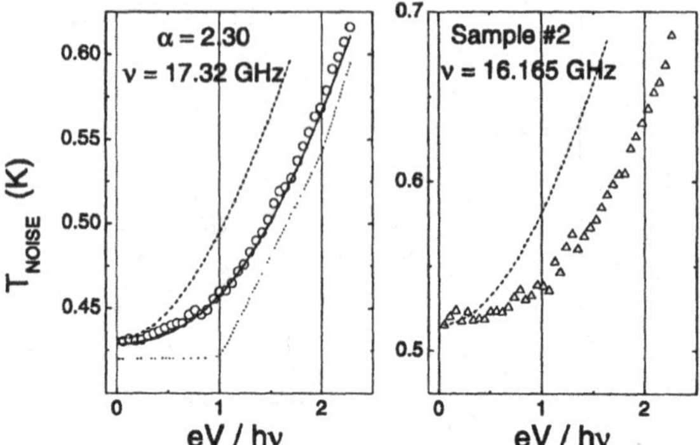

Figure 6. Left graph: Noise temperature as a function of $eV/h\nu$ with 17.32 GHz ac excitation. The measured electron temperature was $229\,mK$ and $\alpha = 2.3$. The dotted line is the expected photo-assisted noise at zero temperature with $\alpha = 2.3$ shifted for comparison. The dashed line is the non photo-assisted shot noise with $T = 430\,mK$. Continuous line is the photo-assisted noise calculated using Eq.(5) without adjustable parameter. Right graph: Noise temperature as a function of $eV/h\nu$ with 16.165 GHz ac excitation for a different sample. The dashed line is the expected non photo-assisted shot noise for $T = 520\,mK$.

power. This provides a calibration of the rf coupling which will be used below [1].

To fully characterize the partition noise of photo-pumped electrons and holes a systematic study as a function of transmission has been performed. The transmissions are measured using simultaneous measurement of the conductance $G = G_0 \sum_n D_n$. Fig.(5) shows the noise temperature variation versus transmission for $\alpha = 2.3$. The first term of Eq.(4) has been subtracted from the data as the transmissions D_n, α and the dependence of the electronic temperature with rf power are known. This allows better comparison with the second term of Eq.(4) which represents the electron hole partition noise. The variation of the noise is clearly proportional to the Fano factor [7, 8] $\sum_n D_n(1-D_n)/\sum_n D_n$ and unambiguously demonstrates the photo-assisted partition noise. Indeed the solid curve is the theoretical comparison with no adjustable parameter. Fig.(5) is the central result of Ref.[3]. We see that the Fano factor can be obtained from non transport photo-assisted shot noise experiments as accurately as the determination from d.c. transport shot noise shown in Fig.1.

A further check that photo-assisted noise is the basic underlying mechanism is to apply both rf and finite bias voltage simultaneously on the QPC. As discussed in previous sections, a singularity is expected in the shot noise variation at $eV = h\nu$

---

[1] Referred to the power P delivered by the rf source before attenuation, we find a proportionality between α and $10^{P(dBm)/20}$ equal to 0.204 ± 0.004 at $17.32\,GHz$ and 1.5 ± 0.1 at $8.73\,GHz$.

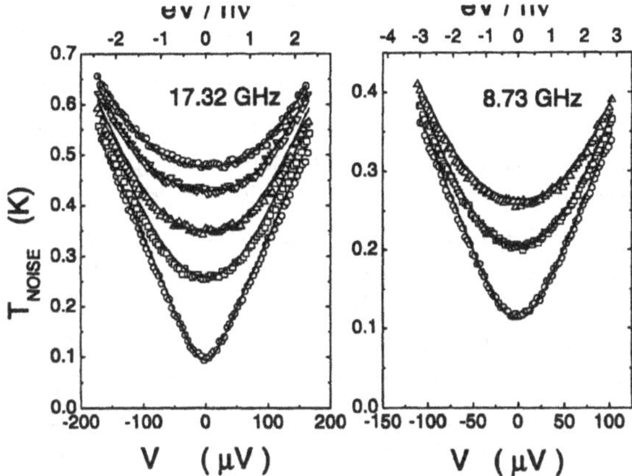

Figure 7. Noise versus bias voltage. The solid lines are the theoretical curves with α and T deduced from the equilibrium noise under rf illumination (see Fig.5). For the 17.32 GHz curves, $\alpha = 0.065$; 1.29; 1.83; 2.30; 2.58 and $T = 94$; 168; 200; 229; $246mK$ and for the 8.73 GHz curves, $\alpha = 0.51$; 1.81; 2.56 and $T = 105.5$; 145; $167.7mK$.

[18, 15, 21, 19]. :

$$T_N = T\frac{\sum_n D_n^2}{\sum_n D_n} + \frac{\sum_n D_n(1-D_n)}{\sum_n D_n}\sum_{\pm}\sum_{l=0}^{+\infty} J_l^2(\alpha)\frac{eV \pm lh\nu}{2k_B}\coth\left(\frac{eV \pm lh\nu}{2k_B T}\right) \tag{5}$$

Fig.(6), left graph, shows total noise measurements versus bias voltage at transmission $1/2$ for the same conditions as Fig.(5) ($\alpha = 2.3, T = 229\,mK$). As we can see on Fig.(6) for bias higher than $h\nu/e$ the noise starts to increase more rapidly, a hallmark of photo-assisted processes. This behavior cannot be attributed to simple thermal rounding, even if we assume that the $V = 0$ noise temperature was corresponding to $430\,mK$ electronic temperature (dashed line). In order to better reveal the expected singularities in the voltage, the dotted curve shows the noise at $T \approx 0$. As all parameters α, T, and D are known, Eq.(5) makes a direct comparison with our data. The agreement is excellent. Even the thermal rounding of the singularity at $eV = h\nu$ is well reproduced. The singularity of noise has been also observed using a different sample with a slightly different coupling and pumping frequency and is displayed on the right graph of Fig.(6).

Finally Fig.(7) shows a set of curves for various rf power values at 17.32 and 8.73 GHz [23]. The curves show that for the two different frequencies, the voltage scale of noise variation is determined by the photon energy quantum $h\nu$ and not by the thermal energy scale. From the regime of nearly pure shot noise to the regime of strongly photo-assisted shot noise all curves compare accurately with theory without any adjustable parameter.

6. Conclusion :

A d.c. current is not necessary to access the Fano factor characterizing the partitioning of electrons and hence the scattering properties of a conductor. In this paper we have discussed the different mechanisms responsible for transport and non-transport shot noise and have particularly considered the case where non-equilibrium was provided using photon irradiation. The results of a recent experiment have been presented. In this experiment, absolute noise measurements on a quantum point contact under rf irradiation have provided a direct demonstration that the quantum partition noise of electrons can be observed when no current flows through a sample. The competition between non-transport photo-assisted processes and normal transport noise processes have been further brought into evidence by the observation of a singularity in the noise derivative for $eV = h\nu$ when applying finite voltage. We hope that this contribution will stimulate new experiments on non-transport shot noise.

References

1. M. J. M. de Jong and C. W. J. Beenakker, edited by L. L. Sohn, L. P. Kouwenhoven, and G. Schön (Kluwer, Dordrecht, 1997).
2. Y. M. Blanter and M. Büttiker, Phys. Rep. **336**, (2000).
3. L.-H. Reydellet, P. Roche, D. C. Glattli, B. Etienne, Y. Jin, cond-mat/0206514, submitted to Phys. Rev. Lett.
4. W. Schottky, Ann. Phys. (Leipzig) **57** (1918) 541.
5. G. B. Lesovik, Pis?ma Zh. Eksp. Teor. Fiz. **49**, 513 (1989) [Sov. Phys. JETP Lett. **49**, 592 (1989)]; M. Büttiker, Phys. Rev. Lett. **65**, 2901 (1990); Th. Martin and R. Landauer, Phys. Rev. B **45**, 1742 (1992).
6. see for example T. Martin and Landauer reference in [5].
7. M. Reznikov *et al.*, Phys. Rev. Lett. **75**, 3340 (1995).
8. A. Kumar *et al.*, Phys. Rev. Lett. **76**, 2778 (1996).
9. B. J. van Wees *et al.*, Phys. Rev. Lett. **60**, 848 (1988); D. A. Wharam *et al.*, J. Phys. C **21**,L209(1988).
10. H. E. van den Brom and J. M. van Ruitenbeek, Phys. Rev. Lett. **82**, 1526 (1999).
11. C. W. J. Beenakker and M. B. uttiker, Phys.Rev.B **46** (1992) 1889.
12. K. E. Nagaev, Phys. Lett. A **169** (1992) 103.
13. A. H. Steinbach, J. M. Martinis, and M. H. Devoret, Phys. Rev. Lett. **76**, 3806 (1996).
14. M. Henny *et al.*, Phys. Rev. B **56**, 2871 (1999).
15. R. J. Schoelkopf *et al.*, Phys. Rev. Lett. **80**, 2437 (1998).
16. M. H. Pedersen, S.A. van Langen, and M. Büttiker, Phys. Rev. B **57**, 1838 (1998)
17. S. Oberholzer, E. V. Sukhorukov, and C. Schönenberger, Nature **415**, 567 (2002).
18. G. B. Lesovik and L. S. Levitov, Phys. Rev. Lett. **72**, 538 (1994).
19. M. H. Pedersen and M. Buttiker, Phys. Rev. B **58**, 12993 (1998).
20. E. V. Sukhorukov and D. Loss, Phys. Rev. B **59**, 13054 (1999).
21. A. A. Kozhevnikov, R. J. Schoelkopf, and D. E. Prober, Phys. Rev. Lett. **84**, 3398 (2000).
22. D. C. Glattli *et al.*, J. Appl. Phys. **81**, 7350 (1997).
23. The weak photocurrent I_{ph} gives an offset in V which is less than $6\mu V$ for $D = 0.5$. This offset is removed.

SHOT NOISE OF COTUNNELING CURRENT

EUGENE SUKHORUKOV
Département de Physique Théorique, Université de Genève, CH-1211 Genève 4, Switzerland.

GUIDO BURKARD and DANIEL LOSS
Department of Physics and Astronomy, University of Basel, Klingelbergstrasse 82, CH-4056 Basel, Switzerland

Abstract. We study the noise of the cotunneling current through one or several tunnel-coupled quantum dots in the Coulomb blockade regime. The various regimes of weak and strong, elastic and inelastic cotunneling are analyzed for quantum-dot systems (QDS) with few-level, nearly-degenerate, and continuous electronic spectra. In the case of weak cotunneling we prove a non-equilibrium fluctuation-dissipation theorem which leads to a universal expression for the noise-to-current ratio (Fano factor). The noise of strong inelastic cotunneling can be super-Poissonian due to switching between QDS states carrying currents of different strengths. The transport through a double-dot (DD) system shows an Aharonov-Bohm effect both in noise and current. In the case of cotunneling through a QDS with a continuous energy spectrum the Fano factor is very close to one.

1. Introduction

In recent years, there has been great interest in the shot noise in mesoscopic systems [1], because it contains additional information about correlations, which is not contained, e.g., in the linear response conductance. The shot noise is characterized by the Fano factor $F = S/eI$, the dimensionless ratio of the zero-frequency noise power S to the average current I. While it assumes the Poissonian value $F = 1$ in the absence of correlations, it becomes suppressed or enhanced when correlations set in as e.g. imposed by the Pauli principle or due to interaction effects. In the present paper we study the shot noise of the cotunneling [2, 3] current. We consider the transport through a quantum-dot system (QDS) in the Coulomb blockade (CB) regime, in which the quantization of charge on the QDS leads to a suppression of the sequential tunneling current except under certain resonant conditions. We consider the transport away from these resonances and

Y. V. Nazarov (ed.), Quantum Noise in Mesoscopic Physics, 149–172.

study the next-order contribution to the current [1] (see Fig. 1). We find that in the weak cotunneling regime, i.e. when the cotunneling rate I/e is small compared to the intrinsic relaxation rate $w_{\rm in}$ of the QDS to its equilibrium state due to the coupling to the environment, $I/e \gg w_{\rm in}$, the zero-frequency noise takes on its Poissonian value, as first obtained for a special case in [6]. This result is generalized here, and we find a universal relation between noise and current for the QDS in the first nonvanishing order in the tunneling perturbation. Because of the universal character of this result Eq. (12) we call it the nonequilibrium fluctuation-dissipation theorem (FDT) [7] in analogy with linear response theory.

One might expect however that the cotunneling, being a two-particle process, may lead to strong correlations in the shot noise and to the deviation of the Fano factor from its Poissonian value $F = 1$. We show in Sec. 4 that this is indeed the case for the regime of strong cotunneling, $I/e \gg w_{\rm in}$. Specifically, for a two-level QDS we predict giant (divergent) super-Poissonian noise [8] (see Sec. 5): The QDS goes into an unstable mode where it switches between states 1 and 2 with (generally) different currents. In Sec. 6 we consider the transport through a double-dot (DD) system as an example to illustrate this effect (see Eq. (37) and Fig. 2). The Fano factor turns out to be a periodic function of the magnetic flux through the DD leading to an Aharonov-Bohm effect in the noise [9]. In the case of weak cotunneling we concentrate on the average current through the DD and find that it shows Aharonov-Bohm oscillations, which are a two-particle effect sensitive to spin entanglement.

Finally, in Sec. 7 we discuss the cotunneling through large QDS with a continuum spectrum. In this case the correlations in the cotunneling current described above do not play an essential role. In the regime of low bias, elastic cotunneling dominates transport,[2] and thus the noise is Poissonian. In the opposite case of large bias, the transport is governed by inelastic cotunneling, and in Sec. 7 we study heating effects which are relevant in this regime.

2. Model system

In general, the QDS can contain several dots, which can be coupled by tunnel junctions, the DD being a particular example [6]. The QDS is assumed to be weakly coupled to external metallic leads which are kept at equilibrium with their associated reservoirs at the chemical potentials μ_l, $l = 1,2$, where the currents I_l can be measured and the average current I through the QDS is defined by Eq. (5). Using a standard tunneling Hamiltonian approach [10], we write

$$H = H_0 + V\,, \quad H_0 = H_L + H_S + H_{\rm int}\,, \tag{1}$$

[1] The majority of papers on the noise of quantum dots consider the sequential tunneling regime, where a classical description ("orthodox" theory) is applicable [4]. In this regime the noise is generally suppressed below its full Poissonian value $F = 1$. This suppression can be interpreted [5] as being a result of the natural correlations imposed by charge conservation.

$$H_L = \sum_{l=1,2} \sum_k \varepsilon_k c_{lk}^\dagger c_{lk} , \quad H_S = \sum_p \varepsilon_p d_p^\dagger d_p , \tag{2}$$

$$V = \sum_{l=1,2} (D_l + D_l^\dagger), \quad D_l = \sum_{k,p} T_{lkp} c_{lk}^\dagger d_p , \tag{3}$$

where the terms H_L and H_S describe the leads and QDS, respectively (with k and p from a complete set of quantum numbers),and tunneling between leads and QDS is described by the perturbation V. The interaction term H_{int} does not need to be specified for our proof of the universality of noise in Sec. 3. The N-electron QDS is in the cotunneling regime where there is a finite energy cost $\Delta_\pm(l, N) > 0$ for the electron tunneling from the Fermi level of the lead l to the QDS $(+)$ and vice versa $(-)$. This energy cost is of the order of the charging energy E_C and much larger than the temperature, $\Delta_\pm(l, N) \sim E_C \gg k_B T$, so that only processes of second order in V are allowed.

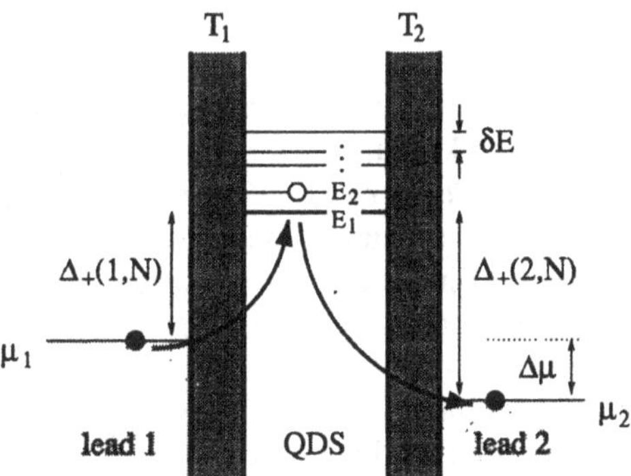

Figure 1. The quantum dot system (QDS) is coupled to two external leads $l = 1, 2$ via tunneling barriers. The tunneling between the QDS and the leads is parametrized by the tunneling amplitudes T_l, where the lead and QDS quantum numbers k and p have been dropped for simplicity, see Eq. (3). The leads are at the chemical potentials $\mu_{1,2}$, with an applied bias $\Delta\mu = \mu_1 - \mu_2$. The eigenstates of the QDS with one added electron ($N + 1$ electrons in total) are indicated by their energies E_1, E_2, etc., with average level-spacing δE. In the cotunneling regime there is a finite energy cost $\Delta_\pm(l, N) > 0$ for the electron tunneling from the Fermi level of the lead l to the QDS $(+)$ and vice versa $(-)$, so that only processes of second order in V (visualized by two arrows) are allowed.

To describe the transport through the QDS we apply standard methods [10] and adiabatically switch on the perturbation V in the distant past, $t = t_0 \to -\infty$. The perturbed state of the system is described by the time-dependent density matrix $\rho(t) = e^{-iH(t-t_0)} \rho_0 e^{iH(t-t_0)}$, with ρ_0 being the grand canonical density matrix of the unperturbed system, $\rho_0 = Z^{-1} e^{-K/k_B T}$, where we set $K = H_0 - \sum_l \mu_l N_l$. Because of tunneling the total number of electrons in each lead $N_l = \sum_k c_{lk}^\dagger c_{lk}$ is no longer conserved. For the outgoing currents $\hat{I}_l = e\dot{N}_l$ we have

$$\hat{I}_l = ei\,[V, N_l] = ei(D_l^\dagger - D_l)\,. \tag{4}$$

The observables of interest are the average current $I \equiv I_2 = -I_1$ through the QDS, and the spectral density of the noise $S_{ll'}(\omega) = \int dt S_{ll'}(t) \exp(i\omega t)$,

$$I_l = \mathrm{Tr}\rho(0)\hat{I}_l, \quad S_{ll'}(t) = \mathrm{Re}\,\mathrm{Tr}\,\rho(0)\delta I_l(t)\delta I_{l'}(0)\,, \tag{5}$$

where $\delta I_l = \hat{I}_l - I_l$. Below we will use the interaction representation where Eq. (5) can be rewritten by replacing $\rho(0) \to \rho_0$ and $\hat{I}_l(t) \to U^\dagger(t)\hat{I}_l(t)U(t)$, with

$$U(t) = T \exp\left[-i\int_{-\infty}^{t} dt'\, V(t')\right]. \tag{6}$$

In this representation, the time dependence of all operators is governed by the unperturbed Hamiltonian H_0.

3. Weak cotunneling: Non-equilibrium fluctuation-dissipation theorem

In this section we prove the universality of noise of tunnel junctions in the weak cotunneling regime $I/e \ll w_{\mathrm{in}}$ keeping the first nonvanishing order in the tunneling Hamiltonian V. Since our final result (12) can be applied to quite general systems out-of-equilibrium we call this result the non-equilibrium fluctuation-dissipation theorem (FDT). In particular, the geometry of the QDS and the interaction H_{int} are completely arbitrary for the discussion of the non-equilibrium FDT in this section.

We note that the two currents $\hat{I}_l$ are not independent, because $[\hat{I}_1, \hat{I}_2] \neq 0$, and thus all correlators $S_{ll'}$ are nontrivial. The charge accumulation on the QDS for a time of order $\Delta_{\mp}^{-1}$ leads to an additional contribution to the noise at finite frequency ω. Thus, we expect that for $\omega \sim \Delta_{\pm}$ the correlators $S_{ll'}$ cannot be expressed through the steady-state current I only and thus I has to be complemented by some other dissipative counterparts, such as differential conductances $G_{ll'}$. On the other hand, at low enough frequency, $\omega \ll \Delta_{\pm}$, the charge conservation on the QDS requires $\delta I_s = (\delta I_2 + \delta I_1)/2 \approx 0$. Below we concentrate on the limit of low frequency and neglect contributions of order of $\omega/\Delta_{\pm}$ to the noise power. In the Appendix we prove that $S_{ss} \sim (\omega/\Delta_{\pm})^2$ (see Eq. (65)), and this allows us to redefine the current and the noise power as $I \equiv I_d = (I_2 - I_1)/2$ and $S(\omega) \equiv S_{dd}(\omega)$. [2] In addition we require that the QDS is in the cotunneling regime, i.e. the temperature is low enough, $k_B T \ll \Delta_{\pm}$, although the bias $\Delta\mu$ is arbitrary as soon as the sequential tunneling to the dot is forbidden, $\Delta_{\pm} > 0$.

---

[2] We note that charge fluctuations, $\delta Q(t) = 2\int_{-\infty}^{t} dt' \delta I_s(t')$, on a QDS are also relevant for device applications such as SET [11]. While we focus on current fluctuations in the present paper, we mention here that in the cotunneling regime the noise power $\langle \delta Q^2 \rangle_\omega$ does not vanish at zero frequency, $\langle \delta Q^2 \rangle_{\omega=0} = 4\omega^{-2} S_{ss}(\omega)|_{\omega\to 0} \neq 0$. Our formalism is also suitable for studying such charge fluctuations; this will be addressed elsewhere.

In this limit the current through a QDS arises due to the direct hopping of an electron from one lead to another (through a virtual state on the dot) with an amplitude which depends on the energy cost $\Delta_\pm$ of a virtual state. Although this process can change the state of the QDS (inelastic cotunneling), the fast energy relaxation in the weak cotunneling regime, $w_{\rm in} \gg I/e$, immediately returns it to the equilibrium state (for the opposite case, see Sec. 4). This allows us to apply a perturbation expansion with respect to tunneling V and to keep only first nonvanishing contributions, which we do next.

It is convenient to introduce the notation $\bar{D}_l(t) \equiv \int_{-\infty}^{t} dt'\, D_l(t')$. We notice that all relevant matrix elements, $\langle N|D_l(t)|N+1\rangle \sim e^{-i\Delta_+ t}$, $\langle N-1|D_l(t)|N\rangle \sim e^{i\Delta_- t}$, are fast oscillating functions of time. Thus, under the above conditions we can write $\bar{D}_l(\infty) = 0$, and even more general, $\int_{-\infty}^{+\infty} dt\, D_l(t) e^{\pm i\omega t} = 0$ (note that we have assumed earlier that $\omega \ll \Delta_\pm$). Using these equalities and the cyclic property of the trace we obtain the following results (for details of the derivation, see Appendix A),

$$I = e \int_{-\infty}^{\infty} dt\, \langle [A^\dagger(t), A(0)] \rangle, \qquad A = D_2 \bar{D}_1^\dagger + D_1^\dagger \bar{D}_2\,, \tag{7}$$

$$S(\omega) = e^2 \int_{-\infty}^{\infty} dt\, \cos(\omega t) \langle \{A^\dagger(t), A(0)\} \rangle\,, \tag{8}$$

where we have dropped a small contribution of order $\omega/\Delta_\pm$ and used the notation $\langle \ldots \rangle = {\rm Tr}\rho_0(\ldots)$.

Next we apply the spectral decomposition to the correlators Eqs. (7) and (8), a similar procedure to that which also leads to the equilibrium fluctuation-dissipation theorem. The crucial observation is that $[H_0, N_l] = 0$, $l = 1, 2$. Therefore, we are allowed to use for our spectral decomposition the basis $|\mathbf{n}\rangle = |E_\mathbf{n}, N_1, N_2\rangle$ of eigenstates of the operator $K = H_0 - \sum_l \mu_l N_l$, which also diagonalizes the grand-canonical density matrix ρ_0, $\rho_\mathbf{n} = \langle \mathbf{n}|\rho_0|\mathbf{n}\rangle = Z^{-1} \exp[-E_\mathbf{n}/k_B T]$. We introduce the spectral function,

$$\mathcal{A}(\omega) = 2\pi \sum_{\mathbf{n},\mathbf{m}} (\rho_\mathbf{n} + \rho_\mathbf{m}) |\langle \mathbf{m}|A|\mathbf{n}\rangle|^2 \delta(\omega + E_\mathbf{n} - E_\mathbf{m})\,, \tag{9}$$

and rewrite Eqs. (7) and (8) in the matrix form in the basis $|\mathbf{n}\rangle$ taking into account that the operator A, which plays the role of the effective cotunneling amplitude, creates (annihilates) an electron in the lead 2 (1) (see Eqs. (3) and (7)). We obtain following expressions

$$I(\Delta\mu) = e\,{\rm th}\left[\frac{\Delta\mu}{2k_B T}\right] \mathcal{A}(\Delta\mu)\,, \tag{10}$$

$$S(\omega, \Delta\mu) = \frac{e^2}{2} \sum_\pm \mathcal{A}(\Delta\mu \pm \omega)\,. \tag{11}$$

We note that because of additional integration over time t in the amplitude A (see Eq. (7)), the spectral density $\mathcal{A}$ depends on μ_1 and μ_2 separately. However, away from the resonances, $\omega \ll \Delta_\pm$, only $\Delta\mu$-dependence is essential, and thus $\mathcal{A}$ can be regarded as being one-parameter function. [3] Comparing Eqs. (10) and (11), we obtain

$$S(\omega, \Delta\mu) = \frac{e}{2} \sum_{\pm} \coth\left[\frac{\Delta\mu \pm \omega}{2k_B T}\right] I(\Delta\mu \pm \omega) \qquad (12)$$

up to small terms on the order of $\omega/\Delta_\pm$. This equation represents our nonequilibrium FDT for the transport through a QDS in the weak cotunneling regime. A special case with $T, \omega = 0$, giving $S = eI$, has been derived earlier [6]. To conclude this section we would like to list again the conditions used in the derivation. The universality of noise to current relation Eq. (12) proven here is valid in the regime in which it is sufficient to keep the first nonvanishing order in the tunneling V which contributes to transport and noise. This means that the QDS is in the weak cotunneling regime with $\omega, k_B T \ll \Delta_\pm$, and $I/e \ll w_{\rm in}$.

4. Strong cotunneling: Correlation correction to noise

In this section we consider the QDS in the strong cotunneling regime, $w_{\rm in} \ll I/e$. Under this assumption the intrinsic relaxation in the QDS is very slow and will in fact be neglected. Thermal equilibration can only take place via coupling to the leads (see Sec. 7). Due to this slow relaxation in the QDS we find that there are non-Poissonian correlations ΔS in the current through the QDS because the QDS has a "memory"; the state of the QDS after the transmission of one electron influences the transmission of the next electron. The microscopic theory of strong cotunneling has been developed in Ref. [5] based on the density-operator formalism and using the projection operator technique. Here we discuss the assumptions and present the results of the theory, equations (14), (15), and (17-19), which are the basis for our further analysis in the Secs. 5 and 6.

First, we assume that the system and bath are coupled only weakly and only via the perturbation V, Eq. (3). The interaction part $H_{\rm int}$ of the unperturbed Hamiltonian H_0, Eq. (1), must therefore be separable into a QDS and a lead part, $H_{\rm int} = H_S^{\rm int} + H_L^{\rm int}$. Moreover, H_0 conserves the number of electrons in the leads, $[H_0, N_l] = 0$, where $N_l = \sum_k c_{lk}^\dagger c_{lk}$. The assumption of weak coupling allows us to keep only the second-order in V contributions to the "golden rule" rates (15) for the Master equation (14).

[3] To be more precise, we neglect small ω-shift of the energy denominators $\Delta_\pm$, which is equivalent to neglecting small terms of order $\omega/\Delta_\pm$ in Eq. (11).

Second, we assume that in the distant past, $t_0 \to -\infty$, the system is in an equilibrium state

$$\rho_0 = \rho_S \otimes \rho_L, \quad \rho_L = \frac{1}{Z_L} e^{-K_L/k_B T}, \tag{13}$$

where $Z_L = \mathrm{Tr}\, \exp[-K_L/k_B T]$, $K_L = H_L - \sum_l \mu_l N_l$, and μ_l is the chemical potential of lead l. Note that both leads are kept at the same temperature T. Physically, the product form of ρ_0 in Eq. (13) describes the absence of correlations between the QDS and the leads in the initial state at t_0. Furthermore, we assume that the initial state ρ_0 is diagonal in the eigenbasis of H_0, i.e. that the initial state is an incoherent mixture of eigenstates of the free Hamiltonian.

Finally, we consider the low-frequency noise, $\omega \ll \Delta_\pm$, i.e. we neglect the accumulation of the charge on the QDS (in the same way as in the Sec. 3). Thus we can write $S_{ll}(\omega) = -S_{l\neq l'}(\omega) \equiv S(\omega)$. This restriction will be lifted in the end of the Sec. 6.1.

We note that the above assumptions limit the generality of the results of present section as compared to those of Sec. 3. On the other hand, they allow us to reduce the problem of the noise calculations to the solution of the Master equation

$$\dot{\rho}_n(t) = \sum_m \left[w_{nm} \rho_m(t) - w_{mn} \rho_n(t) \right], \tag{14}$$

with the stationary state condition $\sum_m (w_{nm} \bar{\rho}_m - w_{mn} \bar{\rho}_n) = 0$. This "classical" master equation describes the dynamics of the QDS, i.e. it describes the rates with which the probabilities ρ_n for the QDS being in state $|n\rangle$ change. The rates $w_{nm} = \sum_{l,l'=1,2} w_{nm}(l', l)$ are the sums of second-order "golden rule" rates

$$w_{nm}(l', l) = 2\pi \sum_{\bar{m},\bar{n}} |\langle \mathbf{n} | (D_l^\dagger, D_{l'}) | \mathbf{m} \rangle|^2 \delta(E_\mathbf{m} - E_\mathbf{n} - \Delta\mu_{ll'}) \rho_{L,\bar{m}}. \tag{15}$$

for all possible cotunneling transitions from lead l to lead l'. In the last expression, $\Delta\mu_{ll'} = \mu_l - \mu_{l'}$ denotes the chemical potential drop between lead l and lead l', and $\rho_{L,\bar{m}} = \langle \bar{m} | \rho_L | \bar{m} \rangle$. We have defined the second order hopping operator

$$(D_l^\dagger, D_{l'}) = D_{l'} \bar{D}_l^\dagger + D_l^\dagger \bar{D}_{l'}, \tag{16}$$

where D_l is given in Eq. (3), and $\bar{D}_l = \int_{-\infty}^0 D_l(t) dt$. Note, that $(D_l^\dagger, D_{l'})$ is the amplitude of cotunneling from the lead l to the lead l' (in particular, we can write $A = (D_1^\dagger, D_2)$, see Eq. (7)). The combined index $\mathbf{m} = (m, \bar{m})$ contains both the QDS index m and the lead index $\bar{m}$. Correspondingly, the basis states used above are $|\mathbf{m}\rangle = |m\rangle|\bar{m}\rangle$ with energy $E_\mathbf{m} = E_m + E_{\bar{m}}$, where $|m\rangle$ is an eigenstate of $H_S + H_S^{\mathrm{int}}$ with energy E_m, and $|\bar{m}\rangle$ is an eigenstate of $H_L + H_L^{\mathrm{int}} - \sum_l \mu_l N_l$ with energy $E_{\bar{m}}$.

For the average current I and the noise power $S(\omega)$ we obtain [5]

$$I = e\sum_{mn} w^I_{nm}\bar{\rho}_m, \quad w^I_{nm} = w_{nm}(2,1) - w_{nm}(1,2), \tag{17}$$

$$S(\omega) = e^2\sum_{mn}[w_{nm}(2,1) + w_{nm}(1,2)]\bar{\rho}_m + \Delta S(\omega), \tag{18}$$

$$\Delta S(\omega) = e^2\sum_{n,m,n',m'} w^I_{nm}\delta\rho_{mn'}(\omega)w^I_{n'm'}\bar{\rho}_{m'}, \tag{19}$$

where $\delta\rho_{nm}(\omega) = \rho_{nm}(\omega) - 2\pi\delta(\omega)\bar{\rho}_n$, and $\bar{\rho}_n$ is the stationary density matrix. Here, $\rho_{nm}(\omega)$ is the Fourier-transformed conditional density matrix, which is obtained from the *symmetrized* solution $\rho_n(t) = \rho_n(-t)$ of the master equation Eq. (14) with the initial condition $\rho_n(0) = \delta_{nm}$.

An explicit result for the noise in this case can be obtained by making further assumptions about the QDS and the coupling to the leads, see the following sections. For the general case, we only estimate ΔS. The current is of the order $I \sim ew$, with w some typical value of the cotunneling rate w_{nm}, and thus $\delta I \sim ew$. The time between switching from one dot-state to another due to cotunneling is approximately $\tau \sim w^{-1}$. The correction ΔS to the Poissonian noise can be estimated as $\Delta S \sim \delta I^2\tau \sim e^2 w$, which is of the same order as the Poissonian contribution $eI \sim e^2 w$. Thus the correction to the Fano factor is of order unity. (Note however, that under certain conditions the Fano factor can diverge, see Secs. 5 and 6.) In contrast to this, we find that for elastic cotunneling the off-diagonal rates vanish, $w_{nm} \propto \delta_{nm}$, and therefore $\delta\rho_{nn} = 0$ and $\Delta S = 0$. Moreover, at zero temperature, either $w_{nn}(2,1)$ or $w_{nn}(1,2)$ must be zero (depending on the sign of the bias $\Delta\mu$). As a consequence, for elastic cotunneling we find Poissonian noise, $F = S(0)/e|I| = 1$.

5. Cotunneling through nearly degenerate states

Suppose the QDS has nearly degenerate states with energies E_n, and level spacing $\delta E_{nm} = E_n - E_m$, which is much smaller than the average level spacing δE. In the regime, $\Delta\mu, k_BT, \delta E_{nm} \ll \delta E$, the only allowed cotunneling processes are the transitions between nearly degenerate states. The noise power is given by Eqs. (18) and (19), and below we calculate the correlation correction to the noise, ΔS. To proceed with our calculation we rewrite Eq. (14) for $\delta\rho(t)$ as a second-order differential equation in matrix form

$$\delta\ddot{\rho}(t) = W^2\delta\rho(t), \quad \delta\rho(0) = 1 - \bar{\rho}, \tag{20}$$

where W is defined as $W_{nm} = w_{nm} - \delta_{nm}\sum_{m'} w_{m'n}$. We solve this equation by Fourier transformation,

$$\delta\rho(\omega) = -\frac{2W}{W^2 + \omega^2 1}, \tag{21}$$

where we have used $W\bar{\rho} = 0$. We substitute $\delta\rho$ from this equation into Eq. (19) and write the result in a compact matrix form,

$$\Delta S(\omega) = -e^2 \sum_{n,m} \left[w^I \frac{2W}{W^2 + \omega^2 1} w^I \bar{\rho} \right]_{nm} . \tag{22}$$

This equation gives the formal solution of the noise problem for nearly degenerate states. As an example we consider a two-level system.

Using the detailed balance equation, $w_{21}\rho_1 = w_{12}\rho_2$, we obtain for the stationary probabilities $\rho_1 = w_{12}/(w_{12} + w_{21})$, and $\rho_2 = w_{21}/(w_{12} + w_{21})$. From Eq. (17) we get

$$I = e \frac{w_{12}(w^I_{11} + w^I_{21}) + w_{21}(w^I_{22} + w^I_{12})}{w_{12} + w_{21}} . \tag{23}$$

A straightforward calculation with the help of Eq. (21) gives for the correction to the Poissonian noise

$$\begin{aligned} \Delta S(\omega) &= \frac{2e^2(w^I_{11} + w^I_{21} - w^I_{22} - w^I_{12})}{(w_{12} + w_{21})[\omega^2 + (w_{12} + w_{21})^2]} \times \\ &\times \left[w^I_{11} w_{12} w_{21} + w^I_{12} w^2_{21} - (1 \leftrightarrow 2) \right] . \end{aligned} \tag{24}$$

In particular, the zero frequency noise $\Delta S(0)$ diverges if the "off-diagonal" rates w_{nm} vanish. This divergence has to be cut at ω, or at the relaxation rate w_{in} due to coupling to the bath (since w_{12} in this case has to be replaced with $w_{12} + w_{\text{in}}$). The physical origin of the divergence is rather transparent: If the off-diagonal rates w_{12}, w_{21} are small, the QDS goes into an unstable state where it switches between states 1 and 2 with different currents in general [12]. The longer the QDS stays in the state 1 or 2 the larger the zero-frequency noise power is. However, if $w^I_{11} + w^I_{21} = w^I_{22} + w^I_{12}$, then $\Delta S(\omega)$ is suppressed to 0. For instance, for the QDS in the spin-degenerate state with an odd number of electrons $\Delta S(\omega) = 0$, since the two states $|\uparrow\rangle$ and $|\downarrow\rangle$ are physically equivalent. The other example of such a suppression of the correlation correction ΔS to noise is given by a multi-level QDS, $\delta E \ll E_C$, where the off-diagonal rates are small compared to the diagonal (elastic) rates [2]. Indeed, since the main contribution to the elastic rates comes from transitions through many virtual states, which do not participate in inelastic cotunneling, they do not depend on the initial conditions, $w^I_{11} = w^I_{22}$, and cancel in the numerator of Eq. (24), while they are still present in the current. Thus the correction $\Delta S/I$ vanishes in this case. Further below in this section we consider a few-level QDS, $\delta E \sim E_C$, where $\Delta S \neq 0$.

To simplify further analysis we consider for a moment the case, where the singularity in the noise is most pronounced, namely, $\omega = 0$ and $|\delta E_{12}| \ll \Delta\mu, k_B T$, so that $w^I_{12} = w^I_{21}$, and $w_{12} = w_{21}$. Then, from Eqs. (23) and (24) we obtain

$$I = \frac{1}{2}(I_1 + I_2) , \quad I_n = e \sum_{m=1,2} w^I_{mn} , \tag{25}$$

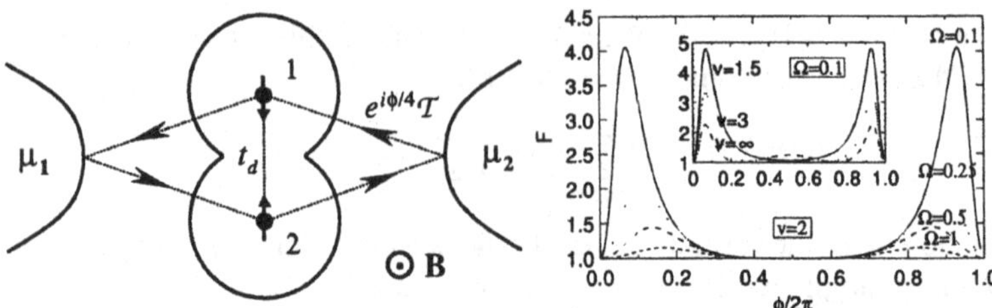

Figure 2. Left: Double-dot (DD) system containing two electrons and being weakly coupled to metallic leads 1, 2, each of which is at the chemical potential μ_1, μ_2. The tunneling amplitudes between dots and leads are denoted by T. The tunneling (t_d) between the dots results in a singlet-triplet splitting $J \sim t_d^2/U$ with the singlet being a ground state. The tunneling path between dots and leads 1 and 2 forms a closed loop (shown by arrows) so that the Aharonov-Bohm phase ϕ will be accumulated by an electron traversing the DD. Right: The Fano factor $F = S(\omega)/I$, with the noise power $S(\omega)$ given in Eqs. (18) and (37), is plotted as a function of the Aharonov-Bohm phase ϕ for the normalized bias $v \equiv \Delta\mu/J = 2$ and for four different normalized frequencies $\Omega \equiv \omega/[G(2\Delta\mu - J)] = 0.1, 0.25, 0.5$, and 1. Inset: the same, but with fixed frequency $\Omega = 0.1$, where the bias v takes the values 1.5, 3, and ∞.

$$\Delta S(0) = \frac{(I_1 - I_2)^2}{4w_{12}}, \tag{26}$$

where I_n is the current through the n-th level of the QDS. Thus in case $|\delta E_{12}| \ll \Delta\mu, k_BT$ the following regimes have to be distinguished: (1) If $k_BT \lesssim \Delta\mu$, then $I_n \propto \Delta\mu$, $w_{12} \propto \Delta\mu$, and thus both, the total current $I = e^{-1}G_D\Delta\mu$, and the total noise $S = FG_D\Delta\mu$ are linear in the bias $\Delta\mu$ (here G_D is the conductance of the QDS). The total shot noise in this regime is super-Poissonian with the Fano factor $F \sim I/(ew_{12}) \gg 1$. (2) In the regime $\Delta\mu \lesssim k_BT \lesssim F^{1/2}\Delta\mu$ the noise correction (26) arises because of the thermal switching the QDS between two states $n = 1, 2$, where the currents are linear in the bias, $I_n \sim G_D\Delta\mu/e$. The rate of switching is $w_{12} \propto k_BT$, and thus $\Delta S \sim FG_D\Delta\mu^2/(k_BT)$. Since $k_BT/\Delta\mu \lesssim F^{1/2}$, the noise correction ΔS is the dominant contribution to the noise, and thus the total noise S can be interpreted as being a thermal telegraph noise [13]. (3) Finally, in the regime $F^{1/2}\Delta\mu \lesssim k_BT$ the first term on the rhs of Eq. (18) is the dominant contribution, and the total noise becomes an equilibrium Nyquist noise, $S = 2G_Dk_BT$.

6. Noise of double-dot system: Two-particle Aharonov-Bohm effect

We notice that for the noise power to be divergent the off-diagonal rates w_{12} and w_{21} have to vanish simultaneously. However, the matrix w_{nm} is not symmetric since the off-diagonal rates depend on the bias in a different way. On the other hand, both rates contain the same matrix element of the cotunneling amplitude $(D_l^\dagger, D_{l'})$, see Eqs. (15) and (16). Although in general this matrix element is not small, it can vanish because of different symmetries of the two states. To illustrate this effect we consider the transport through a double-dot (DD) system (see Ref. [6] for details) as an example. Two leads are equally coupled to two dots in such a way that a closed loop is formed, and the dots are also connected, see Fig. 2. Thus, in a magnetic field the tunneling is described by the Hamiltonian Eq. (3) with

$$D_l = \sum_{s,j} T_{lj} c_{ls}^\dagger d_{js}\,, \qquad l,j = 1,2\,, \tag{27}$$

$$T_{11} = T_{22} = T_{12}^* = T_{21}^* = e^{i\phi/4}T\,, \tag{28}$$

where the last equation expresses the equal coupling of dots and leads and ϕ is the Aharonov-Bohm phase. Each dot contains one electron, and weak tunneling t_d between the dots causes the exchange splitting [14] $J \sim t_d^2/U$ (with U being the on-site repulsion) between one spin singlet and three triplets

$$\begin{aligned} |S\rangle &= \frac{1}{\sqrt{2}}[d_{1\uparrow}^\dagger d_{2\downarrow}^\dagger - d_{1\downarrow}^\dagger d_{2\uparrow}^\dagger]|0\rangle\,, \\ |T_0\rangle &= \frac{1}{\sqrt{2}}[d_{1\uparrow}^\dagger d_{2\downarrow}^\dagger + d_{1\downarrow}^\dagger d_{2\uparrow}^\dagger]|0\rangle\,, \\ |T_+\rangle &= d_{1\uparrow}^\dagger d_{2\uparrow}^\dagger|0\rangle\,, \quad |T_-\rangle = d_{1\downarrow}^\dagger d_{2\downarrow}^\dagger|0\rangle\,. \end{aligned} \tag{29}$$

In the case of zero magnetic field, $\phi = 0$, the tunneling Hamiltonian V is symmetric with respect to the exchange of electrons, $1 \leftrightarrow 2$. Thus the matrix element of the cotunneling transition between the singlet and three triplets $\langle S|V(E - H_0)^{-1}V|T_i\rangle$, $i = 0, \pm$, vanishes because these states have different orbital symmetries. A weak magnetic field breaks the symmetry, contributes to the off-diagonal rates, and thereby reduces noise. Next, we consider weak and strong cotunneling regimes.

6.1. WEAK COTUNNELING

In this regime, $I/e \ll w_{\rm in}$, according to the non-equilibrium FDT (see Sec. 3) the zero-frequency noise contains the same information as the average current (the Fano factor $F = 1$). Therefore, we first concentrate on current. We focus on the regime, $\Delta\mu \gg J$, where inelastic cotunneling [15] occurs with singlet and triplet contributions being different, and where we can neglect the dynamics generated by J compared to the one generated by the bias ("slow spins"). Close to the sequential tunneling peak, $\Delta_- \ll \Delta_+ \sim U$, we keep only the term $D_1^\dagger \bar{D}_2$ in the amplitude (7). After some calculations we obtain

$$I = e^{-1}C(\varphi)G\Delta\mu, \tag{30}$$

$$C(\varphi) = \sum_{s,s'}\left[\langle d_{1s'}^\dagger d_{1s} d_{1s}^\dagger d_{1s'}\rangle + \cos\varphi\langle d_{1s'}^\dagger d_{1s} d_{2s}^\dagger d_{2s'}\rangle\right], \tag{31}$$

where $G = \pi(e\nu T^2/\Delta_-)^2$ is the conductance of a single dot in the cotunneling regime [16], and we assumed Fermi liquid leads with the tunneling density of states ν. Eq. (30) shows that the cotunneling current depends on the properties of the equilibrium state of the DD through the coherence factor $C(\varphi)$ given in (31). The first term in C is the contribution from the topologically trivial tunneling path (phase-incoherent part) which runs from lead 1 through, say, dot 1 to lead 2 and back. The second term (phase-coherent part) in C results from an exchange process of electron 1 with electron 2 via the leads 1 and 2 such that a closed loop is formed enclosing an area A (see Fig. 2). Note that for singlet and triplets the initial and final spin states are the same after such an exchange process. Thus, in the presence of a magnetic field B, an Aharonov-Bohm phase factor $\varphi = ABe/h$ is acquired.

Next, we evaluate $C(\varphi)$ explicitly in the singlet-triplet basis (29) and discuss the applications to the physics of quantum entanglement (see the Ref. [6]). Note that only the singlet $|S\rangle$ and the triplet $|T_0\rangle$ are entangled EPR pairs while the remaining triplets are not (they factorize). Assuming that the DD is in one of these states we obtain the important result

$$C(\varphi) = 2 \mp \cos\varphi \,. \tag{32}$$

Thus, we see that the singlet (upper sign) and the triplets (lower sign) contribute with opposite sign to the phase-coherent part of the current. One has to distinguish, however, carefully the entangled from the non-entangled states. The phase-coherent part of the entangled states is a genuine *two-particle* effect, while the one of the product states cannot be distinguished from a phase-coherent *single-particle* effect. Indeed, this follows from the observation that the phase-coherent part in C factorizes for the product states $T_\pm$ while it does not so for S, T_0. Also, for states such as $|\uparrow\downarrow\rangle$ the coherent part of C vanishes, showing that two different (and fixed) spin states cannot lead to a phase-coherent contribution since we *know* which electron goes which part of the loop.

Finally, we present our results [6] for the high-frequency noise in the quantum range of frequancies, $\omega \sim \Delta_- \ll \Delta_+$, and in the slow-spin regime $\Delta\mu \gg J$. This range of frequancies is beyond the regime of the applicability of the non-equilibrium FDT, and therefore there is no simple relation between the average current and the noise (see the Sec. 3). After lengthy calculations using the perturbation expansion of (6) up to third order in V we obtain

$$\begin{aligned} S(\omega) &= (e\pi\nu T^2)^2\left[X_\omega + X^*_{-\omega}\right], \\ ImX_\omega &= \frac{C(\varphi)}{2\omega}\left[\theta(\mu_1-\omega)-\theta(\mu_2-\omega)\right], \end{aligned} \tag{33}$$

$$\begin{aligned} ReX_\omega &= \frac{C(\varphi)}{2\pi\omega} sign(\mu_1-\mu_2+\omega)\ln\left|\frac{(\mu_1+\omega)(\mu_2-\omega)}{\mu_1\mu_2}\right| \\ &-\frac{1}{2\pi\omega}\left[\theta(\omega-\mu_1)\ln\left|\frac{\mu_2-\omega}{\mu_2}\right| + \theta(\omega-\mu_2)\ln\left|\frac{\mu_1-\omega}{\mu_1}\right|\right], \end{aligned} \tag{34}$$

where $\mu_l = \Delta_-(l)$. Thus the real part of $S(\omega)$ is even in ω, while the imaginary part is odd. A remarkable feature here is that the noise acquires an imaginary (i.e. odd-frequency) part for finite frequencies, in contrast to single-barrier junctions, where Im$S(\omega)$ always vanishes since we have $\delta I_1 = -\delta I_2$ for all times. In double-barrier junctions considered here we find that at small enough bias $\Delta\mu \ll \Delta_- = (\Delta_-(1) + \Delta_-(2))/2$, the odd part, Im$S(\omega)$, given in (33) exhibits two narrow peaks at $\omega = \pm\mu$, which in real time lead to slowly decaying oscillations,

$$S_{odd}(t) = \pi(e\nu T^2)^2 C(\varphi)\frac{\sin(\Delta\mu t/2)}{t\Delta_-}\sin(t\Delta_-). \tag{35}$$

These oscillations again depend on the phase-coherence factor C with the same properties as discussed before. These oscillations can be interpreted as a temporary build-up of a charge-imbalance on the DD during an uncertainty time $\sim 1/\Delta_-$, which results from cotunneling of electrons and an associated time delay between out- and ingoing currents.

6.2. STRONG COTUNNELING

The fact that in the perturbation V all spin indices are traced out helps us to map the four-level system to only two states $|S\rangle$ and $|T\rangle$ classified according to the orbital symmetry (since all triplets are antisymmetric in orbital space). In Appendix B we derive the mapping to a two-level system and calculate the transition rates $w_{nm}(l', l)$ ($n, m = 1$ for a singlet and $n, m = 2$ for all triplets) using Eqs. (15) and (16) with the operators D_l given by Eq. (27). Doing this we obtain the following result

$$\begin{aligned} & w_{nm}(1,1) = w_{nm}(2,2) = w_{nm}(1,2) = 0, \\ & w_{nm} = w^I_{nm} = \frac{\pi}{2}\left(\frac{\nu T^2}{\Delta_-}\right)^2 \\ & \times\left\{\begin{array}{ll} (1+\cos\phi)\Delta\mu & (1-\cos\phi)(\Delta\mu+J) \\ 3(1-\cos\phi)(\Delta\mu-J) & 3(1+\cos\phi)\Delta\mu \end{array}\right\}, \end{aligned} \tag{36}$$

which holds close to the sequential tunneling peak, $\Delta_- \ll \Delta_+ \sim U$ (but still $\Delta_- \gg J, \Delta\mu$), and for $\Delta\mu > J$. We substitute this equation into the Eq. (24) and write the correction $\Delta S(\omega)$ to the Poissonian noise as a function of normalized bias $v = \Delta\mu/J$ and normalized frequency $\Omega = e\omega/[G(2\Delta\mu - J)]$

$$\Delta S(\omega) = 6eGJ\frac{(v^2-1)[1+(v-1)\cos\phi]^2(1-\cos\phi)}{(2v-1)^3[\Omega^2+(1-\cos\phi)^2]}. \tag{37}$$

From this equation it follows that the noise power has singularities as a function of ω for zero magnetic field, and it has singularities at $\phi = 2\pi m$ (where m is integer) as a function of the magnetic field (see Fig. 2). We would like to emphasize that the noise is singular even if the exchange between the dots is weak, $J \ll \Delta\mu$. In the case $\Delta\mu < J$ the transition from the singlet to the triplet is forbidden by conservation of energy, $w_{21}(2,1) = 0$, and we immediately obtain from Eq. (24) that $\Delta S(\omega) = 0$, i.e. the total noise is Poissonian (as it is always the case for elastic cotunneling). In the case of large bias, $\Delta\mu \gg J$, two dots contribute independently to the current $I = 2e^{-1}G\Delta\mu$, and from Eq. (37) we obtain the Fano factor

$$F = \frac{3}{8}\frac{\cos^2\phi(1-\cos\phi)}{\Omega^2 + (1-\cos\phi)^2} + 1, \quad \Delta\mu \gg J. \tag{38}$$

This Fano factor controls the transition to the telegraph noise and then to the equilibrium noise at high temperature, as described above. We notice that if the coupling of the dots to the leads is not equal, then $w_{nm}(l,l) \neq 0$ serves as a cut-off of the singularity in $\Delta S(\omega)$.

Finally, we remark that the Fano factor is a periodic function of the phase ϕ (see Fig. 2); this is nothing but an Aharonov-Bohm effect in the noise of the cotunneling transport through the DD. However, in contrast to the Aharonov-Bohm effect in the cotunneling current through the DD which has been discussed earlier in the Sec. 6.1, the noise effect does not allow us to probe the ground state of the DD, since the DD is already in a mixture of the singlet and three triplet states.

7. Cotunneling through continuum of single-electron states

We consider now the transport through a multi-level QDS with $\delta E \ll E_C$. In the low bias regime, $\Delta\mu \ll (\delta E\, E_C)^{1/2}$, the elastic cotunneling dominates transport [2], and according to the results of Sec. 4 the noise is Poissonian. Here we consider the opposite regime of inelastic cotunneling, $\Delta\mu \gg (\delta E\, E_C)^{1/2}$. Since a large number M of levels participate in transport, we can neglect the correlations which we have studied in Secs. 5 and 6, since they become a $1/M$-effect. Instead, we concentrate on the heating effect, which is not relevant for the 2-level system considered before. The condition for strong cotunneling has to be rewritten in a single-particle form, $\tau_{\rm in} \gg \tau_c$, where $\tau_{\rm in}$ is the single-particle energy relaxation time on the QDS due to the coupling to the environment, and τ_c is the time of the cotunneling transition, which can be estimated as $\tau_c \sim e\nu_D\Delta\mu/I$ (where ν_D is the density of QDS states). Since the energy relaxation rate on the QDS is small, the multiple cotunneling transitions can cause high energy excitations on the dot, and this leads to a nonvanishing backward tunneling, $w_{nm}(1,2) \neq 0$. In the absence of correlations between cotunneling events, Eqs. (17) and (18) can be rewritten in terms of forward and backward tunneling currents I_+ and I_-,

$$I = I_+ - I_-, \quad S = e(I_+ + I_-), \tag{39}$$

$$I_+ = e\sum_{n,m} w_{nm}(2,1)\bar{\rho}_m\,, \quad I_- = e\sum_{n,m} w_{nm}(1,2)\bar{\rho}_m\,, \tag{40}$$

where the transition rates are given by (15).

It is convenient to rewrite the currents $I_\pm$ in a single-particle basis. To do so we substitute the rates Eq. (15) into Eq. (40) and neglect the dependence of the tunneling amplitudes Eq. (3) on the quantum numbers k and p, $T_{lkp} \equiv T_l$, which is a reasonable assumption for QDS with a large number of electrons. Then we define the distribution function on the QDS as

$$f(\varepsilon) = \nu_D^{-1}\sum_p \delta(\varepsilon - \varepsilon_p)\mathrm{Tr}\,\bar{\rho}d_p^\dagger d_p \tag{41}$$

and replace the summation over p with an integration over ε. Doing this we obtain the following expressions for $T = 0$

$$I_\pm = C_\pm \frac{G_1 G_2}{2\pi e^3}\left(\frac{1}{\Delta_+} + \frac{1}{\Delta_-}\right)^2 (\Delta\mu)^3, \tag{42}$$

$$C_\pm = \frac{1}{\Delta\mu^3}\iint d\varepsilon d\varepsilon' \Theta(\varepsilon - \varepsilon' \pm \Delta\mu) f(\varepsilon)[1 - f(\varepsilon')], \tag{43}$$

where $G_{1,2} = \pi e^2 \nu\nu_D |T_{1,2}|^2$ are the tunneling conductances of the barriers 1 and 2, and where we have introduced the function $\Theta(\varepsilon) = \varepsilon\theta(\varepsilon)$ with $\theta(\varepsilon)$ being the step-function. In particular, using the property $\Theta(\varepsilon+\Delta\mu) - \Theta(\varepsilon-\Delta\mu) = \varepsilon + \Delta\mu$ and fixing

$$\int d\varepsilon[f(\varepsilon) - \theta(-\varepsilon)] = 0, \tag{44}$$

(since $I_\pm$ given by Eq. (42) and Eq. (43) do not depend on the shift $\varepsilon \to \varepsilon + const$) we arrive at the following general expression for the cotunneling current

$$I = \Lambda\frac{G_1 G_2}{12\pi e^3}\left(\frac{1}{\Delta_+} + \frac{1}{\Delta_-}\right)^2 (\Delta\mu)^3, \tag{45}$$

$$\Lambda = 1 + 12\Upsilon/(\Delta\mu)^2, \tag{46}$$

$$\Upsilon = \int d\varepsilon\varepsilon[f(\varepsilon) - \theta(-\varepsilon)] \geq 0, \tag{47}$$

where the value $\nu_D\Upsilon$ has the physical meaning of the energy acquired by the QDS due to the cotunneling current through it.

We have deliberately introduced the functions $C_\pm$ in the Eq. (42) to emphasize the fact that if the distribution $f(\varepsilon)$ scales with the bias $\Delta\mu$ (i.e. f is a function of $\varepsilon/\Delta\mu$), then $C_\pm$ become dimensionless universal numbers. Thus both, the prefactor Λ (given by Eq. (46)) in the cotunneling current, and the Fano factor,

$$F = \frac{C_+ + C_-}{C_+ - C_-}, \tag{48}$$

take their universal values, which do not depend on the bias $\Delta\mu$. We consider now such universal regimes. The first example is the case of weak cotunneling, $\tau_{\rm in} \ll \tau_c$, when the QDS is in its ground state, $f(\varepsilon) = \theta(-\varepsilon)$, and the thermal energy of the QDS vanishes, $\Upsilon = 0$. Then $\Lambda = 1$, and Eq. (45) reproduces the results of Ref. [2]. As we have already mentioned, the backward current vanishes, $I_- = 0$, and the Fano factor acquires its full Poissonian value $F = 1$, in agreement with our nonequilibrium FDT proven in Sec. 3. In the limit of strong cotunneling, $\tau_{\rm in} \gg \tau_c$, the energy relaxation on the QDS can be neglected. Depending on the electron-electron scattering time τ_{ee} two cases have to be distinguished: The regime of cold electrons $\tau_{ee} \gg \tau_c$ and regime of hot electrons $\tau_{ee} \ll \tau_c$ on the QDS. Below we discuss both regimes in detail and demonstrate their universality.

7.1. COLD ELECTRONS

In this regime the electron-electron scattering on the QDS can be neglected and the distribution $f(\varepsilon)$ has to be found from the master equation Eq. (14). We multiply this equation by $\nu_D^{-1}\sum_p \delta(\varepsilon - \varepsilon_p)\langle n|d_p^\dagger d_p|n\rangle$, sum over n and use the tunneling rates from Eq. (15). Doing this we obtain the standard stationary kinetic equation which can be written in the following form

$$\int d\varepsilon' \sigma(\varepsilon' - \varepsilon) f(\varepsilon')[1 - f(\varepsilon)] = \int d\varepsilon' \sigma(\varepsilon - \varepsilon') f(\varepsilon)[1 - f(\varepsilon')], \tag{49}$$

$$\sigma(\varepsilon) = 2\lambda\Theta(\varepsilon) + \sum_{\pm} \Theta(\varepsilon \pm \Delta\mu), \tag{50}$$

where $\lambda = (G_1^2 + G_2^2)/(2G_1G_2) \geq 1$ arises from the equilibration rates $w_{mn}(l,l)$. (We assume that if the limits of the integration over energy ε are not specified, then the integral goes from $-\infty$ to $+\infty$.) From the form of this equation we immediately conclude that its solution is a function of $\varepsilon/\Delta\mu$, and thus the cold electron regime is universal as defined in the previous section. It is easy to check that the detailed balance does not hold, and in addition $\sigma(\varepsilon) \neq \sigma(-\varepsilon)$. Thus we face a difficult problem of solving Eq. (49) in its full nonlinear form. Fortunately, there is a way to avoid this problem and to reduce the equation to a linear form which we show next.

We group all nonlinear terms on the rhs of Eq. (49): $\int d\varepsilon' \sigma(\varepsilon' - \varepsilon) f(\varepsilon') = h(\varepsilon) f(\varepsilon)$, where $h(\varepsilon) = \int d\varepsilon' \{\sigma(\varepsilon' - \varepsilon) f(\varepsilon') + \sigma(\varepsilon - \varepsilon')[1 - f(\varepsilon')]\}$. The trick is to rewrite the function $h(\varepsilon)$ in terms of known functions. For doing this we split the integral in $h(\varepsilon)$ into two integrals over $\varepsilon' > 0$ and $\varepsilon' < 0$, and then use Eq. (44) and the property of the kernel $\sigma(\varepsilon) - \sigma(-\varepsilon) = 2(1+\lambda)\varepsilon$ to regroup terms in such a way that $h(\varepsilon)$ does not contain $f(\varepsilon)$ explicitly. Taking into account Eq. (47) we arrive at the following linear integral equation

$$\int d\varepsilon' \sigma(\varepsilon' - \varepsilon) f(\varepsilon') = [(1+\lambda)(\varepsilon^2 + 2\Upsilon) + (\Delta\mu)^2] f(\varepsilon), \tag{51}$$

where the parameter Υ is the only signature of the nonlinearity of Eq. (49).

Since Eq. (51) represents an eigenvalue problem for a linear operator, it can in general have more than one solution. However, there is only one physical solution, which satisfies the conditions

$$0 \le f(\varepsilon) \le 1, \quad f(-\infty) = 1, \quad f(+\infty) = 0. \tag{52}$$

Indeed, using a standard procedure one can show that two solutions of the integral equation (51), f_1 and f_2, corresponding to different parameters $\Upsilon_1 \neq \Upsilon_2$ should be orthogonal, $\int d\varepsilon f_1(\varepsilon) f_2(-\varepsilon) = 0$. This contradicts the conditions Eq. (52). The solution is also unique for the same Υ, i.e. it is not degenerate (for a proof, see the Ref. [5]). From Eq. (49) and conditions Eq. (52) it follows that if $f(\varepsilon)$ is a solution then $1 - f(-\varepsilon)$ also satisfies Eqs. (49) and (52). Since the solution is unique, it has to have the symmetry $f(\varepsilon) = 1 - f(-\varepsilon)$.

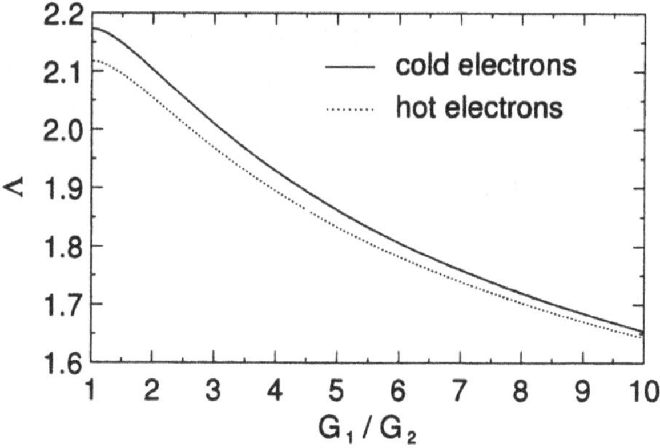

Figure 3. The prefactor Λ in the expression (45) for the cotunneling current characterizes a universal cotunneling transport in the regime of weak cotunneling, $\tau_{\rm in} \ll \tau_c$, ($\Lambda = 1$, see Ref. [2]), and in the regime of strong cotunneling, $\tau_{\rm in} \gg \tau_c$ ($\Lambda > 1$). Here Λ is plotted as a function of G_1/G_2 (same as a function of G_2/G_1) for the strong cotunneling, for the cold-electron case, $\tau_{ee} \gg \tau_c$ (solid line) and for the hot-electron case, $\tau_{ee} \ll \tau_c$ (dotted line). $G_{1,2}$ are the tunneling conductances of a junctions connecting leads 1 and 2 with the QDS.

We solve Eqs. (51) and (52) numerically and use Eqs. (43) and (48) to find that the Fano factor is very close to 1 (it does not exceed the value $F \approx 1.006$). Next we use Eqs. (46) and (47) to calculate the prefactor Λ and plot the result as a function of the ratio of tunneling conductances, G_1/G_2, (Fig. 3, solid line). For equal coupling to the leads, $G_1 = G_2$, the prefactor Λ takes its maximum value 2.173, and thus the cotunneling current is approximately twice as large compared to its value for the case of weak cotunneling, $\tau_{\text{in}} \ll \tau_c$. Λ slowly decreases with increasing asymmetry of coupling and tends to its minimum value $\Lambda = 1$ for the strongly asymmetric coupling case G_1/G_2 or $G_2/G_1 \gg 1$.

7.2. HOT ELECTRONS

In the regime of hot electrons, $\tau_{ee} \ll \tau_c$, the distribution is given by the equilibrium Fermi function $f_F(\varepsilon) = [1 + \exp(\varepsilon/k_B T_e)]^{-1}$, while the electron temperature T_e has to be found self-consistently from the kinetic equation. Eq. (49) has to be modified to take into account electron-electron interactions. This can be done by adding the electron collision integral $I_{ee}(\varepsilon)$ to the rhs of (49). Since the form of the distribution is known we need only the energy balance equation, which can be derived by multiplying the modified equation (49) by ε and integrating it over ε. The contribution from the collision integral $I_{ee}(\varepsilon)$ vanishes, because the electron-electron scattering conserves the energy of the system. Using the symmetry $f_F(\varepsilon) = 1 - f_F(-\varepsilon)$ we arrive at the following equation

$$\iint d\varepsilon d\varepsilon' f_F(\varepsilon')[1 - f_F(\varepsilon)]\sigma(\varepsilon' - \varepsilon)\varepsilon = 0. \tag{53}$$

Next we regroup the terms in this equation such that it contains only integrals of the form $\int_0^\infty d\varepsilon f_F(\varepsilon)(\ldots)$. This allows us to get rid of nonlinear terms, and we arrive at the following equation,

$$\int d\varepsilon \varepsilon^3 [f_F(\varepsilon) - \theta(-\varepsilon)] + 3\Upsilon^2 = \frac{(\Delta\mu)^4}{8(1+\lambda)}, \tag{54}$$

which holds also for the regime of cold electrons. Finally, we calculate the integral in Eq. (54) and express the result in terms of the dimensionless parameter $\alpha = \Delta\mu/k_B T_e$,

$$\alpha = \pi \left[8(1+\lambda)/5\right]^{1/4}. \tag{55}$$

Thus, since the distribution again depends on the ratio $\varepsilon/\Delta\mu$, the hot electron regime is also universal.

The next step is to substitute the Fermi distribution function with the temperature given by Eq. (55) into Eq. (43). We calculate the integrals and arrive at the closed analytical expressions for the values of interest,

$$\Lambda = 1 + \frac{2\pi^2}{\alpha^2} = 1 + \sqrt{\frac{5}{2(1+\lambda)}}, \tag{56}$$

$$F = 1 + \frac{12}{2\pi^2 + \alpha^2} \sum_{n=1}^{\infty} \left[\frac{1}{n^2} + \frac{2}{\alpha n^3}\right] e^{-\alpha n}, \tag{57}$$

where again $\lambda = (G_1^2 + G_2^2)/2G_1 G_2 \geq 1$. It turns out that similar to the case of cold electrons, Sec. 7.1, the Fano factor for hot electrons is very close to 1 (namely, it does not exceed the value $F \approx 1.007$). Therefore, we do not expect that the super-Poissonian noise considered in this section (i.e. the one which is due to heating of a large QDS caused by inelastic cotunneling through it) will be easy to observe in experiments. On the other hand, the transport-induced heating of a large QDS can be observed in the cotunneling current through the prefactor Λ, which according to Eq. (56) takes its maximum value $\Lambda = 1 + \sqrt{5/4} \approx 2.118$ for $G_1 = G_2$ and slowly reaches its minimum value 1 with increasing (or decreasing) the ratio G_1/G_2 (see Fig. 3, dotted line). Surprisingly, the two curves of Λ vs G_1/G_2 for the cold- and hot-electron regimes lie very close, which means that the effect of the electron-electron scattering on the cotunneling transport is rather weak.

8. Conclusions

Here we give a short summary of our results. In Sec. 3, we have derived the non-equilibrium FDT, i.e. the universal relation (12) between the current and the noise, for QDS in the weak cotunneling regime. Taking the limit $T, \omega \to 0$, we show that the noise is Poissonian, i.e. $F = 1$.

In Sec. 4, we present the results of the microscopic theory of strong cotunneling, Ref. [5]: The master equation, Eq. (14), the average current, Eq. (17), and the current correlators, Eqs. (18) and (19), for a QDS system coupled to leads in the strong cotunneling regime $w_{\text{in}} \ll I/e$ at small frequencies, $\omega \ll \Delta_{\pm}$. In contrast to sequential tunneling, where shot noise is either Poissonian ($F = 1$) or suppressed due to charge conservation ($F < 1$), we find that the noise in the inelastic cotunneling regime can be super-Poissonian ($F > 1$), with a correction being as large as the Poissonian noise itself. In the regime of elastic cotunneling $F = 1$.

While the amount of super-Poissonian noise is merely estimated at the end of Sec. 4, the noise of the cotunneling current is calculated for the special case of a QDS with nearly degenerate states, i.e. $\delta E_{nm} \ll \delta E$, in Sec. 5, where we apply our results from Sec. 4. The general solution Eq. (22) is further analyzed for two nearly degenerate levels, with the result Eq. (24). More information is gained in the specific case of a DD coupled to leads considered in Sec. 6, where we determine the average current Eqs. (30-31) and noise Eqs. (33-34) in the weak cotunneling regime and the correlation correction to noise Eq. (37) in the strong cotunneling regime as a function of frequency, bias, and the Aharonov-Bohm phase threading the tunneling loop, finding signatures of the Aharonov-Bohm effect and of the quantum entanglement.

Finally, in Sec. 7, another important situation is studied in detail, the cotunneling through a QDS with a continuous energy spectrum, $\delta E \ll \Delta\mu \ll E_C$. Here, the correlation between tunneling events plays a minor role as a source of super-Poissonian noise, which is now caused by heating effects opening the possibility for tunneling events in the reverse direction and thus to an enhanced noise power. In Eq. (48), we express the Fano factor F in the continuum case in terms of the dimensionless numbers $C_{\pm}$, defined in Eq. (43), which depend on the electronic distribution function $f(\varepsilon)$ in the QDS (in this regime, a description on the single-electron level is appropriate). The current Eq. (45) is expressed in terms of the prefactor Λ, Eq. (46). Both F and Λ are then calculated for different regimes. For weak cotunneling, we immediately find $F = 1$, as anticipated earlier, while for strong cotunneling we distinguish the two regimes of cold ($\tau_{ee} \gg \tau_c$) and hot ($\tau_{ee} \ll \tau_c$) electrons. For both regimes we find that the Fano factor is very close to one, while Λ is given in Fig. 3.

Acknowledgements

This work has been partially supported by the Swiss National Science Foundation.

Appendix A

In this Appendix we present the derivation of Eqs. 7 and 8. In order to simplify the intermediate steps, we use the notation $\bar{O}(t) \equiv \int_{-\infty}^{t} dt' O(t')$ for any operator O, and $O(0) \equiv O$. We notice that, if an operator O is a linear function of operators D_l and $D_l^\dagger$, then $\bar{O}(\infty) = 0$ (see the discussion in Sec. 3). Next, the currents can be represented as the difference and the sum of $\hat{I}_1$ and $\hat{I}_2$,

$$\hat{I}_d = (\hat{I}_2 - \hat{I}_1)/2 = ie(X^\dagger - X)/2 , \tag{58}$$

$$\hat{I}_s = (\hat{I}_1 + \hat{I}_2)/2 = ie(Y^\dagger - Y)/2 , \tag{59}$$

where $X = D_2 + D_1^\dagger$, and $Y = D_1 + D_2$. While for the perturbation we have

$$V = X + X^\dagger = Y + Y^\dagger . \tag{60}$$

First we concentrate on the derivation of Eq. (7) and redefine the average current Eq. (5) as $I = I_d$ (which gives the same result anyway, because the average number of electrons on the QDS does not change $I_s = 0$).

To proceed with our derivation, we make use of Eq. (6) and expand the current up to fourth order in T_{lkp}:

$$I = i \int_{-\infty}^{0} dt \int_{-\infty}^{t} dt' \langle \hat{I}_d V(t) V(t') \bar{V}(t') \rangle - i \int_{-\infty}^{0} dt \langle \bar{V} \hat{I}_d V(t) \bar{V}(t) \rangle + \text{c.c.} \tag{61}$$

Next, we use the cyclic property of trace to shift the time dependence to $\hat{I}_d$. Then we complete the integral over time t and use $\bar{I}_d(\infty) = 0$. This procedure allows us to combine first and second term in Eq. (61),

$$I = -i \int_{-\infty}^{0} dt \langle [\bar{I}_d V + \bar{V} \hat{I}_d] V(t) \bar{V}(t) \rangle + \text{c.c.} \tag{62}$$

Now, using Eqs. (58) and (60) we replace operators in Eq. (62) with X and $X^\dagger$ in two steps: $I = e \int_{-\infty}^{0} dt \langle [\bar{X}^\dagger X^\dagger - \bar{X} X] V(t) \bar{V}(t) \rangle$ + c.c., where some terms cancel exactly. Then we work with $V(t)\bar{V}(t)$ and notice that some terms cancel, because they are linear in c_{lk} and $c_{lk}^\dagger$. Thus we obtain $I = e \int_{-\infty}^{0} dt \langle [\bar{X}^\dagger X^\dagger - \bar{X} X][X^\dagger(t) \bar{X}^\dagger(t) + X(t) \bar{X}(t)] \rangle$ + c.c.. Two terms $\bar{X} X X \bar{X}$ and $\bar{X}^\dagger X^\dagger X^\dagger \bar{X}^\dagger$ describe tunneling of two electrons from the same lead, and therefore they do not contribute to the normal current. We then combine all other terms to extend the integral to $+\infty$,

$$I = e \int_{-\infty}^{\infty} dt \langle \bar{X}^\dagger(t) X^\dagger(t) X \bar{X} - \bar{X} X X^\dagger(t) \bar{X}^\dagger(t) \rangle \tag{63}$$

Finally, we use $\int_{-\infty}^{\infty} dt X(t) \bar{X}(t) = -\int_{-\infty}^{\infty} dt \bar{X}(t) X(t)$ (since $\bar{X}(\infty) = 0$) to get Eq. (7) with $A = X\bar{X}$. Here, again, we drop terms $D_1^\dagger \bar{D}_1^\dagger$ and $D_2 \bar{D}_2$ responsible for tunneling of two electrons from the same lead, and obtain A as in Eq. (7).

Next, we derive Eq. (8) for the noise power. At small frequencies $\omega \ll \Delta_\pm$ fluctuations of I_s are suppressed because of charge conservation (see below), and we can replace $\hat{I}_2$ in the correlator Eq. (5) with $\hat{I}_d$. We expand $S(\omega)$ up to fourth order in T_{lkp}, use $\int_{-\infty}^{+\infty} dt\, \hat{I}_d(t) e^{\pm i\omega t} = 0$, and repeat the steps leading to Eq. (62). Doing this we obtain,

$$S(\omega) = -\int_{-\infty}^{\infty} dt \cos(\omega t) \langle [\bar{V}(t), \hat{I}_d(t)][\bar{V}, \hat{I}_d] \rangle \,. \tag{64}$$

Then, we replace V and $\hat{I}_d$ with X and $X^\dagger$. We again keep only terms relevant for cotunneling, and in addition we neglect terms of order $\omega/\Delta_\pm$ (applying same arguments as before, see Eq. (65)). We then arrive at Eq. (8) with the operator A given by Eq. (7).

Finally, in order to show that fluctuations of I_s are suppressed, we replace $\hat{I}_d$ in Eq. (64) with $\hat{I}_s$, and then use the operators Y and $Y^\dagger$ instead of X and $X^\dagger$. In contrast to Eq. (63) terms such as $\bar{Y}^\dagger Y^\dagger Y \bar{Y}$ do not contribute, because they contain integrals of the form $\int_{-\infty}^{\infty} dt \cos(\omega t) D_l(t) \bar{D}_{l'}(t) = 0$. The only nonzero contribution can be written as

$$S_{ss}(\omega) = \frac{e^2 \omega^2}{4} \int_{-\infty}^{\infty} dt \cos(\omega t) \langle [\bar{Y}^\dagger(t), \bar{Y}(t)][\bar{Y}^\dagger, \bar{Y}] \rangle \,, \tag{65}$$

where we have used integration by parts and the property $\bar{Y}(\infty) = 0$. Compared to Eq. (8) this expression contains an additional integration over t, and thereby it is of order $(\omega/\Delta_{\pm})^2$.

Appendix B

In this Appendix we calculate the transition rates Eq. (15) for a DD coupled to leads with the coupling described by Eqs. (27) and (28) and show that the four-level system in the singlet-triplet basis Eq. (29) can be mapped to a two-level system. For the moment we assume that the indices n and m enumerate the singlet-triplet basis, $n, m = S, T_0, T_+, T_-$. Close to the sequential tunneling peak, $\Delta_- \ll \Delta_+$, we keep only terms of the form $D_l^\dagger \bar{D}_{l'}$. Calculating the trace over the leads explicitly, we obtain at $T = 0$,

$$\begin{aligned} w_{nm}(l', l) &= \frac{\pi\nu^2}{2\Delta_-^2}\,\Theta(\mu_l - \mu_{l'} - \delta E_{nm}) \\ &\quad\times \sum_{j,j'} T_{lj}^* T_{lj'} T_{l'j'}^* T_{l'j} M_{nm}(j, j'), \end{aligned} \tag{66}$$

$$M_{nm}(j, j') = \sum_{s,s'} \langle n|d_{sj}^\dagger d_{s'j}|m\rangle \langle m|d_{s'j'}^\dagger d_{sj'}|n\rangle, \tag{67}$$

with $\Theta(\varepsilon) = \varepsilon\theta(\varepsilon)$, and $\delta E_{nm} = 0, \pm J$, and we have assumed $t_d \ll \Delta_-$.

Since the quantum dots are the same we get $M_{nm}(1,1) = M_{nm}(2,2)$, and $M_{nm}(1,2) = M_{nm}(2,1)$. We calculate these matrix elements in the singlet-triplet basis explicitly,

$$M(1,1) = \frac{1}{2}\begin{pmatrix} 1 & 1 & 1 & 1 \\ 1 & 1 & 1 & 1 \\ 1 & 1 & 2 & 0 \\ 1 & 1 & 0 & 2 \end{pmatrix}, \tag{68}$$

$$M(1,2) = \frac{1}{2}\begin{pmatrix} 1 & -1 & -1 & -1 \\ -1 & 1 & 1 & 1 \\ -1 & 1 & 2 & 0 \\ -1 & 1 & 0 & 2 \end{pmatrix}. \tag{69}$$

Assuming now equal coupling of the form Eq. (28) we find that for $l = l'$ the matrix elements of the singlet-triplet transition vanish (as we have expected, see Sec. 5). On the other hand the triplets are degenerate, i.e. $\delta E_{nm} = 0$ in the triplet sector. Then from Eq. (66) it follows that $w_{nm}(l, l) = 0$. Next, we have $\Theta(\mu_2 - \mu_1 - \delta E_{nm}) = 0$, since for nearly degenerate states we assume $\Delta\mu > |\delta E_{nm}|$, and thus $w_{nm}(1,2) = 0$. Finally, for $w_{nm} = w_{nm}^I = w_{nm}(2,1)$ we obtain,

$$w_{SS} = \frac{\pi}{2}\left(\frac{\nu T^2}{\Delta_-}\right)^2 \Delta\mu(1 + \cos\phi), \tag{70}$$

$$w_{ST} = \frac{\pi}{2}\left(\frac{\nu T^2}{\Delta_-}\right)^2 (\Delta\mu + J)(1-\cos\phi), \tag{71}$$

$$w_{TS} = \frac{\pi}{2}\left(\frac{\nu T^2}{\Delta_-}\right)^2 (\Delta\mu - J)(1-\cos\phi), \tag{72}$$

$$w_{TT} = \frac{\pi}{2}\left(\frac{\nu T^2}{\Delta_-}\right)^2 \Delta\mu \times \begin{pmatrix} 1+\cos\phi & 1+\cos\phi & 1+\cos\phi \\ 1+\cos\phi & 2+2\cos\phi & 0 \\ 1+\cos\phi & 0 & 2+2\cos\phi \end{pmatrix}. \tag{73}$$

Next we prove the mapping to a two-level system. First we notice that because the matrix w_{TT} is symmetric, the detailed balance equation for the stationary state gives $\bar{\rho}_n/\bar{\rho}_m = w_{mn}/w_{nm} = 1$, $n, m \in T$. Thus we can set $\bar{\rho}_n \to \bar{\rho}_2/3$, for $n \in T$. The specific form of the transition matrix Eqs. (70-73) helps us to complete the mapping by setting $(1/3)\sum_{m=2}^4 w_{1m} \to w_{12}$, $\sum_{n=2}^4 w_{n1} \to w_{21}$, and $(1/3)\sum_{n,m=2}^4 w_{nm} \to w_{22}$, so that we get the new transition matrix Eq. (36), while the stationary master equation for the new two-level density matrix does not change its form. If in addition we set $(1/3)\sum_{m=2}^4 \delta\rho_{1m}(t) \to \delta\rho_{12}(t)$, $\sum_{n=2}^4 \delta\rho_{n1}(t) \to \delta\rho_{21}(t)$, and $(1/3)\sum_{n,m=2}^4 \delta\rho_{nm}(t) \to \delta\rho_{22}(t)$, then the master equation Eq. (14) for $\delta\rho_{nm}(t)$ and the initial condition $\delta\rho_{nm}(0) = \delta_{nm} - \bar{\rho}_n$ do not change either. Finally, one can see that under this mapping Eq. (19) for the correction to the noise power $\Delta S(\omega)$ remains unchanged. Thus we have accomplished the mapping of our singlet-triplet system to the two-level system with the new transition matrix given by Eq. (36).

References

1. For a recent review on shot noise, see: Ya. M. Blanter and M. Büttiker, *Shot Noise in Mesoscopic Conductors*, Phys. Rep. **336**, 1 (2000).
2. D. V. Averin and Yu. V. Nazarov, in *Single Charge Tunneling*, eds. H. Grabert and M. H. Devoret, NATO ASI Series B: Physics Vol. 294, (Plenum Press, New York, 1992).
3. D. C. Glattli *et al.*, Z. Phys. B **85**, 375 (1991).
4. For an review, see D. V. Averin and K. K. Likharev, in *Mesoscopic Phenomena in Solids*, edited by B. L. Al'tshuler, P. A. Lee, and R. A. Webb (North-Holland, Amsterdam, 1991).
5. E. V. Sukhorukov, G. Burkard, and D. Loss, Phys. Rev. B **63**, 125315 (2001).
6. D. Loss and E. V. Sukhorukov, Phys. Rev. Lett. **84**, 1035 (2000).
7. Such a non-equilibrium FDT was derived for single barrier junctions long ago by D. Rogovin, and D. J. Scalapino, Ann. Phys. (N. Y.) **86**, 1 (1974).
8. For the super-Poissonian noise in resonant double-barrier structures, see: G. Iannaccone *et al.*, Phys. Rev. Lett. **80**, 1054 (1998); V. V. Kuznetsov *et al.*, Phys. Rev. B **58**, R10159 (1998); Ya. M. Blanter and M. Büttiker, Phys. Rev. B **59**, 10217 (1999).
9. The Aharonov-Bohm effect in the noise of non-interacting electrons in a mesoscopic ring has been discussed in Ref. [1].

10. G. D. Mahan, *Many Particle Physics*, 2nd Ed. (Plenum, New York, 1993).
11. M. H. Devoret, and R. J. Schoelkopf, Nature **406**, 1039 (2000).
12. One could view this as an analog of a whistle effect, where the flow of air (current) is strongly modulated by a bistable state in the whistle, and vice versa. The analogy, however, is not complete, since the current through the QDS is random due to quantum fluctuations.
13. See, e.g., Sh. Kogan, *Electronic Noise and Fluctuations in Solids*, (Cambridge University Press, Cambridge, 1996).
14. G. Burkard, D. Loss, and D. P. DiVincenzo, Phys. Rev. B **59**, 2070 (1999)
15. Note that the Aharonov-Bohm effect is not suppressed by this inelastic cotunneling, since the *entire* cotunneling process involving also leads is elastic: the initial and final states of the *entire* system have the same energy.
16. P. Recher, E. V. Sukhorukov, and D. Loss, Phys. Rev. Lett. **85**, 1962 (2000).

PART TWO

Quantum Measurement and Englanglement

QUBITS AS SPECTROMETERS OF QUANTUM NOISE

R.J. SCHOELKOPF, A.A. CLERK, S.M. GIRVIN, K.W. LEHNERT and M.H. DEVORET
Departments of Applied Physics and Physics
Yale University, PO Box 208284, New Haven, CT 06520-8284

1. Introduction

Electrical engineers and physicists are naturally very interested in the noise of circuits, amplifiers and detectors. This noise has many origins, some of which are completely unavoidable. For example, a dissipative element (a resistor) at finite temperature inevitably generates Johnson noise. Engineers long ago developed spectrum analyzers to measure the intensity of this noise. Roughly speaking, these spectrum analyzers consist of a resonant circuit to select a particular frequency of interest, followed by an amplifier and square law detector (e.g. a diode rectifier) which measures the mean square amplitude of the signal at that frequency.

With the advent of very high frequency electronics operating at low temperatures, we have entered a new regime $\hbar\omega > k_{\mathrm{B}}T$, where quantum mechanics plays an important role and one has to begin to think about *quantum noise* and quantum-limited amplifiers and detectors. This topic is well-studied in the quantum optics community and is also commonplace in the radio astronomy community. It has recently become of importance in connection with quantum computation and the construction of mesoscopic electrical circuits which act like artificial atoms with quantized energy levels. It is also important for understanding the quantum measurement process in mesoscopic systems.

In a classical picture, the intensity of Johnson noise from a resistor vanishes linearly with temperature because thermal fluctuations of the charge carriers cease at zero temperature. One knows from quantum mechanics, however, that there are quantum fluctuations even at zero temperature, due to zero-point motion. Zero-point motion is a notion from quantum mechanics that is frequently misunderstood. One might wonder, for example, whether it is physically possible to use a spectrum analyzer to detect the zero-point motion. The answer is quite definitely

Y.V. Nazarov (ed.), Quantum Noise in Mesoscopic Physics, 175–203.

yes, if we use a quantum system! Consider for example a hydrogen atom in the 2p excited state lying 3/4 of a Rydberg above the 1s ground state. We know that this state is unstable and has a lifetime of only about 1 ns before it decays to the ground state and emits an ultraviolet photon. This spontaneous decay is a natural consequence of the zero-point motion of the electromagnetic fields in the vacuum surrounding the atom. In fact, the rate of spontaneous decay gives a simple way in which to *measure* this zero point motion of the vacuum. Placing the atom in a resonant cavity can modify the strength of the noise at the transition frequency, and this effect can be measured via a change in the decay rate.

At finite temperature, the vacuum will contain blackbody photons which will increase the rate of decay due to stimulated emission and also cause transitions in the reverse direction, 1s $\rightarrow$ 2p, by photon absorption. With these ideas in mind, it is now possible to see how to build a quantum spectrum analyzer.

The remainder of this article is organized as follows. First we describe the general concept of a two-level system as a quantum spectrum analyzer. We next review the Caldeira-Leggett formalism for the modelling of a dissipative circuit element, such as a resistor, and its associated quantum noise. Then, a brief discussion of the single Cooper-pair box, a circuit which behaves as a two-level system or qubit, is given. We then discuss the effects of a dissipative electromagnetic environment on the box, and treat the case of a simple linear, but *nonequilibrium* environment, consisting of a classical tunnel junction which produces shot noise under bias. Finally, we describe a theoretical technique for calculating the properties of a Cooper-pair box coupled to a measurement system, which will be a *nonlinear, nonequilibrium* device, such as a single-electron transistor. Equivalently, this allows one to calculate the full *quantum noise spectrum* of the measurement device. Results of this calculation for the case of a normal SET are presented.

2. Two-level systems as spectrum analyzers

Consider a quantum system (atom or electrical circuit) which has its two lowest energy levels ϵ_0 and ϵ_1 separated by energy $E_{01} = \hbar\omega_{01}$. We suppose for simplicity that all the other levels are far away in energy and can be ignored. The states of any two-level system can be mapped onto the states of a fictitious spin-1/2 particle since such a spin also has only two states in its Hilbert space. With spin up representing the ground state ($|g\rangle$) and spin down representing the excited state ($|e\rangle$), the Hamiltonian is (taking the zero of energy to be the center of gravity of the two levels)

$$H_0 = -\frac{\hbar\omega_{01}}{2}\sigma_z. \tag{1}$$

In keeping with the discussion above, our goal is to see how the rate of 'spin-flip' transitions induced by an external noise source can be used to analyze the spectrum of that noise. Suppose for example that there is a noise source with amplitude $f(t)$ which can cause transitions via the perturbation[1]

$$V = Af(t)\sigma_x, \tag{2}$$

where A is a coupling constant. The variable $f(t)$ represents the noise source. We can temporarily pretend that f is a classical variable, although its quantum operator properties will be forced upon us very soon. For now, only our two-level spectrum analyzer will be treated quantum mechanically.

We assume that the coupling A is under our control and can be made small enough that the noise can be treated in lowest order perturbation theory. We take the state of the two-level system to be

$$|\psi(t)\rangle = \begin{pmatrix} \alpha_g(t) \\ \alpha_e(t) \end{pmatrix}. \tag{3}$$

In the interaction representation, first-order time-dependent perturbation theory gives

$$|\psi_{\rm I}(t)\rangle = |\psi(0)\rangle - \frac{i}{\hbar}\int_0^t d\tau\ \hat{V}(\tau)|\psi(0)\rangle. \tag{4}$$

If we initially prepare the two-level system in its ground state then the amplitude to find it in the excited state at time t is

$$\alpha_e = -\frac{iA}{\hbar}\int_0^t d\tau\ \langle e|\hat{\sigma}_x(\tau)|g\rangle f(\tau) + O(A^2), \tag{5}$$

$$= -\frac{iA}{\hbar}\int_0^t d\tau\ e^{i\omega_{01}\tau} f(\tau) + O(A^2). \tag{6}$$

We can now compute the probability

$$p_e(t) \equiv |\alpha_e|^2 = \frac{A^2}{\hbar^2}\int_0^t\int_0^t d\tau_1 d\tau_2\, e^{-i\omega_{01}(\tau_1-\tau_2)} f(\tau_1)f(\tau_2) + O(A^3) \tag{7}$$

We are actually only interested on the average time evolution of the system

$$\bar{p}_e(t) = \frac{A^2}{\hbar^2}\int_0^t\int_0^t d\tau_1 d\tau_2\, e^{-i\omega_{01}(\tau_1-\tau_2)} \langle f(\tau_1)f(\tau_2)\rangle + O(A^3) \tag{8}$$

We can now perform a change of variables in the integrals, $\tau = \tau_1 - \tau_2$ and $T = (\tau_1 + \tau_2)/2$, and we get

$$\bar{p}_e(t) = \frac{A^2}{\hbar^2}\int_0^t dT \int_{-B(T)}^{B(T)} d\tau\, e^{-i\omega_{01}\tau} \langle f(T+\tau/2)f(T-\tau/2)\rangle + O(A^3) \tag{9}$$

[1] The most general perturbation would also couple to σ_y but we assume that (as is often, though not always, the case) a spin coordinate system can be chosen so that the perturbation only couples to σ_x. Noise coupled to σ_z commutes with the Hamiltonian but is nevertheless important in dephasing coherent superpositions of the two states. We will not discuss such processes here.

where

$$B(T) = T \text{ if } T < t/2$$
$$= t - T \text{ if } T > t/2.$$

Let us now suppose that the noise correlation function is stationary (time translation invariant) and has a finite but small autocorrelation time τ_f. Then for $t \gg \tau_f$ we can set the bound $B(T)$ to infinity in the last integral and write

$$\bar{p}_e(t) = \frac{A^2}{\hbar^2} \int_0^t dT \int_{-\infty}^{\infty} d\tau \, e^{-i\omega_{01}\tau} \langle f(\tau) f(0) \rangle + O(A^3) \tag{10}$$

The integral over τ is effectively a sum of a very large number $N \sim t/\tau_f$ of random terms [2] and hence the value undergoes a random walk as a function of time. Introducing the noise spectral density

$$S_f(\omega) = \int_{-\infty}^{+\infty} d\tau \, e^{i\omega\tau} \langle f(\tau) f(0) \rangle, \tag{11}$$

we find that the probability to be in the excited state increases *linearly* with time,[3]

$$\bar{p}_e(t) = t \frac{A^2}{\hbar^2} S_f(-\omega_{01}) \tag{12}$$

The time derivative of the probability gives the transition rate

$$\Gamma_\uparrow = \frac{A^2}{\hbar^2} S_f(-\omega_{01}) \tag{13}$$

Note that we are taking in this last expression the spectral density on the negative frequency side. If f were a strictly classical source $\langle f(\tau) f(0) \rangle$ would be real and $S_f(-\omega_{01}) = S_f(+\omega_{01})$. However, because as we discuss below f is actually an operator acting on the environmental degrees of freedom, $[f(\tau), f(0)] \neq 0$ and $S_f(-\omega_{01}) \neq S_f(+\omega_{01})$.

Another possible experiment is to prepare the two-level system in its excited state and look at the rate of decay into the ground state. The algebra is identical to that above except that the sign of the frequency is reversed:

$$\Gamma_\downarrow = \frac{A^2}{\hbar^2} S_f(+\omega_{01}). \tag{14}$$

[2] The size of these random terms depends on the variance of f and on the value of $\omega_{01}\tau_f$ For $\omega_{01}\tau_f \gg 1$ the size will be strongly reduced by the rapid phase oscillations of the exponential in the integrand.

[3] Note that for very long times, where there is a significant depletion of the probability of being in the initial state, first-order perturbation theory becomes invalid. However, for sufficiently small A, there is a wide range of times $\tau_f \ll t \ll 1/\Gamma$ for which Eq. 12 is valid. Eqs. 13 and 14 then yield well-defined rates which can be used in a master equation to describe the full dynamics including long times.

We now see that our two-level system does indeed act as a quantum spectrum analyzer for the noise. Operationally, we prepare the system either in its ground state or in its excited state, weakly couple it to the noise source, and after an appropriate interval of time (satisfying the above inequalities) simply measure whether the system is now in its excited state or ground state. Repeating this protocol over and over again, we can find the probability of making a transition, and thereby infer the rate and hence the noise spectral density at positive and negative frequencies. Note that in contrast with a classical spectrum analyzer, we can separate the noise spectral density at positive and negative frequencies from each other since we can separately measure the downward and upward transition rates. Negative frequency noise transfers energy *from the noise source to the spectrometer*. That is, it represents energy emitted by the noise source. Positive frequency noise transfers energy *from the spectrometer to the noise source*.[4] In order to exhibit frequency resolution, $\Delta\omega$, adequate to distinguish these two cases, it is crucial that the two-level quantum spectrometer have sufficient phase coherence so that the linewidth of the transitions satisfies the condition $\omega_{01}/\Delta\omega \geq \max[k_{\mathrm{B}}T/\hbar\omega_{01}, 1]$.

In thermodynamic equilibrium, the transition rates must obey detailed balance $\Gamma_{\downarrow}/\Gamma_{\uparrow} = e^{\beta\hbar\omega_{01}}$ in order to give the correct equilibrium occupancies of the two states of the spectrometer. This implies that the spectral densities obey

$$S_f(+\omega_{01}) = e^{\beta\hbar\omega_{01}} S_f(-\omega_{01}). \tag{15}$$

Without the crucial distinction between positive and negative frequencies, and the resulting difference in rates, one always finds that our two level systems are completely unpolarized. If, however, the noise source is an amplifier or detector biased to be out of equilibrium, no general relation holds.

We now rigorously treat the quantity $f(\tau)$ as quantum operator in the Hilbert space of the noise source. The previous derivation is unchanged, and Eqs. (13,14) are still valid provided that we interpret the angular brackets in Eq. (8) as representing the quantum statistical expectation value for the operator correlation (in the absence of the coupling to the spectrometer)

$$S_f(\omega) = \int_{-\infty}^{+\infty} d\tau\, e^{i\omega\tau} \sum_{\alpha,\gamma} \rho_{\alpha\alpha}\, \langle\alpha|f(\tau)|\gamma\rangle\langle\gamma|f(0)|\alpha\rangle \tag{16}$$

where for simplicity we have assumed that (in the absence of the coupling to the spectrometer) the density matrix is diagonal in the energy eigenbasis and time-

[4] Unfortunately, there are several conventions in existence for describing the noise spectral density. It is common in engineering contexts to use the phrase 'spectral density' to mean $S_f(+\omega) + S_f(-\omega)$. This is convenient in classical problems where the two are equal. In quantum contexts, one sometimes sees the asymmetric part of the noise $S_f(+\omega) - S_f(-\omega)$ referred to as the 'quantum noise.' We feel it is simpler and clearer to simply discuss the spectral density for positive and negative frequencies *separately*, since they have simple physical interpretations and directly relate to measurable quantities. This convention is especially useful in non-equilibrium situations where there is no simple relation between the spectral densities at positive and negative frequencies.

independent (but not necessarily given by the equilibrium expression). This yields the standard quantum mechanical expression for the spectral density

$$S_f(\omega) = \int_{-\infty}^{+\infty} d\tau\, e^{i\omega\tau} \sum_{\alpha,\gamma} \rho_{\alpha\alpha}\, e^{\frac{i}{\hbar}(\epsilon_\alpha - \epsilon_\gamma)t}\, |\langle\alpha|f|\gamma\rangle|^2 \tag{17}$$

$$= 2\pi\hbar \sum_{\alpha,\gamma} \rho_{\alpha\alpha}\, |\langle\alpha|f|\gamma\rangle|^2 \delta(\epsilon_\gamma - \epsilon_\alpha - \hbar\omega). \tag{18}$$

Substitution of this into Eqs. (13,14) we derive the familiar Fermi Golden Rule expressions for the two transition rates.

In standard courses, one is not normally taught that the transition rate of a discrete state into a continuum as described by Fermi's Golden Rule can (and indeed should!) be viewed as resulting from the continuum acting as a quantum noise source. The above derivation hopefully provides a motivation for this interpretation.

One standard model for the continuum is an infinite collection of harmonic oscillators. The electromagnetic continuum in the hydrogen atom case mentioned above is a prototypical example. The vacuum electric field noise coupling to the hydrogen atom has an extremely short autocorrelation time because the range of mode frequencies ω_α (over which the dipole matrix element coupling the atom to the mode electric field $\mathbf{E}_\alpha$ is significant) is extremely large, ranging from many times smaller than the transition frequency to many times larger. Thus the autocorrelation time of the vacuum electric field noise is considerably less than 10^{-15}s, whereas the decay time of the hydrogen 2p state is about 10^{-9}s. Hence the inequalities needed for the validity of our expressions are very easily satisfied.

Of course in the final expression for the transition rate, energy conservation means that only the spectral density at the transition frequency enters. However, in order for the expression to be valid (and in order for the transition rate to be time independent), it is essential that there be a wide range of available photon frequencies so that the vacuum noise has an autocorrelation time much shorter than the inverse of the transition rate.

3. Quantum Noise from a Resistor

Instead of an atom in free space, we might consider a quantum bit capacitively coupled to a transmission line. The transmission line is characterized by an inductance per unit length ℓ and capacitance per unit length c. A semi-infinite transmission line presents a frequency-independent impedance $Z = R_0 = \sqrt{\ell/c}$ at its end and hence acts like an ideal resistor. The dissipation is caused by the fact that currents injected at one end launch waves which travel off to infinity and do not return. Very conveniently, however, the system is simply a large collection of harmonic oscillators (the normal modes) and hence can be readily quantized. This representation of a physical resistor is essentially the one used by Caldeira and

Leggett [1] in their seminal studies of the effects of dissipation on tunneling. The only difference between this model and the vacuum fluctuations in free space discussed above is that the relativistic bosons travel in one dimension and do not carry a polarization label. This changes the density of states as a function of frequency, but has no other essential effect.

The Lagrangian for the system is

$$\mathcal{L} = \int_0^\infty dx \, \frac{\ell}{2} j^2 - \frac{1}{2c} q^2, \tag{19}$$

where $j(x, t)$ is the local current density and $q(x, t)$ is the local charge density. Charge conservation connects these two quantities via the constraint

$$\partial_x j(x, t) + \partial_t q(x, t) = 0. \tag{20}$$

We can solve this constraint by defining a new variable

$$\theta(x, t) \equiv \int_0^x dx' \, q(x', t) \tag{21}$$

in terms of which the current density is $j(x, t) = -\partial_t \theta(x, t)$ and the charge density is $q(x, t) = \partial_x \theta(x, t)$. For any well-behaved function $\theta(x, t)$, the continuity equation is automatically satisfied so there are no dynamical constraints on the θ field. In terms of this field the Lagrangian becomes

$$\mathcal{L} = \int_0^\infty dx \, \frac{\ell}{2} \left(\partial_t \theta\right)^2 - \frac{1}{2c} \left(\partial_x \theta\right)^2 \tag{22}$$

The Euler-Lagrange equation for this Lagrangian is simply the wave equation $v^2 \partial_x^2 \theta - \partial_t^2 \theta = 0$ where the mode velocity is $v = 1/\sqrt{\ell c}$.

From Eq. (21) we can deduce that the proper boundary conditions (in the absence of any coupling to the qubit) for the θ field are $\theta(0, t) = \theta(L, t) = 0$. (We have temporarily made the transmission line have a finite length L.) The normal mode expansion that satisfies these boundary conditions is

$$\theta(x, t) = \sqrt{\frac{2}{L}} \sum_{n=1}^\infty \varphi_n(t) \sin \frac{k_n \pi x}{L}, \tag{23}$$

where φ_n is the normal coordinate and $k_n \equiv \frac{\pi n}{L}$. Substitution of this form into the Lagrangian and carrying out the spatial integration yields a set of independent harmonic oscillators representing the normal modes.

$$\mathcal{L} = \sum_{n=1}^\infty \frac{\ell}{2} (\dot{\varphi}_n)^2 - \frac{1}{2c} k_n^2 \varphi_n^2. \tag{24}$$

From this we can find the momentum p_n canonically conjugate to φ_n and quantize the system to obtain an expression for the voltage at the end of the transmission line in terms of the mode creation and destruction operators

$$V = \sqrt{\frac{2}{L}}\frac{1}{c}\partial_x\theta(0,t) = \frac{1}{c}\sum_{n=1}^{\infty} k_n\sqrt{\frac{\hbar}{2\ell\Omega_n}}(a_n^\dagger + a_n). \tag{25}$$

The spectral density of voltage fluctuations is then found to be

$$S_V(\omega) = 2\pi\frac{2}{L}\sum_{n=1}^{\infty}\frac{\hbar\Omega_n}{2c}\{n_\gamma(\hbar\Omega_n)\delta(\omega+\Omega_n)+[n_\gamma(\hbar\Omega_n)+1]\delta(\omega-\Omega_n)\}, \tag{26}$$

where $n_\gamma(\hbar\omega)$ is the Bose occupancy factor for a photon with energy $\hbar\omega$. Taking the limit $L \to \infty$ and converting the summation to an integral yields

$$S_V(\omega) = 2R_0\hbar|\omega|\{n_\gamma(\hbar|\omega|)\Theta(-\omega) + [n_\gamma(\hbar\omega) + 1]\Theta(\omega)\}, \tag{27}$$

where Θ is the step function. We see immediately that at zero temperature there is no noise at negative frequencies because energy can not be extracted from zero-point motion. However there remains noise at positive frequencies indicating that the vacuum is capable of absorbing energy from the qubit.

A more compact expression for this 'two-sided' spectral density of a resistor is

$$S_V(\omega) = \frac{2R_0\hbar\omega}{1 - e^{-\hbar\omega/k_BT}}, \tag{28}$$

which reduces to the more familiar expressions in various limits. For example, in the classical limit $k_BT \gg \hbar\omega$ the spectral density is equal to the Johnson noise result[5]

$$S_V(\omega) = 2R_0k_BT, \tag{29}$$

which is frequency independent, and in the quantum limit it reduces to

$$S_V(\omega) = 2R_0\hbar\omega\Theta(\omega). \tag{30}$$

Again, the step function tells us that the resistor can only absorb energy, not emit it, at zero temperature.

If we use the engineering convention and add the noise at positive and negative frequencies we obtain

$$S_V(\omega) + S_V(-\omega) = 2R_0\hbar\omega\coth\frac{\hbar\omega}{2k_BT} \tag{31}$$

for the symmetric part of the noise, which appears in the quantum fluctuation-dissipation theorem[2]. The antisymmetric part of the noise is simply

$$S_V(\omega) - S_V(-\omega) = 2R_0\hbar\omega. \tag{32}$$

[5] Note again that in the engineering convention this would be $S_V(\omega) = 4R_0k_BT$.

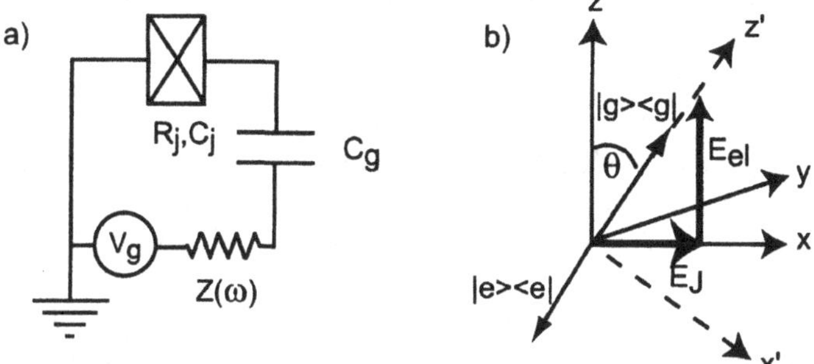

Figure 1. a) Circuit diagram of Cooper-pair box. b) Pseudo-spin representation of the energies of Cooper-pair box. The density matrix for the two pure eigenstates lie along the total effective field, collinear with the z' axis.

This quantum treatment can also be applied to any arbitrary dissipative network[3]. If we have a more complex circuit containing capacitors and inductors, then in all of the above expressions, R_0 should be replaced by $\mathrm{Re}Z(\omega)$ where $Z(\omega)$ is the complex impedance presented to the qubit.

4. The Single Cooper-Pair Box: a Two-Level Quantum Circuit

The Cooper-pair box (CPB) is a simple circuit [4], consisting of a small superconducting "island", connected to a large reservoir via a single small-capacitance Josephson junction, depicted as a box with a cross (Fig. 1). The island is charge biased by applying a voltage (V_g) to a nearby lead, called the gate, which has a small capacitance to the island, C_g. The junction is characterized by its capacitance, C_j, and its tunnel resistance, R_j. At temperatures well below the transition temperature of the superconductor ($T_C \sim 1.5$ K for the usual Al/AlOx/Al junctions), none of the many ($\sim 10^9$) quasiparticle states on the island should be thermally occupied, and the number of Cooper-pairs on the island is the only relevant degree of freedom.

We may then write the Hamiltonian for the box in terms of the states of different numbers of pairs on the island, which are eigenstates of the number operator, $\hat{n}|n\rangle = n|n\rangle$. The box Hamiltonian consists of an electrostatic term, plus a Josephson term describing the coupling of the island to the lead,

$$H = H_{electrostatic} + H_{Josephson} \tag{33}$$

$$= 4E_C \sum_n (n - n_g)^2 |n\rangle\langle n| - \frac{E_J}{2} \sum_n (|n+1\rangle\langle n| + h.c.) \tag{34}$$

The energy scale for the electrostatic interaction is given by the charging energy, $E_C = e^2/2C_\Sigma$, where $C_\Sigma = C_j + C_g$ is the total island capacitance, while the Josephson energy, E_J, is set by the tunnel resistance and the gap of the superconductor,

$$E_J = \frac{h\Delta}{8e^2 R_j} = \frac{\Delta}{8}\frac{R_K}{R_j}. \tag{35}$$

The electrostatic term is easily modulated by changing the voltage on the gate; the quantity $n_g = C_g V_g/2e$ that appears in the Hamiltonian corresponds to the total polarization charge (in units of Cooper pairs) injected into the island by the voltage source.

This Hamiltonian leads to particularly simple behavior in the charge regime, when the electrostatic energy dominates over the Josephson coupling, $4E_C \gg E_J$. In this case we can restrict the discussion to only two charge states, $|n=0\rangle$ and $|n=1\rangle$. For convenience we can reference the energies of the two states to their midpoint, $E_{mid} = 4E_C(1-2n_g)^2$, so that the Hamiltonian now becomes

$$H = \frac{1}{2}\begin{pmatrix} -E_{el} & -E_J \\ -E_J & E_{el} \end{pmatrix} \tag{36}$$

where E_{el} is the electrostatic energy that is now *linear* in the gate charge, $E_{el} = 4E_C(1-2n_g)$. It is also now apparent that the Hamiltonian is identical to that of a fictitious spin-1/2 particle,

$$H = -\frac{E_{el}}{2}\sigma_z - \frac{E_J}{2}\sigma_x, \tag{37}$$

under the influence of two psuedo-magnetic fields, $B_z = E_{el}$ and $B_x = E_J$, as depicted in Fig. 1. In other words, the box is a qubit or two-level system[6]. The state of the system is in general a linear combination of the states $|n=0\rangle$ and $|n=1\rangle$. The state can be depicted using the density matrix, which corresponds to a point on the Bloch sphere, where the north pole (+z-direction) corresponds to $|n=0\rangle$. The ground and excited states of the system will be aligned and anti-aligned with the total fictitious field, i.e. in the $\pm z'$ directions.

It is also apparent from this discussion that the states of the box can be easily manipulated by changing the gate voltage. The energies of the ground and excited states, as a function of n_g, are displayed in Figure 2. The energy difference between the ground and excited bands varies from $4E_C$ at $n_g = 0, 1$, to a minimum at the charge degeneracy point, $n_g = 1/2$. At this point, the Josephson coupling leads to an avoided crossing, and the splitting is E_J.

Also plotted is the expectation value of the number operator, $\langle|\hat{n}|\rangle$, which is proportional to the total charge on the island. In the geometrical picture of Fig. 1b,

[6] Of course, this is an approximation, as there are other charge states ($|n=2\rangle$, etc.) which are possible, but require much higher energy. Even outside the charge regime (i.e. when $E_J \geq 4E_C$) the two lowest levels of the box can be used to realize a qubit [5]. In this case, the two states do not exactly correspond with eigenstates of charge, and matrix elements are more complicated to calculate. Nonetheless, this regime can also be used as an electrical quantum spectrum analyzer.

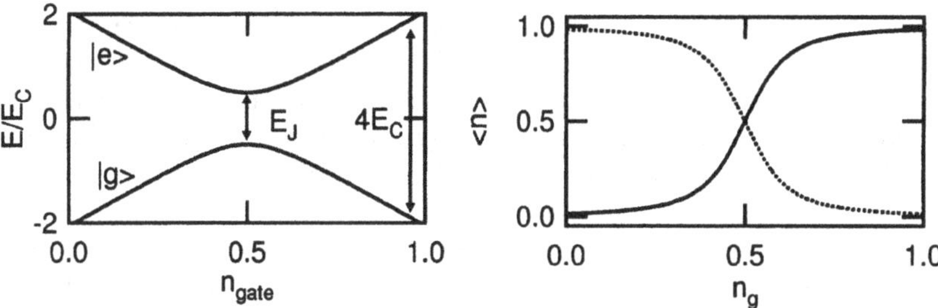

Figure 2. Energies (left) of ground and excited states of a Cooper-pair box with $E_C = E_J$ vs. dimensionless gate charge, $n_g = C_g V_g / 2e$. The expectation value of a charge measurement, $\langle n \rangle$, (right) for the ground (solid line) and excited (dotted line) states vs. n_g.

a measurement of charge ($\hat{n}$) is equivalent to projecting the state on the z-axis, $\hat{n} = \frac{1}{2}(1 - \sigma_z)$. We see that as the gate charge is changed from 0 to 1, the ground state is initially $|n = 0\rangle$, and the character of the ground and excited states interchange on passing through the degeneracy point, leading to the transition between $\langle \hat{n} \rangle$ = 0 and 1, which is broadened by quantum fluctuations (the σ_x coupling). At the degeneracy point, the ground and excited states lie in the $\pm$x-directions, i.e. they are symmetric and antisymmetric combinations of the two charge states. In general, we will denote the ground and excited state of the CPB at a particular gate voltage as $|g\rangle$ and $|e\rangle$, which are given in terms of the charge states by $|g\rangle = \cos(\theta/2)|0\rangle + \sin(\theta/2)|1\rangle$ and $|e\rangle = -\sin(\theta/2)|0\rangle + \cos(\theta/2)|1\rangle$ respectively, where $\theta = \arctan[E_J/E_{el}]$ is a function of the gate voltage.

A nice property of the CPB in this regime is that the various matrix elements can be calculated in a straightforward way. For example, the expectation value of $\hat{n}$ in the ground state, $\langle g|\hat{n}|g\rangle$, is therefore equal to $1/2(1 - \langle g|\sigma_z|g\rangle) = \sin^2(\theta/2)$, from which we can find the ground state charge as shown in Fig. 2. A perturbation in the gate charge, due for example to a fluctuation or change in the applied gate voltage, will lead to a proportional change in the electrostatic energy, or the z-component of the fictitious magnetic field. Such a perturbation will cause both dephasing and transitions between states.

5. General Discussion of CPB Coupled to a Dissipative Environment

In the previous section we described the Hamiltonian and the eigenstates for a Cooper-pair box which is "charge-biased," i.e. controlled with a voltage applied to a gate capacitor C_g, as shown in Fig. 1. In our earlier treatment of the box, the voltage and the dimensionless gate charge, n_g were treated as fixed parameters of the Hamiltonian (c-numbers). In this case, the box's evolution is purely deterministic and conservative. However, it is impossible, even in principle, to control such a voltage with arbitrary precision at all frequencies. In Fig. 1, the idealized source of the gate voltage is drawn in series with an impedance $Z(\omega)$ of the gate lead. Generally, this gate lead will be connected to external wiring (a transmission line), with a typical real impedance comparable to the impedance of free space ($\sim 50\ \Omega$) at the microwave transition frequencies of the box. From the fluctuation-dissipation theorem we know that this impedance will introduce noise on the gate voltage, even at zero temperature.

There are several effects of the voltage noise on the box, or the coupling of our spin-1/2 circuit to the many external degrees of freedom represented by the gate impedance. First, even at zero temperature, we will find a finite excited state lifetime, $T1$, for the box. Second, at finite temperature, we will find a finite polarization of our psuedo-spin, i.e. some steady-state probability to find the spin in its excited state. Finally, the gate noise introduces a random effective field felt by the spin, and a loss of phase coherence for a superpostition state. It is this last effect which is most important in making high-fidelity qubits and performing quantum computations, but it is the first two which depend most explicitly on the quantum nature of the noise. We deal in this manuscript with only these first two features of the box's coupling to the electromagnetic environment, and ignore the dephasing[7]. Of course, the other parameter in the Hamiltonian, the Josephson energy, can in principle fluctuate, especially as in many experiments the box's junction is split into a small SQUID in order to provide external tuning of E_J with an applied flux. We concentrate here only on the voltage noise (fluctuations in the σ_z part of the Hamiltonian) for simplicity.

We begin with a very simple treatment of the dynamics of the two-level system under the influence of the gate voltage noise. We are interested in the ensemble-averaged behavior of our psuedo-spin, which is best described using the density matrix approach, and is detailed in Section 7 on the coupling of the box to a measuring SET. The basic effects, apart from dephasing, however, can be captured simply by examining the probabilities p_g and p_e of finding the box in its ground ($|g\rangle$) or excited ($|e\rangle$) states. The noise of the external environment can drive transitions from ground to excited state and vice-versa, at rates $\Gamma_\uparrow$ and $\Gamma_\downarrow$, respectively. The coupled master equations for these probabilities are

$$\frac{dp_e}{dt} = p_g\Gamma_\uparrow - p_e\Gamma_\downarrow \tag{38}$$

$$\frac{dp_g}{dt} = p_e\Gamma_\downarrow - p_g\Gamma_\uparrow \tag{39}$$

Of course conservation of probability tells us that $p_e + p_g = 1$, so we introduce the polarization of the spin-1/2 system, $P = p_g - p_e$. In steady-state, the detailed balance condition is $p_e\Gamma_\downarrow = p_g\Gamma_\uparrow$. The two rates $\Gamma_\uparrow$ and $\Gamma_\downarrow$ are related by Equations 13 and 14 to the spectral densities of the noise at negative and positive frequencies.

[7] For a nice recent treatment of dephasing in Josephson junctions, see Ref. [6].

We see immediately that if the spectral density is symmetric (classical!), then the rates for transitions up and down are equal, the occupancies of the two states are exactly equal, and the polarization of the psuedo-spin is identically zero. It is the quantum, or antisymmetric, part of the noise which gives the finite polarization of the spin. Even in NMR, where the temperature is large compared to the level splitting ($\hbar\omega_{01} \leq k_B T$), this effect is well-known and crucial, as the small but non-zero polarization is the subject of the field!

Solving for the steady-state polarization, we find

$$P_{ss} = \frac{\Gamma_\downarrow - \Gamma_\uparrow}{\Gamma_\downarrow + \Gamma_\uparrow} = \frac{S(+\omega_{01}) - S(-\omega_{01})}{S(+\omega_{01}) + S(-\omega_{01})} \tag{40}$$

An measurement of the steady-state polarization allows one to observe the amount of asymmetry in the noise, so the two-level system is a *quantum* spectrum analyzer.

If we can create a non-equilibrium polarization, $P = P_{ss} + \Delta P$ (a pure state is not necessary) of our two-level system, we expect it to return to the steady state value. Substituting the modified probabilities $p_e(t) = p_{e_{ss}} - \Delta P(t)/2$ and $p_g(t) = p_{g_{ss}} + \Delta P(t)/2$ into our master equations above, we find an equation for the deviation of the polarization

$$\frac{d(\Delta P(t))}{dt} = -\Delta P(t)(\Gamma_\uparrow + \Gamma_\downarrow). \tag{41}$$

Thus the system decays to its steady-state polarization with the relaxation rate $\Gamma_1 = \Gamma_\uparrow + \Gamma_\downarrow = (A/\hbar)^2[S(-\omega_{01}) + S(+\omega_{01})]$ related to the *total* noise at both positive and negative frequencies. In NMR, the time $1/\Gamma_1$ is referred to as $T1$. In the zero-temperature limit, there is no possibility of the qubit absorbing energy from the environment, so $\Gamma_\uparrow = 0$, and we find full polarization $P = 1$, and a decay of any excited state population at a rate $\Gamma_\downarrow = 1/T1$ which is the spontaneous emission rate.

It is worth emphasizing that a quantum noise source is always characterized by two numbers (at any frequency), related to the positive and negative frequency spectral densities, or to the symmetric and antisymmetric parts of the noise. These two quantities have different effects on a two-level system, introducing a finite polarization and finite excited-state lifetime. Consequently, a measurement of *both* the polarization and $T1$ of a two-level system is needed to fully characterize the quantum noise coupled to the qubit. Such measurements in electrical systems are now possible, and some of us [7] have recently performed such a characterization for the specific case of a CPB coupled to a superconducting single-electron transistor.

Our discussion in this section uses the language of NMR to describe the effects

TABLE I. Different ways to characterize a quantum reservoir.

Fermi Golden Rule	$\Gamma_\uparrow(\omega) = \frac{A^2}{\hbar^2} S_V(+\lvert\omega\rvert)$	$\Gamma_\downarrow(\omega) = \frac{A^2}{\hbar^2} S_V(-\lvert\omega\rvert)$
Fluct.-Diss. Relation	$n_\gamma(\omega) = 2\Gamma_\uparrow/(\Gamma_\downarrow - \Gamma_\uparrow)$	$\mathrm{Re}[Z(\omega)] = \frac{\hbar}{A^2\omega}(\Gamma_\downarrow - \Gamma_\uparrow)$
NMR	$T1 = (\Gamma_\downarrow + \Gamma_\uparrow)^{-1}$	$P = (\Gamma_\downarrow - \Gamma_\uparrow)/(\Gamma_\downarrow + \Gamma_\uparrow)$
Quantum Optics	$B_{Einstein} = \Gamma_\uparrow$	$A_{Einstein} = \Gamma_\downarrow - \Gamma_\uparrow$

on the two-level system. There are, however, several possible protocols[8] for measuring the quantum noise, and several different "basis sets" or measured quantities which describe the noise process or the quantum reservoir to which the two-level system is coupled. Table 1 contains a "translation" between the specific pairs of quantities that are commonly used in different disciplines, and their relation to the positive and negative noise spectral densities. In all cases, though, *two separate numbers* are required to specify the properties of a quantum reservoir.

6. The Box Coupled to an Ohmic Environment

We can now proceed to the effects of a specific dissipative coupling to the Cooper-pair box. The noise on the gate voltage will lead to a fluctuation of the gate charge parameter, n_g, and thus to a fluctuation of the electrostatic energy, i.e. the σ_z term in the Hamiltonian (Eq. 37). Depending on the average value of n_g, this fluctuation will consist of fluctuations which are both longitudinal ($\|$ to σ'_z) and transverse ($\perp$ to σ'_z). To calculate the rates of transitions between the states $|e\rangle$ and $|g\rangle$, we need to find the coupling strength A of this perturbation in the σ'_x direction. Referring to Fig. 1, we see that $\sigma_z = \cos(\theta)\sigma'_z - \sin(\theta)\sigma'_x$. If we let the gate charge now be $n_g(t) = \bar{n_g} + \delta n_g(t)$, we may rewrite the Hamiltonian Eq. 37 in the new eigenbasis as

$$H = -\frac{E_{01}}{2\hbar}\sigma'_z + 4E_C\cos(\theta)\delta n_g(t)\sigma'_z - 4E_C\sin(\theta)\delta n_g(t)\sigma'_x\,. \tag{42}$$

The time-varying term in the σ'_z direction will effectively modulate the transition frequency, $\omega_{01} = E_{01}/\hbar$, and cause dephasing. In terms of the gate voltage noise, $V(t)$, the σ'_x perturbation term has the form $AV(t)\sigma'_x = e\kappa\sin(\theta)V(t)\sigma'_x$, where e is the electron's charge and $\kappa = C_g/C_\Sigma$ is the capacitive coupling. Using Eq. 14, we find

$$\Gamma_\downarrow = \left(\frac{e}{\hbar}\right)^2 \kappa^2 \sin^2(\theta) S_V(+\omega_{01}). \tag{43}$$

[8] The idea of watching the decay from the pure states $|e\rangle$ and $|g\rangle$ to measure $S_V(\pm|\omega_{01}|)$ separately was described in Section 2.

If the environment is effectively at zero temperature ($\hbar\omega_{01} \gg k_B T$), then $S_V(+\omega_{01})=2\hbar\omega_{01}R$, and the quality factor of the transition is

$$Q = \omega_{01}/\Gamma_\downarrow = \frac{1}{\kappa^2 \sin^2\theta} \frac{R_K}{4\pi R}, \tag{44}$$

where $R_K = h/e^2$ is the resistance quantum.

For a finite temperature, we have rates in both directions, and the polarization of the psuedo-spin is given by the ratio of the antisymmetric (Eq. 32) to symmetric (Eq. 31) spectral densities, $P = \mathrm{th}(\hbar\omega_{01}/2k_B T)$, as one expects for any two-level system at temperature T. An example of the average charge state of a Cooper-pair box at finite temperature, and of the polarization and equilibration time T1, are shown in Figure 3. As the gate voltage is varied, the transition frequency of the box changes from a maximum near $n_g = 0$, to a minimum $\omega_{01} = E_J/\hbar$ at the degeneracy point $n_g = 0.5$ and then back again. We see that the states of the box are generally most "fragile" near the avoided crossing. First, the energy splitting is a minimum here, leading to the lowest polarization of the psuedo-spin. Second, because the eigenstates $|g\rangle$ and $|e\rangle$ point in the σ_x direction, the matrix elements for the voltage fluctuations of the environment are maximal, i.e. the noise is orthogonal to the spin. This also implies that the dephasing effects are minimal at this degeneracy point, which offers great advantages for improving the decoherence times [5], but the excited state lifetimes are smallest at this point. One also sees that the lifetimes become large away from the degeneracy, where the electrostatic energy dominates over the Josephson energy, which offers advantages when measuring the charge state. In the limit that E_J could be suppressed to zero during a measurement, the matrix elements (for voltage noise) vanish, and a quantum non-demolition (QND) measurement [8] could be performed. The idea of using the qubit as a quantum spectrum analyzer is precisely the reverse, where we measure the "destruction" in the two-level system (i.e. inelastic transitions caused by the coupling of the states to the environment), in order to learn about the quantum noise spectrum of the fluctuations.

The Cooper-pair box can of course also be used to measure the more interesting spectral densities of *nonequilibrium* devices. The simplest example is to replace the gate resistance by a tunnel junction. If we arrange to bias the junction using, e.g. an inductor, a dc current I and an average dc voltage V can be maintained across the junction. Classically, the current noise of such a tunneling process is frequency independent, $S_I = 2eI$. The voltage noise density presented to the CPB's gate would then be $S_V = 2eIR_T^2$, where R_T is the junction tunnel resistance. In fact, this "white" spectral density can only extend up to frequencies of order $\omega = eV/\hbar$, the maximum energy of electrons tunneling through the junction. The correct high-frequency form of the *symmetrized* noise density was

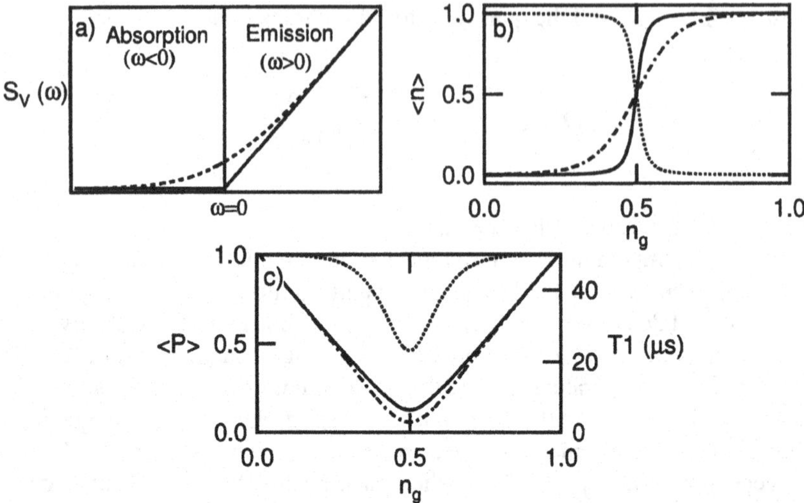

Figure 3. The box coupled to an equilibrium, Ohmic environment, i.e. a resistor. a) Two-sided noise spectral density, of the voltage, $S_V(\omega)$, for a resistor at T=0 (solid line) and finite temperature (dashed line) b) Average charge of box with $E_C = 1K$, $E_J = 0.5K$ when coupled with strength $\kappa = 0.01$ to a resistor with resistance $R = 50\ \Omega$ and temperature $T = 0.5K$. c) Polarization (dotted line) and relaxation time T1 for the same parameters. Full line is the rate of spontaneous emission, i.e. T1 at zero temperature.

calculated by Rogovin and Scalapino [9],

$$S_V(\omega) = R(\hbar\omega + eV)\coth\left[\frac{\hbar\omega + eV}{2k_BT}\right] + R(\hbar\omega - eV)\coth\left[\frac{\hbar\omega - eV}{2k_BT}\right], \tag{45}$$

and was *indirectly* measured in a mesoscopic conductor using a conventional spectrum analyzer [10]. This noise can also be expressed [11] in its two-sided form

$$S_V(\omega) = \frac{(\hbar\omega + eV)R_T}{1 - e^{-\frac{\hbar\omega + eV}{k_BT}}} + \frac{(\hbar\omega - eV)R_T}{1 - e^{-\frac{\hbar\omega - eV}{k_BT}}}, \tag{46}$$

and is displayed in Fig. 4. Notice that the antisymmetric part of this noise is the same as that of the ordinary resistor, $S_V(+\omega) - S_V(-\omega) = 2\hbar\omega R_T$, and is independent of the voltage. Also shown in Fig. 5 is the polarization and relaxation time, T1, of a CPB coupled to a shot noise environment. We see that full polarization is achieved only when $\hbar\omega_{01} \gg eV$. For low transition frequencies, the polarization is inversely proportional to the current through the junction. Aguado and Kouwenhoven [11] have described the use of a double quantum dot as a two-level system to probe this behavior of the shot noise.

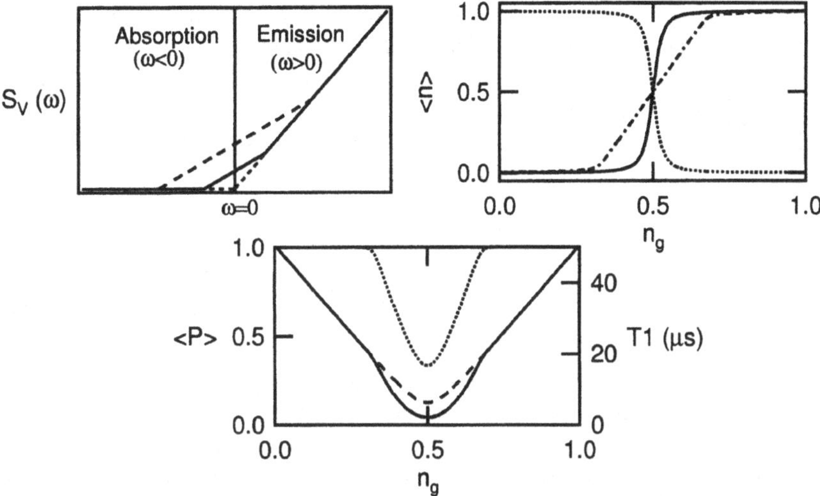

Figure 4. The box coupled to an nonequilibrium, Ohmic environment, i.e. a 50 Ω tunnel junction. a) Two-sided noise spectral density, of the voltage, $S_V(\omega)$, for a junction at T=0.02 K, with zero voltage (dotted line), and increasing bias voltages (solid and dashed lines). b) Average charge of box with $E_C = 1K$, $E_J = 0.5K$ when coupled with strength $\kappa = 0.01$ to a junction biased at $eV = 1.5$ K. c) Polarization (dotted line) and relaxation time T1 for the same parameters (solid) and for T=0, V=0 (dashed line).

Given our discussion so far, it is now interesting to ask what the effects of a real quantum measurement on a quantum circuit will be. A quantum measurement device will in general be neither linear, Ohmic, nor equilibrium. Obviously, if we hope to characterize this measurement process, and to understand what one will observed when the qubit is coupled to the noise processes of the measuring device, we will need to calculate the full quantum (two-sided!) noise spectrum of the amplifier or detector.

7. Single-Electron Transistor Coupled to a Two-Level System

We have seen in previous sections that a two-level system (TLS) may be used as a "spectrum analyzer" to measure quantum noise. Here, we use this technique to *theoretically* calculate quantum noise. Instead of simply studying the "noisy" system of interest in isolation, one can study a composite system consisting of the "noisy" system coupled to a TLS; by calculating the relaxation and excitation rates of the TLS, one can efficiently calculate the quantum noise of interest[9]. We demonstrate the usefulness of this technique by outlining a calculation of the quantum charge noise of a single electron transistor (SET). This is an important example, as when an SET is used as an electrometer, it is this noise which determines the measurement backaction.

The SET consists of a metallic island attached via tunnel junctions to source and drain reservoirs. It is described by the Hamiltonian:

$$H_{SET} = \sum_{k,\alpha=L,R,I} (\varepsilon_k - \mu_\alpha)\, c^\dagger_{k\alpha} c_{k\alpha} + E_C(n-\mathcal{N})^2 + H_T \tag{47}$$

$$H_T = t \sum_{k,q,\alpha=L,R} \left(F^\dagger c^\dagger_{kI} c_{q\alpha} + h.c. \right) \tag{48}$$

The first term in H_{SET} describes the kinetic energy of electrons in the leads ($\alpha = L, R$) and on the island ($\alpha = I$). The second term is the Coulomb charging energy which depends on n, the number of excess electrons on the island. This interaction term can be tuned by changing the voltage on a nearby gate electrode which is capacitively coupled to the island; $\mathcal{N}$ represents the dimensionless value of this voltage. Finally, H_T describes the tunneling of electrons through the two SET tunnel junctions; the conductance of each junction (in units of e^2/h) is given by $g = 4\pi^2 t^2 \nu_0^2$, with ν_0 being the density of states. $F^\dagger$ is an auxiliary operator which increases n by one: $\left[n, F^\dagger\right] = F^\dagger$. For simplicity, we assume that the two junctions of the SET are completely symmetric (i.e. equal junction capacitances and dimensionless conductances).

Throughout this section, we will be interested in the regime of sequential tunneling in the SET, where transport involves energy-conserving transitions between two charge states of the SET island, say $n = 0$ and $n = 1$. These transitions are described by simple rates, which can be derived via Fermi's Golden rule:

$$\Gamma^\alpha_{n\pm1,n} = \gamma([\Delta E]^\alpha_{n\pm1,n}) \tag{49}$$

$$\gamma(\Delta E) = \frac{g\Delta E/h}{1 - e^{-\Delta E/(k_B T)}} \tag{50}$$

$$\Delta E^\alpha_{n\pm1,n} = \mp 2E_C\left(n \pm \frac{1}{2} - \mathcal{N}\right) \pm \left(\frac{1}{2} - \delta_{\alpha,R}\right) eV_{DS} \tag{51}$$

$\Gamma^\alpha_{n\pm1,n}$ is the tunneling rate from the charge state n to $n \pm 1$ through junction α; ΔE is the energy gained in making the tunneling transition, and includes contributions both from the drain-source voltage V_{DS} and from the charging energy. Sequential tunneling is the dominant transport mechanism when the junction

[9] Note that the spirit of our approach is similar to that employed in the theory of full counting statistics [12]. There too one attaches an auxiliary spin 1/2 to the scattering system of interest, and studies the dynamics of the fully coupled system to obtain the statistics of charge transfer in the scatterer.

conductances are small (i.e. $g/(2\pi) \ll 1$), and the dimensionless gate voltage $\mathcal{N}$ is not too far away from a charge degeneracy point. Sequential tunneling is the most important regime for measurement applications, as it yields the largest SET currents.

At low temperatures, only tunnel events which follow the voltage are possible. There are thus there only two relevant rates: $n = 0 \to 1$ transitions occur through the left junction at a rate Γ^L_{10}, while $n = 1 \to 0$ occur through the right junction at a rate Γ^R_{01}. The average current will be given by:

$$\langle I \rangle = e\bar{\Gamma} \equiv e\frac{\Gamma^L_{10}\Gamma^R_{01}}{\Gamma^L_{10} + \Gamma^R_{01}} \tag{52}$$

We are interested in calculating $S_Q(\omega)$, the quantum noise associated with fluctuations of the charge on the central island of the SET. It is defined as:

$$S_Q(\omega) = \int_{-\infty}^{\infty} dt \langle n(t) n(0) \rangle e^{-i\omega t} \tag{53}$$

Note that we can equivalently think of S_Q as describing the voltage noise of the island, as $V_{island} = en/C_\Sigma$, where C_Σ is the total capacitance of the island. In two limits, the form of the island charge noise can be anticipated. For $\omega \to 0$, the noise will correspond to classical telegraph noise– the island charge n fluctuates between the values 0 and 1, with Poisson-distributed waiting times determined by the rates Γ^L_{10} and Γ^R_{01}. We thus expect a symmetric, Lorentzian form [13] for the noise at low frequencies:

$$S_Q(\omega) \to \frac{2\bar{\Gamma}}{\omega^2 + (\Gamma^L_{10} + \Gamma^R_{01})^2} \qquad (\omega \ll E_C) \tag{54}$$

For large frequencies $|\omega| \gg E_C$, we expect that correlations due to the charging energy will have no influence on the noise. The system will effectively look like two tunnel junctions in parallel, and we can use the results of Sec. 6 for the corresponding voltage noise. Noting that each junction effectively consists of a resistor and capacitor in parallel, we have at zero temperature:

$$\begin{aligned} S_Q(\omega) &= \frac{C_\Sigma^2}{e^2} \times S_V(\omega) \to \frac{C_\Sigma^2}{e^2}\left[2\hbar\omega \mathrm{Re}\, Z_{tot}\Theta(\omega)\right] \qquad (|\omega| \gg E_C) \\ &= 4\left(\frac{g}{2\pi}\right)\frac{\omega}{\left(\frac{g}{2\pi}\frac{4E_C}{\hbar}\right)^2 + \omega^2}\Theta(\omega) \end{aligned} \tag{55}$$

Note that $S_Q(\omega)$ decays as $1/\omega$ at large frequencies, whereas Eq. (54) for classical telegraph noise decays as $1/\omega^2$.

Given these two limiting forms, the question now becomes one of how the SET interpolates between them. One might expect that the two results should simply be added in quadrature, but as with combining thermal and quantum noise (see Sec. 6), this is approach is too simple. A completely quantum mechanical way of calculating the noise for any frequency is needed. This was recently accomplished by Johansson *et. al* [14] using an extension of a technique developed by Schöller and Schön [15]. Here, we re-derive their results using the coupled system approach outlined above. This method is physically motivated and allows for a heuristic interpretation of the final result.

8. SET Coupled to a Qubit

We now consider a system where the SET is coupled to a two-level system (i.e. a qubit), with a coupling Hamiltonian which can induce transitions in the TLS. Using spin operators to describe the qubit, and assuming operation at the degeneracy point for simplicity, where the transitions are fastest[10], we have:

$$H = H_{SET} - \frac{1}{2}\Omega\sigma_x + A\sigma_z n, \tag{56}$$

where Ω is the qubit splitting frequency[11], and A is the coupling strength. We can define the rates $\Gamma_\uparrow$ and $\Gamma_\downarrow$ which are, respectively, the rate at which the qubit is excited by the SET, and the rate at which the qubit is relaxed by the SET. For a weak coupling ($A \to 0$), one has (c.f. Eq. 14,13 in Sec. 2):

$$\Gamma_{\downarrow/\uparrow} = \frac{A^2}{\hbar} S_Q(\pm\Omega) \tag{57}$$

Eq. (57) tells us that if we know the rates $\Gamma_\uparrow$ and $\Gamma_\downarrow$ for a weakly coupled system at an arbitrary splitting frequency Ω, we know the quantum noise $S_Q(\Omega)$ at all frequencies. This is the essence of the technique previously described, in which a qubit acts as a quantum spectrum analyzer of noise. Here, we mimic this approach theoretically by obtaining $\Gamma_\uparrow$ and $\Gamma_\downarrow$ from a *direct* analysis of the coupled system in the limit of weak coupling ($A \to 0$). The object of interest is the reduced density matrix ρ which describes both the charge n of the transistor island *and* the state of the qubit. We are interested in two quantities. First, what is the stationary state of the qubit? The stationary populations of the two qubit states (which are determined from the time-independent solution for ρ) will tell us the polarization of the qubit, and the amount of asymmetry in the noise (c.f. Eq.40). Second, how

[10] This amounts to maximizing the "destruction" due to the SET's noise, and the case where $\theta = \pi/2$, the qubit eigenstates are in the $\sigma'_z = \sigma_x$ direction, and the SET's perturbation is in the $-\sigma'_x = \sigma_z$ direction (c.f. Eq. 42). After the noise of the SET is found, we can then recalculate the effects on the qubit at various n_g or values of θ by including the modified matrix elements in the coupling coefficient, A.

[11] Henceforth we use Ω for the transition frequency, instead of the previous notation $\omega_{01} = E_{01}/\hbar$, for compactness.

quickly do the qubit populations relax to their stationary value? This relaxation will be described by a time-dependent solution of ρ characterized by a mixing rate Γ_{mix} which is the *sum* of $\Gamma_\uparrow$ and $\Gamma_\downarrow$ (c.f. Eq. 41). From these two results we can solve for the individual values of $\Gamma_{\uparrow/\downarrow}$.

To deal with the dynamics of ρ, we make use of the fact that sequential tunneling processes are completely described by lowest-order perturbation theory in the tunneling Hamiltonian H_T. Keeping only second order terms (there are no non-vanishing first order terms), one obtains the following standard evolution equation in the interaction picture:

$$\frac{d}{dt}\rho(t) = -\frac{1}{\hbar}\int_{-\infty}^{t} dt' \langle [H_T(t), [H_T(t'), \rho(t') \otimes \rho_F]] \rangle \tag{58}$$

The angular brackets denote the trace over the single-particle degrees of freedom in the SET leads and island; ρ_F is the equilibrium density matrix corresponding to the state of these degrees of freedom in the absence of tunneling.[12] Note that a similar density matrix analysis of a qubit coupled to a SET was recently discussed by Makhlin *et. al* [16]; unlike the present case, these authors restricted attention to a vanishingly small splitting frequency Ω.

To make progress with Eq. (58), we make a Markov approximation, which involves replacing $\rho(t')$ on the right-hand side with $\rho(t)$. This is permissible as we are interested in the *slow* dynamics of ρ. We want to find both the stationary solution of ρ, for which the Markov approximation is exact, and the mixing mode, a mode whose time dependence is $\propto e^{-(\Gamma_\uparrow+\Gamma_\downarrow)t}$. This mode is also arbitrarily slow in the weak coupling limit $A \to 0$ of interest. Finding the stationary mode and the mixing mode correspond to evaluating the polarization and T1 of the qubit (c.f. Eq. 40 and Eq. 41), as was shown earlier for the master equation of the *probabilities* in Section 5. Note that the Markov approximation should be made in the *Schrödinger* picture, as it is in the Schrödinger picture that ρ will be nearly stationary (i.e. all oscillations associated with the qubit splitting frequency Ω will be damped out in the long-time limit).

Evaluation of Eq. (58) in the Markov approximation results in the appearance of rates which are generalizations of those given in Eq. (49). Now, however, these rates depend on the initial and final state of the qubit– tunneling transitions can simultaneously change both the charge state of the SET island *and* the state of the qubit. The resulting equation is most easily presented if we write the reduced density matrix ρ in the basis of eigenstates at zero tunneling. For each value of island charge n, there is a different qubit Hamiltonian, and correspondingly a different a qubit ground state $|g_n\rangle$ and excited state $|e_n\rangle$. When a tunneling event occurs in the SET, there is a sudden change in the qubit Hamiltonian. As the qubit ground and excited states at different values of n are *not* orthogonal, tunneling

[12] In the diagrammatic language of Ref. [15], Eq. (58) is equivalent to keeping all $(H_T)^2$ terms in the self-energy of the Keldysh propagator governing the evolution of ρ.

transitions in the SET are able to cause "shake-up" transitions in the qubit. In the limit $A \to 0$, the relevant matrix overlaps are given by:

$$\langle g_m | g_n \rangle = 1 - \frac{1}{2}\left(\frac{A(m-n)}{\Omega}\right)^2 = \langle e_m | e_n \rangle \tag{59}$$

$$\langle e_m | g_n \rangle = \frac{A(m-n)}{\Omega} \tag{60}$$

Defining the frequency dependent rate $\Gamma_{n\pm1,n}(\omega)$ as:

$$\Gamma_{n\pm1,n}(\omega) \equiv \sum_{\alpha=L,R} \gamma([\Delta E]^{\alpha}_{n\pm1,n} + \hbar\omega), \tag{61}$$

where ΔE and $\gamma(\Delta E)$ are defined in Eqs. (51) and (50), the required tunnel rates take the form:

$$\Gamma_{m,n} \equiv \Gamma_{m,n}(0) \qquad \Gamma^{\pm}_{m,n} \equiv \Gamma_{m,n}(\pm\Omega) \tag{62}$$

The Γ^+ rates correspond to tunneling events where the qubit is simultaneously relaxed, and thus there is an additional energy Ω available for tunneling. For large Ω, tunneling processes which are normally energetically forbidden can occur if they are accompanied by qubit relaxation. Similarly, the Γ^- rates describe tunnelling events where the qubit is simultaneously excited, with the consequence that there is less energy available for tunneling.

Returning to the evolution equation Eq. (58), note that we do not need to track elements of ρ which are off-diagonal in the island charge index n– there is no coherence between different charge states, as tunneling events necessarily create an electron-hole excitation. Further, if we focus on small qubit frequencies, we may continue to restrict attention to only $n = 0$ and $n = 1$ (i.e. Ω is not large enough to "turn on" tunneling processes which are normally energetically forbidden). Thus, there are 8 relevant matrix elements of ρ– for each of the four qubit density matrix elements (i.e. gg, ee, ge, eg), there are two possible island charge states. We combine these elements into a vector $\rho = (\rho_{gg}, \rho_{ee}, \rho_{ge}, \rho_{eg})$, where $\rho_{gg} = \left(\langle 0, g_0|\rho|0, g_0\rangle, \langle 1, g_1|\rho|1, g_1\rangle\right)$, etc. Organizing the resulting evolution equation in powers of the coupling A, we obtain in the Schrödinger picture:

$$\frac{d}{dt}\rho = (\mathbf{\Lambda_0} + \frac{A}{\Omega}\mathbf{\Lambda_1} + \frac{A^2}{\Omega^2}\mathbf{\Lambda_2} + ...)\rho \tag{63}$$

We discuss the significance of the matrices $\mathbf{\Lambda_j}$ in what follows.

The 8×8 matrix $\mathbf{\Lambda_0}$ describes the evolution of the system at zero coupling:

$$\mathbf{\Lambda_0} = \begin{pmatrix} M & 0 & 0 & 0 \\ 0 & M & 0 & 0 \\ 0 & 0 & +i\Omega + M' & 0 \\ 0 & 0 & 0 & -i\Omega + M' \end{pmatrix}, \tag{64}$$

with the 2×2 matrices M and M' being defined by:

$$M = \begin{pmatrix} -\Gamma_{10} & \Gamma_{01} \\ \Gamma_{10} & -\Gamma_{01} \end{pmatrix} \qquad M' = \frac{1}{2} \begin{pmatrix} -(\Gamma_{10}^{+} + \Gamma_{10}^{-}) & \Gamma_{01}^{+} + \Gamma_{01}^{-} \\ \Gamma_{10}^{+} + \Gamma_{10}^{-} & -(\Gamma_{01}^{+} + \Gamma_{01}^{-}) \end{pmatrix} \tag{65}$$

At zero coupling there are no transitions between different qubit states, and hence $\mathbf{\Lambda_0}$ has a block-diagonal form. There are two independent stationary solutions of Eq. (63) at $A = 0$ (i.e. two zero eigenvectors of $\mathbf{\Lambda_0}$), which correspond to being either in the qubit ground or qubit excited state:

$$\mathbf{z}_g = (p_0, p_1, 0, 0, 0, 0, 0, 0)\,, \qquad \mathbf{z}_e = (0, 0, p_0, p_1, 0, 0, 0, 0)\,. \tag{66}$$

(p_0, p_1) are the stationary probabilities of being in the $n = 0$ or $n = 1$ charge states:

$$(p_0, p_1) = \left(\frac{\Gamma_{01}}{\Gamma_{01} + \Gamma_{10}}, \frac{\Gamma_{10}}{\Gamma_{01} + \Gamma_{10}} \right) \tag{67}$$

The existence of two zero-modes is directly related to the fact that at zero coupling ($A = 0$), the probabilities to be in the qubit ground and excited state are individually conserved.

At non-zero coupling, the matrices $\mathbf{\Lambda_1}$ and $\mathbf{\Lambda_2}$ appearing in Eq. (63) generate transitions between different qubit states. The matrix $\mathbf{\Lambda_2}$ directly couples ρ_{gg} and ρ_{ee}, while $\mathbf{\Lambda_1}$ couples ρ_{gg} and ρ_{ee} to the off-diagonal blocks ρ_{ge} and ρ_{eg}. The effect of these matrices will be to break the degeneracy of the two zero modes of Eq. (63) existing at $A = 0$. After this degeneracy is broken, there will still be one zero mode ρ_0, describing the stationary state of the *coupled* system (the existence of a stationary solution is guaranteed by the conservation of probability). For weak coupling, the qubit density matrix obtained from ρ_0 will be diagonal in the basis $\{|g_{\langle n \rangle}\rangle, |e_{\langle n \rangle}\rangle\}$, which corresponds to the average SET charge $\langle n \rangle = p_1$. The ratio of the occupancies of these two qubit states will yield the ratio between the relaxation rate $\Gamma_\downarrow$ and the excitation rate $\Gamma_\uparrow$. In addition, there will also be a slow, time-dependent mode of Eq. (63) arising from breaking the degeneracy of the two $A = 0$ zero modes. This time-dependent mode will describe how a linear combination of z_g and z_e relaxes to the true stationary state, and will have an eigenvalue $\lambda = -\Gamma_\uparrow - \Gamma_\downarrow$, i.e. the mixing rate.

Thus, we need to do degenerate second order perturbation theory in the coupling A to obtain the relaxation and excitation rates $\Gamma_\downarrow$ and $\Gamma_\uparrow$. The only subtlety here is that the matrix M is not Hermitian, implying that it has distinct right and left eigenvectors. Letting $\tilde{\mathbf{z}}$ represent the left eigenvector of $\mathbf{\Lambda_0}$ corresponding to the right eigenvector $\mathbf{z}$, we define the projector matrix $\mathbf{P}$ as:

$$\mathbf{P} = |z_g\rangle\langle\tilde{z}_g| + |z_e\rangle\langle\tilde{z}_e|, \tag{68}$$

and let $\mathbf{P}_\perp$ denote $1 - \mathbf{P}$. As usual, degenerate second order perturbation theory requires diagonalizing the perturbation in the space of the degenerate eigenvectors.

We are thus led to look at the matrix Q, defined as:

$$Q = \frac{A^2}{\Omega^2}\left(\mathbf{P}\mathbf{\Lambda_2}\mathbf{P} + \mathbf{P}\mathbf{\Lambda_1}\mathbf{P}_\perp\left[-\mathbf{\Lambda_0}\right]^{-1}\mathbf{P}_\perp\mathbf{\Lambda_1}\mathbf{P}\right) \tag{69}$$

From the definition of Q, we may immediately identify the rates $\Gamma_\uparrow$ and $\Gamma_\downarrow$:

$$\Gamma_\uparrow = \langle \tilde{z}_e|Q|z_g\rangle \qquad \Gamma_\downarrow = \langle \tilde{z}_g|Q|z_e\rangle \tag{70}$$

We thus see how the rates $\Gamma_{\uparrow,\downarrow}$ arise in the present approach– they are related to breaking the degeneracy between two zero-modes (stationary solutions) which exist at zero coupling. Note that there are two distinct contributions to $\Gamma_{\uparrow,\downarrow}$, coming from the two terms in the matrix Q: a "direct" contribution involving $\mathbf{\Lambda_2}$ and an "interference" contribution involving $\mathbf{\Lambda_1}$ acting twice. These two terms have a different physical interpretation, as will become clear.

Let us first consider the rate $\Gamma_\uparrow$, which describes how noise in the SET causes ground to excited state transitions in the qubit. For this rate, our approximation of only keeping two charge states will be valid for all splitting frequencies Ω. To evaluate the "direct" contribution to this rate, which involves the first term in the matrix Q, note that the relevant part of $\mathbf{\Lambda_2}$ has the expected form:

$$\mathbf{\Lambda_2}|_{ee,gg} = \begin{pmatrix} 0 & \Gamma_{01}^- \\ \Gamma_{10}^- & 0 \end{pmatrix} \tag{71}$$

i.e. it consists of tunnel rates which correspond to having given up an energy Ω to the qubit. Using Eqs. (69) and (70), we find:

$$\Gamma_\uparrow|_{direct} = \left(\frac{A}{\Omega}\right)^2\left(p_0\Gamma_{10}^- + p_1\Gamma_{01}^-\right) = \left(\frac{A}{\Omega}\right)^2\left(\frac{\Gamma_{01}\Gamma_{10}^- + \Gamma_{10}\Gamma_{01}^-}{\Gamma_{10}+\Gamma_{01}}\right) \tag{72}$$

The direct contribution to $\Gamma_\uparrow$ has a very simple form: for each charge state $n = 0, 1$, add the rate to tunnel out of n while exciting the qubit, weighted by both the probability to be in state n, and the overlap between ground and excited states (i.e. $(A/\Omega)^2$). This is very similar to how one typically calculates the current for a SET: one adds up the current associated with each charge state (i.e. a difference of rates), weighted by the occupancy of the state. The direct contribution to $\Gamma_\uparrow$ neglects any possible coherence between successive excitation events; as a result, it fails to recover the classical expression of Eq. (54) in the small-Ω limit.

We now consider the "interference" contribution to $\Gamma_\uparrow$ coming from the second term in the expression for matrix Q (c.f. Eq. (69)). After some algebra, we obtain the following for the interference contribution to $\Gamma_\uparrow$:

$$\Gamma_\uparrow|_{int} = -\frac{2A^2}{\Omega^2}\left(p_0\Gamma_{10}^- + p_1\Gamma_{01}^-\right)\frac{(\Gamma_\Sigma)^2}{\Omega^2+(\Gamma_\Sigma)^2} \tag{73}$$

where:

$$\Gamma_\Sigma \equiv \frac{\Gamma_{10}^- + \Gamma_{10}^+ + \Gamma_{01}^- + \Gamma_{01}^+}{2} \tag{74}$$

This contribution is purely negative, and is only significant (relative to the direct contribution) at low frequencies $\Omega < \Gamma$. We can interpret this equation as describing the interference between *two* consecutive excitation events. For example, consider the first term in Eq. (73). This describes a process where a SET initially in the charge state $n = 0$ undergoes a tunnel event to the $n = 1$ state, creating a superposition of qubit ground and excited states. At some later time the SET relaxes to the stationary distribution (p_0, p_1) of the charge states, again partially exciting the qubit. Letting Δt represent the time between these two events, we have the approximate sequence:

$$\begin{aligned} |0, g_0\rangle \; &\rightarrow \Gamma_{10}^- \;\; |1, g_1\rangle + \alpha|1, e_1\rangle \\ &\rightarrow \Delta t \;\; e^{i\Omega\Delta t/2}|1, g_1\rangle + e^{-i\Omega\Delta t/2}\alpha|1, e_1\rangle \qquad (75) \\ &\rightarrow \Gamma_\Sigma \;\; \left(e^{i\Omega\Delta t/2}\beta - e^{-i\Omega\Delta t/2}\alpha\right)|0, e_0\rangle + \ldots \qquad (76) \end{aligned}$$

Here, α is the amplitude associated with qubit excitation having occurred during the first ($n = 0 \rightarrow 1$) tunnel event, while β is the amplitude associated with excitation occurring during the second ($n = 1 \rightarrow 0$) tunneling. These amplitudes will be given by the corresponding matrix overlap elements:

$$\alpha = \langle e_1|g_0\rangle \simeq \frac{A}{\Omega} \qquad \beta = \langle e_0|g_1\rangle \simeq -\frac{A}{\Omega} \tag{77}$$

In the final state after the two tunnelings (Eq. (76)), there are two terms in the amplitude of the state $|0, e_0\rangle$, corresponding to the fact that qubit excitation could have occurred in either the first or the second tunnel event. To get a rate for this double excitation event, we should take the modulus squared of the final $|0, e_0\rangle$ state amplitude, then multiply by the occupancy of the initial state (p_0) and the rate of the first tunnel event (Γ_{10}). The interference term in the resulting expression takes the form:

$$\Gamma_\uparrow|_{int} = (p_0\Gamma_{10}) \times 2\mathrm{Re}\left(\alpha^*\beta e^{i\Omega\Delta t}\right) = -(p_0\Gamma_{10})\frac{2A^2}{\Omega^2}\cos(\Omega\Delta t) \tag{78}$$

The above expression is a function of the time Δt between the first and second tunnel events. This time is determined by the fact that the intermediate superposition state (Eq. (75)) corresponds to a non-stationary distribution of charge on the SET island, and will decay via tunneling to the stationary distribution (p_0, p_1) at a rate Γ_Σ. Taking this decay to be Poissonian, and averaging over Δt, we obtain:

$$\Gamma_\uparrow|_{int} = -(p_0\Gamma_{10})\frac{2A^2}{\Omega^2}\frac{(\Gamma_\Sigma)^2}{\Omega^2 + (\Gamma_\Sigma)^2} \tag{79}$$

This is precisely the first term in Eq. (73); the second term can be obtained in the same way, by now considering a situation where the SET is initially in the $n = 1$ charge state. As claimed, $\Gamma_\uparrow|_{int}$ corresponds to the interference between two consecutive excitation events. The negative sign of this contribution can be directly traced to the matrix overlap elements (c.f. Eq. 77). Also, we see that the suppression of the interference term at large Ω results from phase randomization occurring during the delay time between the two excitation events.

Returning to the total noise, we combine Eq. (73) with the direct contribution Eq. (72) to $\Gamma_\uparrow$; comparing against Eq. (57), we obtain the final expression for $S_Q(\Omega)$ at all *negative* frequencies:

$$S_Q(-|\Omega|) = \frac{p_0\Gamma_{10}^- + p_1\Gamma_{01}^-}{\Omega^2 + \frac{1}{4}\left(\Gamma_{10}^- + \Gamma_{10}^+ + \Gamma_{01}^- + \Gamma_{01}^+\right)^2} \tag{80}$$

Note for large $|\Omega|$ (i.e. $|\Omega| > \max(\Delta E_{01}^\alpha, \Delta E_{10}^\alpha) \simeq V_{DS}/2$), $S_Q(-\Omega)$ will vanish identically at zero temperature. Physically, this cutoff corresponds to the largest amount of energy the SET can give up to the qubit during a single tunnel event; giving up more energy would suppress the event completely (i.e. the tunnel rates have a step-function form at zero temperature, c.f. Eq. (50)). If one included higher order processes in the tunneling (i.e. went beyond sequential tunneling), correlated tunneling events involving the full voltage drop over both junctions, V_{DS}, would move this cutoff to higher values of absolute frequency.

We now turn to the relaxation rate $\Gamma_\downarrow$, and hence the positive frequency parts of S_Q. The calculation proceeds exactly as that for $\Gamma_\uparrow$, the only modification being that one now needs to include the charge states $n = 2$ and $n = -1$, as the SET could absorb enough energy from the qubit to make transitions to these states possible. We can combine the result for $\Gamma_\downarrow$ with Eq. (80) to obtain a single, compact expression for the noise at all frequencies first obtained by Johansson *et. al* [14]: [13]

$$S_Q(\omega) = \frac{p_0\left[\Gamma_{10}(\omega) + \Gamma_{-1,0}(\omega)\right] + p_1\left[\Gamma_{01}(\omega) + \Gamma_{21}(\omega)\right]}{\omega^2 + \frac{1}{4}\left[\Gamma_{10}(\omega) + \Gamma_{10}(-\omega) + \Gamma_{01}(\omega) + \Gamma_{01}(-\omega)\right]^2} \tag{81}$$

Shown in Figure 5 is the symmetrized noise $S_Q(\omega) + S_Q(-\omega)$ for typical SET parameters. One can clearly see abrupt changes in the slope of this curve; each of these kinks corresponds to a threshold frequency at which a given tunneling process either turns on or turns off. For comparison, curves corresponding to classical telegraph noise and to the uncorrelated noise of two tunnel junctions are also shown. At low frequencies the symmetrized true noise matches the classical

[13] Eq. (81) ignores additional order $g/(2\pi)$ terms which arise in the denominator at positive frequencies large enough to turn on tunneling to higher charge states; such terms are clearly negligible in the sequential tunneling regime due to the smallness of $g/(2\pi)$.

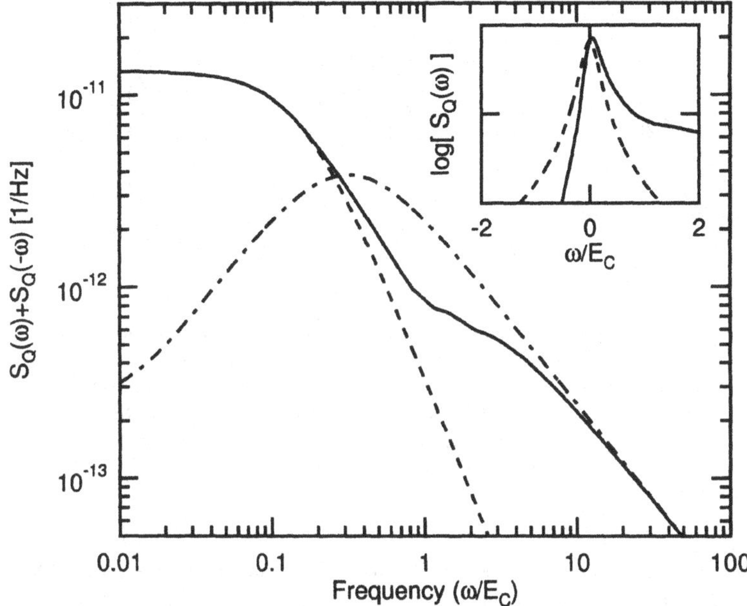

Figure 5. Symmetrized SET charge noise as a function of frequency, for typical SET parameters: $g = 1$, $E_C/k_B = 2\text{K}$, $\mathcal{N} = 0.33$, $eV_{DS} = E_C$, and $T = 20\text{m}K$. The dashed line is the classical telegraph noise (Eq. (54)), while the dot-dashed line is the noise of two parallel tunnel junctions (Eq. (55)). Inset: full (non-symmetrized) quantum noise for identical SET parameters; the dashed line is the symmetric classical telegraph noise.

curve; for higher frequencies, it lies *above* the classical curve but *below* the curve corresponding to the uncorrelated case. The inset of this figure shows the both the negative and positive frequency parts of $S_Q(\omega)$.

It is easy to check that in the limit $\omega \to 0$, Eq. (81) recovers the classical telegraph expression of Eq. (54). In the high-frequency, zero temperature limit, one can also see that Eq. (81) approaches the uncorrelated result of Eq. (55) from *below*:

$$S_Q(\omega) \to \Theta(\omega) \frac{4\left(\frac{g}{2\pi}\right)\omega\left(1 - \frac{E_C}{2\hbar\omega}\right)}{\omega^2 + \frac{g^2}{\pi^2}\omega^2} \to \Theta(\omega)\frac{2g}{\pi\omega} \tag{82}$$

Note that at high frequencies, it is only the "direct" terms which contribute to the noise– the interference contribution is not important in the limit of uncorrelated tunneling. The fact that the noise approaches the high frequency limit from below results from the tendency of charging energy induced correlations (which are present for a finite ω/E_C) to suppress fluctuations of n, and thus suppress the noise. Note that the interpolation between the low and high frequency limits here

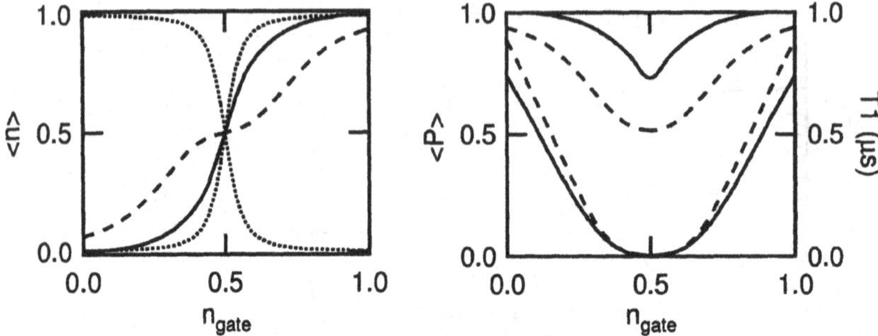

Figure 6. a) Average charge state of a Cooper pair box coupled to a SET, as a function of box gate voltage, using identical SET parameters as above. The box parameters are $E_C/k_B = 0.5$K, $E_J/k_B = 0.25$K and the coupling constant is $\kappa = 0.04$. We also include relaxation effects due to a 10% coupling to a 50Ω environment. The dashed curve corresponds to assuming the SET produces classical telegraph noise, the solid curve corresponds to using the full quantum noise of the SET, and the dashed-dot curve is the box ground state. b) The relaxation time T_1 for the same system, as a function of box gate voltage.

is very different than, e.g., interpolating between thermal noise and zero-point fluctuation noise in a tunnel junction. In the latter case, one is effectively *combining* two sources of noise; here, one is simply turning off correlations brought on by the charging energy by increasing ω.

Finally, shown in Figure 6 is the average charge state of a Cooper pair box coupled to a SET with identical parameters to that in Fig. 5. We have also included here the relaxation effects of the environment, modelled as in Sec. 6 as a 50Ω impedance. Note that even near the box degeneracy point, there are large deviations between the result obtained using the full quantum noise of the SET and that obtained from using only classical telegraph noise. In Fig. 6b, we show the relaxation rate T_1 for the same system. Note that the differences between using the full quantum noise and the classical expression are not so evident here.

9. Summary

In this article, we have emphasized the need to discuss quantum noise processes using their two-sided spectral densities. Because of the quantum nature of noise, the positive and negative frequencies are generally unequal, in order to account for spontaneous emission. A two-level system was shown to be an ideal spectrum analyzer for probing the quantum nature of a noise process or reservoir. With the advent of real electrical circuits which behave as coherent two-level systems (e.g., [5],[7]), we can now build and use *quantum electrical* spectrum analyzers. We also described the use of a qubit as a *theoretical* tool, by following the evolution of the density matrix of a TLS coupled to the noise-producing system of interest. This technique appears to be quite powerful, as it can yield analytical results for the full quantum noise spectrum of a wide variety of devices, including the superconducting SET [17]. The distinction between the classical noise and the quantum noise, found in this way, leads to dramatically different predictions (c.f. Fig. 6 and Ref. [17]) for continuous measurements of qubits with an SET. The "coupled-system" calculational approach also allows predictions of the dephasing by the measurement, the performance relative to the Heisenberg uncertainty limit [13], the fidelity of single-shot measurements of the qubit states, and the effects of strong coupling to the qubit. The combined theoretical and experimental advances raise many interesting possibilities for testing our understanding of quantum measurement theory with mesoscopic devices.

Acknowledgements

The authors acknowledge the generous support of this work by the NSA and ARDA under ARO contract ARO-43387-PH-QC (RS,SG), by the NSF under DMR-0196503 & DMR-0084501 (AC,SG), the David and Lucile Packard Foundation (RS), and the W.M. Keck Foundation.

References

1. A.O. Caldeira and A.J. Leggett, Ann. Phys. **149**, 374 (1983).
2. H.B. Callen and T.A. Welton, Phys. Rev. **83**, 34 (1951).
3. M.H. Devoret, in *Quantum Fluctuations*, S. Reynaud, E. Giacobino and J. Zinn-Justin, eds. (Elsevier, Amsterdam, 1995).
4. V. Bouchiat, D. Vion, P. Joyez, D. Esteve and M. Devoret, Physica Scripta **T76**, 165 (1998).
5. D. Vion, A. Aassime, A. Cottet, P. Joyez, H. Pothier, C. Urbina, D. Esteve, and M.H. Devoret, Science **296**, 886 (2002).
6. J.M. Martinis *et al.*, submitted to Phys. Rev. B, 2002.
7. K.W. Lehnert, K. Bladh, L.F. Spietz, D. Gunnarson, D.I. Schuster, P. Delsing, and R.J. Schoelkopf, submitted to Phys. Rev. Lett., 2002.
8. V.B. Braginsky and F. Ya. Khalili, "Quantum Measurement," (Cambridge University Press, New York, 1992).)
9. A.J. Dahm *et al.*, Phys. Rev. Lett. **22**, 1416 (1969); D. Rogovin and D.J. Scalapino, Ann. Phys. **86**, 1 (1974).
10. R.J. Schoelkopf, P.J. Burke, A.A. Kozhevnikov, D.E. Prober, and M.J. Rooks, Phys. Rev. Lett. **78**, 3370 (1997).
11. R. Aguado and L.P. Kouwenhoven, Phys. Rev. Lett. **84**, 1986 (2000).
12. L. S. Levitov, H. Lee and G. B. Lesovik, J. Math. Phys. 37, 4845 (1996).
13. M.H. Devoret and R.J. Schoelkopf, Nature **406**, 1039 (2000).
14. G. Johansson, Andreas Käck, and Göran Wendin, Phys. Rev. Lett. **88**, 046802 (2002).
15. H. Schoeller and G. Schön, Phys. Rev. **B**, 18436 (1994).
16. Y. Makhlin, B. Shnirman and G. Schön, Phys. Rev. Lett. **85**, 4578 (2000); *ibid.*, Rev. Mod. Phys. **73**, 357 (2001).
17. A. A. Clerk, S. M. Girvin, A. K. Nguyen and A. D. Stone, Phys. Rev. Lett., in press.

NOISY QUANTUM MEASUREMENT OF SOLID-STATE QUBITS: BAYESIAN APPROACH

A. N. KOROTKOV
Dept. of Electrical Engineering, University of California, Riverside, CA 92521, USA

Abstract. We discuss a recently developed formalism which describes the quantum evolution of a solid-state qubit due to its continuous measurement. In contrast to the conventional ensemble-averaged formalism, it takes into account the measurement record and therefore is able to consider individual realizations of the measurement process. The formalism provides testable experimental predictions and can be used for the analysis of a quantum feedback control of solid-state qubits. We also discuss generalization of the Bayesian formalism to the continuous measurement of entangled qubits.

1. Introduction

Bayesian approach to the problem of continuous quantum measurement is a relatively new subject in solid-state mesoscopics, even though this approach has a long history [1, 2] as a general quantum framework and is rather well developed, for example, in quantum optics [3] (for more references, see Ref. [4]) The main problem considered in this paper is a very simple question: *what is the evolution of a quantum two-level system (qubit) during the process of its measurement by a solid-state detector* (Fig. 1)? In spite of the question simplicity, the answer is not that trivial.

The textbook "orthodox" quantum mechanics [5] says that the measurement should instantly collapse the qubit state, so that after the measurement the qubit state is either $|1\rangle$ or $|2\rangle$, depending on the measurement outcome. [The measurement basis is obviously defined by the detector; in particular, it is a charge basis for the examples of Figs. 1(a) and 1(b).] Such answer is sufficient for typical optical experiments when the measurement is instantaneous (a scintillator flash or a photocounter click). However, for typical solid-state setups (as well as for some more advanced setups in quantum optics [3]) the instantaneous collapse is not a sufficient answer. In particular, in the examples of Fig. 1 typically the detector is weakly coupled to the qubit, so the measurement process can take a significant

Y. V. Nazarov (ed.), Quantum Noise in Mesoscopic Physics, 205–228.

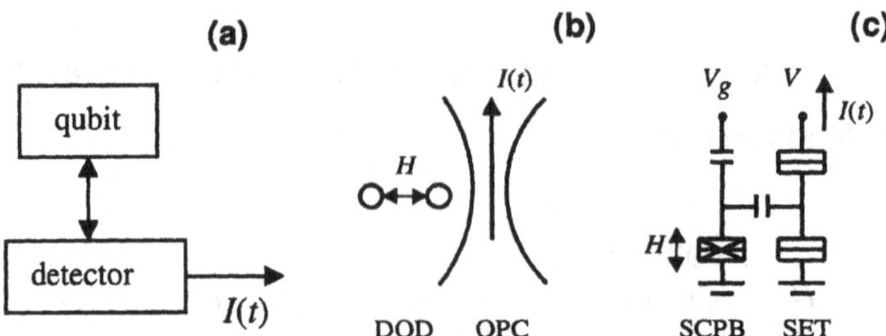

Figure 1. (a) General schematic of a continuously measured solid-state qubit and two particular realizations of the setup: (b) a qubit made of double quantum dot (DQD) measured by a quantum point contact (QPC) and (c) a qubit based on single-Cooper-pair box (SCPB) measured by a single-electron transistor (SET).

time and therefore the collapse should be considered as a continuous process. The notion of a continuous evolution due to measurement is well accepted in the solid-state community and is usually considered within the framework of the Leggett's formalism [6, 7]. This formalism gives the decoherence-based answer to the question posed above. It says that the nondiagonal matrix elements of the qubit density matrix (obtained by tracing over the detector degrees of freedom) gradually decay to zero, while the diagonal matrix elements do not evolve (assuming that the qubit does not oscillate by itself, $H = 0$, where H describes the tunneling between $|1\rangle$ and $|2\rangle$). So, after the completed measurement we have an incoherent mixture of the states $|1\rangle$ and $|2\rangle$.

Let us notice that these two answers to our question obviously *contradict* each other and the "orthodox" answer cannot be obtained as some limiting case of the decoherence answer (since decoherence does not lead to localization into one definite state). The resolution of the apparent contradiction is simple: two approaches consider different objects. The decoherence approach describes the *average evolution of the ensemble* of qubits, while the "orthodox" quantum mechanics is designed to treat a *single quantum system*. This difference also explains the inability of the decoherence formalism to take the measurement outcome into account.

Obviously, it is desirable to have a formalism which would combine advantages of the two approaches and describe the *continuous measurement of a single qubit*. Then the "orthodox" result would be a limiting case for very fast (and "strong") measurement, while the decoherence result could be obtained by an ensemble averaging. The Bayesian formalism [4, 8] which is the subject of this paper has been developed exactly for that purpose (some extensions of the Bayesian formalism will be discussed later). Notice that the formalism has been also reproduced in a somewhat different language (using the terminology of quantum trajectories, quantum jumps, and quantum state diffusion) by another group [9]. It is important to stress that the Bayesian approach is not a phenomenological formalism which just correctly describes two previously known cases. It claims the description of a real and experimentally verifiable evolution of a single qubit in a process of measurement.

Simply speaking, the Bayesian formalism gives the following answer to the question posed above (for $H = 0$). During the measurement process the diagonal matrix elements of the qubit density matrix evolve according to the classical Bayes formula [10, 11] which takes into account the noisy detector output [$I(t)$ in Fig. 1] and describes a gradual qubit localization into one of the states $|1\rangle$ or $|2\rangle$, depending on $I(t)$. The evolution of nondiagonal matrix elements can be easily calculated using somewhat surprising result that a good (ideal) detector preserves the purity of the qubit state, so that the decoherence is actually just a consequence of averaging over different detector outcomes $I(t)$ for different members of the ensemble. (Nonideal detectors also produce some amount of qubit decoherence, which is calculated within the formalism.)

Notice that in the case of an ideal detector, our result can actually be considered as a simple consequence of the so-called Quantum Bayes Theorem (we borrow this name from the book on quantum noise by Gardiner [12], even though it is not a theorem in a mathematical sense). However, the application of this "theorem" is not always straightforward, so instead of applying it as an ansatz, we derive the Bayesian formalism for particular measurement setups, starting from the textbook quantum mechanics.

It is difficult to avoid philosophical questions discussing a problem related to quantum measurements. In brief, philosophy of the Bayesian approach is exactly the philosophy of the "orthodox" quantum mechanics. A minor technical difference is that instead of assuming instantaneous information on measurement result corresponding to instantaneous "orthodox" collapse, we consider a more realistic case of continuous information flow.

Finally, let us mention that the problem of solid-state qubit evolution due to continuous measurement was recently a subject of theoretical study by many groups (see, e.g. [13–18]). However, most of them assumed ensemble averaging and so obtained results different from the Bayesian results (except the Australian group [9, 19, 20] which also studies single realizations of the measurement process).

2. Simple model

Even though Bayesian approach is applicable to a broad range of measurement setups, let us start with a particularly simple setup [Fig. 1(b)] consisting of a double quantum dot occupied by a single electron, the position of which is measured by a low-transparency Quantum Point Contact (QPC) or (which is almost the same) by just a tunnel junction [Fig. 2(a)]. Basically following the model of Ref. [13] we assume that the detector barrier height depends on the location of the electron

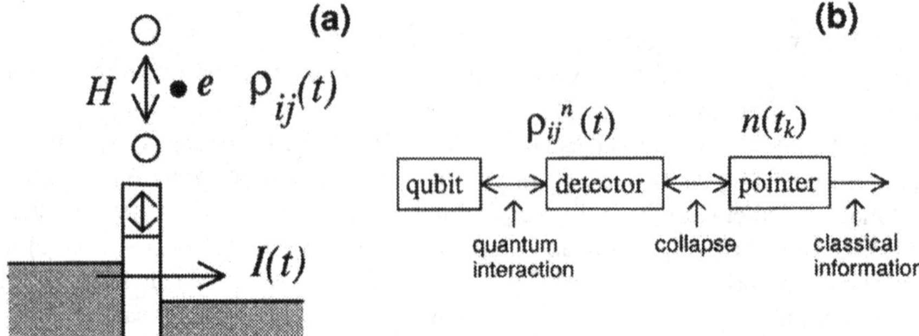

Figure 2. (a) Tunnel junction as a detector of the double-quantum-dot qubit. The electron location in the DQD affects the detector barrier height. The noisy current $I(t)$ (detector output) reflects the evolution of the qubit density matrix $\rho_{ij}(t)$. (b) Idea of the Bayesian formalism derivation via Bloch equations. The number n of electrons passed through the detector is periodocally collapsed (forced to choose a definite value) at moments t_k.

in either dot 1 or 2; then the current through the tunnel junction (which is the detector output) is sensitive to the electron location.

The Hamiltonian of the system,

$$\mathcal{H} = \mathcal{H}_{QB} + \mathcal{H}_{DET} + \mathcal{H}_{INT}, \tag{1}$$

consists of terms describing the double-dot qubit, the detector, and their interaction. The qubit Hamiltonian,

$$\mathcal{H}_{QB} = \frac{\varepsilon}{2}\,(c_2^\dagger c_2 - c_1^\dagger c_1) + H\,(c_1^\dagger c_2 + c_2^\dagger c_1), \tag{2}$$

is characterized by the energy asymmetry ε between two dots and the tunneling strength H (we assume real H without loss of generality). The detector and interaction Hamiltonians can be written as

$$\mathcal{H}_{DET} = \sum_l E_l a_l^\dagger a_l + \sum_r E_r a_r^\dagger a_r + \sum_{l,r} T(a_r^\dagger a_l + a_l^\dagger a_r),$$
$$\mathcal{H}_{INT} = \sum_{l,r} \frac{\Delta T}{2}\,(c_1^\dagger c_1 - c_2^\dagger c_2)(a_r^\dagger a_l + a_l^\dagger a_r), \tag{3}$$

where both T and ΔT are assumed real and their dependence on the states in electrodes (l, r) is neglected. For simplicity we assume zero temperature (Bayesian formalism at finite temperatures has been considered in Refs. [4, 9, 21, 22]). If the electron occupies dot 1, then the average current through the detector is $I_1 = 2\pi(T + \Delta T/2)^2 \rho_l \rho_r e^2 V/\hbar$ (V is the voltage across the tunnel junction and $\rho_{l,r}$ are the densities of states in the electrodes), while if the measured electron is in the dot 2, the average current is $I_2 = 2\pi(T - \Delta T/2)^2 \rho_l \rho_r e^2 V/\hbar$.

The difference between the currents,

$$\Delta I \equiv I_1 - I_2, \tag{4}$$

determines the detector response to the electron position. Because of the detector shot noise, the two states cannot be distinguished instantaneously and the signal-to-noise ratio (S/N) gradually improves with the increase of the measurement duration. The S/N becomes close to unity after the "measurement" time

$$\tau_m = \frac{(\sqrt{S_1} + \sqrt{S_2})^2}{2(\Delta I)^2}, \tag{5}$$

where the spectral densities S_1 and S_2 of the detector shot noise for states $|1\rangle$ and $|2\rangle$ are given by the Schottky formula,

$$S_{1,2} = 2eI_{1,2}. \tag{6}$$

[Actually, Eq. (5) exactly corresponds to S/N=1 for $S_1 = S_2$, while for S/N$\neq$1 it gives the proper asymptotic scaling at $t \gg \tau_m$.] To avoid an explicit account of the detector quantum noise we will consider only processes at frequencies $\omega \ll eV/\hbar$; in particular, we assume $\max(\hbar\tau_m^{-1}, |\varepsilon|, |H|) \ll eV$.

Notice that due to electron charge discreteness and stochastic nature of tunneling, the total number $n(t)$ of electrons passed through the detector is sometimes a more convenient magnitude to work with than the current $I(t) = e\, dn(t)/dt$. In particular, we will use $n(t)$ instead of $I(t)$ for the Bayesian formalism derivation in the next section.

3. Derivation of the Bayesian formalism via "Bloch" equations

The "conventional" ensemble-averaged equations for the qubit density matrix $\rho_{ij}(t)$,

$$\dot{\rho}_{11} = -\dot{\rho}_{22} = -2\,\frac{H}{\hbar}\,\mathrm{Im}\,\rho_{12}, \tag{7}$$

$$\dot{\rho}_{12} = \mathrm{i}\,\frac{\varepsilon}{\hbar}\,\rho_{12} + \mathrm{i}\,\frac{H}{\hbar}\,(\rho_{11} - \rho_{22}) - \Gamma_d\,\rho_{12}, \tag{8}$$

do not take into account any information about the detector outcome and describe the effect of continuous measurement by the ensemble decoherence rate [13]

$$\Gamma_d = \frac{(\sqrt{I_1} - \sqrt{I_2})^2}{2e}. \tag{9}$$

(Notice a relation $\Gamma_d\tau_m = 1/2$; as will be seen later, this means that the detector is ideal).

Equations (7)–(8) imply tracing over all detector degrees of freedom, including the measurement outcome. An important step towards taking into account the measurement record was a derivation [13] of "Bloch" equations for the density matrix $\rho_{ij}^n(t)$ which is divided into components with different number n of electrons passed through the detector:

$$\dot{\rho}_{11}^n = -\frac{I_1}{e}\,\rho_{11}^n + \frac{I_1}{e}\,\rho_{11}^{n-1} - 2\,\frac{H}{\hbar}\,\mathrm{Im}\,\rho_{12}^n\,, \tag{10}$$

$$\dot{\rho}_{22}^n = -\frac{I_2}{e}\,\rho_{22}^n + \frac{I_2}{e}\,\rho_{22}^{n-1} + 2\,\frac{H}{\hbar}\,\mathrm{Im}\,\rho_{12}^n\,, \tag{11}$$

$$\dot{\rho}_{12}^n = \mathrm{i}\,\frac{\varepsilon}{\hbar}\,\rho_{12}^n + \mathrm{i}\,\frac{H}{\hbar}\,(\rho_{11}^n - \rho_{22}^n) - \frac{I_1+I_2}{2e}\,\rho_{12}^n + \frac{\sqrt{I_1 I_2}}{e}\,\rho_{12}^{n-1}\,. \tag{12}$$

Eqs. (7)–(9) can be obtained from the Bloch equations using summation over n and relation $\rho_{ij} = \sum_n \rho_{ij}^n$. (Absence of nondiagonal in n matrix elements ρ_{ij}^{nm} is related to the assumption of large detector voltage [13].)

Despite the Bloch equations carry the total number n of electrons passed through the detector, they cannot take into account the whole measurement record $n(t)$ for a particular realization of measurement process. We should make a simple but important step for that: *we should introduce a sufficiently frequent collapse of $n(t)$ corresponding to a particular realization of the measurement record* [4]. This idea is illustrated in Fig. 2(b). Including "detector" into the quantum part of the setup, we anyway have to deal with a classical information, so we introduce a classical "pointer" which periodically (at times t_k) forces the system "qubit+detector" to choose a definite value for $n(t_k)$. An obvious drawback of such construction is that it is absolutely not clear what should be a sequence of t_k (in other words, how strongly the detector and pointer should be coupled). In general, the frequency of this collapse can depend on the physical parameters of interaction between the measurement stage included in the "detector" Hamiltonian and the next stage. The only obvious fact is that if in an experiment we can *record* $n(t)$ with some frequency, then the collapse should take place at least not less frequent. Still it is unclear how much more frequent. Fortunately, for a model described in the previous section the results do not depend on the choice of t_k if $\Delta t_k \equiv t_k - t_{k-1}$ are sufficiently small, so the natural choice is $\Delta t_k \to 0$.

Technically the procedure is the following. During the time between t_{k-1} and t_k the "qubit+detector" evolves according to the Bloch equations (10)–(12), while at time t_k the number n is collapsed in the "orthodox" way [5]. This means that the probability $P(n)$ of getting some $n(t_k)$ is equal to

$$P(n) = \rho_{11}^n(t_k) + \rho_{22}^n(t_k), \tag{13}$$

while after a particular n_k is picked, the density matrix ρ_{ij}^n is immediately updated (collapsed):

$$\rho_{ij}^n(t_k+0) = \delta_{n,n_k}\,\rho_{ij}(t_k+0), \tag{14}$$

$$\rho_{ij}(t_k+0) = \frac{\rho_{ij}^{n_k}(t_k-0)}{\rho_{11}^{n_k}(t_k-0)+\rho_{22}^{n_k}(t_k-0)}, \tag{15}$$

where δ_{n,n_k} is the Kronecker symbol. After that the evolution is again described by Eqs. (10)–(12) with n shifted by n_k until the next collapse occurs at $t = t_{k+1}$, and so on. *This procedure is the main step* in the derivation of the Bayesian formalism.

Let us discuss now the relation of this procedure to the classical Bayes theorem [10, 11] which says that a posteriori probability $P(B_i|A)$ of a hypothesis B_i after learning an information A (B_i form a complete set of mutually exclusive hypotheses) is equal to

$$P(B_i|A) = \frac{P(B_i)P(A|B_i)}{\sum_k P(B_k)P(A|B_k)} \tag{16}$$

where $P(B_i)$ is a priori probability of the hypothesis B_i (before learning information A) and $P(A|B_i)$ is the conditional probability of the event A under hypothesis B_i.

Assuming for a moment $H = 0$ and $\varepsilon = 0$ in the qubit Hamiltonian (so that the qubit evolution is due to measurement only), it is easy to find [4] that Eqs. (10)–(12) and our procedure (14)–(15) lead to the qubit evolution as

$$\rho_{11}(t_k) = \frac{\rho_{11}(t_{k-1})P_1(\Delta n_k)}{\rho_{11}(t_{k-1})P_1(\Delta n_k)+\rho_{22}(t_{k-1})P_2(\Delta n_k)}, \tag{17}$$

$$\rho_{22}(t_k) = \frac{\rho_{22}(t_{k-1})P_2(\Delta n_k)}{\rho_{11}(t_{k-1})P_1(\Delta n_k)+\rho_{22}(t_{k-1})P_2(\Delta n_k)}, \tag{18}$$

$$\rho_{12}(t_k) = \rho_{12}(t_{k-1})\frac{[\rho_{11}(t_k)\rho_{22}(t_k)]^{1/2}}{[\rho_{11}(t_{k-1})\rho_{22}(t_{k-1})]^{1/2}}, \tag{19}$$

where $\Delta n_k = n_k - n_{k-1}$ is the number of electrons passed through the detector during time Δt_k and

$$P_i(n) = \frac{(I_i\Delta t_k/e)^n}{n!}\exp(-I_i\Delta t_k/e) \tag{20}$$

is the classical Poisson distribution for this number assuming either qubit state $|1\rangle$ or $|2\rangle$. One can see that the diagonal matrix elements ρ_{ii} *exactly obey the classical Bayes formula* (16), i.e. *as if* the qubit is really either in the state $|1\rangle$ or $|2\rangle$, but not in both simultaneously. Actually, this fact is not much surprising because at least in some sense ρ_{ii} are the probabilities. Equation (19) is a little more surprising and says that the measurement preserves the degree of qubit purity $\rho_{12}/(\rho_{11}\rho_{22})^{1/2}$; for instance, *a pure state remains pure during the whole measurement process.*

After introducing the main procedure (14)–(15), further derivation of the Bayesian formalism is pretty simple and depends on whether we want to consider finite detector response, $|\Delta I| \sim I_0 \equiv (I_1 + I_2)/2$ or a weak response, $|\Delta I| \ll I_0$. In the first case each event of tunneling through the detector carries significant information and significantly affects the qubit state, so a reasonable "experimental" setup implies recording the time of each tunneling event. Then during the time periods when no electrons are passing through the detector, the evolution is essentially described by the Bloch equations (10)–(12) with $n = 0$, while the frequent collapses $[\Delta t_k \ll \min(e/I_1, e/I_2, \hbar/H, \hbar/\varepsilon)]$ just restore the density matrix normalization, leading to the continuous (but not unitary!) qubit evolution [4, 9]:

$$\dot{\rho}_{11} = -\dot{\rho}_{22} = -2\frac{H}{\hbar}\,\mathrm{Im}\,\rho_{12} - \frac{\Delta I}{e}\rho_{11}\rho_{22}, \tag{21}$$

$$\dot{\rho}_{12} = \frac{i\varepsilon}{\hbar}\rho_{12} + \frac{iH}{\hbar}(\rho_{11} - \rho_{22}) + \frac{\Delta I}{2e}(\rho_{11} - \rho_{22})\rho_{12}. \tag{22}$$

However, at moments when one electron passes through the detector, the qubit state changes abruptly (corresponding to $\Delta n_k = 1$ and $\Delta t_k \to 0$ in Eqs. (17)–(19)):

$$\rho_{11}(t+0) = \frac{I_1\rho_{11}(t-0)}{I_1\rho_{11}(t-0) + I_2\rho_{22}(t-0)}, \tag{23}$$

$$\rho_{22}(t+0) = 1 - \rho_{11}(t+0), \tag{24}$$

$$\rho_{12}(t+0) = \rho_{12}(t-0)\left[\frac{\rho_{11}(t+0)\,\rho_{22}(t+0)}{\rho_{11}(t-0)\,\rho_{22}(t-0)}\right]^{1/2}. \tag{25}$$

These abrupt changes are usually called "quantum jumps" [9]. Notice that for $I_1 > I_2$ the jumps always shift the qubit state closer to $|1\rangle$ (because detector tunneling is "more likely" for state $|1\rangle$), while continuous nonunitary evolution shifts the state towards $|2\rangle$. On average the evolution is still given by conventional Eqs. (7)–(9).

The case of a weak detector response, $|\Delta I| \ll I_0$, is more realistic from the experimental point of view. In this case the measurement time τ_m as well as the ensemble decoherence time Γ_d^{-1} are much longer than the average time e/I_0 between electron passages in the detector. If the tunneling in the qubit is also sufficiently slow, $\hbar/H \gg e/I_0$, we can completely disregard individual events in the detector and consider the detector current $I(t)$ as quasicontinuous. Then Eqs. (17)–(18) for the evolution due to measurement only (neglecting H and ε) transform into equations which again have clear Bayesian interpretation:

$$\rho_{11}(t+\tau) = \frac{\rho_{11}(t)P_1(\overline{I})}{\rho_{11}(t)P_1(\overline{I}) + \rho_{22}(t)P_2(\overline{I})}, \tag{26}$$

$$\rho_{22}(t+\tau) = \frac{\rho_{22}(t)P_2(\overline{I})}{\rho_{11}(t)P_1(\overline{I}) + \rho_{22}(t)P_2(\overline{I})}, \tag{27}$$

where

$$\overline{I} \equiv \frac{1}{\tau} \int_{t}^{t+\tau} I(t')\, dt' \tag{28}$$

is the detector current averaged over the time interval $(t, t+\tau)$ and $P_i(\overline{I})$ are its classical Gaussian probability distributions for two qubit states:

$$P_i(\overline{I}) = \frac{1}{(2\pi D)^{1/2}} \exp[-\frac{(\overline{I} - I_i)^2}{2D}], \;\; D = S_0/2\tau, \tag{29}$$

(here $S_0 = 2eI_0$ is the detector noise), while Eq. (19) essentially does not change.

Differentiating Eqs. (26), (27), and (19) over time and including additional evolution due to H and ε, we obtain the *main equations of the Bayesian formalism*:

$$\dot{\rho}_{11} = -\dot{\rho}_{22} = -2\frac{H}{\hbar}\, \mathrm{Im}\, \rho_{12} + \rho_{11}\rho_{22} \frac{2\Delta I}{S_0} [I(t) - I_0], \tag{30}$$

$$\dot{\rho}_{12} = \mathrm{i}\frac{\varepsilon}{\hbar}\rho_{12} + \mathrm{i}\frac{H}{\hbar}(\rho_{11} - \rho_{22}) - (\rho_{11} - \rho_{22})\frac{\Delta I}{S_0}[I(t) - I_0]\, \rho_{12}. \tag{31}$$

In each realization of measurement the noisy detector outcome $I(t)$ is different; however, for each realization we can precisely monitor the evolution of the qubit density matrix, plugging experimental $I(t)$ into Eqs. (30)–(31). Let us stress again that these equations show the absence of any qubit decoherence during the process of measurement. Pure initial state remains pure; moreover, initially mixed state gradually purifies during the measurement if $H \neq 0$ [4, 8]. The gradual state purification is essentially due to acquiring more and more information about the qubit state from the measurement record.

While the qubit state does not decohere in each individual realization of the measurement, different members of the ensemble evolve differently because of random $I(t)$. Averaging Eqs. (30)–(31) over random $I(t)$ and using the relation [which follows from Eqs. (13), (17)–(18), and (20)]

$$I(t) - I_0 = \frac{\Delta I}{2}(\rho_{11} - \rho_{22}) + \xi(t), \tag{32}$$

where $\xi(t)$ is a zero-correlated ("white") random process with the same spectral density as the detector noise, $S_\xi = S_0$, we obtain conventional Eqs. (7)–(8). Therefore, the ensemble-averaged decoherence in our model is just a consequence of averaging over the measurement outcome (similar conclusion is also valid in the finite response case).

Notice that since $I(t)$ contains the white noise contribution, Eqs. (30)–(31) are nonlinear stochastic differential equations [23] and dealing with them requires a special care. The problem is that their analysis depends on the choice of the derivative definition. Two mainly used definitions are the symmetric derivative: $\dot{\rho}(t) \equiv \lim_{\tau\to 0}[\rho(t+\tau/2) - \rho(t-\tau/2)]/\tau$ which leads to the so-called

Stratonovich interpretation of the stochastic differential equations, and the forward derivative: $\dot{\rho}(t) \equiv \lim_{\tau\to 0}[\rho(t+\tau)-\rho(t)]/\tau$ (Itô interpretation). Usual calculus rules remain valid only in the Stratonovich form [23], so the physical intuition works better when using Stratonovich definition. However, Itô interpretation allows simple averaging over the noise and because of that is usually preferred by mathematicians. Since we derived Eqs. (30)–(31) by a simple first-order differentiation, we automatically obtained them in the Stratonovich form (keeping the second-order terms in the expansion, we can obtain different equations, depending on the definition of the derivative). Since sometimes Itô form is more preferable, let us translate Bayesian equations into Itô form using the following general rule [23]. For an arbitrary system of equations

$$\dot{x}_i(t) = G_i(\mathbf{x},t) + F_i(\mathbf{x},t)\,\xi(t) \tag{33}$$

in Stratonovich interpretation, the corresponding Itô equation which has the same solution is

$$\dot{x}_i(t) = G_i(\mathbf{x},t) + \frac{S_\xi}{4}\sum_k \frac{\partial F_i(\mathbf{x},t)}{\partial x_k}\, F_k(\mathbf{x},t) + F_i(\mathbf{x},t)\,\xi(t)\,, \tag{34}$$

where $x_i(t)$ are the components of the vector $\mathbf{x}(t)$, G_i and F_i are arbitrary functions, and the constant S_ξ is the spectral density of the white noise process $\xi(t)$. Applying this transformation to Eqs. (30)–(31), we get the following equations in Itô interpretation:

$$\dot{\rho}_{11} = -\dot{\rho}_{22} = -2\frac{H}{\hbar}\,\mathrm{Im}\,\rho_{12} + \rho_{11}\rho_{22}\,\frac{2\Delta I}{S_0}\,\xi(t)\,, \tag{35}$$

$$\dot{\rho}_{12} = \mathrm{i}\frac{\varepsilon}{\hbar}\,\rho_{12} + \mathrm{i}\frac{H}{\hbar}\,(\rho_{11}-\rho_{22}) - (\rho_{11}-\rho_{22})\frac{\Delta I}{S_0}\,\rho_{12}\,\xi(t) - \frac{(\Delta I)^2}{4S_0}\,\rho_{12}\,, \tag{36}$$

while the relation between pure noise $\xi(t)$ and the current $I(t)$ is still given by Eq. (32). Notice that the last term in Eq. (36) does not actually mean the single qubit decoherence (pure state remains pure), but is just a feature of the Itô form [it directly corresponds to the ensemble decoherence after averaging over $\xi(t)$].

4. Derivation of the formalism via correspondence principle

Another derivation [8] of the Bayesian formalism for a single qubit can be based on the logical use of the correspondence principle [5], classical Bayes formula, and results of the conventional ensemble-averaged formalism. Even though this way lacks some advantages of the "microscopic" derivation discussed in the previous section, it can be applied to a broader class of solid-state detectors, in particular, to the finite-transparency quantum point contact and (with some extension) to the single-electron transistor and SQUID. In this section we will assume a double-dot qubit measured by a finite-transparency QPC and treat the detector current $I(t)$ as a quasicontinuous noisy signal that implies weak detector response, $|\Delta I| \ll I_0$.

Let us start with a completely *classical* case when the electron is actually located in one of two dots and does not move, but we do not know in which one, so the measurement gradually reveals the actual electron location. This is a well studied problem of the probability theory. The measurement process can be described as an evolution of probabilities (we still call them ρ_{11} and ρ_{22}) which reflect our knowledge about the system state. Then for arbitrary τ (which can be comparable to τ_m) the current $\overline{I}$ averaged over time interval $(t, t+\tau)$ [see Eq. (28)] has the probability distribution

$$P(\overline{I}) = \rho_{11}(t)P_1(\overline{I}) + \rho_{22}(t)P_2(\overline{I}), \tag{37}$$

where P_i are given by Eq. (29) and depend on the detector white noise spectral density S_0 which should not necessarily satisfy Schottky formula. After the measurement during time τ the information about the system state has increased and the probabilities ρ_{11} and ρ_{22} should be updated using the measurement result $\overline{I}$ and the Bayes formula (26)–(27). This completely describes the classical measurement process.

The next step in the derivation is an important assumption: in the quantum case with $H = 0$ the evolution of ρ_{11} and ρ_{22} is still given by Eqs. (26)–(27) because there is no possibility to distinguish between classical and quantum cases, performing only this kind of measurement. Even though this assumption is quite obvious, it is not derived formally but should rather be regarded as a *consequence of the correspondence principle.*

The correspondence with classical measurement cannot describe the evolution of ρ_{12}; however, there is still an upper limit: $|\rho_{12}| \leq [\rho_{11}\rho_{22}]^{1/2}$. Surprisingly, this inequality is sufficient for exact calculation of $\rho_{12}(t)$ in the case of a QPC as a detector (we still assume $H = 0$). Averaging the inequality over all possible detector outputs $\overline{I}$ using distribution (37) we get the inequality

$$\begin{aligned} |\langle \rho_{12}(t+\tau)\rangle| \leq \langle |\rho_{12}(t+\tau)|\rangle \leq \langle [\rho_{11}(t+\tau)\rho_{22}(t+\tau)]^{1/2}\rangle \\ = [\rho_{11}(t)\rho_{22}(t)]^{1/2} \exp[-(\Delta I)^2\tau/4S_0] \end{aligned} \tag{38}$$

[here the decaying exponent is a consequence of changing ρ_{11} and ρ_{22} that reduces their average product]. On the other hand, from the conventional approach we know [16, 18, 24, 25] that the ensemble-averaged qubit decoherence rate caused by a QPC is equal to $\Gamma_d = (\Delta I)^2/4S_0$, where $S_0 = 2eI_0(1-\mathcal{T})$ and $\mathcal{T}$ is the QPC transparency. This means that inequality (38) actually *reaches its upper bound.* This is possible *only if* in each realization of the measurement process an initially pure density matrix $\rho_{ij}(t)$ stays pure all the time, $|\rho_{12}|^2 = \rho_{11}\rho_{22}$. Moreover, since the phase of $\rho_{12}(t+\tau)$ should be the same for all realizations (to ensure $|\langle \rho_{12}(t+\tau)\rangle| = \langle|\rho_{12}(t+\tau)|\rangle$), the only possibility in absence of a detector-induced shift of ε is

$$\frac{\rho_{12}(t+\tau)}{[\rho_{11}(t+\tau)\rho_{22}(t+\tau)]^{1/2}} = \frac{\rho_{12}(t)}{[\rho_{11}(t)\rho_{22}(t)]^{1/2}} \exp(\mathrm{i}\varepsilon\tau/\hbar) \tag{39}$$

(if the coupling with detector shifts ε, we just have to use the shifted value).

As the next step of the derivation, let us allow an arbitrary mixed initial state of the qubit (but still $H = 0$). It can always be represented as a mixture of two states with the same diagonal matrix elements, one of which is pure, while the other state does not have nondiagonal matrix elements. Since nondiagonal matrix elements for the latter state cannot appear in the process of measurement and since the evolution of the diagonal matrix elements is equal for both states, one can conclude that Eq. (39) remains valid, i.e. for mixed states the degree of purity is conserved (gradual purification does not occur at $H = 0$). The final step of the formalism derivation is differentiating Eqs. (26), (27), and (39) over time and adding (in a simple way) the evolution due to H.

In this way we reproduce Eqs. (30)–(31). However, as seen from the derivation, now they are applicable to a *broader class of detectors* (which includes the finite-transparency QPC) for which $\Gamma_d = (\Delta I)^2/4S_0$. This relation can also be expressed as $\Gamma_d\tau_m = 1/2$ since $\tau_m = 2S_0/(\Delta I)^2$ for a weakly responding detector. As will be discussed in the next section, this is a condition of an ideal quantum detector.

5. Nonideal detectors

The relation $\Gamma_d\tau_m = 1/2$ which is valid for the models of a tunnel junction and QPC as detectors, basically says that the *larger output noise S_0 of a detector leads to a smaller backaction* characterized by ensemble decoherence Γ_d. This is quite expected from quantum mechanical point of view (the faster we get information, the faster we should collapse the measured state). However, it is obviously not necessarily the case for an arbitrary solid-state detector; for example, the increase of output noise S_0 can be due to later stages of signal amplification, which do not affect Γ_d. In other words, it is easy to imagine a bad detector which produces a lot of both output and backaction noises.

To take into account an extra detector noise, we can phenomenologically add a dephasing rate γ_d into the Bayesian equations:

$$\dot{\rho}_{11} = -\dot{\rho}_{22} = -2\frac{H}{\hbar}\,\mathrm{Im}\,\rho_{12} + \rho_{11}\rho_{22}\frac{2\Delta I}{S_0}\,[I(t) - I_0], \tag{40}$$

$$\dot{\rho}_{12} = \mathrm{i}\frac{\varepsilon}{\hbar}\rho_{12} + \mathrm{i}\frac{H}{\hbar}(\rho_{11}-\rho_{22}) - (\rho_{11}-\rho_{22})\frac{\Delta I}{S_0}\,[I(t)-I_0]\,\rho_{12} - \gamma_d\rho_{12}. \tag{41}$$

This obviously increases the ensemble decoherence rate:

$$\Gamma_d = \frac{(\Delta I)^2}{4S_0} + \gamma_d. \tag{42}$$

A natural definition of a detector ideality (quantum efficiency) in this case is

$$\eta \equiv 1 - \frac{\gamma_d}{\Gamma_d} = \frac{1}{2\Gamma_d \tau_m}. \tag{43}$$

An upper limit for η is 100% because of a fundamental limitation

$$\Gamma_d \tau_m \geq 1/2, \tag{44}$$

which is a by-product of the Bayesian derivation for the case of quasicontinuous detector current and small difference between noises S_1 and S_2 (so that $S_1 = S_2 = S_0$) – see inequality (38). [In the case of a detector with $S_1 \neq S_2$ and possibility to observe each passing electron, Eq. (44) remains valid; however, a meaningful model would imply Poisson statistics (20) and therefore Schottky formula for the detector noise.]

The extra dephasing γ_d in Eq. (41) can be interpreted [8] as an effect of extra environment or (mathematically) as due to a second detector "in parallel", the output of which is not read out (then we have to average over possible outputs). It can be also interpreted as an effect of extra noise S_{add} at the detector output, $S_0 = (\Delta I)^2/4\Gamma_d + S_{add}$. In this latter case one can argue that the qubit is actually in a pure state and the evolution of the diagonal matrix elements is actually different from what is given by Eq. (40), because the measured current $I(t)$ is not the "actual" detector current. Yes, we would know the exact pure state if our amplifiers did not produce extra noise S_{add}; however, since we do not have access to the "actual" detector current, we should perform averaging over the extra noise. It is easy to show that such averaged qubit density matrix (which is a density matrix "for us") still satisfy Eqs. (40)–(41).

Introduction of the detector ideality η allows us to consider a continuous transition from the conventional ensemble-averaged result (7)–(8) to the "pure" Bayesian result (30)–(31). The effect of a pure environment can be considered as a measurement with an extremely bad detector, $\eta = 0$. Technically it corresponds to $\Delta I = 0$ in Eqs. (40)–(41), transforming them into conventional equations. The case of a detector with very small efficiency, $\eta \ll 1$, can be treated in two steps: first, we analyze the effect of the decoherence term ($\gamma_d \approx \Gamma_d$), and then we use the classical (still Bayesian) analysis to relate the qubit density matrix and the measurement outcome. So, only for good detectors with the efficiency η comparable to unity, the quantum Bayesian approach discussed in this paper is really necessary. Some people could argue that it is so difficult to create a solid-state detector with good quantum efficiency, that the Bayesian formalism is useless at the present-day level of technology. However, actually at present such detectors are becoming available. For example, the analysis of experimental data of the recent "which path" experiment [26] shows that their QPC had a quantum efficiency quite close to 100%.

The account of the detector nonideality by introducing extra decoherence rate γ_d into Eq. (41) implicitly assumes the absence of a direct correlation between the output detector noise and the backaction noise affecting the qubit energy asymmetry ε. However, such correlation is a typical situation, for example, for a single-electron transistor as a detector [27]. In this case the knowledge of the noisy detector output $I(t)$ gives some information about the probable backaction noise "trajectory" $\varepsilon(t)$ which can be used to improve our knowledge of the qubit state. Compensation for the most probable trajectory $\varepsilon(t)$ leads to the improved Bayesian equations for the SET in which Eq. (41) is replaced with

$$\begin{aligned}\dot{\rho}_{12} = {} & \mathrm{i}\frac{\varepsilon}{\hbar}\rho_{12} + \mathrm{i}\frac{H}{\hbar}(\rho_{11}-\rho_{22}) - (\rho_{11}-\rho_{22})\frac{\Delta I}{S_0}\left[I(t)-I_0\right]\rho_{12} \\ & + \mathrm{i}\,K\left[I(t)-I_0\right]\rho_{12} - \tilde{\gamma}_d\rho_{12},\end{aligned} \tag{45}$$

where $K = (d\varepsilon/d\varphi)S_{I\varphi}/S_0\hbar$, φ is the electric potential of the SET central electrode, and $S_{I\varphi}$ is the mutual low-frequency spectral density between the current noise and φ noise. [Strictly speaking, $S_{I\varphi}$ in our notation is only the usual real part of the mutual spectral density, which reflects the detector "asymmetry", while the imaginary part can formally describe the detector response [16]. Also notice that since the small shift of the SET operating point for two localized qubit states in general affects the energy ε, it should be defined self-consistently in Eq. (45).]

The dephasing rate $\tilde{\gamma}_d$ should now satisfy equation

$$\tilde{\gamma}_d = \Gamma_d - \frac{(\Delta I)^2}{4S_0} - \frac{K^2 S_0}{4} \tag{46}$$

to correspond to the the ensemble-averaged dynamics still described by Eqs. (7)–(8).

The obvious inequality $\tilde{\gamma}_d \geq 0$ (in the opposite case the relation $|\rho_{12}|^2 \leq \rho_{11}\rho_{22}$ would be violated in a single realization of measurement) imposes a lower bound for the ensemble decoherence rate Γ_d:

$$\Gamma_d \geq \frac{(\Delta I)^2}{4S_0} + \frac{K^2 S_0}{4}, \tag{47}$$

which is stronger than the inequality $\Gamma_d\tau_m \geq 1/2$.

Inequality (47) can be easily expressed in terms of the energy sensitivity of an SET. Let us define the output energy sensitivity as $\epsilon_I \equiv (dI/dq)^{-2}S_I/2C$ where C is the total SET island capacitance, dI/dq is the response to the externally induced charge q, and we have changed the notation $S_I \equiv S_0$ for a more symmetric look of the formulas. Notice that ϵ_I has the same dimension as $\hbar$. Similarly, let us characterize the backaction noise intensity by $\epsilon_\varphi \equiv CS_\varphi/2$ and the correlation between two noises by the magnitude $\epsilon_{I\varphi} \equiv (dI/dq)^{-1}S_{I\varphi}/2$. Since in absence of other decoherence sources $\Gamma_d = S_\varphi(C\Delta E/2e\hbar)^2$, where ΔE is the energy coupling between qubit and single-electron transistor [4, 17], and using also the reciprocity property $\Delta q = C\Delta E/e = d\varepsilon/d\varphi$, we can rewrite Eq. (47) as

$$\epsilon \equiv (\epsilon_I \epsilon_\varphi - \epsilon_{I\varphi}^2)^{1/2} \geq \hbar/2. \tag{48}$$

This is a result known for 20 years [28] for SQUIDs (see also [16, 29–33]).

When the limit $\epsilon = \hbar/2$ is achieved, the decoherence rate

$$\tilde{\gamma}_d = \frac{(\Delta I)^2}{4S_I}\left[\frac{\epsilon_I\epsilon_\varphi - \epsilon_{I\varphi}^2}{(\hbar/2)^2} - 1\right] \tag{49}$$

in Eq. (45) vanishes, $\tilde{\gamma}_d = 0$. In this sense the detector is ideal, $\tilde{\eta} = 1$, where

$$\tilde{\eta} \equiv 1 - \frac{\tilde{\gamma}_d}{\Gamma_d} = \frac{\hbar^2(dI/dq)^2}{S_I S_\varphi} + \frac{(S_{I\varphi})^2}{S_I S_\varphi}, \tag{50}$$

even though it can be a nonideal detector ($\eta < 1$) by the previous definition, $\eta = \hbar^2(dI/dq)^2/S_I S_\varphi$.

Another possible definition of the detector efficiency in this case is

$$\tilde{\eta}_2 \equiv \frac{(\hbar/2)^2}{\epsilon_I\epsilon_\varphi - \epsilon_{I\varphi}^2} = \frac{\hbar^2(dI/dq)^2}{S_I S_\varphi - S_{I\varphi}^2}. \tag{51}$$

Notice a simple relation,

$$\eta = \tilde{\eta} = \tilde{\eta}_2 = \frac{(\hbar/2)^2}{\epsilon_I\epsilon_\varphi} = \frac{1}{2\Gamma_d\tau_m}, \tag{52}$$

in absence of correlation between noises of $\varphi(t)$ and $I(t)$, $(S_{I\varphi})^2 \ll S_I S_\varphi$.

Even though Eqs. (45)–(52) were derived for the SET as a detector, it is rather obvious that they are applicable to virtually any solid-state detector with continuous output (for a dc SQUID the current output should obviously be replaced by the voltage output). In particular, the conclusion that reaching the quantum-limited total energy sensitivity $\epsilon = \hbar/2$ is equivalent to the detector ideality, is quite general.

As we already discussed, the tunnel junction and QPC at zero temperature (actually, for small temperatures $\beta^{-1} \ll eV$) are theoretically ideal quantum detectors. The fact that a SQUID can reach the limit of an ideal detector follows from the results of Ref. [28]. A normal state SET is not a good quantum detector ($\eta \ll 1$) at usual operating points above the Coulomb Blockade threshold [4, 17]. However, its quantum efficiency improves when we go closer to the threshold

[4, 29] and becomes much better when the operating point is in the cotunneling range (below the threshold), in which case the limit of an ideal detector can be achieved [30, 31]. Superconducting SET is generally better than normal SET as a quantum-limited detector and can approach 100% ideality in the supercurrent regime [32] as well as in the double Josephson-plus-quasiparticle regime [33] (possibly a threshold of a quasiparticle current is also a good operating point in this sense; however, this regime has been studied only for the current so far [34], but not for the noise). Finally, the resonant-tunneling SET [35] can reach ideality factor $\tilde{\eta}_2 = 3/4$ at large bias and complete ideality, $\tilde{\eta} = \tilde{\eta}_2 = 1$, in the small-bias limit.

6. Some experimental predictions

6.1. DIRECT EXPERIMENTS

The Bayesian equations tell us that we can monitor the qubit evolution in a single realization of the measurement process using the record of the noisy detector output. In particular, for an ideal detector *we can monitor the qubit wavefunction* (except the overall phase) if the initial qubit state is pure or, for a mixed initial state, after monitoring for a sufficiently long time so that the gradual purification has enough time to produce a practically pure state.

This prediction (and hence, the validity of the Bayesian equations) can in principle be tested experimentally. For example [8], let us first prepare the double-dot in the symmetric coherent state, $\rho_{11} = \rho_{22} = |\rho_{12}| = 1/2$, make $H = 0$ (raise the barrier), and begin measurement with a QPC [Fig. 1(b)]. According to our formalism, after some time τ (the most interesting case is $\tau \sim \tau_m$) the qubit state remains pure but becomes asymmetric ($\rho_{11} \neq \rho_{22}$) and can be calculated with Eqs. (30) and (31). To prove this, an experimentalist can use the knowledge of the wavefunction to move the electron "coherently" into the first dot with 100% probability. (Notice that if the qubit is in a mixed state, no unitary transformation can end up in the state $|1\rangle$ with certainty.) For instance, experimentalist switches off the detector at $t = \tau$, reduces the barrier (to get finite H), and creates the energy difference $\varepsilon = [(1 - 4|\rho_{12}(\tau)|^2)^{1/2} - 1] H \mathrm{Re}\rho_{12}(\tau)/|\rho_{12}(\tau)|^2$; then after the time period $\Delta t = [\pi - \arcsin(\mathrm{Im}\rho_{12}(\tau)\, \hbar\Omega/H)]/\Omega$ the "whole" electron will be moved into the first dot, that can be checked by the detector switched on again. [Here $\Omega \equiv (4H^2 + \varepsilon^2)^{1/2}/\hbar$ is the frequency of unperturbed coherent ("Rabi") oscillations of the qubit.]

Another experimental idea [8] is to demonstrate the gradual purification of the double-dot density matrix. Let us start with a completely mixed (unknown) state ($\rho_{11} = \rho_{22} = 1/2$, $\rho_{12} = 0$) of the double-dot qubit with finite H. Then using the detector output $I(t)$ and Eqs. (30)–(31) an experimentalist gradually gets more and more knowledge about the randomly evolving qubit state (gradual purification), eventually ending up with almost pure wavefunction with precisely known *phase* of Rabi oscillations (we are not talking about the wavefunction phase, but about the phase of diagonal matrix elements oscillations). The final check of the wavefunction can be similar to that described in the previous paragraph. It can be even simpler, since with the knowledge of the phase of oscillations it is easy to stop the evolution by raising the barrier when the electron is in the first dot with certainty. Notice that if fast real-time calculations are not available, the moment of raising the barrier can be random, while lucky cases can be selected later from the record of $I(t)$.

Direct experiments of this kind as well as experiments on quantum feedback control and on Bayesian measurement of entangled qubits (discussed in sections 7 and 8), are still too difficult for realization at the present-day level of technology. In the next two subsections we will discuss experiments which seem to be realizable (though very hard) at present.

6.2. SPECTRAL DENSITY OF THE DETECTOR CURRENT

Naively thinking, a qubit with $H \neq 0$ should perform coherent (Rabi) oscillations with frequency Ω and these oscillations should lead to an oscillating contribution of the detector current $I(t)$. On the other hand, conventional Eqs. (7)–(8) seem naively to imply that the qubit eventually reaches a stationary state and no oscillations should be present in $I(t)$ after a sufficiently long observation. So, it is interesting to find what is the actual spectral density of the detector current $S_I(\omega)$ [it is easier to measure this quantity experimentally, than to record $I(t)$].

The Bayesian formalism predicts [21, 19, 22] the presence of the spectral peak at the Rabi frequency Ω, however, the height of this peak cannot be larger than 4 times the noise pedestal. In particular for a symmetric qubit ($\varepsilon = 0$)

$$\frac{S_I(\omega)}{S_0} = 1 + \frac{4\eta}{(\omega/\Omega)^2 + (\omega^2 - \Omega^2)^2/\Omega^2\Gamma_d^2} . \tag{53}$$

Actually, an experimental confirmation of this formula would not be a direct verification of the Bayesian formalism, since Eq. (53) can be also obtained by other methods, including the master equation method [25, 21, 16] and the method based on the Bloch equations [22].

6.3. BELL-TYPE EXPERIMENT

Another experiment which also seems to be much easier than the direct experiments but can unambiguously test the Bayesian formalism, is a Bell-type experiment in which one qubit is measured by two detectors [36]. An idea (Fig. 3) is to prepare the qubit in a coherent state $(|1\rangle + |2\rangle)/\sqrt{2}$, then to switch on the first detector (A) for a relatively short time τ_A (so that the measurement is only partially completed), and to switch on the second detector (B) a little later. If the

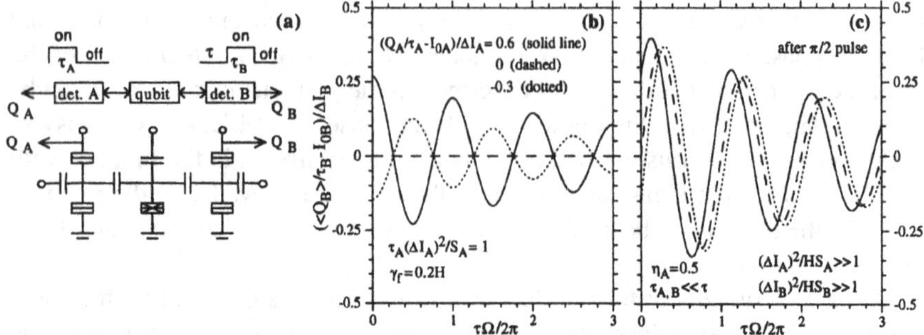

Figure 3. (a) Schematic of the proposed Bell-type correlation experiment [36], in which a SCPB qubit is measured by two SETs during short time periods τ_A and τ_B shifted in time by τ. The first measurement leads to an incomplete collapse of the qubit initial state $(|1\rangle + |2\rangle)/\sqrt{2}$ and affects the result of the second measurement. (b) The average result $\langle Q_B \rangle$ of the second measurement for a selected result Q_A of the first measurement. The sign and amplitude of Rabi oscillations depend on Q_A, reflecting the change of the diagonal matrix elements of the qubit density matrix. (c) same as (b) if $\pi/2$ pulse is applied immediately after the first measurement. Now the phase of oscillations depends on Q_A. The full-swing oscillations (with amplitude of 0.5 in the ideal case) indicate a pure qubit state after the first measurement.

first measurement changes the qubit state according to the Bayesian formalism, then the second measurement can check this change. An output from a single run of the measurement are two charges Q_A and Q_B passed through two detectors. Performing the experiment many times and analyzing the correlation between Q_A and Q_B, one can recover the effect of the first measurement on the qubit state [36] (to check the change of the nondiagonal matrix element it is necessary to apply a $\pi/2$ pulse right after the first measurement). The main advantage of this Bell-type experiment in comparison with the direct Bayesian experiments is that the wide bandwidth for the output signal is not necessary; instead, it is traded for the wide bandwidth of two input lines (switching detectors on and off), which is much easier to realize experimentally.

7. Quantum feedback control of a qubit

The Bayesian formalism can be used as a basis for the design and analysis of a quantum feedback control of a solid-state qubit. As an example, such feedback control can maintain for arbitrary long time the desired phase of a qubit Rabi oscillations, synchronizing them with a classical reference oscillator, even in presence of dephasing environment [4, 37]. The overall idea is very close to a classical feedback loop [Fig. 4(a)]. The oscillating qubit evolution is monitored by a weakly coupled detector ($C \equiv \hbar(\Delta I)^2/S_0 H < 1$), the phase $\phi(t)$ of actual Rabi oscillations is compared with the desired phase $\phi_0(t)$, and the difference signal $\Delta\phi$ is used to control the qubit barrier height. If qubit is slightly behind the

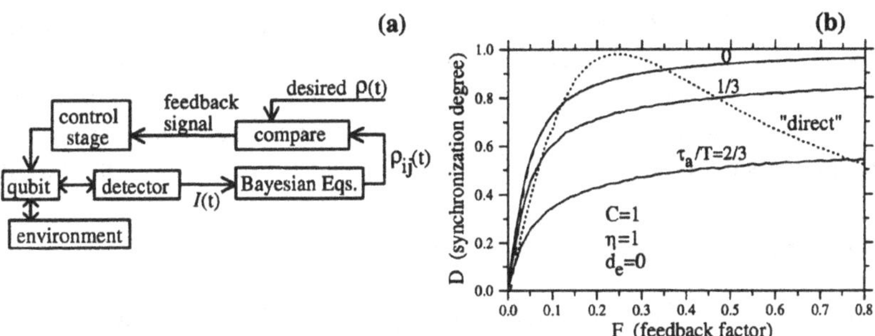

Figure 4. (a) Schematic of the quantum feedback loop maintaining the Rabi oscillations of a qubit by synchronizing them with a classical harmonic signal. (b) Solid lines: the synchronization degree D as a function of the feedback factor F for several values of available bandwidth τ_a^{-1}. While synchronization can approach 100% for wide bandwidth, it worsens when τ_a becomes comparable to the oscillation period $T = 2\pi/\Omega$. Dashed line (alsmost coinciding with the upper solid line): analytical result $D = \exp(-\mathcal{C}/32F)$. Dotted line: synchronization degree for a direct feedback with $\tau_a = T/10$. (From Ref. [37].)

desired phase, then H is decreased, so the oscillations run faster to catch up; if the qubit is ahead of proper phase, H is increased. It is natural to use a linear control: $H_{fb} = H(1 - F \times \Delta\phi)$, where F is a dimensionless feedback factor.

The only difference of this loop from a classical feedback is that even weakly coupled detector disturbs the qubit oscillations. However, the induced fluctuations of the oscillation phase are slow, and the information obtained from the detector happens to be enough to monitor the phase fluctuations and compensate them. The quantitative analysis [37] shows that in a limit of good synchronization and absence of extra environment the qubit correlation function $K_z(\tau) \equiv \langle z(t)z(t+\tau)\rangle$ (here $z \equiv \rho_{11} - \rho_{22}$) is given by

$$K_z(\tau) = \frac{\cos\Omega\tau}{2} \exp\left[\frac{\mathcal{C}}{16F}\left(e^{-2FH\tau/\hbar} - 1\right)\right], \tag{54}$$

and does not decay to zero at $\tau \to \infty$. Correspondingly, the degree of the qubit synchronization, $D \equiv 2\langle\rho\rho_d\rangle - 1$ (here ρ_d is the desired density matrix corresponding to ideal oscillations) is found to be $D = \exp(-\mathcal{C}/32F)$ and approaches 100% at $F \gg \mathcal{C}$.

The quality of the qubit oscillations synchronization decreases with the decrease of available feedback bandwidth τ_a^{-1} [Fig. 4(b)]. It also decreases when the qubit is dephased by an extra environment. For a weak dephasing rate γ_e we found numerically [37] a dependence $D_{max} \simeq 1 - 0.5d_e$ where $d_e \equiv \gamma_e/[(\Delta I)^2/4S_0]$. This means that the feedback loop can efficiently suppress the qubit dephasing due to coupling to the environment if this coupling is much weaker than the coupling to a nearly ideal detector.

Besides the linear feedback $H_{fb} = H(1 - F \times \Delta\phi)$, we have also studied the "direct" feedback $H_{fb}(t)/H - 1 = F\{2[I(t) - I_0]/\Delta I - \cos\Omega t\}\sin\Omega t$ and found that it can also provide a good phase synchronization if F/C is close to 1/4 [Fig. 4(b)]. The direct feedback is much easier for an experimental realization because it does not require solving the Bayesian equations (40)–(41) in real time.

8. Bayesian measurement of entangled qubits

The Bayesian formalism has been generalized to a continuous quantum measurement of entangled qubits in Ref. [38]. Suppose a detector is coupled to N entangled qubits. In the "measurement" basis there are 2^N states $|i\rangle$ of the qubits which correspond to up to 2^N different dc current levels I_i of the detector (some of these currents can coincide, for example, if two or more qubits are coupled equally strong to the detector). It has been shown that the generalization of Eqs. (40)–(41) for this case is [38]

$$\dot{\rho}_{ij} = \frac{-i}{\hbar}[H_{qb}, \rho]_{ij} + \rho_{ij}\frac{1}{S_0}\sum_k \rho_{kk}\left[\left(I(t) - \frac{I_k + I_i}{2}\right)(I_i - I_k) \right.$$
$$\left. + \left(I(t) - \frac{I_k + I_j}{2}\right)(I_j - I_k)\right] - \gamma_{ij}\rho_{ij}, \tag{55}$$

where the first term describes the unitary evolution due to the Hamiltonian of qubits H_{qb} and

$$\gamma_{ij} = (\eta^{-1} - 1)(I_i - I_j)^2/4S_0, \tag{56}$$

while Eq. (32) is replaced by

$$I(t) = \sum_i \rho_{ii}(t) I_i + \xi(t). \tag{57}$$

Notice that there is no mutual decoherence ($\gamma_{ij} = 0$) between states $|i\rangle$ and $|j\rangle$ even for a nonideal detector if the corresponding classical currents coincide, $I_i = I_j$. This is because the detector noise cannot destroy the coherence between states which are equally coupled to the detector.

Translating Eq. (55) from Stratonovich form into Itô form, we get

$$\dot{\rho}_{ij} = \frac{-i}{\hbar}[H_{qb}, \rho]_{ij} + \rho_{ij}\frac{1}{S_0}\left(I(t) - \sum_k \rho_{kk} I_k\right)\left(I_i + I_j - 2\sum_k \rho_{kk} I_k\right)$$
$$- \left(\gamma_{ij} + \frac{(I_i - I_j)^2}{4S_0}\right)\rho_{ij}, \tag{58}$$

while in the ensemble-averaged equations the second term of Eq. (58) (which depends on $I(t)$) is averaged to zero.

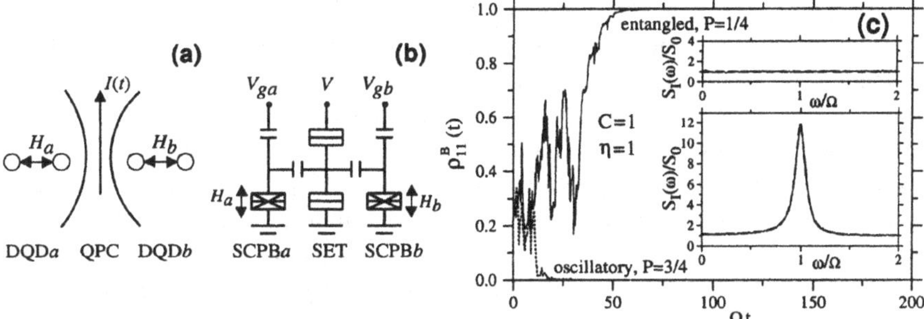

Figure 5. (a) Two qubits made of double quantum dots measured by an equally coupled quantum point contact. (b) Similar setup made of single-Cooper-pair boxes measured by a single-electron transistor. (c) Two Monte Carlo realizations of the two-qubit state evolution starting from the fully mixed state for a symmetric setup (ρ_{11}^B is the diagonal component of the two-qubit density matrix, corresponding to the Bell state $(|\uparrow_a\downarrow_b\rangle - |\downarrow_a\uparrow_b\rangle)/\sqrt{2}$). With probability 1/4 the qubits become fully entangled, $\rho_{11}^B \to 1$ ("spontaneous entanglement"); then the detector output is a pure noise (upper inset). With probability 3/4 the state is gradually collapsed into the orthogonal subspace, $\rho_{11}^B \to 0$; then the detector signal shows a spectral peak at the Rabi frequency Ω with the peak-to-pedestal ratio of 32/3. (From Ref. [39].)

These Bayesian equations have been applied in Ref. [39] to the analysis of a simple setup (Fig. 5) in which a detector is equally coupled to two similar qubits (both qubits are symmetric, $\varepsilon_a = \varepsilon_b = 0$, and do not interact directly with each other). An interesting effect has been found in the case when the Rabi frequencies $\Omega_a = 2H_a/\hbar$ and $\Omega_b = 2H_b/\hbar$ exactly coincide. Then there are two possible scenarios of the two-qubit evolution due to measurement, starting from a general mixed state. Either qubits become fully entangled collapsing into the Bell state $(|\uparrow_a\downarrow_b\rangle - |\downarrow_a\uparrow_b\rangle)/\sqrt{2}$ (we call this process spontaneous entanglement), or the state falls into the orthogonal subspace of the two-qubit Hilbert space. Experimentally these two scenarios can be distinguished by different spectral density $S_I(\omega)$ of the detector current [Fig. 5(c)]. In the case of Bell state, $S_I(\omega)$ is just the flat noise S_0 of the detector because the signals from two qubits compensate each other, while in the other scenario $S_I(\omega)$ has a spectral peak at the Rabi frequency, which height is equal to $32\eta/3$ for a weakly coupled detector ($\mathcal{C}_a = \mathcal{C}_b < 1$) [39].

The probabilities of two scenarios are equal to the contributions of two subspaces in the initial state $\rho(0)$; for the case of fully mixed initial state they are equal to 1/4 and 3/4, respectively. The considered setup can obviously be used for a *preparation of the Bell state* without direct interaction between two qubits. Notice that if the state collapsed into the orthogonal subspace, we can apply some noise which affects ε_a and/or ε_b and therefore mixes the two-qubit density matrix, and try the measurement again. In this way the probability $1 - (3/4)^M$ to obtain the Bell state can be made arbitrary close to 100% by allowing sufficiently large number M of attempts.

In actual experiment the symmetry of the setup cannot be made exact. In this case the Bell state and the oscillating state are not infinitely stable and there will be switching between them. The calculations show [39] that the switching rate $\Gamma_{B\to O}$ from the Bell state into the oscillating state is equal to $\Gamma_{B\to O} = (\Delta\Omega)^2/2\Gamma_d$ due to slightly different Rabi frequencies $[\Gamma_d = \eta^{-1}(\Delta I)^2/4S_0]$, $\Gamma_{B\to O} = (\Delta C/C)^2\Gamma_d/8$ due to slightly different coupling, and $\Gamma_{B\to O} = (\gamma_a + \gamma_b)/2$ due to an extra environment acting on two qubits separately. The rate of the return switching is 3 times smaller, $\Gamma_{O\to B} = \Gamma_{B\to O}/3$. Notice that in this case the averaged height of the Rabi spectral peak is equal to $8\eta S_0$, which is exactly twice as much as for a single qubit.

Even though such experiment on spontaneous entanglement is still extremely difficult for a realization, it should be noted that for the observation of the phenomenon the detector quantum efficiency η should not necessarily be close to 100%; it should only be large enough to allow distinguishing the Rabi spectral peak with the peak-to-pedestal ratio of $32\eta/3$.

9. Discussion

In this paper we have reviewed the basic derivation and some applications of the Bayesian approach to continuous quantum measurement of solid-state qubits. Even though this is a new subject for the solid-state community, many similar formalisms have been developed in other fields of quantum physics. Generally, this type of approach which takes into account the measurement outcome, is usually called selective or conditional quantum measurement. However, there is a rather broad variety of formalisms and their interpretations within the approach (for example, see reviews [40–43]). Some of key words related to this subject are: quantum trajectories, quantum state diffusion, quantum jumps, weak measurements, stochastic evolution of the wavefunction, stochastic Schrödinger equation, complex Hamiltonian, restricted path integral, quantum Bayes theorem, etc. The approach of conditional quantum measurements is relatively well developed in quantum optics. In particular, the optical quantum feedback control has been well studied theoretically (see, e.g. [44–48]) and was recently realized experimentally [49]. In relation to continuous quantum measurement of single systems, the quantum optics seems to be about 10 years ahead of the solid-state physics. However, the interest to this problem in the solid-state community has significantly increased after the "which path" experiments [26, 50]. Quite possibly it will be a rapidly growing field in the nearest future, especially because of its direct relation to the solid-state quantum computing.

In this paper we have discussed two solid-state experiments which seem to be realizable (though very difficult) today. First, it would be interesting to measure the spectral density of the detector current when the measured qubit performs coherent oscillations, and compare experimental results with the theoretical prediction that the spectral peak in the best case is 4 times higher than the noise pedestal. Second, the Bell-type correlation experiment with one qubit measured by two detectors would be able to verify that the qubit state remains pure during the whole measurement process and show the possibility of monitoring the qubit evolution precisely. This would be the first step towards realization of the quantum feedback control of solid-state qubits. A continuous monitoring of entangled qubits would be another very interesting experiment. Hopefully, the rapid progress of solid-state technology will make these experiments possible in a reasonably near future.

The author thanks D. Averin, E. Buks, G. Milburn, and R. Ruskov for useful discussions. The work was supported by NSA and ARDA under ARO grant DAAD19-01-1-0491.

References

1. Davies, E. B. (1976) *Quantum Theory of Open Systems* (Academic, London).
2. Kraus, K. (1983) *States, Effects, and Operations: Fundamental Notions of Quantum Theory* (Springer, Berlin).
3. Wiseman, H. M. and Milburn, G. J. (1993) *Phys. Rev. A* **45**, 1652–1666.
4. Korotkov, A. N. (2001) *Phys. Rev. B* **63**, 115403.
5. von Neumann, J., (1955) *Mathematical Foundations of Quantum Mechanics* (Princeton Univ. Press, Princeton, 1955); Messiah, A. (1961) *Quantum Mechanics* (North-Holland Publishing, Amsterdam).
6. Caldeira, A. O. and Leggett, A. J. (1983) *Ann. Phys.* (N.Y.) **149**, 374–456.
7. Zurek, W. H. (1991) *Phys. Today* **44** (10), 36–44.
8. Korotkov, A. N. (1998) quant-ph/9808026; (1999) *Phys. Rev. B* **60**, 5737–5742.
9. Goan, H.-S., Milburn, G. J., Wiseman, H. M. and Sun, H. B. (2001) *Phys. Rev. B* **63**, 125326.
10. Bayes, T. (1763) *Philos. Trans. R. Soc. London* **53**, 370–418; Laplace, P. S. (1812) *Théorie analytique des probabilités* (Ve Courcier, Paris).
11. Borel, E. (1965) *Elements of the Theory of Probability* (Prentice-Hall, Englewood Cliffs, NJ).
12. Gardiner, C. W. (1991) *Quantum noise* (Springer, Heidelberg), Chap. 2.2.
13. Gurvitz, S. A. (1997) *Phys. Rev. B* **56**, 15215–15223.
14. Gurvitz, S. A. (1998) cond-mat/9808058.
15. Stodolsky, L. (1999) *Phys. Lett. B* **459**, 193–200.
16. Averin, D. V. (2000) cond-mat/0004364.
17. Makhlin, Y., Schön, G. and Shnirman, A. (2001) *Rev. Mod. Phys.* **73**, 357–400.
18. Hackenbroich, G. (2001) *Phys. Rep.* **343**, 464–538.
19. Goan, H.-S. and Milburn, G. J. (2001) *Phys. Rev. B* **64**, 235307.
20. Goan, H.-S. (2002) cond-mat/0205582.
21. Korotkov, A. N. (2001) *Phys. Rev. B* **63**, 085312.
22. Ruskov. R. and Korotkov, A. N. (2002) cond-mat/0202303.
23. Øksendal, B. (1998) *Stochastic Differential Equations* (Springer, Berlin).
24. Aleiner, I. L., Wingreen, N. S. Meir, Y. (1997) *Phys. Rev. Lett.* **79**, 3740–3743.
25. Korotkov, A. N. and Averin, D. V. (2001) *Phys. Rev. B* **64**, 165310.
26. Buks, E., Schuster, R., Heiblum, M., Mahalu, D. and Umansky, V. (1998) *Nature* **391**, 871–874.
27. Korotkov, A. N. (1994) *Phys. Rev. B* **49**, 10381–10391.

28. Danilov, V. V., Likharev, K. K. and Zorin, A. B. (1983) *IEEE Trans. Magn.* MAG-**19**, 572–575.
29. Devoret, M. H. and Schoelkopf, R. J. (2000) *Nature* **406**, 1039–1046.
30. Averin, D. V. (2000) cond-mat/0010052.
31. Maassen van den Brink, A. (2002) *Europhys. Lett.* **58**, 562–568.
32. Zorin, A. B. (1996) *Phys. Rev. Lett.* **76**, 4408–4411.
33. Clerk, A. A., Girvin, S. M., Nguyen, A. K. and Stone, A. D. (2002) cond-mat/0203338.
34. Averin, D. V., Korotkov, A. N., Manninen, A. J. and Pekola, J. P. (1997) *Phys. Rev. Lett.* **78**, 4821–4824.
35. Averin, D. V. (2000) *Fortschr. Phys.* **48**, 1055–1074.
36. Korotkov, A. N. (2001) *Phys. Rev. B* **64**, 193407.
37. Ruskov, R. and Korotkov, A. N. (2002) *Phys. Rev. B* **66**, 041401.
38. Korotkov, A. N. (2002) *Phys. Rev. A* **65**, 052304.
39. Ruskov. R. and Korotkov, A. N. (2002) cond-mat/0206396.
40. Carmichael, H. J. (1993) *An Open System Approach to Quantum Optics*, Lecture notes in physics (Springer, Berlin).
41. Plenio, M. B. and Knight, P. L. (1998) *Rev. Mod. Phys.* **70**, 101–144.
42. Mensky, M. B. (1998) *Phys. Usp.* **41**, 923–940; quant-ph/9812017.
43. Presilla, C., Onofrio, R. and Tambini, U. (1996) *Ann. Phys.* (N.Y.) **248**, 95–121.
44. Caves, C. M. and Milburn, G. J. (1987) *Phys. Rev. A* **36**, 5543–5555.
45. Wiseman, H. M. and Milburn, G. J. (1993) *Phys. Rev. Lett.* **70**, 548–551.
46. Tombesi, P. and Vitali, D. (1995) *Phys. Rev. A* **51**, 4913–4917.
47. Doherty, A. C., Habib, S., Jacobs, K., Mabuchi, H. and Tan, S. M. (2000) *Phys. Rev. A* **62**, 012105.
48. Wiseman, H. M., Mancini, S. and Wang, J. (2002) *Phys. Rev. A* **66**, 013807.
49. Armen, M. A., Au, J. K., Stockton, J. K., Doherty, A. C. and Mabuchi, H. (2002) quant-ph/0204005.
50. Sprinzak, D., Buks, E., Heiblum, M. and Shtrikman, H. (2000) *Phys. Rev. Lett.* **84**, 5820–5823.

LINEAR QUANTUM MEASUREMENTS

D.V. AVERIN
*Department of Physics and Astronomy,
Stony Brook University, SUNY
Stony Brook, New York 11794-3800, U.S.A.*

Abstract.
Linear response theory describes quantum measurement with an arbitrary detector weakly coupled to a measured system. This description produces generic quantitative relation characterizing the detector that is analogous to the fluctuation-dissipation theorem for equilibrium systems. The detector characteristic obtained in this way shows how effective is the trade-off between the back-action dephasing and information acquisition by the detector.

1. Introduction

If one puts aside philosophical questions created by the perceived counterintuitive features of the quantum mechanics that are frequently discussed in the literature on the quantum measurement problem (an entry point to this literature is provided, e.g., by the collections of papers [1–3] or monographs [4, 5]) it is easy to see that the physics of quantum measurements is fairly well understood by now at least on a qualitative level. The process of quantum measurement is dynamic interaction between a microscopic quantum system and a macroscopic detector that establishes correlations between the states of these systems. Although it is impossible to give a universal definition of "macroscopic" in this context, a reasonable working definition is that the macroscopic detector is a system with negligible quantum fluctuations. Such systems are quite abundant in nature and can be described quantitatively within the framework of quantum mechanics.

Since the description of quantum measurement process as an interaction between microscopic and macroscopic systems is quite broad, it is of interest to see whether there are any universal quantitative features of this process that are independent of specific physical realization of the detectors and the measured system. The purpose of this work is to show that one such universal description of the measurement process can be obtained from the linear response theory. This was first pointed out in [6], and this work gives concise and more accurate presentation of

Y.V. Nazarov (ed.), Quantum Noise in Mesoscopic Physics, 229–239.

this result. One example of the area where a universal description of the quantum measurement process can be applied usefully is mesoscopic quantum dynamics of solid-state, in particular Josephson-junction, qubits. Recent widespread interest in the development of solid-state qubits for quantum information processing brought with it the discussion of a large number of different detectors for measurement of quantum dynamics of these qubits. These detectors include quantum point contacts [7–11], normal-metal [12–15] and superconducting [16, 17] SET transistors operated in different regimes, resonant tunneling structures [18], generic mesoscopic conductors [19], SQUIDs [20]. With such a variety of detectors (the list of references above is by no means complete), a possibility of giving a quantitative description of some detector properties independently of its physical realization is obviously very helpful for understanding of the process of quantum measurement.

2. Measurement model and basic relations

As was discussed in the introduction, the most essential feature of the model of the quantum measurement process considered in this work is its universality. The model applies to measurement with an arbitrary detector which have to satisfy only some basic conditions. Schematics of this model is shown in Fig. 1 and includes the measured system and the detector with the Hamiltonians H_S and H_D, respectively. The detector and system are coupled through an interaction Hamiltonian H_{int} which almost without any loss of generality can be written as the product of some system "coordinate" x and the detector "force" f:

$$H_{int} = xf\,. \tag{1}$$

The operator x acts as the observable measured by the detector. In principle, one could imagine a situation when a detector couples to some system through a combination of the several product terms like (1), but the detector sensitive to several different dynamic variables of the system would be quite unusual.

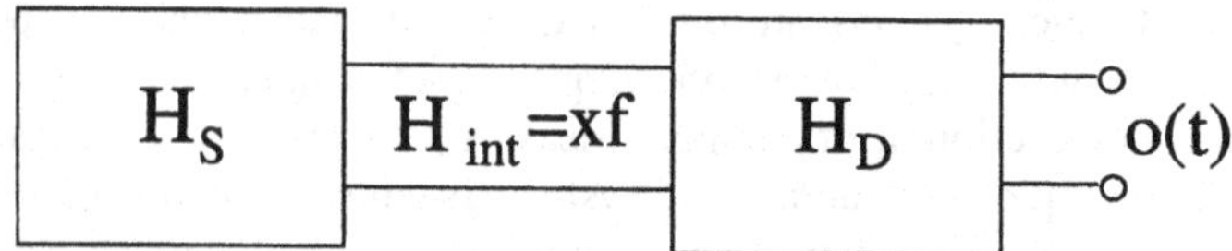

Figure 1. Schematics of the generic quantum measurement process. The detector couples to the measured system via the product of the measured observable x and the detector force f and variation of its state is reflected in the output variable o with essentially classical dynamics.

The general structure of the Hamiltonian of the system:

$$H = H_S + H_D + H_{int}\,, \tag{2}$$

is similar to the Hamiltonians studied in the discussions of the dissipative quantum dynamics based on the "system+reservoir" models (see, e.g., [21]) if the detector is identified with the reservoir. This analogy is not accidental, and many features of the dynamics of the measurement process are similar to the dissipative quantum dynamics of open systems. For instance, as will also be discussed below, coupling to the detector leads to the dephasing of the measured system, similar to the reservoir-induced dephasing.

There are, however, important differencies between the two situations. The main difference is that in the case of the dissipative quantum dynamics induced by a reservoir, changes in the reservoir caused by the system of interest are assumed to be unobservable, while the detector as a matter of principle should have an output observable $o(t)$ that provides some information about its state. To represent the complete measurement process, $o(t)$ should behave as a classical variable, i.e. magnitude of quantum fluctuations in it should be much smaller than the classical component. This condition implies that there is another difference between the detector and the dissipative reservoir.The detector should be able to convert weak quantum input signal $x(t)$ into the classical output signal $o(t)$ that is large on the scale of quantum fluctuations. Such a transformation requires amplification and can not be achieved in an equilibrium system, i.e., the detector, in contrast to the reservoir has to be driven out of the equilibrium.

In the case of the dissipative reservoirs, it is well known that the linear response theory is a powerful tool that produces a general quantitative statement, fluctuation-dissipation theorem, about the reservoir-induced relaxation. The theorem is independent of the specific microscopic model of the system. The analogy between the measurement and the environment-induced relaxation leads naturally to the question whether some general relations characterizing quantum measurement can be derived from the linear response theory. As we will see below, the answer to this question is yes, and the linear response theory applied to the measurement problem enables one to derive generic restriction on the energy sensitivity of a general linear quantum detector, characterizing how close the detector is to being quantum-limited, i.e., how close the detector-induced back-action dephasing in the measured system is to the fundamental limit set by the quantum mechanics. While such a restriction on the energy sensitivity was known before in several specific detector models, linear response theory establishes it for an arbitrary detector weakly coupled to the measured system.

For the linear response theory to be applicable to the measurement set-up shown in Fig. 1, and to provide some meaningful information, the detector has to satisfy certain conditions. Two most important are:

- the detector is weakly coupled to the measured system so that its response to variations in this system is linear,
- the detector is in the stationary state.

Since the detector can not be in equilibrium, the second item on this list generalizes the equilibrium condition of the standard linear-response theory. Both of these conditions are not very restrictive, and are satisfied in many experimental situations.

Linearity of the response is not sufficient to specify the detector properties completely, since the linear response coefficient can have an arbitrary frequency dependence, i.e. one needs to assume some spectral characteristics of the detector. Here we consider the simplest case when the response time of the detector is much shorter that the characteristic time of the dynamics of the measured system. This means that the response coefficient λ is constant in the relevant frequency range. Although this assumption is less fundamental that the previous two, and in principle can be avoided, it simplifies the discussion considerably.

Quantitatively, linearity of the detector response means that dynamics of the detector operators can be treated in the lowest order of the perturbation theory in the detector-system coupling H_{int}. Expanding the evolution operator $U(t) = e^{-iHt/\hbar}$ with the total Hamiltonian (2) upto the first order in H_{int} we see that the detector output $o(t)$ can be written as:

$$o(t) = q(t) + \frac{i}{\hbar}\int^{t} d\tau [f(\tau), q(t)] x(\tau) \, . \tag{3}$$

The first term $q(t) \equiv e^{iH_D t/\hbar} o(0) e^{-iHt/\hbar}$ in this expression can be interpreted as the output noise of the detector that exists even in the absence of the input signal, while the second term represents the detector's linear response to measured system. In zeroth order in coupling, the detector and the system are independent, so the averages over their states in eq. (3) can be taken separately. Tracing out the detector variables in (3) we obtain the regular expression for the linear response, extended to the situation when the "driving force" $x(t)$ is a quantum operator. With the adopted assumption of instantaneous detector response:

$$\frac{i}{\hbar}\langle [f(t), q(t+\tau)] \rangle_D = \lambda \delta(\tau - 0) \tag{4}$$

eq. (3) reduces to:

$$\langle o(t) \rangle_D = \lambda x(t) \, . \tag{5}$$

The infinitesimal shift in the argument of the δ-function represents small but finite response time of the detector, and is needed to resolve the ambiguity in eq. (3). In equation (4), we used the assumption (introduced above) that the detector is in the stationary state and the q–f correlator depends therefore only on the difference of the time arguments, and $\langle \ldots \rangle_D$ denotes the average of the stationary detector density matrix. Since $q(t)$ represents the output noise, we took $\langle q(t) \rangle_D = 0$.

Similarly, if the detector produces some non-vanishing average force $\langle f(t) \rangle_D \equiv F \neq 0$, it is natural to include it in the system Hamiltonian H_S, and one can therefore always assume that $\langle f(t) \rangle_D = 0$. The dynamics of the coupling force f generated by the detector is characterized then by the correlation function $\langle ff(t)\rangle_D$ and physically can be viewed as "back-action" noise that dephases the measured system. Following the assumption of the instantaneous response (4) we also take both the output noise and the back-action noise to be δ-correlated:

$$\langle q(t+\tau)q(t)\rangle_D = 2\pi S_q\delta(\tau)\,, \quad \langle f(t+\tau)f(t)\rangle_D = 2\pi S_f\delta(\tau)\,. \tag{6}$$

Here we introduced spectral densities of the output (S_q) and back-action (S_f) noise that are constant in the relevant low frequency range. Besides the absence of characteristic internal frequency of the detector, eqs. (6) also imply that dynamics of both $q(t)$ and $f(t)$ is essentially classical, since the frequency-dependent part of the spectrum that corresponds to zero-point fluctuations gives negligible contribution to the correlators (6).

3. Back-action dephasing versus acquisition of information

Qualitatively, measurement dynamics defined by the total Hamiltonian (2) and assumptions (4) and (6) about the detector characteristics, includes two related processes. One is the dephasing and relaxation or excitation of the measured system by the back-action force $f(t)$. This part of the measurement is very similar to the dissipative quantum dynamics of a system weakly coupled to a reservoir. The second process is acquisition of information about the state of the measured system reflected in the evolution of the detector output $o(t)$. On the quantitative level, measurement dynamics depends on the Hamiltonian H_S of the measured system and the operator structure of the measured observable x.

Probably the simplest situation which can be used to illustrate the two qualitative aspects of the measurement dynamics is the case of the stationary system, $H_S = 0$, with an observable x that has several discrete eigenstates $|j\rangle$: $x|j\rangle = x_j|j\rangle$ and eigenvalues x_j. Since under the assumption (6) the back-action force $f(t)$ can be treated as a classical random force generated by the detector, one can write the time-evolution equation directly for the density matrix ρ of the measured system. From the Hamiltonian (2), this equation in the basis of x-eigenstates is:

$$\dot{\rho}_{j,j'} = \frac{-i}{\hbar}(x_j - x_{j'})f(t)\rho_{j,j'}\,.$$

Solution of this equation averaged over different realizations of $f(t)$ using eq. (6), gives:

$$\rho_{j,j'}(t) = \rho_{j,j'}(0)e^{-\Gamma_d t}\,, \quad \Gamma_d \equiv \pi(x_j - x_{j'})^2 S_f/\hbar^2\,. \tag{7}$$

¿From this equation one can see that when the system is prepared in the initial state which is a superposition of the eigenstates of x, i.e. $\rho_{j,j'}$ has non-vanishing off-diagonal elements, interaction with the detector dephases the system and suppresses these off-diagonal elements. The suppression does not happen only if the states j and j' are degenerate with respect to x: $x_j = x_{j'}$, so that the detector does not distinguish them. Equation (7) shows also that the rate Γ of such a back-action dephasing is determined by the spectral density of the back-action force $f(t)$.

The dephasing (7) leads to diagonalization of the density matrix in the basis of the measured observable and coincides with the simplest instance of the environment-induced dephasing of an open quantum system. From the measurement perspective, it describes the dynamic side of the "wave-function collapse", since when the measurement is completed, the measured system should find itself in one of the eigenstates of the observable being measured, i.e. x. The density matrix diagonal in the x representation corresponds precisely to such a situation. This interpretation of the back-action dephasing is strengthened by the relation between the dephasing rate Γ and the rate of information acquisition - see, e.g., [22]. Indeed, from this perspective, in the example of the preceding paragraph, the detector has to distinguish different eigenstates of the observable x and to provide information in which eigenstate the system finds itself. According to the linear response relation (5), the difference between two eigenvalues of x, $x_j - x_{j'}$, translates into the difference $\delta o = \lambda(x_j - x_{j'})$ of the dc values of the detector output o. The characteristic time τ_m on which this difference in o can be distinguished in the presence of the output noise is determined by the low-frequency spectral density S_q of this noise. Averaging the δ-correlated noise over the time interval Δt leaves characteristic noise amplitude $\Delta o = (2\pi S_q/\Delta t)^{1/2}$. The characteristic time τ_m of acquisition of the information about the dc value of the detector output, and therefore, about the eigenstate of the measured observable, is determined by the condition that the noise amplitude is reduced at least to half the distance δo between the signal level. This condition gives

$$\tau_m = 8\pi S_q/[\lambda(x_j - x_{j'})]^2 \,, \tag{8}$$

and shows that the relation between the back-action dephasing rate and the "measurement time" τ_m is independent of the eigenvalue difference $x_j - x_{j'}$, and depends only on the linear-response parameters of the detector:

$$\tau_m \Gamma_d = 8(\pi/\hbar\lambda)^2 S_q S_f \,. \tag{9}$$

Now the final step is to show that the linear response theory relates parameters λ, S_q, S_f, in such a way that the dephasing rate is fundamentally linked to the measurement time:

$$\tau_m \Gamma_d \geq 1/2 \,. \tag{10}$$

To see this, we start by taking Fourier transform of eq. (4):

$$-i\frac{\hbar\lambda}{2\pi} = S_{fq}(\omega) - S_{fq}^*(-\omega), \quad S_{fq}(\omega) = \frac{1}{2\pi}\int d\tau e^{i\omega\tau}\langle fq(\tau)\rangle_D \,. \tag{11}$$

Expanding expression for the correlator $S_{fq}(\omega)$ in the basis of the energy eigenstates $|\varepsilon\rangle$ of the detector we obtain:

$$S_{fq}(\omega) = \int d\varepsilon \rho_D(\varepsilon)\nu(\varepsilon)\nu(\varepsilon - \hbar\omega)\langle\varepsilon|f|\varepsilon - \hbar\omega\rangle\langle\varepsilon - \hbar\omega|q|\varepsilon\rangle \,, \qquad (12)$$

where $\nu(\varepsilon)$ is the density of the detector energy states, and since the detector was assumed to be in the stationary state, its density matrix $\rho_D(\varepsilon)$ in the energy basis is diagonal.

Making use of the expression (12) and similar expressions for noise spectral densities, we can establish an inequality relating them. To obtain it, we note that eq. (12) can be viewed from a somewhat artificial, but useful perspective as a scalar product of two functions of energy ε, $\langle\varepsilon - \hbar\omega|q|\varepsilon\rangle$ and $\langle\varepsilon - \hbar\omega|f|\varepsilon\rangle$. Since both the density of states and the probability $\rho_D(\varepsilon)$ are positive, the scalar product defined by eq. (12) satisfies all usual requirements of the scalar product. Using the standard notation for this product, we can write (12) simply as

$$S_{fq}(\omega) =< f|q > \, . \qquad (13)$$

Spectral densities of the back-action and the output noise can be also expressed in terms of this scalar product. The total spectral densities S_q, S_f are defined as

$$S_q(\omega) = (S_{qq}(\omega) + S_{qq}(-\omega))/2, \quad S_{qq}(\omega) = \frac{1}{2\pi}\int d\tau e^{i\omega\tau}\langle qq(\tau)\rangle_D \,, \qquad (14)$$

with similar equations for $S_f(\omega)$. Writing the correlators $S_{qq}(\omega)$ and $S_{ff}(\omega)$ in the same way as $S_{fq}(\omega)$ (12) in the basis of the detector energy eigenstates one sees immediately that they again can be expressed very simply in the language of the scalar product:

$$S_{qq}(\omega) =< q|q > \,, \quad S_{ff}(\omega) =< f|f > \, . \qquad (15)$$

Schwarz inequality for the introduced scalar product gives, when applied to eqs. (13) and (15):

$$S_{ff}(\omega)S_{qq}(\omega) \geq \mid S_{fq}(\omega) \mid^2 \, . \qquad (16)$$

Finally, combining eq. (11), inequality (16), similar inequality for components of the correlators at frequency $-\omega$, and our assumption that the noise spectral densities $S_q(\omega)$ and $S_f(\omega)$ are constant at low frequencies, we see that

$$\lambda = -4\pi \mathrm{Im} S_{fq}/\hbar \,,$$

and therefore

$$|\lambda| \leq \frac{4\pi}{\hbar}[S_f S_q - (\mathrm{Re} S_{fq})^2]^{1/2} \,, \qquad (17)$$

where all spectral densities and the response coefficient are now taken at low frequencies. This inequality is the main result of application of the linear response theory to the measurement problem.

One consequence of the inequality (17) is the proof of the relation (10) between the back-action dephasing rate and the measurement time. It follows directly from eq. (17) that when the detector is "symmetric", i.e. Re$S_{fq} = 0$, so that there are no classical correlations between the back-action noise and the output noise, inequality (17) transforms eq. (9) directly into the (10). In this derivation of (10), this inequality appears to be the consequence of some relation between the correlators appearing in the linear response theory. From the perspective of the measurement problem it can be interpreted in a much broader sense. Inequality (10) shows that in general the dephasing of the measured system by a detector can be arbitrary strong without providing any information on the system. However, acquisition of information about the state of the system creates some minimum dephasing that suppresses coherence between the eigenstates of the measured observable. The fact that such a minimum exists is dictated by the basic principle of quantum mechanics which requires the successful measurement of an observable to localize the system in one of the eigenstates of this observable. If the detector is such that it is causing only this minimum back-action dephasing dictated by quantum mechanics, it is typically referred to as "quantum-limited" or "ideal" detector.

Interplay between information acquisition by the detector and back-action dephasing of the measured system has a simple form of inequality (10) only in the situation of measurement of a static system. When the system Hamiltonian H_S is non-vanishing, this interplay is affected by the dynamics of the system and in general manifests itself less directly. One studied example of this [23, 24] is the measurement of a *two-state system*. Coherent quantum oscillations in this system are transformed by measurement into classical oscillations of the detector output. In this case, the trade-off between the back-action dephasing and information acquisition is reflected in the height of the oscillations peak S_m in the output spectrum relative to the output noise S_q. Inequality (10) limits then the peak height: $S_m/S_q \leq 4$, with eaquality reached by measurement with a symmetric quantum limited detector.

Another system which has been studied in this context is *harmonic oscillator* [25, 6]. Similarly to the two-state system, dynamics of the measurement process is reflected in this case in the frequency spectrum of the detector output, which contains the detector output noise and a peak at the oscillator frequency. The trade-off between the detector back-action and information gain becomes relevant if one asks a question what is the minimum contribution of the detector noise to the spectrum. Qualitatively, there are two ways in which the detector noise contributes to the spectrum. The detector output noise gives direct contribution to the spectrum, while the back-action noise contributes to the output spectrum indirectly, by inducing additional oscillations at the detector input that are transformed to the output by the detector response. Minimization of the total noise contribution (reduced to the detector input and normalized to the zero-point spectrum of the oscillator) shows that the minimum total noise is given by the "energy sensitivity" ϵ:

$$\epsilon = \frac{2\pi}{|\lambda|}[S_f S_q - (\mathrm{Re} S_{qf})^2]^{1/2} \,. \tag{18}$$

Energy sensitivity has the dimension of action, and relation (17) between the parameters of the linear response theory translates into the following limitation on ϵ:

$$\epsilon \geq \hbar/2 \,. \tag{19}$$

This inequality can be qualitatively expressed as a statement that the detector, when measuring a harmonic oscillator, adds at least half an excitation quantum of this oscillator to the measured signal.

Equation (19) was obtained for the first time in a different physical context of linear amplification of electromagnetic radiation, where this interpretation has direct meaning – see, e.g., the papers on quantum noise in linear amplifies reprinted in [1], also [26] and references therein. The magnitude of the signal in this context is described naturally by the number of photons in a mode and amplification is characterized by the dimensionless photon gain: the ration of these numbers in the output and input modes. Relation between such a linear amplifier and the linear detector considered in this work is established by the fact that in the limit of large gain, the input signal of the amplifier which in general can be quantum (i.e., contain only few photons) is transformed into the output signal which contains number of photons much larger than one and is in this sense classical. This means that the amplifier in this regime acts essentially as a detector, since after amplification of the signal to the classical level, it can be dealt with without any restrictions.

The last point to be made here concerns the role of the classical correlations between the output and back-action noises of the detector that are described by the nonvanishing real part of the spectral density S_{fq}. Comparison of eq. (9) for the trade-off between the measurement time and the back-action dephasing, and eq. (19) for the energy sensitivity shows that they behave differently in the case of non-vanishing $\mathrm{Re} S_{fq}$. While the energy sensitivity (19) can reach its quantum limit even in asymmetric detectors with $\mathrm{Re} S_{fq} \neq 0$, the $\tau_m \Gamma_d$ product (9) in this situation can only be larger than its quantum limit $1/2$. The origin of this discrepancy is that in asymmetric detectors the output noise contains some information about the back-action noise which should be utilized to reach the optimum quantum-limited performance of the detector. In the simple treatment that leads to inequality (10), this information is lost, the fact that increases the dephasing beyond the minimum required by the quantum mechanics. In the case of measurement of the harmonic oscillator, the noise minimization procedure implicitly makes use of this information, since the precise frequency where the noise is minimum depends on the correlation strength $\text{Re}S_{fq}$. In the case of information/dephasing trade-off, the information provided by the correlations contained in non-vanishing $\text{Re}S_{fq}$ can also be used to approach the quantum limit of dephasing. Conceptually (and probably also practically) the simplest way to use the information contained in the output noise of the detector to reduce the back-action dephasing is to apply the output noise with an appropriate transfer coefficient to the measured system together with the back-action noise. Although physically this procedure might require a rather complicated set-up that actually applies part of the detector output to the measured system, formally this is achieved by simply redefining the back-action force of the detector:

$$f \rightarrow f - (\text{Re}S_{fq}/S_q)o \equiv f' .$$

It is straightforward to see that the rate Γ_d' of back-action dephasing created by the redefined force f' in measurement of a static system can indeed be equal to the quantum minimum of $1/2\tau_m$ even for asymmetric detectors with $\text{Re}S_{fq} \neq 0$, if the relation (17) between the linear response properties of this detector is satisfied as equality. The same conclusion can be reached for measurement of the two-state systems [22]. This provides one more example of the fact that many physical characteristics of the the measurement process can be understood conveniently as dynamics of the information.

References

1. *Quantum theory and measurement*, Ed. by J.A. Wheeler and W.H. Zurek (Princeton Univ. Press, 1983).
2. *Quantum optics, experimental gravity, and measurement theory*, Ed. by P. Meystre and M.O. Scully (Plenum, NY, 1983).
3. *Sixty two years of uncertainty: historical, philosophical, and physical inquiries into the foundations of quantum mechanics*, Ed. by A.I. Miller (Plenum, NY, 1990).
4. V.B. Braginsky and F.Ya. Khalili, *Quantum measurement*, (Cambridge, 1992).
5. M.B. Mensky, *Quantum measurement and decoherence: models and phenomenology* (Kluwer, Dordrecht, 2000).
6. D.V. Averin, in: "*Exploring the quantum/classical frontier: recent advances in macroscopic quantum phenomena*", Ed. by J.R. Friedman and S. Han, (Nova Publishes, Hauppauge, NY, 2002), p. 441; cond-mat/0004364.
7. M. Field, C.G. Smith, M. Pepper, D.A. Ritchie, J.E.F. Frost, G.A.C. Jones, and D.G. Hasko, Phys. Rev. Lett. **70**, 1311 (1993).
8. M. Kataoka, C.J.B. Ford, G. Faini, D. Mailly, M.Y. Simmons, D.R. Mace, C.-T. Liang, and D. A. Ritchie, Phys. Rev. Lett. **83**, 160 (1999).
9. D. Sprinzak, E. Buks, M. Heiblum, and H. Shtrikman, Phys. Rev. Lett. **84**, 5820 (2000).
10. S.A. Gurvitz, Phys. Rev. B **56**, 15215 (1997).
11. A.N. Korotkov, Phys. Rev. B **60**, 5737 (1999).
12. Yu. Makhlin, G. Schön, and A. Shnirman, Phys. Rev. Lett. **85**, 4578 (2000).
13. D.V. Averin, in: "*Macroscopic Quantum Coherence and Quantum Computing*" Ed. by D.V. Averin, B. Ruggiero, and P. Silvestrini, (Kluwer, 2001), p. 399; cond-mat/0010052.

14. M.H. Devoret and R.J. Schoelkopf, Nature **406**, 1039 (2000).
15. G. Johansson, A. Käck, and G. Wendin, Phys. Rev. Lett. **88**, 046802 (2002).
16. A.B. Zorin, Phys. Rev. Lett. **76**, 4408 (1996).
17. A.A. Clerk, S.M. Girvin, A.K. Nguyen, and A.D. Stone, Phys. Rev. Lett. **89**, 176804 (2002).
18. D.V. Averin, Fortschrit. der Physik **48**, 1055 (2000).
19. S. Pilgram and M. Büttiker Phys. Rev. Lett. **89**, 200401 (2002).
20. S.-X. Li, Y. Yu, Y. Zhang, W. Qiu, S. Han, and Z. Wang, Phys. Rev. Lett. **89**, 098301 (2002).
21. U. Weiss, *Quantum dissipative systems*, (World Scientific, 1999).
22. A.N. Korortkov, cond-mat/0209629.
23. A.N. Korotkov and D.V. Averin, Phys. Rev. B **64**, 165310 (2001).
24. L.N. Bulaevskii, M. Hruska, G. Ortiz, cond-mat/0212049.
25. V.V. Danilov, K.K. Likharev, and A.B. Zorin, IEEE Trans. Magn. **19**, 572 (1983).
26. C.M: Caves, Phys. Rev. D **26**, 1817 (1982).

SHOT NOISE FOR ENTANGLED AND SPIN-POLARIZED ELECTRONS

J. C. EGUES†, P. RECHER, D. S. SARAGA, V. N. GOLOVACH, G. BURKARD, E. V. SUKHORUKOV, AND D. LOSS
Department of Physics and Astronomy, University of Basel, Klingelbergstrasse 82, CH-4056 Basel, Switzerland

Abstract. We review our recent contributions on shot noise for entangled electrons and spin-polarized currents in novel mesoscopic geometries. We first discuss some of our recent proposals for electron entanglers involving a superconductor coupled to a double dot in the Coulomb blockade regime, a superconductor tunnel-coupled to Luttinger-liquid leads, and a triple-dot setup coupled to Fermi leads. We briefly survey some of the available possibilities for spin-polarized sources. We use the scattering approach to calculate current and shot noise for spin-polarized currents and entangled/unentangled electron pairs in a novel beam-splitter geometry with a *local* Rashba spin-orbit (s-o) interaction in the incoming leads. For single-moded incoming leads, we find *continuous* bunching and antibunching behaviors for the *entangled* pairs – triplet and singlet – as a function of the Rashba rotation angle. In addition, we find that unentangled triplets and the entangled one exhibit distinct shot noise; this should allow their identification via noise measurements. Shot noise for spin-polarized currents shows sizable oscillations as a function of the Rashba phase. This happens only for electrons injected perpendicular to the Rashba rotation axis; spin-polarized carriers along the Rashba axis are noiseless. The Rashba coupling constant α is directly related to the Fano factor and could be extracted via noise measurements. For incoming leads with s-o induced interband-coupled channels, we find an additional spin rotation for electrons with energies near the crossing of the bands where interband coupling is relevant. This gives rise to an additional modulation of the noise for both electron pairs and spin-polarized currents. Finally, we briefly discuss shot noise for a double dot near the Kondo regime.

1. Introduction

Fluctuations of the current away from its average usually contain supplementary information, not provided by average-current measurements alone. This is particularly true in the non-linear response regime where these quantities are not related via the fluctuation-dissipation theorem. At zero temperature, non-equilibrium current noise is due to the discreteness of the electron charge and is termed shot noise. This dynamic noise was first investigated by Schottky in connection with

† PERMANENT ADDRESS: DEPARTMENT OF PHYSICS AND INFORMATICS, UNIVERSITY OF SÃO PAULO AT SÃO CARLOS, 13560-970 SÃO CARLOS/SP, BRAZIL.

Y. V. Nazarov (ed.), Quantum Noise in Mesoscopic Physics, 241–274.

thermionic emission [1]. Quantum shot noise has reached its come of age in the past decade or so and constitutes now an indispensable tool to probe mesoscopic transport [2]; in particular, the role of fundamental correlations such as those imposed by quantum statistics.

More recently, shot noise has been investigated in connection with transport of entangled [3]–[8] and spin polarized electrons [4], [8]–[11] and has proved to be a useful probe for both entanglement and spin-polarized transport. Entanglement [12] is perhaps one of the most intriguing features of quantum mechanics since it involves the concept of non-locality. Two-particle entanglement is the simplest conceivable form of entanglement. Yet, these Einstein-Podolsky-Rosen (EPR) pairs play a fundamental role in potentially revolutionary implementations of quantum computation, communication, and information processing [13]. In this context, such a pair represents two qubits in an entangled state. The generation and detection of EPR pairs of photons has already been accomplished. On the other hand, research involving two-particle entanglement of massive particles (e.g. electrons) in a solid-state matrix is still in its infancy, with a few proposals for its physical implementation; some of these involve quantum-dot setups as sources of mobile spin-entangled electrons [4], [14], [15]. Spin-polarized transport [16], [17], [18], on the other hand, is a crucial ingredient in semiconductor spintronics where the spin (and/or possibly the charge) of the carriers play the relevant role in a device operation. To date, robust spin injection has been achieved in Mn-based semiconductor layers (*pin* diode structures) [16]. High-efficiency spin injection in other semiconductor systems such as hybrid ferromagnetic/semiconductor junctions is still challenging.

It is clear that the ability to create, transport, coherently manipulate, and detect entangled electrons and spin-polarized currents in mesoscopic systems is highly desirable. Here we review some of our recent works [4], [7], [8], [14], [15], [19], [20] addressing some of these issues and others in connection with noise. Shot noise provides an additional probe in these novel transport settings. We first address the production of mobile entangled electron pairs (Sec. 2). We discuss three proposals involving a superconductor coupled to two dots [14], a superconductor coupled to Luttinger-liquid leads [19], and a triple-dot arrangement [15]. Our detailed analysis of these "entanglers" does not reveal any intrinsic limitation to their experimental feasibility. We also mention some of the available sources of spin-polarized electrons (Sec. 3). Ballistic spin filtering with spin-selective semimagnetic tunnel barriers [17] and quantum dots as spin filters [18] are also briefly discussed.

We investigate transport of entangled and spin polarized electrons in a beam-splitter (four-port) configuration [21], [22] with a local Rashba spin-orbit interaction in the incoming leads [23], Fig. (1). A local Rashba term provides a convenient way to coherently spin-rotate electrons as they traverse quasi one-dimensional channels, as was first pointed out by Datta and Das [24]. Within the scattering formalism [2], we calculate shot noise for both entangled and spin-polarized electrons.

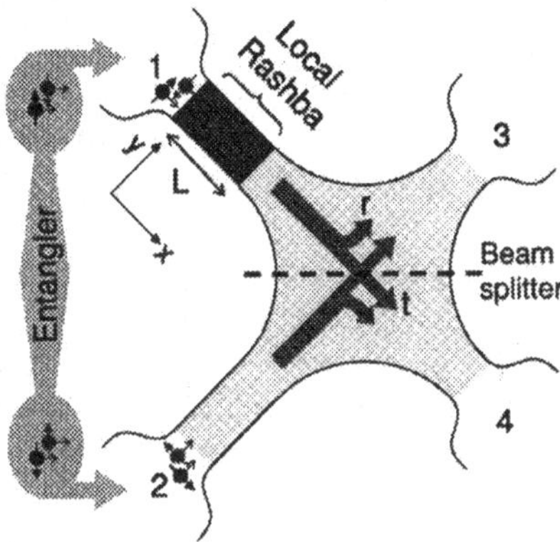

Figure 1. Novel electron beam-splitter geometry with a local Rashba s-o interaction in lead 1. An entangler or a spin-polarized electron source (not shown) inject either entangled pairs or spin-polarized carriers into leads 1 and 2. The portion of the entangled pairs (or the spin-polarized carriers) traversing lead 1 undergoes a Rashba-induced spin rotation. This *continuously* changes the symmetry of the *spin* part of the pair wave function. Adapted from Ref. [8] .

For entangled electrons, shot noise is particularly relevant as a probe for fundamental two-particle interference. More specifically, shot noise (charge noise) directly probes the orbital symmetry of the EPR pair wave function. However, the symmetry of the orbital degree of freedom ("the charge") is intrinsically tied to that of the spin part of the pair wave function via the Pauli principle. That is, the total electron-pair wave function is antisymmetric thus imposing a fundamental connection between the spin and orbital parts of the pair wave function. Hence charge noise measurements probe in fact the spin symmetry of the pair. Moreover, if one can alter the spin state of the pair (say, via some proper coherent spin rotation) this will definitely influence shot-noise measurements. This is precisely what we find here for singlet and triplet pairs.

The coherent local Rashba spin rotation in one of the incoming leads of our setup, continuously alters the (spin) symmetry of the pair wave function thus giving rise to sizable shot noise oscillations as a function of the Rashba phase. Noise measurements in our novel beam-splitter should allow one to distinguish entangled triplets from singlets and entangled triplets from the unentangled ones, through their Rashba phase. Entangled pairs display continuous bunching or antibunching behavior. In addition, triplets (entangled or not) defined along different quantization axes (x, y, or z) exhibit distinctive noise, thus allowing the detection of their spin polarization via charge noise measurements.

Shot noise for spin-polarized currents also probes effects imposed by the Pauli principle through the Fermi functions in the leads. These currents also exhibit Rashba-induced oscillations for spin polarizations perpendicular to the Rashba rotation axis. We find zero shot noise for spin-polarized carriers with polarizations along the Rashba axis and for unpolarized injection. Moreover, the Rashba-induced modulations of the Fano factor for both entangled and spin-polarized electrons offer a direct way to extract the s-o coupling constant via noise measurements.

We also consider incoming leads with two transverse channels. In the presence of a weak s-o induced interchannel coupling, we find an additional spin rotation due to the coherent transfer of carriers between the coupled channels in lead 1. This extra rotation gives rise to further modulation of the shot noise characteristics for both entangled and spin-polarized currents; this happens only for carriers with energies near the band crossings in lead 1. Finally, we briefly discuss shot noise for transport through a double dot near the Kondo regime [20].

2. Sources of mobile spin-entangled electrons

A challenge in mesoscopic physics is the experimental realization of an electron "entangler" – a device creating mobile entangled electrons which are spatially separated. Indeed, these are essential for quantum communication schemes and experimental tests of quantum non-locality with massive particles. First, one should note that entanglement is rather the rule than the exception in nature, as it arises naturally from Fermi statistics. For instance, the ground state of a helium atom is the spin singlet $|\uparrow\downarrow\rangle - |\downarrow\uparrow\rangle$. Similarly, one finds a singlet in the ground state of a quantum dot with two electrons. These "artificial atoms" [25] are very attractive for manipulations at the single electron level, as they possess tunable parameters and allow coupling to mesoscopic leads – contrary to real atoms. However, such "local" entangled singlets are not readily useful for quantum computation and communication, as these require control over each individual electron as well as non-local correlations. An improvement in this direction is given by two coupled quantum dots with a single electron in each dot [26], where the spin-entangled electrons are already spatially separated by strong on-site Coulomb repulsion (like in a hydrogen molecule). In this setup, one could create mobile entangled electrons by simultaneously lowering the tunnel barriers coupling each dot to separate leads. Another natural source of spin entanglement can be found in superconductors, as these contain Cooper pairs with singlet spin wave functions. It was first shown in Ref. [27] how a non-local entangled state is created in two uncoupled quantum dots when coupled to the same superconductor. In a non-equilibrium situation, the Cooper pairs can be extracted to normal leads by Andreev tunnelling, thus creating a flow of entangled pairs [14],[19],[28]–[31].

A crucial requirement for an entangler is to create *spatially separated* entangled electrons; hence one must avoid whole entangled pairs entering the same lead. As will be shown below, energy conservation is an efficient mechanism for the suppression of undesired channels. For this, interactions can play a decisive role. For instance, one can use Coulomb repulsion in quantum dots [14],[15] or in Luttinger liquids [19],[28]. Finally, we mention recent entangler proposals using leads with narrow bandwidth [32] and/or generic quantum interference effects [33]. In the following, we discuss our proposals towards the realization of an entangler that produces mobile non-local singlets [34]. We set $\hbar = 1$ in this section.

2.1. SUPERCONDUCTOR-BASED ELECTRON ENTANGLERS

Here we envision a *non-equilibrium* situation in which the electrons of a Cooper pair tunnel coherently by means of an Andreev tunnelling event from a SC to two separate normal leads, one electron per lead. Due to an applied bias voltage, the electron pairs can move into the leads thus giving rise to mobile spin entanglement. Note that an (unentangled) single-particle current is strongly suppressed by energy conservation as long as both the temperature and the bias are much smaller than the superconducting gap. In the following we review two proposals where we exploit the repulsive Coulomb charging energy between the two spin-entangled electrons in order to separate them so that the residual current in the leads is carried by non-local singlets. We show that such entanglers meet all requirements for subsequent detection of spin-entangled electrons via noise measurements (charge measurement, see Secs. 5 and 8) or via spin-projection measurements (Bell-type measurement, see Sec. 3.3).

2.1.1. *Andreev entangler with quantum dots*

The proposed entangler setup (see Fig. 2) consists of a SC with chemical potential μ_S which is weakly coupled to two quantum dots (QDs) in the Coulomb blockade regime [25]. These QDs are in turn weakly coupled to outgoing Fermi liquid leads, held at the same chemical potential μ_l. A bias voltage $\Delta\mu = \mu_S - \mu_l$ is applied between the SC and the leads. The tunnelling amplitudes between the SC and the dots, and dots and leads, are denoted by T_{SD} and T_{DL}, respectively (see Fig. 2). The two intermediate QDs in the Coulomb blockade regime have chemical potentials ϵ_1 and ϵ_2, respectively. These can be tuned via external gate voltages, such that the tunnelling of two electrons via different dots into different leads is resonant for $\epsilon_1 + \epsilon_2 = 2\mu_S$ [35]. As it turns out [14], this two-particle resonance is suppressed for the tunnelling of two electrons via the same dot into the same lead by the on-site repulsion U of the dots and/or the superconducting

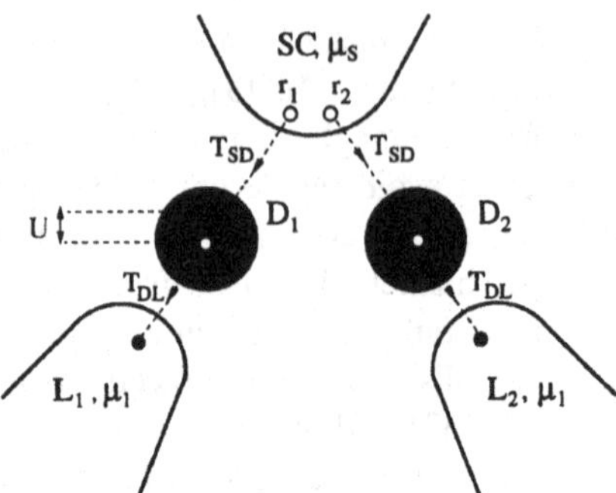

Figure 2. The entangler setup. Two spin-entangled electrons forming a Cooper pair tunnel with amplitude T_{SD} from points $\mathbf{r}_1$ and $\mathbf{r}_2$ of the superconductor, SC, to two dots, D_1 and D_2, by means of Andreev tunnelling. The dots are tunnel-coupled to normal Fermi liquid leads L_1 and L_2, with tunnelling amplitude T_{DL}. The superconductor and leads are kept at chemical potentials μ_S and μ_l, respectively. Adapted from [14].

gap Δ. Next, we specify the parameter regime of interest here in which the initial spin entanglement of a Cooper pair in the SC is successfully transported to the leads.

Besides the fact that single-electron tunnelling and tunnelling of two electrons via the same dot should be excluded, we also have to suppress transport of electrons which are already on the QDs. This could lead to effective spin-flips on the QDs, which would destroy the spin entanglement of the two electrons tunnelling into the Fermi leads. A further source of unwanted spin-flips on the QDs is provided by its coupling to the Fermi liquid leads via particle-hole excitations in the leads. The QDs can be treated each as one localized spin-degenerate level as long as the mean level spacing $\delta\epsilon$ of the dots exceeds both the bias voltage $\Delta\mu$ and the temperature k_BT. In addition, we require that each QD contains an even number of electrons with a spin-singlet ground state. A more detailed analysis of such a parameter regime is given in [14] and is stated here

$$\Delta, U, \delta\epsilon > \Delta\mu > \gamma_l, k_BT, \text{and } \gamma_l > \gamma_S. \tag{1}$$

In (1) the rates for tunnelling of an electron from the SC to the QDs and from the QDs to the Fermi leads are given by $\gamma_S = 2\pi\nu_S|T_{SD}|^2$ and $\gamma_l = 2\pi\nu_l|T_{DL}|^2$, respectively, with ν_S and ν_l being the corresponding electron density of states per spin at the Fermi level. We consider asymmetric barriers $\gamma_l > \gamma_s$ in order to exclude correlations between subsequent Cooper pairs on the QDs. We work at the particular interesting resonance $\epsilon_1, \epsilon_2 \simeq \mu_S$, where the injection of the electrons into different leads takes place at the same orbital energy. This is a crucial requirement for the subsequent detection of entanglement via noise [4, 8]. In this regime, we have calculated and compared the stationary charge current of two spin-entangled electrons for two competing transport channels in a T-matrix approach.

The ratio of the desired current for two electrons tunnelling into *different* leads (I_1) to the unwanted current for two electrons into the *same* lead (I_2) is [14]

$$\frac{I_1}{I_2} = \frac{4\mathcal{E}^2}{\gamma^2}\left[\frac{\sin(k_F\delta r)}{k_F\delta r}\right]^2 e^{-2\delta r/\pi\xi}, \qquad \frac{1}{\mathcal{E}} = \frac{1}{\pi\Delta} + \frac{1}{U}, \tag{2}$$

where $\gamma = \gamma_1 + \gamma_2$. The current I_1 becomes exponentially suppressed with increasing distance $\delta r = |\mathbf{r}_1 - \mathbf{r}_2|$ between the tunnelling points on the SC, on a scale given by the superconducting coherence length ξ which is the size of a Cooper pair. This does not pose a severe restriction for conventional s-wave materials with ξ typically being on the order of μm. In the relevant case $\delta r < \xi$ the suppression is only polynomial $\propto 1/(k_F\delta r)^2$, with k_F being the Fermi wave vector in the SC. On the other hand, we see that the effect of the QDs consists in the suppression factor $(\gamma/\mathcal{E})^2$ for tunnelling into the same lead [36]. Thus, in addition to Eq. (1) we have to impose the condition $k_F\delta r < \mathcal{E}/\gamma$, which can be satisfied for small dots with $\mathcal{E}/\gamma \sim 100$ and $k_F^{-1} \sim 1\,\mathcal{A}$. As an experimental probe to test if the two spin-entangled electrons indeed separate and tunnel to different leads we suggest to join the two leads 1 and 2 to form an Aharonov-Bohm loop. In such a setup the different tunnelling paths of an Andreev process from the SC via the dots to the leads can interfere. As a result, the measured current as a function of the applied magnetic flux ϕ threading the loop contains a phase coherent part I_{AB} which consists of oscillations with periods h/e and $h/2e$ [14]

$$I_{AB} \sim \sqrt{8I_1I_2}\cos(\phi/\phi_0) + I_2\cos(2\phi/\phi_0), \tag{3}$$

with $\phi_0 = h/e$ being the single-electron flux quantum. The ratio of the two contributions scales like $\sqrt{I_1/I_2}$ which suggest that by decreasing I_2 (e.g. by increasing U) the $h/2e$ oscillations should vanish faster than the h/e ones.

We note that the efficiency as well as the absolute rate for the desired injection of two electrons into different leads can even be enhanced by using lower dimensional SCs [19, 37] . In two dimensions (2D) we find that $I_1 \propto 1/k_F\delta r$ for large $k_F\delta r$, and in one dimension (1D) there is no suppression of the current and only an oscillatory behavior in $k_F\delta r$ is found. A 2D-SC can be realized by using a SC on top of a two-dimensional electron gas (2DEG) [38], where superconducting correlations are induced via the proximity effect in the 2DEG. In 1D, superconductivity was found in ropes of single-walled carbon nanotubes [39].

Finally, we note that the coherent injection of Cooper pairs by an Andreev process allows the detection of individual spin-entangled electron pairs in the leads. The delay time τ_{delay} between the two electrons of a pair is given by $1/\Delta$, whereas the separation in time of subsequent pairs is given approximately by $\tau_{\text{pairs}} \sim 2e/I_1 \sim \gamma_l/\gamma_S^2$ (up to geometrical factors) [14]. For $\gamma_S \sim \gamma_l/10 \sim 1\mu$eV and $\Delta \sim 1$meV we obtain that the delay time $\tau_{\text{delay}} \sim 1/\Delta \sim 1$ps is much smaller than the delivery time τ_{pairs} per entangled pair $2e/I_1 \sim 40$ns. Such a time separation is indeed necessary in order to detect individual pairs of spin-entangled electrons.

2.1.2. *Andreev entangler with Luttinger-liquid leads*

Next we discuss a setup with an s-wave SC weakly coupled to the center (bulk) of two separate one-dimensional leads (quantum wires) 1,2 (see Fig. 3) which exhibit Luttinger liquid (LL) behavior, such as carbon nanotubes [40–42]. The leads are assumed to be infinitely extended and are described by conventional LL-theory [44].

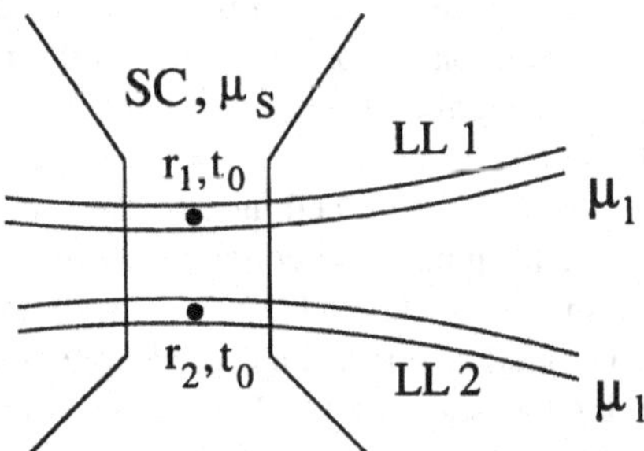

Figure 3. Two quantum wires 1,2, with chemical potential μ_l and described as infinitely long Luttinger liquids (LLs), are deposited on top of an s-wave superconductor (SC) with chemical potential μ_S. The electrons of a Cooper pair can tunnel by means of an Andreev process from two points $\mathbf{r}_1$ and $\mathbf{r}_2$ on the SC to the center (bulk) of the two quantum wires 1 and 2, respectively, with tunnelling amplitude t_0. Adapted from [19].

Interacting electrons in one dimension lack the existence of quasi particles like they exist in a Fermi liquid and instead the low energy excitations are collective charge and spin modes. In the absence of backscattering interaction the velocities of the charge and spin excitations are given by $u_\rho = v_F/K_\rho$ for the charge and $u_\sigma = v_F$ for the spin, where v_F is the Fermi velocity and $K_\rho < 1$ for repulsive interaction between electrons ($K_\rho = 1$ corresponds to a 1D-Fermi gas). As a consequence of this non-Fermi liquid behavior, tunnelling into a LL is strongly suppressed at low energies. Therefore one should expect additional interaction effects in a two-particle tunnelling event (Andreev process) of a Cooper pair from the SC to the leads. We find that strong LL-correlations result in an additional suppression for tunnelling of two coherent electrons into the *same* LL compared to single electron tunnelling into a LL if the applied bias voltage μ between the SC and the two leads is much smaller than the energy gap Δ of the SC.

To quantify the effectiveness of such an entangler, we calculate the current for the two competing processes of tunnelling into different leads (I_1) and into the same lead (I_2) in lowest order via a tunnelling Hamiltonian approach. Again we account for a finite distance separation δr between the two exit points on the SC when the two electrons of a Cooper pair tunnel to different leads. For the current I_1 of the desired pair-split process we obtain, in leading order in μ/Δ and at zero temperature [19, 37]

$$I_1 = \frac{I_1^0}{\Gamma(2\gamma_\rho + 2)} \frac{v_F}{u_\rho} \left[\frac{2\Lambda\mu}{u_\rho}\right]^{2\gamma_\rho}, \quad I_1^0 = \pi e \gamma^2 \mu F_d[\delta r], \tag{4}$$

where $\Gamma(x)$ is the Gamma function and Λ is a short distance cut-off on the order of the lattice spacing in the LL and $\gamma = 4\pi\nu_S\nu_l|t_0|^2$ is the dimensionless tunnel conductance per spin with t_0 being the bare tunnelling amplitude for electrons to tunnel from the SC to the LL-leads (see Fig. 3). The electron density of states per spin at the Fermi level for the SC and the LL-leads are denoted by ν_S and ν_l, respectively. The current I_1 has its characteristic non-linear form $I_1 \propto \mu^{2\gamma_\rho+1}$ with $\gamma_\rho = (K_\rho + K_\rho^{-1})/4 - 1/2 > 0$ being the exponent for tunnelling into the bulk of a *single* LL. The factor $F_d[\delta r]$ in (4) depends on the geometry of the device and is given here again by $F_d[\delta r] = [\sin(k_F\delta r)/k_F\delta r]^2 \exp(-2\delta r/\pi\xi)$ for the case of a 3D-SC. In complete analogy to subsection 2.1.1 the power law suppression in $k_F\delta r$ gets weaker in lower dimensions.

This result should be compared with the unwanted transport channel where two electrons of a Cooper pair tunnel into the same lead 1 or 2 but with $\delta r = 0$. We find that such processes are indeed suppressed by strong LL-correlations if $\mu < \Delta$. The result for the current ratio I_2/I_1 in leading order in μ/Δ and for zero temperature is [19, 37]

$$\frac{I_2}{I_1} = F_d^{-1}[\delta r] \sum_{b=\pm 1} A_b \left(\frac{2\mu}{\Delta}\right)^{2\gamma_{\rho b}}, \quad \gamma_{\rho+} = \gamma_\rho, \; \gamma_{\rho-} = \gamma_\rho + (1 - K_\rho)/2, \tag{5}$$

where A_b is an interaction dependent constant [45]. The result (5) shows that the current I_2 for injection of two electrons into the same lead is suppressed compared to I_1 by a factor of $(2\mu/\Delta)^{2\gamma_{\rho+}}$, if both electrons are injected into the same branch (left or right movers), or by $(2\mu/\Delta)^{2\gamma_{\rho-}}$ if the two electrons travel in different directions [46]. The suppression of the current I_2 by $1/\Delta$ reflects the two-particle correlation effect in the LL, when the electrons tunnel into the same lead. The larger Δ, the shorter the delay time is between the arrivals of the two partner electrons of a Cooper pair, and, in turn, the more the second electron tunnelling into the same lead will feel the existence of the first one which is already present in the LL. This behavior is similar to the Coulomb blockade effect in QDs, see subsection 2.1.1. Concrete realizations of LL-behavior is found in metallic carbon nanotubes with similar exponents as derived here [41, 42]. In metallic single-walled carbon nanotubes $K_\rho \sim 0.2$ [40] which corresponds to $2\gamma_\rho \sim 1.6$. This suggests the rough estimate $(2\mu/\Delta) < 1/k_F\delta r$ for the entangler to be efficient. As a consequence, voltages in the range $k_BT < \mu < 100\mu\text{eV}$ are required for

$\delta r \sim$ nm and $\Delta \sim 1$meV. In addition, nanotubes were reported to be very good spin conductors [43] with estimated spin-flip scattering lengths of the order of μm [28].

We remark that in order to use the beam-splitter setup to detect spin entanglement via noise the two LL-leads can be coupled further to Fermi liquid leads. In such a setup the LL-leads then would act as QDs [47]. Another way to prove spin entanglement is to carry out spin-dependent current-current correlation measurements between the two LLs. Such spin dependent currents can be measured e.g. via spin filters (Sec. 3).

2.2. TRIPLE-QUANTUM DOT ENTANGLER

In this proposal [15], the pair of spin-entangled electrons is provided by the ground state of a single quantum dot D_C with an even number of electrons, which is the spin-singlet [48]; see Fig. 4. In the Coulomb blockade regime [25], electron interactions in each dot create a large charging energy U that provides the energy filtering necessary for the suppression of the non-entangled currents. These arise either from the escape of the pair to the same lead, or from the transport of a single electron. The idea is to create a resonance for the joint transport of the two electrons from D_C to secondary quantum dots D_L and D_R, similarly to the resonance described in Sec. 2.1.1 . For this, we need the condition $\epsilon_L + \epsilon_R = 2\epsilon_C$, where ϵ_L and ϵ_R are the energy levels of the available state in D_L and D_R, and $2\epsilon_C$ is the total energy of the two electrons in D_C. On the other hand, the transport of a single electron from D_C to D_L or D_R is suppressed by the energy mismatch $\epsilon_C \pm U \neq \epsilon_L, \epsilon_R$, where $\epsilon_C \pm U$ is the energy of the $2^{\text{nd}}/1^{\text{st}}$ electron in D_C [49].

We describe the incoherent sequential tunneling between the leads and the dots in terms of a master equation [50] for the density matrix ρ of the triple-dot system

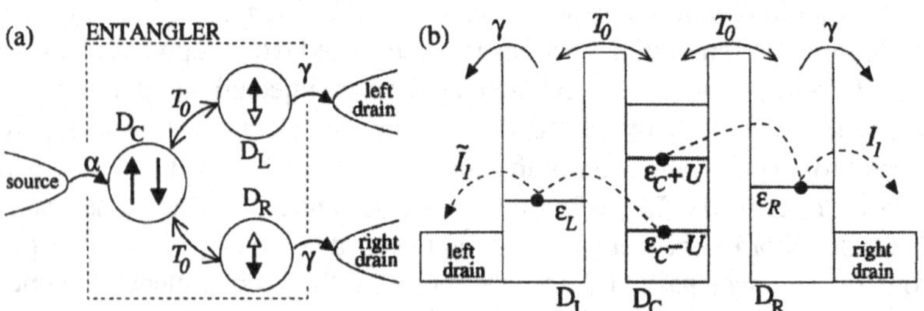

Figure 4. (a) Setup of the triple quantum dot entangler. Three leads are coupled to three quantum dots in the Coulomb blockade regime. The central dot D_C can accept 0, 1 or 2 electrons provided from the source lead with rate α; with 2 electrons, its ground state is the spin singlet. The tunnelling amplitudes T_0 describe the coherent tunnelling between D_C and the secondary dots D_L and D_R, which can only accept 0 or 1 electron. Each electron can finally tunnel out to the drain leads with a rate γ. (b) *Single-particle* energy level diagram. The dashed arrows represent the single-electron currents I_1 and $\tilde{I}_1$. Adapted from [15].

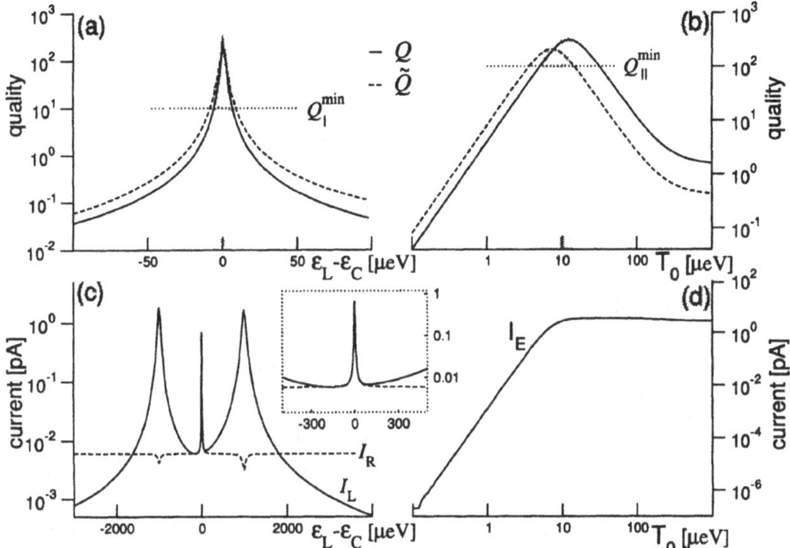

Figure 5. Quality and current of the entangler, with the parameters $\alpha = 0.1, \gamma = 1, T_0 = 10, U = 1000$ in μeV. (a) Quality Q and $\widetilde{Q}$, around the resonance at $\epsilon_L - \epsilon_C = 0$ where the entangled current dominates. In gray, the width of the resonance defined by $Q, \widetilde{Q} > Q_{\rm I}^{\rm min} = 10$ is $|\epsilon_L - \epsilon_C| < 6\,\mu$eV, as predicted by Eq.(6). (b) Q and $\widetilde{Q}$ as a function of T_0 at resonance ($\epsilon_L = \epsilon_C$). In gray, the region where the quality of the entangler is $Q, \widetilde{Q} > Q_{\rm II}^{\rm min} = 100$ corresponding to Eq. (7). (c) Entangled and non-entangled current in the left (I_L) and in the right (I_R) drain leads. The inset shows the resonance in a larger scale. (d) Saturation of the entangled current I_E. Adapted from [15].

(valid for $k_{\rm B}T > \gamma$). The stationary solution of the master equation is found with MAPLE, and is used to define stationary currents. Besides the entangled current I_E coming from the *joint* transport of the electrons from D_C to D_L and D_R, the solitary escape of one electron of the singlet can create a non-entangled current I_1, as it could allow a new electron coming from the source lead to form a new spin-singlet with the remaining electron. Another non-entangled current $\tilde{I}_1$ can be present if only one electron is transported across the triple-dot system; see Fig. 4(b). The definition of entangler *qualities* $Q = I_E/I_1$ and $\tilde{Q} = I_E/\tilde{I}_1$ enables us to check the suppression of these non-entangled currents.

In Fig. 5 we present results in the case where $\epsilon_R = \epsilon_C$. This gives a two-electron resonance at $\epsilon_L = \epsilon_C = \epsilon_R$, and create mobile entangled electrons with the same orbital energy, as required in the beam-splitter setup to allow entanglement detection [4], [8]. The exact analytical expressions are extremely lengthy, but we can get precise conditions for an efficient entangler regime by performing a Taylor expansion in terms of α, γ, T_0 (defined in Fig. 4). Introducing the conditions $Q, \tilde{Q} > Q_{\rm I}^{\rm min}$ away from resonance ($\epsilon_L \neq \epsilon_C$) and $Q, \tilde{Q} > Q_{\rm II}^{\rm min}$ at resonance ($\epsilon_L = \epsilon_C$), we obtain the conditions [15]

$$|\epsilon_L - \epsilon_C| < 2T_0/\sqrt{Q_{\rm I}^{\rm min}}\,, \tag{6}$$

$$\gamma\sqrt{Q_{\rm II}^{\rm min}/8} < T_0 < U\sqrt{4\alpha/\gamma Q_{\rm II}^{\rm min}}. \tag{7}$$

We need a large U for the energy suppression of the one-electron transport, and $\gamma \ll T_0$ because the joint transport is a higher-order process in T_0. The current saturates to $I_E \to e\alpha$ when $T_0^4 \gg U^2\gamma\alpha/32$ [see Fig. 5(d)] when the bottleneck process is the tunneling of the electrons from the source lead to the central dot. We see in (c) that equal currents in the left and right drain lead, $I_L - I_R$, are characteristic of the resonance $\epsilon_L = \epsilon_C$, which provides an experimental procedure to locate the efficient regime.

Taking realistic parameters for quantum dots [25, 51] such as $I_E = 20$ pA, $\alpha = 0.1\ \mu$eV and $U = 1$ meV, we obtain a maximum entangler quality $Q_{\rm II}^{\rm min} = 100$ at resonance, and a finite width $|\epsilon_L - \epsilon_C| \simeq 6\ \mu$eV where the quality is at least $Q_{\rm I}^{\rm min} = 10$. Note that one must avoid resonances with excited levels which could favour the undesired non-entangled one-electron transport. For this, one can either tune the excited levels away by applying a magnetic field, or require a large energy levels spacing $\Delta\epsilon_i \simeq 2U$, which can be found in vertical quantum dots or carbon nanotubes [25]. We can estimate the relevant timescales by simple arguments. The entangled pairs are delivered every $\tau_{\rm pairs} \simeq 2/\alpha \simeq 13\,$ns. The average separation between two entangled electrons within one pair is given by the time-energy uncertainty relation: $\tau_{\rm delay} \simeq 1/U \simeq 0.6\,$ps, while their maximal separation is given by the variance of the exponential decay law of the escape into the leads: $\tau_{\rm max} \simeq 1/\gamma \simeq 0.6\,$ns. Note that $\tau_{\rm delay}$ and $\tau_{\rm max}$ are both well below reported spin decoherence times (in bulk) of 100 ns [52].

3. Spin-polarized electron sources

Here we briefly mention some of the possibilities for spin-polarized electron sources possibly relevant as feeding Fermi-liquid reservoirs to our beam-splitter configuration. Even though we are concerned here with mesoscopic *coherent* transport, we emphasize that the electron sources themselves can be diffusive or ballistic.

Currently, there is a great deal of interest in the problem of spin injection in hybrid mesoscopic structures. At the simplest level we can say that the "Holy Grail" here is essentially the ability to spin inject *and* detect spin-polarized charge flow across interfaces. The possibility of controlling and manipulating the degree of spin polarization of the flow is highly desirable. This would enable novel spintronic devices with flexible/controllable functionalities.

Recently, many different experimental possibilities for spin injection/detection have been considered: (i) all-optical [53] and (ii) all-electrical [54], [55] spin injection and detection in semiconductors and metal devices, respectively, and (iii) electric injection with optical detection in hybrid (Mn-based) ferromagnetic/non-magnetic and paramagnetic/non-magnetic semiconductor *pin* diodes [16]. For an account of the experimental efforts currently underway in the field of spin-polarized transport, we refer the reader to Ref. [13]. Below we focus on our proposals for spin filtering with a semimagnetic tunnel barrier [17] and a quantum dot [18]. These can, in principle, provide alternative schemes for spin injection into our beam splitter.

3.1. QUANTUM SPIN FILTERING

Ballistic Mn-based tunnel junctions [17] offer an interesting possibility for generating spin-polarized currents. Here the s-d interaction in the paramagnetic layer gives rise to a spin-dependent potential. An optimal design can yield high barriers for spin-up and vanishingly small barriers for spin-down electrons. Hence, a highly spin-selective tunnel barrier can be achieved in the presence of an external magnetic field. Note that here *ballistic spin filtering* – due to the blocking of one spin component of the electron flow – is the relevant mechanism for producing a spin-polarized current. Earlier calculations have shown that full spin polarizations are attainable in ZnSe/ZnMnSe spin filters [17].

3.2. QUANTUM DOTS AS SPIN FILTERS

Spin polarized currents can also be generated by a quantum dot [18]. In the Coulomb blockade regime with Fermi-liquid leads, it can be operated as an efficient spin-filter [56] at the single electron level. A magnetic field lifts the spin degeneracy in the dot while its effect is negligible [57] in the leads. As a consequence, only one spin direction can pass through the quantum dot from the source to the drain. The transport of the opposite spin is suppressed by energy conservation and singlet-triplet splitting. This filtering effect can be enhanced by using materials with different g-factors for the dot and the lead. To increase the current signal, one could also use an array of quantum dots, e.g. self-assembled dots.

3.3. SPIN FILTERS FOR SPIN DETECTION AND BELL INEQUALITIES

Besides being a source of spin-polarized currents, such spin filters (with or without spin-polarized sources [18],[58]) could be used to measure electron spin, as they convert spin information into charge: the transmitted charge current depends on the spin direction of the incoming electrons [26]. Such filters could probe the degree of polarization of the incoming leads. In addition, Bell inequalities measurements could be performed with such devices [59, 60].

4. Scattering formalism: basics

Current. In a multi-probe configuration with incoming and outgoing leads related via the scattering matrix $\mathbf{s}_{\gamma\beta}$, the current operator in lead γ within the Landauer-Büttiker [61] approach is given by

$$\hat{I}_\gamma(t) = \frac{e}{h}\sum_{\alpha\beta}\int d\varepsilon d\varepsilon' e^{i(\varepsilon-\varepsilon')t/\hbar}\mathbf{a}^\dagger_\alpha(\varepsilon)\mathbf{A}_{\alpha,\beta}(\gamma;\varepsilon,\varepsilon')\mathbf{a}_\beta(\varepsilon'),$$

$$\mathbf{A}_{\alpha\beta}(\gamma;\varepsilon,\varepsilon') = \delta_{\gamma\alpha}\delta_{\gamma\beta}\mathbf{1} - \mathbf{s}^\dagger_{\gamma\alpha}(\varepsilon)\mathbf{s}_{\gamma\beta}(\varepsilon'), \tag{8}$$

where we have defined the two-component object $\mathbf{a}^\dagger_\alpha(\varepsilon) = (a^\dagger_{\alpha,\uparrow}(\varepsilon), a^\dagger_{\alpha,\downarrow}(\varepsilon))$ with $a^\dagger_{\alpha,\sigma}(\varepsilon)$ denoting the usual fermionic creation operator for an electron with energy ε and spin component $\sigma =\uparrow,\downarrow$ in lead α. Here the spin components σ are along a properly defined quantization axis (e.g., x, y or z).

Noise. Let $\delta\hat{I}_\gamma(t) = \hat{I}_\gamma(t) - \langle I\rangle$ denote the current-fluctuation operator at time t in lead γ ($\langle I\rangle$: average current). We define noise between leads γ and μ in a multi-terminal system by the average power spectral density of the symmetrized current-fluctuation autocorrelation function [62]

$$S_{\gamma\mu}(\omega) = \frac{1}{2}\int \langle \delta\hat{I}_\gamma(t)\delta\hat{I}_\mu(t') + \delta\hat{I}_\mu(t')\delta\hat{I}_\gamma(t)\rangle e^{i\omega t} dt. \tag{9}$$

The angle brackets in Eq. (9) denote either an ensemble average or an expectation value between relevant pairwise electron states. We focus on noise at zero temperatures. In this regime, the current noise is solely due to the discreteness of the electron charge and is termed *shot noise*.

4.1. SCATTERING MATRIX

Electron beam splitter. This device consists of four quasi one-dimensional leads (point contacts) electrostatically defined on top of a 2DEG [21], [22]. An extra "finger gate" in the central part of the device acts as a potential barrier for electrons traversing the system, i.e., a "beam splitter". That is, an impinging electron from, say, lead 1 has probability amplitudes r to be reflected into lead 3 and t to be transmitted into lead 4.

Beam splitter s *matrix.* The transmission processes at the beam splitter can be suitably described in the language of the scattering theory: $s_{13} = s_{31} = r$ and $s_{14} = s_{41} = t$; similarly, $s_{23} = s_{32} = t$ and $s_{24} = s_{42} = r$, see Fig. 6. We also neglect backscattering into the incoming leads, $s_{12} = s_{34} = s_{\alpha\alpha} = 0$. Note that the beam splitter s matrix is spin independent; this no longer holds in the presence of a spin-orbit interaction. We also assume that the amplitudes r and t are energy independent. The unitarity of **s** implies $|r|^2 + |t|^2 = 1$ and $\mathrm{Re}(r^*t) = 0$. Below we use the above scattering matrix to evaluate noise.

5. Noise of entangled electron pairs: earlier results

Singlet and triplets. Let us assume that an entangler is now "coupled" to the beam-splitter device so as to inject entangled (and unentangled) electron pairs into the incoming leads, Fig. 6. This will certainly require some challenging lithographic patterning and/or elaborate gating structures.

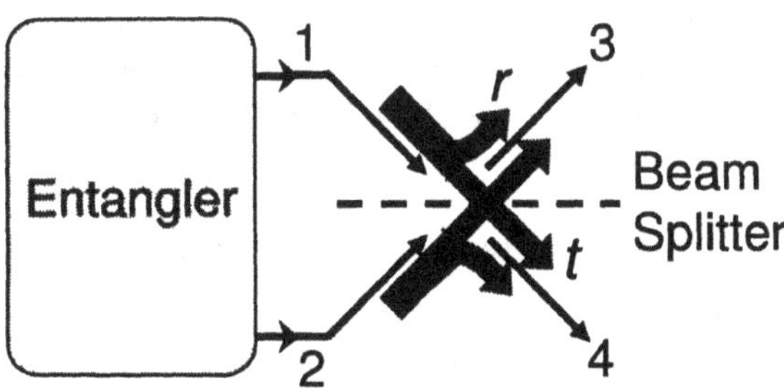

Figure 6. Electron entangler coupled to a beam splitter device. "Entangler" here represents one of the proposed setups of Sec. 2 or some other arrangement providing both triplet and singlet pairs via proper level tuning with gate electrodes. Adapted from Ref. [4].

Let us consider the following two-electron states

$$|S\rangle = \frac{1}{\sqrt{2}}\left[a^\dagger_{1\uparrow}(\varepsilon_1)a^\dagger_{2\downarrow}(\varepsilon_2) - a^\dagger_{1\downarrow}(\varepsilon_1)a^\dagger_{2\uparrow}(\varepsilon_2)\right]|0\rangle, \tag{10}$$

$$|Te\rangle = \frac{1}{\sqrt{2}}\left[a^\dagger_{1\uparrow}(\varepsilon_1)a^\dagger_{2\downarrow}(\varepsilon_2) + a^\dagger_{1\downarrow}(\varepsilon_1)a^\dagger_{2\uparrow}(\varepsilon_2)\right]|0\rangle, \tag{11}$$

and

$$|Tu_\sigma\rangle = a^\dagger_{1\sigma}(\varepsilon_1)a^\dagger_{2\sigma}(\varepsilon_2)|0\rangle, \qquad \sigma = \uparrow, \downarrow . \tag{12}$$

The above states correspond to the singlet $|S\rangle$, the entangled triplet $|Te\rangle$, and the unentangled triplets $|Tu_\sigma\rangle$, respectively, injected electron pairs. Note that $|0\rangle$ denotes the "lead vacuum", i.e., an empty lead or a Fermi sea. Here we follow Ref. [4] and assume that the injected pairs have *discrete* energies $\varepsilon_{1,2}$.

To determine the average current and shot noise for electron pairs we have to calculate the expectation value of the noise two-particle states in Eqs. (10)-(12). In the limit of zero bias, zero temperature, and zero frequency, we find [4]

$$S_{33}^{S/Te,u_\sigma} = \frac{2e^2}{h\nu}T(1-T)(1 \pm \delta_{\varepsilon_1,\varepsilon_2}), \tag{13}$$

for the shot noise in lead 3 for singlet (upper sign) and triplets (lower sign) with $T \equiv |t|^2$ (transmission coefficient). The corresponding currents in lead 3 are $I_3^{S,Te,u_\sigma} = I = \frac{e}{h\nu}$. Note the density of states factor ν in Eqs. (13) arising from the discrete spectrum used [63].

Bunching and antibunching. For $\varepsilon_1 = \varepsilon_2$ the Fano factors corresponding to the shot noise in Eq. (13) are $F^S = S_{33}^S/eI = 4T(1-T)$, for the singlet and $F^{Te,u_\sigma} = 0$, for all three triplets. Interestingly, the Fano factor for a singlet pair is enhanced by a factor of two as compared to the Fano factor $2T(1-T)$ for a *single* uncorrelated electron beam [64] impinging on the beam splitter; the Fano factor for the triplets is suppressed with respect to this uncorrelated case. This enhancement of F^S and suppression of F^{Te,u_σ} is due to *bunching* and *antibunching*, respectively, of electrons in the outgoing leads. This result offers the possibility of distinguishing singlet from triplet states via noise measurements (triplets cannot be distinguished among themselves here; a further ingredient is needed for this, e.g., a local Rashba interaction in one of the incoming leads).

6. Electron transport in the presence of a *local* Rashba s-o interaction

The central idea here is to use the gate-controlled Rashba coupling to rotate the electron spins [24] traversing the Rashba-active region (lead 1 of the beam splitter), thus altering in a controllable way the resulting transport properties of the system. Below we first discuss the effects of the Rashba s-o interaction in one-dimensional systems; the incoming leads are essentially quasi one-dimensional wires, i.e, “quantum point contacts”. A local Rashba interaction can in principle be realized with an additional gating structure (top and back gates [65]).

We focus on wires with one and two transverse channels [66]. This latter case allows us to study the effects of s-o induced interband coupling on both current and shot noise.

6.1. RASHBA WIRES WITH UNCOUPLED TRANSVERSE CHANNELS

6.1.1. *Hamiltonian, eigenenergies and eigenvectors*

The Rashba spin-orbit interaction is present in low-dimensional systems with *structural* inversion asymmetry. Roughly speaking, this interaction arises from the gradient of the confining potential (“triangular shape”) at the interface between two different materials [67]. For a non-interacting one-dimensional wire with *uncoupled* transverse channels, the electron Hamiltonian in the presence of the Rashba coupling α reads [68]

$$H_n = -\frac{\hbar^2}{2m^*}\partial_x^2 + \epsilon_n + i\alpha\sigma_y\partial_x. \tag{14}$$

In Eq. (14) $\partial_x \equiv \partial/\partial x$, σ_y is the Pauli matrix, m^* is the electron effective mass, and ϵ_n is the bottom of the nth-channel energy band in absence of s-o interaction. For an infinite-barrier transverse confinement of width w, $\epsilon_n = n^2\pi^2\hbar^2/(2mw^2)$.

The Hamiltonian in (14) yields the usual set of Rashba bands [69]

$$\varepsilon_s^n = \hbar^2(k - sk_R)^2/2m^* + \epsilon_n - \epsilon_R, \qquad s = \pm \tag{15}$$

where $k_R = m^*\alpha/\hbar^2$ and $\epsilon_R = \hbar^2 k_R^2/2m^* = m^*\alpha^2/2\hbar^2$ ("Rashba energy"). The corresponding wave functions are eigenvectors of σ_y with the orbital part being a plane wave times the transverse-channel wave function. Figure 7 shows that the parabolic bands are shifted sideways due to the Rashba interaction. Note that these bands are still identified by a unique spin index $s = \pm$ which in our convention corresponds to the eigenspinors $|\mp\rangle \sim |\uparrow\rangle \mp |\downarrow\rangle$ of σ_y.

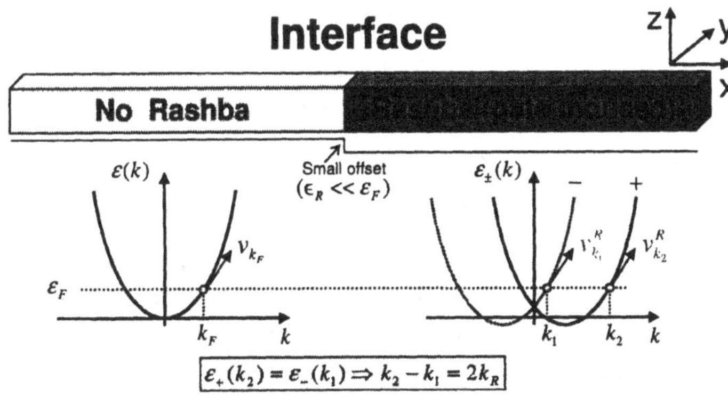

Figure 7. Schematic of a portion of a gate-induced no-Rashba/Rashba "interface" and its corresponding band structure. Note the small band offset arising solely from the mismatch ϵ_R.

6.1.2. *Boundary conditions and spin injection*

Here we assume a unity transmission across the interface [70] depicted in Fig. 7. For a spin-up electron with wave vector k_F entering the Rashba region at $x = 0$, we have the following boundary conditions for the wave function and its derivative [69, 71]

$$|\uparrow\rangle e^{ik_F x}|_{x\to 0^-} = \frac{1}{\sqrt{2}}[|+\rangle e^{ik_2 x} + |-\rangle e^{ik_1 x}]_{x\to 0^+}, \tag{16}$$

and

$$|\uparrow\rangle v_{k_F} e^{ik_F x}|_{x\to 0^-} = \frac{1}{\sqrt{2}}[|+\rangle v_{k_2}^R e^{ik_2 x} + |-\rangle v_{k_1}^R e^{ik_1 x}]_{x\to 0^+}, \tag{17}$$

with the Fermi and Rashba group velocities defined by $v_{k_F} = \hbar k_F/m^*$, $v_{k_1}^R = \frac{\hbar}{m^*}(k_1 + k_R)$, and $v_{k_2}^R = \frac{\hbar}{m^*}(k_2 - k_R)$. The wave vectors k_1 and k_2 are defined by the "horizontal" intersections with the Rashba bands $\varepsilon_-(k_1) = \varepsilon_+(k_2)$, see Fig. 7. This results in the condition $k_2 - k_1 = 2k_R$ which implies that the Rashba group velocities are the same at these points: $v_{k_1}^R = v_{k_2}^R$. Equation (17) is satisfied provided that [69]

$$v_{k_F} = \frac{1}{2}(v_{k_1}^R + v_{k_2}^R) = \frac{\hbar}{m^*}\sqrt{\frac{2m^*}{\hbar^2}(\varepsilon_F + \epsilon_R)}, \tag{18}$$

where the last equality follows from conservation of energy, $\varepsilon_-(k_1) = \varepsilon_+(k_2) = \varepsilon_F$. Note that the group velocity of the incoming spin-up electron is completely "transferred" to the Rashba states at the interface.

Spin-rotated state at $x = L$. For an incoming spin-up electron, we have at the exit of the Rashba region the spin-rotated state

$$\psi_{\uparrow,L} = \frac{1}{\sqrt{2}}[|+\rangle e^{ik_2 L} + |-\rangle e^{ik_1 L}], \tag{19}$$

which is consistent with the boundary conditions (16) and (17). After some straightforward manipulations (and using $k_2 - k_1 = 2k_R$), we find

$$\psi_{\uparrow,L} = \begin{pmatrix} \cos\theta_R/2 \\ \sin\theta_R/2 \end{pmatrix} e^{i(k_1+k_R)L}, \tag{20}$$

with the usual Rashba angle $\theta_R = 2m^*\alpha L/\hbar^2$ [24, 72]. A similar expression holds for an incoming spin-down electron. Note that the boundary conditions at $x = L$ are trivially satisfied since we assume unity transmission. The overall phase of the spinor in Eq. (20) is irrelevant for our purposes; we shall drop it from now on.

6.1.3. *Rashba spin rotator*

From the results of the previous section we can now define a unitary operator which describes the action of the Rashba-active region on any incoming spinor

$$\mathbf{U_R} = \begin{pmatrix} \cos\theta_R/2 & -\sin\theta_R/2 \\ \sin\theta_R/2 & \cos\theta_R/2 \end{pmatrix}. \tag{21}$$

Note that all uncoupled transverse channels are described by the same unitary operator $\mathbf{U_R}$. The above unitary operator allows us to incorporate the s-o induced precession effect straightforwardly into the scattering formalism (Sec. 7).

6.2. RASHBA WIRE WITH TWO COUPLED TRANSVERSE CHANNELS

The Rashba s-o interaction also induces a coupling between the bands described in the previous section. Here we extend our analysis to the case of two *weakly* coupled Rashba bands.

6.2.1. *Exact and approximate energy bands*

Projecting the two-dimensional Rashba Hamiltonian [68] onto the basis of the two lowest uncoupled Rashba states, we obtain the quasi one-dimensional Hamiltonian [72]

$$H = \begin{bmatrix} \varepsilon_+^a(k) & 0 & 0 & -\alpha d \\ 0 & \varepsilon_-^a(k) & \alpha d & 0 \\ 0 & \alpha d & \varepsilon_+^b(k) & 0 \\ -\alpha d & 0 & 0 & \varepsilon_-^b(k) \end{bmatrix}, \tag{22}$$

where the interband coupling matrix element is $d \equiv \langle \phi_a(y) | \partial / \partial y | \phi_b(y) \rangle$ and $\phi_n(y)$ is the transverse channel wave function. Here we label the uncoupled Rashba states by $n = a, b$ [73]. The Hamiltonian above gives rise to two sets of parabolic Rashba bands for zero interband coupling $d = 0$. These bands are sketched in Fig. 8 (thin lines). Note that the uncoupled Rashba bands cross. For positive k vectors the crossing is at $k_c = (\epsilon_b - \epsilon_a)/2\alpha$. For non-zero interband coupling $d \neq 0$ the bands anti-cross near k_c (see thick lines); this follows from a straightforward diagonalization of the 4x4 matrix in Eq. (22). We are interested here in the *weak* interband coupling limit. In addition, we consider electron energies near the crossing; away from the crossing the bands are essentially uncoupled and the problem reduces to that of the previous section. In what follows, we adopt a perturbative description for the energy bands near k_c which allows us to obtain analytical results.

"Nearly-free electron bands". In analogy to the usual nearly-free electron approach in solids [74], we restrict the diagonalization of Eq. (22) to the 2x2 central block which corresponds to the degenerate Rashba states crossing at k_c

$$\tilde{H} = \begin{bmatrix} \varepsilon_-^a(k) & \alpha d \\ \alpha d & \varepsilon_+^b(k) \end{bmatrix}. \tag{23}$$

To lowest order we find

$$\varepsilon_\pm^{\text{approx}}(k) = \frac{\hbar^2 k^2}{2m} + \frac{1}{2}\epsilon_b + \frac{1}{2}\epsilon_a \pm \alpha d. \tag{24}$$

The corresponding eigenvectors are the usual linear combination of the *zeroth order* degenerate states at the crossing

$$|\psi_\pm\rangle = \frac{1}{\sqrt{2}} \left[|-\rangle_a \pm |+\rangle_b \right], \tag{25}$$

where the ket sub-indices denote the respective (uncoupled) Rashba channel [for simplicity, we omit the orbital part of the wave functions in (25)].

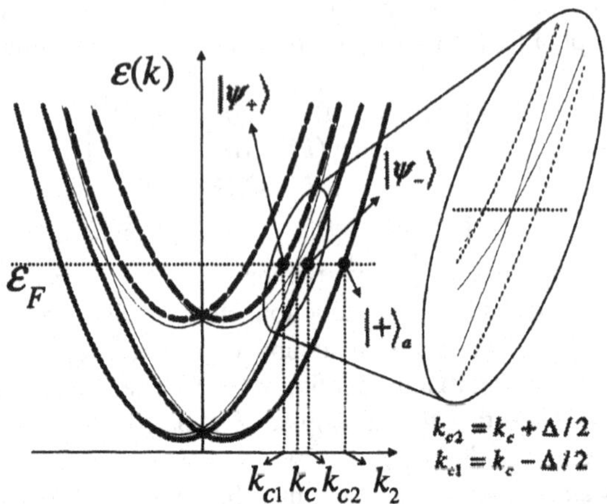

Figure 8. Band structure for a wire with two sets of Rashba bands. Both the uncoupled (thin lines) and the interband-coupled (thick solid and dashed lines) are shown. The uncoupled Rashba bands cross at k_c. Spin-orbit induced interband coupling gives rise to anti crossing of the bands near k_c. Inset: blowup of the region near the crossing. The nearly-free electron bands [perturbative approach, Eq. (24)] describe quite well the exact dispersions near the crossing (cf. dotted and solid + dashed lines in the inset). The solid circles ("intersections") indicate the relevant k points for spin injection [Eq. (27)]; their corresponding zeroth-order eigenvectors [Eq. (25)] are also indicated.

6.2.2. *Boundary conditions and spin injection near the crossing*

Here we extend the analysis in Sec. 6.1.2 to the case of two interband-coupled bands. We first determine the k points corresponding to the "horizontal intersections" near the crossing at k_c, i.e., k_{c1} and k_{c2}, see Fig. 8. We need these points since incoming spin-up electrons will be primarily injected into those states (and also into k_2, conservation of energy). By defining $k_{c1} = k_c - \Delta/2$ and $k_{c2} = k_c + \Delta/2$ and then imposing $\varepsilon_+^{\text{approx}}(k_{c1}) = \varepsilon_-^{\text{approx}}(k_{c2})$ (assumed $\sim \varepsilon_F$) we find,

$$\Delta = \frac{2m\alpha d}{\hbar^2 k_c} = 2\frac{k_R}{k_c}d. \tag{26}$$

For a spin-up electron in the lowest wire state in the "no-Rashba" region (channel a), we can again write at $x = 0$ [70]

$$| \uparrow \rangle e^{ikx}|_{x\to 0^-} = \frac{1}{\sqrt{2}}\left\{\frac{1}{\sqrt{2}}\left[|\psi_+\rangle e^{ik_{c1}x} + |\psi_-\rangle e^{ik_{c2}x}\right] + |+\rangle_a e^{ik_2 x}\right\}_{x\to 0^+}, \tag{27}$$

in analogy to Eq. (16). Note that we only need to include three intersection points in the above "expansion" since the incoming spin-up electron is in channel a.

Equation (27) satisfies the continuity of the wave function. The boundary condition for the derivative of the wave function is also satisfied provide that $\Delta/4 \ll k_F$. This condition is readily fulfilled for realistic parameters (Sec. 8.3). Hence, fully spin-polarized injection into the Rashba region is still possible in the presence of a weak interband coupling. Here we are considering a fully spin-polarized injector so that the intrinsic limitation due to the "conductivity mismatch" [75] is not a factor.

Generalized spin-rotated state at $x = L$. Here again we can easily determine the form of the state at the exit of the Rashba region. For an incoming spin-up electron in the lowest band of the wire, we find

$$\Psi_{\uparrow,L} = \frac{1}{2} e^{i(k_c+k_R)L} \begin{pmatrix} \cos(\theta_d/2)e^{-i\theta_R/2} + e^{i\theta_R/2} \\ -i\cos(\theta_d/2)e^{-i\theta_R/2} + ie^{i\theta_R/2} \\ -i\sin(\theta_d/2)e^{-i\theta_R/2} \\ \sin(\theta_d/2)e^{-i\theta_R/2} \end{pmatrix}. \tag{28}$$

A similar state holds for a spin-down incoming electron. The state (28) satisfies the boundary conditions at $x = L$ (again, provided that $\Delta \ll 4k_F$. Equation (28) essentially tells us that a weak s-o interband coupling gives rise to an additional spin rotation (besides θ_R) described by the mixing angle $\theta_d = \theta_R d/k_c$. This extra modulation enhances spin control in a Datta-Das spin-transistor geometry. In Ref. [72] we show that the spin-resolved current in this case is

$$I_{\uparrow,\downarrow} = \frac{e}{h} eV[1 \pm \cos(\theta_d/2)\cos\theta_R], \tag{29}$$

where V is the source-drain bias.

7. Novel Beam-splitter geometry with a local Rashba interaction

Figure 1 shows an schematic of our proposed beam-splitter geometry with a local Rashba-active region of length L in lead 1. Below we discuss its scattering matrix in the absence of interband coupling. In this case, each set of Rashba bands can be treated independently.

Combined s *matrices.* An electron entering the system through port 1, first undergoes a unitary Rashba rotation $\mathbf{U_R}$ in lead 1 then reaches the beam splitter which either reflects the electron into lead 3 or transmits it into lead 4. This happens for electrons injected into either the first or the second set of uncoupled Rashba bands. Since the Rashba spin rotation is unitary, we can combine the relevant matrix elements of the beam-splitter s matrix, connecting leads 1 and 3 ($s_{14} = s_{41}$) and 1 and 4 ($s_{14} = s_{41}$), with the Rashba rotation matrix $\mathbf{U_R}$ thus obtaining effective *spin-dependent* 2×2 matrices of the form $\mathbf{s_{13}^R} = \mathbf{s_{31}^R} = s_{13}\mathbf{U_R}$ A similar definition holds for $\mathbf{s_{14}^R} = s_{41}\mathbf{U_R} = \mathbf{s_{41}^R}$. Note also that $\mathbf{s_{23}} = \mathbf{s_{32}} = t\mathbf{1}$ and $\mathbf{s_{24}} = \mathbf{s_{42}} = r\mathbf{1}$ since no Rashba coupling is present in lead 2. All the other matrix elements are zero. Hence the new effective beam-splitter s matrix which incorporates the effect of the Rashba interaction in lead 1 reads

$$\mathbf{s} = \begin{pmatrix} 0 & 0 & \mathbf{s}_{13}^{\mathrm{R}} & \mathbf{s}_{14}^{\mathrm{R}} \\ 0 & 0 & \mathbf{s}_{23} & \mathbf{s}_{24} \\ \mathbf{s}_{31}^{\mathrm{R}} & \mathbf{s}_{32} & 0 & 0 \\ \mathbf{s}_{41}^{\mathrm{R}} & \mathbf{s}_{42} & 0 & 0 \end{pmatrix}. \tag{30}$$

Note that incorporating the s-o effects directly into the beam-splitter scattering matrix makes it spin dependent. The Rashba interaction does not introduce any noise in lead 1. This is so because the electron transmission coefficient through lead 1 is essentially unity [70]; a quantum point contact is noiseless for unity transmission.

Coupled Rashba bands. The interband-coupled case can, in principle, be treated similarly. However, we follow a different simpler route to determine the shot noise in this case. We discuss this in more detail in Sec. 8.1.2.

8. Noise of entangled and spin-polarized electrons in the presence of a local Rashba spin-orbit interaction

Starting from the noise definition in (Eq. 9), we briefly outline here the derivation of noise expressions for pairwise electron states (entangled and unentangled) and spin-polarized electrons (Secs. 2 and 3). For each of these two cases, we present results with and without s-o induced interband coupling.

8.1. SHOT NOISE FOR SINGLET AND TRIPLETS

8.1.1. *Uncoupled Rashba bands: single modulation θ_R*

To determine noise, we calculate the expectation value of the noise operator (Eq. 9) between pairwise electron states. We have derived shot noise expressions for both singlet and triplet states for a generic *spin-dependent* s matrix. Our results quite generally show that unentangled triplets and the entangled triplet display distinctive shot noise for spin-dependent scattering matrices. Below we present shot noise formulas for the specific case of interest here; namely, the beam-splitter scattering matrix in the presence of a local Rashba term [Eq. (30)]. In this case, for singlet and triplets defined along different quantization axes ($\hat{x}$ and $\hat{z}$ are equivalent directions perpendicular to the Rashba rotation axis $-\hat{y}$), we find

$$S_{33}^{S}(\theta_R) = \frac{2e^2}{h\nu}T(1-T)[1+\cos(\theta_R)\delta_{\varepsilon_1,\varepsilon_2}], \tag{31}$$

$$S_{33}^{Te_y}(\theta_R) = \frac{2e^2}{h\nu}T(1-T)[1-\cos(\theta_R)\delta_{\varepsilon_1,\varepsilon_2}], \tag{32}$$

$$S_{33}^{Te_z}(\theta_R) = S_{s-o}^{Tu_y}(\theta_R) = \frac{2e^2}{h\nu} T(1-T)(1-\delta_{\varepsilon_1,\varepsilon_2}), \tag{33}$$

and

$$S_{33}^{Tu_\uparrow}(\theta_R) = S_{33}^{Tu_\downarrow}(\theta_R) = \frac{2e^2}{h\nu} T(1-T)[1-\cos^2(\theta_R/2)\delta_{\varepsilon_1,\varepsilon_2}]. \tag{34}$$

Equations (32)–(34) clearly show that entangled and unentangled triplets present distinct noise as a functions of the Rashba phase. Note that for $\theta_R = 0$, we regain the formulas in Sec. 5.

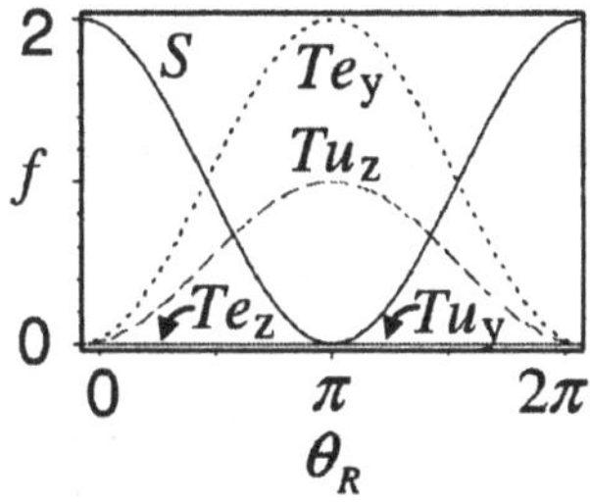

Figure 9. "Reduced" Fano factor f as a function of the Rashba phase for singlet and triplets along different quantization axes. Note that singlet and entangled triplet states show *continuous* bunching and antibunching behaviors as θ_R is increased. Unentangled triplets display distinctive noise for a given polarization and for different polarizations. Adapted from Ref. [8].

Figure 9 shows the "reduced" Fano factor $f = F/2T(1-T)$, $F = S_{33}/eI$ (here $I = e/h\nu$), as a function of the Rashba angle θ_R for the noise expressions (31)–(34). We clearly see that singlet and triplet pairs exhibit distinct shot noise in the presence of the s-o interaction. The singlet S and entangled (along the Rashba rotation axis $\hat{y}$) triplet Te_y pairs acquire an oscillating phase in lead 1 thus originating intermediate degrees of bunching/antibunching (solid and dotted lines, respectively). Triplet states (entangled and unentangled) display distinctive noise as a function of the Rashba phase, e.g., Te_y is noisy and Tu_y is noiseless. Hence entangled and unentangled triplets can also be distinguished via noise measurements. Note that for $\theta_R = 0$ all three triplets exhibit identically zero noise [see Eq.(13)].

8.1.2. *Interband-coupled Rashba bands: additional modulation* θ_d

Here we determine noise for injected pairs with energies near the crossing $\varepsilon(k_c)$ using an alternate scheme. We calculate the relevant expectation values of the noise by using pairwise states defined from the generalized spin-rotated state in Eq. (28) and its spin-down counterpart. Since these states already incorporate all the relevant effects (Rashba rotation and interband mixing), we can calculate noise by using the "bare" beam splitter matrix elements, generalized to account for two channels. The beam-splitter does not mix transverse channels; hence this extension is trivial, i.e., block diagonal in the channel indices. This approach was first developed in Ref. [7].

Rashba-evolved pairwise electron states. The portion of an electron-pair wave function "propagating" in lead 1 undergoes the effects of the Rashba interaction: ordinary precession θ_R and additional rotation θ_d. Using Eq. (28) (and its spin-down counterpart) we find the following states

$$\begin{aligned} |S/Te_z\rangle_L \;=\; & \frac{1}{2}[\cos(\theta_d/2)e^{-i\theta_R/2}+e^{i\theta_R/2}]\frac{|\uparrow\downarrow\rangle_{aa}\mp|\downarrow\uparrow\rangle_{aa}}{\sqrt{2}}+ \\ & \frac{1}{2}[-i\cos(\theta_d/2)e^{-i\theta_R/2}+ie^{i\theta_R/2}]\frac{|\downarrow\downarrow\rangle_{aa}\pm|\uparrow\uparrow\rangle_{aa}}{\sqrt{2}}+ \\ & \frac{1}{2}[-i\sin(\theta_d/2)e^{-i\theta_R/2}]\frac{|\uparrow\downarrow\rangle_{ba}\pm|\downarrow\uparrow\rangle_{ba}}{\sqrt{2}}+ \\ & \frac{1}{2}[\sin(\theta_d/2)e^{-i\theta_R/2}]\frac{|\downarrow\downarrow\rangle_{ba}\mp|\uparrow\uparrow\rangle_{ba}}{\sqrt{2}}. \end{aligned} \tag{35}$$

The notation $|Te_z\rangle_L$ and $|S\rangle_L$ emphasizes the type of injected pairs (singlets or triplets at $x=0$) propagating through the length L of the Rashba-active region in lead 1. Similar expressions hold for $|Tu_{\uparrow,\downarrow}\rangle_L$. In addition, we use the shorthand notation $|\downarrow\uparrow\rangle_{ba}\equiv|\downarrow_{1b}\uparrow_{2a}\rangle$, denoting a pair with one electron in channel b of lead 1 and another in channel a of lead 2. Here we consider incoming pairs with $\hat{z}$ polarizations only. Despite the seemingly complex structure of the above pairwise states, they follow quite straightforwardly from the general state $\Psi_{\uparrow,L}$ in (28) (and its counterpart $\Psi_{\downarrow,L}$). For instance, the unentangled triplet $|Tu_\uparrow\rangle_L$ is obtained from the tensor product between $\Psi_{\uparrow,L}$ [which describes as electron crossing lead 1 (initially spin up and in channel a)] and a spin-up state in channel a of lead 2: $|Tu_\uparrow\rangle_L=|\Psi_{\uparrow,L}\rangle\bigotimes|\uparrow\rangle_{2a}$.

Noise. We can now use the above states to determine shot noise at the zero frequency, zero temperature, and zero applied bias. Using the shot-noise results of Sec. 5 (trivially generalized for two channels), we find for the noise in lead 3

$$\begin{aligned} S_{33}^{Tu_\uparrow}(\theta_R,\theta_d) \;=\; & S_{33}^{Tu_\downarrow}(\theta_R,\theta_d)=\frac{2e^2}{h\nu}T(1-T)\times \\ & \left[1-\frac{1}{2}\left(1+\cos\frac{\theta_d}{2}\cos\theta_R-\frac{1}{2}\sin^2\frac{\theta_d}{2}\right)\delta_{\varepsilon_1,\varepsilon_2}\right], \end{aligned} \tag{36}$$

$$S_{33}^{Te_z}(\theta_R,\theta_d) = \frac{2e^2}{h\nu}T(1-T)\left[1-\frac{1}{2}\left(\cos^2\frac{\theta_d}{2}+1\right)\delta_{\varepsilon_1,\varepsilon_2}\right], \tag{37}$$

and

$$S_{33}^{S}(\theta_R,\theta_d) = \frac{2e^2}{h\nu}T(1-T)\left[1+\left(\cos\frac{\theta_d}{2}\cos\theta_R\right)\delta_{\varepsilon_1,\varepsilon_2}\right]. \tag{38}$$

Equations (36)-(38) describe shot noise only for injected pairs with energies near the crossing, say, within αd of $\varepsilon(k_c)$. Away from the crossing or for $d = 0$, the above expressions reduce to those of Sec. 8.1.1. We can also define "reduced" Fano factors as before; the interband mixing angle θ_d further modulates the Fano factors. For conciseness, we present the angular dependence of the Fano factors in the next section.

8.2. SHOT NOISE FOR SPIN-POLARIZED ELECTRONS

We have derived a general shot noise formula for the case of spin-polarized sources by performing the ensemble average in Eq. (9) over appropriate thermal reservoirs. The resulting expression corresponds to the standard Landauer-Büttiker formula for noise with spin-dependent s matrices. Below we present results for the specific beam-splitter s matrix in (30).

8.2.1. *Uncoupled-band case: single modulation θ_R*

For incoming leads with a degree of spin polarization p and for the scattering matrix (30), we find at zero temperatures

$$S_{33}^{p}(\theta_R) = 2eIT(1-T)p\sin^2\frac{\theta_R}{2}, \tag{39}$$

where $I = 2e^2V/[h(1+p)]$ is the average current in lead 3. The "reduced" Fano factor corresponding to Eq. (39) is $f_p = p\sin^2(\theta_R/2)$. Figure 10 shows f_p as a function of the Rashba angle θ_R. For spin polarized injection along the Rashba rotation axis $(-\hat{y})$ no noise results in lead 3. This is a consequence of the Pauli exclusion principle in the leads. Spin-polarized currents with polarization perpendicular to the Rashba axis exhibit sizable oscillations as a function of θ_R. Full shot noise is obtained for $\theta_R = \pi$ since the spin polarization of the incoming flow is completely reversed within lead 1.

Probing/detecting spin-polarized currents. Since *unpolarized* incoming beams in lead 1 and 2 yield zero shot noise in lead 3, the results shown in Fig. 10 provide us with an interesting way to *detect* spin-polarized currents via their noise. In addition, noise measurements should also allow one to probe the direction of the spin-polarization of the injected current.

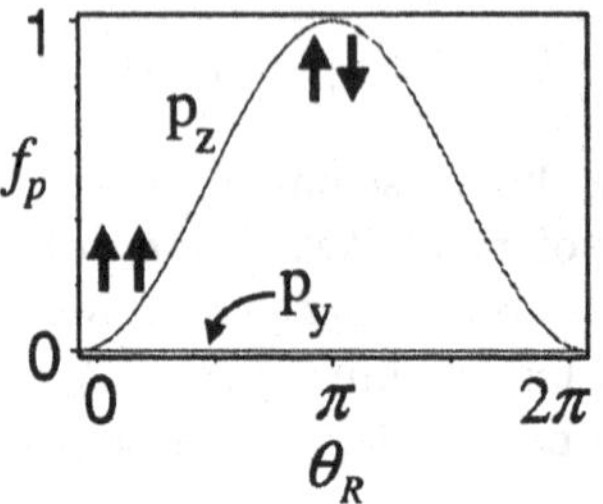

Figure 10. "Reduced" Fano factor for fully spin-polarized ($p = 1$) incoming beams in leads 1 and 2 as a function of the Rashba phase. Polarizations along two distinct quantization axes are shown (p_z and p_z). For spin injection along the Rashba rotation axis ($-\hat{y}$), no precession occurs in lead 1 and shot noise is identically zero (Pauli principle). Spin-polarized carriers injected along $\hat{z}$ undergo precession and hence exhibit shot noise. Adapted from Ref. [8].

Measuring the s-o coupling. We can express the s-o coupling constant in terms of the reduced Fano factor. For a fully spin-polarized beam ($p = 1$), we have

$$\alpha = \frac{\hbar^2}{m^* L} \arcsin \sqrt{f_p}. \tag{40}$$

Equation (40) provides a direct means of extracting the Rashba s-o coupling α via shot noise measurements. We can also obtain a similar expression for α from the unentangled triplet noise formula (34).

8.2.2. *Interband-coupled case: extra modulation θ_d*

The calculation in the previous section can be extended to the interband-coupled case for electrons impinging near the anti crossing of the bands [$\sim \varepsilon(k_c)$]. Here we present a simple "back-of-the-envelope" derivation of the the shot noise for the fully spin-polarized current case ($p = 1$) from that of the spin-up unentangled triplet Eq. (36). Here we imagine that the spectrum of the triplet $Tu_\uparrow$ forms now a continuum and integrate its noise expression (after making $\varepsilon_1 = \varepsilon_2$) over some energy range to obtain the noise of a spin-polarized current. Assuming T constant in the range $(\varepsilon_F, \varepsilon_F + eV)$, we find to linear order in eV

$$S_{33}^{\uparrow}(\theta_R, \theta_d) = eIT(1-T)\left(1 - \cos\frac{\theta_d}{2}\cos\theta_R + \frac{1}{2}\sin^2\frac{\theta_d}{2}\right). \tag{41}$$

Figures 11(a) and 11(b) illustrate the angular dependencies of the reduced Fano factors for both the spin-polarized case Eq. (41) and that of the singlet Eq.

(38). Note that the further modulation θ_d due to interband mixing can drastically change the noise for both spin-polarized and entangled electrons. For the singlet pairs, for instance, it can completely reverse the bunching/antibunching features. Hence further control is gained via θ_d which can, in principle, be tuned independently of θ_R (see Sec. 8.3).

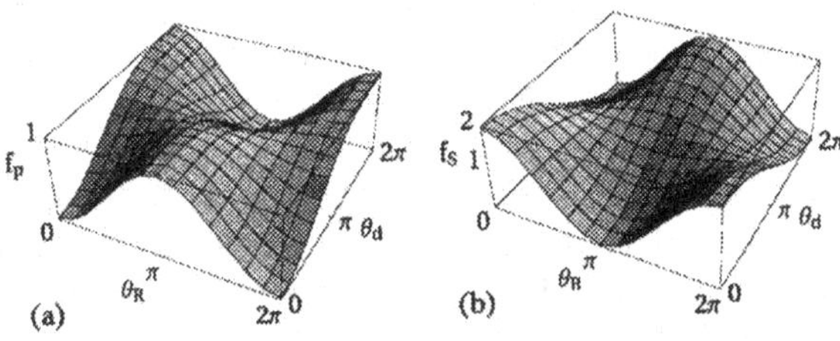

Figure 11. Reduced Fano factors $f = f_p$ (a) and f_S (b), for fully spin-polarized ($p = 1$, $\hat{z}$ direction) incoming electrons and for singlet pairs, respectively, as a function of the Rashba angle θ_R and the interband mixing angle θ_d. The additional phase θ_d can significantly alter the noise characteristics.

8.3. REALISTIC PARAMETERS: ESTIMATES FOR θ_R AND θ_D.

We conclude this section by presenting some estimates for the relevant spin-rotation angles θ_R and θ_d for realistic system parameters. Let us assume, for the sake of concreteness, an infinite confining potential of width w. In this case, the transverse wire modes in absence of the Rashba interaction are quantized with energies $\epsilon_n = \hbar^2\pi^2 n^2/(2m^* w^2)$. Let us now set $\epsilon_b - \epsilon_a = 3\hbar^2\pi^2/(2m^* w^2) = 16\epsilon_R$ which is a "reasonable guess". Since $\epsilon_R = m^*\alpha^2/2\hbar^2$, we find $\alpha = (\sqrt{3}\pi/4)$ $\hbar^2/m^* w^2$=3.45 $\times$ 10^{-11} eVm [65] (which yields $\epsilon_R \sim 0.39$ meV) for $m^* = 0.05 m_0$ and $w = 60$ nm. For the above choice of parameters, the energy at the crossing is $\epsilon_-^a(k_c) = \epsilon_+^b(k_c) = \epsilon(k_c) = 24\epsilon_R \sim 9.36$ meV. Electrons with energies around this value are affected by the s-o interband coupling, i.e., they undergo the additional spin rotation θ_d. The relevant wave vector at the crossing is $k_c = 8\epsilon_R/\alpha$. Assuming the $L = 69$ nm for the length of the Rashba channel, we find $\theta_R = \pi$ and $\theta_d = \theta_R d/k_c \sim \pi/2$ since $d/k_c = 2/(3k_R w)$ and $k_R w = \sqrt{3}\pi/4 \sim 4/3$ for $\epsilon_b - \epsilon_a = 16\epsilon_R$ which implies $d/k_c \sim 0.5$. The preceding estimates are conservative. We should point out that both θ_R and θ_d can, in principle, be varied independently via side gates. It should also be possible to "over rotate" θ_R (say, by using a larger L) and hence increase θ_d. As a final point we note that $\Delta/4k_F \sim 0.05 \ll 1$ [k_F is obtained by making

$\varepsilon_F = \hbar^2 k_F^2/2m^* = \epsilon(k_c)$] which assures the validity of the boundary condition for the velocity operator.

9. Relevant issues and outlook

Relevant time scales. Typical parameters for a finite-size electron beam splitter (tunnel coupled to reservoirs) defined on a GaAs 2DEG are: a device size $L_0 \sim 1$ μm, a Fermi velocity in the range $v_F \sim 10^4 - 10^5$ m/s and an orbital coherence length of ~ 1 μm [21]. These values lead to traversal times $\tau_t = L_0/v_F$ in the range $\sim 10 - 100$ ps; these are lower bounds for the actual dwell time $\tau_{\text{dwell}} \sim 1/\gamma_R$ of the electrons in the beam splitter, where γ_R is the tunnelling rate from the leads of the beam splitter to the reservoirs. Hence the electrons keep their orbital coherence across the beam-splitter at low temperatures. Moreover, long spin dephasing times in semiconductors ($\sim$ 100 ns for bulk GaAs [52]) should allow the propagation of entangled electrons without loss of spin coherence.

For the noise calculation with entangled/unentangled pairs, we have assumed discrete energy levels in the incoming leads. A "particle-in-a-box" estimate of the level spacing $\delta\varepsilon$ due to longitudinal quantization of the beam splitter leads yields $\delta\varepsilon \sim \hbar v_F/L_0 \sim 0.01 - 0.1$ meV. The relevant broadening of these levels is given by the coupling $\gamma_R \ll \delta\varepsilon$, which justifies the discrete level assumption. Here we take $\gamma_R \lesssim \gamma \sim 1$ μeV, where γ is the tunnelling rate from the entangler to the beam splitter (Sec. 2). In addition, the stationary state description we use requires that the electrons have enough time to "fill in" the extended states in the beam splitter before they leave to the reservoirs: $\tau_{\text{dwell}} \gtrsim \tau_{\text{inj}} \sim 1/\gamma$. Here τ_{inj} is the injection time from the entangler to the beam splitter. To have well separated pairs of entangled electrons, we also need $\tau_{\text{delay}} < \tau_{\text{pairs}} \sim$ ns (Sec. 2), where τ_{delay} ($\sim$ ps) is the time delay between two entangled electrons of the same pair, and τ_{pairs} ($\sim$ ns) is the time separation between two subsequent pairs.

Interactions in the beam splitter. For entangled electrons it would be advantageous to reduce electron-electron interaction in the beam splitter, which is the main source of orbital decoherence at low temperatures. This could be achieved by depleting the electron sea in the beam splitter, e.g., by using the lowest channel in a quantum point contact. A further possibility is to use a superconductor for the beam splitter [76]. A superconductor would have the advantage that the entangled electrons could be injected into the empty quasiparticle states right at their chemical potential. Because of the large gap Δ between these states and the condensate, the injected electron cannot exchange energy (nor spin) with the underlying condensate of the superconductor.

An alternative way to detect entangled pairs would be to use a superconductor as an analyzer: arriving entangled (spin-singlet) pairs can enter the superconductor whereas any triplet state is not allowed. Thus, the current of entangled pairs is larger than otherwise.

10. Noise of a double QD near the Kondo regime

Spin-flip processes in a spin $1/2$ quantum dot attached to leads result in a renormalization of the single-particle transmission coefficient T, giving rise to the Kondo effect [77] below the Kondo temperature T_K. Theoretical studies on shot noise in this system are available [78]–[80], and show that the noise S obeys qualitatively the same formula as for noninteracting electrons but with a renormalized T. Here, we consider a system where the spin fluctuations (that are enhanced near the Kondo regime) strongly affect the noise, resulting in some cases in super-Poissonian noise – a result which cannot be obtained from the "non-interacting" formula.

We consider two lateral quantum dots (DD), connected in series between two metallic leads via tunnel contacts, see inset of Fig. 12*a*. The dots are tuned into the Coulomb blockade regime, each dot having a spin $1/2$ ground state. The low energy sector of the DD consists of a singlet $|S\rangle$ and a triplet $|T\rangle \equiv \{|T_+\rangle, |T_0\rangle, |T_-\rangle\}$, with the singlet-triplet splitting K. The Kondo effect in this system has been studied extensively [20]–[84]. Two peculiar features in the linear conductance G have been found: a peak in G *vs* the inter-dot tunnel coupling t_H (see Fig. 12*a*), revealing the non-Fermi-liquid critical point of the two-impurity Kondo model (2IKM) [85]; and a peak in G *vs* an applied perpendicular magnetic field B (see Fig. 12*b*), as a result of the singlet-triplet Kondo effect at $K = 0$ [20].

The problem of shot noise in DDs with Kondo effect is rather involved. Here we propose a phenomenological approach. For bias $\Delta\mu \gg T_K, K$, the scattering problem can be formulated in terms of the following scattering matrix

$$\begin{aligned} \mathrm{s} = & \begin{pmatrix} r_S & t_S \\ t_S & r_S \end{pmatrix} |S\rangle\langle S| + \begin{pmatrix} r_T & t_T \\ t_T & r_T \end{pmatrix} |T\rangle\langle T| \\ & + \begin{pmatrix} r_{TS} & t_{TS} \\ t_{TS} & r_{TS} \end{pmatrix} |T\rangle\langle S| + \begin{pmatrix} r_{ST} & t_{ST} \\ t_{ST} & r_{ST} \end{pmatrix} |S\rangle\langle T|, \end{aligned} \tag{42}$$

where $t_{i(j)}$ and $r_{i(j)}$ are the transmission and reflection amplitudes. The spin fluctuations in the DD cause fluctuations in the transmission through the DD. The dominant mechanism is qualitatively described by the following stochastic model

$$f(t) = [f_1(t)\,(1 - F(t)) + f_2(t)F(t)]\left(1 - \left|\dot{F}(t)\right|\right) + f_3(t)\left|\dot{F}(t)\right|, \tag{43}$$

where $f_i(t) = 0, 1$ is a white noise ($i = 1, 2, 3$) with $\langle f_i(t)\rangle = \bar{f}_i$ and $\langle f_i(t) f_i(0)\rangle - \bar{f}_i^2 = \bar{f}_i(1 - \bar{f}_i)\delta(t/\Delta t)$, and $F(t) = 0, 1$ is a telegraph noise with $\bar{F} = \beta/(1+\beta)$ and $\langle F(t)F(0)\rangle - \bar{F}^2 = \beta\exp(-ct)/(1+\beta)^2$, for $t \geq 0$. In this model, the time t is discretized in intervals of $\Delta t = h/2\Delta\mu$. The derivative $\dot{F}(t)$ takes values $0, \pm 1$. The function $f_{1(2)}(t)$ describes tunnelling through the DD, with the DD staying in the singlet (triplet) state, while $f_3(t)$ describes tunnelling accompanied by the DD transition between singlet and triplet. The relation to formula (42) is given by: $\bar{f}_1 = |t_S|^2 = 1 - |r_S|^2$, $\bar{f}_2 = |t_T|^2 = 1 - |r_T|^2$, and $f_3 = |t_{ST}|^2/(|t_{ST}|^2 + |r_{ST}|^2) = |t_{TS}|^2/(|t_{TS}|^2 + |r_{TS}|^2)$. The telegraph noise is described by two parameters: $\beta = w_{12}/w_{21}$ and $c = w_{12} + w_{21}$, where w_{ij} is the probability to go from i to j.

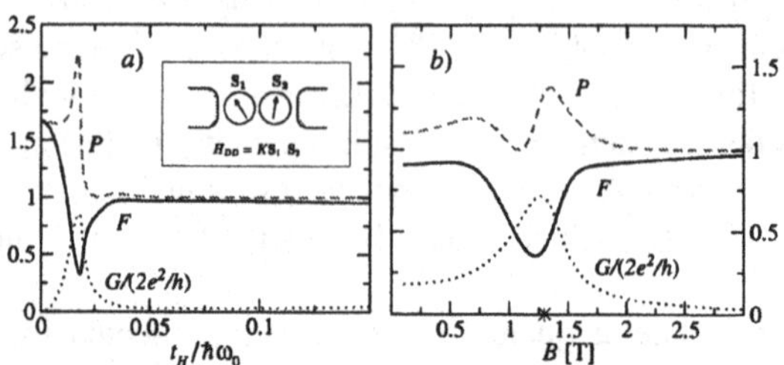

Figure 12. *a)* Linear conductance G (dotted line), Fano factor (solid line), and the factor P (dashed line), in vicinity of the 2IKM critical point. Inset: DD setup. *b)* Similar to (*a*), but in the vicinity of the singlet-triplet Kondo effect ("*" denotes $K = 0$).

The quantity of interest is the Fano factor $F = S/e|I|$. For a single-channel non-interacting system, one has $F = 1 - \mathrm{T}$. In order to show the effect of interaction, we introduce the factor $P = F/(1 - \mathrm{T})$. The noise power at zero frequency is then given by $S = 2eI_{\mathrm{imp}}\mathrm{T}(1 - \mathrm{T})P$, where $I_{\mathrm{imp}} = 2e\Delta\mu/h$. For the average transmission probability we obtain

$$\mathrm{T} \equiv \langle f \rangle = \frac{\bar{f}_1 + \beta \bar{f}_2}{1 + \beta} + \frac{\beta c \Delta t}{(1+\beta)^2}\left(2\bar{f}_3 - \bar{f}_1 - \bar{f}_2\right). \tag{44}$$

The noise can be calculated as $S = 2eI_{\mathrm{imp}}S_f$, with $S_f = \mathrm{T}(1-\mathrm{T}) + \Delta S_f$, where

$$\begin{aligned}\Delta S_f = & \frac{2\beta}{(1-q)(1+\beta)^2}\left\{q(\bar{f}_1 - \bar{f}_2)^2 + \frac{c\Delta t(\bar{f}_1 - \bar{f}_2)}{(1+\beta)}\times\right. \\ & [\bar{f}_3(\beta - 1)(q+1) + \bar{f}_1(1 - \beta q) + \bar{f}_2(q - \beta)] + \\ & \left.\frac{(c\Delta t)^2}{4}\left[(2\bar{f}_3 - \bar{f}_1 - \bar{f}_2)^2 - (\bar{f}_1 - \bar{f}_2)^2\right]\right\},\end{aligned} \tag{45}$$

with $q = \exp(-c\Delta t)$. The factor P is then given by $P = 1 + \Delta S_f/(\mathrm{T} - \mathrm{T}^2)$. Deviations of P from $P = 1$ show the effect of interactions in the DD. We plot the Fano factor and the factor P for a DD on Fig. 12. The results show that the spin fluctuations affect the shot noise in the regions where $K \lesssim T_K$. A peculiar feature in P is found both at the 2IKM critical point (Fig. 12*a*) and at the point of the singlet-triplet Kondo effect (Fig. 12*b*).

For $\Delta\mu \ll T_K$ the DD spin is screened, and correlations between two electrons passing through the DD occur only via virtual excitations of the Kondo state. The shot noise is expected to qualitatively obey the non-interacting formula with the renormalized T.

11. Summary

We presented our recent works on shot noise for spin-entangled electrons and spin-polarized currents in novel beam splitter geometries. After a detailed description of various schemes ("entanglers") to produce entangled spin states, we calculated shot noise within the scattering approach for a beam splitter with and without a local s-o interaction in the incoming leads. We find that the s-o interaction significantly alters the noise. Entangled/unentangled pairs and spin-polarized currents show sizable shot noise oscillations as a function of the Rashba phase. Interestingly, we find an additional phase modulation due to s-o induced interband coupling in leads with two channels. Shot noise measurements should allow the identification/characterization of both entangled and unentangled pairs as well as spin-polarized currents. Finally, we find that the s-o coupling constant α is directly related to the Fano factor; this offers an alternative means of extracting α via noise.

This work was supported by NCCR Nanoscience, the Swiss NSF, DARPA, and ARO.

References

1. W. Schottky, Ann. Phys. 57 (1918) 541.
2. Ya. M. Blanter and M. Büttiker, Phys. Rep. **336**, 1 (2000).
3. D. Loss and E.V. Sukhorukov, Phys. Rev. Lett. **84**, 1035 (2000), cond-mat/9907129.
4. G. Burkard, D. Loss, and E.V. Sukhorukov, Phys. Rev. B **61**, R16303 (2000), cond-mat/9906071. For an early account see D. P. DiVincenzo and D. Loss, J. Magn. Magn. Mat. **200**, 202 (1999), cond-mat/9901137.
5. W. D. Oliver *et al.*, in *Quantum Mesoscopic Phenomena and Mesoscopic Devices in Microelectronics*, vol. 559 of NATO ASI Series C: Mathematical and Physical Sciences, eds. I. O. Kulik and R. Ellialtioglu (Kluwer, Dordrecht, 2000), pp. 457-466.
6. F. Taddei and R. Fazio, Phys. Rev. B **65**, 075317 (2002).
7. J. C. Egues, G. Burkard, and D. Loss, to appear in the Journal of Superconductivity; cond-mat/0207392.
8. J. C. Egues, G. Burkard, and D. Loss, Phys. Rev. Lett. **89**, 176401 (2002); cond-mat/0204639.
9. B. R. Bulka *et al.* Phys. Rev. B **60**, 12246 (1999).
10. F. G. Brito, J. F. Estanislau, and J. C. Egues, J. Magn. Magn. Mat. **226-230**, 457 (2001).
11. F. M. Souza, J. C. Egues, and A. P. Jauho, cond-mat/0209263.
12. J. J. Sakurai, *Modern Quantum Mechanics*, San Fu Tuan, Ed., (Addison-Wesley, New York,1994); (Ch. 3, p. 223). See also J. I. Cirac, Nature **413**, 375 (2001).
13. *Semiconductor Spintronics and Quantum Computation*, Eds. D. D. Awschalom, D. Loss, and N. Samarth (Springer, Berlin, 2002).
14. P. Recher, E.V. Sukhorukov, and D. Loss, Phys. Rev. B **63**, 165314 (2001); cond-mat/0009452.
15. D. S. Saraga and D. Loss, cond-mat/0205553.
16. R. Fiederling *et al.*, Nature **402**, 787 (1999); Y. Ohno *et al.*, Nature **402**, 790 (1999).
17. See J. C. Egues Phys. Rev. Lett. **80**, 4578 (1998) and J. C. Egues *et al.* Phys. Rev. B **64**, 195319 (2001) for *ballistic* spin filtering in semimagnetic heterostruc ures.
18. P. Recher, E. V. Sukhorukov, and D. Loss, Phys. Rev. Lett. **85**, 1962 (2000), cond-mat/0003089.
19. P. Recher and D. Loss, Phys. Rev. B **65**, 165327 (2002), cond-mat/0112298.

20. V.N. Golovach and D. Loss, cond-mat/0109155.
21. R. C. Liu *et al.*, Nature (London), **391**, 263 (1998).
22. M. Henny *et al.*, Science **284**, 296 (1999); W. D. Oliver *et al.*, Science **284**, 299 (1999). See also M. Büttiker, Science **284**, 275 (1999).
23. G. Feve *et al.* (cond-mat/0108021) also investigate transport in a beam splitter configuration. These authors assume a "global" s-o interaction and formulate the scattering approach using Rashba states in single-moded leads.
24. S. Datta and B. Das, Appl. Phys. Lett. **56**, 665 (1990).
25. L.P. Kouwenhoven, G. Schön, L.L. Sohn, Mesoscopic Electron Transport, NATO ASI Series E: Applied Sciences-Vol.345, 1997, Kluwer Academic Publishers, Amsterdam.
26. D. Loss and D. P. DiVincenzo, Phys. Rev. A **57**, 120 (1998), cond-mat/9701055.
27. M.-S. Choi, C. Bruder, and D. Loss, Phys. Rev. B **62**, 13569 (2000); cond-mat/0001011.
28. C. Bena, S. Vishveshwara, L. Balents, and M.P.A. Fisher, Phys. Rev. Lett. **89**, 037901 (2002).
29. G.B. Lesovik, T. Martin, and G. Blatter, Eur. Phys. J. B **24**, 287 (2001).
30. R. Mélin, cond-mat/0105073.
31. V. Bouchiat *et al.*, cond-mat/0206005.
32. W.D. Oliver, F. Yamaguchi, and Y. Yamamoto, Phys. Rev. Lett. **88**, 037901 (2002).
33. S. Bose and D. Home, Phys. Rev. Lett. **88**, 050401 (2002).
34. In principle, an entangler producing entangled triplets $|\uparrow\downarrow\rangle + |\downarrow\uparrow\rangle$ or *orbital* entanglement would also be desirable.
35. This condition reflects energy conservation in the Andreev tunnelling event from the SC to the two QDs.
36. This reduction factor of the current I_2 compared to the resonant current I_1 reflects the energy cost in the virtual states when two electrons tunnel via the same QD into the same Fermi lead and are given by U and/or Δ. Since the lifetime broadenings γ_1 and γ_2 of the two QDs 1 and 2 are small compared to U and Δ such processes are suppressed.
37. P. Recher and D. Loss, Journal of Superconductivity: Incorporating Novel Magnetism 15 (1): 49-65, February 2002; cond-mat/0205484.
38. A.F. Volkov, P.H.C. Magne, B.J. van Wees, and T.M. Klapwijk, Physica C **242**, 261 (1995).
39. M. Kociak, A.Yu. Kasumov, S. Guron, B. Reulet, I.I. Khodos, Yu.B. Gorbatov, V.T. Volkov, L. Vaccarini, and H. Bouchiat, Phys. Rev. Lett. **86**, 2416 (2001).
40. M. Bockrath *et al.*, *Nature* **397**, 598 (1999).
41. R. Egger and A. Gogolin, Phys. Rev. Lett. **79**, 5082 (1997); R. Egger, Phys. Rev. Lett. **83**, 5547 (1999).
42. C. Kane, L. Balents, and M.P.A. Fisher, Phys. Rev. Lett. **79**, 5086 (1997).
43. L. Balents and R. Egger, Phys. Rev. B, **64** 035310 (2001).
44. For a review see e.g. H.J. Schulz, G. Cuniberti, and P. Pieri, cond-mat/9807366; or J. von Delft and H. Schoeller, Annalen der Physik, Vol. **4**, 225-305 (1998).
45. The interaction dependent constants A_b are of order one for not too strong interaction between electrons in the LL but are decreasing when interaction in the LL-leads is increased [19]. Therefore in the case of substantially strong interaction as it is present in metallic carbon nanotubes, the pre-factors A_b can help in addition to suppress I_2.
46. Since $\gamma_{\rho-} > \gamma_{\rho+}$, it is more probable that two electrons coming from the same Cooper pair travel in the same direction than into different directions when injected into the same LL-lead.
47. In order to have exclusively singlet states as an input for the beamsplitter setup, it is important that the LL-leads return to their spin ground-state after the injected electrons have tunnelled out again into the Fermi leads. For an infinite LL, spin excitations are gapless and therefore an arbitrary small bias voltage μ between the SC and the Fermi liquids gives rise to spin excitations in the LL. However, for a realistic finite size LL (e.g. a nanotube), spin excitations are gapped on an energy scale $\sim \hbar v_F/L$, where L is the length of the LL. Therefore, if $k_BT, \mu < \hbar v_F/L$

only singlets can leave the LL again to the Fermi leads, since the total spin of the system has to be conserved. For metallic carbon nanotubes, the Fermi velocity is $\sim 10^6$m/s, which gives an excitation gap of the order of a few meV for $L \sim \mu$m; this is large enough for our regime of interest.

48. A singlet-triplet transition for the ground state of a quantum dot can be driven by a magnetic field; see S. Tarucha *et al.*, Phys. Rev. Lett. **84**, 2485 (2000).
49. This symmetric setup of the charging energy U is obtained when the gate voltages are tuned such that the total Coulomb charging energies in D_C are equal with zero or two electrons.

50. K. Blum, *Density Matrix Theory and Applications* (Plenum, New York, 1996).
51. T.H. Oosterkamp *et al.*, Nature (London) **395**, 873 (1998); T. Fujisawa *et al.*, Science **282**, 932 (1998).
52. J.M. Kikkawa and D.D. Awschalom, Phys. Rev. Lett. **80**, 4313 (1998).
53. I. Malajovich, J. M. Kikkawa, D. D. Awschalom, J. J. Berry, and D. D. Awschalom, Phys. Rev. Lett. **84**, 1015 (2000); I. Malajovich, J. J. Berry, N. Samarth, and D. D. Awschalom, Nature **411**, 770 (2001).
54. M. Johnsson and R. H. Silsbee, Phys. Rev. Lett. **55**, 1790 (1985); M. Johnsson and R. H. Silsbee, Phys. Rev. B **37**, 5326 (1988); M. Johnsson and R. H. Silsbee, Phys. Rev. B **37**, 5712 (1988).
55. F. J. Jedema, A. T. Filip, and B. J. van Wees, Nature **410**, 345 (2001); F. J. Jedema, H. B. Heersche, J. J. A. Baselmans, and B. J. van Wees, Nature **416**, 713 (2002).

56. In addition, for fully spin-polarized leads the device can act as a single spin memory with read-in and read-out capabilities if the dot is subjected to a ESR source.
57. This is true as long as the Zeeman splitting in the leads is much smaller than their Fermi energies.

58. H.-A. Engel and D. Loss, Phys. Rev. B **65**, 195321 (2002), cond-mat/0109470.
59. S. Kawabata, J. Phy. Soc. Jpn. **70**, 1210 (2001).
60. N.M. Chtchelkatchev, G. Blatter, G.B. Lesovik, and T. Martin, cond-mat/0112094.

61. M. Büttiker, Phys. Rev. B **46**, 12485 (1992); Th. Martin and R. Landauer, Phys. Rev. B **45**, 1742 (1992). For a recent comprehensive review on shot noise, see Ref. [2].
62. Our noise definition here differs by a factor of two from that in the review article by Blanter and Büttiker (Ref. [2]); these authors define their power spectral density of the noise with a coefficient two in front (see definition following Eq. (49) and footnote 4 in Ref. [2]). We use a standard Fourier transform (no factor of two in front) to define the noise spectral density.
63. For a discrete energy spectrum we need to insert a density-of-states factor ν in the current and noise definitions; see Ref. [4].
64. Note that the uncorrelated-beam case here refers to a beam splitter configuration with only one of the incoming leads "open". This is an important point since a beam splitter is noiseless for (unpolarized) uncorrelated beams in both incoming leads.

65. G. Engels *et al.* Phys. Rev. B **55**, R1958 (1997); J. Nitta *et al.*, Phys. Rev. Lett. **78**, 1335 (1997); D. Grundler Phys. Rev. Lett. **84**, 6074 (2000); Y. sato *et al.* J. Appl. Phys. **89**, 8017 (2001).
66. A. V. Moroz and C. H. W. Barnes, Phys. Rev. B **60**, 14272 (1999); F. Mireles and G. Kirczenow, *ibid.* **64**, 024426 (2001); M. Governale and U. Zülicke, Phys. Rev. B **66** 073311 (2002).
67. G. Lommer *et al.*, Phys. Rev. Lett. **60**, 728 (1988), G. L. Chen *et al.*, Phys. Rev. B **47**, 4084(R) (1993), E. A. de Andrada e Silva *et al.*, Phys. Rev. B **50**, 8523 (1994), and F. G. Pikus and G. E. Pikus Phys. Rev. B **51**, 16928 (1995).
68. Yu. A. Bychkov and E. I. Rashba, JETP Lett. **39**, 78 (1984).
69. L. W. Molenkamp *et al.*, Phys. Rev. B **64**, R121202 (2001); M. H. Larsen *et al.*, *ibid.* **66**, 033304 (2002).

70. The Rashba-active region in lead 1 is (supposed to be) *electrostatically* induced. This implies that there is no band-gap mismatch between the Rashba region and the adjacent regions in lead 1 due to materials differences. There is, however, a small mismatch arising from the Rashba energy ϵ_R; this is the amount the Rashba bands are shifted down with respect to the bands in the absence of s-o orbit in the channel. Since typically $\epsilon_R \ll \varepsilon_F$, we find that the transmission is indeed very close to unity (see estimate in Ref. [8]).
71. Note that the velocity operator is not diagonal in the presence of the Rashba interaction.
72. J. C. Egues, G. Burkard, and D. Loss, cond-mat/0209692.
73. In the absence of the s-o interaction, we assume the wire has two sets of spin-degenerate parabolic bands for each k vector. In the presence of s-o interaction but neglecting s-o induced interband coupling, there is a one-to-one correspondence between the parabolic bands with no spin orbit and the Rashba bands; hence they can both be labelled by the same indices.
74. N. W. Ashcroft and N. D. Mermin, *Solid State Physics*, Ch. 9. (Holt, Rinehart, and Winston, New York, 1976).
75. G. Schmidt, D. Ferrand, L. W. Molenkamp, A. T. Filip, and B. J. van Wees, Phys. Rev. B **62**, R4790 (2000).
76. L. P. Kouwenhoven, *private communication.*
77. L. I. Glazman and M.E. Raikh, JETP Lett. **47**, 452 (1988); T. K. Ng and P. A. Lee, Phys. Rev. Lett. **61**, 1768 (1988).
78. Y. Meir and A. Golub, Phys. Rev. Lett. **88**, 116802 (2002).
79. F. Yamaguchi and K. Kawamura, Physica B **227**, 116 (1996).
80. A. Schiller and S. Hershfield, Phys. Rev. B **58**, 14978 (1998).
81. G. Burkard, D. Loss, and D.P. DiVincenzo, Phys. Rev. B **59**, 2070 (1999), cond-mat/9808026.
82. W. Izumida and O. Sakai, Phys. Rev. B **62**, 10260 (2000).
83. A. Georges and Y. Meir, Phys. Rev. Lett. **82**, 3508 (1999).
84. T. Aono and M. Eto, Phys. Rev. B **63**, 125327 (2001).
85. I. Affleck, A. W. W. Ludwig, and B. A. Jones, Phys. Rev. B **52**, 9528 (1995).

THE GENERATION AND DETECTION OF SINGLE AND ENTANGLED ELECTRONS IN MESOSCOPIC 2DEG SYSTEMS

W. D. OLIVER*, G. Feve, N. Y. Kim, F. Yamaguchi and Y. Yamamoto †
Quantum Entanglement Project, ICORP, JST
Departments of Electrical Engineering and Applied Physics
E. L. Ginzton Laboratory, Stanford, CA 94305

1. Introduction

The study of higher-order correlation functions of electrons and composite particles in mesoscopic systems has opened the field of quantum electron optics [1]. Since the 1950's, probing higher-order correlation functions in photon and atom-cavity systems has led to a more complete understanding of the fundamental quantum statistical [2, 3] and quantum mechanical phenomena resulting from entanglement [4–6], including violations of Bell's inequality [7–9], quantum nondemolition measurements [10, 11], and teleportation [12–14]. Many current-current correlation experiments have demonstrated that the quantum optical effects manifest in second-order correlation functions can also be observed clearly in mesoscopic electron systems [15–24]. As with photon quantum optics [25–27], this has led naturally to the investigation of higher-order correlation functions beyond the second order, the search for generating functions, and full counting statistics for particles in mesoscopic systems [28–34].

Higher-order correlation functions are particularly useful in probing the nonclassical nature of entanglement. The Einstein-Podolsky-Rosen (EPR) state [4–6] is an interesting example of two-particle entanglement that can be used in secure quantum communication protocols [35, 36], quantum information processing [37], and fundamental tests of quantum mechanics [7]. Although entanglement

* email: woliver@stanford.edu

† also at NTT Basic Research Laboratories, 3-1 Morinosato-Wakamiya Atsugi, Kanagawa, 243-01 Japan

Y. V. Nazarov (ed.), Quantum Noise in Mesoscopic Physics, 275–296.

with ions [38] and between atom and cavity field modes [39, 40] has been demonstrated, to our knowledge, there have yet to be any experimental demonstrations specifically utilizing EPR-type entangled electrons. Recently, there have been several proposals to generate entangled electrons [41–51]. There have also been several proposals to detect the entanglement through bunching/anti-bunching experiments [52–54], teleportation experiments [55], and tests of Bell's inequality [56–59]. In addition, there has been work related to the characterization of entanglement with electrons or, more generally, fermions [42, 60].

In this article, we will review our recent efforts aimed at the generation and detection of electron entanglement in 2DEG systems. We will begin on the detection side, reviewing our quantum electron optics toolbox: an electron intensity interferometer (Hanbury Brown and Twiss (HBT) experiment [2]) and an electron collision analyzer (Hong-Ou-Mandel experiment [3]). We extend their application to generalized states and sources with generalized quantum statisitics; for example, we characterize a source of entangled electrons in a proposed electron bunching and anti-bunching experiment. Since these experiments are proposed for 2DEG systems, we consider next the effect of the Rashba spin-orbit (SO) coupling on such experiments by deriving a coherent scattering formalism that is applicable to both single and entangled electron states. We then consider the generation of entangled electrons via a quantum dot in a manner analogous to $\chi^{(3)}$ four-wave mixing with photons.

The transport experiments discussed in this paper occur in ballistic systems. The inelastic phonon scattering and the elastic ionized impurity scattering lengths are much longer than the characteristic size of the system at cryogenic temperatures (typically 1.5 K in our experiments), and so the electrons are coherent across the entire transport device [21, 23, 54]. The screening length (typically 5 nm) is assumed to be much smaller than the Fermi wavelength (typically 40 nm), so a non-interacting quasi-particle picture is assumed and Coulomb interaction is neglected. Consequently, the interactions related to quantum-interference will occur in this device, while those related to the Coulomb charge can be ignored. We further assume ideal thermal reservoirs, independent transport channels, and transmission probabilities independent of the applied bias voltage. This approach directly follows the coherent scattering formalism in the low-frequency, low-temperature limit: $\hbar\omega << k_B\Theta << eV_{ds} << E_F$ [61, 62].

2. Quantum Optics Toolbox

There are at least six fundamental elements to the quantum optics toolbox that we can use to investigate quantum optics with electrons. These are shown below with their photon counterparts. The first four have been realized experimentally, while the remaining two have yet to be demonstrated.

Electron Quantum Optics	*Photon Quantum Optics*
quantum point contact	waveguide, single-mode source
spatial beamsplitter	spatial beamsplitter
intensity interferometer	Hanbury Brown, Twiss experiment
collision analyzer	Hong, Ou, Mandel experiment
phase shifter	half wave plate
polarization beamsplitter	polarization beamsplitter

2.1. QUANTUM POINT CONTACTS

Quantum point contacts (QPC) act as single-mode, Fermi-degenerate sources at cryogenic temperatures. For our purposes, they are tunable waveguides for electrons, allowing a means to selectively deplete a degenerate, single, transverse mode of electrons. For the first transverse mode accommodated by the QPC, the transmission probability q for the associated longitudinal modes (in particular, those in the net-transport energy window $[E_{\rm F}, E_{\rm F} + eV_{\rm ds}]$, where $V_{\rm ds}$ is the drain source voltage) can be tuned in the range $0 \leq q \leq 1$. The conductance G correspondingly can be tuned in the range $0 \leq G \leq G_{\rm Q}$, where we define the spin-degenerate quantum unit $G_{\rm Q} \equiv 2e^2/h$ and $G = qG_{\rm Q}$. The unilateral power spectral density of the non-equilibrium transport noise in the low-frequency limit, $S = 2eV_{\rm ds}G_{\rm Q}q(1-q) = 2eI_{\rm ds}(1-q)$, where $I_{\rm ds}$ is the drain source current. This correspondingly falls in the range $2eI_{\rm ds} \geq S \geq 0$ (full shot noise to noiseless). The Fano factor $F \equiv S/2eI_{\rm ds}$ is a metric that compares the measured noise to full shot noise within the measurement bandwidth, and it correspondingly falls in the range $1 \geq F \geq 0$ [15, 16, 21, 61–70].

2.2. SPATIAL BEAMSPLITTER

A (spatial) beamsplitter operates on the spatial or orbital part of the particle wave-function. It is described by a unitary scattering matrix relating the incoming and outgoing modes; for example, (see Fig. 2.1a),

$$\begin{pmatrix} \hat{b}_{3\uparrow} \\ \hat{b}_{3\downarrow} \\ \hat{b}_{4\uparrow} \\ \hat{b}_{4\downarrow} \end{pmatrix} = \underbrace{\begin{pmatrix} r & 0 & t & 0 \\ 0 & r & 0 & t \\ t & 0 & r & 0 \\ 0 & t & 0 & r \end{pmatrix}}_{S} \begin{pmatrix} \hat{a}_{1\uparrow} \\ \hat{a}_{1\downarrow} \\ \hat{a}_{2\uparrow} \\ \hat{a}_{2\downarrow} \end{pmatrix}. \tag{1}$$

The unitarity of the S matrix requires that $|r|^2 + |t|^2 \equiv R + T = 1$ and $rt^* + tr^* = 0$. In practice, an incoming particle impinges on the beamsplitter and, ideally, has

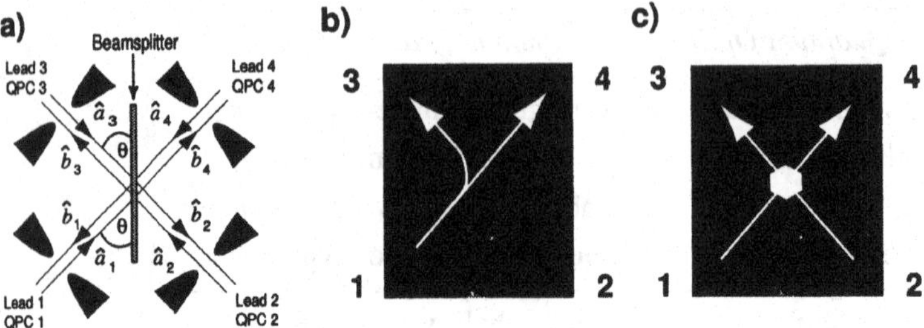

Figure 1. a) Beamsplitter model. Four QPCs define the beamsplitter. $\hat{a}_i$ and $\hat{b}_i$ are annihilation operators for the incoming and outgoing modes respectively. b) Electron micrograph of the device used in the electron HBT experiment. Arrows denote the electron flow from the QPC source at port 1 to the output ports 3 and 4. The normalized cross-covariance is measured between ports 3 and 4. c) The same type of device is used in the electron collision experiment. Arrows show inputs at ports 1 and 2, collision at the beamsplitter, and outputs at ports 3 and 4. The noise is measured at either output port 3 or 4.

a probability T ($R = 1 - T$) to be transmitted (reflected). The devices used in Refs. [23] (Fig 2.1b) and [21] (Fig 2.1c) are examples of an experimental electron beamsplitter. Each port of the electron beamsplitter comprises a QPC, defined electrostatically by negative voltages applied to Schottky gates and by an etched trench [21]. At cryogenic temperatures, the input QPCs at leads 1 and 2 serve as Fermi-degenerate, single-mode electron sources, and are typically biased at [21, 54] or below [23, 54] the first conductance plateau. The output QPCs at leads 3 and 4 are typically biased "wide open" to collect all electrons from the beamsplitter device without reflection back towards the beamsplitter. The beamsplitter itself is also defined electrostatically by a narrow Schottky gate in close proximity to the 2DEG to provide a sharp potential barrier [21]. A small voltage is applied to tune the beamsplitter transmission probability T.

The beamsplitter is considered "linear;" it operates on single modes independently, without mixing energies. It transmits and reflects single particles, regardless of their quantum statistics, leading to standard partition noise. However, two-particle quantum mechanical interference effects between identical particles [21, 3] or entangled states [53, 54, 56, 71] that collide at the beamsplitter will tend to enhance or suppress this noise through particle bunching and anti-bunching. This is not a particular two-particle effect of the beamsplitter; the beamsplitter still operates on the incoming modes independently. Rather, it is the quantum interference between identical modes, as manifest in their final-state enhancement and suppression, that causes the noise enhancement and suppression.

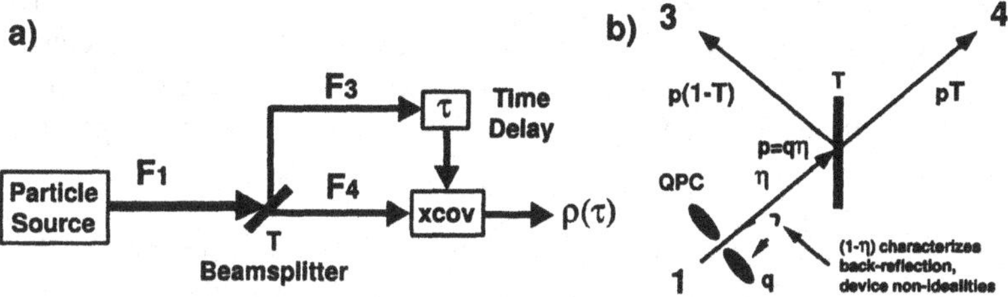

Figure 2. a) Cross-covariance schematic. The Fano factor of the source, F_1, can be determined from the cross-covariance, $\rho(\tau)$. b) Probability model for HBT-type intensity interferometer. The QPC transmission probability is q, η accounts for device non-idealities, $p = \eta q$ is the probability the source successfully sends an electron to the beamsplitter, and T is the beamsplitter transmission probability.

2.3. INTENSITY INTERFEROMETER

In principle, one can measure the current noise of a particle source (beamsplitter, QPC, interferometer, *etc.*) by placing a noise detector immediately after it. This is usually done in mesoscopic systems using an amplifier followed by a square-law device and an averager to get a quantity proportional to the noise power within the measurement bandwidth. However, a careful calibration is needed to interpret the resulting value, because the transfer function accounting for the noise detection circuit is usually not known *a priori*. One possibility is to put a calibrated noise source in parallel with one's device, provided it does not hamper the operation of the device. Another possibility is to characterize the detection system without the device. However, upon replacing the device, the impedance seen by the noise detection circuit will change, both due to the device itself and, more importantly, the change in parasitic reactance. This can alter the bandwidth and/or the gain of the noise detection circuit, potentially skewing the calibration.

Another approach is to physically place a beamsplitter after the noise source and measure the normalized cross-covariance of the output fluxes as shown in figure 2.3a. The normalized current cross-covariance is defined as

$$\rho(\tau) = \frac{\langle \Delta I_3(t) \Delta I_4(t+\tau) \rangle}{\langle \Delta I_3^2 \rangle^{\frac{1}{2}} \langle \Delta I_4^2 \rangle^{\frac{1}{2}}}, \tag{2}$$

where I_i is the particle current at port i and τ is the relative delay time between the beamsplitter outputs. By definition, this is the cross-correlation of the output current fluctuations $\Delta I_3(t)$ and $\Delta I_4(t+\tau)$, normalized by the geometric mean of the output current variances. (Note: one can alternatively consider $N_i = I_i/eB$, the particle number within the measurement bandwidth B.) One can uniquely determine the source Fano factor F_1 from the normalized cross-covariance for a

$T = 1/2$ beamsplitter with zero delay, $\rho(\tau = 0) = (F_1 - 1)/(F_1 + 1)$. The normalized cross-covariance is positive for $F_1 > 1$ (super-Poisson distributed noise), negative for $0 \leq F_1 < 1$ (sub-Poisson distributed noise), and zero for $F_1 = 1$ (classical, Poisson distributed noise). The advantages of this approach are:

1. For matched electronics between the two outputs, the transfer function of the experimental detection system is effectively normalized out of the expression.
2. One experimentally measures the normalized cross-covariance, a one-to-one function in the Fano factor of an arbitrarily complex source.
3. No independent shot-noise calibration is required to interpret the experimental result.

The positive cross-covariance of a partitioned photon flux from a thermal photon reservoir was demonstrated in 1956 by Hanbury Brown and Twiss [2]. This intensity interferometry technique has subsequently become an important tool for probing the statistics of photons generated by various types of sources [72–74]. The negative cross-covariance of a partitioned electron current from a thermal fermion source was measured in the ballistic regime [23] using the beamsplitter device in Fig. 2.1b for twenty different source Fano factors in the range $0.17 \leq F_1 \leq 0.47$, spanning negative cross-covariances in the range $-0.74 \leq \rho(0) \leq -0.33$. A similar experiment was performed simultaneously and independently in the quantum Hall regime for the Fano factor $F = 0$ [22] and at four different source Fano factors [75]. Recently, an HBT-type experiment has also been observed with electrons in free space using a tungsten field emitter as the electron source [76].

Although intensity interferometry has been utilized in the quantum optics community for nearly 50 years, there are several interesting and subtle aspects to this technique. These aspects may be the origin, at least in part, of some misunderstanding related to the interpretation of the Hanbury Brown and Twiss-type experiment with electrons. These aspects include, but are not limited to, the following four issues:

1. The role of the beamsplitter in the measured anti-correlation;
2. The meaning of anti-bunching and the role of the measurement bandwidth;
3. The similarities and differences between a continuous-wave HBT and a coincidence counting HBT experiment;
4. The meaning of the τ dependence of the cross-covariance (Fig. 3 of Ref. [23]) and the role of the measurement bandwidth.

Below, we briefly present and clarify these issues. We then address directly an erroneous criticism of our work [77].

2.3.1. *Role of the Beamsplitter*

The role of the beamsplitter can be seen from the model shown in Fig. 2.3b. The source successfully sends a particle to the beamsplitter with probability $p = \eta q$, where q is the transmission probability through the quantum point contact and η characterizes the non-ideal back-reflection from the beamsplitter back through the QPC [23, 54]. The beamsplitter has a transmission probability T and a reflection probability $\bar{T} \equiv 1 - T$. A straightforward application of probability theory yields:

$$\rho(\tau = 0) = -\left[\frac{pT \cdot p\bar{T}}{(1 - pT)(1 - p\bar{T})}\right]^{\frac{1}{2}} = \frac{F_1 - 1}{\sqrt{F_1 + \frac{\bar{T}}{T}}\sqrt{F_1 + \frac{T}{\bar{T}}}}, \tag{3}$$

where $F_1 = (1 - p)$ is the Fano factor of the source, and mode independence has been assumed. The coherent scattering formalism leads to an identical result [61]. As stated above, we see that the cross-covariance $\rho(0)$ is one-to-one in the Fano factor of the source. In the limit $T = 1/2$, this result simplifies to

$$\rho(\tau = 0) = -\frac{p}{2 - p} = \frac{F_1 - 1}{F_1 + 1}. \tag{4}$$

A measurement of the cross-covariance in the $T = 1/2$ limit depends only on the statistics of the source, not on the properties of the beamsplitter.

One might think that the negative cross-covariance measured in Refs. [22, 23, 75] is due to the beamsplitter transmitting or reflecting each individual particle, as opposed to being a property of the source. However, this line of reasoning is fallacious in general. In an intensity interferometry experiment,

1. The beamsplitter statistics (*e.g.* the transmission probability) are known and fixed *a priori*; the goal is to determine the statistics (*e.g.* the Fano factor) of an unknown source from the measured cross-covariance;
2. The beamsplitter *always* transmits or reflects each individual particle, whether from a thermal boson, classical, or fermion source; yet, one experimentally measures a positive, null, and negative cross-covariance respectively.

Therefore, it is not a property of the beamsplitter that determines the sign of the cross-covariance in an HBT set-up. Rather, the sign of the cross-covariance is determined by the change in the input particle flux with time, and this is a statistical property of the source. A thermal photon source emits particles with super-Poisson statistics ($F > 1$) leading to a positive cross-covariance; a thermal fermion source emits particles with sub-Poisson statistics ($0 < F < 1$) leading to a negative cross-covariance. For a classical Poisson source ($F = 1$), the cross-covariance is zero. This is because a Poisson process that undergoes random partitioning remains a Poisson process, and so there is no correlation between the output fluctuations.

An intuitive understanding of the HBT measurement is the following. Noise originates from two points in this system: the source and the beamsplitter. The source noise is described by the source Fano factor F_1. The beamsplitter partitions the particles from the source, thereby introducing additional partition noise. A measurement of the cross-covariance is *effectively* a comparison of these two noises: source noise to partition noise. In isolation, standard partition noise is half shot-noise. In fact, the fluctuation in number difference between the two outputs is full shot noise [78]. This indicates, at least intuitively, how the shot noise self-calibration can occur in such a system.

Beyond simple intuition, the HBT set-up can be analyzed exactly as an aggregate "source-plus-partition" process. In quantum optics, this process is described in part by the so-called Burgess variance theorem [79, 78, 23, 24]. In probability theory, it is known generally as the "random sum of random variables" problem. See, for example, Ref. [23, 54, 80, 79] for more details.

2.3.2. *Anti-Bunching and the Measurement Bandwidth*

The HBT experiment measures a negative cross-covariance, and this is a direct indication of electron anti-bunching. Anti-bunching is defined as the tendency for particles to arrive separately, with sub-Poissonian fluctuations. It is *not* necessarily a statement that all particles arrive one at a time in a perfectly regulated manner, just as bunching does not mean that all the particles arrive at once and never come again. Rather, bunching and anti-bunching are always statements regarding the statistics of the particle flux.

In quantum optics, the degree of bunching/anti-bunching is usually described in terms of the second-order coherence function, $g^{(2)}(\tau)$, [25]

$$g^{(2)}(\tau) = \frac{\langle I_3(t)I_4(t+\tau)\rangle}{\langle I_3\rangle \langle I_4\rangle} = 1 + \rho(\tau)\frac{\langle \Delta I_3^2\rangle^{\frac{1}{2}} \langle \Delta I_4^2\rangle^{\frac{1}{2}}}{\langle I_3\rangle \langle I_4\rangle}. \tag{5}$$

The range of the second-order coherence function is $0 \leq g^{(2)}(\tau) \leq \infty$. Bunching corresponds to $g^{(2)}(0) > 1$, anti-bunching corresponds to $g^{(2)}(0) < 1$, and $g^{(2)}(0) = 1$ is the classical Poisson limit. It is clear that $g^{(2)}(\tau) < 1$ for any $\rho(\tau) < 0$. Since the HBT experiment with electrons measures negative cross-covariance, it demonstrates anti-bunching.

The role of the measurement bandwidth now becomes clear as well. The suppression of the $g^{(2)}(\tau)$ function below its classical value becomes stronger and sharper in τ as the bandwidth becomes larger [23, 25]. This is because the measurement of $(\langle \Delta I_2^2\rangle^{\frac{1}{2}} \langle \Delta I_3^2\rangle^{\frac{1}{2}})/(\langle I_2\rangle \langle I_3\rangle)$ scales as the measurement bandwidth (for bandwidths smaller than the correlation and/or arrival frequency). This is intuitive, since the measurement bandwidth effectively determines a time-window over which one estimates the statistics of the particle flux. If the measurement bandwidth is low compared with the arrival frequency of the particles, the measurement time-window is large and one effectively estimates the degree of bunching/anti-bunching based on a snapshot of many particles. The averaging time (an additional bandwidth) then effectively determines how many of these snapshots are considered in the final estimate of the statistics.

The above picture was the case in the original HBT continuous-wave experiment with photons [2], which used a 3-27 MHz measurement bandwidth for a 10 THz thermal photon source [81]. It was also the case for the electron HBT experiment in Ref. [23] which used a 2-10 MHz measurement bandwidth for a 0.1-1.0 THz thermal electron source. Based solely on this, both experiments measure roughly the same degree of bunching/anti-bunching. However, in the photon experiment, the high degree of loss further reduced the expected number of photon coincidences by a factor of 10000, thereby reducing their correlation and the resulting estimated degree of bunching [81]. The electron version does not suffer from such losses due to charge conservation. Therefore, the estimated degree of bunching/anti-bunching was in fact larger in the electron HBT experiment [23].

2.3.3. *Continuous-wave and coincidence counting versions of the HBT experiment*

The original HBT experiments with photons [2] and electrons [22, 23] were of the continuous-wave type explained above. Subsequently, coincidence counting techniques were also used to implement intensity interferometry. Both continuous-wave and coincidence counting techniques can, in principle, reveal the correlation time of the source provided the detectors used are sufficiently ideal. That is, both approaches are limited by the detectors.

Detectors oftentimes have a low measurement bandwidth compared with the particle source and/or long recovery times. One way to overcome slow detectors, in principle, is to attenuate the source beam, thereby reducing the mode degeneracy and arrival rate of the particles. The advantage of the coincidence counting technique lies in the fact that $g^{(2)}$ is independent of the mode degeneracy. Therefore, provided one can maintain an adequate signal-to-noise ratio, "ideal" coincidence counting can make accurate $g^{(2)}$ measurements despite low-degeneracy and slow detectors. This is why "ideal" coincidence counting techniques are sometimes realized in pulsed experiments with very low degeneracy (highly attenuated) beams [82, 83]. The penalty is that a Fano factor measurement becomes Poisson limited due to the low-degeneracy. We further note that detector resolution, sometimes called time-jitter, cannot be overcome through attenuation.

On the other hand, one can certainly coincidence count in the "non-ideal" limit. This occurs in practice when the source bandwidth remains too large, the detectors too slow, and/or the signal-to-noise too poor to be overcome by attenuation alone. In this case, multiple counts are recorded, but only a fraction are related to the bunching/anti-bunching effect. This limits the estimated degree of bunching/anti-bunching that can be measured (just as the above continuous wave version is limited with a low bandwidth detector). The recent observation of free-space electron anti-correlations was in this coincidence counting limit; the

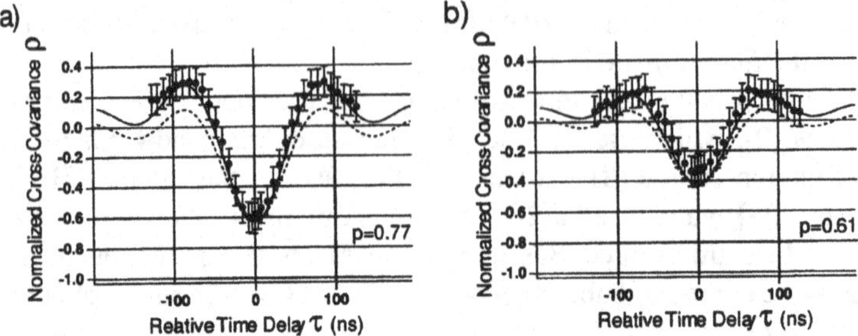

Figure 3. Normalized cross-covariance as a function of the relative delay time for a) $p = 0.77$ b) $p = 0.61$. In both graphs, the circles are the measured data points. The solid line is a simulation of the device and detection circuit for the actual measurement bandwidth 2-10 MHz. The dashed line is for dc-10 MHz, exhibiting no side-lobe dc-offset.

detector bandwidth was a factor 10^{-3} smaller than the electron source bandwidth, leading to mostly random coincidences with a small reduction in the coincidence rate due to the electron anti-bunching [76].

Both the semiconductor [22, 23, 75] and the free-field [76] HBT experiments with electrons unequivocally measure a reduction of the $g^{(2)}$ function and, thereby, demonstrate anti-bunching of electrons. The degree of reduction is limited by the detectors in both cases. The difference between these two approaches is the mode degeneracy: it was of order unity in the semiconductor experiments [22, 23, 75], whereas it was was of order 10^{-4} in the free-field experiment [76].

For more information on the early versions of the continuous-wave and the coincidence counting HBT-type experiments, see Refs. [2, 72, 84, 81, 85–88].

2.3.4. *Normalized Cross-Covariance* vs. τ

The relative delay-time dependence of the normalized cross-covariance, $\rho(\tau)$, was measured for each of the twenty source Fano factors; four of these graphs appeared in Ref. [23]. Two examples are reproduced here in Fig. 2.3.4, and they show the measured normalized cross covariance as a function of the relative delay time τ for two particular values of the QPC transmission probability p (see Fig. 2.3). The minimum value near $\tau = 0$ is the normalized cross-covariance $\rho(0)$ discussed previously. The characteristic shape of the experimental cross-covariance is a direct consequence of the bandpass filter (2-10 MHz) used in the measurement circuit. Ideally, a broadband measurement would contain frequency components from dc to a high-frequency cutoff of at least 100 GHz (inverse of the electron emission time), yielding a cross-covariance with a sharp, delta-like anti-correlation of a width corresponding to this electron interarrival time. However, cutting off all frequency components higher than 10 MHz creates a sinc-like oscillatory behavior with a center lobe full width at half maximum FWHM $\approx$ 100 ns; this is the inverse of the cutoff frequency 10 MHz. Removing the low frequency components below 2 MHz introduces a slow, dc-offset modulation with period $\approx$ 500 ns, which is manifest in Fig. 2.3.4 as an apparent side-lobe dc offset. The solid line in Fig. 2.3.4 is a simulation of the device and measurement circuit considering the actual measurement bandwidth 2-10 MHz. We also include here a dashed line corresponding to a simulation that considers a bandwidth of 0-10 MHz, and, thus, there is no side-lobe dc offset.

The relative delay-time dependence of the cross-covariance conveys three important experimental facts.

1. A variation of τ allows the tuning and matching of the electrical path lengths of the two arms of the interferometer;
2. The FWHM does not change, even though the input QPC is tuned over a wide range of transmission probabilities and, thereby, a wide range of resistances. This is a direct indication that the measurement bandwidth did not change despite the change in resistance.
3. The characteristic sinc-like oscillation is consistent with the low-frequency limit (white noise limit) of the coherent scattering formalism [61], the theory used to describe this experiment.

2.3.5. *Response to Criticism of Gavish, Levinson and Imry*

Finally, we address a specific criticism [77] of our Hanbury Brown and Twiss-type experiment [23]. It is asserted in Ref. [77] that a mesoscopic device at temperature $\Theta = 0$ is attached to a passive detector comprising an LC circuit [89, 90], also at temperature $\Theta = 0$. The authors state that since a passive detector can only absorb energy, only the excess noise is measured.

We consider the detector model used in Ref. [77] to be incomplete. There is a bias V applied to the device, and so there is a net flow of energy into the system. However, it is not clear how this energy is absorbed and dissipated, how a measurement is made (either a dissipative or QND measurement), or how the back-action of the dissipation and/or measurement may effect a $\Theta = 0$ system. In addition, the authors compare their theory to the experimental demonstration of the intensity interferometer in Ref. [23], neglecting to incorporate the modulation/synchronous-detection technique and the measurement bandwidth into their model. Because their model is incomplete, they draw erroneous conclusions:

- The authors state that the cross-covariance vs. τ in Ref. [23] has a 100 ns FWHM and is inconsistent with their broadband theory.

  - *The experimental modulation/synchronous-detection technique does, in fact, measure the excess noise discussed by the authors in Ref. [77].*

- *The experimental result is consistent with a theory that incorporates the measurement bandwidth.*

– The authors in Ref. [77] further question the validity of the so-called "vacuum beam" model used in Refs. [21–23].

- *It is well known that such an interpretation is valid for the net transport in the the low-frequency ($\omega \to 0$) limit, and invalid outside this limit [64, 68, 69, 91–94]. The experimental measurement bandwidths used in Refs. [21–23] are clearly in the low frequency limit and, therefore, the model is valid.*

2.4. COLLISION ANALYZER

The Hanbury Brown and Twiss-type intensity interferometry experiment is a useful tool to probe the noise of a given source. If that source is also known to be a degenerate thermal reservoir, then one can also conclude the type of particle, fermion or boson, from the cross-covariance. However, this is not true for arbitrary noise sources. For example, a single-photon turnstile device is an engineered source of quiet, single-file photons [95, 82, 83]. An HBT-type measurement of this source would yield a *negative* cross-covariance since the source is quiet, *i.e.*, $F_1 = 0$. To determine directly the statistical nature of quantum particles, one should investigate the two-particle contributions to the noise [61, 62]. Such terms arise in the collision of particles [21, 3].

The electron collision analyzer in its original form comprised the collision of two identical electrons at the electron beamsplitter (see Fig. 2.1c), resulting in electron anti-bunching [21]. The photon complement was a collision of identical photons at a beamsplitter, the so-called Hong, Ou, Mandel experiment, resulting in photon bunching [3]. The quantum statistical nature of the constituent particles that undergo the collision determines the symmetry of the orbital wavefunction on which the beamsplitter operates. This in turn determines the two-particle contributions to the noise at the output ports after collision. This concept can be extended to the analysis and characterization of any incoming two-particle state, including entangled states.

For example, we consider here the simple case of two identical bosons (fermions). The particles share the same polarization (spin). The symmetrization postulate states that these quantum particles must be symmetric (anti-symmetric) under particle exchange. This directly leads to the following state (see Fig. 2.1c):

$$|\psi_{in}\rangle = \frac{1}{\sqrt{2}}(|1_A 2_B\rangle \pm |2_A 1_B\rangle) \otimes |\sigma_A \sigma_B\rangle, \tag{6}$$

where A and B are the particle labels, 1 and 2 correspond to the beamsplitter input port modes at the same energy E, and σ corresponds to the polarization (spin). The orbital state is necessarily symmetric (anti-symmetric) for bosons (fermions), since the spin state is symmetric. A spatial beamsplitter operates on the orbital state only, and the output state assuming $T = 1/2$ is [1]

$$|\psi_{\text{out,boson}}\rangle = \frac{1}{\sqrt{2}}(|3_A 3_B\rangle + |4_A 4_B\rangle) \otimes |\sigma_A \sigma_B\rangle \quad (7)$$

$$|\psi_{\text{out,fermion}}\rangle = \frac{1}{\sqrt{2}}(|3_A 4_B\rangle - |4_A 3_B\rangle) \otimes |\sigma_A \sigma_B\rangle, \quad (8)$$

where we have omitted overall phase factors. It is clear that identical photons (fermions) will bunch (anti-bunch) upon collision. We can further state that the positive (negative) symmetry of the incoming orbital state determines the output bunching (anti-bunching). (*c.f.*, for the entangled states discussed below, this general statement is also true, provided the particles arrive from different sides of the beamsplitter [56].) Note that two classical particles would independently scatter into a statistical mixture of the anti-bunched and bunched outputs. In all three cases, the average particle number at an output is one. However, the Fano factor after collision distinguishes the three cases: $F_3 = 1/2$ for classical particles, $F_3 = 1$ for bosons (bunching), and $F_3 = 0$ for fermions (anti-bunching).

2.5. PHASE SHIFTER AND POLARIZATION BEAMSPLITTER

The phase shifter and polarization beamsplitter have yet to be demonstrated experimentally. However, there are a few ideas on how to implement them in ballistic 2DEG systems.

The phase shifter might be realized through a dysprosium micromagnet located near arm 1 of the interferometer in Fig. 3. A gate could shorten or lengthen the path of the electrons through this local magnetic field [56]. Alternatively, a phase shift might be realized using a local Rashba spin-orbit effect [96, 97]. A gate could be used to tune the strength of the local Rashba interaction and/or to alter the path length of the electrons through this local interaction region.

A polarization beamsplitter might be achievable using a strong local magnetic field from, for example, a dysprosium micromagnet, causing a large Zeeman splitting. This approach might be realistic provided the polarization beamsplitter were the last element in the interferometer. This is the case in the proposed Bell's inequality experiment in Ref. [56]. Another idea would be to forego the magnetic field and use a quantum point contact biased at the 0.7 structure, or, more ideally, the 0.5 structure using a back-gate [98]. It remains to be demonstrated that interactions in such a QPC lead to a polarization-based beamsplitting.

3. Bunching and Anti-bunching Experiment

The concept of orbital symmetry and scattering noise can be extended to the characterization of electron entangled states [53, 54, 56], as was demonstrated with photon entangled states [71], through a bunching/anti-bunching experiment (a form of Bell-state analyzer). In Ref. [54], we assume the existence of an electron entangler that emits a non-local spin-singlet state (the spin-triplet with different spins will also work) into a two-particle interferometer in a noiseless, regulated manner (Fig. 3),

$$|\Psi_{in}\rangle = \frac{1}{2}\left(|1_A 2_B\rangle + |2_A 1_B\rangle\right) \otimes \left(|\uparrow_A \downarrow_B\rangle - |\downarrow_A \uparrow_B\rangle\right), \tag{9}$$

where 1 and 2 denote the arms of the interferometer and A and B are the particle labels. A spin-selective phase shift [54] is applied, for example, to the spin-up electrons in the arm 1 of the interferometer. This effectively allows us to tune from the spin-singlet to the spin-triplet as a function of the phase. The entangled state interferes with itself at the collision analyzer, and we use an HBT intensity interferometer to measure the noise in output arm 3 of the output state,

$$\begin{aligned} |\Psi_{out}\rangle = \frac{1}{4}\Big[& i\left(e^{i\phi}+1\right)\left(|3_A 3_B\rangle + |4_A 4_B\rangle\right) \otimes \left(|\uparrow_A \downarrow_B\rangle - |\downarrow_A \uparrow_B\rangle\right) + \\ & \left(e^{i\phi}-1\right)\left(|3_A 4_B\rangle - |4_A 3_B\rangle\right) \otimes \left(|\uparrow_A \downarrow_B\rangle + |\downarrow_A \uparrow_B\rangle\right)\Big], \end{aligned} \tag{10}$$

to within an overall phase factor. The two-electron output can be shifted from bunching to anti-bunching as a function of the phase shift ϕ. The HBT intensity interferometer then measures a cross-covariance corresponding to the output Fano factor,

$$F(\phi) = \frac{1+\rho(\tau=0,\phi)}{1-\rho(\tau=0,\phi)} = \cos^2\left(\frac{\phi}{2}\right). \tag{11}$$

The Fano factor oscillates between $F = 1$ (two-particle bunching, twice standard partition noise) and $F = 0$ (two-particle anti-bunching).

4. The Rashba Effect within the Coherent Scattering Formalism

The bunching/anti-bunching experiment discussed in section 3 assumes a non-interacting particle picture. The entangled state generated by the entangler in Fig. 3 is not altered during its flight towards the phase shifter and beamsplitter along the arms of the interferometers (it is, of course, altered by the phase shifter and beamsplitters). However, it is known that spin-orbit coupling can occur in 2DEG systems [99–101], due to the bulk inversion asymmetry (Dresselhaus effect) [102] and the structure inversion asymmetry (Rashba effect) [103]. Any coupling between the spin and orbital subspaces could alter the electron EPR-state and/or the experimental results based on this state. For example, in the absence of spin-orbit coupling, the spin and orbital subspaces are orthogonal and their symmetries are related only through the anti-symmetrization postulate for

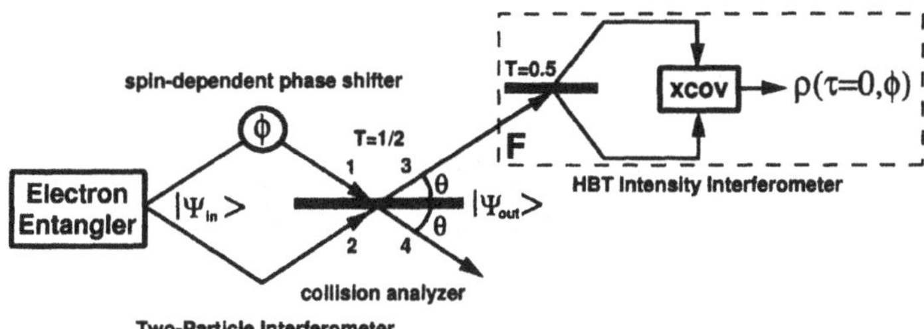

Figure 4. Electron bunching and anti-bunching experiment. An electron entangler emits a spin-singlet state into a two-particle interferometer. A phase shift ϕ is applied to spin-up electrons in arm 1. The Fano factor of the interferometer output after collision is measured using an HBT-type intensity interferometer.

fermions. As discussed in the previous section, this relationship is utilized in the bunching/anti-bunching experiment: a spin phase-shifter is used to alter the spin-subspace symmetry, thereby altering the orbital subspace symmetry on which the spatial beamsplitter will operate, leading to electron bunching and anti-bunching. Clearly, the presence of spin-orbit coupling could effect the EPR-state itself and/or the experimental technique used to analyze it.

As a first attempt to address the spin-orbit coupling issue, we consider the Rashba spin-orbit coupling within the coherent scattering formalism, and its effect on scattering experiments such as the bunching/anti-bunching experiment [104]. The Rashba Hamiltonian is

$$\hat{H}_{SO} = \frac{\alpha}{\hbar}(\hat{\sigma} \times \hat{p})_z = i\alpha(\hat{\sigma}_y \frac{\partial}{\partial_x} - \hat{\sigma}_x \frac{\partial}{\partial_y}), \tag{12}$$

where α is the Rashba parameter that characterizes the aggregate strength of the various spin-orbit coupling mechanisms that are linear in the momentum, and transport occurs in the x-y plane. It takes values in the range 1 to 10×10^{-10} eV cm for a large variety of materials including GaAs/$Al_xGa_{1-x}As$ [105], depending on the shape of the confining well. To simplify the presentation, we consider α to be independent of the in-plane momentum.

The 2DEG free-particle and Rashba Hamiltonians together can be diagonalized; one can determine the stationary states and energy dispersion relation for free electrons subjected to the Rashba spin-orbit interaction. In short, these new stationary states reflect the surface inversion asymmetry, phenomenologically an electric field in the z-direction perpendicular to the 2DEG. An electron moving through this electric field feels a virtual magnetic field perpendicular to the direction of motion, as shown in Fig. 4a. The "spin" σ of the electron then lies parallel ($\sigma = -$) or anti-parallel ($\sigma = +$) to this magnetic field. Thus, as the electron scatters to different leads with different orientations, its direction of motion changes

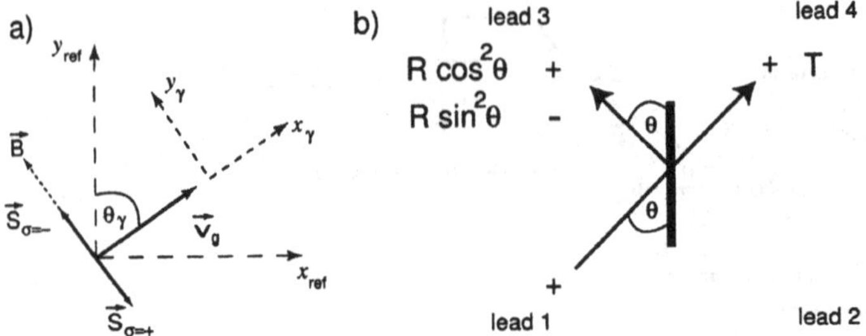

Figure 5. a) An electron spin-orbit coupling state travels with group velocity v_g along lead γ which makes an angle θ_γ with respect to the reference frame. Due to the surface inversion asymmetry, the electron feels a virtual magnetic field perpendicular to v_g. The spin-orbit label σ is the quantum number for "spin", and its direction is aligned ($\sigma = -$) or anti-aligned ($\sigma = +$) to this field. b) A beamsplitter scatters an incoming $\sigma = +$ electron. A transmitted electron does not change direction, and remains a $\sigma = +$ state with probability T. The reflected electron changes direction, and is projected onto both $\sigma = +$ and $\sigma = -$ states with probabilities $R\cos^2\theta$ and $R\sin^2\theta$ respectively, according to the lead angle θ.

and, therefore, the "spin" direction will also change. This results in a lead angular dependence of the Rashba stationary states [104].

The Rashba stationary states can be used to derive a multi-terminal coherent scattering formalism that connects the spin-orbit coupling states in different leads through a scattering matrix [104], as is done for the non-interacting electron case [61]. For example, the current operator takes on the following form,

$$\hat{I}_\alpha(t) = \frac{e}{h}\sum_{\rho}\sum_{\sigma\sigma'}\sum_{\beta\gamma}\int dEdE'e^{\frac{i(E-E')t}{\hbar}}\ \hat{a}^{\dagger}_{\beta\sigma}(E)A^{\alpha\rho}_{\beta\gamma\sigma\sigma'}(E,E')\hat{a}_{\gamma\sigma'}(E'),$$

$$A^{\alpha\rho}_{\beta\gamma\sigma\sigma'}(E,E') = \delta_{\alpha\beta}\delta_{\alpha\gamma}\delta_{\rho\sigma}\delta_{\rho\sigma'} - S^{*\alpha\rho}_{\beta\sigma}(E,\theta_\alpha,\theta_\beta)S^{\alpha\rho}_{\gamma\sigma'}(E',\theta_\alpha,\theta_\gamma), \tag{13}$$

where E denotes the energy, (α,β,γ) denote the leads of the device, (σ,ρ) denote the spin-orbit label (a quantum number for the "spin" in the spin-orbit basis), and the matrix $A^{\alpha\rho}_{\beta\gamma\sigma\sigma'}(E,E')$ is seen to be dependent on the lead orientations through the angles $\theta_\alpha,\theta_\beta$ in the scattering matrix $S^{\alpha\rho}_{\beta\sigma}(E,\theta_\alpha,\theta_\beta)$. The form of the current operator is identical to that in Ref. [61], except that the electron spin is replaced by the spin-orbit label σ and there is a lead angular dependence. For the beamsplitter device shown in Figs. 2.1a, 3, 4b, the scattering matrix reflecting this angular dependence is found to be [104],

$$\begin{pmatrix} \hat{b}_{3+} \\ \hat{b}_{3-} \\ \hat{b}_{4+} \\ \hat{b}_{4-} \end{pmatrix} = \underbrace{\begin{pmatrix} r\cos\theta & i\,r\sin\theta & t & 0 \\ i\,r\sin\theta & r\cos\theta & 0 & t \\ t & 0 & r\cos\theta & -i\,r\sin\theta \\ 0 & t & -i\,r\sin\theta & r\cos\theta \end{pmatrix}}_{S} \begin{pmatrix} \hat{a}_{1+} \\ \hat{a}_{1-} \\ \hat{a}_{2+} \\ \hat{a}_{2-} \end{pmatrix}, \tag{14}$$

where unitarity again requires $|r|^2 + |t|^2 \equiv R + T = 1$ and $rt^* + tr^* = 0$. Compared with Eq. 1, we find off-diagonal elements in the reflection submatrices of Eq. 14. The behavior of this scattering matrix is illustrated in Fig. 4b. For example, a $\sigma = +$ state is injected in lead 1 and scattered by a beamsplitter into leads 3 and 4. The transmitted electron does not change direction, so its "spin" remains $\sigma = +$ with probability T. However, the reflected electron does change direction, and the initial "spin" $\sigma = +$ is scattered to the Rashba states in lead 3: $\sigma = +$ with probability $R\cos^2\theta$ and $\sigma = -$ with probability $R\sin^2\theta$. Thus, we find a probability $R\sin^2\theta$ of a so-called "spin-flip" in the scattering process. It is shown in Ref. [104] that this "spin-flip" effect leads to a new means of experimentally measuring the Rashba parameter α in a collision of spin-polarized electrons.

We consider here the effect of the Rashba spin-orbit coupling on the degree of bunching in the bunching/anti-bunching experiment discussed in Section 3. There it was shown that applying a spin-dependent phase shift ϕ to one arm of the interferometer with a spin-singlet state input (see Fig. 3) results in an oscillation of the Fano factor between ideal bunching and ideal anti-bunching (see Eq. 11). Here, we consider a spin-singlet state in the σ basis. In Ref. [104], it is shown that spin-orbit coupling reduces the degree of bunching according to the lead angle θ of the beamsplitter,

$$F(\phi, \theta) = \cos^2\theta \cos^2\left(\frac{\phi}{2}\right). \tag{15}$$

This result can be understood in the following way. The bunching effect occurs for spin-singlet states that collide at the beamsplitter. However, the probability that the Rashba spin-orbit coupling leaves the "spin" unchanged upon scattering goes as $\cos^2\theta$ (for this case of two-particle collision, it is exactly $\cos^2\theta$ [104]). Thus, the degree of bunching is scaled by the extent to which the singlet state remains unaltered. In other words, as the purity of the singlet state is "diluted", the bunching effect is reduced. What is learned from this exercise is that one can reduce the destructive effect of the Rashba spin-orbit coupling on scattering experiments by designing beamsplitters with shallow angles θ.

We note here that a Fano factor reduction was also found in Ref. [96, 97] through a local Rashba inter-band coupling. There also, the reduction is due to a dilution of the original singlet state, although the mechanism for the dilution effect is distinct. These two Fano factor reduction mechanisms help to illustrate the fragility of entanglement and entanglement experiments.

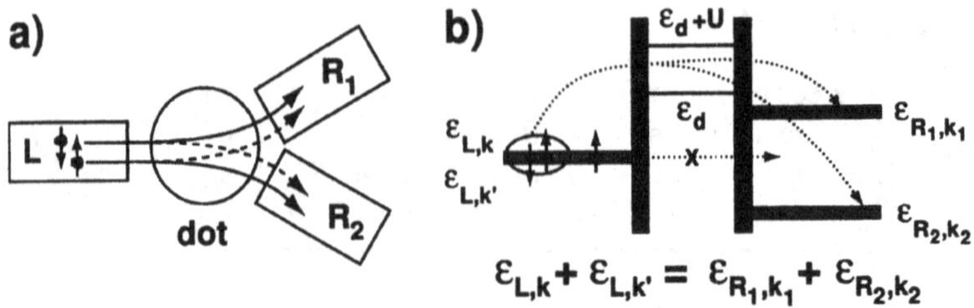

Figure 6. a) Three port quantum dot entangler. b) Energy band diagram with non-degenerate energy leads which act as energy filters.

5. Electron Entanglement *via* a Quantum Dot

Until this point, we have assumed the existence of an electron entangler, a device that emits electrons into a non-local spin-singlet state. There have been several proposals to entangle electrons in a variety of systems [41–51]. Here, we consider electron entanglement *via* a quantum dot in a manner analogous to photon entanglement generated by a $\chi^{(3)}$ parametric amplifier [47].

The electron entangler comprises a three-port quantum dot operating in the coherent tunneling regime [106, 107]. The dot is coupled to one input and two output leads as shown in Fig. 6a, with an energy band diagram shown in Fig. 6b. The lead and dot energy levels are arranged such that the dot is empty. There are two key concepts to the successful operation of this entangler. The first concept is that the leads are non-degenerate and of relatively narrow width in energy, thus acting as "energy filters". Single electron tunneling does not conserve energy and is forbidden. However, two-electron co-tunneling events can conserve energy, sending one electron each to leads R_1 and R_2. This co-tunneling process is the lowest-order contribution to the tunneling current. The second concept is that double occupancy of the dot incurs an on-site Coulomb energy, mediating the electron entanglement as a nonlinear medium does for photon entanglement.

As detailed in Ref. [47], the system is analyzed using the Anderson Hamiltonian with an on-site Coulomb energy term U. We consider only a single, spin-degenerate energy level for the dot, and there are no single electron excitations within the dot.

$$\begin{aligned} \hat{H}_{\text{And}} &= \hat{H}_0 + \hat{V}, \\ \hat{H}_0 &= \sum_{\eta,k,\sigma} \varepsilon_{\eta,k} \hat{a}^\dagger_{\eta,k,\sigma} \hat{a}_{\eta,k,\sigma} + \sum_{\sigma} \varepsilon_d \hat{c}^\dagger_\sigma \hat{c}_\sigma + U \hat{n}_\uparrow \hat{n}_\downarrow, \\ \hat{V} &= \sum_{\eta,k,\sigma} \left(V_\eta \hat{a}^\dagger_{\eta,k,\sigma} \hat{c}_\sigma + h.c. \right), \end{aligned} \tag{16}$$

where $\eta \in \{L, R_1, R_2\}$ is the lead label, k is the lead electron momentum, $\sigma \in \{\uparrow, \downarrow\}$ is the electron spin, V_η is the overlap matrix element between the dot and the lead states, $\hat{a}$ ($\hat{a}^\dagger$) is the annihilation (creation) operator for the lead electrons, $\hat{c}$ ($\hat{c}^\dagger$) is the annihilation (creation) operator for the dot electrons, and $\hat{n}_\sigma \equiv \hat{c}_\sigma^\dagger \hat{c}_\sigma$ is the dot electron number operator. The dot energy levels, ε_d and $\varepsilon_d + U$ shown in Fig. 6b, are taken to be off-resonance with the leads. The left lead energy is below its quasi-Fermi level so that the lead is full of electrons. The right leads are empty. We use a T-matrix formalism [108, 109] to calculate the transition amplitude $\langle\phi_f|\hat{T}(\varepsilon_i)|\phi_i\rangle$ between an initial state $|\phi_i\rangle$ and a final state $|\phi_f\rangle$ to order $\mathcal{O}(V^4)$, where $\hat{T}(\varepsilon_i) = \hat{V} + \hat{V}\frac{1}{\varepsilon_i - \hat{H}_0}\hat{T}(\varepsilon_i)$ is the transition operator. As discussed in Ref. [47], the initial state is taken to be a selection of any two electrons with spins σ and σ' from the left lead, $|\phi_i\rangle = \hat{a}^\dagger_{L,k,\sigma}\hat{a}^\dagger_{L,k',\sigma'}|0\rangle$. The final states are the singlet and triplet states at the output, $|s\rangle, |t\rangle \equiv (1/\sqrt{2})\left(\hat{a}^\dagger_{R_1\uparrow}\hat{a}^\dagger_{R_2\downarrow} \mp \hat{a}^\dagger_{R_1\downarrow}\hat{a}^\dagger_{R_2\uparrow}\right)|0\rangle$, and the same-spin states, $|\uparrow\uparrow\rangle, |\downarrow\downarrow\rangle$: $|\phi_f\rangle = |s\rangle, |t\rangle, |\uparrow\uparrow\rangle, |\downarrow\downarrow\rangle$. The resulting transition amplitudes are,

$$\langle s|\hat{T}|\phi_i\rangle = \sqrt{2}\frac{V_L^{*2}V_{R_1}V_{R_2}2E_L}{(E_L^2 - \Delta_R^2)(E_L^2 - \Delta_L^2)}\frac{U}{2E_L - U}, \tag{17}$$

$$\langle t|\hat{T}|\phi_i\rangle = 0 \tag{18}$$

$$\langle\uparrow\uparrow|\hat{T}|\phi_i\rangle = \langle\downarrow\downarrow|\hat{T}|\phi_i\rangle = 0, \tag{19}$$

where we have taken $\varepsilon_d = 0$, $E_L \equiv \frac{1}{2}(\varepsilon_{L,k} + \varepsilon_{L,k'}) = \frac{1}{2}(\varepsilon_{R_1,k_1} + \varepsilon_{R_2,k_2})$, $\Delta_L \equiv \frac{1}{2}(\varepsilon_{L,k} - \varepsilon_{L,k'})$, and $\Delta_R \equiv \frac{1}{2}(\varepsilon_{R_1,k_1} - \varepsilon_{R_2,k_2})$.

Only the singlet state amplitude remains at the output of the entangler, and it scales with a U-dependent function. In the limit that $U = 0$, the singlet state vanishes, emphasizing the necessity of the non-linear interaction. The results can be understood in the following way. The Anderson Hamiltonian in Eq. 16 conserves spin. The singlet portion of the two-electron input state is effectively filtered to the output side of the dot, while the triplet portion of the two-electron input remains on the input side. As explained in Ref. [47], the reason is due to Fermi statistics (including the Pauli principle) and energy conservation. The non-linear interaction U allows for Coulomb-mediated wave mixing [110]; the optical analogue is a $\chi^{(3)}$ parametric amplifier, a four-wave mixing process in which two input photons interact within the $\chi^{(3)}$ nonlinear medium and may generate an entangled pair of output photons. Experimentally, the empty dot limit might be achievable with semiconductors using side-gating [111] or carbon nanotubes [112]. Realization of the energy filters might be achieved using additional quantum dots [45] or superlattice structures [113].

6. Conclusion

In conclusion, we have discussed the generation and detection of single and entangled electron states in mesoscopic 2DEG systems. We presented the HBT-type intensity interferometer and the collision analyzer, experiments originally used for the detection single- and two-particle contributions to the noise due to single electron states (independent modes) produced by quantum point contacts. We extended their application to the detection of generalized states, in particular, the entangled electron states in a bunching/anti-bunching experiment. Finally, we considered the generation of entangled spin-singlet states *via* a quantum dot.

References

1. Y. M. Blanter and M. Büttiker, Physics Reports **336**, 1 (2000).
2. R. Hanbury Brown and R. Q. Twiss, Nature **177**, 27 (1956).
3. C. K. Hong, Z. Y. Ou, and L. Mandel, Phys. Rev. Lett. **59**, 2044 (1987).
4. A. Einstein, B. Podolsky, and N. Rosen, Phys. Rev. **47**, 777 (1935).
5. N. Bohr, Phys. Rev. **48**, 696 (1935).
6. D. Bohm, *Quantum Theory*, Constable, London (1954).
7. J. S. Bell, Physics **1**, 195 (1964).
8. A. Aspect, P. Grangier, and G. Roger, Phys. Rev. Lett. **47**, 460 (1981).
9. G. Weihs, *et al.*, Phys. Rev. Lett. **81**, 5039 (1998).
10. P. Grangier, J. A. Levenson, and J.-P. Poizat, Nature **396**, 537 (1998).
11. G. Nogues, *et al.*, Nature **400**, 239 (1999).
12. D. Bouwmeester, *et al.*, Nature **390**, 575 (1997).
13. D. Boschi, *et al.*, Phys. Rev. Lett. **80**, 1121 (1998).
14. A. Furusawa, *et al.*, Science **282**, 706 (1998).
15. M. Reznikov, M. Heiblum, H. Shtrikman, and D. Mahalu, Phys. Rev. Lett. **75**, 3340 (1995).
16. A. Kumar, *et al.*, Phys. Rev. Lett. **76**, 2778 (1996).
17. A. H. Steinbach, J. M. Martinis, and M. H. Devoret, Phys. Rev. Lett. **76**, 3806 (1996).
18. R. J. Schoelkopf, *et al.*, Phys. Rev. Lett. **78**, 3370 (1997).
19. L. Saminadayar, D. C. Glattli, Y. Jin, and B. Etienne, Phys. Rev. Lett. **79**, 2526 (1997).
20. R. de Picciotto, *et al.*, Nature **389**, 162 (1997).
21. R. C. Liu, B. Odom, Y. Yamamoto, and S. Tarucha, Nature **391**, 263 (1998).
22. M. Henny, *et al.*, Science **284**, 296 (1999).
23. W. D. Oliver, J. Kim, R. C. Liu, and Y. Yamamoto, Science **284**, 299 (1999).
24. E. Comforti, *et al.*, Nature **416**, 515 (2002).
25. L. Mandel and E. Wolf, *Optical Coherence and Quantum Optics*, Cambridge University Press, New York (1995).
26. Y. Yamamoto and A. İmamoğlu, *Mesoscopic Quantum Optics*, John Wiley and Sons, New York (1999).
27. C. W. J. Beenakker and H. Schomerus, Phys. Rev. Lett. **86**, 700 (2001).
28. L. S. Levitov and G. B. Lesovik, JETP **58**, 230 (1993).
29. D. A. Ivanov and L. S. Levitov, JETP **58**, 461 (1993).
30. B. A. Muzykantskii and D. E. Khmelnitskii, Phys. Rev. B **50**, 3982 (1994).
31. H. Lee, L. S. Levitov, and A. Y. Yakovets, Phys. Rev. B **51**, 4079 (1995).
32. W. Belzig and Y. V. Nazarov, Phys. Rev. Lett. **87**, 067006 (2001).
33. W. Belzig and Y. V. Nazarov, Phys. Rev. Lett. **87**, 197006 (2001).

34. Y. V. Nazarov and D. A. Bagrets, Phys. Rev. Lett. **88**, 196801 (2002).
35. C. H. Bennett and G. Brassard, in *Proceedings of the IEEE Conference on Computers, Systems, and Signal Processing, Bangalore, India*, p. 174, IEEE, New York (1984).
36. A. K. Ekert, Phys. Rev. Lett. **67**, 661 (1991).
37. C. H. Bennett and D. P. DiVincenzo, Nature **404**, 247 (2000).
38. M. A. Rowe, *et al.*, Nature **409**, 791 (2001).
39. E. Hagley, *et al.*, Phys. Rev. Lett. **79**, 1 (1997).
40. X. Maître, *et al.*, Phys. Rev. Lett. **79**, 769 (1997).
41. D. Loss and D. P. DiVincenzo, Phys. Rev. A **57**, 120 (1998).
42. J. Schliemann, D. Loss, and A. H. MacDonald, Phys. Rev. B **63**, 085311 (2001).
43. G. Burkard, D. Loss, and D. P. DiVincenzo, Phys. Rev. B **59**, 2070 (1999).
44. C. H. W. Barnes, J. M. Shilton, and A. M. Robinson, Phys. Rev. B **62**, 8410 (2000).
45. P. Recher, E. V. Sukhorukov, and D. Loss, Phys. Rev. B **63**, 165314 (2001).
46. G. B. Lesovik, T. Martin, and G. Blatter, Eur. Phys. J. B **24**, 287 (2001).
47. W. D. Oliver, F. Yamaguchi, and Y. Yamamoto, Phys. Rev. Lett. **88**, 037901 (2002).
48. P. Recher and D. Loss, Phys. Rev. B **65**, 165327 (2002).
49. D. S. Saraga and D. Loss, cond-mat/0205553 (2002).
50. V. Bouchiat, *et al.*, cond-mat/0206005 (2002).
51. C. Bena, S. Vishveshwara, L. Balents, and M. P. A. Fisher, Phys. Rev. Lett. **89**, 037901 (2002).
52. D. Loss and E. V. Sukhorukov, Phys. Rev. Lett. **84**, 1035 (2000).
53. G. Burkard, D. Loss, and E. V. Sukhorukov, Phys. Rev. B **61**, R16303 (2000).
54. W. D. Oliver, *et al.*, in I. O. Kulik and R. Ellialtioglu (Eds.), *Quantum Mesoscopic Phenomena and Mesoscopic Devices in Microelectroncs*, pp. xxx–xxx, Kluwer Academic Publishers, Dordrecht (2000).
55. O. Sauret, D. Feinberg, and T. Martin, cond-mat/0203215 (2002).
56. X. Maître, W. D. Oliver, and Y. Yamamoto, Physica E **6**, 301 (2000).
57. A. Bertoni, *et al.*, Phys. Rev. Lett. **84**, 5912 (2000).
58. R. Ionicioiu, P. Zanardi, and F. Rossi, Phys. Rev. A **63**, 050101(R) (2001).
59. N. M. Chtchelkatchev, G. Blatter, G. B. Lesovik, and T. Martin, cond-mat/0112094 (2001).
60. J. Schliemann, *et al.*, Phys. Rev. A **64**, 022303 (2001).
61. M. Büttiker, Phys. Rev. B **46**, 12485 (1992).
62. T. Martin and R. Landauer, Phys. Rev. B **45**, 1742 (1992).
63. R. Landauer, Phil. Mag. **21**, 863 (1970).
64. V. A. Khlus, Sov. Phys. JETP **66**, 1243 (1987).
65. B. J. Van Wees, *et al.*, Phys. Rev. Lett. **60**, 848 (1988).
66. D. A. Wharam, *et al.*, J. Phys. C **21**, L209 (1988).
67. A. Szafer and A. D. Stone, Phys. Rev. Lett. **62**, 300 (1989).
68. G. B. Lesovik, JETP Lett. **49**, 592 (1989).
69. M. Büttiker, Phys. Rev. Lett. **65**, 2901 (1990).
70. B. Yurke and G. P. Kochanski, Phys. Rev. B **41**, 8184 (1990).
71. M. Michler, K. Mattle, H. Weinfurter, and A. Zeilinger, Phys. Rev. A **53**, R1209 (1996).
72. G. A. Rebka and R. V. Pound, Nature **180**, 1035 (1957).
73. H. J. Kimble, M. Dagenais, and L. Mandel, Phys. Rev. Lett. **39**, 691 (1977).
74. F. Diedrich and H. Walther, Phys. Rev. Lett. **58**, 203 (1987).
75. S. Oberholzer, *et al.*, Physica E **6**, 314 (2000).
76. H. Kiesel, A. Renz, and F. Hasselbach, Nature **418**, 392 (2002).
77. U. Gavish, Y. Levinson, and Y. Imry, Phys. Rev. Lett. **87**, 216807 (2001).
78. R. C. Liu and Y. Yamamoto, Phys. Rev. B **49**, 10520 (1994).
79. R. E. Burgess, Discussions of the Faraday Society **28**, 151 (1959).
80. A. W. Drake, *Fundamentals of Applied Probability Theory*, McGraw-Hill, New York (1967).

81. R. Hanbury Brown and R. Q. Twiss, Nature **178**, 1447 (1956).
82. P. Michler, *et al.*, Science **290**, 2282 (2000).
83. C. Santori, *et al.*, Phys. Rev. Lett. **86**, 1502 (2001).
84. E. Brannon and H. I. S. Ferguson, Nature **178**, 481 (1956).
85. E. M. Purcell, Nature **178**, 1449 (1956).
86. R. Q. Twiss and R. Hanbury Brown, Nature **179**, 1128 (1957).
87. R. Q. Twiss, A. G. Little, and R. Hanbury Brown, Nature **180**, 324 (1957).
88. R. Q. Twiss and A. G. Little, Aust. J. Phys. **12**, 77 (1959).
89. G. Lesovik and R. Loosen, JETP Lett. **65**, 295 (1997).
90. U. Gavish, Y. Levinson, and Y. Imry, Phys. Rev. B **62**, 10637 (2000).
91. M. Büttiker, Phys. Rev. B **45**, 3807 (1992).
92. S. R. E. Yang, Solid State Comm. **81**, 375 (1992).
93. R. C. Liu and Y. Yamamoto, Phys. Rev. B **50**, 17411 (1994).
94. M. J. M. de Jong and C. W. J. Beenakker, in L. L. Sohn, L. P. Kouwenhoven, and G. Schön (Eds.), *Mesoscopic Electron Transport*, pp. 225–258, Kluwer, The Netherlands (1997).
95. J. Kim, O. Benson, H. Kan, and Y. Yamamoto, Nature **397**, 500 (1999).
96. J. C. Egues, G. Burkard, and D. Loss, cond-mat/0204639 (2002).
97. J. C. Egues, G. Burkard, and D. Loss, cond-mat-0207392 (2002).
98. S. Nuttinck, *et al.*, Jpn. J. Appl. Phys. **39**, L655 (2000).
99. B. Das, S. Datta, and R. Reifenberger, Phys. Rev. B **41**, 8278 (1990).
100. J. Luo, H. Munekata, F. F. Fang, and P. J. Stiles, Phys. Rev. B **41**, 7685 (1990).
101. G. L. Chen, *et al.*, Phys. Rev. B **47**, 4084 (1993).
102. G. Dresselhaus, Phys. Rev. **100**, 580 (1955).
103. E. I. Rashba, Sov. Phys. Solid State **2**, 1109 (1960).
104. G. Feve, W. D. Oliver, M. Aranzana, and Y. Yamamoto, Phys. Rev. B (2002).
105. T. Hassenkam, *et al.*, Phys. Rev. B **55**, 9298 (1997).
106. D. V. Averin and Y. V. Nazarov, in H. Grabert and M. H. Devoret (Eds.), *Single Charge Tunneling: Coulomb Blockade Phenomena in Nanostructures, Vol. 294*, Plenum Press, New York (1992).
107. E. V. Sukhorukov, G. Burkard, and D. Loss, Phys. Rev. B **63**, 125315 (2001).
108. L. S. Rodberg and R. M. Thaler, *Introduction to the Quantum Theory of Scattering*, Academic Press, New York (1967).
109. G. D. Mahan, *Many Particle Physics*, Plenum Press, New York (1990).
110. M. Kitagawa and M. Ueda, Phys. Rev. Lett. **67**, 1852 (1991).
111. S. Tarucha, *et al.*, Phys. Rev. Lett. **77**, 3613 (1996).
112. H. W. C. Postma, *et al.*, Science **293**, 76 (2001).
113. M. Higashiwaki, *et al.*, Jpn. J. Appl. Phys. **35**, L606 (1996).

WHAT QUANTITY IS MEASURED IN AN EXCESS NOISE EXPERIMENT?

U. GAVISH, Y. IMRY, Y. LEVINSON
Weizmann Institute of Science
76100 Rehovot, Israel

AND

B. YURKE
Bell Laboratories
07974 Murray Hill New Jersey

1. Introduction

Consider a measurement in which the current coming out of a mesoscopic sample is filtered around a frequency $\Omega > 0$, then amplified, measured, squared, then this process is repeated many times and the results are averaged. The final result of such a procedure is called the *current fluctuations*, the *power spectrum*, or the *current noise*, at frequency Ω.

Often, two such measurements are performed on the same system: the first while it is driven out of equilibrium (e.g., by applying a DC voltage [1] or electromagnetic radiation [2],[3] to it) and the second in equilibrium (the voltage source is turned off, i.e., the power supply becomes a short). The *excess noise* is defined as the difference in the noise between the first and the second measurement. The present work analyzes what quantity one should calculate in order to predict excess noise.

In the rest of the introduction some basic concepts in amplification theory are introduced. In Sec. 2 the measurement procedure is defined. In Sec. 3 we analyze the classical case and in sections 4 and 5 the quantum one - when $\hbar\Omega$ is comparable with or larger than the temperature, the voltage, or the RF radiation frequency applied to the sample. Finally, a possible verification of the results is presented.

Y.V. Nazarov (ed.), Quantum Noise in Mesoscopic Physics, 297–311.

Our main result can be stated as follows. The result of a noise measurement depends on the particular instrumentation used in the setup - what type of amplifier and detector are used, what are their temperatures, etc. It is generally, and usually, neither the fourier transform of the current correlator in the sample, nor its symmetrized version. However, the *excess* noise is instrumentation-independent, and is equal to the amplifier-gain squared times the difference in the fourier transform of the current correlator, in and out of equilibrium. This difference has a clear physical meaning: it is the difference in the power emitted *from* the sample and *into* the filter, in and out of equilibrium.

1.1. COSINE AND SINE COMPONENTS OF TIME DEPENDENT FUNCTIONS

A real (and well behaved) function, $I(t)$, can be fourier-represented as:

$$I(t) = \frac{1}{2\pi} \int_0^{\infty} d\omega \left[I(\omega) e^{-i\omega t} + I^*(\omega) e^{i\omega t} \right]. \tag{1}$$

where $I(\omega) = \int_{-\infty}^{\infty} dt I(t) e^{i\omega t}$. If $I(\omega)$ is negligible outside a narrow bandwidth Δ_f around a center frequency $\Omega > 0$, $\Delta_f \ll \Omega$, then it is useful to write $I(t)$ in the form [4]:

$$I(t) = I_c(t) \cos \Omega t + I_s(t) \sin \Omega t, \tag{2}$$

where are $I_c(t)$ and $I_s(t)$ are real and slowly varying - they have fourier-components only at frequencies smaller than Δ_f. $I_c(t)$ and $I_s(t)$ are called the *cosine* and *sine* components of $I(t)$. Defining the time average of $f(t)$ as:

$$\overline{f(t)} \equiv \frac{1}{T_0} \lim_{T_0 \to \infty} \int_t^{t+T_0} dt' f(t'), \tag{3}$$

then (if $\Delta_f \ll \Omega$) Eq. (2) implies: $\overline{I^2(t)} = \frac{1}{2}\overline{I_c^2(t)} + \frac{1}{2}\overline{I_s^2(t)}$. In a *stationary state*, the choice of the origin of time does not affect average quantities and therefore using and Eq. (2) for $t = 0$ and $t = \pi/(2\Omega)$ one gets:

$$\overline{I^2(t)} = \overline{I_c^2(t)} = \overline{I_s^2(t)}. \tag{4}$$

Moreover, in a stationary state time averaging may be replaced by averaging over realizations. We choose these realizations as repeated measurement on the same system at $N \gg 1$ different times, t_n, which are distributed over a whole time interval, T_0, which is longer than other time scales in the systems: $\hbar/eV$, $\hbar/(k_B T)$, Ω^{-1}, and Δ_f^{-1} (in practice, usually the largest time is Δ_f^{-1} and therefore it is enough to require that the measurements are distributed over a time interval longer than it.). Thus, we can replace the time average: $\overline{f(t)} \to \langle f(t) \rangle = \frac{1}{N} \sum_{n=1}^{N} f(t_n)$.

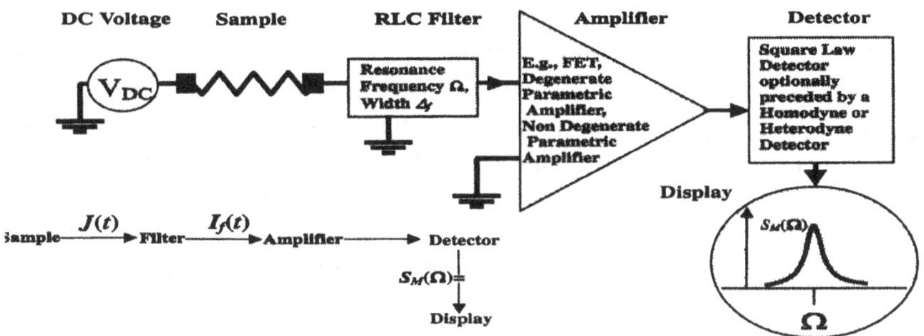

Figure 1. Noise Measurement Procedure

1.2. LINEAR AMPLIFIERS

Below, we shall consider only *linear* amplifiers. An amplifier is linear if the current coming out of it, $I_a(t)$, is related to the one entering it, $I_f(t)$, by:

$$I_a(t) = G_1 I_{f,c}(t) \cos \Omega t + G_2 I_{f,s}(t) \sin \Omega t, \tag{5}$$

where $I_{f,c}(t)$ and $I_{f,s}(t)$ are the cosine and sine components of $I_f(t)$. and where at least one of the numbers G_1 and G_2 is large. If $G_1 = G_2$ the amplifier is called *phase-insensitive*. A phase insensitive amplifier does what one would naively expect from an amplifier - it just multiplies the incoming signal by a large number. If $G_1 \neq G_2$ the amplifier is called *phase sensitive* and it affects the two components of the incoming signal differently. An important special case is when $G_i \gg 1 \gg G_j$, $i \neq j$, where the amplifier amplifies only one of the components while disposing of the other.

Phase sensitive and insensitive amplifiers have similar classical behaviors, but may have different technical advantages. However, when quantum effects are important they differ fundamentally due to the limitations Heisenberg principle puts on them - see Ref. [4] and the Appendix.

In noise measurement in mesoscopic systems phase insensitive amplifiers are commonly used (e.g., in setups that include a field effect transistor) but we consider below also the phase sensitive case such as the degenerate parametric amplifiers [4] because their quite developed technology (that was used, for example, in order to enable sensitive detection in experimental gravitational physics [5]) seems to be less familiar in the mesoscopic community and also in order to demonstrate the universality of our result.

2. Noise and excess noise measurement procedures

2.1. NOISE MEASUREMENT

In a typical noise measurement (shown in Fig. 1) at frequency Ω the current $J(t)$ that flows out of the sample (which is assumed to be in a stationary state, but not necessarily in equilibrium,) is filtered with an RLC circuit around a resonance frequency $\Omega = (LC)^{-\frac{1}{2}}$. The current coming out of the filter, $I_f(t)$, is amplified, measured at some arbitrary time point, t, and the result of this measurement, $I_a(t)$, is squared (say, by a square law detector). This measurement is then repeated at $N \gg 1$ different times, $\{t_n\}$, and the results are averaged and divided by the filter bandwidth, Δ_f, giving a number which we shall denote by $S_M(\Omega)$,

$$S_M(\Omega) = \frac{1}{N\Delta_f} \sum_{n=1}^{N} I_a^2(t_n). \tag{6}$$

Finally, a voltage proportional to $S_M(\omega)$ is sent out to drive a display.

The explanation for why the above measurement is, at least in the classical case, a measurement of the current spectrum is given in the next section. Meanwhile we make the following comments:

1. The above procedure is not a measurement of a function of time. It is merely a series of *independent* samplings of a stationary process. Though many setups perform these samplings at times which are separated by a *constant* time interval, we stress that this is *not* essential (and actually may create confusion): the result will be the same even if the measurement is performed at *random times* as explained below Eq. (4)).

2. In some setups it is convenient to convert the signal to a low-frequency one and measure only the sine or cosine components, for example, by mixing the signal with a local oscillator (as is done in Heterodyne and Homodyne detection), that is, multiplying it by a pure sine or a cosine and averaging the result over time before squaring it. Such a mixing may introduce additional noise that should be taken into account. Numerical factors that may multiply the signal as a result of such a procedure are then cancelled by, e.g., calibration of the setup with respect to a source of noise with a known power spectrum such as a resistor in thermal equilibrium.

If the system, the filter and the amplifier are all in a stationary state, then according to Eq. (4) such a procedure yields the same $S_M(\Omega)$ as in the case of measuring and squaring the whole signal.

We shall always assume that the sample and the filter are in a stationary state, but we shall not necessarily assume that this is the case for the amplifier. It is typically the case for semiconductor amplifiers such as the field effect transistors used in noise measurement in mesoscopic systems but it is *not* the case in several types of parametric amplifiers [4].

2.2. EXCESS NOISE MEASUREMENT

In an excess noise measurement one subtracts the noise measured when the system is in equilibrium from that which is measured when the same system is driven out of equilibrium:

$$S_{M,excess}(\Omega) = S_{M,noneq}(\Omega) - S_{M,eq}(\Omega). \tag{7}$$

The excess noise is useful when one is interested in looking into the changes in the system which are due to driving it out of equilibrium. It is also useful when a particular setup (amplifier temperature and type, etc) affects the measurement by introducing an *additional* noise which is independent of the sample state, so by taking the difference between the two noise powers one can get rid of the instrumentation-dependent noise power.

Equilibrium properties, can not be described by the excess noise since by definition it vanishes in equilibrium.

Since in most cases mesoscopic samples are driven out of equilibrium by an external DC voltage, V, we shall consider the quantity:

$$S_{M,excess}(\Omega) = S_{M,V}(\Omega) - S_{M,0}(\Omega), \tag{8}$$

however, other means (e.g., by application of external radiation [2],[3]) can be used to drive the system out of equilibrium.

2.3. STATEMENT OF THE PROBLEM

What quantity should one calculate in order to predict $S_{M,excess}(\Omega)$? Will this quantity depend on the properties of the sample only, or also on the particular experimental setup?

To answer these questions we first consider the classical case.

3. Which quantity is measured in a classical noise measurement?

3.1. THE CLASSICAL CASE WITHOUT AMPLIFICATION

Consider a current $I(t)$ flowing in a system which is in a stationary state. Consider a long time interval $T_0 \gg \omega$ and define the *restricted fourier transform* of $I(t)$,

$$I_{T_0}(\omega) = \int_{-\frac{T_0}{2}}^{\frac{T_0}{2}} dt e^{i\omega t} I(t), \tag{9}$$

the *power spectrum* of $I(t)$,

$$S_I(\omega) = \lim_{T_0 \to \infty} \frac{|I_{T_0}(\omega)|^2}{T_0}. \tag{10}$$

and the *correlator* of $I(t)$,

$$c(\tau) = c(-\tau) = \overline{I(0)I(\tau)}. \tag{11}$$

The Wiener-Khinchin theorem [6] states that

$$S(\omega) = S(-\omega) = \int_{-\infty}^{\infty} d\tau e^{-i\omega\tau} c(\tau), \tag{12}$$

$$c(\tau) = \frac{1}{\pi} \int_{0}^{\infty} d\omega cos\omega\tau S_I(\omega). \tag{13}$$

and specifically also that,

$$c(0) = \overline{I^2(t)} = \frac{1}{\pi} \int_{0}^{\infty} d\omega S_I(\omega). \tag{14}$$

By its definition, a filter greatly reduces the fourier components at frequencies further than a band width Δ_f away from its center frequency, Ω . Therefore, assuming a regular behavior of $J_{T_0}(\omega)$, and a small Δ_f, the restricted transform of the current coming out of that filter, $I_{f,T_0}(\omega)$, is related to that of the incoming current, $J_{T_0}(\omega)$, by

$$\begin{aligned} I_{f,T_0}(\omega) &= \gamma J_{T_0}(\Omega) \quad & |\omega - \Omega| \lesssim \Delta_f, \\ I_{f,T_0}(\omega) &= 0 & |\omega - \Omega| \gtrsim \Delta_f, \end{aligned} \tag{15}$$

where γ is constant. An example of a filter is shown in Fig. 2. For the moment we do not consider the possibility that the filter adds its own thermal noise to $I_f(t)$. Thus, according to their definitions, the power spectrum of $I_f(t)$ and $J(t)$ are related by:

$$\begin{aligned} S_f(\omega) &= \gamma^2 S_J(\Omega) \quad & |\omega - \Omega| \lesssim \Delta_f, \\ S_f(\omega) &= 0 & |\omega - \Omega| \gtrsim \Delta_f. \end{aligned} \tag{16}$$

Applying Eq. (14) to $I_f(t)$, Eq. (12) to $J(t)$, and using Eq. (16) one gets

$$\frac{1}{\Delta_f} \overline{I_f^2(t)} = \gamma^2 S_J(\Omega) = \gamma^2 \int_{-\infty}^{\infty} d\tau e^{i\Omega\tau} \overline{J(0)J(\tau)}. \tag{17}$$

Eq. (17) clarifies what might be a confusing feature of a noise measurement with a filter: taking the square of the current coming out of the filter at one time yields information on the correlation in the current in the sample at two different times.

The average energy stored in an RLC filter is [7]:

$$\langle E_f \rangle = L \langle I_f^2 \rangle. \tag{18}$$

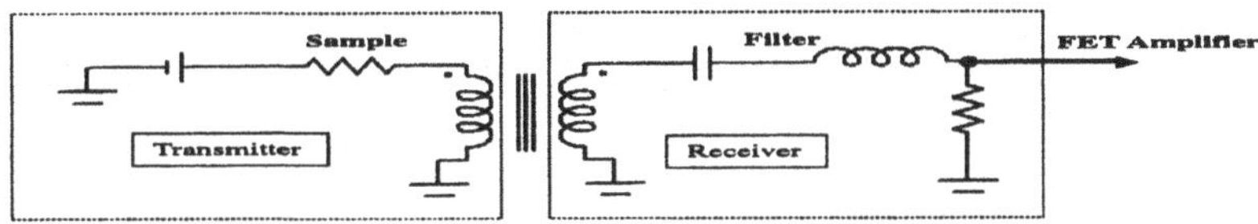

Figure 2. Inductive coupling to an RLC filter

For later purposes we note that this equation is valid also in a stationary quantum state as a consequence of the virial theorem [8]. Since, for the moment, the amplification stage is ignored we have $I_f(t) = I_a(t)$. Therefore, making use of the definition of the measured noise, Eq. (6), we see that

$$S_M^{(\text{no ampl})}(\Omega) = \frac{1}{\Delta_f L}\langle E_f \rangle = \gamma^2 \int_{-\infty}^{\infty} d\tau e^{i\Omega\tau} \langle J(0)J(\tau)\rangle. \tag{19}$$

3.2. THE CLASSICAL CASE WITH AMPLIFICATION

We proceed now to include the amplification stage. In order to avoid complications due to the differences between different types of amplifiers we assume here that only the cosine component of the output current, $I_{a,c}(t)$, is measured, squared and averaged. Such an assumption enables us to assign a single gain, G, to the amplifier (whether it is phase sensitive or not) that multiplies the incoming signal. Thus, $I_f(t)$ enters the amplifier and $I_a(t)$ comes out of it. Then, the component $I_{a,c}(t) = GI_{f,c}(t)$ is measured and squared. But $\overline{I_{f,c}^2} = \overline{I_f^2}$ and therefore by Eqs. (17)-(19) one has:

$$S_M(\Omega) = \frac{G^2}{\Delta_f L}\langle E_f \rangle = \bar{G}^2 \int_{-\infty}^{\infty} d\tau e^{i\Omega\tau} \langle J(0)J(\tau)\rangle. \tag{20}$$

where $\bar{G} = \gamma G$.

3.3. THERMAL NOISE PRODUCED BY THE SETUP

Even when considered classically, the measurement setup can perform according to Eq. (5) and (20) only as long as its components are operating at temperatures which are low compared with the signal power spectrum. At higher temperatures, one should take into account the thermal noise, $S_{N,T}(\omega)$, produced by these components and add it to the output signal. We shall not discuss the particular form of this noise, except for mentioning that in equilibrium and at low frequencies its contribution is $k_B T$ (the Nyquist-Johnson noise) times the amplifier-gain squared, and that it is independent of the input, i.e., of the state of the sample. Thus, we write:

$$S_M(\Omega) = \frac{G^2}{\Delta_f L}\langle E_f\rangle + S_{N,T}(\Omega) = G^2 \int_{-\infty}^{\infty} d\tau e^{i\Omega\tau}\langle J(0)J(\tau)\rangle + S_{N,T}(\Omega) \tag{21}$$

where $S_{N,T}(\Omega)$ is defined as the noise measured with no input.
Comments:

1. Other types of noise such (e.g., $1/f$ noise) may occur inside the setup components. However, unlike the noise required by thermodynamics (and in the quantum case also that required by the Heisenberg principle), these may in principle be eliminated and thus are not considered here.

2. Adding a setup noise which is assumed to be independent of the input, is justified by assuming that the statistical distributions of the state of the total system is a product of that of the amplifier (and the detector) and the filter+sample and that the coupling between those parts is weak. In the quantum case the assumption is that the density matrix of the system is a product of that of the amplifier (and the detector) and the sample+filter. For more details see the appendix.

3.4. CLASSICAL EXCESS NOISE

According to Eqs. (8), (21), the excess noise in a classical measurement is:

$$S_{M,excess}(\Omega) = \frac{G^2}{\Delta_f L}\langle E_f\rangle_{excess} = \bar{G}^2 \int_{-\infty}^{\infty} d\tau e^{i\Omega\tau}\langle J(0)J(\tau)\rangle_{excess} \tag{22}$$

where

$$\langle E_f\rangle_{excess} = \langle E_f\rangle_V - \langle E_f\rangle_0, \tag{23}$$

and

$$\langle J(0)J(\tau)\rangle_{excess} = \langle J(0)J(\tau)\rangle_V - \langle J(0)J(\tau)\rangle_0, \tag{24}$$

are the differences in the filter energies and the correlators in and out of equilibrium. Eqs. (21) and (22) tell us what are the measured quantities in noise and excess noise measurements in classical situations, i.e., when quantum effects can be neglected. They show that although the measured noise, Eq. (21), is setup-dependent because the term $S_{N,T}(\Omega)$ depends on the setup type and temperature, the measured excess noise is not.

Eq. (22) also gives a simple physical picture to the excess noise: when a voltage is applied to the sample and drives it out of equilibrium, the current (or charge) fluctuations in the sample change (typically, they increase), and interact with the charges or currents in the RLC circuit (e.g., through capacitive or inductive coupling) causing an increase in the energy flow from the sample into the circuit in a similar way to that in which the current fluctuations in an antenna of a classical transmitter emit energy into a receiver (Fig. 2). Part of this energy is accumulated in the capacitor and the inductor and part is dissipated in the resistor. Eventually the system arrives at a stationary (though not an equilibrium) state where the filter energy is higher than before. The measured excess noise is simply this increase in the filter energy multiplied by the amplifier gain.

Having obtained a detailed picture of the classical noise measurement we are now ready to analyze the quantum case.

4. Which quantity is measured in quantum noise measurement?

When the measured frequency is higher than the temperature of the sample or the setup, quantum effects become important. One may then consider replacing the current $J(t)$, and the average over realizations $\langle J(0)J(\tau)\rangle$, in Eqs. (21) and (22) by, respectively, the Heisenberg current operator of the electrons in the sample, $\hat{J}(t)$, and the expectation value of the product of operators, $\hat{J}(0)\hat{J}(\tau)$, in the quantum stationary state of the system. However, in attempting to do so one immediately encounters the following two questions (which are answered below):

1. The current operator does not commute with itself at different times (the product $\hat{J}(0)\hat{J}(\tau)$ is not Hermitian) and therefore it is not clear in which order, $\hat{J}(0)\hat{J}(\tau)$ or $\hat{J}(\tau)\hat{J}(0)$ the product should be written or whether it should be replaced by its *symmetrized* version $(\hat{J}(0)\hat{J}(\tau) + \hat{J}(\tau)\hat{J}(0))/2$, (which is Hermitian) as is customarily suggested in text books [9].

2. What are the properties of the setup noise (the analog $S_{N,T}(\Omega)$) in the quantum case, and specifically, does it vanish in the limit of zero temperature as was the case in the classical regime?

4.1. THE QUANTUM CASE WITHOUT AMPLIFICATION

Consider first a mesoscopic system, e.g., a ballistic quantum point contact between two ohmic contacts, in which a DC voltage is applied to the left contact and the current flowing out of the right one interacts with the current in an RLC circuit (modelled by an harmonic oscillator with a small damping) through an inductive coupling of the form (see Fig. 2):

$$\alpha\hat{J}(t)\hat{I}_f(t).$$

The above system was considered in Refs. [10] and [11] (in the limit of small α and Δ_f). It was shown, that as a result of this interaction there is an energy flow between the electronic system and the filter. The current fluctuations in the electronic system excite the harmonic modes of the filter in a similar way to that in which current fluctuations in an antenna excite the photon modes in the electromagnetic field of the vacuum . As a result of switching the *interaction* on adiabatically while keeping the DC voltage constant, (unlike what is considered above where the *voltage* is switched on), the energy of the filter was found to increase by an amount of [10]-[11]:

$$\delta\langle E_f\rangle = L(\langle\hat{I}_f^2(t)\rangle_\alpha - \langle\hat{I}_f^2(t)\rangle_{\alpha=0}) =$$
$$\gamma^2\left[(N+1)S_J(\Omega) - NS_J(-\Omega)\right] = \gamma^2\left[S_J(\Omega) - 2N\hbar\Omega G_d(\Omega)\right], \qquad (25)$$

where $S_J(\Omega)$ is the transform of the *nonsymmetrized* quantum correlator,

$$S_J(\Omega) = \int_{-\infty}^{\infty} d\tau c^{i\Omega\tau}\langle\hat{J}(0)\hat{J}(\tau)\rangle, \qquad (26)$$

where N is the average number of quanta in the oscillator, and $G_d(\Omega)$ is the differential conductance [12]. γ^2 is equal to α^2 times some multiplicative factors (which anyhow cancel in the setup calibration). $S_J(\Omega)$ has a physical meaning: it is proportional to the fermi-golden rule emission-rate of quanta of $\hbar\Omega$ from the sample into the filter [11]. Similarly, $S_J(-\Omega)$ is proportional to the absorption rate. Thus, Eq. (25) has a simple physical meaning: the change in the filter energy correspond to the spontaneous emission from the sample plus the net energy flow due to the difference between the induced emission and absorption.

In order to obtain the change in the filter energy due to the application of the voltage, one has to calculate the difference between finite and zero voltage: $\langle E_f\rangle_{excess} = \delta\langle E_f\rangle_V - \delta\langle E_f\rangle_0$. One obtains:

$$\langle E_f\rangle_{excess} = \gamma^2\int_{-\infty}^{\infty} d\tau e^{i\Omega\tau}\langle\hat{J}(0)\hat{J}(\tau)\rangle_{excess}, \qquad (27)$$

where it was assumed that the voltage-dependence of the differential conductance is weak. Eq. (27) expresses the change in the filter energy when the sample is in and out of equilibrium. We are now ready to proceed to the final step and take the amplification into account.

4.2. THE QUANTUM CASE WITH AMPLIFICATION

Consider the current $\hat{I}_f(t)$ in an RLC circuit which serves as the input signal of a linear amplifier. The quantum theory of linear amplifiers [4] specifies what limitations the Heisenberg principle puts on their performances. The limitations relevant to our case can be summarized as follows:

An amplifier that amplifies both sine and cosine components of the input signal *must* add noise, which we will denote by $S_{N,Q}(\Omega)$, to the measured signal, and this noise does not vanish at zero temperature. Therefore, a phase insensitive amplifier must add noise to the measured signal. However, the added noise is not necessarily distributed evenly between the two components. The amount of noise which is added, say, to the cosine component, depends on the particular amplification setup and temperature. The fluctuation of the current coming out of a phase insensitive amplifier are given by the form [4]:

$$S_M(\Omega) = \Delta_f^{-1}\langle \hat{I}_a^2(t)\rangle = G^2\Delta_f^{-1}\langle \hat{I}_f^2(t)\rangle + S_{N,Q}(\Omega) \tag{28}$$

where $S_{N,Q}(\Omega)$ depends only on the properties and state of the amplifier and detector while $\langle \hat{I}_f^2(t)\rangle$ depends only on those of the filter. Similarly, the fluctuation of the current coming out of a phase sensitive amplifier (which can be, ideally, noiseless and which amplifies only the cosine component and disposes of the sine component) are given by the form:

$$S_M(\Omega) = \Delta_f^{-1}\langle \hat{I}_{a,c}^2(t)\rangle = G^2\Delta_f^{-1}\langle \hat{I}_{f,c}^2(t)\rangle. \tag{29}$$

A brief review on the origin of the additive form of the right hand side of Eq. (28) is given in the appendix.

5. Excess quantum noise measurement

Since the filter is in a stationary state, one has $\langle \hat{I}_f^2\rangle = \langle \hat{I}_{f,c}^2\rangle = \frac{1}{L}\langle E_f\rangle$ and therefore, Eqs. (18), (28) and (29) yield:

$$S_{M,excess}(\Omega) = G^2(L\Delta_f)^{-1}\langle E_f\rangle_{excess}. \tag{30}$$

Eqs. (27) and (30) imply

$$S_{M,excess}(\Omega) = \bar{G}^2\int_{-\infty}^{\infty} d\tau e^{i\Omega\tau}\langle \hat{J}(0)\hat{J}(\tau)\rangle_{excess}, \tag{31}$$

where $\bar{G} = \gamma G$.

Eq. (31) is our main result. It shows that in order to predict the result of an excess noise measurement one should calculate the correlators (no symmetrization is required) in and out of equilibrium and take the difference between them. It also shows that there will be no contribution from the zero-point fluctuations since $S(\Omega > 0)$ (the *emission* spectrum) does not contain such contribution - the zero point fluctuations can not emit energy.

When the *sample* (but not necessarily the setup) is at zero temperature the correlator vanishes in equilibrium since the sample can not emit anything. Therefore:

$$S_{M,excess}(\Omega) = G^2\int_{-\infty}^{\infty} d\tau e^{i\Omega\tau}\langle \hat{J}(0)\hat{J}(\tau)\rangle, \qquad k_BT_s = 0. \tag{32}$$

We note that the equality in Eq. (31) is not term-by-term. The excess noise is equal to the excess of the correlator but the noise by itself is not equal to the correlator by itself since according to Eq. (28) the former depends also on the setup properties while the latter depends only on the sample properties.

The derivation of Eq. (31) is valid for all linear amplifiers used in noise measurements in mesoscopic systems and the parametric ones which are analyzed in Ref. [4]. The condition that the conductance G_d remains constant when the DC voltage is turned on, that was used in deriving this equation, may be understood by considering the zero temperature case: the excess noise is the power flow of energy from the sample into the filter. In order to enable an efficient measurement of this power an impedance matching is needed between the sample, the transmission lines and the filter. If the sample conductance is very different in its equilibrium and nonequilibrium states, then, initially good impedance-matching with the detector that enables an efficient power flow in equilibrium, means a bad impedance matching out of equilibrium with an inefficient power flow. In such a case one should correct Eq. (31) by taking into account the different impedance ratios in and out of equilibrium.

6. Suggested verifications of the theory

A straightforward way to verify Eq. (32) is to measure the excess noise in a single-channel ballistic quantum point contact at high frequencies, $\hbar\Omega \gtrsim eV$. For such a system the nonsymmetrized correlator is given by (see Ref. [14] for the symmetrized version, and [15] and [13] for the nonsymmetrized one):

$$S(\Omega, T_s, V) = \frac{e^2}{h}|t|^2(1-|t|^2)\sum_{\epsilon=\pm 1} F(\hbar\Omega + \epsilon eV) + \frac{e^2}{h}|t|^4 F(\hbar\Omega) \qquad (33)$$

where $F(x) = x(e^{x/k_B T_s} - 1)^{-1}$, T_s is the sample temperature and $|t|^2$ is the transmission of the channel. According to our theory, such an excess noise measurement would yield $S(\Omega, T_s, V) - S(\Omega, T_s, 0)$ for any amplifier type or temperature, while without taking the excess the result will generally depend substantially on type of the setup and its temperature. In particular, for $T_s \ll eV, \hbar\Omega$, the excess noise measurement will yield $S(\Omega, 0, V)$, i.e., the nonsymmetrized correlator (with no contribution from the zero point fluctuations) while without taking the excess the result will generally differ from both the non-symmetrized and the symmetrized correlators and will depend on the particular setup.

7. Appendix. Noise added in amplification

7.1. REQUIREMENTS FROM QUANTUM LINEAR AMPLIFIER OUTPUT

Consider an RLC circuit (see e.g., Fig. 2), which we shall call 'the input', connected into a linear amplifier. Let $\hat{I}_f(t)$ be the Heisenberg operator of the current in the input. This operator acts on the degrees of freedom of the circuit and therefore its expectation values are determined when the circuit state is given. Let $\hat{I}_a(t)$ be the Heisenberg operator of the current at the output port of the amplifier. In general, this operator acts on both the input and the amplifier degrees of freedom. For an ideal linear amplifier:

$$\hat{I}_a(t) = \hat{I}_{a,c}(t)\cos\Omega t + \hat{I}_{a,s}(t)\sin\Omega t = G_1\hat{I}_{f,c}(t)\cos\Omega t + G_2\hat{I}_{f,s}(t)\sin\Omega t. \quad (34)$$

Note that in such an ideal case the output operator, $\hat{I}_a(t)$, acts only on the degrees of freedom of the input (and not any of the setup) - the input state determines completely the output independently of the amplifier state. However, the Heisenberg principle limits on the possibility of realizing such a device: it requires the sine and the cosine components to obey a minimum uncertainty relation (see Eq. (2.19) in Ref. [4]). Suppose one constructs a minimum-uncertainty state for $\hat{I}_{a,c}(t)$ and $\hat{I}_{a,s}(t)$, so that they can be measured simultaneously at the maximum accuracy permitted without violating the Heisenberg principle. If $G_1G_2 > 1$ (as is the case for a phase insensitive amplifier where $G_1 = G_2 \gg 1$) then $\hat{I}_{f,c}(t)$ and $\hat{I}_{f,s}(t)$ could have been measured simultaneously up to an accuracy which is G_1G_2 times better than allowed simply by measuring $\hat{I}_{a,c}(t)$ and $\hat{I}_{a,s}(t)$, and then applying Eq. (34). Therefore, an amplifier with $G_1G_2 > 1$ and specifically a phase insensitive amplifier is forbidden in quantum mechanics.

One way to overcome these limitations is to build an amplifier in which, for example, $G_1 \gg 1$, $G_1G_2 = 1$, that is, a phase sensitive amplifier that amplifies the cosine component and diminishes the sine component so that in the limit of $G_1 = G \to \infty$ we can write:

$$\hat{I}_a(t) = G\hat{I}_{f,c}(t)\cos\Omega t \quad (35)$$

An example for such a device is the degenerate parametric amplifier [4]. Another way to overcome the above limitations is to allow $G_1G_2 > 1$, (and in particular $G_1 = G_2 = G \gg 1$,) but add to the right hand side of Eq. (34) an additional term that will operates on the amplifier degrees of freedom so that:

$$\hat{I}_a(t) = G\hat{I}_f(t) + \hat{I}_{N,Q}(t), \quad (36)$$

where $\hat{I}_{N,Q}(t)$ is an operator acting on the amplifier degrees of freedom such as the electronic state in a field effect transistor or the idler resistor state in a non-degenerate parametric amplifier.

7.2. INDEPENDENCE OF THE AMPLIFIER NOISE OF THE SAMPLE STATE

Let us assume that the input is in a stationary state in which the average currents vanish and take the expectation square of the output current, as is done in a noise measurement. In order to do so we should specify the state of the system or more generally, the density matrix. Here we make an important assumption that the total density matrix of the system, ρ_s is a product of the amplifier density matrix, ρ_a, and the input one, ρ_f: ρ_a and ρ_f, $\rho_s = \rho_f \rho_a$. Such an assumption is justified, e.g., when the interaction between the amplifier and the input is small compared to their coupling with the thermal baths that determine their temperatures. Then using Eq. (35) and averaging over time much larger than Ω^{-1} one has

$$\langle \hat{I}_a^2(t) \rangle_s = \frac{1}{2} G^2 \langle \hat{I}_{f,c}^2(t) \rangle_{f,a} = \frac{1}{2} G^2 \langle \hat{I}_{f,c}^2(t) \rangle_f, \tag{37}$$

where $\langle A \rangle_x = Tr \rho_x A$. This relationship is identical to Eq. (29) except for the factor 1/2 which appeared due to different definitions of the gain. In the phase insensitive case one gets from Eq. (36)

$$\begin{aligned} \langle \hat{I}_a^2(t) \rangle_s &= G^2 \left[\langle \hat{I}_f^2(t) \rangle_f + \langle \hat{I}_f^2(t) \rangle_a + 2 \langle \hat{I}_f(t) \rangle_f \langle \hat{I}_a(t) \rangle_a \right] \\ &= G^2 \langle \hat{I}_f^2(t) \rangle_f + S_{N,Q}(\Omega) \Delta_f \end{aligned} \tag{38}$$

where $S_{N,Q}(\Omega)\Delta_f = G^2 \langle \hat{I}_f^2(t) \rangle_a$, as in Eq. (28). Eqs. (37) and (38) shows that the amplifier noise is *additive* to the input noise (in Eq. (37) it is trivially zero), at least as long as one assumes that the density matrix of the system can be factorized as described above.

8. Acknowledgment

E. Comforti, R. dePiccioto, C. Glattli, M. Heiblum, W. Oliver, D. Prober, M. Reznikov, U. Sivan and A. Yacoby are warmly thanked for many discussions, and for providing detailed descriptions of their experimental setups. U. G. would like to express special thanks to B. Doucot for his support and for many very helpful discussions.

References

1. M. Reznikov, M. Heiblum, H. Shtrikman, and D. Mahalu, Phys. Rev. Lett. **75**, 3340 (1995); A. Kumar, L. Saminadayar, D. C. Glattli, Y. Jin and B. Etienne, Phys. Rev. Lett. **76**, 2778 (1996); R. dePiccioto et al. Nature, **389**, 162 (1997); L. Saminadayar et al. Phys. Rev. Lett. **79**, 2526 (1997).
2. R.J. Schoelkopf, A.A. Kozhevnikov, D.E.Prober, M.J. Rooks, Phys. Rev. Lett. 80, 2437 (1998)
3. L.-H. Reydellet, P. Roche, D. C. Glattli, B. Etienne and Y. Jin, cond-mat/0206514 v2 3 Jul 2002.

4. B. Yurke and J. S. Denker, Phys. Rev. A. **29**, 1419 (1984).
5. See, e.g., K. S. Thorne et al. in *Sources of Gravitational Radiation*, L. Smarr (ed.), Cambridge University Press, London, p. 49 (1979).
6. E.g., see C. W. Gardiner, *Handbook of Stochastic Methods*, 2nd ed., Sec. 1.4.2, p. 15, Springer-Verlag, (1990)
7. See, e.g., E. M. Purcell, *Electricity and Magnetism*, Sec. 8.5, Mcgraw-Hill (1965).
8. E.g., see under 'Virial Theorem' in C. Cohen-Tannoudji, B. Diu and F. Laloe, *Quantum Mechanics*, 2nd ed., ch. 3, p. 344 (1977).
9. See e.g., L. D. Landau and E. M. Lifshitz, *Statistical Physics Part 1*, 3rd ed., Sec. 118, p. 360, Eq. 118.4, Butterworth Heinemann (1997).
10. G. Lesovik and R. Loosen, JETP Lett., **65**, 295 (1997).
11. U. Gavish, Y. Levinson and Y. Imry, Phys. Rev. **B 62** R10637 (2000). Y. Levinson and Y. Imry, Proc. of Nato ASI "Mesoscope 99" in Ankara/Antalya, Kluwer 1999.
12. In deriving the right equality in Eq. (25) the fluctuation-dissipation theorem (or generelized Kubo formula) was used *out of equilibrium*. The theorem provides a relation between the negative and positive frequency brances of the correlator. See Eq. (21) in Sec. 4 of Ref. [13].
13. U. Gavish, Y. Levinson and Y. Imry, Quantum noise, Detailed Balance and Kubo Formula in Nonequilibrium Systems, in *Electronic Correlations: from Meso- to Nano-Physics*, Proc. of the XXXVI Rencontres de Moriond, March 2001, T. Martin, G. Montambaux and J. T. T. Vân, eds., EDP Sciences (2001), p. 243.
14. S.-R. Eric Yang, solid state communications, **81**, 375, (1992).
15. R. Aguado and L. P. Kouwenhoven, Phys. Rev. Lett., **84**, 1986 (2000)

NOISE CORRELATIONS, ENTANGLEMENT, AND BELL INEQUALITIES

T. MARTIN AND A. CREPIEUX
CPT et Université de la Méditerranée
Case 907, 13288 Marseille

N. CHTCHELKATCHEV
L.D. Landau Institute for Theoretical Physics
Kosygina Str. 2, 117940 Moscow

1. Introduction

In condensed matter physics, the interactions between the constituents of the system are typically known. The building blocks are mere electrons, protons and neutrons, their collective behavior has been shown to lead to a variety of astonishing phenomena. Classic examples of such quantum correlated systems are superconductivity [1], superfluidity [2] and the fractional quantum Hall effect (FQHE) [3]. In these instances, the departure from usual behavior is often symptomatic of the presence of a non trivial ground state: a ground state which cannot be described by a systematic application of perturbation theory on the noninteracting system.

Investigations on such ground states naturally lead to that of the elementary excitations of the system. One typically probes the system with an external interaction which triggers the population of excited states. The subsequent measurement of the thermodynamical properties then provides some crucial information. From a different angle, transport measurements deal with open systems connected to reservoirs. There are several ways to approach the issue of (quasi)-particle correlations using transport experiments. One can chose to study a system where interactions are most explicit, such as the FQHE, otherwise, to look for statistical interactions in Fermi and Bose systems. Noise being a multi-particle diagnosis, it can be employed to study both issues.

The aim of this chapter is to describe two situations where positive noise correlations can be directly monitored using a transport experiment, either with

Y.V. Nazarov (ed.), Quantum Noise in Mesoscopic Physics, 313–335.

a superconductor or with a correlated electron system. To be more precise, the present text reflects the presentations made by the three authors during the Delft NATO workshop. Bell inequalities and quantum mechanical non-locality with electrons injected from a superconductor will be addressed first [4]. Next, noise correlations will be computed in a carbon nanotube where electrons are injected in the bulk from a STM tip [5]. The first topic is the result of an ongoing collaboration with G. Lesovik and G. Blatter over the years. The unifying theme is that in both branched quantum circuits, entanglement is explicit and can be illustrated via noise correlations. Entanglement can be achieved either for pairs of electrons in the case of superconductor sources connected to Fermi liquid leads, or alternatively for pairs of quasiparticle excitations of the correlated electron fluid.

A normal metal fork attached to a superconductor can exhibit positive correlations [6, 7], which had been attributed primarily to photonic systems in the seminal Hanbury-Brown and Twiss experiment [8]. They arise when the source of particles is a superconductor [6, 9–14] or they also occur in systems with floating voltage probes [15]. Here, evanescent Cooper pairs can be emitted on the normal side, due to the proximity effect [16]. These Cooper pairs can either decay in one given lead, which gives a negative contribution to noise correlations, or may split at the junction on the normal side with its two constituent electrons propagating in different leads. This latter effect constitutes a justification for positive noise correlations. If filters, such as quantum dots, are added to the leads, such a mechanism generates delocalized, entangled electron pairs [7, 17]. In order to exit from the superconductor, a Cooper pair has to be split between the two normal leads. This provides a solid state analog of Einstein-Podolsky-Rosen (EPR) states which were proposed to demonstrate the non-local nature of quantum mechanics [18]. Photon entanglement has triggered the proposition of new information processing schemes based on quantum mechanics for quantum cryptography or for teleportation [19]. Concrete proposals for quantum information prossessing devices based on electron transport and electron interactions in condensed matter have been recently presented. [7, 17, 20, 21]. Here we will analyze whether a non-locality test can be conceived for electrons propagating in quantum wave guides. It is indeed possible to perform the exact analog of a Bell inequalities violation [22] for photons in a condensed matter system [4].

As cited above, positive correlation do not necessarily require a superconducting source of electrons. A particular geometry using a one dimensional correlated electron liquid can be studied for the same purposes. In particular, it allows to probe directly the underlying charges of the collective excitations in the Luttinger liquid. In order to make contact with experiments, the setup consists of a nanotube whose bulk is contacted by an STM tip which injects electrons by tunneling, while both extremities of the nanotube collect the current. The current, the noise and the noise correlations are computed, and the effective charges are determined by comparison with the Schottky formula. For an “infinite” nanotube, the striking result is that noise correlations contribute to second order in the electron tunneling, in sharp contrast with a fermionic system which requires fourth order. The noise correlations are then positive, because the tunneling electron wave function is split in two counter propagating modes of the collective excitations in the nanotube.

The transport properties of quasiparticles will be addressed from first principles. The understanding of interactions – statistical or otherwise – in noise correlations experiments will be approached here from the point of view of scattering theory and of the Keldysh technique [23].

2. Hanbury–Brown and Twiss correlations

Particularly interesting is the role of electronic correlations in quantum transport. Correlations can have several causes. First, they may originate from the interactions between the particles themselves. Second, correlations are generated by a measurement which involves two or more particles. In the latter case, non-classical correlations may occur solely because of the bosonic or fermionic statistics of particles, with or without interactions. The measurement of noise – the Fourier transform of the current–current correlation function – constitutes a two particle measurement, as implied in the average of the two current operators:

$$S_{\alpha\beta}(\omega=0)=\frac{1}{2}\int dt\,\left(\langle I_\alpha(t)I_\beta(0)+I_\beta(0)I_\alpha(t)\rangle-2\langle I_\alpha\rangle\langle I_\beta\rangle\right)\,. \qquad (1)$$

Here I_α is the current operator in reservoir α, and the time arguments on the average currents have been dropped, assuming a stationary regime.

Consider the case of photons propagating in vacuum: the archetype of a weakly interacting boson system. It was shown [8] that when a photon beam is extracted from a thermal source such as a mercury arc lamp, the intensity correlations measured in two separated photo-multipliers are always positive. On average, each photon scattering state emanating from the source can be populated by several photons at a time – due to the bunching property of bosons. As a result when a photon is detected in one of the photo-multipliers, it is likely to be correlated with another detection in the other photo-tube. The positive correlations can be considered as a diagnosis of the statistics of the carriers performed with a quantum transport experiment.

2.1. NOISE CORRELATIONS IN NORMAL METALS

What should be the equivalent test for electrons ? A beam of electrons can be viewed as a train of wave packets, each of which is populated at most by two electrons with opposite spins. If the beam is fully occupied, negative correlations are expected because the measurement of an electron in one detector is accompanied by the absence of a detection in the other one, as depicted in Fig. 2.1a. It was understood recently [24] that if the electrons propagate in a quantum wire with few lateral modes, maximal occupancy could be reached, and the anti-correlation signal would then be substantial.

Consider the device drawn in Fig. 2.1a : electrons emanating from reservoir 3 have a probability amplitude $s_{1(2)3}$ to end up in reservoir $1(2)$. For simplicity we assume that each lead is connected to a single electron channel. The scattering matrix describing this multi-terminal system is hermitian because of current conservation. On general grounds, it is possible to derive the following sum rule for the autocorrelation noise and the noise correlations [25]:

$$\sum_{\alpha} S_{\alpha\alpha}(\omega = 0) + \sum_{\alpha\neq\beta} S_{\alpha\beta}(\omega = 0) = 0 \,. \tag{2}$$

Note that this sum rule only holds for normal conductors, but it not valid for conductors which involve superconducting reservoirs. The scattering theory of electron transport [23] then specifies how to compute the current and the noise correlations between the two branches. The current in lead α is given by

$$\langle I_\alpha \rangle = \frac{e}{h} \int dE \left(\mathrm{Tr}[\mathbf{1}_\alpha - s^\dagger_{\alpha\alpha} s_{\alpha\alpha}] f_\alpha(E) - \sum_{\beta\neq\alpha} s^\dagger_{\alpha\beta} s_{\alpha\beta} f_\beta(E) \right) , \tag{3}$$

with $s_{\alpha\beta}$ the amplitude to go from reservoir β to reservoir α, $\mathbf{1}_\alpha$ is the identity matrix for lead α, and f_α the associated Fermi function. The zero frequency noise correlations in the zero temperature limit are found in general to be [25]:

$$S_{\alpha\beta} = \frac{e^2}{h} \sum_{\gamma\neq\delta} \int dE \, \mathrm{Tr}[s^\dagger_{\alpha\gamma} s_{\alpha\delta} s^\dagger_{\beta\delta} s_{\beta\gamma}] f_\gamma(E)[1 - f_\delta(E)] \,. \tag{4}$$

Note that when considering the tunnel limit, the lowest non vanishing contribution is of fourth order in the tunneling amplitude $\Gamma \sim s_{\alpha\beta}$ for fermionic particles. Applying the above results to the three-terminal situation depicted in Fig. 2.1 in the presence of a symmetric voltage bias V between 3 and $1(2)$ so that no flow occurs between 1 and 2:

$$S_{12}(0) = -\frac{2e^3|V|}{h} |s_{13} s_{23}|^2 \,. \tag{5}$$

These electronic noise correlations were measured recently [26] by two groups working either in the integral quantum Hall effect regime or in the ballistic regime, with beam splitters designed with metallic gates.

Negative correlations here are most natural, because the injection of electrons is made from a degenerate Fermi gas. Yet there exist situations where they can be positive in a fermionic system.

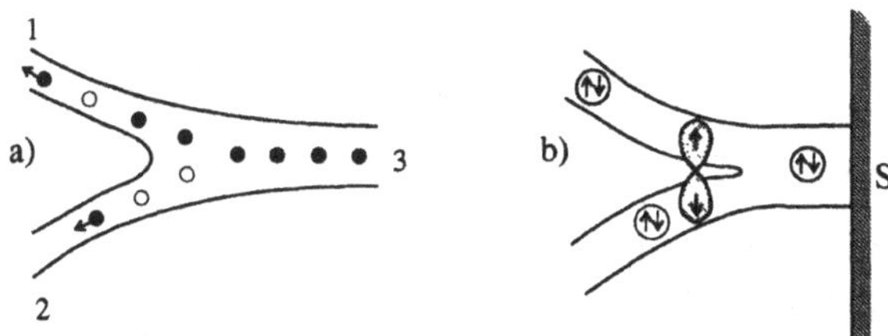

Figure 1. a) Hanbury-Brown and Twiss geometry in a normal metal fork with electrons injected from 3 and collected in reservoirs 1 and 2. Occupied (empty) electron wave packet states are identified as black (white) dots. b) Hanbury-Brown and Twiss geometry in a superconductor–normal metal fork. Cooper pairs are emitted from the superconductor, and the two constituent electrons can either propagate in the same lead, or propagate in an entangled state in both leads.

2.2. FORK GEOMETRY WITH A SUPERCONDUCTOR SOURCE

If the reservoir which injects electrons in the fork is a superconductor as in Fig. 2.1b, both positive and negative noise correlations are possible [6]. Charge transfer between the injector and the two collectors 1 and 2 is then specified by the Andreev scattering process, where an electron is reflected as a hole. Positive correlations are linked to the proximity effect, as superconducting correlations (Cooper pairs) leak in the two normal leads. Depending on the nature of the junction in Fig. 2.1b, it may be more favorable for a pair to be distributed among the two arms than for a pair to enter a lead as a whole. The detection of an electron in 1 is then accompanied by the detection of an electron in 2, giving a positive correlation signal.

In the scattering theory for normal–superconductor (NS) systems [27], the fermion operators which enter the current operator are given in terms of the quasiparticle states using the Bogolubov transformation:

$$\psi_\sigma(x) = \sum_n \left(u_n(x)\gamma_{n\,\sigma} - \sigma v_n^*(x)\gamma_{n\,-\sigma}^\dagger \right) . \tag{6}$$

Here, $\gamma_{n\,\sigma}^\dagger$ are quasiparticle creation operators, $n = (i, \alpha, E)$ contains information on the reservoir (i) from which the particle ($\alpha = e, h$) is incident with energy E and σ labels the spin.

The contraction of these two operators gives the distribution function of the particles injected from each reservoir, which for a potential bias V are: $f_{ie} \equiv f(E - eV)$ for electrons incoming from i, similarly $f_{ih} \equiv f(E + eV)$ for holes, and $f_{i\alpha} = f(E)$ for both types of quasiparticles injected from the superconductor (f is the Fermi–Dirac distribution). The solution of the Bogolubov–de Gennes equations provide the electron and hole wave functions describing scattering states α (particle) and i (lead) are expressed in terms of the elements $s_{ij\alpha\beta}$ of the S–matrix which describes the whole NS ensemble:

$$u_{i\alpha}(x_j) = [\delta_{ij}\delta_{\alpha e}e^{ik_+ x_j} + s_{jie\alpha}e^{-ik_+ x_j}]/\sqrt{v_+}\,, \quad (7)$$
$$v_{i\alpha}(x_j) = [\delta_{ij}\delta_{\alpha h}e^{-ik_- x_j} + s_{jih\alpha}e^{ik_- x_j}]/\sqrt{v_-}\,, \quad (8)$$

where x_j denotes the position in normal lead j and $k_\pm$ $(v_\pm)$ are the usual momenta (velocities) of the two branches.

Specializing now to the NS junction connected to a beam splitter (inset of Fig. 2.2), 6×6 matrix elements are sufficient to describe all scattering processes. At zero temperature, the noise correlations between the two normal reservoirs simplify to:

$$S_{12}(0) = \frac{2e^2}{h}\int_0^{eV} dE \sum_{i=1,2}$$
$$\times\Big[\sum_{j=1,2}(s^*_{1iee}s_{1jeh} - s^*_{1ihe}s_{1jhh})\left(s^*_{2jeh}s_{2iee} - s^*_{2jhh}s_{2ihe}\right)$$
$$+ \sum_{\alpha=e,h}(s^*_{1iee}s_{14e\alpha} - s^*_{1ihe}s_{14h\alpha})(s^*_{24e\alpha}s_{2iee} - s^*_{24h\alpha}s_{2ihe})\Big]\,, \quad (9)$$

where the subscript 4 denotes the superconducting lead. The first term represents normal and Andreev reflection processes, while the second term invokes the transmission of quasiparticles through the NS boundary. It was noted [28] that in the pure Andreev regime the noise correlations vanish when the junction contains no disorder: electron (holes) incoming from 1 and 2 are simply converted into holes (electrons) after bouncing off the NS interface. The central issue, whether changes in the transparency can induce changes in the sign of the correlations, is now addressed.

We now consider only the subgap or Andreev regime, were $eV \ll \Delta$, the superconducting gap, for which a simple model for a disordered NS junction is readily available. The junction is composed of four distinct regions (see inset Fig. 2.2). The interface between 3 (normal) and 4 (superconductor) exhibits only Andreev reflection, with scattering amplitude for electrons into holes $r_A = \gamma\exp(-i\phi)$ (the phase of $\gamma = \exp[-i\arccos(E/\Delta)] \simeq -i$ is the Andreev phase and ϕ is the phase of the superconductor). Next, 3 is connected to two reservoirs 1 and 2 by a beam splitter which is parameterized by a single parameter $0 < \epsilon < 1/2$ identical to that of Ref. [29]: the splitter is symmetric, its scattering matrix coefficients are real, and transmission between 3 and the reservoirs is maximal when $\epsilon = 1/2$, and vanishes at $\epsilon = 0$.

Performing the energy integrals in Eq. (9):

$$S_{12}(\epsilon) = \frac{2e^2}{h}eV\frac{\varepsilon^2}{2(1-\varepsilon)^4}\left(-\varepsilon^2 - 2\varepsilon + 1\right)\,. \quad (10)$$

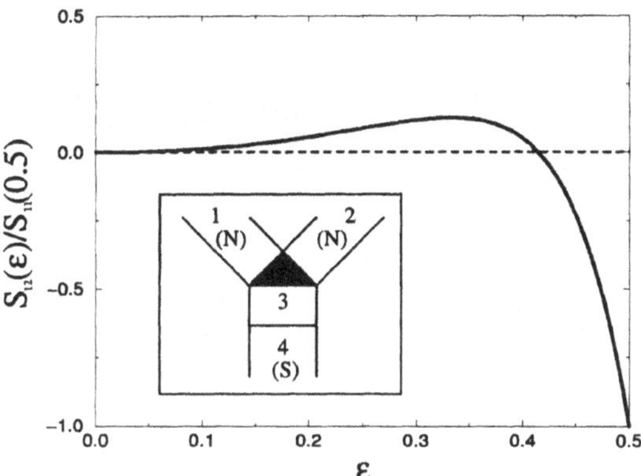

Figure 2. Noise correlation between the two normal reservoirs of the device (inset), as a function of the transmission probability of the beam splitter, showing both positive and negative correlations. Inset: the device consists of a superconductor (4)–normal (3) interface which is connected by a beam splitter (shaded triangle) to reservoirs (1) and (2). $\epsilon = 0.5$ corresponds to maximal transmission.

The noise correlations vanish at $\epsilon = 0$ and $\epsilon = \sqrt{2} - 1$. The correlations (Fig. 2.2) are positive (bosonic) for $0 < \epsilon < \sqrt{2} - 1$ and negative (fermionic) for $\sqrt{2}-1 < \epsilon < 1/2$. At maximal transmission into the normal reservoirs ($\epsilon = 1/2$), the correlations normalized to the noise in 1 (or 2) give the negative minimal value: electrons and holes do not interfere and propagate independently into the normal reservoirs. It is then expected to obtain the signature of a purely fermionic system. When the transmission ϵ is decreased, multiple Andreev processes start taking some importance. Further reducing the beam splitter transmission allows to balance the contribution of split Cooper pairs with that of Cooper pairs entering the leads as a whole. Note that inclusion of disorder or/and additional leads has been discussed by many authors, using either the scattering approach [9, 12, 14] or circuit theory [10] to treat the diffusive limit. To summarize, both positive and negative noise correlations are possible there, and such results are discussed in detail in the contributions of [30] and [31] contained in this volume. In particular, the possibility for positive noise cross-correlation is reduced for asymmetric multichannel conductors. Although sample specific, the scattering theory results of Refs. [6, 14] are found to be rather robust in the presence of disorder.

2.3. FILTERING SPIN/ENERGY

Applying spin or energy filters to the normal arms 1 and 2 (Fig. 2.3), it is possible to generate positive correlations only [7]. For electron emanating from a superconductors, it is possible to project either the spin or the energy with an appropriate

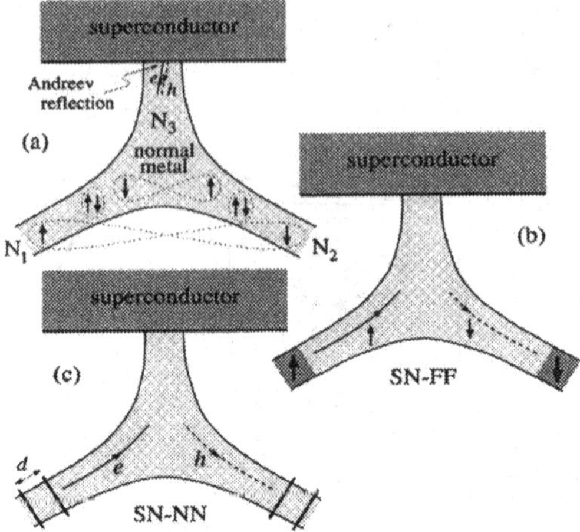

Figure 3. Normal-metal–superconductor (NS) junction with normal-metal leads arranged in a fork geometry. (a) Without filters, entangled pairs of quasi-particles (Cooper pairs) injected in N_3 propagate into leads N_1 or N_2 either as a whole or one by one. The ferromagnetic filters in setup (b) separates the entangled spins, while the energy filters in (c) separate electron- and hole quasi-particles.

filter, without perturbing the entanglement of the remaining degree of freedom (energy or spin). Energy filters, which are more appropriate towards a comparison with photon experiments, will have resonant energies symmetric above and below the superconductor chemical potential which serve to select electrons (holes) in leads $1(2)$. The positive correlation signal then reads:

$$S_{12}(0) = \frac{e^2}{h} \sum_{\zeta} \int_0^{e|V|} d\varepsilon \mathcal{T}_{\zeta}^A(\varepsilon)[1 - \mathcal{T}_{\zeta}^A(\varepsilon)] \,, \tag{11}$$

where the index $\zeta = h, \sigma, 2$, ($\sigma = \uparrow, \downarrow$) identifies the incoming hole state for energy filters (positive energy electrons with arbitrary spin are injected in lead 1 here). $\zeta = h, \uparrow, 1$ $(h, \downarrow, 2)$ applies for spin filters (spin up electrons – with positive energy – emerging from the superconductor are selected in lead 1). $\mathcal{T}_{\zeta}^A$ is then the corresponding (reverse) crossed-Andreev reflection probability [32] for each type of setup: the energy (spin) degree of freedom is frozen, while the spin (energy) degree of freedom is unspecified. $eV < 0$ insures that the constituent electrons of a Cooper pair from the superconductor are emitted into the leads without suffering from the Pauli exclusion principle. Moreover, because of such filters, the propagation of a Cooper pair as a whole in a given lead is prohibited. Note the similarity with the quantum noise suppression mentioned above. This is no accident: by adding constraints to our system, it has become a two terminal device, such that the noise correlations between the two arms are identical to the noise in one arm:

$$S_{11}(\omega = 0) = S_{12}(\omega = 0)\ . \tag{12}$$

The positive correlation and the perfect locking between the auto and cross correlations provide a serious symptom of entanglement. One can speculate (however without a rigorous proof yet) that the wave function which describes the two electron state in the case of spin filters reads:

$$|\Phi^{\rm spin}_{\varepsilon,\sigma}\rangle = \alpha|\varepsilon,\sigma;-\varepsilon,-\sigma\rangle + \beta|-\varepsilon,\sigma;\varepsilon,-\sigma\rangle\ , \tag{13}$$

where the first (second) argument in $|\phi_1;\phi_2\rangle$ refers to the quasi-particle state in lead 1 (2) evaluated behind the filters, ε is the energy and σ is a spin index. The coefficients α and β can be tuned by external parameters, e.g., a magnetic field. Note that by projecting the spin degrees of freedom in each lead, the spin entanglement is destroyed, but energy degrees of freedom are still entangled, and can help provide a measurement of quantum mechanical non locality nevertheless. A measurement of energy ε in lead 1 (with a quantum dot) projects the wave function so that the energy $-\varepsilon$ has to occur in lead 2. On the other hand, energy filters do preserve spin entanglement, and are appropriate to make a Bell test (see below). In this case the two electron wavefunction takes the form:

$$|\Phi^{\rm energy}_{\varepsilon,\sigma}\rangle = \alpha|\varepsilon,\sigma;-\varepsilon,-\sigma\rangle + \beta|\varepsilon,-\sigma;-\varepsilon,\sigma\rangle\ . \tag{14}$$

Electrons emanating from the energy filters (coherent quantum dots) could be analyzed provided a measurement can be performed on the spin of the outgoing electrons with ferromagnetic leads.

2.4. TUNNELING APPROACH TO ENTANGLEMENT

We recall a perturbative argument which supports the claim that two electrons originating from the same Cooper pair are entangled. Consider a system composed of two quantum dots (energies $E_{1,2}$) next to a superconductor. An energy diagram is depicted in Fig. 2.4. The electron states in the superconductor are specified by the BCS wave function $|\Psi_{BCS}\rangle = \prod_k(u_k + v_k c^\dagger_{k\uparrow}c^\dagger_{-k\downarrow})|0\rangle$. Tunneling to the dots is described by a single electron hopping Hamiltonian:

$$H_T = \sum_{k\sigma}[t_{1k}c^\dagger_{1\sigma} + t_{2k}c^\dagger_{2\sigma}]c_{k\sigma} + h.c.\ , \tag{15}$$

with $c^\dagger_{k\sigma}$ creates an electron with spin σ. Now let assume that the transfer Hamiltonian acts on a single Cooper pair.

Using the T-matrix to lowest (2nd) order, the wave function contribution of the two particle state with one electron in each dot reads:

$$|\delta\Psi_{12}\rangle \ = \ H_T\frac{1}{i\eta - H_0}H_T|\Psi_{BCS}\rangle$$

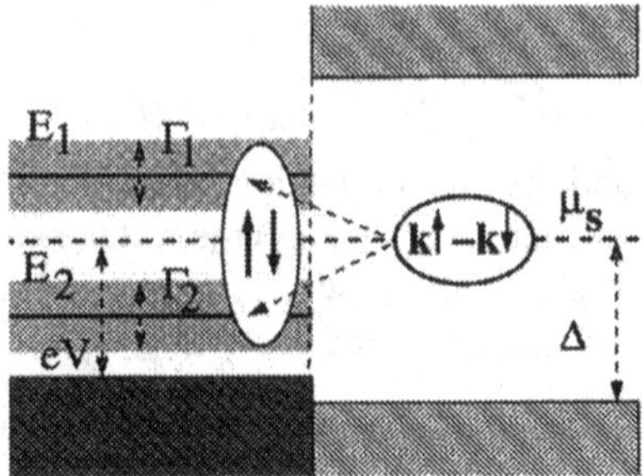

Figure 4. Transfer of a Cooper pair on two quantum energy levels $E_{1,2}$ with a finite width $\Gamma_{1,2}$. The superconductor is located on the right hand side. The transfer of a Cooper pair gives an entangled state in the dots because it implies the creation and destruction of the same quasiparticle in the superconductor. The source drain voltage eV for measuring noise correlations is indicated.

$$= \sum_k v_k u_k t_{1k} t_{2k} \left(\frac{1}{i\eta - E_k - E_1} + \frac{1}{i\eta - E_k - E_2} \right) \times [c^\dagger_{1\uparrow} c^\dagger_{2\downarrow} - c^\dagger_{1\downarrow} c^\dagger_{2\uparrow}] |\Psi_{BCS}\rangle \,, \tag{16}$$

where E_k is the energy of a Bogolubov quasiparticle. The state of Eq. (16) has entangled spin degrees of freedom. This is clearly a result of the spin symmetry of the tunneling Hamiltonian. Given the nature of the correlated electron state in the superconductor in terms of Cooper pairs, H_T can only produce singlet states in the dots.

3. Bell inequalities with electrons

In photon experiments, entanglement is identified by a violation of Bell inequalities (BI) – which are obtained with a hidden variable theory. But in the case of photons, the BIs have been tested using photo-detectors measuring coincidence rates [33]. Counting quasi-particles one-by-one in coincidence measurements is difficult to achieve in solid-state systems where stationary currents and noise are the natural observables [27]. Performing BI checks in condensed matter systems may appear to be a mere generalization of the corresponding checks achieved for photons [34]. Here, the BIs are re-formulated in terms of current-current cross-correlators (noise correlations) [4].

In order to derive Bell inequalities, we consider that a source provides two streams of particles (labeled 1 and 2) as in Fig. 3a injecting quasi-particles into two arms labelled by indices 1 and 2. Filter $F^d_{1(2)}$ are transparent for electrons spin-polarized along the direction $\mathbf{a}(\mathbf{b})$. Assuming separability and locality [22]

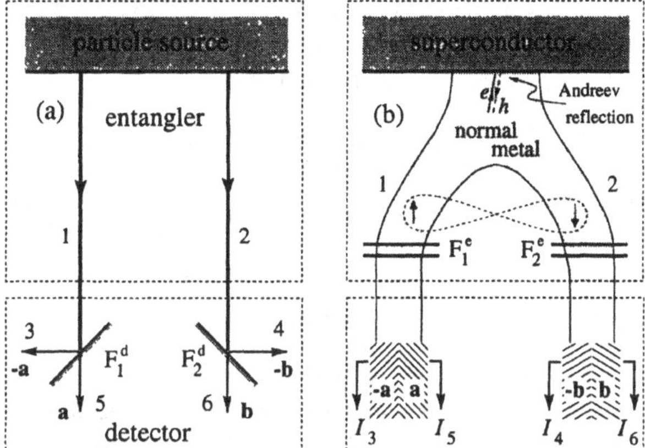

Figure 5. a) Schematic setup for the measurement of Bell inequalities: a source emits particles into leads 1 and 2. The detector measures the correlation between beams labelled with odd and even numbers. Filters $F^d_{1(2)}$ select the spin: particles with polarization along the direction $\pm\mathbf{a}(\pm\mathbf{b})$ are transmitted through filter $F^d_{1(2)}$ into lead 5 and 3 (6 and 4). b) Solid state implementation, with superconducting source emitting Cooper pairs into the leads. Filters $F^e_{1,2}$ (e.g., Fabry-Perot double barrier structures or quantum dots) prevent Cooper pairs from entering a single lead. Ferromagnets with orientations $\pm\mathbf{a}$, $\pm\,\mathbf{b}$ play the role of the filters $F^d_{1(2)}$ in a); they are transparent for electrons with spin aligned along their magnetization axis.

the density matrix for joint events in the leads α, β is chosen to be:

$$\rho = \int d\lambda f(\lambda)\rho_\alpha(\lambda) \otimes \rho_\beta(\lambda)\,, \tag{17}$$

where the lead index α is even and β is odd (or vice-versa); the distribution function $f(\lambda)$ is positive. $\rho_\alpha(\lambda)$ are standard density matrices for a given lead, which are hermitian. The total density matrix ρ is the most general density matrix one can build for the source/detector system assuming no entanglement and only local correlations [35].

Consider the current operator $I_\alpha(t)$ in lead $\alpha = 1, \dots, 6$ (see Fig. 3) and the associated particle number operator $N_\alpha(t,\tau) = \int_t^{t+\tau} I_\alpha(t')dt'$. Particle-number correlators are defined as:

$$\langle N_\alpha(t,\tau)N_\beta(t,\tau)\rangle_\rho = \int d\lambda f(\lambda)\langle N_\alpha(t,\tau)\rangle_\lambda\langle N_\beta(t,\tau)\rangle_\lambda\,, \tag{18}$$

with indices α/β odd/even or even/odd. The average $\langle N_\alpha(t,\tau)\rangle_\lambda$ depends on the state of the system in the interval $[t, t+\tau]$. An average over large time periods is introduced in addition to averaging over λ, e.g.,

$$\langle N_\alpha(\tau)N_\beta(\tau)\rangle \equiv \frac{1}{2T}\int_{-T}^{T} dt\langle N_\alpha(t,\tau)N_\beta(t,\tau)\rangle_\rho\,, \tag{19}$$

where $T/\tau \rightarrow \infty$ (a similar definition applies to $\langle N_\alpha(\tau)\rangle$). Particle number fluctuations are written as $\delta N_\alpha(t,\tau) \equiv N_\alpha(t,\tau) - \langle N_\alpha(\tau)\rangle$.

Let x, x', y, y', X, Y be real numbers such that:

$$|x/X|, |x'/X|, |y/Y|, |y'/Y| < 1 . \tag{20}$$

Then $-2XY \leq xy - xy' + x'y + x'y' \leq 2XY$. Define accordingly:

$$x = \langle N_5(t,\tau)\rangle_\lambda - \langle N_3(t,\tau)\rangle_\lambda \, , \; x' = \langle N_{5'}(t,\tau)\rangle_\lambda - \langle N_{3'}(t,\tau)\rangle_\lambda \, , \tag{21}$$
$$y = \langle N_6(t,\tau)\rangle_\lambda - \langle N_4(t,\tau)\rangle_\lambda \, , \; y' = \langle N_{6'}(t,\tau)\rangle_\lambda - \langle N_{4'}(t,\tau)\rangle_\lambda \, , \tag{22}$$

where the subscripts with a 'prime' indicate a different direction of spin-selection in the detector's filter (e.g., let **a** denote the direction of the electron spins in lead 5 ($-$**a** in lead 3), then the subscript $5'$ in Eqs. (21) and (22) means that the electron spins in lead 5 are polarized along **a**$'$ (along $-$**a**$'$ in the lead 3). The quantities X, Y are defined as

$$X \; = \; \langle N_5(t,\tau)\rangle_\lambda + \langle N_{3(}(t,\tau)\rangle_\lambda = \langle N_1(t,\tau)\rangle_\lambda \, , \tag{23}$$
$$Y \; = \; \langle N_6(t,\tau)\rangle_\lambda + \langle N_4(t,\tau)\rangle_\lambda = \langle N_2(t,\tau)\rangle_\lambda \, , \tag{24}$$

where primed quantities ($5'$, $3'$), ($6'$, $4'$) also apply. The Bell inequality follows after appropriate averaging:

$$|F(\mathbf{a},\mathbf{b}) - F(\mathbf{a},\mathbf{b}') + F(\mathbf{a}',\mathbf{b}) + F(\mathbf{a}',\mathbf{b}')| \leq 2 \, , \tag{25}$$
$$F(\mathbf{a},\mathbf{b}) = \frac{\langle [N_1(\mathbf{a},t) - N_1(-\mathbf{a},t)][N_2(\mathbf{b},t) - N_2(-\mathbf{b},t)]\rangle}{\langle [N_1(\mathbf{a},t) + N_1(-\mathbf{a},t)][N_2(\mathbf{b},t) + N_2(-\mathbf{b},t)]\rangle} \, , \tag{26}$$

with **a**, **b** the polarizations of the filters $F_{1(2)}$ (electrons spin-polarized along **a** (**b**) can go through filter $F_{1(2)}$ from lead 1(2) into lead 5(6)). This is the quantity we want to test, using a quantum mechanical theory of electron transport. Here it will be written in terms of noise correlators, as particle number correlators at equal time can be expressed in general as a function of the finite frequency noise cross-correlations. Assuming short times (see below), one obtains $\langle N_\alpha(\tau) N_\beta(\tau)\rangle \approx \langle I_\alpha\rangle\langle I_\beta\rangle\tau^2 + \tau S_{\alpha\beta}(\omega = 0)$ where $\langle I_\alpha\rangle$ is the average current in the lead α and $S_{\alpha\beta}$ denotes the shot noise. One then gets:

$$F(\mathbf{a},\mathbf{b}) = \frac{S_{56} - S_{54} - S_{36} + S_{34} + \Lambda_-}{S_{56} + S_{54} + S_{36} + S_{34} + \Lambda_+} \, , \tag{27}$$

with $\Lambda_\pm = \tau(\langle I_5\rangle \pm \langle I_3\rangle)(\langle I_6\rangle \pm \langle I_4\rangle)$. Consider now the solid-state analog of the Bell-device as sketched in Fig. 3b where the particle source is a superconductor (S). The test of the Bell inequality (25) requires information about the dependence of the noise on the mutual orientations of the magnetizations $\pm$**a** and $\pm$**b** of the ferromagnetic spin-filters.

Consider an example of the solid-state analog of the Bell-device [Fig. 1(b)] where the particle source is a superconductor. The chemical potential of the superconductor is larger than that of the leads, which means that electrons are flowing out of the superconductor. Two normal leads 1 and 2 are attached to it in a fork geometry [7, 17] and the filters $\mathrm{F}^{\mathrm{e}}_{1,2}$ enforce the energy splitting of the injected pairs. $\mathrm{F}^{\mathrm{d}}_{1,2}$-filters play the role of spin-selective beam-splitters in the detector. Quasi-particles injected into lead 1 and spin-polarized along the magnetization $\mathbf{a}$ enter the ferromagnet 5 and contribute to the current I_5, while quasi-particles with the opposite polarization contribute to the current I_3. For a biased superconductor with grounded normal leads, we find in the tunneling limit the noise

$$S_{\alpha\beta} = e\sin^2\left(\frac{\theta_{\alpha\beta}}{2}\right)\int_0^{|eV|} d\varepsilon \mathcal{T}^A(\varepsilon)\,, \tag{28}$$

which integral also represents the current in a given lead (we have dropped the subscript in $\mathcal{T}^A(\varepsilon)$ assuming the two channels are symmetric). Here $\alpha = 3, 5$, $\beta = 4, 6$ or vice versa; $\theta_{\alpha\beta}$ denotes the angle between the magnetization of leads α and β, e.g., $\cos(\theta_{56}) = \mathbf{a}\cdot\mathbf{b}$, and $\cos(\theta_{54}) = \mathbf{a}\cdot(-\mathbf{b})$. Below, we need configurations with different settings $\mathbf{a}$ and $\mathbf{b}$ and we define the angle $\theta_{\mathbf{ab}} \equiv \theta_{56}$. V is the bias of the superconductor.

The Λ-terms in Eq. (27) can be dropped if $\langle I_\alpha\rangle\tau \ll 1$, $\alpha = 3, \ldots, 6$, which corresponds to the assumption that only one Cooper pair is present on average. The resulting BIs Eqs. (25)-(27) then neither depend on τ nor on the average current but only on the shot-noise, and $F = -\cos(\theta_{\mathbf{ab}})$; the left hand side of Eq. (25) has a maximum when $\theta_{\mathbf{ab}} = \theta_{\mathbf{a'b}} = \theta_{\mathbf{a'b'}} = \pi/4$ and $\theta_{\mathbf{ab'}} = 3\theta_{\mathbf{ab}}$. With this choice of angles the BI Eq.(25) is *violated*, thus pointing to the nonlocal correlations between electrons in the leads 1,2 [see Fig. 3(b)].

If the filters have a width Γ the current is of order $e\mathcal{T}^A\Gamma/h$ and the condition for neglecting the reducible correlators becomes $\tau \ll \hbar/\Gamma\mathcal{T}^A$. On the other hand, in order to insure that no electron exchange between 1 and 2 one requires $\tau \ll \tau_{\mathrm{tr}}/\mathcal{T}^A$ (τ_{tr} is the time of flight from detector 1 to 2). The conditions for BI violation require very small currents, because of the specification that one entangled pair at a time is in the system. Yet it is necessary to probe noise cross correlations of these same small currents. The noise experiments which we propose here are closely related to coincidence measurements in quantum optics. [33]

If we allow the filters to have a finite line width, which could reach the energy splitting of the pair, the violation of BI can still occur, although violation is not maximal. Moreover, when the source of electron is a normal source, it is possible to show that in "standard device geometries", where single electron physics is at play, Bell inequalities are not violated: in this situation the term $\Lambda_\pm$ dominates over the noise correlation contribution. Nevertheless, it would be possible in practice to violate Bell inequalities if the normal source itself, composed of quantum dots as suggested in Ref. [36, 37], could generate entangled electron states as the result of electron-electron interactions.

Note that there are other inequalities which test entanglement for particles sources with the number of terminals two and larger than (see, e.g., Ref. [35]); tests of such inequalities can be implemented in a similar manner as discussed above. For example, one can use Clauser-Horne (CH) inequalities [38] (instead of Bell Inequalities) to test the entanglement in the solid-state systems shown in the Fig. 3. The derivation of CH-inequality is similar to that of BI; it is based on the lemma: if x, x', y, y', X, Y are real numbers such that $x, x' \in [0, X]$ and $y, y' \in [0, Y]$. The following inequality then holds:

$$-XY \leq xy - xy' + x'y + x'y' - Yx' - Xy \leq 0\,. \tag{29}$$

The definitions for x, x', y, y' are the same as in Eqs.(21) and (22), but $X = \langle N_5^{(0)}(t,\tau)\rangle_\lambda$, $Y = \langle N_6^{(0)}(t,\tau)\rangle_\lambda$, where $\langle N_6^{(0)}(t,\tau)\rangle_\lambda$ is the number of particles coming into the arm of the detector $\alpha = 5, 6$ when it doesn't include the spin-filter[38, 39]. Finally we get the CH-inequality in a similar manner as the Bell inequality, with the same assumption of small times:

$$S_{56}(\mathbf{a},\mathbf{b}) - S_{56}(\mathbf{a},\mathbf{b}') + S_{56}(\mathbf{a}',\mathbf{b}) + S_{56}(\mathbf{a}',\mathbf{b}') - S_{56}(\mathbf{a}',-) - S_{56}(-,\mathbf{b}) \leq 0\,, \tag{30}$$

where $S_{56}(\mathbf{a}',-)$ is the shot noise when there is no spin-filter on the way of the particles coming from the source into the the detector arm 6. In the tunneling limit $S_{56}(\mathbf{a}',\mathbf{b})/S_{56}(\mathbf{a}',-) = \sin^2(\theta_{56}/2)$ and Eq. (20) gives S_{56}. CH-inequalities are maximally violated when $\theta_{\mathbf{a},\mathbf{b}} = \theta_{\mathbf{a}',\mathbf{b}'} = \pi/2$, $\theta_{\mathbf{a},\mathbf{b}'} = \pi/4$, and $\theta_{\mathbf{a}',\mathbf{b}} = 3\pi/4$ [this choice of angles is different than in BI case]; then the left-hand side of Eq. (30) is $(\sqrt{2}-1)/2$.

CH-inequalities have one working advantage compared to BIs: Eq. (30) includes only correlations between terminals 5 and 6, the number of correlators in Eq. (30) is decreased compared to the BI's. Moreover, in the case of CH the spin-filters of the detector can include only one ferromagnet in each arm rather than two as in Fig. 5b.

4. Electron injection in the bulk of a nanotube

So far, noise was computed for non–interacting systems. If one puts aside the fact that electrons are converted into holes the calculation of noise in NS systems is similar to the normal case. Recall now the classic results for the FQHE. The long wave length edge excitations along a quantum Hall bar can be described by a Luttinger liquid [40]. Backscattering can be induced by bringing together two counter–propagating edges using a point contact. In the absence of impurities or backscattering, the maximal edge current is $I_M = \nu e^2/h$, while for weak backscattering, the current voltage characteristic is highly non linear for Laughlin

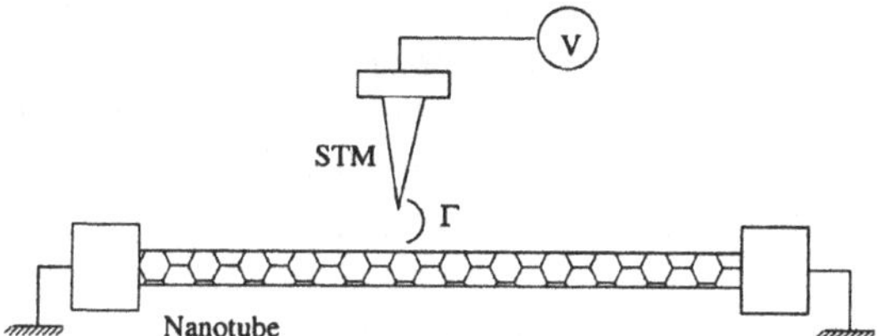

Figure 6. Schematic configuration of the nanotube–STM device: electrons are injected from the tip at $x = 0$: current is measured at both nanotube ends, which are set to the ground

fractions, i.e. $\langle I_B \rangle \sim V^{2\nu-1}$ ($\langle I_B \rangle$ is the average backscattering current and V is the voltage bias between the two edges). A two terminal noise measurement performed on a gated mesoscopic device in this regime provides a direct link to the quasiparticle charge. Quasiparticles are scattered from one edge to the other one by one, so the usual Schottky formula $S_B = 2e^* \langle I_B \rangle$ which relates the zero frequency backscattering noise to the average current flowing between the two edges applies [41], with an effective carrier charge $e^* = \nu e$ contains the electron filling factor [42–44]. This fractional charge was measured recently by several groups [45, 46]. Statistical interaction between quasiparticle have been addressed theoretically [47] in an Hanbury-Brown and Twiss geometry.

Attention is now turning towards conductors which are essentially free of defects and which have one–dimensional character. Carbon nanotubes can have metallic behavior, with two propagating modes at the Fermi level, and constitute good candidates to study Luttinger liquid behavior. In particular, their tunnel density of states – and thus their $I(V)$ characteristics – is known to have a power law behavior [48–50].

Nanotube Luttinger liquids are non-chiral in nature, so a straightforward transposition of the results obtained for chiral edge systems is not obvious. Nevertheless, non–chiral Luttinger liquids also have underlying chiral fields [51, 52]. Such chiral fields correspond to excitations with anomalous (non-integer) charge, which has eluded detection so far.

The transport geometry (Fig. 4) implies: tunneling from the tip (normal or ferromagnetic metal) to the nanotube, and subsequent propagation of collective excitations along the nanotube. In the absence of tunneling, the Hamiltonian is thus simply the sum of the nanotube Hamiltonian, described by a two mode Luttinger liquid, together with the tip Hamiltonian. Using the standard conventions [53], the operator describing an electron with spin σ moving along the direction r, from mode α is specified in terms of a bosonic field:

$$\Psi_{r\alpha\sigma}(x,t) = \frac{1}{\sqrt{2\pi a}} e^{i\alpha k_F x + i r q_F x + i\sqrt{\frac{\pi}{2}} \sum_{j\delta} h_{\alpha\sigma j\delta}(\phi_{j\delta}(x,t) + r\theta_{j\delta}(x,t))}, \tag{31}$$

with a a short distance cutoff, k_F the Fermi momentum, q_F the momentum mismatch associated with the two modes, and the convention $r = \pm$, $\alpha = \pm$ and $\sigma = \pm$ are chosen for the direction of propagation, for the nanotube branch, and for the spin orientation. The non-chiral Luttinger liquid bosonic fields $\theta_{j\delta}$ and $\phi_{j\delta}$, with $j\delta \in \{c+, c-, s+, s-\}$ identifying the charge/spin and total/relative fields, have been introduced. The coefficients are defined as $h_{\alpha\sigma c+} = 1$, $h_{\alpha\sigma c-} = \alpha$, $h_{\alpha\sigma s+} = \sigma$ et $h_{\alpha\sigma s-} = \alpha\sigma$. The Hamiltonian which describes the collective excitations in the nanotube has the standard form:

$$H = \frac{1}{2}\sum_{j\delta}\int_{-\infty}^{\infty} dx \left(v_{j\delta}(x) K_{j\delta}(x) (\partial_x \phi_{j\delta}(x,t))^2 + \frac{v_{j\delta}(x)}{K_{j\delta}(x)} (\partial_x \theta_{j\delta}(x,t))^2 \right) \quad (32)$$

with an interaction parameter $K_{j\delta}(x)$ and velocity $v_{j\delta}(x)$ which allows to address both homogeneous and inhomogeneous Luttinger liquids.

For the STM tip, one assumes for simplicity that only one electronic mode couples to the nanotube, so it can be described by a semi-infinite Luttinger liquid (with interaction parameter $K = 1$) for simplicity. For the sake of generality, we allow the two spin components of the tip fields to have different Fermi velocities u_F^σ, which allows to treat the case of a ferromagnetic metal. The fermion operator at the tip location $x = 0$ is then:

$$c_\sigma(t) = \frac{1}{\sqrt{2\pi a}} e^{i\tilde{\varphi}_\sigma(t)} . \quad (33)$$

Here, $\tilde{\varphi}_\sigma$ is the chiral Luttinger liquid field, whose Keldysh Green's function (here at $x = 0$) is given in [54].

The tunneling Hamiltonian is a standard hopping term:

$$H_T(t) = \sum_{\varepsilon r\alpha\sigma} \Gamma^{(\varepsilon)}(t) [\Psi^\dagger_{r\alpha\sigma}(0,t) c_\sigma(t)]^{(\varepsilon)} . \quad (34)$$

Here the superscript (ε) leaves either the operators in bracket unchanged ($\varepsilon = +$), or transforms them into– their hermitian conjugate ($\varepsilon = -$). The voltage bias between the tip and the nanotube is included using the Peierls substitution: the hopping amplitude $\Gamma^{(\varepsilon)}$ acquires a time dependent phase $\exp(i\varepsilon\omega_0 t)$, with the bias voltage identified as $V = \hbar\omega_0/e$. We use the convention $\hbar \to 1$. Similarly, one can define the tunneling current:

$$I_T(t) = ie \sum_{\varepsilon r\alpha\sigma} \varepsilon \Gamma^{(\varepsilon)}_{r\alpha\sigma}(t) [\Psi^\dagger_{r\alpha\sigma}(0,t) c_\sigma(t)]^{(\varepsilon)} . \quad (35)$$

For this problem which implies propagation of excitations along the nanotube it is also necessary to compute the (total) charge current using the bosonized fields Eq. (31):

$$I_\rho(x,t) = 2ev_F\sqrt{\frac{2}{\pi}}\partial_x \phi_{c+}(x,t) . \quad (36)$$

Note that the contribution from terms containing $2k_F$ oscillations has been dropped.

The Keldysh technique is employed to compute the non-equilibrium currents $\langle I_T\rangle$ and $\langle I_\rho(x)\rangle$, and noises S_T and $S_\rho(x,x')$ to second order in Γ. The contribution of the nanotube fields and of the tip are regrouped into two time ordered products. Each can be related to an correlator of several exponentiated bosonic fields. Such correlators are readily expressed in terms of the Keldysh Green's functions $G^{\theta\theta}_{j\delta}$, $G^{\phi\phi}_{j\delta}$, $G^{\theta\phi}_{j\delta}$, $G^{\phi\theta}_{j\delta}$, associated with the fields $\theta_{j\delta}$ and $\phi_{j\delta}$, as well as the tip Green's function g_σ. The following results apply to both homogeneous and inhomogeneous Luttinger liquids:

$$\begin{aligned}
\langle I_\rho(x)\rangle = &-\frac{ev_F\Gamma^2}{2\pi^2a^2}\sum_{\eta\eta_1r_1\sigma_1}\int_{-\infty}^{+\infty}d\tau\,\sin(\omega_0\tau)e^{2\pi g_{\sigma_1(\eta_1-\eta_1)}(\tau)}\\
&\times e^{\frac{\pi}{2}\sum_{j\delta}(G^{\phi\phi}_{j\delta(\eta_1-\eta_1)}(0,0,\tau)+G^{\theta\theta}_{j\delta(\eta_1-\eta_1)}(0,0,\tau))}\\
&\times e^{\frac{\pi r_1}{2}\sum_{j\delta}(G^{\phi\theta}_{j\delta(\eta_1-\eta_1)}(0,0,\tau)+G^{\theta\phi}_{j\delta(\eta_1-\eta_1)}(0,0,\tau))}\\
&\times\int_{-\infty}^{+\infty}d\tau'\partial_x\Big(G^{\phi\phi}_{c+(\eta\eta_1)}(x,0,\tau')-G^{\phi\phi}_{c+(\eta-\eta_1)}(x,0,\tau')\\
&\qquad+r_1G^{\phi\theta}_{c+(\eta\eta_1)}(x,0,\tau')-r_1G^{\phi\theta}_{c+(\eta-\eta_1)}(x,0,\tau')\Big)\ ,
\end{aligned}\tag{37}$$

$$\begin{aligned}
S_\rho(x,x',\omega=0) = &-\frac{e^2v_F^2\Gamma^2}{(\pi a)^2}\sum_{\eta\eta_1r_1\sigma_1}\int_{-\infty}^{+\infty}d\tau\,\cos(\omega_0\tau)e^{2\pi g_{\sigma_1(\eta_1-\eta_1)}(\tau)}\\
&\times e^{\frac{\pi}{2}\sum_{j\delta}(G^{\phi\phi}_{j\delta(\eta_1-\eta_1)}(0,0,\tau)+G^{\theta\theta}_{j\delta(\eta_1-\eta_1)}(0,0,\tau))}\\
&\times e^{\frac{\pi r_1}{2}\sum_{j\delta}(G^{\phi\theta}_{j\delta(\eta_1-\eta_1)}(0,0,\tau)+G^{\theta\phi}_{j\delta(\eta_1-\eta_1)}(0,0,\tau))}\\
&\times\int_{-\infty}^{+\infty}d\tau_1\partial_x\Big(G^{\phi\phi}_{c+(\eta\eta_1)}(x,0,\tau_1)+r_1G^{\phi\theta}_{c+(\eta\eta_1)}(x,0,\tau_1)\\
&\qquad-G^{\phi\phi}_{c+(\eta-\eta_1)}(x,0,\tau_1)+r_1G^{\phi\theta}_{c+(\eta-\eta_1)}(x,0,\tau_1)\Big)\\
&\times\int_{-\infty}^{+\infty}d\tau_2\partial_{x'}\Big(G^{\phi\phi}_{c+(-\eta\eta_1)}(x',0,\tau_2)+r_1G^{\phi\theta}_{c+(-\eta\eta_1)}(x',0,\tau_2)\\
&\qquad-G^{\phi\phi}_{c+(-\eta-\eta_1)}(x',0,\tau_2)+r_1G^{\phi\theta}_{c+(-\eta-\eta_1)}(x',0,\tau_2)\Big)\ .
\end{aligned}\tag{38}$$

Here, $\eta,\eta_1,\eta_2=\pm$ are indices which specify on which branch of the Keldysh contour the times τ,τ_1,τ_2 and 0 are attached.

5. Nanotube noise correlations and effective charges

An accepted diagnosis to detect effective or anomalous charges is to compare the noise with the associated current with the Schottky formula in mind. Consider an infinite, homogeneous nanotube characterized by interaction parameters $K^N_{j\delta}$. The current reads:

$$\langle I_\rho(x)\rangle = \frac{e\Gamma^2}{\pi a}\Big(\sum_\sigma \frac{1}{u_F^\sigma}\Big)\frac{sgn(\omega_0)|\omega_0|^\mu}{\Gamma(\mu+1)}\Big(\frac{a}{v_F}\Big)^\mu sgn(x)\,, \tag{39}$$

$$\langle I_T(x)\rangle = 2|\langle I_\rho(x)\rangle|\,, \tag{40}$$

where Γ is the Gamma function. We thus obtain a non-linear dependence on voltage $|\omega_0|^\mu$ when interactions are present, with exponent $\mu = \sum_{j\delta}(K^N_{j\delta} + (K^N_{j\delta})^{-1})/8$. A striking result is that despite the fact that electrons are tunneling from the STM tip to the bulk of the nanotube, the zero frequency current fluctuations are proportional to the current (for $x' = x >> a$) with an anomalous effective charge:

$$S_\rho(x,x,\omega=0) = \frac{1+(K^N_{c+})^2}{2}e|\langle I_\rho(x)\rangle|\,, \tag{41}$$

$$S_T = e\langle I_T\rangle\,. \tag{42}$$

More can be learned from a measurement of the noise correlations. Indeed, our geometry can be considered as a Hanbury-Brown and Twiss [8] correlation device. Such experiments have now been completed for photons and more recently for electrons in quantum waveguides. Here the interesting aspect is that electronic excitations do not represent the right eigenmodes of the nanotube. For $x = -x' >> a$ the noise cross-correlations read:

$$S_\rho(x,-x,\omega=0) = -\frac{1-(K^N_{c+})^2}{2}e|\langle I_\rho(x)\rangle|\,. \tag{43}$$

Note that the prefactors in Eqs. (41) and (43) can readily be interpreted using the language of Ref. [51, 52]. According to these works, a tunneling event to the bulk of a nanotube is accompanied by the propagation of two counter-propagating charges $Q_\pm = (1 \pm K^N_{c+})/2$ in opposite directions. Each charge is as likely to go right or left.

The current noise and noise correlations can be interpreted as an average over the two types of excitations:

$$S_\rho(x,x) \sim \frac{(Q_+^2+Q_-^2)}{2} = \frac{1+(K^N_{c+})^2}{4}\,, \tag{44}$$

$$S_\rho(x,-x) \sim -Q_+Q_- = -\frac{1-(K^N_{c+})^2}{4}\,. \tag{45}$$

The current operator for the nanotube charge is measured in the same positive x direction at both extremities of the nanotube. If one measures the current away from the electron source (the STM), which corresponds to the standard convention for multi-terminal systems, such correlations become positive. Here the noise correlations have the added particularity that they occur to second order in a perturbative tunneling calculation. Strictly positive noise correlations are known to occur in superconducting-normal systems with filters, with applications toward entanglement [7, 20, 55]. As noted before, positive correlations do not constitute a rigorous proof for entanglement. In the present case, one single electron is injected, but it enters a correlated system where electrons are not "welcome" as the eigenstates of the nanotube consist of a coherent superposition of bosonic modes. It therefore has to be split into left and right excitations, unless one imposes one dimensional Fermi liquid leads ($K^L_{j\delta} = 1$). Here, we are dealing with entanglement between collective excitations of the Luttinger liquid.

A drawing where the two types of charges "flow away" from the tip while propagating along the nanotube is depicted in the lower part of Fig. 5. Both charges $Q_\pm$ are equally likely to go right or left, and they are emitted as a pair with opposite labels. Written in terms of the chiral quasiparticle fields, the addition of an electron at $x = 0$ with given spin σ on a nanotube in the ground state $|O_{LL}\rangle$ gives:

$$\sum_{r\alpha} \Psi^\dagger_{r\alpha\sigma}(0)|O_{LL}\rangle = \frac{1}{\sqrt{2\pi a}} \sum_{\alpha} \prod_{j\delta} \Big\{ [\tilde{\psi}^\dagger_{j\delta+}(0)]^{Q_{j\delta+}} [\tilde{\psi}^\dagger_{j\delta-}(0)]^{Q_{j\delta-}} + [\tilde{\psi}^\dagger_{j\delta+}(0)]^{Q_{j\delta-}} [\tilde{\psi}^\dagger_{j\delta-}(0)]^{Q_{j\delta+}} \Big\} |O_{LL}\rangle \, , \tag{46}$$

where for each sector (charge/spin, total/relative mode) the charges $Q_{j\delta\pm} = (1 \pm K^N_{j\delta})/2$ have been introduced, and chiral fractional operators are defined as:

$$\tilde{\psi}_{j\delta\pm}(x) = \exp\left[i\sqrt{\frac{\pi}{2K^N_{j\delta}}} h_{\alpha\sigma j\delta} \tilde{\varphi}^\pm_{j\delta}(x) \right] , \tag{47}$$

with $\tilde{\varphi}^r_{j\delta}$ the chiral bosonic fields of the (nonchiral) Luttinger liquid which time evolution is simply obtained with the substitution $\tilde{\varphi}^r_{j\delta}(x) \to \tilde{\varphi}^r_{j\delta}(x - rt)$. Consequently, quantum mechanical non-locality is quite explicit here. This entanglement is the direct consequence of the correlated state of the Luttinger liquid: the addition of an electron does not yield an eigenstate of the Luttinger liquid Hamiltonian. Electrons must be decomposed into specific modes, which happen to propagate in opposite directions. It differs significantly from its analogs which use superconductors [7, 14, 17, 20, 55].

When considering for instance the total charge sector $j\delta = c+$, the wave function is similar to a triplet spin state (a symmetric combination of "up" and

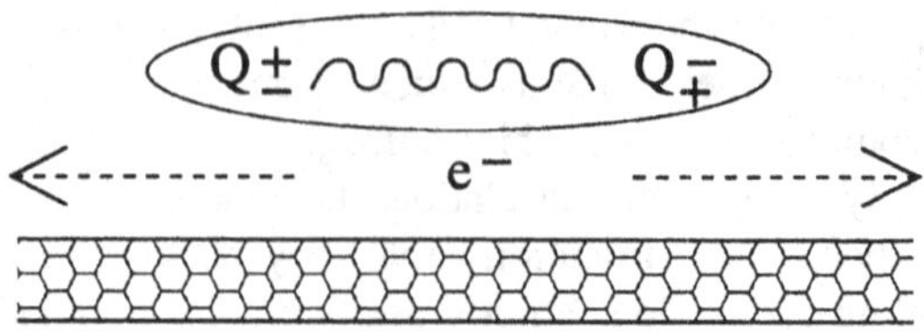

Figure 7. An additional electron injected in the bulk of the nanotube gives rise to right and left moving chiral excitations which have entangled charge degrees of freedom.

"down" states, or "plus" and "minus" charges) for electrons, with the electrons being replaced by chiral quasiparticle operators. Yet, each chiral field $\tilde{\varphi}^r_{j\delta}$ can be written as a linear superposition of boson creation and annihilation operators, and these bosonic fields appear in an exponential. This expresses entanglement between "many–boson" states.

We mention briefly the effect of one-dimensional Fermi liquid contacts. They can be included as in Ref. [56] by connecting both nanotube ends to Luttinger liquids with interaction parameters $K^L_{j\delta} = 1$. Standard results proper to Fermi liquid systems are then recovered. To a first approximation, the Luttinger liquid parameters of the nanotube disappear from transport quantities (current or noise) when such quantities are evaluated in the leads. However, higher order corrections in the voltage still carry a dependence on $K^N_{j\delta}$. The charge noise follows a Schottky formula $S_\rho = e\langle I_\rho\rangle$, and the noise correlations to order Γ^2 vanish – for fermions, the leading order being in Γ^4.

6. Conclusion

Hanbury–Brown and Twiss geometries provide a physical test of mesoscopic transport. They can be used to check the bosonic/fermionic statistics of the carriers, or alternatively to generate entangled streams of particles. Information about statistics is necessarily contained in quantum measurements which involve two particles or more: here the zero frequency noise correlations play the role of the intensity correlator in the early quantum optics experiment of Ref. [8].

A general form of BI-tests in solid-state systems has been proposed, which is formulated in terms of current-current cross-correlators (noise correlations), the natural observables in the stationary transport regime of a solid state device. For a superconducting source injecting correlated pairs into a normal-metal fork completed with appropriate filters [7, 17], the analysis of such BIs shows that this device is a source of entangled electrons with opposite spins when the fork is weakly coupled to a superconductor.

The possibility of a Bell test for this superconductor-normal "fork" devices puts electronic entanglement in condensed matter systems on a firm footing. It is now appropriate to imagine practical information processing devices which exploit this entanglement, in a similar manner as in quantum optics. In particular, a proposal for electron teleportation consisting of 5 quantum dots (2 superconducting dots and 3 normal dots) together with a superconducting circuit was presented at the workshop. The spin state of an electron can be transfered to another dot without direct matter transfer, and the teleportation sequence is selected with the electrostatic interactions between the dots. Details of this proposal are presented elsewhere [57].

A diagnosis for detecting the chiral excitations of a Luttinger liquid nanotube has been presented, which is based on the knowledge of low frequency current fluctuation spectrum in the nanotube. Typical transport calculations either address the propagation in a nanotube, or compute tunneling I(V) characteristics. Here, both is necessary to obtain the quasiparticle charges. Also, both the noise autocorrelation and the noise cross-correlations are needed to identify the charges $Q_\pm$.

Granted, this relies on the assumption that one dimensional Fermi liquid leads are avoided. Multiple scattering at the contacts [58] may allow to preserve the contribution of the noise cross-correlations to second order in the tunneling amplitude. In particular, special circumstances such as embedded contacts [14], transport quantities seem not be renormalized by the contact parameters. This type of geometry could in fact be tested to analyze the type of contacts which one has between the nanotube and its connections. If the ratio $S_\rho(x, -x, \omega_0)/\langle I_\rho(x)\rangle$ does not depend on the tunneling distance (log Γ), this is a clear indication that contacts do not affect this quasiparticle entanglement.

Acknowledgements

The contribution on the Bell inequality tests for electrons was performed in collaboration with G. Blatter and G. Lesovik [4]; the contribution on noise correlations in nanotubes was achieved together with R. Guyon and P. Devillard [5]. Discussions and collaborations with J. Torrès, I. Safi and D. Feinberg on normal-superconducting systems, Luttinger liquids and teleportation are also gratefully acknowledged.

References

1. P.-G. de Gennes, *Superconductivity of Metals and Alloys*, Addison Wesley, 1989; J. R. Schrieffer, *Theory of Superconductivity*, Benjamin, 1964.
2. I. M. Khalatnikov and P. C. Hohenberg, *An introduction to the Theory of Superfluidity*, Advanced Book Classics, 2000.
3. R. B. Laughlin, Rev. Mod. Phys. **71**, 863 (1999); H.L. Stormer, *ibid*, 875 (1999).
4. N. Chtchelkatchev, G. Blatter, G. B. Lesovik, and T. Martin cond-mat/0112094.
5. A. Crépieux, R. Guyon, P. Devillard, and T. Martin, cond-mat/0209291.

6. J. Torrès and T. Martin, Eur. Phys. J. B **12**, 319 (1999).
7. G.B. Lesovik, T. Martin, and G. Blatter, Eur. Phys. J. B **24**, 287 (2001).
8. R. Hanbury-Brown and Q. R. Twiss, Nature **177**, 27 (1956).
9. T. Gramespacher and M. Büttiker Phys. Rev. B **61**, 8125 (2000).
10. J. Börlin, W. Belzig, and C. Bruder Phys. Rev. Lett. **88**, 197001 (2002).
11. M. Schechter, Y. Imry, and Y. Levinson Phys. Rev. B **64**, 224513 (2001).
12. F. Taddei and R. Fazio, Phys. Rev. B **65**, 134522 (2002).
13. P. Samuelsson and M. Büttiker, Phys. Rev. Lett. **89**, 046601 (2002).
14. V. Bouchiat, N. Chtchelkatchev, D. Feinberg, G.B. Lesovik, T. Martin, and J. Torres, cond-mat/0206005.
15. C. Texier and M. Büttiker, Phys. Rev. B **62**, 7454 (2000).
16. V.T. Petrashov *et al.*, Phys. Rev. Lett. **70**, 347 (1993); *ibid.* **74**, 5268 (1995); A. Dimoulas *et al.*, *ibid.* **74**, 602 (1995); H. Courtois *et al.*, *ibid.* **76**, 130 (1996); F.B. Müller-Allinger *et al.*, *ibid.* **84**, 3161 (2000).
17. P. Recher , E.V. Sukhorukov, and D. Loss, Phys. Rev. B **63**, 165314 (2001).
18. E. Schrödinger, Naturwissenschaften **23**, 807 (1935); *ibid.* **23**, 823 (1935); *ibid.* **23**, 844 (1935); A. Einstein, B. Podolsky, and N. Rosen, Phys. Rev. Lett. **47**, 777 (1935).
19. D. Bouwmeester, A. Ekert, and A. Zeilinger, *The Physics of Quantum Information: Quantum Cryptography, Quantum Teleportation, Quantum Computation* (Springer-Verlag, Berlin, 2000).
20. P. Recher and D. Loss, Phys. Rev. B **65**, 165327 (2002).
21. D. Loss and D. P. DiVicenzo, Phys. Rev. A **57**, 120 (1998).
22. J.S. Bell, Physics (Long Island City, N.Y.) **1**, 195 (1965); J.S. Bell, Rev. Mod. Phys. **38**, 447 (1966).
23. S. Datta, *Electronic Transport in Mesoscopic systems* (Cambridge University Press 1996); Y. Imry, *Introduction to Mesoscopic Physics* (Oxford University Press, Oxford, 1997).
24. T. Martin and R. Landauer, Phys. Rev. B **45**, 1742 (1992).
25. M. Büttiker, Phys. Rev. Lett. **65**, 2901 (1990); Phys. Rev. B **46**, 12485 (1992).
26. M. Henny *et al.*, Science **296**, 284 (1999); W. Oliver *et al.*, Science **296**, 299 (1999).
27. M.J.M. de Jong and C.W.J. Beenakker, Phys. Rev. B **49**, 16070 (1994); B.A. Muzykantskii and D.E. Khmelnitskii, *ibid.* **50**, 3982 (1994); M.P. Anantram and S. Datta, *ibid.* **53**, 16 390 (1996); G. Lesovik, T. Martin, and J. Torrès, *ibid.* **60**, 11935 (1999).
28. T. Martin, Phys. Lett. A **220**, 137 (1996).
29. Y. Gefen, J. Imry, and R. Landauer, Phys. Rev. Lett. **52**, 139 (1984); M. Büttiker, Y. Imry, and M. Ya. Azbel, Phys. Rev. A **30**, 1982 (1984).
30. M. Büttiker, NATO workshop on noise, Delft, Y. Nazarov and Y. Blanter eds. (Kluwer 2002).
31. W. Belzig *et al.*, NATO workshop on noise, Delft, Y. Nazarov and Y. Blanter eds. (Kluwer 2002).
32. R. Melin and D. Feinberg, Eur. Phys. J. B **26**, 101 (2002).
33. A. Aspect, J. Dalibard, and G. Roger, Phys. Rev. Lett. **49**, 1804 (1982).
34. R. Ionicioiu, P. Zanardi, and F. Rossi, Phys. Rev. A **63**, 050101 (2001).
35. R. Werner and M. Wolf, quant-ph/0107093.
36. W.D. Oliver, F. Yamaguchi, and Y. Yamamoto, Phys. Rev. Lett. **88**, 037901 (2002).
37. D.S. Saraga and D. Loss cond-mat/0205553.
38. J.F. Clauser and M.A. Horne, Phys. Rev. D **10**, 526 (1974).
39. Z.Y. Ou and L. Mandel, Phys. Rev. Lett. **61**, 50 (1988).
40. X.G. Wen, Int. J. Mod. Phys. B **6**, 1711 (1992); Adv. Phys. **44**, 405 (1995).
41. W. Schottky, Ann. Phys. (Leipzig) **57**, 541 (1918).
42. C. L. Kane and M.P.A. Fisher, Phys. Rev. Lett. **72**, 724 (1994).
43. C. de C. Chamon, D. E. Freed, and X.G. Wen, Phys. Rev. B **51**, 2363 (1995-II).

44. P. Fendley, A. W. W. Ludwig, and H. Saleur, Phys. Rev. Lett. **75**, 2196 (1995).
45. L. Saminadayar, D.C. Glattli, Y. Jin, and B. Etienne, Phys. Rev. Lett. **79**, 2526 (1997).
46. R. de-Picciotto *et al.*, Nature **389**, 162 (1997).
47. I. Safi, P. Devillard, and T. Martin, Phys. Rev. Lett. **86**, 4628 (2001).
48. M. Bockrath, D. H. Cobden, A. Rinzler, R.E. Smalley, L. Balents, and P. McEuen, Nature **397**, 598 (1999).
49. C. Kane, L. Balents, and M. P. A. Fisher, Phys. Rev. Lett. **79**, 5086 (1997).
50. R. Egger, A. Bachtold, M. Fuhrer, M. Bockrath, D. Cobden, and P. McEuen, in *Interacting Electrons in Nanostructures*, edited by R. Haug and H. Schoeller (Springer, 2001).
51. I. Safi, Ann. Phys. Fr. **22**, 463 (1997).
52. K.-V. Pham, M. Gabay and P. Lederer, Phys. Rev. B **61**, 16397 (2000).
53. R. Egger and A. Gogolin, Eur. Phys. J. B **3**, 781 (1998).
54. C. Chamon and D. E. Freed, Phys. Rev. B **60**, 1842-1853 (1999).
55. C. Bena *et al.*, Phys. Rev. Lett. **89**, 037901 (2002).
56. I. Safi and H. Schulz, Phys. Rev. B **52**, 17040 (1995); D. Maslov and M. Stone, Phys. Rev. B **52**, 5539 (1995); V.V. Ponomarenko, Phys. Rev. B **52**, 8666 (1995).
57. O. Sauret, D. Feinberg, and T. Martin, cond-mat/0203215.
58. K.-I. Imura, K.-V. Pham, P. Lederer, and F. Piéchon, Phys. Rev. B **66**, 035313 (2002).

NOISE IN THE SINGLE ELECTRON TRANSISTOR AND ITS BACK ACTION DURING MEASUREMENT

G. JOHANSSON
Institut für Theoretische Festkörperphysik
Universität Karlsruhe, D-76128 Karlsruhe, Germany

P. DELSING, K. BLADH, D. GUNNARSSON, T. DUTY, A. KÄCK, G. WENDIN
Microtechnology Center at Chalmers MC2,
Department of Microelectronics and Nanoscience,
Chalmers University of Technology and Göteborg University, S-41296

A. AASSIME
Service de Physique de l'Etat Condensé, CEA-Saclay, F-91191 Gif-sur-Yvette, France

Abstract. Single electron transistors (SETs) are very sensitive electrometers and they can be used in a range of applications. In this paper we give an introduction to the SET and present a full quantum mechanical calculation of how noise is generated in the SET over the full frequency range, including a new formula for the quantum current noise. The calculation agrees well with the shot noise result in the low frequency limit, and with the Nyquist noise in the high frequency limit. We discuss how the SET and in particular the radio-frequency SET can be used to read out charge based qubits such as the single Cooper pair box. We also discuss the backaction which the SET will have on the qubit. The back action is determined by the spectral power of voltage fluctuations on the SET island. We will mainly treat the normal state SET but many of the results are also valid for superconducting SETs.

1. Introduction

The single-electron transistor (SET) is known as a highly sensitive electrometer [1, 2] based on the Coulomb blockade [3, 4]. Electrons tunnel one by one through two small-capacitance tunnel junctions, with capacitances C_L and C_R. The gate, with capacitance C_g, is used to modulate the generated current, and the island is also capacitively coupled to the system to be measured through C_c. The charging energy $E_C = e^2/2C_\Sigma$, $C_\Sigma = C_L + C_R + C_g + C_c$, associated with a single electron prevents sequential tunneling through the island at voltages below a threshold

Y. V. Nazarov (ed.), Quantum Noise in Mesoscopic Physics, 337–355.

V_t, which can be controlled by applying a voltage V_g to the gate. This Coulomb blockade is effective at temperatures $T < E_C/k_B$ and for junction resistances larger than the resistance quantum $R_K = h/e^2 \approx 25.8\text{k}\Omega$.

SETs can be made using several different technologies, for example they can be fabricated from metallic (often aluminum based) systems, from quantum dots in GaAs and silicon, or from carbon nanotubes.

Often the SETs are operated at low temperatures, however room temperature operation has been demonstrated in several cases[5–7].

Both the ultimate sensitivity and the back action of SET during measurement is determined by the noise in the SET. By understanding the noise in the SET we can optimize the use of the SET for each application.

1.1. RF-SET

The conventional SET, based on measuring either the current or voltage across the transistor, has suffered from the relatively large output resistance R of the transistor. For the typical resistance values of $100\,\text{k}\Omega$ and cable capacitance of $C \sim 1\text{nF}$, the corresponding RC time limits the bandwidth to a few kHz.

With the invention of the radio-frequency SET (RF-SET)[8] the SET was made fast and very sensitive. By connecting the SET very close to a cold amplifier the upper frequency limit was improved to about 1 MHz[9, 10]. With the invention of the RF-SET[8] frequencies above 100 MHz could be reached.

The operation principle of the RF-SET is similar to that of a radar. A weak RF-signal (the carrier) is launched via a directional coupler and a bias tee, towards the tank circuit in which the SET is embedded. The reflected signal depends critically on the dissipation in the tankcircuit, and thus in the SET. This means that the carrier is modulated by the gate signal. The reflected signal is amplified by a cold amplifier and a number warm amplifiers. The reflected signal can be analyzed either in the frequency domain or in the time domain. Typical carrier frequencies are in the range of 0.3-2 GHz.

Charge sensitivities of $1.2 \cdot 10^{-5}$ and $3.2 \cdot 10^{-6}$ $e/\sqrt{\text{Hz}}$ have been reached for signal frequencies of 100 and 2 MHz, respectively[8, 11]. The RF-SET can be used as a readout device in applications from very sensitive charge meters and current standards[4] in which electrons are counted or pumped one by one, to read out of quantum bits[12–15] or to work as photon detectors [16].

2. SET - orthodox theory

Consider a small metallic SET island coupled via low transparency tunnel barriers to two external leads, and coupled capacitively to an object to be measured. In Fig. 1 we have used the example of a Single Cooper-pair Box (SCB)[17] which may serve as a qubit in a quantum computer. The SCB is controlled by a voltage source V_g^{qb}.

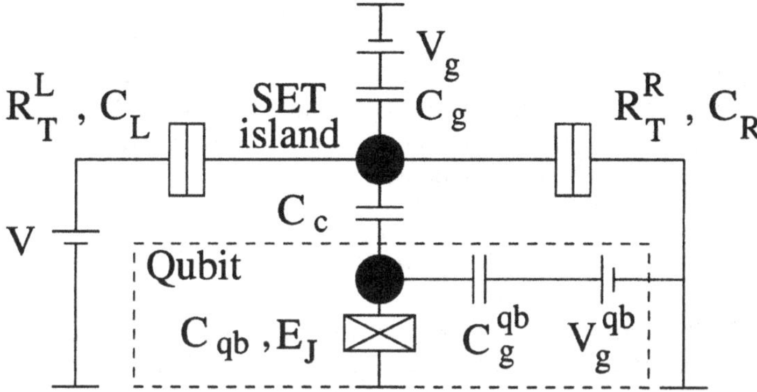

Figure 1. Schematic figure of the SET capacitively coupled to a Single Cooper-pair Box (SCB).

Following orthodox SET theory[2, 3] we use the integer number of electrons on the SET island (N) as a basis for describing its dynamics. This is motivated by the low transparency junctions. The charge on the island is the sum of the electrons, the background charge, and the charge induced by the the voltages on the three island capacitances. Electrostatics gives for the Coulomb charging energy E_{ch} and island potential V_I:

$$\begin{aligned} E_{ch}(N) &= E_C(N-n_x)^2, \ E_C = \frac{e^2}{2C_\Sigma}, \\ V_I(N) &= \alpha_L V - \frac{e(N-n_x)}{C_\Sigma}, \ \alpha_L = \frac{C_R}{C_\Sigma}, \ \alpha_R = \frac{C_L}{C_\Sigma}, \end{aligned} \tag{1}$$

where $n_x e = V_g^{qb} C_g^{qb}$ is the charge induced on the gate capacitance C_g^{qb} by the external voltage source V_g^{qb}. $\alpha_{L/R}$ describes the capacitive voltage division over the L/R junction, were we have neglected the small gate and coupling capacitances $\{C_c, C_g\} \ll \{C_L, C_R\}$.

The orthodox theory neglects mainly three things: the effect of the electromagnetic environment of the SET [18], higher order tunneling processes, i.e. cotunneling [19], and that for high frequency dynamics the transition rates are frequency dependent. As long as the electromagnetic environment has a low impedance compared to the quantum resistance $R_Q = h/4e^2 \approx 6.45\text{k}\Omega$ (which is often the case for SETs) the corrections due to the environment are small. The same is true for the cotunneling corrections as long as the resistance of the SET is large compared to R_Q. To take into account the frequency dependence of the transition rates, including the energy exchange with the measured system, is the main objective of the quantum theory presented in section 3.2.

2.1. TRANSITION RATES

The dynamics of the SET consists of stochastic transitions between the charge states, i.e. by electrons randomly tunneling on and off the island. In orthodox SET theory the rates for the different transitions are given by the Golden Rule rates for tunneling through the left and right tunnel junctions

$$\Gamma^{L/R}(E_{if}) = \frac{R_K}{R_T^{L/R}} \frac{E_{if}}{h} \frac{1}{1 - e^{-E_{if}/k_B T}}, \tag{2}$$

where E_{if} is the energy difference between the initial and final state and the Bose function appears from the convolution of the two Fermi functions for the filled initial states and the empty final states in the leads. E_{if} has two terms; one is the change in charging energy, and the other is the work done by the voltage bias.

To be specific we now consider the electrons to gain energy from the voltage source by tunneling from left to right. Then the rates $\Gamma_{n\pm}^{L/R}$ for transitions from charge state n to $n \pm 1$ across the left/right junction are

$$\begin{aligned} \Gamma_{n+}^{L} &= \Gamma^{L}(E_{n(n+1)} + eV_L), \; \Gamma_{n+}^{R} = \Gamma^{R}(E_{n(n+1)} - eV_R), \\ \Gamma_{n-}^{L} &= \Gamma^{L}(E_{n(n-1)} - eV_L), \; \Gamma_{n-}^{R} = \Gamma^{R}(E_{n(n-1)} + eV_R), \end{aligned} \tag{3}$$

where $E_{mn} = E_{ch}(m) - E_{ch}(n)$ and $V_{L/R} = \alpha_{L/R} V$ is the voltage drop over the L/R junction. We denote the sum of rates taking the SET island from n to $n \pm 1$ with $\Sigma_{n\pm} = \Gamma_{n\pm}^{L} + \Gamma_{n\pm}^{R}$.

2.2. MASTER EQUATION AND STEADY-STATE

During a specific measurement, i.e. a realisation of the stochastic process, the number of (extra) electrons on the SET island as a function of time is piecewise constant, with steplike changes due to instantaneous tunnel events (see Fig. 2a-b). Mathematically one may describe the dynamics of this process averaged over a large number of measurements. Introducing the probability $P_n(t)$ of finding the SET in charge state n at time t we may write down a master equation

$$\partial_t \bar{P}(t) = \hat{\Sigma} \cdot \bar{P}(t), \tag{4}$$

where $\bar{P} = [\ldots P_{-1}(t)\; P_0(t)\; P_1(t) \ldots]^T$ is a column vector and $\hat{\Sigma}$ is the tridiagonal transition rate matrix with off-diagonal elements $(\hat{\Sigma})_{(n\pm1)n} = \Sigma_{n\pm}$, and to ensure probability conservation $(\hat{\Sigma})_{nn} = -(\Sigma_{n+} + \Sigma_{n-})$. The solution to Eq. (4) may be written $\bar{P}(t) = \hat{\Pi}(t) \cdot \bar{P}(0)$, introducing the time-evolution operator (matrix) $\hat{\Pi}(t) = e^{\hat{\Sigma}t}$. To compute the matrix exponent one needs to diagonalize $\hat{\Sigma}$. The steady-state corresponds to the eigenvector of $\hat{\Sigma}$ with eigenvalue zero, and the other eigenvalues determine rates for exponential decay of the corresponding eigenvector.

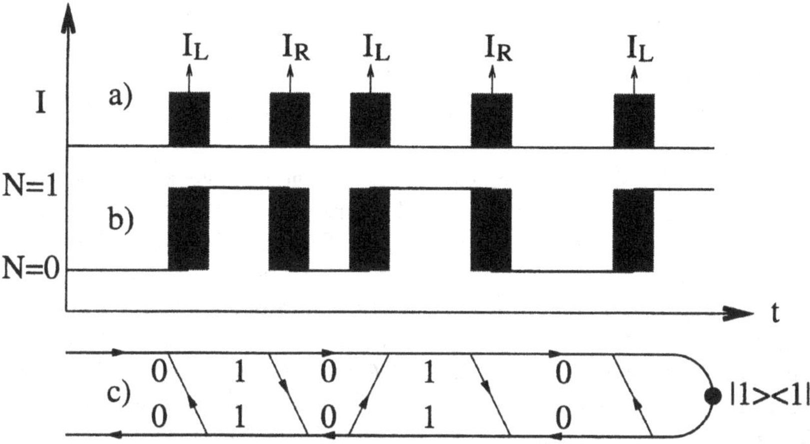

Figure 2. In orthodox theory the transitions between charge states in the SET are considered to be instantaneous. For a specific realisation (measurement) this results in: a) The currents through the left/right junctions I_L/I_R consist of a train of delta-peaks. b) The charge on the island $N(t)$ is a telegraph signal. In order to calculate the transition rates one uses the Golden Rule approximation, which includes integration over a timescale of h/E_{if}, where E_{if} is the energy gain in the transition. This is indicated as grey areas in figures a) and b). c) In the real-time Keldysh formalism this timescale of tunneling reappears as a timescale on which the density matrix is off-diagonal.

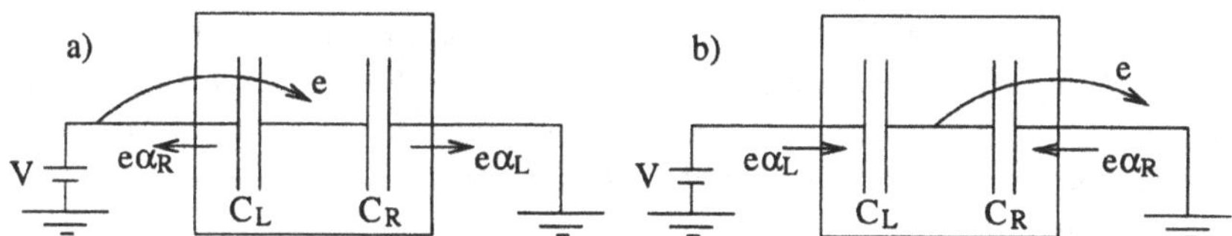

Figure 3. When an electron tunnels from/to the leads to/from the island the charges on the capacitances will redistribute. The charge transport across the boundary of a region containing the SET island plus its nearby capacitances is shown for a) an electron tunneling onto the island from the left lead, and b) an electron tunneling from the island into the right lead. $e\alpha_{L/R}$ represents a displacement charge.

2.3. CHARGE, VOLTAGE AND CURRENT

In order to calculate physical quantities we introduce operators for the number of (excess) electrons on the island ($\hat{N}$) and the tunnel currents across the left ($\hat{I}_L$) and right ($\hat{I}_R$) junctions,

$$(\hat{N})_{nn} = n, \quad (\hat{I}_L)_{(n\pm1)n} = \pm e\Gamma^L_{n\pm}, \quad (\hat{I}_R)_{(n\pm1)n} = \mp e\Gamma^R_{n\pm}, \tag{5}$$

noting only the non-zero elements, i.e. $\hat{N}$ is diagonal, and $\hat{I}_{L/R}$ are both tridiagonal with zeros on the main diagonal. The steady-state expectation value O^{st} of an operator $\hat{O}$ is given by

$$\begin{aligned} O^{st} &= \langle\hat{O}\rangle_{st} \equiv [1\ldots 1]\cdot\hat{O}\cdot\bar{P}^{st}, \;\; \bar{P}^{st} = \hat{\Pi}(\infty)\cdot\bar{P}(0) \Rightarrow \\ N^{st} &= \sum_n nP_n^{st}, I_L^{st} = e\sum_n P_n^{st}(\Gamma_{n+}^L - \Gamma_{n-}^L), I_R^{st} = e\sum_n P_n^{st}(\Gamma_{n-}^R - \Gamma_{n+}^R), \end{aligned} \tag{6}$$

which defines the meaning of the brackets $\langle\,\rangle_{st}$. One gets the voltage of the SET island by multiplying the number of extra electrons N by e/C_Σ, and adding the gate bias dependent constant according to Eq. (1). One should also note that each tunnel event in the SET is followed by a fast redistribution of charge on the nearby capacitances, see Fig. 3. Thus a tunnel event in one junction will create a displacement current also in the opposite lead. The operator $\hat{I}$ for the externally measurable current is given by[20–24] $\hat{I} = \alpha_L\hat{I}_L + \alpha_R\hat{I}_R$, (see Eq. (1)). For the steady-state current we have $I^{st} = I_L^{st} = I_R^{st}$, using the detailed balance[25] $P_n^{st}\Sigma_{n+} = P_{n+1}^{st}\Sigma_{(n+1)-}$, but for finite frequency properties the difference between tunnel currents and the externally measurable current becomes important.

3. Noise - Fluctuations and Correlations

The SET dynamics is noisy due to its stochastic nature. The fluctuations in the current through the SET determine the measurement time (t_m) needed to separate the dc currents corresponding to different charges at the input of the SET. The fluctuations of the charge on the SET island induce a fluctuating voltage on the capacitance coupling to the measured system (see Fig. 1), which may be disturbed. This effect is called the back-action of the measurement, i.e. the meter acting back on the measured system.

The fluctuations in the SET can be thought of in terms of two contributions, one from the shot noise of the sequential tunneling described by orthodox theory, and one from the quantum fluctuations. At low frequency, the shot noise will dominate, as described by orthodox SET theory in section 3.1 below. At high frequencies the SET bias may be neglected and the quantum fluctuations, i.e. Nyquist noise, will dominate. The Nyquist voltage noise is given by the impedance of the SET island to ground[26], i.e. from the two junctions in parallell ($Z_p(\omega)$), while the current noise is given by the impedance through the SET, i.e. from the two junctions in series ($Z_s(\omega)$):

$$\begin{aligned} S_{VV}(\omega) &= 2\hbar\omega\mathrm{Re}\{Z_p(\omega)\} \rightarrow \frac{e^2}{C_\Sigma^2}\frac{R_K}{\pi\omega}\left[\frac{1}{R_T^L}+\frac{1}{R_T^R}\right] + O(\omega^{-2}), \\ S_{II}(\omega) &= 2\hbar\omega\frac{\mathrm{Re}\{Z_s(\omega)\}}{|Z_s(\omega)|^2} \rightarrow 2\hbar\omega\left[\frac{\alpha_L^2}{R_T^L}+\frac{\alpha_R^2}{R_T^R}\right] + O(1), \end{aligned} \tag{7}$$

given to leading order in ω^{-1}.

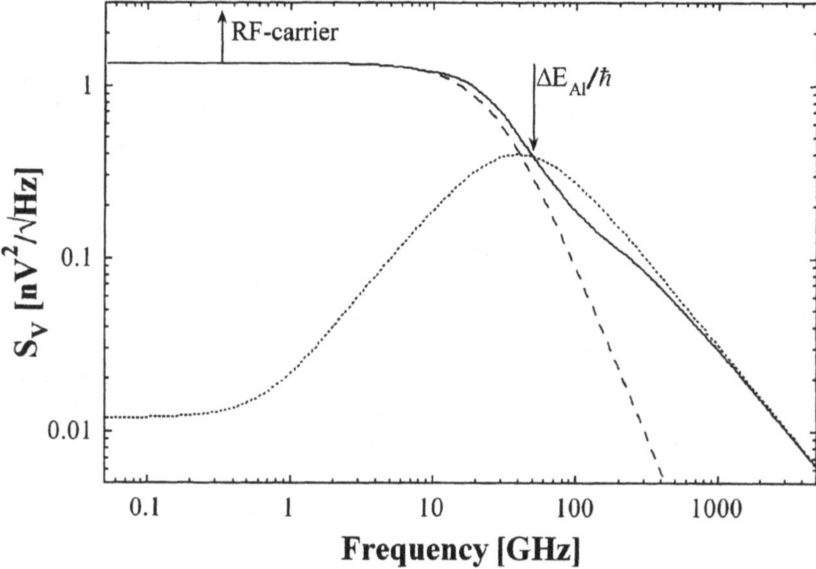

Figure 4. Comparison of the calculations of the symmetrized spectral density of voltage fluctuations of the SET island using the full quantum mechanical calculation (full line), the classical shot noise (dashed line), and the quantum fluctuations assuming a linear SET impedance (dotted line). Parameters for sample #1 described in Table 3, are used in all calculations

To obtain the spectrum of fluctuations for intermediate frequencies it is necessary to solve the full quantum problem, which was done in [27] for the voltage noise. The result of this calculation and the comparison to the shot noise and quantum fluctuation results in the two limits, are shown in Fig. 3 for sample # 1 described in Table 3. As can be seen, the result of the full calculation coincides with the shot noise and the quantum noise in the low and high frequency limits respectively.

In addition to this spectrum, the RF excitation gives a component at f_{RF}, and the non-linearity of the IV-characteristics gives an additional component at $3f_{RF}$. However these frequencies are much lower than the relevant mixing frequency $\Delta E/\hbar$ for the SCB-qubit described in section 5.

3.1. NOISE IN THE SET: ORTHODOX THEORY

Since orthodox SET theory gives the correct low-frequency limits for the noise, and is also the natural reference for discussing the quantum theory, we will describe this in some detail. Together with section 2 the following discussion is sufficient for calculating the orthodox theory noise, including an arbitrary number of charge states. In section 4 we will go through the analytically solvable case of two charge states.

We are interested in the fluctuation of charge and current around their average values. Therefore we define fluctuation operators, $\delta\hat{O} = \hat{O} - O^{st}1$, where 1 is the unit matrix with the same dimension as $\hat{O}$ and O^{st} is the steady-state expectation value of the operator. We may then write the following expressions for the fluctuation correlation function ($\tau > 0$)

$$K_{AB}(\tau) = \langle \delta\hat{A} \cdot \Pi(\tau) \cdot \delta\hat{B} \rangle_{st}, \;\; K_{AB}(-\tau) = K_{BA}(\tau). \tag{8}$$

The master equation (4) is dissipative and therefore irreversible in time, which explains the need for the special negative τ definition. Special care has to be taken for the autocorrelation of a current pulse, i.e. the shot-noise. Since the first current operator instantaneously changes the state of the system, the second current operator does not operate on the same state, and thus the shot-noise is not included. Using some arbitrary representation for the δ-function current pulse on the tunneling time-scale $E_{if}/\hbar$ one gets $\int \delta(t-t_0)\delta(t)dt = \delta(t_0)$ which gives

$$K_{I_rI_r}(\tau) \to K_{I_rI_r}(\tau) + e\delta(\tau) \sum_n P_n^{st}(\Gamma_{n+}^r + \Gamma_{n-}^r), \;\; r \in \{Ł, R\}. \tag{9}$$

We will also need the spectral densities of the fluctuations, i.e. the Fourier transforms of the correlation functions

$$S_{AB}(\omega) = \int_{-\infty}^{\infty} e^{-i\omega\tau} K_{AB}(\tau) d\tau = \int_{0}^{\infty} e^{-i\omega\tau} K_{AB}(\tau) + e^{i\omega\tau} K_{BA}(\tau) d\tau. \tag{10}$$

By e.g. diagonalizing $\hat{\Sigma}$ one finds that the needed Laplace transform of $\hat{\Pi}(\tau)$ is given by Eq. (13), just replacing $\hat{\Sigma}(\omega) \to \hat{\Sigma}$.

Here we use the unsymmetrized definitions of the correlation functions in order to separate noise where the tunnel processes absorb the energy $|\hbar\omega|$ (positive frequencies) from noise where the tunnel processes emit the energy $|\hbar\omega|$ (negative frequencies), when we go to the quantum expressions in the next section. When this separation is not needed one may use the symmetrized definition $S_{AB}^{sym}(\omega) = \int_{-\infty}^{\infty} e^{-i\omega\tau}(K_{AB}(\tau) + K_{BA}(\tau))d\tau = S_{AB}(\omega) + S_{AB}(-\omega)$. In the classical calculations $S_{II}(\omega) = S_{II}(-\omega)$; therefore the symmetric definition only gives a factor of 2 needed to recover the usual expression for the shot-noise across a single junction $S_{II}^{sym}(\omega) = 2eI^{st}$[28].

Apart from using unsymmetrised correlation functions, the formalism for the orthodox theory presented in this section is equivalent to, and inspired by, the work of Korotkov[22].

3.2. NOISE IN THE SET: QUANTUM THEORY RESULTS

The master equation (4) contains only dissipative transitions between SET charge states. The fact that the SET might be in a coherent state like $|N=0\rangle + |N=1\rangle$ was only taken into account in deriving the tunneling rates. This Golden Rule derivation includes an integration over the timescale $t_{GR} = h/E_{if}$, and the main approximation behind Eq. (4) is that this timescale is short compared to the dynamics you describe. For steady-state or low-frequency properties this is fulfilled since the tunneling rates include the small tunnel conductance, i.e. $1/\Sigma \sim t_{GR}R_K/R_T \ll t_{GR}$. The shot-noise term had to be handled separately since it corresponds to correlations on the time-scale t_{GR}. For high frequency dynamics, i.e. $\hbar\omega \sim E_{if}$, the approximation breaks down and one has to consider the effects of quantum coherence.

One way to include quantum coherence is to use the real-time diagrammatic Keldysh approach described by Schoeller and Schön[29]. In the sequential tunneling approximation the low-frequency results coincide with orthodox theory[22], and the diagrams used may also be compared with the time evolution of the state of the SET island, see Fig. 2c. We will not further describe the method here, only state the results in such a way that the finite frequency noise, including arbitarily many charge states, may be calculated.

3.2.1. *Inelastic transition rates*

When we take into account that the SET can absorb or emit energy the transition rates are modified. First of all the tunnel rates become frequency dependent, since the noise energy $|\hbar\omega|$ should be created or absorbed in a tunnel event. The rates

$$\Gamma^L_{n\pm}(\omega) = \frac{1}{2}\Gamma^L(E_{n(n\pm1)} \pm eV_L + \hbar\omega),\ \ \Gamma^R_{n\pm}(\omega) = \frac{1}{2}\Gamma^R(E_{n(n\pm1)} \mp eV_R + \hbar\omega), \tag{11}$$

are here defined so that the transition is facilitated by positive $\hbar\omega$, i.e. the tunnel event absorbs energy for positive $\hbar\omega$. The definition of the $\hat{\Sigma}$-matrix below Eq. (4) is still valid replacing the rates with the sum of positive and negative frequency rates

$$\Sigma^{L/R}_{n\pm}(\omega) = \Gamma^{L/R}_{n\pm}(\omega) + \Gamma^{L/R}_{n\pm}(-\omega),\ \ \Sigma_{n\pm}(\omega) = \Sigma^L_{n\pm}(\omega) + \Sigma^R_{n\pm}(-\omega). \tag{12}$$

Note that in the zero-frequency limit the orthodox transition rates are recovered, i.e. $\Sigma_{n\pm}(0) = \Sigma_{n\pm}$. The time-evolution is evaluated by Laplace-transformation, and the Laplace transform of the time-evolution operator obeys the following Dyson type of equation[31]

$$\hat{\Pi}(\omega) = \frac{i}{\omega}\left[\hat{1} - \frac{i}{\omega}\hat{\Sigma}(\omega)\right]^{-1}, \tag{13}$$

where $\hat{1}$ is the unit matrix with the same dimension as the $\hat{\Sigma}$-matrix, determined by the number of relevant charge states.

3.2.2. *Finite Frequency Noise Expressions*

We now present the noise expressions valid at finite frequency. The charge noise is given by

$$S_{NN}(\omega) = 2Re\left[[1\dots 1]\cdot \hat{N}\cdot \hat{\Pi}(\omega)\cdot \hat{N}_1(\omega)\cdot \bar{P}^{st}\right], \tag{14}$$

where the charge operator now is tridiagonal and frequency dependent when it stands in the first position ($\hat{N}_1$),

$$(\hat{N}_1)_{(n\pm 1)n} = \mp i\Gamma_{n\pm}(\omega))/\omega,\ \ (\hat{N}_1)_{nn} = -[(\hat{N}_1)_{(n-1)n} + (\hat{N}_1)_{(n+1)n}]. \tag{15}$$

The current noise is given by

$$S_{II}(\omega) = S_{II}^{shot}(\omega) + [1\dots 1]\cdot \hat{I}_2(\omega)\cdot \left[\hat{\Pi}(\omega) + \hat{\Pi}(-\omega)\right]\cdot \hat{I}_1(\omega)\cdot \bar{P}^{st}, \tag{16}$$

where the shot-noise contribution is

$$S_{II}^{shot}(\omega) = e^2 \sum_{n,r\in\{L,R\}} 2P_n^{st}\alpha_r^2\left[\Gamma_{n+}^r(\omega) + \Gamma_{n-}^r(\omega)\right], \tag{17}$$

and the current operators are both tridiagonal matrices with non-zero entries

$$\begin{aligned} (\hat{I}_{L1})_{(n\pm 1)n} &= \pm e\left[\Gamma_{n\pm}^L(0) + \Gamma_{n\pm}^L(\omega)\right], \\ (\hat{I}_{L1})_{nn} &= e[\Gamma_{n+}^L(0) - \Gamma_{n+}^L(\omega)] - e[\Gamma_{n-}^L(0) - \Gamma_{n-}^L(\omega)], \\ (\hat{I}_{L2})_{(n\pm 1)n} &= \pm e\Sigma_{n\pm}^L(\omega),\ \ (\hat{I}_{L2})_{nn} = 0. \end{aligned} \tag{18}$$

The operators for the right junction are constructed by changing $L \to R$ and multiplying by -1. The total current operator is still $\hat{I}_n = \alpha_L \hat{I}_{Ln} + \alpha_R \hat{I}_{Rn}$.

We see that also the current and charge operators aquire a frequency dependence and that the first and second operator have different form, due to correct ordering along the Keldysh contour.

3.3. DIFFERENT LIMITS OF THE QUANTUM EXPRESSIONS

An example of $S_{VV}(\omega)$ and $S_{II}(\omega)$ is shown in Fig. 5, with the orthodox results as comparison. In the limit $\omega \to 0$ the orthodox results are recovered. The quantum corrections in the low-frequency regime are discussed in detail for the two-level system in section 4.2. The noise at large negative frequencies, which correspond to noise where the SET emits large energies, tend to zero. In our approximation the voltage noise is analytically zero when $P_n^{st}\Gamma_{n\pm}(\omega) = 0$ for all n, i.e. when no inelastic tunneling events are allowed from the steady-state. The shot-noise part of the current noise behaves similarly, and the correction has the same order of magnitude as the cotunneling shot-noise, which we do not take into account.

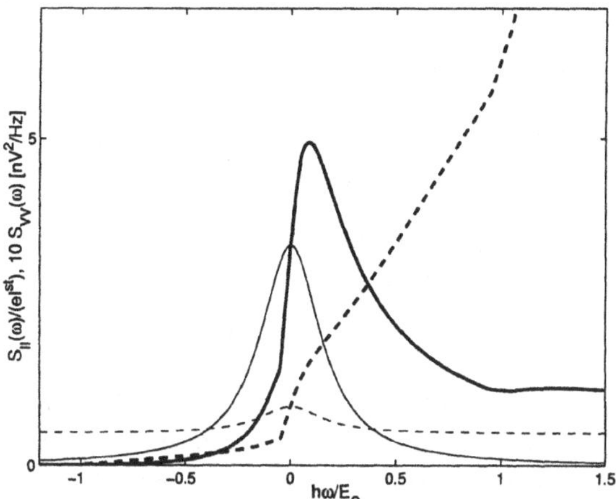

Figure 5. Calculated $S_{VV}(\omega)$ (bold solid line) and $S_{II}(\omega)$ (bold dashed line) for a symmetric SET at zero temperature with $R_T^L = R_T^R = R_K$ and $E_C = 3\,k_B$. The thin lines are the results from orthodox theory. The SET is biased at $n_x = 0.25$, and slightly above the Coulomb threshold $eV = 1.1E_C$, with $I^{st} \approx 0.5\text{nA}$. Both $S_{VV}(\omega)$ and $S_{II}(\omega)$ go to zero at $\hbar\omega = 1.05E_C$ which correspond to the maximal extractable energy.

Therefore, to the accuracy of our approximation, the current noise vanishes when the shot-noise vanishes, i.e. when $P_n^{st}\Gamma_{n\pm}(\omega) = 0$ for all n.

In the high positive frequency limit the spectral noise density of the SET should be independent of the bias and be dominated by the Nyquist noise in Eq. (7). In this limit all rates entering the quantum noise formulas simplify to

$$\Gamma_{n\pm}^{L/R}(\omega) \to \Gamma^{L/R} \equiv \frac{R_K}{2R_T^{L/R}}\frac{\omega}{2\pi} + O(1) = \frac{\hbar\omega}{2e^2 R_T^{L/R}} + O(1), \tag{19}$$

where $O(1)$ is a bias-dependent constant. Using $S_{VV}(\omega) = e^2/C_\Sigma^2\, S_{NN}(\omega)$ we thus arrive at the following high-frequency quantum noise expressions

$$S_{VV}(\omega) \approx \frac{2e^2}{\omega^2 C_\Sigma^2} 2(\Gamma^L + \Gamma^R), \;\; S_{II}(\omega) \approx S_{II}^{shot}(\omega) \approx 4e^2(\alpha_L^2\Gamma^L + \alpha_R^2\Gamma^R), \tag{20}$$

where we used that $\sum_n P_n^{st} = 1$. One may easily check that this agrees with Eq. (7).

4. Two level approximation for the SET

As an explicit example we now show the analytically solvable case with low bias and temperature, such that only the two lowest energy charge states, say $N = 0$ and $N = 1$, are occupied, i.e. the only non-zero rates are Σ_{0+} and Σ_{1-}.

4.1. ORTHODOX THEORY

The master equation Eq. (4) then simplifies to

$$\partial_t \begin{pmatrix} P_0(t) \\ P_1(t) \end{pmatrix} = \begin{pmatrix} -\Sigma_{0+} & \Sigma_{1-} \\ \Sigma_{0+} & -\Sigma_{1-} \end{pmatrix} \begin{pmatrix} P_0(t) \\ P_1(t) \end{pmatrix}. \tag{21}$$

with the solution given by the time-evolution operator

$$\hat{\Pi}(t) = \begin{pmatrix} P_0^{st} & P_0^{st} \\ P_1^{st} & P_1^{st} \end{pmatrix} + \begin{pmatrix} P_1^{st} & -P_0^{st} \\ -P_1^{st} & P_0^{st} \end{pmatrix} e^{-\Sigma t} \; (t > 0), \tag{22}$$

where we defined the sum rate $\Sigma = \Sigma_{0+} + \Sigma_{1-}$, and the steady-state occupation probabilities $P_0^{st} = \Sigma_{1-}/\Sigma$ and $P_1^{st} = \Sigma_{0+}/\Sigma$. The first term in the time evolution matrix (operator) $\Pi(t)$ gives the steady-state solution, and the second term shows simple exponential relaxation with a single rate Σ towards the steady state. The charge and tunnel current operators are

$$\hat{N} = \begin{pmatrix} 0 & 0 \\ 0 & 1 \end{pmatrix}, \; \hat{I}_L = e\Sigma_{0+} \begin{pmatrix} 0 & 0 \\ 1 & 0 \end{pmatrix}, \; \hat{I}_R = e\Sigma_{1-} \begin{pmatrix} 0 & 1 \\ 0 & 0 \end{pmatrix}. \tag{23}$$

The steady-state properties and correlation functions are

$$\begin{aligned}
&N^{st} = P_1^{st}, \; I_L^{st} = I_R^{st} = P_0^{st}\Sigma_{0+} = P_1^{st}\Sigma_{1-}, \\
&K_{NN}(\tau) = P_0^{st} P_1^{st} e^{-\Sigma\tau}, \; S_{NN}(\omega) = \frac{2I^{st}}{\omega^2 + \Sigma^2} = \frac{P_0^{st}\Sigma_{0+} + P_1^{st}\Sigma_{1-}}{\omega^2 + \Sigma^2}, \\
&K_{II}(\tau) = \\
&= eI^{st}[(\alpha_R^2 + \alpha_L^2)\delta(\tau) + (\alpha_L\Sigma_{0+} - \alpha_R\Sigma_{1-})(\alpha_R\Sigma_{0+} - \alpha_L\Sigma_{1-})\frac{e^{-\Sigma\tau}}{\Sigma}], \\
&S_{II}(\omega) = eI^{st}[(\alpha_R^2 + \alpha_L^2) + 2\frac{(\alpha_L\Sigma_{0+} - \alpha_R\Sigma_{1-})(\alpha_R\Sigma_{0+} - \alpha_L\Sigma_{1-})}{\Sigma^2 + \omega^2}] = \\
&= eI^{st}\frac{\Sigma_{0+}^2 + \Sigma_{1-}^2 + \omega^2(\alpha_R^2 + \alpha_L^2)}{\Sigma^2 + \omega^2}.
\end{aligned} \tag{24}$$

Notice that both charge- and current-noise are proportional to the steady-state current.

4.2. LOW FREQUENCY QUANTUM THEORY

In the finite but low-frequency regime, i.e. when still only $\Gamma_{0+}^L(\omega)$ and $\Gamma_{1-}^R(\omega)$ are non-zero, the quantum expressions simplify to:

$$S_{NN}(\omega) = \frac{P_0^{st} 2\Gamma_{0+}^L(\omega) + P_1^{st} 2\Gamma_{1-}^R(\omega)}{\omega^2 + (\Sigma_{0+}(\omega) + \Sigma_{1-}(\omega))^2},$$

$$S_{II}(\omega) = 2e^2\left[P_0^{st}\alpha_L^2\Gamma_{0+}^L + P_1^{st}\alpha_R^2\Gamma_{1-}^R\right] + \frac{\alpha_L\Sigma_{0+} - \alpha_R\Sigma_{1-}}{\omega^2 + (\Sigma_{0+} + \Sigma_{1-})^2} \times$$
$$\times\left[eI^{st}(\Sigma_{0+} - \Sigma_{1-}) + 2e^2(\Sigma_{0+} + \Sigma_{1-})(P_1^{st}\alpha_R\Gamma_{1-}^R - P_0^{st}\alpha_L\Gamma_{0+}^L)\right], \tag{25}$$

where the frequency dependence of $\Sigma_{n\pm}(\omega)$ and $\Gamma_{n\pm}^{L/R}(\omega)$ has been suppressed for brevity. At zero temperature, with symmetric junctions $C_L = C_R$ and $R_T^L = R_T^R = R_T$, the expressions simplify further:

$$\begin{aligned} S_{NN}(\omega) &= \frac{1}{e}\left[I^{st} + \frac{\hbar\omega}{eR_T}\right]\frac{1}{(\Sigma_{0+}(0) + \Sigma_{1-}(0))^2 + \omega^2}, \\ S_{II}(\omega) &= \frac{e}{2}\left[I^{st} + \frac{\hbar\omega}{2eR_T}\right]\left[1 + \frac{(\Sigma_{0+}(0) - \Sigma_{1-}(0))^2}{(\Sigma_{0+}(0) + \Sigma_{1-}(0))^2 + \omega^2}\right]. \end{aligned} \tag{26}$$

We find that the classically derived expressions (Eq. 24), symmetric in $\pm\omega$, get an asymmetric quantum correction linear in ω. The correction is added to the steady-state current and is proportional to $\hbar\omega/eR_T$, which would be the current through one junction with voltage bias $V = \hbar\omega/e$.

Thus the important difference between the orthodox result and the full quantum result is that we get an asymmetry in the noise spectrum between the positive and negative frequencies in the full quantum result. This becomes important for example in a qubit measurement since it will drastically change the occupation of the two levels in the qubit. In the next sections we will discuss such a qubit, and how the noise from the SET will affect the measurement of the qubit.

5. The Qubit

The qubit we consider is made up of the two lowest lying energy levels in a single Cooper-pair box (SCB) [17]. An SCB is a small superconducting island, with charging energy $E_C^{qb} = e^2/2(C_{qb}+C_g^{qb}+C_c)$, coupled to a superconducting reservoir via a Josephson junction with Josephson energy E_J. In order to have a good qubit the following inequalities have to be fulfilled: $\Delta_s > E_C^{qb} \gg E_J \gg k_BT$, where Δ_s is the superconducting energy gap and T is the temperature. The low temperature is required to prevent thermal excitations and the high superconducting gap is needed to suppress quasiparticle tunneling. For suitable values of the gate voltage (close to $n_g = 1/2$) the box can be described by the following two-level Hamiltonian [17, 15]

$$H_q^{qb} = -\frac{4E_C^{qb}}{2}(1 - 2n_g)\hat{\sigma}_z - \frac{E_J}{2}\hat{\sigma}_x \tag{27}$$

written in the charge basis $\langle\uparrow| = \langle n = 0| \equiv (1\ 0), \langle\downarrow| = \langle n = 1| \equiv (0\ 1)$, where n is the number of extra Cooper-pairs on the island, $\hat{\sigma}_{x,z}$ are the Pauli matrices, and $n_g = C_g V_g/2e$ is the number of gate-induced Cooper-pairs. By changing the gate voltage the eigenstates of the qubit can be tuned from being almost pure charge states to a superposition of charge states. The eigenstates of the system written in the charge basis are

$$\begin{aligned} |0\rangle &= \cos(\eta/2)|\uparrow\rangle + \sin(\eta/2)|\downarrow\rangle \\ |1\rangle &= -\sin(\eta/2)|\uparrow\rangle + \cos(\eta/2)|\downarrow\rangle, \end{aligned} \tag{28}$$

where $\eta = \arctan(E_J/4E_{qb}(1-2n_g))$ is the mixing angle. The energy difference between the two states is $\Delta E = \sqrt{(4E_{qb})^2(1-2n_q)^2 + E_J^2}$ and the average charge of the eigenstates is

$$Q_0 = 2e\langle 0|\downarrow\rangle\langle\downarrow|0\rangle = 2e\sin^2(\eta/2), \quad Q_1 = 2e\langle 1|\downarrow\rangle\langle\downarrow|1\rangle = 2e\cos^2(\eta/2). \tag{29}$$

6. SET Charge Measurement of Qubit

We assume that the qubit is in the state $c_0(0)|0\rangle + c_1(0)|1\rangle$ before a measurement. A perfect charge measurment, i.e. qubit read-out, will now give the charge Q_0 or Q_1 with probability $|c_0|^2$ and $|c_1|^2$ respectively. The two qubit states correspond to slightly different SET gate voltages, and therefore to two slightly different sets of transition rates, and two different steady-state currents I_0^{st} and I_1^{st}. The current fluctuates so there is a finite measurement time t_m needed to separate I_0^{st} from I_1^{st}[15]

$$t_m^{theory} = \frac{(I_1^{st} - I_0^{st})^2}{8S_{II}(0)} \approx \frac{e^2 R_T C_\Sigma}{\Delta Q^2}. \tag{30}$$

where $\Delta Q = (Q_1 - Q_0)\cdot C_c/C_{qb}$ is the charge difference seen by the SET, and where for the last estimate have we used a symmetric SET biased slightly above the Coulomb threshold at $n_x = 0.25$. The experimentally determined charge sensitivity δQ gives the measurement time according to

$$t_m^{exp} = \left(\frac{2\delta Q}{\Delta Q}\right)^2, \tag{31}$$

and by comparing Eq. (30) and Eq. (31) we may deduce a rough theoretical estimate for the charge sensitivity δQ^{theory}. For the parameters of sample # 1 and # 2 listed in Table 3 we find $\delta Q^{theory} = e\sqrt{R_T C_\Sigma}/2 \sim 1.5\cdot 10^{-6} e/\sqrt{Hz}$, which further substantiates that the measurements are indeed almost shot noise limited.

7. SET Back Action on the Qubit

When performing a measurement on a qubit, the measurement necessarily dephases the qubit. However, there can also be transitions between the two qubit states which destroys the information that the read-out system tries to measure. This mixing occurs due to charge fluctuations of the SET island $S_{NN}(\omega)$, which creates a fluctuating charge on the coupling capacitance C_c, which is equivalent to a fluctuating qubit gate charge $n_g \to n_g + \delta n_g(t)$. The characteristic time for this mixing process is the time t_{mix}. In the weak coupling limit $C_c \ll \{C_\Sigma, C_{qb}\}$, which is relevant for the qubit measurement setup, we may use pertubation theory to evaluate t_{mix}.

The fluctuating term in the qubit Hamiltonian, written in the charge basis is

$$\delta H_q^{qb}(t) = \frac{4E_{qb}}{2} 2\delta n_g(t)\hat{\sigma}_z = E_I \delta n(t)\hat{\sigma}_z, \;\; E_I = \frac{2e^2 C_c}{C_{qb} C_\Sigma}, \tag{32}$$

where $e\,\delta n(t)$ represents the fluctuations of charge on the SET-island, and where we have omitted the term quadratic in $\delta n_g(t)$. We also defined an electrostatic SET-qubit interaction energy E_I. In the eigenbasis, the full qubit Hamiltonian now reads

$$\delta H_e(t) = -\frac{\Delta E}{2}\hat{\sigma}_z + E_I \delta n(t)(\cos(\eta)\hat{\sigma}_z + \sin(\eta)\hat{\sigma}_x). \tag{33}$$

The $\cos(\eta)$ term leads to dephasing, and the $\sin(\eta)$ term to interlevel transitions (level mixing and relaxation).

Weak coupling means $E_I \ll \{E_C, E_C^{qb}\}$ and we may use standard second order time-dependent pertubation theory[30] to see the effect of the charge fluctuations. Assuming that the qubit is in a pure coherent state $c_0(0)|0\rangle + c_1(0)|1\rangle$ at time $t = 0$, we may express the effect of the fluctuating SET charge in terms of the time-evolution of the quantities $P_0^{qb}(t) = |c_0(t)|^2, P_1^{qb}(t) = |c_1(t)|^2$ and $\xi(t) = |c_0(t)c_1^*(t)|$. In terms of the qubit density-matrix, $P_{0,1}^{qb}(t)$ are the diagonal elements, determining the occupation of respective state, and $\xi(t)$ is the magnitude of the off-diagonal elements, describing the quantum coherence between the states. We find that the qubit occupation probabilities obey a similar master equation as the two-level SET (Eq. (21)), where now the transisition rates are determined by the spectral density of the charge fluctuations at the frequency corresponding to the qubit level splitting:

$$\Sigma_{0+}^{qb} = \frac{E_I^2 \sin^2\eta}{\hbar^2} S_{NN}(\Delta E/\hbar), \;\; \Sigma_{1-}^{qb} = \frac{E_I^2 \sin^2\eta}{\hbar^2} S_{NN}(-\Delta E/\hbar), \tag{34}$$

where $S_{NN}(\omega)$ is the asymmetric charge (number) noise spectral density defined in Eq. (10). The qubit therefore relaxes exponentially on the timescale t_{mix}, where

$$t_{mix}^{-1} = \Sigma_{qb} = \Sigma_{0+}^{qb} + \Sigma_{1-}^{qb}, \tag{35}$$

towards the steady-state $P_0^{qb,st} = \Sigma_{1-}^{qb}/\Sigma^{qb}$, $P_1^{qb,st} = \Sigma_{0+}^{qb}/\Sigma^{qb}$. One may note that the quantum corrections in Eq. (26) cancel in the expression $S_{NN}(\omega) + S_{NN}(-\omega)$ determining the mixing time.

Due to the charge fluctuations the quantum coherence decays exponentially

$$\xi(t) = \xi(0)e^{-t/\tau_\varphi}, \quad \tau_\varphi^{-1} = 2\frac{E_I^2 \cos^2 \eta}{\hbar^2} S_{NN}(0) + \frac{\Sigma^{qb}}{2}, \tag{36}$$

where τ_φ is the timescale for dephasing.

7.1. SIGNAL TO NOISE RATIO IN QUBIT READ OUT

Now we can use the measured data for samples #1 and #2 to calculate both t_m and t_{mix}, and thus we can also get the expected signal-to-noise ratio for a single-shot measurement, which is simply given by $SNR_{SS} = \sqrt{t_{mix}/t_m}$. The spectral densities $S_{VV}(f)$ for the two samples at the optimum charge sensitivity, at a current of 6.7 and 8.0 nA for samples #1 and #2, respectively, are displayed in Fig. 6. As can be seen both through Eq. (34) and in Fig. 6, the mixing time increases strongly with increasing ΔE. In our case ΔE is limited by the superconducting energy gap Δ, which for aluminum films corresponds to about 2.5 K. Using niobium as the qubit material would substantially increase the mixing time. If we assume that E_C^{qb} and thus also ΔE can be scaled with Δ and that the coupling C_c/C_{qb} is kept constant, mixing times of several ms can be reached. In that case, other sources than the SET noise would most probably dominate the mixing. The results for the samples #1 and #2 are summarized in Table 3, where a coupling $C_c/C_{qb} = 0.01$ is assumed.

7.2. COULOMB STAIRCASE

One may use the SET to measure the so-called Coulomb staircase, i.e. the average charge of the qubit as a function of gate voltage. In an ideal situation with no energy availible from an external source, at zero temperature, the qubit would

TABLE III.

SET Sample	Qubit material	δQ e/$\sqrt{\text{Hz}}$	ΔE [K]	$S_V(\omega)$ [nV^2/Hz]	t_m [μs]	t_{mix} [μs]	SNR
1	Al	6.3	2.4	0.29	0.40	8.6	4.6
1	Nb	6.3	15.5	0.056	0.40	1860	68
2	Al	3.2	2.4	0.39	0.10	6.4	8.0
2	Nb	3.2	15.5	0.080	0.10	1300	114

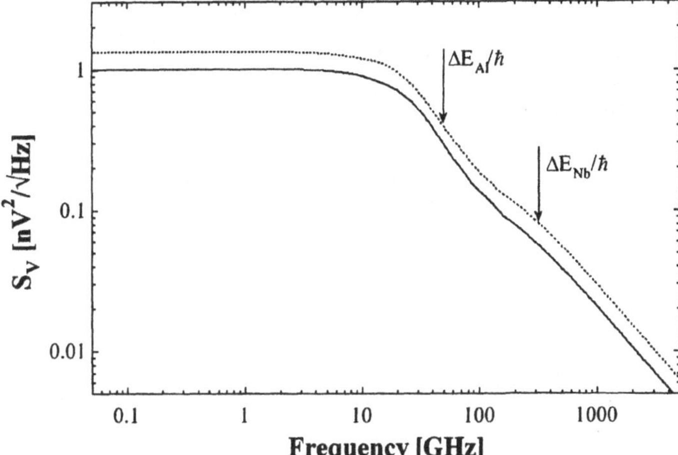

Figure 6. Calculated $S_{VV}^{sym}(f)$ for samples #1(full line) and #2(dashed line). The arrows show the energy separations of the two qubit states for an aluminum and a niobium qubit, respectively.

follow the ground state adiabatically and the charge would increment in steps of $2e$ at $n_g = n + 0.5$, n integer. These steps are not perfectly sharp because of the Josephson energy mixing the charge states[17]. Now assuming that the qubit equilibrium is determined by the SET back action the charge measured should instead be $P_0^{qb,st}Q_0 + P_1^{qb,st}Q_1$. The steps are now rounded further[32] due to the finite probability for the qubit to be excited, see Fig. 7. The quantum asymmetry of the noise-spectrum, indicating the difference of qubit excitation and relaxation, is vital to recover the correct Coulomb staircase. Straighforward use of orthodox theory would predict an equal population of the qubit states.

Acknowledgements

One of the authors would like to acknowledge fruitful discussions with Yuriy Makhlin and Alexander Shnirman. This work was supported by the Swedish grant agencies NFR/VR and SSF, the Wallenberg and Göran Gustavsson foundations, and by the SQUBIT project of the IST-FET programme of the EC.

References

1. T. A. Fulton and G. J. Dolan, Phys. Rev. Lett. **59**, 109 (1987)
2. K. K.Likharev, IEEE Trans. Mag. **23**, 1142 (1987)
3. D. V. Averin and K. K. Likharev: in *Mesoscopic Phenomena in Solids*, eds. B.L. Altshuler, P.A. Lee and R.A. Webb (North-Holland, Amsterdam, 1991) p. 173.
4. H. Grabert and M. H. Devoret, NATO Adv. Study Inst. Ser., Ser. B **279**, 74 (1992).

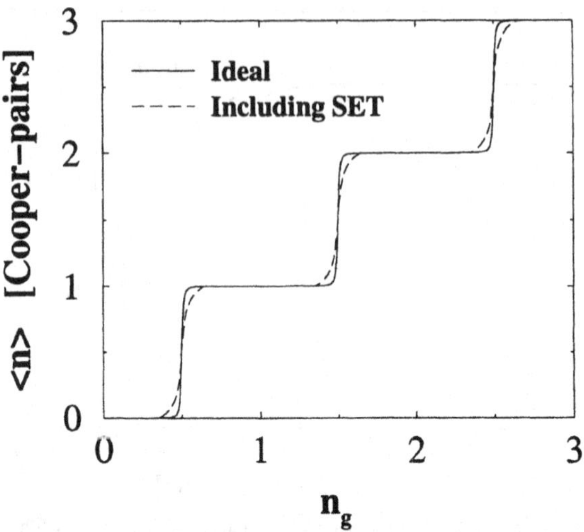

Figure 7. The calculated Coulomb staircase of an SCB as measured by the SET, assuming that the SET backaction determine the steady-state of the qubit.

5. K. Matsumoto, M. Ishii, K. Segawa, Y. Oka, B. J. Vartanian and J. S. Harris, Appl. Phys. Lett. **68**, 35 (1996)
6. Yu. A. Pashkin, Y. Nakamura, and J. S. Tsai Appl. Phys. Lett. **76**, 2256 (2000)
7. T. W. Kim, D. C. Choo, J. H. Shim, and S. O. Kang Appl. Phys. Lett., **80**, 2168 (2002)
8. R. J.Schoelkopf, P. Wahlgren, A. A. Kozhevnikov, P. Delsing, and D. E. Prober, Science **280**, 1238 (1998)
9. J. Pettersson, P. Wahlgren, P. Delsing, D. B. Haviland, T. Claeson, N. Rorsman and H. Zirath, Phys.l Rev. B **53**, R13272, (1996)
10. S. L Pohlen, R. J. Fitzgerald, J. M. Hergenrother, and M. Tinkham, Appl. Phys. Lett. 74, 2884-2886 (1999)
11. A. Aassime, D.Gunnarsson, K. Bladh, P. Delsing, and R. Schoelkopf, Appl. Phys. Lett. **79**, 4031 (2001)
12. A. Aassime, G. Johansson, G. Wendin, P. Delsing, and R. Schoelkopf, Phys. Rev. Lett. **86**, 3376 (2001)
13. M. J. Lea, P. G. Frayne, and Yu. Mukharsky, Fortschr. Phys. **48**, 1109 (2000)
14. B. E. Kane, N. S. McAlpine, A. S. Dzurak, R. G. Clark, G. J. Milburn, He. Bi. Sun and Howard Wiseman, Phys. Rev. B, **61**, 2961 (2000)
15. Y. Makhlin, G. Schön, and A. Shnirman Rev. Mod. Phys., **73**, 357 (2001)
16. T. R. Stevenson,A. Aassime, P. Delsing, R. Schoelkopf, K. Segall, and C. M. Stahle, IEEE Transactions On Applied Superconductivity **11**, 692 (2001)
17. V. Bouchiat, D. Vion, P. Joyez, D. Esteve and M. H. Devoret, Phys. Scripta **T76**, 165 (1998)
18. G L Ingold, Yu V Nazarov in: NATO ASI Series B: Physics, vol. 294, Single Charge Tunneling: Coulomb Blockade Phenomena in Nanostructures, Eds. H Grabert, M Devoret, p. 21-106
19. L. J. Geerligs, D. V. Averin, and J. E. Mooij Phys. Rev. Lett. **65**, 3037 (1990)
20. W. Shockley, Journal of Applied Physics, **9**, 635 (1938)
21. Ya. M. Blanter and M. Buttiker, Phys. Rep. **336**, 1 (2000)

22. A. N. Korotkov, Phys. Rev. B, **59**, 10 381 (1994)
23. L. Fedichkin and V. V'yurkov, Appl. Phys. Lett., **64** 2535 (1994)
24. U. Hanke, Yu. M. Galperin, and K. A. Chao, Physical Review B **50**, 1595 (1994)
25. G. R. Grimmett, and D. R. Stirzaker, *Probability and Random Processes*, (Clarendon Press, Oxford, 1992), 245
26. H. B. Callen and T. A. Welton, Phys. Rev., **83**, 34 (1951)
27. G. Johansson, A. Käck and G. Wendin, Phys. Rev. Lett. **88**, 046802 (2002)
28. W. Schottky, Annals of Physics (Leipzig) **57** 541 (1918)
29. H. Schoeller and G. Schön, Phys. Rev. B, **50**, 18 436 (1994).
30. J. J. Sakurai, *Modern Quantum Mechanics*, (Addison-Wesley, 1985)
31. F. Dyson, Phys. Rev. **75**, 1736 (1949)
32. Y. V. Nazarov, J. Low Temp. Phys. **90** 77 (1993)

DEPHASING AND RENORMALIZATION IN QUANTUM TWO-LEVEL SYSTEMS

ALEXANDER SHNIRMAN
Institut für Theoretische Festkörperphysik, Universität Karlsruhe, 76128 Karlsruhe, Germany

GERD SCHÖN
Institut für Theoretische Festkörperphysik, Universität Karlsruhe, 76128 Karlsruhe, Germany; and Forschungszentrum Karlsruhe, Institut für Nanotechnologie, 76021 Karlsruhe, Germany

Abstract.

Motivated by fundamental questions about the loss of phase coherence at low temperature we consider relaxation, dephasing and renormalization effects in quantum two-level systems which are coupled to a dissipative environment. We observe that experimental conditions, e.g., details of the initial state preparation, determine to which extent the environment leads to dephasing or to renormalization effects. We analyze an exactly solvable limit where the relation between both can be demonstrated explicitly. We also study the effects of dephasing and renormalization on response functions.

1. Introduction

The dynamics of quantum two-level systems has always been at the focus of interest, but recently attracted increased attention because of the prospects of quantum state engineering and related low-temperature experiments. A crucial requirement for many of these concepts is the preservation of phase coherence in the presence of a noisy environment. Typically one lacks a detailed microscopic description of the noise source, but frequently it is sufficient to model the environment by a bath of harmonic oscillators, with frequency spectrum adjusted to reproduce the observed power spectrum. The resulting 'spin-boson' models have been studied in the literature, in particular the one with bilinear coupling and 'Ohmic' spectrum, but spin-boson models with different power spectra appear equally interesting in view of several experiments. In this article we, therefore, analyze spin-boson model with general power spectra with respect to relaxation, dephasing and renormalization processes at low temperatures. We study the de-

Y. V. Nazarov (ed.), Quantum Noise in Mesoscopic Physics, 357–370.

phasing of nonequilibrium initial states. We show how the state preparation affects the effective high-frequency cut-off and separates renormalization from dephasing effects. We also demonstrate how dephasing processes influence low-temperature linear response and correlation functions.

2. Spin-boson model

In this section we review the theory and properties of spin-boson models, which have been studied extensively before (see the reviews [4, 11]). A quantum two-level system is modeled by a spin degree of freedom in a magnetic field. It is coupled linearly to an oscillator bath representing the environment. The total Hamiltonian is

$$\mathcal{H} = \mathcal{H}_{\rm ctrl} + \sigma_z \sum_j c_j (a_j + a_j^\dagger) + \mathcal{H}_{\rm b} \,, \tag{1}$$

where the controlled part is $\mathcal{H}_{\rm ctrl} = -\frac{1}{2}B_z\,\sigma_z - \frac{1}{2}B_x\,\sigma_x = -\frac{1}{2}\Delta E\,(\cos\theta\;\sigma_z + \sin\theta\;\sigma_x)$, the oscillator bath is described by $\mathcal{H}_{\rm b} = \sum_j \hbar\omega_j\, a_j^\dagger a_j$, and the bath 'force' operator $X = \sum_j c_j(a_j + a_j^\dagger)$ is assumed to couple linearly to σ_z. For later convenience $\mathcal{H}_{\rm ctrl}$ has also been written in terms of a mixing angle $\theta = \tan^{-1}(B_x/B_z)$, depending on the direction of the magnetic field, and the energy splitting of the eigenstates, $\Delta E = \sqrt{B_x^2 + B_z^2}$.

In thermal equilibrium the Fourier transform of the symmetrized correlation function of the force operator is given by

$$S_X(\omega) \equiv \langle [X(t), X(t')]_+ \rangle_\omega = 2\hbar J(\omega) \coth \frac{\hbar\omega}{2k_{\rm B}T} \,. \tag{2}$$

Here the bath spectral density has been introduced, defined by $J(\omega) \equiv \frac{\pi}{\hbar}\sum_j c_j^2\,\delta(\omega - \omega_j)$. At low frequencies it typically has a power-law form up to a high-frequency cut-off ω_c,

$$J(\omega) = \frac{\pi}{2}\,\hbar\alpha\,\omega_0^{1-s}\omega^s\Theta(\omega_c - \omega) \,. \tag{3}$$

Generally one distinguishes Ohmic ($s = 1$), sub-Ohmic ($s < 1$), and super-Ohmic ($s > 1$) spectra. In Eq. (3) an additional frequency scale ω_0 has been introduced. Since it appears only in the combination $\alpha\omega_0^{1-s}$ it is arbitrary and in Ref. [4] has been chosen equal to the high-frequency cut-off. Here we prefer to distinguish both frequencies, since the cut-off ω_c will play an important role in what follows.

The spin-boson model has been studied mostly for the case where an Ohmic bath is coupled linearly to the spin. One reason is that linear damping, proportional to the velocity, is encountered frequently in real systems. Another is that suitable systems with Ohmic damping show a quantum phase transition at a critical strength of the dissipation, $\alpha_{\rm cr} \sim 1$. On the other hand, in the context of quantum-state engineering we should concentrate on systems with weak damping, but allow for general spectra of the fluctuations.

Spin-boson models with sub-Ohmic damping ($0 < s < 1$) have been considered earlier [4, 11] but did not attract much attention. It was argued that sub-Ohmic dissipation would totally suppress coherence, transitions between the states of the two-level system would happen only at finite temperatures and would be incoherent. At zero temperature the system should be localized in one of the eigenstates of σ_z, since the bath renormalizes the off-diagonal part of the Hamiltonian B_x to zero. This scenario is correct for intermediate to strong damping. It is, however, not correct for weak damping. Indeed the 'NIBA' approximation developed in Ref. [4] fails in the weak-coupling limit for transverse noise, while a more accurate renormalization procedure [3] predicts damped coherent behavior. In the context of quantum-state engineering we are interested in precisely this *coherent sub-Ohmic* regime. We will demonstrate a simple criterion which allows to define a border between coherent and incoherent regimes.

A further reason to study the sub-Ohmic case is that it allows us to mimic the universally observed $1/f$ noise. For instance, a bath with $s = 0$ and $J(\omega) = (\pi/2)\alpha\hbar\omega_0$ produces at low frequencies, $\hbar\omega \ll k_B T$, a $1/f$ noise spectrum $S_X(\omega) = E_{1/f}^2/|\omega|$ with $E_{1/f}^2 = 2\pi\alpha\hbar\omega_0 k_B T$. Since frequently nonequilibrium sources are responsible for the $1/f$ noise, the temperature here should be regarded as a fitting parameter rather than a thermodynamic quantity.

Below we will also consider the super-Ohmic case ($s > 1$) as it allows us to study renormalization effects clearly.

3. Relaxation and dephasing

We first consider the Ohmic case in the weak damping regime, $\alpha \ll 1$. Still we distinguish two regimes: the 'Hamiltonian-dominated' regime, which is realized when ΔE is large enough, and the 'noise-dominated' regime, which is realized, e.g., at degeneracy points where $\Delta E \to 0$. The exact border between both regimes will be specified below.

In the Hamiltonian-dominated regime it is natural to describe the evolution of the system in the eigenstates of $\mathcal{H}_{\rm ctrl}$, which are

$$|0\rangle = \cos\frac{\theta}{2}|\uparrow\rangle + \sin\frac{\theta}{2}|\downarrow\rangle \quad \text{and} \quad |1\rangle = -\sin\frac{\theta}{2}|\uparrow\rangle + \cos\frac{\theta}{2}|\downarrow\rangle\,. \quad (4)$$

Denoting by τ_x and τ_z the Pauli matrices in the eigenbasis, we have

$$\mathcal{H} = -\frac{1}{2}\Delta E\,\tau_z + (\sin\theta\,\tau_x + \cos\theta\,\tau_z)\,X + \mathcal{H}_{\rm b}\,. \quad (5)$$

Two different time scales describe the evolution in the spin-boson model. The first, the dephasing time scale τ_φ, characterizes the decay of the off-diagonal elements of the qubit's reduced density matrix $\hat\rho(t)$ in the preferred eigenbasis (4), or, equivalently of the expectation values of the operators $\tau_\pm \equiv (1/2)(\tau_x \pm i\tau_y)$. Frequently dephasing processes lead to an exponential long-time dependence,

$$\langle \tau_\pm(t) \rangle \equiv \mathrm{tr}\left[\tau_\pm \hat{\rho}(t)\right] \propto \langle \tau_\pm(0) \rangle \, e^{\mp i\Delta E t/\hbar} \, e^{-t/\tau_\varphi} \,, \tag{6}$$

but other decay laws occur as well and will be discussed below. The second, the relaxation time scale τ_{relax}, characterizes how diagonal entries tend to their equilibrium values,

$$\langle \tau_z(t) \rangle - \langle \tau_z(\infty) \rangle \propto e^{-t/\tau_{\mathrm{relax}}} \,, \tag{7}$$

where $\langle \tau_z(\infty) \rangle = \mathrm{th}(\Delta E/2k_{\mathrm{B}}T)$.

In Refs. [4, 11] the dephasing and relaxation times were evaluated in a path-integral technique. In the regime $\alpha \ll 1$ it is easier to employ the perturbative (diagrammatic) technique developed in Ref. [9] and the standard Bloch-Redfield approximation. The rates are

$$\Gamma_{\mathrm{relax}} \equiv \tau_{\mathrm{relax}}^{-1} = \frac{1}{\hbar^2} \sin^2\theta \; S_X\left(\omega = \Delta E/\hbar\right) \,, \tag{8}$$

$$\Gamma_\varphi \equiv \tau_\varphi^{-1} = \frac{1}{2}\,\Gamma_{\mathrm{relax}} + \frac{1}{\hbar^2}\cos^2\theta \, S_X(\omega = 0) \,, \tag{9}$$

where $S_X(\omega) = \pi\alpha\hbar^2\omega\coth(\hbar\omega/2k_{\mathrm{B}}T)$. We observe that only the 'transverse' part of the fluctuating field X, coupling to τ_x and proportional to $\sin\theta$, induces transitions between the eigenstates (4) of the unperturbed system. Thus the relaxation rate[1] (8) is proportional to $\sin^2\theta$. The 'longitudinal' part of the coupling of X to τ_z, which is proportional to $\cos\theta$, does not induce relaxation processes. It does, however, contribute to dephasing since it leads to fluctuations of the eigenenergies and, thus, to a random relative phase between the two eigenstates. This is the origin of the 'pure' dephasing contribution to Eq. (9), which is proportional to $\cos^2\theta$. We rewrite Eq. (9) as $\Gamma_\varphi = \frac{1}{2}\Gamma_{\mathrm{relax}} + \cos^2\theta\,\Gamma_\varphi^*$, where $\Gamma_\varphi^* = S_X(\omega = 0)/\hbar^2 = 2\pi\alpha k_{\mathrm{B}}T/\hbar$ is the pure dephasing rate.

The pure dephasing rate Γ_φ^* characterizes the strength of the dissipative part of the Hamiltonian, while ΔE characterizes the coherent part. The Hamiltonian-dominated regime is realized when $\Delta E \gg \Gamma_\varphi^*$, while the noise-dominated regime is realized in the opposite case.

In the noise-dominated regime, $\Delta E \ll \Gamma_\varphi^*$, the coupling to the bath is the dominant part of the total Hamiltonian. Therefore, it is more convenient to discuss the problem in the eigenbasis of the observable σ_z to which the bath is coupled. The spin can tunnel incoherently between the two eigenstates of σ_z. One can

[1] The equilibration is due to two processes, excitation $|0\rangle \to |1\rangle$ and relaxation $|1\rangle \to |0\rangle$, with rates $\Gamma_{+/-} \propto \langle X(t)X(t')\rangle_{\omega=\pm\Delta E/\hbar}$. Both rates are related by a detailed balance condition, and the equilibrium value $\langle \tau_z(\infty) \rangle$ depends on both. On the other hand, Γ_{relax} is determined by the sum of both rates, i.e., the symmetrized noise power spectrum S_X.

again employ the perturbative analysis [9], but use directly the Markov instead of the Bloch-Redfield approximation. The resulting rates are given by

$$\begin{aligned}\Gamma_{\text{relax}} &= B_x^2/\Gamma_\varphi^* = B_x^2/(2\pi\hbar\alpha k_{\text{B}}T) \\ \Gamma_\varphi &= \Gamma_\varphi^* = 2\pi\hbar\alpha k_{\text{B}}T\,.\end{aligned} \tag{10}$$

In this regime the dephasing is much faster than the relaxation. In fact, as a function of the coupling strength α the dephasing and relaxation rates evolve in opposite directions. The α-dependence of the relaxation rate is an indication of the Zeno (watchdog) effect [2]: the environment frequently 'observes' the state of the spin, thus preventing it from tunneling.

4. Longitudinal coupling, exact solution for factorized initial conditions

The last forms of Eqs. (8) and (9) express the two rates in terms of the noise power spectrum at the relevant frequencies. These are the level spacing of the two-state system and zero frequency, respectively. The expressions apply in the weak-coupling limit for spectra which are regular at these frequencies. For the relaxation rate (8) the generalization to sub- and super-Ohmic cases merely requires substituting the relevant $S_X(\omega = \Delta E/\hbar)$. However, this does not work for the pure dephasing contribution. Indeed $S_X(\omega = 0)$ is infinite in the sub-Ohmic regime, while it vanishes for the super-Ohmic case and the Ohmic case at $T = 0$. As we will see this does not imply infinitely fast or slow dephasing in these cases. To analyze these cases we study the exactly soluble model of longitudinal coupling, $\theta = 0$.

Dephasing processes are contained in the time evolution of the quantity $\langle\tau_+(t)\rangle$ obtained after tracing out the bath. This quantity can be evaluated analytically for $\theta = 0$ (when $\tau_+ = \sigma_+$) for an initial state which is described by a factorized density matrix $\hat\rho(t=0) = \hat\rho_{\text{spin}} \otimes \hat\rho_{\text{bath}}$ (which implies that the two-level system and bath are initially disentangled). In this case the Hamiltonian $\mathcal{H} = -\frac{1}{2}\Delta E\,\sigma_z + \sigma_z X + \mathcal{H}_{\text{b}}$ is diagonalized by a unitary transformation by $U \equiv \exp\left(-i\sigma_z\Phi/2\right)$ where the bath operator Φ is defined as

$$\Phi \equiv i\sum_j \frac{2c_j}{\hbar\omega_j}(a_j^\dagger - a_j)\,. \tag{11}$$

The 'polaron transformation' yields $\tilde{\mathcal{H}} = U\mathcal{H}U^{-1} = -\frac{1}{2}\Delta E\,\sigma_z + \mathcal{H}_{\text{b}}$. It has a clear physical meaning: the oscillators are shifted in a direction depending on the state of the spin. Next we observe that the operator σ_+ is transformed as $\tilde\sigma_+ = U\sigma_+U^{-1} = e^{-i\Phi}\sigma_+$, and the observable of interest can be expressed as

$$\langle\sigma_+(t)\rangle = \text{Tr}\left[\hat\rho(t=0)\sigma_+(t)\right] = \text{Tr}\left[U\hat\rho(t=0)U^{-1}e^{-i\Phi(t)}\sigma_+\right]\,. \tag{12}$$

The time evolution of $\Phi(t) = e^{i\mathcal{H}_b t}\Phi e^{-i\mathcal{H}_b t}$ is governed by the bare bath Hamiltonian. After some algebra, using the fact that the initial density matrix is factorized, we obtain $\langle\sigma_+(t)\rangle \equiv \mathcal{P}(t)\, e^{-i\Delta E t}\, \langle\sigma_+(0)\rangle$ where

$$\mathcal{P}(t) = \mathrm{Tr}\left[e^{i\Phi(0)/2}e^{-i\Phi(t)}e^{i\Phi(0)/2}\hat{\rho}_{\mathrm{bath}}\right] . \tag{13}$$

The expression (13) applies for any initial state of the bath as long as it is factorized from the spin. In particular, we can assume that the spin was initially (for $t \le 0$) kept in the state $|\uparrow\rangle$ and the bath had relaxed to the thermal equilibrium distribution for this spin value: $\hat{\rho}_{\mathrm{bath}} = \hat{\rho}_\uparrow \equiv Z_\uparrow^{-1}e^{-\beta\mathcal{H}_\uparrow}$, where $\mathcal{H}_\uparrow = \mathcal{H}_b + \sum_j c_j(a_j + a_j^\dagger)$. In this case we can rewrite the density matrix as $\hat{\rho}_{\mathrm{bath}} = e^{i\Phi/2}\hat{\rho}_b e^{-i\Phi/2}$, where the density matrix of the decoupled bath is given by $\hat{\rho}_b \equiv Z_b^{-1}e^{-\beta\mathcal{H}_b}$, and the function $\mathcal{P}(t)$ reduces to

$$\mathcal{P}(t) \to P(t) \equiv \mathrm{Tr}\left(e^{-i\Phi(t)}\, e^{i\Phi}\, \hat{\rho}_b\right) . \tag{14}$$

This expression (with Fourier transform $P(E)$) has been studied extensively in the literature [4, 8, 7, 6, 1]. It can be expressed as $P(t) = \exp K(t)$, where

$$K(t) = \frac{4}{\pi\hbar}\int_0^\infty d\omega\, \frac{J(\omega)}{\omega^2}\left[\coth\left(\frac{\hbar\omega}{2k_BT}\right)(\cos\omega t - 1) - i\sin\omega t\right] . \tag{15}$$

For an Ohmic bath at non-zero temperature and not too short times, $t > \hbar/k_BT$, it reduces to $\mathrm{Re}\,K(t) \approx -S_X(\omega = 0)t/\hbar^2 = -2\pi\,\alpha\,k_BT\,t/\hbar$, consistent with Eq. (9) in the limit $\theta = 0$.

For $1/\omega_c < t < \hbar/k_BT$, and thus for all times at $T = 0$, one still finds a decay of $\langle\sigma_+(t)\rangle$ governed by $\mathrm{Re}\ K(t) \approx -2\ \alpha\ \ln(\omega_c t)$, which implies a power-law decay

$$\langle\sigma_+(t)\rangle = (\omega_c t)^{-2\alpha}e^{-i\Delta Et/\hbar}\langle\sigma_+(0)\rangle . \tag{16}$$

Thus even at $T = 0$, when $S_X(\omega = 0) = 0$, the off-diagonal elements of the density matrix decay in time. All oscillators up to the high-frequency cut-off ω_c contribute to this decay. The physical meaning of this result will be discussed later. We can also define a cross-over temperature T^* below which the power-law decay dominates over the subsequent exponential one. A criterion is that at $t = \hbar/k_BT^*$ the short-time power-law decay has reduced the off-diagonal components already substantially, i.e $\langle\sigma_+(\hbar/k_BT^*)\rangle = 1/e$. This happens at the temperature $kT^* = \hbar\omega_c\exp(-1/2\alpha)$. Thus the dephasing rate is $\Gamma_\varphi^* = k_BT^*/\hbar$ for $T < T^*$ and $\Gamma_\varphi^* = 2\pi\alpha k_BT/\hbar$ for $\alpha T > T^*$.

For a sub-Ohmic bath with $0 < s < 1$ due to the high density of low-frequency oscillators exponential dephasing is observed even for short times, $|\langle\sigma_+(t)\rangle| \propto \exp[-\alpha(\omega_0 t)^{1-s}]$ for $t < \hbar/k_BT$, as well as for longer times, $|\langle\sigma_+(t)\rangle| \propto \exp[-\alpha Tt\,(\omega_0 t)^{1-s}]$ for $t > \hbar/k_BT$. In the exponents of the short- and long-time decay laws we have omitted factors which are of order one, except if s is close to either 1 or

0, in which case a more careful treatment is required. The dephasing rates resulting from the decay laws are $\Gamma^*_\varphi \propto T^* = \alpha^{1/(1-s)}\omega_0$ (cf. Ref. [10]) for $T < T^*$ and $\Gamma^*_\varphi \propto (\alpha T/\omega_0)^{1/(2-s)}\omega_0$ for $T > T^*$. Again the crossover temperature T^* marks the boundary between the regimes where either the initial, temperature-independent decay or the subsequent decay at $t > \hbar/k_B T$ is more important.

These results allow us to further clarify the question of coherent vs. incoherent behavior in the sub-Ohmic regime. It is known from earlier work [4, 11] that the dynamics of the spin is incoherent even at $T = 0$ if $\Delta E \ll \alpha^{1/(1-s)}\hbar\omega_0$ (only the transverse case $\theta = \pi/2$ was considered). This condition implies $\Delta E \ll \Gamma^*_\varphi$, i.e., it marks the noise-dominated regime. We are mostly interested in the opposite, Hamiltonian-dominated regime, $\Delta E \gg \hbar\Gamma^*_\varphi$. The NIBA approximation used in Ref. [4] fails in this limit, while a more accurate RG study [3] predicts damped coherent behavior.

Finally we discuss the super-Ohmic regime, $s > 1$. In this case, within a short time of order ω_c^{-1}, the exponent Re $K(t)$ increases to a finite value Re $K(\infty) = -\alpha(\omega_c/\omega_0)^{s-1}$ and then remains constant for $\omega_c^{-1} < t < \hbar/k_B T$ (we again omit factors of order one and note that the limit $s \to 1$ requires more care). This implies an initial reduction of the off-diagonal element $|\langle\sigma_+(t)\rangle|$ followed by a saturation at $|\langle\sigma_+(t)\rangle| \propto \exp[-\alpha(\omega_c/\omega_0)^{s-1}]$. For $t > \hbar/k_B T$ an exponential decay develops, $|\langle\sigma_+(t)\rangle| \propto \exp[-\alpha\, Tt\,(\omega_0 t)^{1-s}]$, but only if $s < 2$. This decay is always dominant and, thus, there is no crossover in this case, i.e., $T^* = 0$. For $s \geq 2$ there is almost no additional decay.

The results obtained for different bath spectra are summarized in Table 4.

5. Preparation effects

In the previous section we considered specific initial conditions with a factorized density matrix. The bath was prepared in an equilibrium state characterized by temperature T, while the spin state was arbitrary. The state of the total system is, thus, a nonequilibrium one and dephasing is to be expected even at zero bath temperature. In this section we investigate what initial conditions may arise in real experiments.

The fully factorized initial state just described can in principle be prepared by the following Gedanken experiment: The spin is forced, e.g., by a strong external field, to be in a fixed state, say $|\uparrow\rangle$. The bath, which is kept coupled to the spin, relaxes to the equilibrium state of the Hamiltonian $\mathcal{H}_\uparrow$, e.g., at $T = 0$ to the ground state $|g_\uparrow\rangle$ of $\mathcal{H}_\uparrow$. Then, at $t = 0$, a *sudden* pulse of the external field is applied to change the spin state, e.g., to a superposition $\frac{1}{\sqrt{2}}(|\uparrow\rangle + |\downarrow\rangle)$. Since the bath has no time to respond, the resulting state is $|i\rangle = \frac{1}{\sqrt{2}}(|\uparrow\rangle + |\downarrow\rangle)\otimes|g_\uparrow\rangle$. Both components of this initial wave function now evolve in time according to the Hamiltonian (1). The first, $|\uparrow\rangle \otimes |g_\uparrow\rangle$, which

TABLE IV. Decay of $|\langle\sigma_+(t)\rangle|$ in time for different bath spectra.

• Ohmic: $J(\omega) = \frac{\pi}{2}\alpha\omega\Theta(\omega_c - \omega)$:

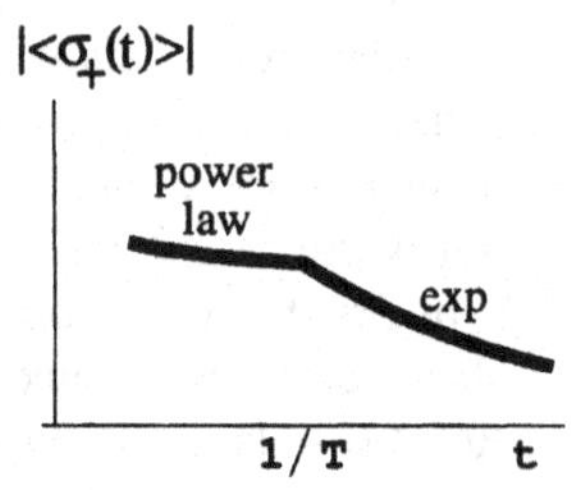

for $\omega_c^{-1} \ll t \ll T^{-1}$ $\quad |\langle\sigma_+(t)\rangle| \approx (\omega_c t)^{-2\alpha}$
for $\qquad t \gg T^{-1}$ $\quad |\langle\sigma_+(t)\rangle| \approx e^{-2\pi\alpha T t}$

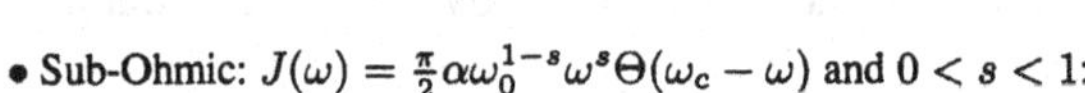

• Sub-Ohmic: $J(\omega) = \frac{\pi}{2}\alpha\omega_0^{1-s}\omega^s\Theta(\omega_c - \omega)$ and $0 < s < 1$:

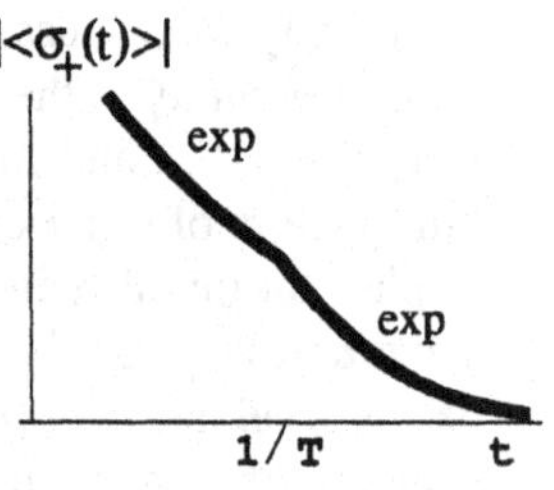

for $\omega_c^{-1} \ll t \ll T^{-1}$ $\quad |\langle\sigma_+(t)\rangle| \approx e^{-\alpha(\omega_0 t)^{1-s}}$
for $\qquad t \gg T^{-1}$ $\quad |\langle\sigma_+(t)\rangle| \approx e^{-\alpha T t\,(\omega_0 t)^{1-s}}$

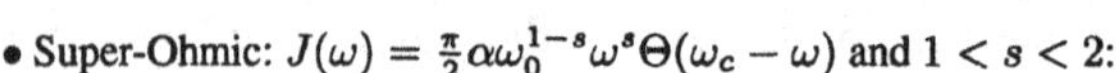

• Super-Ohmic: $J(\omega) = \frac{\pi}{2}\alpha\omega_0^{1-s}\omega^s\Theta(\omega_c - \omega)$ and $1 < s < 2$:

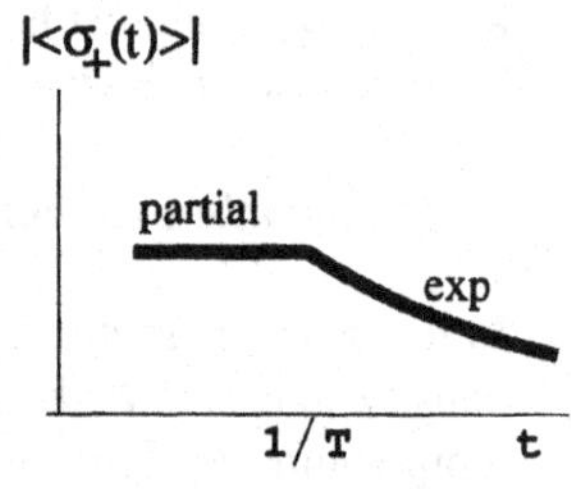

for $\omega_c^{-1} \ll t \ll T^{-1}$ $\quad |\langle\sigma_+(t)\rangle| \approx e^{-\alpha(\omega_c/\omega_0)^{s-1}}$
for $\qquad t \gg T^{-1}$ $\quad |\langle\sigma_+(t)\rangle| \approx e^{-\alpha T t\,(\omega_0 t)^{1-s}}$

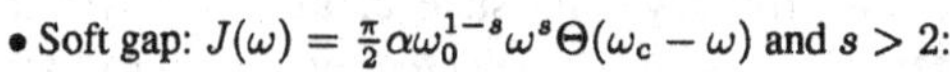

• Soft gap: $J(\omega) = \frac{\pi}{2}\alpha\omega_0^{1-s}\omega^s\Theta(\omega_c - \omega)$ and $s > 2$:

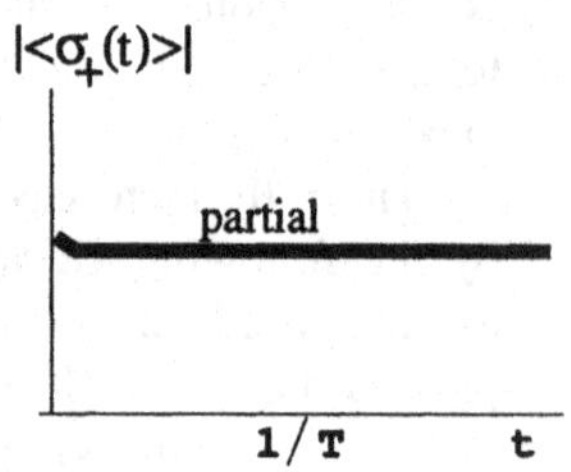

for $\omega_c^{-1} \ll t$ $\quad |\langle\sigma_+(t)\rangle| \approx e^{-\alpha(\omega_c/\omega_0)^{s-1}}$

is an eigenstate of (1), acquires only a trivial phase factor. The time evolution of the second component is more involved. Up to a phase factor it is given by $|\downarrow\rangle\otimes\exp(-i\mathcal{H}_\downarrow t/\hbar)|\,\mathrm{g}_\uparrow\rangle$ where $\mathcal{H}_\downarrow \equiv \mathcal{H}_\mathrm{b} - \sum_j c_j(a_j + a_j^\dagger)$. As the state $|\,\mathrm{g}_\uparrow\rangle$ is not an eigenstate of $\mathcal{H}_\downarrow$, entanglement between the spin and the bath develops, and the coherence between the components of the spin's state is reduced by the factor $|\langle \mathrm{g}_\uparrow|\exp(-i\mathcal{H}_\downarrow t/\hbar)|\mathrm{g}_\uparrow\rangle| = |\langle \mathrm{g}_0|\exp(-i\Phi(t))\exp(i\Phi)|\mathrm{g}_0\rangle| = |P_{\omega_c}(t, T = 0)| < 1$. The function $P(t)$ was defined in Eq. (14). The subscript ω_c is added to indicate the value of the high-frequency cut-off of the bath which will play an important role in what follows.

In a real experiment of the type discussed the preparation pulse takes a finite time, τ_p. For instance, the $(\pi/2)_x$-pulse which transforms the state $|\uparrow\rangle \rightarrow \frac{1}{\sqrt{2}}(|\uparrow\rangle + |\downarrow\rangle)$, can be accomplished by putting $B_z = 0$ and $B_x = \hbar\omega_p$ for a time span $\tau_p = \pi/2\omega_p$. During this time the bath oscillators partially adjust to the changing spin state. The oscillators with (high) frequencies, $\omega_j \gg \omega_p$, follow the spin adiabatically. In contrast, the oscillators with low frequency, $\omega_j \ll \omega_p$, do not change their state. Assuming that the oscillators can be split into these two groups, we see that just after the $(\pi/2)_x$-pulse the state of the system is $\frac{1}{\sqrt{2}}\left(|\uparrow\rangle\otimes|\mathrm{g}_\uparrow^\mathrm{h}\rangle + |\downarrow\rangle\otimes|\mathrm{g}_\downarrow^\mathrm{h}\rangle\right)\otimes|\mathrm{g}_\uparrow^\mathrm{l}\rangle$ where the superscripts 'h' and 'l' refer to high- and low-frequency oscillators, respectively. Thus, we arrive at an initial state with only the low-frequency oscillators factorized from the spin. For the off-diagonal element of the density matrix we obtain

$$|\langle\sigma_+(t)\rangle| = Z(\omega_c, \omega_p)|P_{\omega_p}(t)|\,, \tag{17}$$

where $Z(\omega_c, \omega_p) \equiv |\langle g_\uparrow^\mathrm{h}|g_\downarrow^\mathrm{h}\rangle|$ and $P_{\omega_p}(t)$ is given by the same expressions as before, except that the high-frequency cut-off is reduced to ω_p.

The high frequency oscillators still contribute to the reduction of $|\langle\sigma_+(t)\rangle|$ – via the factor $Z(\omega_c, \omega_p)$ – however, this effect is reversible. To illustrate this we consider the following continuation of the experiment. After the $(\pi/2)$ pulse we allow for a free evolution of the system during time t with magnetic field $B_z = \Delta E$ along z axis. Then we apply a $(-\pi/2)$ pulse and measure σ_z. Without dissipation the result would be $\langle\sigma_z\rangle = \cos(\Delta E t)$. With dissipation the state of the system after time t is

$$\frac{1}{\sqrt{2}}\left(e^{i\Delta Et/2}|\uparrow\rangle\otimes|\mathrm{g}_\uparrow^\mathrm{h}\rangle\otimes|\mathrm{g}_\uparrow^\mathrm{l}\rangle + e^{-i\Delta Et/2}|\downarrow\rangle\otimes|\mathrm{g}_\downarrow^\mathrm{h}\rangle\otimes e^{-i\mathcal{H}_\downarrow t/\hbar}|\mathrm{g}_\uparrow^\mathrm{l}\rangle\right)\,. \tag{18}$$

After the $(-\pi/2)$ pulse (also of width $\pi/2\omega_p$) we obtain the following state:

$$\begin{aligned}&\frac{1}{2}|\uparrow\rangle\otimes|\mathrm{g}_\uparrow^\mathrm{h}\rangle\otimes\left(e^{i\Delta Et/2}|\mathrm{g}_\uparrow^\mathrm{l}\rangle + e^{-i\Delta Et/2}e^{-i\mathcal{H}_\downarrow t/\hbar}|\mathrm{g}_\uparrow^\mathrm{l}\rangle\right) + \\ &\frac{1}{2}|\downarrow\rangle\otimes|\mathrm{g}_\downarrow^\mathrm{h}\rangle\otimes\left(-\,e^{i\Delta Et/2}|\mathrm{g}_\uparrow^\mathrm{l}\rangle + e^{-i\Delta Et/2}e^{-i\mathcal{H}_\downarrow t/\hbar}|\mathrm{g}_\uparrow^\mathrm{l}\rangle\right)\,.\end{aligned} \tag{19}$$

From this we finally get $\langle\sigma_z\rangle = \mathrm{Re}\left[P_{\omega_p}(t)e^{-i\Delta Et}\right]$. Thus the amplitude of the coherent oscillations of σ_z is reduced only by the factor $|P_{\omega_p}(t)|$ associated with slow oscillators. The high frequency factor $Z(\omega_c, \omega_p)$ does not appear. To interpret this result we note that we could have discussed the experiment in terms of renormalized spins $|\tilde{\uparrow}\rangle \equiv |\uparrow\rangle|g^{\mathrm{h}}_\uparrow\rangle$ and $|\tilde{\downarrow}\rangle \equiv |\downarrow\rangle|g^{\mathrm{h}}_\downarrow\rangle$, and assuming that the high frequency cutoff of the bath is ω_p.

It is interesting to compare further the two scenarios with instantaneous and finite-time preparation further. The time evolution after the instantaneous preparation is governed by $P_{\omega_c}(t)$. For $T \ll \omega_p$ and $t \gg 1/\omega_p$ and arbitrary spectral density $J(\omega)$ it satisfies the following relation: $P_{\omega_c}(t) = Z^2(\omega_c, \omega_p)P_{\omega_p}(t)$, which follows from $\langle \mathrm{g}^{\mathrm{h}}_\uparrow|\exp(-i\mathcal{H}_\downarrow t/\hbar)|\mathrm{g}^{\mathrm{h}}_\uparrow\rangle \to |\langle g^{\mathrm{h}}_\uparrow|g^{\mathrm{h}}_\downarrow\rangle|^2$ for $t \gg 1/\omega_p$. Thus for the instantaneous preparation the reduction due to the high frequency oscillators is equal to $Z^2(\omega_c, \omega_p)$, while a look at the finite-time preparation result (17) shows that in this case the reduction is weaker, given by a single power of $Z(\omega_c, \omega_p)$ only. Moreover, in the slow preparation experiment the factor $Z(\omega_c, \omega_p)$ originates from the overlap of two 'simple' wave functions, $|\mathrm{g}^{\mathrm{h}}_\uparrow\rangle$ and $|\mathrm{g}^{\mathrm{h}}_\downarrow\rangle$, which can be further adiabatically manipulated, as described above, and this reduction can be recovered. This effect is to be interpreted as a renormalization. On the other hand, for the instantaneous preparation the high frequency contribution to the dephasing originates from the overlap of the states $|\mathrm{g}^{\mathrm{h}}_\uparrow\rangle$and $e^{-i\mathcal{H}_\downarrow t/\hbar}|\mathrm{g}^{\mathrm{h}}_\uparrow\rangle$. The latter is a complicated excited state of the bath with many nonzero amplitudes evolving with different frequencies. There is no simple (macroscopic) way to reverse the dephasing associated with this state. Thus we observe that the time scale of the manipulating pulses determines the border between the oscillators responsible for dephasing and the oscillators responsible for renormalization.

6. Response functions

In the limit $\theta = 0$ we can also calculate exactly the linear response of $\tau_x = \sigma_x$ to a weak magnetic field in the x-direction, $B_x(t)$:

$$\chi_{xx}(t) = \frac{i}{\hbar}\,\Theta(t)\langle\tau_x(t)\tau_x(0) - \tau_x(0)\tau_x(t)\rangle\,. \tag{20}$$

Using the equilibrium density matrix

$$\hat{\rho}^{\mathrm{eq}} = (1 + e^{-\beta\Delta E})^{-1}\left[|\uparrow\rangle\langle\uparrow| \otimes \hat{\rho}_\uparrow + e^{-\beta\Delta E}|\downarrow\rangle\langle\downarrow| \otimes \hat{\rho}_\downarrow\right]\,, \tag{21}$$

where $\hat{\rho}_\uparrow \propto \exp(-\beta\mathcal{H}_\uparrow)$ is the bath density matrix adjusted to the spin state $|\uparrow\rangle$, and similar for $\hat{\rho}_\downarrow$, we obtain the susceptibility

$$\chi_{xx}(t) = -\frac{2\hbar^{-1}\Theta(t)}{1 + e^{-\beta\Delta E}}\,\mathrm{Im}\left[P_{\omega_c}(t)e^{-i\Delta Et} + e^{-\beta\Delta E}P_{\omega_c}(t)e^{i\Delta Et}\right]\,. \tag{22}$$

Thus, to calculate the response function one has to use the full factor $P_{\omega_c}(t)$, corresponding to the instantaneous preparation of an initial state. This can be understood by looking at the Kubo formula (20). The operator $\tau_x(0)$ flips only the bare spin without touching the oscillators, as if an infinitely sharp $(\pi/2)$ pulse was applied.

The imaginary part of the Fourier transform of $\chi(t)$, which describes dissipation, is

$$\chi''_{xx}(\omega) = \frac{1}{2(1+e^{-\beta\Delta E})}\left[P(\hbar\omega - \Delta E) + e^{-\beta\Delta E}P(\hbar\omega + \Delta E)\right] - ...(-\omega)\,. \tag{23}$$

At $T = 0$ and positive values of ω we use the expression for $P(E)$ from Ref. [1] to obtain

$$\chi''_{xx}(\omega) = \frac{1}{2}P(\hbar\omega - \Delta E) = \Theta(\hbar\omega - \Delta E)\frac{e^{-2\gamma\alpha}(\hbar\omega_c)^{-2\alpha}}{2\Gamma(2\alpha)}(\hbar\omega - \Delta E)^{2\alpha-1}\,. \tag{24}$$

We observe that the dissipative part χ''_{xx} has a gap ΔE, which corresponds to the minimal energy needed to flip the spin, and a power-law behavior as ω approaches the threshold. This behavior of $\chi''_{xx}(\omega)$ is known from the *orthogonality catastrophe* scenario [5]. It implies that the ground state of the oscillator bath for different spin states, $|g_\uparrow\rangle$ and $|g_\downarrow\rangle$, are *macroscopically* orthogonal. In particular, for an Ohmic bath we recover the behavior typical for the problem of X-ray absorption in metals [5].

As $\chi''(\omega)$ characterizes the dissipation in the system (absorption of energy from the perturbing magnetic field) it is interesting to understand the respective roles of high- and low-frequency oscillators. We use the spectral decomposition for χ'' at $T = 0$,

$$\chi''(\omega) = \pi\sum_n |\langle 0|\tau_x|n\rangle|^2\left[\delta(\omega - E_n) - \delta(\omega + E_n)\right]\,, \tag{25}$$

where n denotes exact eigenstates of the system. These are $|\uparrow\rangle|n_\uparrow\rangle$ and $|\downarrow\rangle|n_\downarrow\rangle$, where $|n_\uparrow\rangle$ and $|n_\uparrow\rangle$ denote the excited (multi-oscillator) states of the Hamiltonians $\mathcal{H}_\uparrow$ and $\mathcal{H}_\downarrow$. The ground state is $|\uparrow\rangle|g_\uparrow\rangle$ and the only excited states that contribute to $\chi''(\omega)$ are $|\downarrow\rangle|n_\downarrow\rangle$ with $\mathcal{H}_\downarrow|n_\downarrow\rangle = (\omega - \Delta E)|n_\downarrow\rangle$. This means that all the oscillators with frequencies $\omega_j > \omega - \Delta E$ have to be in the ground state. Therefore we obtain for $\omega_p > \omega - \Delta E$

$$\chi''_{\omega_c}(\omega) = Z^2(\omega_c, \omega_p)\chi''_{\omega_p}(\omega)\,. \tag{26}$$

To interpret this result we generalize the coupling to the magnetic field by introducing a g factor: $H_{\text{int}} = -(g/2)\delta B_x(t)\sigma_x$. Then, if the applied magnetic field can be independently measured, the observable quantity corresponding, e.g., to the energy absorption is

$$\chi_{xx}(t) = g^2\frac{i}{\hbar}\,\Theta(t)\langle[\sigma_x(t), \sigma_x(0)]\rangle\,. \tag{27}$$

Thus, Eq. (26) tells us that by measuring the response of the spin at frequencies $\omega < \omega_p + \Delta E$ we cannot distinguish between a model with upper cutoff ω_c and $g = 1$ and a model with cutoff ω_p and $g = Z(\omega_c, \omega_p)$. This is the usual situation in the renormalization group context. Thus, again, we note that high-frequency oscillators are naturally associated with renormalization effects.

We collect the results for χ'' for various spectra in Table 5. The results are shown for temperatures lower and higher than the crossover temperature introduced in Section 4.

7. Summary

The examples presented above show that for a quantum two-state system with a non-equilibrium initial state, described by a factorized initial density matrix, dephasing persists down to zero bath temperature. An Ohmic environment leads to a power-law dephasing at $T = 0$, while a sub-Ohmic bath yields exponential dephasing. The reason is that the factorized initial state, even with the bath in the ground state of the bath Hamiltonian, is actually a superposition of many excited states of the total coupled system. In a real experiment only a part of the environment, the oscillators with low frequencies, can be prepared factorized from the two level system. These oscillators still lead to dephasing, whereas the high-frequency oscillators lead to renormalization effects. The examples demonstrate that experimental conditions, e.g., details of the system's initial state preparation, determine which part of the environment contributes to dephasing and which part leads to renormalization. The finite preparation time $\sim 1/\omega_p$ also introduces a natural high-frequency cutoff in the description of dephasing effects. We have further demonstrated that dephasing and renormalization effects influence the response functions of the two level system. We noted that they exhibit features known for the orthogonality catastrophe, including a power-law divergence above a threshold.

8. Acknowledgments

We thank Y. Makhlin for valuable contributions to the present work and M. Büttiker, M.H. Devoret, D. Es:eve, Y. Gefen, D. Golubev, Y. Imry, D. Loss, A.D. Mirlin, A. Rosch, U. Weiss, R. Whitney, and A.D. Zaikin for stimulating discussions. The work is part of the EU IST Project SQUBIT and of the **CFN** (Center for Functional Nanostructures) which is supported by the DFG (German Science Foundation).

TABLE V. Response functions for different bath spectra.

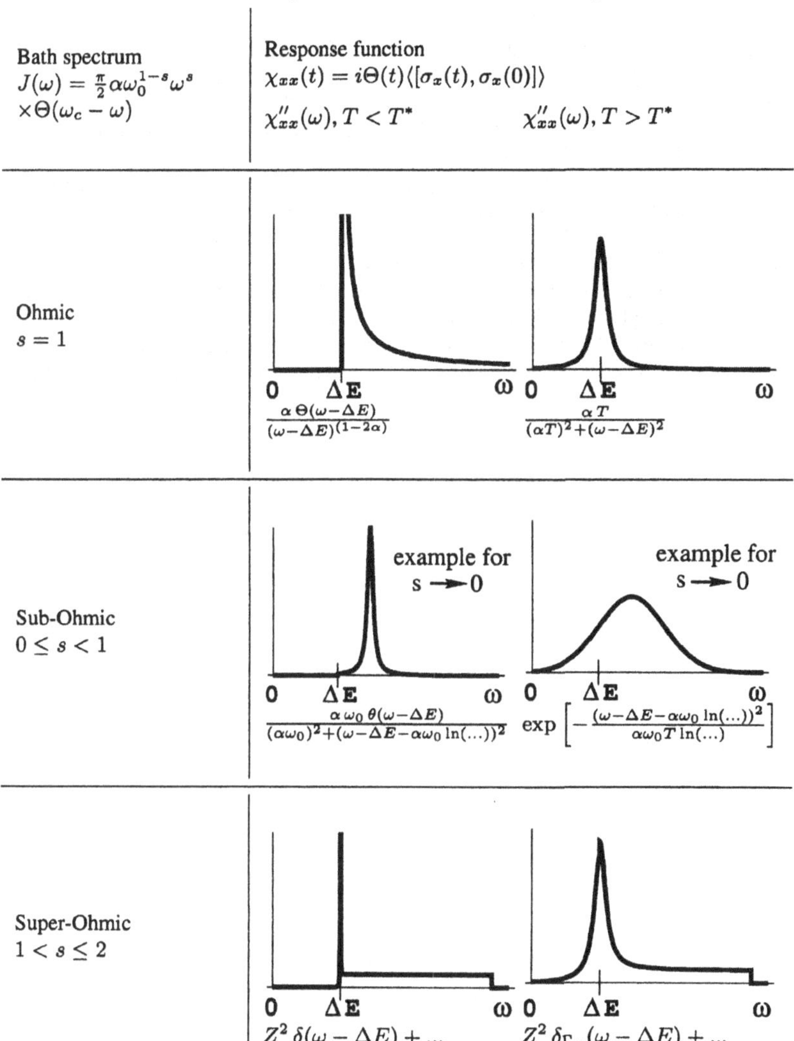

References

1. Devoret, M. H., D. Esteve, H. Grabert, G. L. Ingold, and H. Pothier: 1990, 'Effect of the electromagnetic environment on the Coulomb blockade in ultrasmall tunnel junctions'. *Phys. Rev. Lett.* **64**, 1824.
2. Harris, R. A. and L. Stodolsky: 1982, 'Two state systems in media and Turing's paradox'. *Phys. Lett. B* **116**, 464.

3. Kehrein, S. K. and A. Mielke: 1996, 'On the spin-boson model with a sub-Ohmic bath'. *Phys. Lett. A* **219**, 313.
4. Leggett, A., S. Chakravarty, A. Dorsey, M. Fisher, A. Garg, and W. Zwerger: 1987, 'Dynamics of the dissipative two-state system'. *Rev. Mod. Phys.* **59**, 1.
5. Mahan, G. D.: 1990, *Many-Particle physics*. New York: Plenum Press, 2nd edition.
6. Yu. V. Nazarov: 1989, 'Anomalous current-voltage characteristics of tunnel junctions'. *Sov. Phys. JETP* **68**, 561.
7. Odintsov, A. A.: 1988, 'Effect of dissipation on the characteristics of small-area tunnel junctions: application of the polaron model'. *Sov. Phys. JETP* **67**, 1265.
8. Panyukov, S. V. and A. D. Zaikin: 1988, 'Quantum fluctuations and quantum dynamics of small Josephson junctions'. *J. Low Temp. Phys.* **73**, 1.
9. Schoeller, H. and G. Schön: 1994, 'Mesoscopic quantum transport: resonant tunneling in the presence of strong Coulomb interaction'. *Phys. Rev. B* **50**, 18436.
10. Unruh, W. G.: 1995, 'Maintaining coherence in quantum computers'. *Phys. Rev. A* **51**, 992.
11. Weiss, U.: 1999, *Quantum dissipative systems*. Singapore: World Scientific, 2nd edition.

PART THREE

Full Counting Statistics

THE STATISTICAL THEORY OF MESOSCOPIC NOISE

A short review

L.S. LEVITOV
Department of Physics, Massachusetts Institute of Technology, 77 Massachusetts Ave, Cambridge, MA 02139

1. Introduction

The measurement performed by optical detectors, such as photon counters, is extended in the time domain, which makes it sensitive to temporal correlations of photons [1]. It has been known long ago in the theory of photodetection [2] that understanding photon counting distribution is essentially a problem of many-particle statistics. Similar considerations apply to the electrical noise measurement, which is fundamentally different from photodetection in that the electrons are not destroyed but just counted. The noise measurement, very much like photodetection, is a sensitive probe of temporal correlations between different transmitted electrons.

Fermi correlations in the electron noise were originally studied by Lesovik [3] (see also Ref. [4]) in a point contact, and then by Büttiker [5] in multiterminal systems, and by Beenakker and Büttiker [6] in mesoscopic conductors. Kane and Fisher proposed to employ the shot noise for detecting fractional quasiparticles in a Quantum Hall edge system [7]. Subsequent theoretical developments are summarized in a recent review [8].

Experimental studies of the shot noise, after first measurements in a point contact by Reznikov et al. [9] and Kumar et al. [10], focused on the quantum Hall regime. The fractional charges $e/3$ and $e/5$ were observed [11–13] at incompressible Landau level filling (see also recent work on noise at intermediate filling [14]). The shot noise in a mesoscopic conductor was observed by Steinbach et al. [15] and Schoelkopf et al. [16], who also studied noise in photo-assisted phase-coherent mesoscopic transport [17].

Y.V. Nazarov (ed.), Quantum Noise in Mesoscopic Physics, 373–396.

In this article we discuss counting statistics of electric noise and consider the probability distribution of charge transmitted in a fixed time interval [18, 19]. This distribution provides detailed information about current fluctuations. The counting statistics have been studied for the DC transport of free fermions [18, 20], the photo-assisted transport [21], the parametrically driven transport [19, 22], and in the mesoscopic regime [23] (also, see a review [24]). Nazarov developed Keldysh formalism for the counting statistics problem and applied it to mesoscopic transport in a weak localization regime [25] and, together with Bagrets, in a multiterminal geometry [26]. Charge doubling due to Andreev scattering in NS junctions was considered by Muzykantskii and Khmelnitskii [27], and in mesoscopic NS systems by Belzig and Nazarov [28]. Andreev and Kamenev [29] studied the problem of mesoscopic pumping in view of the results of Ref. [19]. Taddei and Fazio discussed counting statistics of entangled electron sources [30]. Statistics of transport in a Coulomb blockade regime was studied by Bagrets and Nazarov [31]. Photon statistics was considered by Beenakker and Schomerus [32] and Kindermann et al. [33]. The problem of back influence of a charge detector on current fluctuations in the context of counting statistics measurement was studied by Nazarov and Kindermann [37].

The possibility of measuring counting statistics using a fast charge integrator scheme was considered recently [34]. From the measured distribution all moments of charge fluctuations can be calculated and, conversely, the knowledge of all moments is in principle sufficient for recovering the full distribution. However, due to the central limit theorem, high moments are probably difficult to access experimentally. Therefore recent literature focused primarily on the third moment. It was found that the third moment obeys a generalized Schottky relation which holds in the tunneling regime at both high and low temperature, but involves a temperature-dependent Fano factor in the mesoscopic regime [22, 35]. Gutman and Gefen [35] studied the third moment using a sigma model approach, while Nagaev [36] demonstrated that all moments are correctly reproduced by an extension of the Boltzmann-Langevin kinetic equation.

In this article, after introducing the counting distribution (Sec.2), we review its microscopic definition based on a passive charge detector (Sec.2.1). In Sec.2.2 we study the statistics of tunneling in a generic many-body system. From the microscopic approach of Sec.2.1 we derive a bidirectional Poisson distribution for tunneling current, obtain a Schottky-like relation for the third moment and discuss its robustness. Then we briefly discuss the relation of the counting statistics theory and the theory of photo-detection (Sec.2.3). Then we proceed to the problem of mesoscopic transport. In Sec.3.1 we review the results on the DC transport and the derivation of a functional determinant formula for the counting distribution generating function. In Sec.3.3 we review the work on the AC transport statistics, and then consider the problem of mesoscopic pumping (Sec.3.4). The counting statistics for generic pumping strategy at weak pumping is given by a bidirectional

Poisson distribution. We show that the Fano factor varies between 0 and 1 as a function of the pumping fields phase difference.

2. General approach

The transmitted charge distribution can be characterized [18, 19] by electron counting probabilities p_n, usually accumulated in a generating function[1]

$$\chi(\lambda) = \sum e^{in\lambda} p_n \,. \tag{1}$$

The function $\chi(\lambda)$ is 2π-periodic in the *counting field* λ and has the property $\chi(0) = 1$ which follows from the probability normalization $\sum_n p_n = 1$. The term "counting field" will be motivated in Sec.2.1, where a microscopic definition of $\chi(\lambda)$ is discussed in which λ appears as a field that couples current fluctuations to a charge detector.

The generating function (1) is particularly well suited for characterizing statistics of the distribution p_n. The so-called *irreducible correlators* $\langle\langle \delta n^k \rangle\rangle$ (also known as *cummulants*) are expressed in terms of $\chi(\lambda)$ as

$$\ln \chi(\lambda) = \sum_{k=1}^{\infty} m_k \frac{(i\lambda)^k}{k!} \,, \qquad m_k \equiv \langle\langle \delta n^k \rangle\rangle \tag{2}$$

The first two correlators in (2) give the mean and the variance:

$$m_1 = \overline{n}, \qquad m_2 = \overline{\delta n^2} = \overline{n^2} - \overline{n}^2, \tag{3}$$

where $\overline{f(n)}$ stands for $\sum_n f(n) p_n$. The third correlator[2]

$$m_3 = \langle\langle \delta n^3 \rangle\rangle \equiv \overline{\delta n^3} = \overline{(n - \overline{n})^3} \tag{4}$$

characterizes the asymmetry (or skewness) of the distribution p_n.

To illustrate the notion of a generating function, let us consider a Poisson process. It describes charge transport at very low transmission, with subsequent transmission events uncorrelated. For the Poisson distribution

$$p_k = \begin{cases} e^{-\bar{n}}\, \bar{n}^k / k! & k \geq 0 \\ 0 & k < 0 \end{cases} \quad \text{and} \quad \chi(\lambda) = \exp\left((e^{i\lambda} - 1) \bar{n} \right) \tag{5}$$

with $\bar{n} = It/e$ the average number of particles transmitted during the time interval t, where I is the time-averaged current and e is elementary charge. Comparing (5)

[1] The function $\chi(\lambda)$ is also called a characteristic function [1].

[2] The relation between cummulants and correlators is generally more complicated than Eq.(4) for the third cummulant. For example, $m_4 = \langle\langle \delta q^4 \rangle\rangle = \overline{\delta q^4} - 3\left(\overline{\delta q^2}\right)^2$.

with (2), one finds that all cummulants of the Poisson distribution are identical: $m_k = \bar{n}$.

Another useful example is binomial statistics. A binomial distribution arises when a fixed number N of independent attempts to transmit particles is made, each attempt successful or unsuccessful with probabilities p and $q = 1 - p$. The probability to transmit k particles in this case is determined by the number of k successful outcomes, equal to $C_N^k = N!/(N-k)!k!$ each. The probability distribution and the generating function in this case are

$$p_k = C_N^k p^k q^{N-k} \quad \text{and} \quad \chi(\lambda) = \left(pe^{i\lambda} + q\right)^N \tag{6}$$

The cummulants of the binomial distribution can be found from (2) by expanding $\ln \chi$ in λ:

$$m_1 = pN, \quad m_2 = pqN, \quad m_3 = pq(q-p)N, \quad ... \tag{7}$$

The binomial distribution (6) describes counting distribution of DC current noise for a single channel scatterer, such as point contact, at zero temperature (Sec.3.1).

Statistically independent processes result in a generating function given by a product of generating functions for constituing processes: $\chi(\lambda) = \chi_1(\lambda)...\chi_k(\lambda)$. For example, consider *a biderectional Poisson distribution* defined as a mixture of two independent Poisson processes transmitting particles in opposite directions with the rates $\bar{n}$ and $\bar{n}'$. In this case,

$$\chi_{2P}(\lambda) = \exp\left((e^{i\lambda} - 1)\bar{n}\right) \cdot \exp\left((e^{-i\lambda} - 1)\bar{n}'\right) \tag{8}$$

In Sec.2.2 we use the distribution (8) to describe statistics of tunneling current. In Sec.3.4 we show that it describes noise a mesoscopic pump.

2.1. A MICROSCOPIC REPRESENTATION OF $\chi(\lambda)$

Here we discuss a microscopic definition of counting statistics for a physical system. It will be shown that the generating function can be written as a certain partition function in Keldysh field theory. Adopting an inductive approach, we shall start with a specific model of a current detector [20, 24]. We obtain $\chi(\lambda)$ for a particular current measurement scheme, and then argue that the result is universal.

In a realistic noise measurement, e.g. in a mesoscopic wire or a point contact, the current fluctuations are not detected directly. Instead, the measurement is performed on the electromagnetic fluctuations (basically, voltage noise) induced by current fluctuations in the system. The electromagnetic fluctuations have to be amplified before detection. The conversion of underlying microscopic fluctuations due to fermions (electrons, fractional charges, etc.) into fluctuations of bosons (photons) is crucial, since Bose fields can be amplified without compromising noise statistics, while Fermi statistics does not allow amplification.

Our goal is a microscopic picture of current fluctuations, rather than a description of realistic measurements. Thus we choose a *gedanken* measurement scheme well suited for that purpose. Consider a spin $1/2$ placed near an electron system and magnetically coupled to the electric current. We restrict the coupling to the spin z component, so that the system in the presence of the spin is described by $\mathcal{H}(\mathbf{q}, \mathbf{p} - \mathbf{a}\sigma_3)$, where $\mathcal{H}(\mathbf{q}, \mathbf{p})$ is the electron Hamiltonian and $\mathbf{a}(r)$ is the spin vector potential scaled by e/c.

The scheme of current detection using the dynamics of a spin coupled to current can be motivated quasiclassically. A spin coupled to a time-dependent classical current $I(t)$ by the interaction $\mathcal{H} = \frac{1}{2}\lambda\sigma_3 I(t)$ will precess at the rate proportional to current, which turns the spin into an analog galvanometer. Indeed, if the spin-current coupling is turned on at $t = 0$, the spin will start precessing around the z axis with the precession angle $\theta(t) = \lambda \int_0^t I(t')dt'$ equal to the transmitted charge times λ. The coupling constant λ, so far arbitrary, will be associated with counting field below.

In a fully quantum-mechanical problem, the spin evolution can be obtained from $i\dot{\sigma} = [\sigma, \mathcal{H}]$. Since the spin-current coupling Hamiltonian commutes with σ_3, the spin dynamics can be found explicitly. For that we consider the transverse spin components $\sigma_\pm \equiv \sigma_1 \pm i\sigma_2$ and write the evolution equation $i\dot{\sigma}_\pm = [\sigma_\pm, \mathcal{H}]$ in the form

$$i\dot{\sigma}_+ = \sigma_+ P_\downarrow \mathcal{H}(\mathbf{q}, \mathbf{p} + \mathbf{a}) - \mathcal{H}(\mathbf{q}, \mathbf{p} - \mathbf{a}) P_\uparrow \sigma_+ \,, \tag{9}$$

$$i\dot{\sigma}_- = \sigma_- P_\uparrow \mathcal{H}(\mathbf{q}, \mathbf{p} - \mathbf{a}) - \mathcal{H}(\mathbf{q}, \mathbf{p} + \mathbf{a}) P_\downarrow \sigma_- \,, \tag{10}$$

with $P_{\uparrow,\downarrow} = \frac{1}{2}(1 + \sigma_3)$ the up and down spin projectors. Here we used the raising/lowering properties of the operators $\sigma_\pm$ and replaced σ_3 by 1 to the left of σ_+ and by -1 to the right of σ_+ (and similarly for the σ_- equation).

We consider a measurement which is performed during time interval $0 < t' < t$, i.e. start with a spin decoupled from the electron system, $\mathbf{a}_{t<0} = 0$, maintain a finite coupling during $0 < t' < t$, and then make it zero again at $t' > t$. The expectation value of the transverse spin component at the time t, found by integrating Eqs. (9),(10), is

$$\langle \sigma_+(t) \rangle \;=\; \langle e^{i\mathcal{H}(\mathbf{q}, \mathbf{p} - \mathbf{a})t} e^{-i\mathcal{H}(\mathbf{q}, \mathbf{p} + \mathbf{a})t} \rangle_{\rm el} \, \langle \sigma_+(0) \rangle_{\rm spin} \,, \tag{11}$$

while $\langle \sigma_-(t) \rangle = \langle \sigma_+(t) \rangle^*$. Note that the result of the coupled spin and current evolution factors into a product of quantities that depend separately on electron dynamics and on the initial state of the spin, as indicated by the subscripts.

The effect of current on spin precession is described by the dependence of the first term in Eq.(11) on the gauge field $\mathbf{a}$. To make contact with the quasiclassical discussion above, let us expand $\mathcal{H}$ in $\mathbf{a}$, assuming it to be small. The result is

$\mathcal{H}(\mathbf{q}, \mathbf{p} \pm \mathbf{a}) = \mathcal{H}(\mathbf{q}, \mathbf{p}) \pm \mathbf{aj}$, where $\mathbf{j}$ is electric current. Substituting this back in Eq.(11) and passing to the interaction representation with respect to the Hamiltonian of fermions uncoupled from the spin, we rewrite the average $\langle ... \rangle_{\rm el}$ in (11) as

$$\left\langle \mathrm{Texp}\left(i \int_t^0 \int \mathbf{aj}(t') d^3r\, dt' \right) \mathrm{Texp}\left(-i \int_0^t \int \mathbf{aj}(t') d^3r\, dt' \right) \right\rangle_{\rm el} \tag{12}$$

Let us consider a specific form of $\mathbf{a}$, taking it to be a pure gauge, $\oint \mathbf{a} d\mathbf{l} = 0$, within the electron system, and nonzero near a particular surface (e.g. a δ–function on the surface). For a classical current, ignoring noncommutativity of current operators at different times, the expression (12) becomes

$$\left\langle e^{-i\theta(t)} \right\rangle_{\rm el} \quad \text{with} \quad \theta(t) = \lambda \int_0^t I(t') dt'. \tag{13}$$

Here $I(t)$ is the total current through the surface and $\lambda = -2 \int \mathbf{a} d\mathbf{l}$, where the integral is taken across the surface. The form of Eq.(13) agrees with what one expects for the precession phase factor averaged over classical current fluctuations. For n electrons transmitted through the system during the measurement time, the precession angle is $\theta(t) = \lambda n$. Thus one can relate the average in (13) with the transmitted charge distribution as

$$\left\langle e^{-i\theta(t)} \right\rangle_{\rm el} = \sum_n e^{i\lambda n} p_n \tag{14}$$

The relation with the spin precession in this case can be seen more clearly by combining the result (14) with Eq.(11), which gives

$$\langle \sigma_+(t) \rangle = \sum_n p_n \left(e^{i\lambda n} \langle \sigma_+(0) \rangle_{\rm spin} \right), \tag{15}$$

This way of writing the result of spin evolution confirms the expected relationship between the charge counting probability distribution and the distribution of the spin precession angles.

This discussion clarifies the meaning of the quantity $\langle ... \rangle_{\rm el}$ in Eq.(11), linking it with the counting distribution generating function. Motivated by this, we use this quantity to give a microscopic definition of counting statistics. We rewrite the quantity $\langle ... \rangle_{\rm el}$ as a Keldysh partition function

$$\chi(\lambda) = \left\langle \mathrm{T_K} \exp\left(-i \int_{C_{0,t}} \hat{\mathcal{H}}_\lambda(t') dt' \right) \right\rangle_{\rm el} \tag{16}$$

with the integral taken over the Keldysh time contour $C_{0,t} \equiv [0 \to t \to 0]$, first forward and then backward in time. The counting field λ is related to the gauge field $\mathbf{a}$ via $\lambda = \mp\frac{1}{2} \int \mathbf{a} d\mathbf{l}$, where the integration path goes across the region where scattering takes place and noise is generated (e.g. across the barrier in the point contact). The sign $\mp$ indicates that the field $\mathbf{a}$ is antisymmetric on the upper and lower parts of the Keldysh contour. Because of that, even though $\mathbf{a}$ resembles in many ways an ordinary electromagnetic gauge field (allowing for gauge transformations, etc.), it has no such meaning. We emphasize that $\mathbf{a}$ is really an auxiliary field describing coupling with a *virtual* measurement device, such as the spin $1/2$ above.

The microscopic formula (16), originating from the analysis of a coupling with spin $1/2$, is in fact adequate for any ideal "passive charge detector" without internal dynamics. We shall use this formula below to obtain counting statistics for several physical situations of interest, including tunneling and mesoscopic transport.

This still leaves some questions about universality and limitations of Eq.(16). Nazarov and Kindermann [37] considered a more general scheme of charge detection and recovered the expression (16). Although this is reassuring, Ref. [37] concludes that the detector back action is inevitable. Thus it is still desirable to study more realistic models of noise detection that include conversion of microscopic current noise into electromagnetic field (photons) as well as an amplifier. The electron-to-photon noise conversion was partially addressed by Beenakker and Schomerus [32].

It is also of interest to compare the back action effects in different models. We argue that the above scheme is likely to describe noise measurement with the least back action, since coupling to the precessing spin $1/2$ affects only the phase of electron forward scattering amplitude, without changing scattering probabilities.

2.2. STATISTICS OF THE TUNNELING CURRENT

The problem of the tunneling current noise provides a simple test for the microscopic formula (16). The starting point of our analysis will be the tunneling Hamiltonian $\hat{\mathcal{H}} = \hat{\mathcal{H}}_1 + \hat{\mathcal{H}}_2 + \hat{V}$, where $\hat{\mathcal{H}}_{1,2}$ describe the leads and $\hat{V} = \hat{J}_{12} + \hat{J}_{21}$ is the tunneling operator. The specific form of the operators $\hat{J}_{12}$, $\hat{J}_{21}$ that describe tunneling of a quasiparticle between the leads will not be important for the most of our discussion. Both the discussion and the results for the tunneling current statistics obtained in this section are valid for a generic interacting many-body system.

The counting field $\mathbf{a}$ in this case can be taken localized on the barrier, entering the Hamiltonian through the phase factors $\exp(\pm i \int \mathbf{a} d\mathbf{l}) = \exp(\pm i\lambda/2)$ of the operators $\hat{J}_{12}$, $\hat{J}_{21}$. The tunneling operator then is

$$\hat{V}_\lambda = e^{\frac{i}{2}\lambda(t)} \hat{J}_{12}(t) + e^{-\frac{i}{2}\lambda(t)} \hat{J}_{21}(t) \tag{17}$$

Here $\lambda(t) = \pm\lambda$ is antisymmetric on the Keldysh contour $C_{0,t}$.

In what follows we compute $\chi(\lambda)$ and establish a relation with the Kubo theorem for the tunneling current [39]. For that, we perform the usual gauge transformation turning the bias voltage into the tunneling operator phase factor as $\hat{J}_{12} \to \hat{J}_{12}e^{-ieVt}$, $\hat{J}_{21} \to \hat{J}_{21}e^{ieVt}$. Passing to the Keldysh interaction representation, we write

$$\chi(\lambda) = \left\langle \mathrm{T_K} \exp\left(-i \int_{C_{0,t}} \hat{V}_{\lambda(t')}(t')dt'\right)\right\rangle \tag{18}$$

Diagrammatically, the partition function (18) is a sum of linked cluster diagrams with appropriate combinatorial factors. To the lowest order in the tunneling operators $\hat{J}_{12}$, $\hat{J}_{21}$ we only need to consider linked clusters of the second order. This gives

$$\chi(\lambda) = e^{W(\lambda)}, \quad W(\lambda) = -\frac{1}{2}\iint_{C_{0,t}} \left\langle \mathrm{T_K} \hat{V}_{\lambda(t')}(t')\hat{V}_{\lambda(t'')}(t'')\right\rangle dt'dt'' \tag{19}$$

This result is correct for the measurement time t much larger than the correlation time in the contacts that determines the characteristic time separation $t' - t''$ at which the correlator in (19) decays.

There are several different contributions to the integral in (19), arising from t' and t'' taken on the forward or backward parts of the contour $C_{0,t}$. Evaluating them separately, we obtain

$$\begin{aligned} W(\lambda) = & \int_0^t\int_0^t \left\langle \hat{V}_{-\lambda}(t')\hat{V}_{\lambda}(t'')\right\rangle dt''dt' \\ & - \int_0^t\int_0^{t'} \left\langle \hat{V}_{\lambda}(t')\hat{V}_{\lambda}(t'')\right\rangle dt''dt' - \int_0^t\int_{t'}^{t} \left\langle \hat{V}_{-\lambda}(t')\hat{V}_{-\lambda}(t'')\right\rangle dt''dt' \end{aligned} \tag{20}$$

We substitute expression (17) in Eq.(20) and average by pairing $\hat{J}_{12}$ with $\hat{J}_{21}$. This gives

$$W(\lambda) = (e^{i\lambda}-1)N_{1\to 2}(t) + (e^{-i\lambda}-1)N_{2\to 1}(t) \tag{21}$$

with

$$N_{1\to 2} = \int_0^t\int_0^t \langle \hat{J}_{21}(t')\hat{J}_{12}(t'')\rangle\, dt'dt'', \tag{22}$$

$$N_{2\to 1} = \int_0^t\int_0^t \langle \hat{J}_{12}(t')\hat{J}_{21}(t'')\rangle\, dt'dt'' \tag{23}$$

where $N_{j\to k}(t) = g_{jk}t$ is the mean charge number transmitted from the contact j to the contact k in a time t. Exponentiating Eq.(21) gives nothing but the bidirectional Poisson distribution $\chi_{2P}(\lambda)$ defined by Eq.(8) with the transition rates given by $\bar{n} = N_{1\to 2} = g_{12}t$, $\bar{n}' = N_{2\to 1} = g_{21}t$, respectively.

Eq.(8) yields interesting relations between different statistics of the distribution. The cummulants $\langle\langle \delta n^k \rangle\rangle$, obtained by expanding $\ln \chi_{2P}(\lambda)$ in λ, are

$$m_k = \langle\langle \delta n^k \rangle\rangle = \begin{cases} (g_{12} - g_{21})t, & \text{k } odd \\ (g_{12} + g_{21})t, & \text{k } even \end{cases} \tag{24}$$

Setting $k = 1, 2$ we express $g_{12} \pm g_{21}$ through the time-averaged current and the low frequency noise spectral density[3]:

$$g_{12} - g_{21} = I/q_0, \quad g_{12} + g_{21} = \mathcal{S}_0/2q_0^2 \tag{25}$$

with q_0 the tunneling charge. Of special interest for us will be the third correlator of the transmitted charge

$$\langle\langle \delta q^3 \rangle\rangle \equiv \overline{\left(\delta q - \overline{\delta q}\right)^3} \tag{26}$$

For this correlator Eq. (24) gives $\langle\langle \delta q^3 \rangle\rangle = C_3 t$ with the coefficient C_3 ("spectral power") related to the current I as

$$C_3 \equiv \langle\langle \delta q^3 \rangle\rangle / t = q_0^2 I \tag{27}$$

We note that the relation (27) holds for the distribution (8) at any ratio $(g_{12} - g_{21})/(g_{12} + g_{21})$ of the mean transmitted charge to the variance.

The quantities (22), (23) have several general properties. First, by writing the expectation values (22), (23) in a basis of exact microscopic states and using the detailed balance relation, we obtain

$$N_{1\to 2}/N_{2\to 1} \equiv g_{12}/g_{21} = \exp(eV/k_{\rm B}T) \tag{28}$$

where V is the voltage applied between the contacts. Using this result to calculate the ratio of the first and second cummulants, Eq.(25), we have $(g_{12} - g_{21})/(g_{12} + g_{21}) = \mathrm{th}(eV/k_{\rm B}T)$. This gives the noise-current relation

$$\mathcal{S}_0 = 2q_0 \coth(eV/k_{\rm B}T)\, I \tag{29}$$

that holds for arbitrary $eV/k_{\rm B}T$. This relation was pointed out by Sukhorukov and Loss [38].

Also, one can establish a relation of the quantities (22), (23) with the Kubo theorem. We consider the tunneling current operator

$$\hat{\mathcal{I}}(t) = -iq_0 \left(\hat{J}_{12}(t) - \hat{J}_{21}(t)\right). \tag{30}$$

From the Kubo theorem for the tunneling current [39], the mean integrated current $\int_0^t \langle \hat{\mathcal{I}}(t') \rangle dt'$ is

$$q_0 \int_0^t \int_0^t \left\langle \left[\hat{J}_{21}(t'), \hat{J}_{12}(t'')\right] \right\rangle dt' dt'' = q_0 \left(N_{1\to 2} - N_{2\to 1}\right) \tag{31}$$

[3] The spectral density of the noise is defined through the symmetrized current correlator as $\mathcal{S}_\omega = \int \overline{\left\{\delta I(t), \delta I(0)\right\}_+} e^{i\omega t} dt$. At $\omega = 0$, one can thus write $\mathcal{S}_0$ in terms of the variance of charge $q(\tau) = \int_t^{t+\tau} I(t')dt'$ transmitted during a long time τ as $\mathcal{S}_0 = \frac{2}{\tau}\overline{\delta q^2(\tau)}$.

By writing $N_{j\to k} = g_{jk}t$, we confirm the first relation (25). To obtain the second relation (25) we consider the variance of the charge transmitted in time t, given by $\langle\langle \delta q^2 \rangle\rangle = q_0^2 \int_0^t\int_0^t \langle \{\hat{\mathcal{I}}(t'), \hat{\mathcal{I}}(t'')\}_+ \rangle\, dt'dt''$. This integral can be rewritten as

$$\int_0^t\int_0^t \left\langle \left\{ \hat{J}_{12}(t'), \hat{J}_{21}(t'') \right\}_+ \right\rangle dt'dt'' = N_{1\to 2} + N_{2\to 1} \tag{32}$$

which immediately leads to the second relation (25).

We conclude that the tunneling current statistics, described by Eq.(8), are simpler than in a generic system. The current-noise relation, typically known in a generic system only at equilibrium (Nyquist) and in the fully out-of-equilibrium (Schottky) regimes, for the tunneling current is given by Eq.(29) at arbitrary eV/k_BT.

In contrast, the relation (27) obeyed by the third correlator (4) is completely insensitive to the crossover between the Nyquist and Schottky noise regimes. The meaning of Eq.(27) is similar to that of the Schottky formula $S_0 = 2\langle\langle \delta q^2 \rangle\rangle = 2q_0 I$. However, the Schottky current-noise relation is valid only when charge flow is unidirectional, i.e. at low temperatures $k_BT \ll eV$, since $g_{12}/g_{21} = \exp(eV/k_BT)$, while Eq.(27) holds at any eV/k_BT.

In experiment, when the current-noise relation is used to determine the tunneling quasiparticle charge q_0 from the tunneling current noise, it is crucial to maintain low temperature $k_BT \ll eV$. The requirement of a cold sample at a relatively high bias voltage is the origin of a well known difficulty in the noise measurement. In contrast, the relation (27) is not constrained by any requirement on sample temperature.

This property of the third moment, if confirmed experimentally, may prove to be quite useful for measuring quasiparticle charge. In particular, this applies to the situations when the $I - V$ characteristic is strongly nonlinear, when it is usually difficult to unambiguously interpret the noise *versus* current dependence as a shot noise effect or as a result of thermal noise generated by non-linear conductance. This appears to be a completely general problem pertinent to any interacting system. Namely, in the systems such as Luttinger liquids, the $I - V$ nonlinearities arise at $eV \geq k_BT$. However, it is exactly this voltage that has to be applied for measuring the shot noise in the Schottky regime.

Finally, we note that the temperature-independence of the third moment is specific for the tunneling problem. In other situations, such a point contact or a mesoscopic system, the third moment is not universal [34–36].

2.3. A RELATION TO THE THEORY OF PHOTODETECTION

We have not specified the statistics of tunneling particles in the above discussion, since everything said so far is applicable for bosons as well as for fermions. To illustrate this, here we discuss the relation of the above to the theory of photon counting [1, 2]. A system of photons interacting with atoms in a photon detector can be described by a Hamiltonian of the form $\mathcal{H} = \mathcal{H}_{\rm p} + \mathcal{H}_{\rm a} + V$, where $\mathcal{H}_{\rm p}$ describes free electromagnetic field, $\mathcal{H}_{\rm a}$ is the Hamiltonian of atoms in the detector, and

$$V = \sum_{j,k} \left(u_{j,k} e^{\frac{i}{2}\lambda} b_j^\dagger a_k + u_{j,k}^* e^{-\frac{i}{2}\lambda} a_k^\dagger b_j \right) \tag{33}$$

describes the interaction of photons with the atoms, i.e. the process of photon absorption and atom excitation. Here a_k are the canonical Bose operators of photon modes, labeled by k, and b_j are the operators describing excitation of the atoms. Since the operator V transfers excitations between the field and the atom systems, it can be interpreted as a "tunneling operator." (The only difference is in the unidirectional character of "current" due to the coupling V, since photons can be only absorbed in the detector but not created.) This analogy allows one to use the formalism of Sec.2.1 to study photon counting, and for that purpose we added a counting field in (33) (compare to Eq.(17)).

Given all that, the generating function for photons has the form (18) which we rewrite to show an explicit dependence on the measurement time:

$$\chi_t(\lambda) = \left\langle \mathcal{U}_{-\lambda}^{-1}(t)\,\mathcal{U}_\lambda(t) \right\rangle, \qquad \mathcal{U}_\lambda(t) = \mathrm{T}\exp\left(-i\int_0^t \hat{V}_\lambda(t')dt'\right), \tag{34}$$

The task of evaluating the partition function (34) is simplified by the weakness of the photon-atom coupling. This means that each atom is excited during the counting time t with a very small probability. The expression (34) can thus be evaluated by taking into account the interaction of a photon with each of the atoms only to the lowest order. This is also similar to the tunneling problem.

However, at this stage the similarity with tunneling ends, since photon coherence time can be much longer than the measurement time t. The method of Sec.2.2, based on the linked cluster expansion of $\ln\chi$, should be modified to account for the long coherence times. Another complication is that the photon density matrix is not specified, since we are not limiting the discussion to thermal photon sources.

We handle the partition function (34) by averaging over atoms, while keeping the photon variables free. As explained above, only pairwise averages of atoms' operators are needed. We write them as $\langle b_j^\dagger(t) b_{j'}(t')\rangle = 0$, $\langle b_j(t) b_{j'}^\dagger(t')\rangle = \tau_j\delta(t-t')\delta_{jj'}$, where τ_j is a constant of the order of the excitation time of an atom, and the δ-function is actually a function of the width $\simeq \tau_j$. (Typically, τ_j is a very short, microscopic time.)

Turning to the calculation, let us consider the difference $\chi_{t+\Delta}(\lambda) - \chi_t(\lambda)$, with the time increment Δ large compared to τ_j, but much smaller than the characteristic photon coherence time. Expanding $\mathcal{U}_\lambda(t+\Delta)$ to the second order in Δ, we write it as

$$\left(1 - i\int_t^{t+\Delta} V_\lambda(t')dt' - \frac{1}{2}\int_t^{t+\Delta}\int_t^{t'} V_\lambda(t')V_\lambda(t'')dt'dt''\right)\mathcal{U}_\lambda(t) \tag{35}$$

Substituting this in Eq.(34) along with a similar expression for $\mathcal{U}_{-\lambda}^{-1}(t+\Delta)$, and averaging over the atoms as described above, we obtain

$$\partial_t \chi_t(\lambda) = (\chi_{t+\Delta}(\lambda) - \chi_t(\lambda))/\Delta = \sum_k \eta_k (e^{i\lambda} - 1) \left\langle \mathcal{U}_{-\lambda}^{-1}(t) a_k^\dagger a_k \mathcal{U}_\lambda(t) \right\rangle \quad (36)$$

with $\eta_k = \sum_j \tau_j |u_{j,k}|^2$ the detector efficiency parameters. The solution of Eq.(36) has the form well known in optics [1, 2] :

$$\chi_t(\lambda) = \prod_k \chi_t^{(k)}(\lambda), \quad \chi_t^{(k)}(\lambda) = \left\langle : \exp\left(\eta_k t (e^{i\lambda} - 1) a_k^\dagger a_k\right) : \right\rangle_k, \quad (37)$$

where : ... : is the normal ordering symbol and $\langle ... \rangle_k$ is the averaging over the photon density matrix. The product in (37) indicates that the counting distributions for different electromagnetic modes are statistically independent. The normal ordering physically means that each photon, after having been detected, is absorbed in the detector and destroyed.

From Eq.(37), the counting probability of m photons in one mode is

$$p_m^{(k)} = \frac{(\eta_k t)^m}{m!} \langle : (a_k^\dagger a_k)^m e^{-\eta_k t a_k^\dagger a_k} : \rangle_k . \quad (38)$$

Eqs.(37,38) is the central result of the theory of photon counting [1]. Particularly interesting is the case of a coherent photon state $|z\rangle$, $a|z\rangle = z\,|z\rangle$, with complex z, corresponding to the radiation field of an ideal laser. In this case Eq.(38) yields the Poisson distribution $p_m = e^{-Jt}(Jt)^m/m!$, $J = \eta |z|^2$, which describes the so-called minimally bunched light sources.

3. Counting statistics of mesoscopic transport

Here we consider the problem of counting statistics in a mesoscopic transport. From now on we adopt the noninteracting particle approximation and use the scattering approach [40], in which the system is characterized by a single particle scattering matrix. Depending on the nature of the problem, the matrix can be stationary or time-dependent. Even for noninteracting particles the problem of counting statistics remains nontrivial due to correlations between different particles arising from Fermi statistics.

The counting statistics can be analyzed using the microscopic formula (16). However, there is a more efficient way of handling the noninteracting problem. One can obtain a formula for the generating function $\chi(\lambda)$ in terms of a functional determinant that involves the scattering matrix and the density matrix of reservoirs. Then for each particular problem one must analyze and evaluate an appropriate determinant. Although functional determinants can be nontrivial to deal with, this approach is still much simpler than the one based directly on Eq.(16).

3.1. STATISTICS OF THE DC TRANSPORT

He we discuss the problem of time-independent scattering. We consider a conductor with m scattering channels describing states within one or several current leads. The scattering is elastic and will be characterized by a $m \times m$ matrix S. Although in applications so far the 2×2 matrices (i.e. the problems with two channels) have been more common than larger matrices, the general determinant structure of $\chi(\lambda)$ will be revealed only for matrices of arbitrary size m.

In this case one can obtain $\chi(\lambda)$ from a quasiclassical argument. For elastic scattering, particles with different energies contribute to counting statistics independently, and thus one can "symbolically" write

$$\chi(\lambda) = \prod_{\epsilon} \chi_\epsilon(\lambda)\,, \quad \text{i.e.} \quad \chi(\lambda) = \exp\left(t \int \ln \chi_\epsilon(\lambda) \frac{d\epsilon}{2\pi\hbar}\right)\,, \tag{39}$$

where $\chi_\epsilon(\lambda)$ is the contribution of particles with energy ϵ. The factor $2\pi\hbar$ is written based on the quasiclassical phase space volume normalization, $d\mathcal{V} = d\epsilon dt/2\pi\hbar$. The quantity $\chi_\epsilon(\lambda)$ depends on the scattering matrix S and on the energy distribution $n_i(\epsilon)$ in the channels.

To obtain $\chi_\epsilon(\lambda)$ we introduce a vector of counting fields λ_j, $j = 1, ..., m$, one for each channel, and consider all possible multi-particle scattering processes at fixed energy. The processes can involve any number $k \leq m$ of particles each coming out of one of the m channels and being scattered into another channel. Since the particles are indistinguishable fermions, no two particles can share an incoming or outgoing channel. One can then write $\chi_\epsilon(\lambda)$ as a sum over all different multiparticle scattering processes:

$$\chi_\epsilon(\lambda) = \sum_{i_1,...,i_k,j_1,...,j_k} e^{\frac{i}{2}(\lambda_{i_1}+...+\lambda_{i_k}-\lambda_{j_1}-...-\lambda_{j_k})} P_{i_1,...,i_k\,|\,j_1,...,j_k}\,, \tag{40}$$

where the rate of k particles transition from channels $i_1, ..., i_k$ into channels $j_1, ..., j_k$ is given by

$$P_{i_1,...,i_k\,|\,j_1,...,j_k} = \left|S^{j_1,...,j_k}_{i_1,...,i_k}\right|^2 \prod_{i \neq i_\alpha} (1 - n_i(\epsilon)) \prod_{i = i_\alpha} n_i(\epsilon)\,. \tag{41}$$

Here $S^{j_1,...,j_k}_{i_1,...,i_k}$ is an antisymmetrized product of k single particle amplitudes, which is nothing but the minor of the matrix S with rows $j_1, ..., j_k$ and columns $i_1, ..., i_k$. The product of n_i and $1 - n_i$ gives the probability to have k particles come out of the channels $i_1, ..., i_k$.

An important insight in the structure of the expression (40) can be obtained by noting that it has a form of a determinant:

$$\chi_\epsilon(\lambda) = \det\left(\hat{1} - \hat{n}_\epsilon + \hat{n}_\epsilon S^{-1}_{-\lambda} S_\lambda\right)\,. \tag{42}$$

Here $\hat{n}_\epsilon$ is a diagonal $m \times m$ matrix of channel occupancy at energy ϵ and the matrix S_λ has matrix elements $e^{\frac{i}{4}(\lambda_j - \lambda_i)} S_{ji}$ with the counting field λ_j in the phase factors. To demonstrate that the expressions (40) and (42) are identical one has to expand the determinant (42) and go through a bit of matrix algebra. The formula (42) is particularly useful because, as we shall see below, it can be generalized to a time-dependent problem.

Let us now focus on the simplest case $m = 2$ which describes transport in a point contact, with the two channels corresponding to current leads. The 2×2 scattering matrix S contains reflection and transmission amplitudes. In this case, since there are only six terms in Eq.(40), the determinant formula (42) is not necessary. In this case Eq.(40) takes the form

$$\begin{aligned} \chi_\epsilon(\lambda) &= (1 - n_1)(1 - n_2) + (|S_{11}|^2 + e^{\frac{i}{2}(\lambda_2 - \lambda_1)}|S_{21}|^2) n_1 (1 - n_2) \\ &+ (|S_{22}|^2 + e^{\frac{i}{2}(\lambda_1 - \lambda_2)}|S_{12}|^2) n_2 (1 - n_1) + |\det S|^2 n_1 n_2 \,, \end{aligned} \quad (43)$$

where the energy dependence of $n_j(\epsilon)$ is suppressed. By using the unitarity relations $|S_{1i}|^2 + |S_{2i}|^2 = 1$, $|\det S| = 1$, Eq.(43) can be simplified:

$$\chi_\epsilon(\lambda) = 1 + p(e^{i\lambda} - 1)) n_1 (1 - n_2) + p(e^{-i\lambda} - 1)) n_2 (1 - n_1) \quad (44)$$

Here $p = |S_{21}|^2 = |S_{12}|^2$ is the transmission coefficient and $\lambda = \lambda_2 - \lambda_1$. (We denote transmission by p instead of the traditional t to avoid confusion with the measurement time.)

To obtain the full counting statistics integrated over all energies, one has to specify the energy distribution in the leads and use Eq.(39). We consider a barrier with energy-independent transmission and the leads at temperature T biased by voltage V. Then $n_{1,2} = n_F(\epsilon \mp eV/2)$ with n_F the Fermi function.

At $T = 0$, since $n_F(\epsilon)$ takes values 0 and 1, for $V > 0$ we have

$$\chi_\epsilon(\lambda) = \begin{cases} e^{i\lambda} p + 1 - p \,, & |\epsilon| < \frac{1}{2} eV \\ 1 \,, & |\epsilon| > \frac{1}{2} eV \end{cases} \quad (45)$$

Doing the integral in Eq.(39) we obtain a binomial distribution $\chi_N(\lambda) = (e^{i\lambda} p + 1 - p)^N$ of the form (6) with the number of attempts $N(t) = eVt/2\pi\hbar$.

This means that, in agreement with intuition, in the energy window eV the transport is just the single particle transmission and reflection, while the states with energies in the Fermi sea, populated in both reservoirs, are noiseless. (At $V < 0$ the result is similar, with $e^{i\lambda}$ replaced by $e^{-i\lambda}$, which corresponds to the DC current sign reversal.)

We note that the noninteger number of attempts $N(t) = eVt/2\pi\hbar$ is an artifact of a quasiclassical calculation. In a more careful analysis the number of attempts is characterized by a narrow distribution P_N peaked at $\overline{N} = N(t)$, and the generating function is a weighted sum $\sum_N P_N \chi_N(\lambda)$. Since the peak width is a sublinear function of the measurement time t (in fact, $\overline{\delta N^2} \propto \ln t$), the statistics to the leading order in t are correctly described by the binomial distribution.

One can also consider the problem at arbitrary $k_{\rm B}T/eV$. The integral in Eq.(39), although less trivial, can still be carried out analytically, giving

$$\chi(\lambda) = \exp\left(-u_+ u_- N_T\right) , \quad N_T = t\, k_{\rm B}T/2\pi\hbar, \tag{46}$$

where

$$u_\pm = v \pm {\rm ch}^{-1}(p\,{\rm ch}(v + i\lambda) + (1-p)\,{\rm ch}\, v), \quad v = eV/2k_{\rm B}T\,. \tag{47}$$

At low temperature $k_{\rm B}T \ll eV$, the expression (46) reproduces the binomial statistics. At low voltage $eV \ll k_{\rm B}T$ (or high temperature) Eq.(46) gives the counting statistics of the equilibrium Nyquist noise:

$$\chi(\lambda) = e^{-\lambda_*^2 N_T} , \qquad \sin(\lambda_*/2) = p^{1/2} \sin(\lambda/2)\,. \tag{48}$$

Interestingly, even at equilibrium the noise is non-gaussian, except for a special case of fully transmitting system, $p = 1$, $\lambda_* = \lambda$, when it is gaussian.

3.2. STATISTICS OF TIME-DEPENDENT SCATTERING

The time-dependent scattering problem describes photon-assisted transport. There are two groups of practically interesting problems: the AC-driven systems with static scattering potential, such as tunneling barriers or point contacts in the presence of a microwave field [17, 41], and the electron pumps with time-dependent scattering potential controlled externally, e.g. by gate voltages [42].

Typically, the time of individual particle transit through the scattering region is much shorter than the period at which the system is driven. This situation is described, in the instantaneous scattering approximation, by a time-dependent scattering matrix $S(t)$ that characterizes single particle scattering at time t. The question of interest is how Fermi statistics of many-body scattering states affects the counting statistics.

One can construct a theory of counting statistics of time-dependent scattering [19] by generalizing the results of Sec.3.1 for the statistics of a generic time-independent scattering. In particular, the determinant formula (42), along with (39), allows a straightforward extension to the time-dependent case. In that, the generating function $\chi(\lambda)$ acquires a form of a functional determinant.

Let us consider a scattering matrix $S(t)$ varying periodically in time with the external (pumping) frequency Ω. The analysis is most simple in the frequency representation [19], in which the scattering operator S has off-diagonal matrix elements $S_{\omega',\omega}$ with a discrete frequency change $\omega' - \omega = n\Omega$. In this approach the energy axis is divided into intervals $n\Omega < \omega < (n{+}1)\Omega$ and each such interval is treated as a separate conduction channel. In doing so it is convenient (and some times necessary) to assign a separate counting field λ_n to each frequency channel, so that the counting field may acquire frequency channel index in addition to the conduction channel dependence discussed above (see (42)).

Since the scattering operator conserves energy modulo multiple of $\hbar\Omega$, the scattering is *elastic* in the extended channel representation, which allows to employ the method of Sec.3.1. Now, we note that the form of the determinant in Eq.(42) is not particularly sensitive to the size of the scattering matrix. Thus one can use it even when the number of channels is infinite, provided that the determinant remains well defined. This procedure brings (42) to the form of a determinant of a matrix with an infinite number of rows and columns. This matrix is then truncated at very high and low frequencies, eliminating empty states and the states deep in the Fermi sea which do not contribute to noise.

Finally, we note that the product rule (39) for $\chi(\lambda)$ at all energies is consistent with the determinant structure, since scattering processes at energies different modulo $n\Omega$ are decoupled. This allows to keep the answer for $\chi(\lambda)$ in the form of the determinant (42), where now the scattering operator S is considered in the entire frequency domain, rather than at discrete frequencies $\omega + n\Omega$. The resulting functional determinant has a simple form in the time representation:

$$\chi(\lambda) = \det\left(\hat{1} + n(t,t')\left(\hat{T}_\lambda(t) - \hat{1}\right)\right), \quad \hat{T}_\lambda(t) = S^\dagger_{-\lambda}(t) S_\lambda(t), \tag{49}$$

where $(S_\lambda)_{ji} = e^{\frac{i}{4}(\lambda_j - \lambda_i)} S_{ji}$ as above, and $\hat{n}$ is the density matrix of reservoirs. The operator $\hat{n}$, diagonal in the channel index, is given by

$$n_j(t,t') = \int n_j(\hbar\omega)\, e^{i\omega(t'-t)}\, d\omega/2\pi \tag{50}$$

In general $\hat{n}$ depends on the energy distribution parameters, such as temperature and bias voltage. In equilibrium, at finite T and V, by taking Fourier transform of a Fermi function $n_F(\epsilon - eV)$, one obtains

$$n(t,t') = \frac{e^{-i\frac{e}{\hbar}V(t-t')}}{2\beta\,\mathrm{sh}(\pi(t-t'+i\delta)/\beta)}, \quad \beta = \hbar/k_B T \tag{51}$$

In equilibrium, at $T = 0$, $V = 0$, this gives $n(t,t') = i/(2\pi(t-t'+i\delta))$. The result (49) holds for an arbitrary (even nonequilibrium) energy distribution in reservoirs.

The functional determinant of an infinite matrix (49) should be handled carefully. One can show that, in a mathematical sense, the quantity (49) is well defined. For the states with energies deep in the Fermi sea, $\hat{n} = 1$ and, since $\det\left(\hat{T}_\lambda(t)\right) = 1$ due to unitarity of S, these states do not contribute to the determinant (49). Similarly, since $\hat{n} = 0$ for the states with very high energy, these states also do not affect the determinant. Effectively, the determinant is controlled by a group of states near the Fermi level, in agreement with intuition about transport in a driven system. The absence of ultraviolet divergences allows one to go freely between different representations, e.g. to switch from the frequency domain to the time domain, which facilitates calculations [24, 21].

The above derivation of the formula (49) based on a generalization of the result (42) for time-independent scattering might seem not entirely rigorous. A more mathematically sound derivation that starts directly from the microscopic expression (16) was proposed recently by Klich [43].

3.3. CASE STUDIES

Here we briefly review the time-dependent scattering problems for which the counting statistics have been studied. From several examples for which $\chi(\lambda)$ has been obtained it appears that the problem does not allow a general solution. Instead, the problem can be handled only for suitably chosen form of the time dependence $S(t)$.

In Ref. [19] a two channel problem was considered with $S(t)$ of the form

$$S(\tau) \equiv \begin{pmatrix} r & t' \\ t & r' \end{pmatrix} = \begin{pmatrix} B + be^{-i\Omega\tau} & \bar{A} + \bar{a}e^{i\Omega\tau} \\ A + ae^{-i\Omega\tau} & -\bar{B} - \bar{b}e^{i\Omega\tau} \end{pmatrix} \tag{52}$$

which is unitary for $|A|^2 + |a|^2 + |B|^2 + |b|^2 = 1$, $A\bar{a} + B\bar{b} = 0$. The problem was solved by using the extended channel representation in the frequency domain, in which each frequency interval $n\Omega < \omega < (n+1)\Omega$ is treated as a separate scattering channel, as discussed above.

For the reservoirs at zero temperature and without bias voltage the charge distribution for m pumping cycles is described by

$$\chi(\lambda) = \left(1 + p_1(e^{i\lambda} - 1) + p_2(e^{-i\lambda} - 1)\right)^m \tag{53}$$

with $p_1 = |a|^4/(|a|^2 + |b|^2)$ and $p_2 = |b|^4/(|a|^2 + |b|^2)$. This result means that at each pumping cycle one electron is pumped in one direction with probability p_1, or in the opposite direction with probability p_2, or no charge is pumped with probability $1 - p_1 - p_2$. The multiplicative dependence of $\chi(\lambda)$ on the number of pumping cycles m means that the outcomes of different cycles are statistically independent. One can thus view (53) as a generalization of the binomial distribution (6).

The problem (52) was also studied in Ref. [19] at a finite bias voltage, when the counting distribution is not as simple as (53). To describe the result, for a given bias voltage V we find an integer n such that $nf < \frac{e}{h}V \le (n+1)f$, where $f = \Omega/2\pi$ is the cyclic frequency in (52). Then for a long measurement time $t \gg \Omega^{-1}$ the counting distribution is

$$\chi(\lambda) = \chi_n^{N_>}(\lambda) \cdot \chi_{n+1}^{N_<}(\lambda), \tag{54}$$

where $N_< = (\frac{e}{\hbar}V - nf)\,t$, $N_> = ((n+1)f - \frac{e}{\hbar}V)\,t$, and the functions $\chi_n(\lambda)$ are polynomials in $e^{\pm i\lambda}$ of finite degree. The form of $\chi_n(\lambda)$ depends on A, B, a, and b (we refer to Ref. [19] for details).

The product rule (54) means that the cummulants of the distribution $\chi(\lambda)$ depend on V in a piecewise linear way, $m_k(V) = N_> m_k^{(n)} + N_< m_k^{(n+1)}$, with cusp-like singularities at $eV = nhf \equiv n\hbar\Omega$. These singularities are generic for the noise in photo-assisted phase-coherent transport [44, 17, 41].

Another time-dependent problem for which solution can be obtained in a closed form is mesoscopic transport in the presence of an AC voltage [24]. The scatterer in this case is a time-independent 2×2 matrix, while the voltage $V(t)$ enters in the phase factors of the density matrix in (49):

$$n_{1,2}(t,t') = e^{\pm\frac{i}{2}(\varphi(t')-\varphi(t))} n_{1,2}^{(0)}(t,t'), \quad \dot\varphi(t) = \frac{e}{\hbar}V(t) \tag{55}$$

(compare this with the formula (51) for constant bias voltage).

The counting distribution (49) for a family of such problems has been studied in Ref. [21]. It was noted earlier [45] that noise is minimized at fixed transmitted charge for a special form of time-dependent voltage:

$$V(t) = \frac{h}{e} \sum_{k=1,\ldots,m} \frac{2\tau_k}{(t-t_k)^2 + \tau_k^2} \tag{56}$$

Each of the Lorentzian voltage pulses (56) corresponds to a 2π phase change in $\varphi(t)$. Interestingly, the noise-minimizing pulses (56) have large degeneracy: they produce noise which is insensitive to the pulses' widths τ_k and peak positions t_k. This calls for an interpretation of the pulses (56) as independent attempts to transmit charge. Not surprisingly, the counting statistics for such pulses was found to be binomial:

$$\chi(\lambda) = (1 + t(e^{i\lambda} - 1))^m \tag{57}$$

with t the transmission constant. The lowest possible noise for a current pumped by voltage pulses is thus equal to that of a DC current with the same transmitted charge.

The method of Ref. [21] also allows to find the distribution for an arbitrary sum of the pulses (56) with alternating signs. For example, two opposite pulses

$$V(t) = \frac{h}{e}\left(\frac{2\tau_1}{(t-t_1)^2 + \tau_1^2} - \frac{2\tau_2}{(t-t_2)^2 + \tau_2^2}\right) \tag{58}$$

give rise to the counting distribution

$$\chi(\lambda) = 1 - 2F + F(e^{i\lambda} + e^{-i\lambda}), \quad F = t(1-t)\left|\frac{z_1^* - z_2}{z_1 - z_2}\right|^2 \tag{59}$$

with $z_{1,2} = t_{1,2} + i\tau_{1,2}$. The quantity $|...|^2$ is a measure of pulses' overlap in time, varying between 0 for a full overlap and 1 for no overlap. For nonoverlapping pulses $\chi(\lambda)$ factors as $(t\,e^{i\lambda} + 1 - t)(t\,e^{-i\lambda} + 1 - t)$, in agreement with the interpretation of a binomial distribution for independent attempts.

3.4. MESOSCOPIC PUMPING

A DC current in a mesoscopic system, such as an open quantum dot, can be induced by pumping, i.e. by modulating its area, shape, or other parameters [46, 48, 47]. After pumping was demonstrated experimentally [42], it came into the focus of mesoscopic literature (for references we refer to [49]). In particular, Brouwer made an interesting observation that time averaged pumped current is a purely geometric property of the path in the scattering matrix parameter space, insensitive to path parameterization.

Transport through a mesoscopic is described [40] by a scattering matrix S which depends on externally driven parameters and varies cyclically with time. The matrix $S(t)$ defines a path in the space of all scattering matrices. For a system with m scattering channels, the matrix space is the group $U(m) = SU(m)\times U(1)$. In an experiment one can, in principle, realize any path in the space of scattering matrices.

Counting distribution for a parametrically driven open system was discussed by Andreev and Kamenev who adapted the results [19] obtained for specific pumping cycles (see Sec.3.3). However, since the relation between the path in the scattering matrix space and the external pumping parameters is generally unknown, only the results valid for generic paths are of interest in this problem.

Here we consider the weak pumping regime, when the path $S(t)$ is a sufficiently small, but otherwise arbitrary loop, and show that in this case the counting distribution is universal [22], having the form of bidirectional Poisson distribution (8). From that, we obtain the dependence of the noise on the amplitude and relative phase of the voltages driving the pump.

Before turning to the calculation, we discuss general dependence of counting statistics on the path in matrix space. Different paths $S(t)$, in principle, give rise to different current and noise. However, there is a remarkable property of invariance with respect to group shifts. Any two paths,

$$S(t) \quad \text{and} \quad S'(t) = S(t)S_0, \tag{60}$$

where S_0 is a time-independent matrix in $U(m)$, give rise to the same counting statistics at zero temperature. We note that only the right shifts of the form (60) leave counting statistics invariant, whereas the left shifts generally change it. One can explain the result (60) qualitatively as follows. The change of scattering matrix, $S(t) \to S'(t) = S(t)S_0$, is equivalent to replacing states in the *incoming* scattering channels by their superpositions $\psi^\alpha = S_{0\beta}^{\alpha}\psi^\beta$. At zero temperature, however, Fermi reservoirs are noiseless and also such are any their superpositions. Correlation between superposition states of noiseless reservoirs is negligible, while current fluctuations arise only during the time-dependent scattering.

Therefore, noise statistics remain unchanged. A simple formal proof of the result (60) is given below.

For a weak pumping field we shall evaluate (49) in the time domain by expanding $\ln\det(...)$ in powers of δS and keeping non-vanishing terms of lowest order. In doing so, however, we preserve full functional dependence on λ which gives all moments of counting statistics. We write $S(t) = e^{A(t)}S^{(0)}$ with antihermitian $A(t)$ representing small perturbation, $\mathrm{tr}A^{\dagger}A \ll 1$. Here $S^{(0)}$ is scattering matrix of the system in the absence of pumping. Substituting this into (49) one obtains

$$\hat{T}_\lambda(t) \equiv \hat{T}^{(0)}_\lambda + \delta T_\lambda(t) = S^{(0)\dagger}_{-\lambda} e^{-A_{-\lambda}(t)} e^{A_\lambda(t)} S^{(0)}_\lambda \tag{61}$$

with $\hat{T}^{(0)}_\lambda = S^{(0)\dagger}_{-\lambda} S^{(0)}_\lambda$ and $A_\lambda(t) = e^{i\frac{\lambda}{4}\sigma_3} A(t) e^{-i\frac{\lambda}{4}\sigma_3}$. Now, we expand (49):

$$\ln\chi(\lambda) = \ln\det Q_0 + \mathrm{tr}R - \frac{1}{2}\mathrm{tr}R^2 + \frac{1}{3}\mathrm{tr}R^3 - ... \tag{62}$$

where $Q_0 = 1 + \hat{n}(\hat{T}^{(0)}_\lambda - 1)$ and $R = Q_0^{-1}\hat{n}\delta T_\lambda$. At zero temperature, from $\hat{n}^2 = \hat{n}$ it follows that $\det Q_0 = 1$ and $R = S^{(0)\,-1}_\lambda \hat{n}\left(e^{-A_{-\lambda}(t)} e^{A_\lambda(t)} - 1\right) S^{(0)}_\lambda$. Therefore,

$$\ln\chi(\lambda) = \mathrm{tr}\;\hat{n}\hat{M} - \frac{1}{2}\mathrm{tr}(\hat{n}\hat{M})^2 + \frac{1}{3}\mathrm{tr}(\hat{n}\hat{M})^3 - ... \tag{63}$$

where $\hat{M} = e^{-A_{-\lambda}(t)} e^{A_\lambda(t)} - 1$. Note that at this stage there is no dependence left on the constant matrix $S^{(0)}$, which proves the invariance under the group shifts (60).

We need to expand (63) in powers of the pumping field, which amounts to taking the lowest order terms of the expansion in powers of the matrix $A(t)$. One can check that the two $\mathcal{O}(A)$ terms arising from the first term in (63) vanish. The $\mathcal{O}(A^2)$ terms arise from the first and second term in (63) and have the form

$$\ln\chi = \frac{1}{2}\mathrm{tr}\left(\hat{n}\left(A^2_{-\lambda} + A^2_\lambda - 2A_{-\lambda}A_\lambda\right)\right) - \frac{1}{2}\mathrm{tr}(\hat{n}B_\lambda)^2 \tag{64}$$

with $B_\lambda(t) = A_\lambda(t) - A_{-\lambda}(t)$. At zero temperature, by using $\hat{n}^2 = \hat{n}$, one can bring (64) to the form

$$\frac{1}{2}\mathrm{tr}\;(\hat{n}\,[A_\lambda, A_{-\lambda}]) + \frac{1}{2}\left(\mathrm{tr}\left(\hat{n}^2 B^2_\lambda\right) - \mathrm{tr}(\hat{n}B_\lambda)^2\right) \tag{65}$$

The first term of (65) has to be regularized in the Schwinger anomaly fashion, by splitting points, $t', t'' = t \pm \epsilon/2$, which gives

$$\frac{1}{2}\oint n(t',t'')\mathrm{tr}\;(A_{-\lambda}(t'')A_\lambda(t') - A_\lambda(t'')A_{-\lambda}(t'))\,dt \tag{66}$$

Averaging over small ϵ can be achieved either by inserting in (66) additional integrals over t', t'', or simply by replacing $A_\lambda(t) \to \frac{1}{2}(A_\lambda(t) + A_\lambda(t'))$, etc. After taking the limit $\epsilon \to 0$, Eq.(66) becomes

$$\frac{i}{8\pi} \oint \mathrm{tr}\,(A_{-\lambda}\partial_t A_\lambda - A_\lambda \partial_t A_{-\lambda})\, dt \tag{67}$$

The second term of (65) can be written as

$$\frac{1}{4(2\pi)^2} \oint\oint \frac{\mathrm{tr}\,(B_\lambda(t) - B_\lambda(t'))^2}{(t-t')^2} dt dt' \tag{68}$$

Now, we decompose $A = a_0 + z + z^\dagger$, so that $[\sigma_3, a_0] = 0$, $[\sigma_3, z] = -2z$, $\left[\sigma_3, z^\dagger\right] = 2z^\dagger$. Then $A_\lambda \equiv e^{-i\frac{\lambda}{4}\sigma_3} A e^{i\frac{\lambda}{4}\sigma_3} = a_0 + e^{i\frac{\lambda}{2}} z^\dagger + e^{-i\frac{\lambda}{2}} z$, $B_\lambda = \left(e^{i\frac{\lambda}{2}} - e^{-i\frac{\lambda}{2}}\right) W$, $W \equiv z^\dagger - z$. Substituting this into (67) and (68) one finds that in terms of $W(t)$ these two expressions become[4]

$$\frac{\sin\lambda}{8\pi} \oint \mathrm{tr}\,([\sigma_3, W]\,\partial_t W)\, dt \tag{69}$$

and

$$\frac{(1-\cos\lambda)}{2(2\pi)^2} \oint\oint \frac{\mathrm{tr}\,(W(t) - W(t'))^2}{(t-t')^2} dt dt' \tag{70}$$

Hence the counting distribution for one pumping cycle is

$$\chi(\lambda) = \exp\left(u(e^{i\lambda} - 1) + v(e^{-i\lambda} - 1)\right) \tag{71}$$

with the transmitted charge average $I = e(u - v)$ and variance $J = e^2(u + v)$, related to the noise spectral density by $S_0 = J\Omega/\pi$ (see Sec.2.2).

The parameters u and v in (71) can be expressed through $z(t)$ and $z^\dagger(t)$ as follows. Let us write $z(t)$ as $z_+(t) + z_-(t)$, where $z_+(t)$ and $z_-(t)$ contain only positive or negative Fourier harmonics, respectively. Then

$$u = \frac{i}{4\pi} \oint \mathrm{tr}\left(z_-^\dagger \partial_t z_+ - z_+ \partial_t z_-^\dagger\right) dt = \sum_{\omega>0} \omega\, \mathrm{tr} z_{-\omega}^\dagger z_\omega, \tag{72}$$

$$v = \frac{i}{4\pi} \oint \mathrm{tr}\left(z_- \partial_t z_+^\dagger - z_+^\dagger \partial_t z_-\right) dt = -\sum_{\omega<0} \omega\, \mathrm{tr} z_{-\omega}^\dagger z_\omega \tag{73}$$

Note that $u \geq 0$ and $v \geq 0$. It is straightforward to show that (69) equals $i\sin\lambda(u - v)$, whereas (70) equals $(\cos\lambda - 1)(u + v)$.

[4] Eq.(69) is essentially identical to the result obtained by Brouwer for the average pumped current [48]. The integral in (69) is invariant under reparameterization, and thus has a purely geometric character determined by the contour $S(t)$ in $U(m)$.

Now we consider a single channel pump, $S(t) \in U(2)$. In this case, z and $z^\dagger$ are complex numbers. For harmonic driving signal, without loss of generality, one can write

$$z(t) = z_1 V_1 \cos(\Omega t + \theta) + z_2 V_2 \cos(\Omega t), \tag{74}$$

where $V_{1,2}$ are pumping signal amplitudes, and complex parameters $z_{1,2}$ depend on microscopic details. From (72) we find the Poisson rates

$$u = \frac{1}{4}\left|z_1 V_1 e^{i\theta} + z_2 V_2\right|^2, \; v = \frac{1}{4}\left|z_1 V_1 e^{-i\theta} + z_2 V_2\right|^2 \tag{75}$$

Note that u and v vanish at particular signal amplitudes ratio V_1/V_2 and phase θ. Once the two Poisson processes (71) are reduced to one, the current to noise ratio gives elementary charge, $I/J = \pm e^{-1}$. This happens at the extrema of I/J as a function of $w = (V_1/V_2)e^{i\theta}$, for (75) reached at $w = -z_2/z_1, \; -\bar{z}_2/\bar{z}_1$.

Reducing the counting statistics (71) to purely poissonian by varying pumping parameters is possible, in principle, for any number of channels n. However, since the number of parameters to be tuned is $2n^2$, this method is practical perhaps only for small channel numbers. Although the method is demonstrated for non-interacting fermions, we argue that it can be applied to interacting systems as well. Poisson statistics results from the absence of correlations of subsequently transmitted particles, which must be the case in a generic system, interacting or noninteracting, at small pumping current. Using the dependence of the rates u, v on the driving signal to maximize I/J, i.e. to eliminate one of the two Poisson processes (71), one could then obtain the charge quantum in the standard way as $e = J/I$.

To summarize the results of this section, in the weak pumping regime the distribution of charge transmitted per cycle is of bidirectional Poisson form (8), i.e., it is fully characterized by only two parameters, average current and noise. The current to noise ratio I/J, scaled by elementary charge, varies between 1 and -1, depending on the relation between driving signals phases and amplitudes. Thus the quantity $\max(|I|/J)$ gives the inverse of elementary charge *without any fitting parameters*. Polianski et al. [49] recently studied the dependence of I/J on the mesoscopic scattering ensemble parameters, and found that, within the random matrix theory, the nearly extremal values close to ± 1 can be reached with finite probability. This may permit to use the noise in a pump to measure quasiparticle charge in open systems, such as Luttinger liquids.

References

1. Mandel, L. and Wolf, E. (1995) *Optical Coherence and Quantum Optics*, Cambridge University Press, Cambridge
2. Glauber, R.J. (1963) Photon Correlations, *Phys. Rev. Lett.* **10**, pp. 84–86; The Quantum Theory of Optical Coherence, *Phys. Rev.* **130**, pp. 2529–2539

3. Lesovik, G.B. (1989) Excess Quantum Shot Noise in 2D Ballistic Point Contacts, *JETP Lett.*, **49**, pp. 592–594
4. Khlus, V.A. (1987) Current and Voltage Fluctuations in Microjunctions of Normal and Superconducting Metals, *Sov. Phys. JETP*, **66**, pp. 1243–1249
5. Büttiker, M. (1990) Scattering Theory of Thermal and Excess Noise in Open Conductors, *Phys. Rev. Lett.*, **65**, pp. 2901–2904
6. Beenakker, C.W.J. and Büttiker, M. (1992) Suppression of shot noise in metallic diffusive conductors, *Phys. Rev. B*, **46**, pp. 1889–1892
7. Kane, C.L. and Fisher, M.P.A. (1994) Nonequilibrium Noise and Fractional Charge in the Quantum Hall Effect, *Phys. Rev. Lett.*, **72**, pp. 724–727
8. Blanter, Ya.M. and Büttiker, M. (2000) Shot Noise in Mesoscopic Conductors, *Phys. Rep.* **336**, pp. 2-166
9. Reznikov, M., Heiblum, M., Shtrikman, H. and Mahalu, D. (1995) Temporal Correlation of Electrons: Suppression of Shot Noise in a Ballistic Quantum Point Contact, *Phys. Rev. Lett.*, **75**, pp. 3340–3343
10. Kumar, A., Saminadayar, L., Glattli, D.C., Jin, Y. and Etienne, B. (1996) Experimental Test of the Quantum Shot Noise Reduction Theory, *Phys. Rev. Lett.*, **76**, pp. 2778–2781
11. de-Picciotto, R., Reznikov, M., Heiblum, M., Umansky, V. Bunin, G. and Mahalu, D. (1997) Direct Observation of a Fractional Charge, *Nature*, **389**, pp. 162–164
12. Saminadayar, L., Glattli, D.C., Jin, Y. and Etienne, B. (1997) Observation of the *e/3* Fractionally Charged Laughlin Quasiparticles, *Phys. Rev. Lett.*, **79**, pp. 2526–2529
13. Reznikov, M., de-Picciotto, R., Griffiths, T.G., Heiblum, M. and Umansky, V. (1999) Observation of Quasiparticles with One-Fifth of an Electron's Charge, *Nature*, **399**, pp. 238–241
14. Griffiths, T.G., Comforti, E., Heiblum, M., Stern, A., Umansky, V. (2000) Evolution of Quasiparticle Charge in the Fractional Quantum Hall Regime, *Phys. Rev. Lett.*, **85**, pp. 3918–3921
15. Steinbach, A.H., Martinis, J.M., Devoret, M.H. (1996) Observation of Hot-Electron Shot Noise in a Metallic Resistor, *Phys. Rev. Lett.*, **76**, pp. 3806-3809
16. Schoelkopf, R.J., Burke, P.J., Kozhevnikov, A.A. and Prober, D.E. (1997) Frequency dependence of shot noise in a diffusive mesoscopic conductor, *Phys. Rev. Lett.*, **78**, pp. 3370–3373
17. Schoelkopf, R.J., Kozhevnikov, A.A., Prober, D.E. and Rooks, M. (1998) Observation of "Photon-Assisted" Shot Noise in a Phase-Coherent Conductor, *Phys. Rev. Lett.*, **80**, pp. 2437–2440
18. Levitov, L.S. and Lesovik, G.B. (1993) Charge Distribution in Quantum Shot Noise, *JETP Lett.*, **58**, pp. 230–235
19. Ivanov, D.A. and Levitov, L.S. (1993) Statistics of Charge Fluctuations in Quantum Transport in an Alternating Field, *JETP Lett.*, **58**, pp. 461–468
20. Levitov, L.S. and Lesovik, G.B. (1994) Quantum Measurement in Electric Circuit, cond-mat/9401004, unpublished
21. Ivanov, D.A., Lee, H.W., Levitov, L.S. (1997) Coherent States of Alternating Current, *Phys. Rev.*, **B56**, pp. 6839–6850
22. Levitov, L.S. (2001) Counting Statistics of Charge Pumping in an Open System, cond-mat/0103617, unpublished
23. Lee, H.W., Levitov, L.S. and Yakovets, A.Yu. (1995) Universal Statistics of Transport in Disordered Conductors" *Phys. Rev. B*, **51**, pp. 4079–4083
24. Levitov, L.S., Lee, H.W. and Lesovik, G.B. (1996) Electron counting statistics and coherent states of electric current," *J. of Math. Phys.*, **37**, pp. 4845–4866
25. Nazarov, Yu.V. (1999) Universalities of Weak Localization, *Ann. Phys. (Leipzig)*, **8**, SI-193; see also: cond-mat/9908143
26. Nazarov, Yu.V. and Bagrets, D.A. (2002) Circuit Theory for Full Counting Statistics in Multi-Terminal Circuits, *Phys. Rev. Lett.*, **88**, pp. 196801-196804

27. Muzykantskii, B.A. and Khmelnitskii, D.E. (1994) Quantum Shot Noise in a NS Point Contact, *Phys. Rev. B*, **50**, pp. 3982–5483
28. Belzig, W. and Nazarov, Yu.V. (2001) Full Current Statistics in Diffusive Normal-Superconductor Structures, *Phys. Rev. Lett.*, **87**, pp. 067006–067009
29. Andreev, A.V. and Kamenev, A. (2000) Counting Statistics of an Adiabatic Pump, *Phys. Rev. Lett.*, **85**, pp. 1294–1297
30. Taddei, F. and Fazio, R. (2002) Counting Statistics for Entangled Electrons, *Phys. Rev. B*, **65**, pp. 075317–075323
31. Bagrets, D.A. and Nazarov, Yu.V. (2002) Full Counting Statistics of Charge Transfer in Coulomb Blockade Systems, cond-mat/0207624
32. Beenakker, C.W.J. and Schomerus, H. (2001) Counting Statistics of Photons Produced by Electronic Shot Noise, *Phys. Rev. Lett.* **86**, pp. 700–703
33. Kindermann, M., Nazarov, Yu.V. and Beenakker, C.W.J. (2002) Manipulation of photon statistics of highly degenerate chaotic radiation, *Phys. Rev. Lett.*, **88**, pp. 063601–063604
34. Levitov, L.S. and Reznikov, M. (2001) Electron Shot Noise Beyond the Second Moment cond-mat/0111057, unpublished
35. Gutman, D.B. and Gefen, Y. (2002) Shot Noise at High Temperatures, cond-mat/0201007
36. Nagaev, K.E. (2002) Cascade Boltzmann-Langevin Approach to Higher-Order Current Correlations in Diffusive Metal Contacts, cond-mat/0203503
37. Nazarov, Yu.V. and Kindermann, M. (2001) Full Counting Statistics of a General Quantum Mechanical Variable, cond-mat/0107133
38. Sukhorukov, E. V. and Loss, D. (2001) Shot Noise of Weak Cotunneling Current: Non-Equilibrium Fluctuation-Dissipation Theorem, in: "Electronic Correlations: From Meso- to Nano-Physics", edited by G. Montambaux and T. Martin, XXXVI Rencontres de Moriond; also: cond-mat/0106307
39. Mahan, G.D. (1981) *Many-Particle Physics*, Sec. 9.3. Plenum Press, New York
40. Beenakker, C.W.J. (1997) Random-Matrix Theory of Quantum Transport, *Rev. Mod. Phys.* **69**, pp. 731–808
41. Glattli D.C., Jin, Y., Reydellet, L.-H. and Roshe, P. (2002) Photo-Assisted Noise in Quantum Point Contacts, this volume
42. Switkes, M., Marcus, C.M., Campman, K. and Gossard, A.C. (1999) An Adiabatic Quantum Electron Pump, *Science* **283**, pp. 1905–1908
43. Klich, I. (2002) Full Counting Statistics, this volume
44. Lesovik, G.B. and Levitov, L.S. (1994) Noise in an AC Biased Junction. Non-stationary Aharonov-Bohm Effect, *Phys. Rev. Lett.*, **72**, pp. 538-541
45. Lee, H.W. and Levitov, L.S. (1995) Estimate of Minimal Noise in a Quantum Conductor, preprint cond-mat/9507011, unpublished
46. Spivak, B. Zhou, F. and Beal Monod, M.T. (1995) Mesoscopic Mechanisms of the Photo-voltaic Effect and Microwave Absorption in Granular Metals, *Phys. Rev. B* **51**, pp. 13226–13230
47. Altshuler, B.L. and Glazman, L.I. (1999) Condensed matter physics – Pumping electrons, *Science* **283**, pp. 1864–1865
48. Brouwer, P.W. (1998) Scattering approach to parametric pumping, Phys. Rev. **B58**, pp. 10135–10138
49. Polianski, M. L., Vavilov, M.G., Brouwer, P.W. (2002) Noise through Quantum Pumps, *Phys. Rev. B* **65**, pp. 245314–245322

AN ELEMENTARY DERIVATION OF LEVITOV'S FORMULA

A short review

I. KLICH
Technion
Department of Physics, 32000 Haifa, Israel.

The field of quantum noise has been rapidly developing in recent years, with the growing possibilities in precision measurements [1], and interest in mesoscopic systems as well as in technological applications of physical effects at the micrometer and nanometer scales.

Of particular interest is the study of the statistics of charge transport between materials coupled through a contact or through a time dependent scatterer. The full statistics of charge transport was studied in a series of works by Levitov et al. [2, 3]. This approach yielded interesting results, and in particular, they where able to express the full counting statistics in terms of a determinant of a single particle operator. Several aspects of Levitov's formula (4) where discussed in following papers [4, 5].

Our aim in this paper is to present a novel derivation of the original Levitov formula. This is done by proving a trace formula (8), which relates certain traces in Fock space to single particle determinants. Using the present approach we find in addition several generalizations, such as a corresponding formula for Bosons.

1. The full counting statistics

The typical setting is the following: consider particle reservoirs, with given temperatures and chemical potentials, which are separated at time zero, and are evolving by a second quantized hamiltonian H_0. At some time the reservoirs are coupled, through a scattering region, and evolve by a new, time dependent hamiltonian. After a time T they are decoupled again and one is interested in the statistics of charge transported from side to side, i.e. to compute $< Q >, < Q^2 >$ and higher moments. The role of the third moment was recently discussed in [6].

Y. V. Nazarov (ed.), Quantum Noise in Mesoscopic Physics, 397–402.

The full statistics of charge transfer may be conveniently represented by the characteristic function of the (charge transport) probability distribution function, defined by:

$$\chi(\lambda_1, ...,; T) = \sum_{\alpha,\beta} P(\alpha(t=0), \beta(t=T)) e^{iq\sum_i \lambda_i(\beta_i - \alpha_i)} \quad (1)$$

Here the summation is over all states $\alpha = (\alpha_1, ...), \beta = (\beta_1, ...)$ labelling the Fock space in the occupation number representation: for fermions these are vectors of zeros and ones and for bosons vectors with integer coefficients, where α_i is the number of particles occupying the single particle state i in α. $P(\alpha(t=0), \beta(t=T))$ is the probability that we started in state α at time 0 and finished in a state β at time T. Thus terms $(\alpha_i - \beta_i)$ appearing in the exponent are just the change in the number of particles occupying the single particle state i. And the parameters λ_i are introduced in the standard manner to calculate different moments. By taking derivatives of χ with respect to λ_i, one can calculate arbitrary moments of the charge accumulation in state i. For example,

$$< Q_i >= -i\partial_{\lambda_i} \log \chi|_{\lambda_1,...=0} \quad (2)$$

And

$$(\Delta Q_i)^2 =< Q_i^2 - < Q_i >^2 >= -\partial^2_{\lambda_i} \log \chi|_{\lambda_1,...=0}. \quad (3)$$

It is also possible to compute the moments of the charge which is transferred to a particular reservoir by taking the derivative with respect to λ after setting $\lambda_i = \lambda$ for all states i belonging to the desired reservoir. $\chi(\lambda)$ may be interpreted as the reaction of the system to coupling with a classical field λ which measures the number of electrons on each side.

For adiabatic change, and short scattering time Levitov et al [2], obtained the following expression for χ:

$$\chi(\lambda) = \det(1 + n(S^\dagger e^{iq\lambda} S e^{-iq\lambda} - 1)) \quad (4)$$

Where n is the occupation number operator and S is the scattering matrix. As was remarked [2], this expression requires careful understanding and regularization. In the following we derive this formula in a new manner which, we hope will allow a convenient way to address these issues.

In order to proceed we first write χ as a trace in Fock space:

$$\begin{aligned} &\chi(\lambda, T) = \quad (5) \\ &\sum_{\alpha,\beta} < \alpha|\rho_0|\alpha > | < \alpha|\mathbb{U}^\dagger|\beta > |^2 e^{iq\sum_i \lambda_i(\beta_i - \alpha_i)} \\ &= \mathrm{Tr}(\rho_0 \mathbb{U}^\dagger e^{iq\sum \lambda_i a_i^\dagger a_i} \mathbb{U} e^{-iq\sum \lambda_i a_i^\dagger a_i}) \end{aligned}$$

Where ρ_0 is the density matrix at the initial time ($t = 0$), $\mathbb{U}$ is the evolution (in Fock space) from time $t = 0$ to time $t = T$ and $a_i^\dagger, a_i$ are the creation and annihilation operators for a given one particle state i. Here it is assumed that the occupation number basis is chosen such that the initial time density matrix ρ_0 is diagonal in it, which implies that the states α are eigenstates of the initial Hamiltonian, and measurement of charge in a specific state is meaningful.

Next, we define the second quantized version of a single particle operator A (i.e. an operator on the single particle Hilbert space) to be the Fock space operator:

$$\Gamma(A) = \sum < i|A|j > a_i^\dagger a_j. \tag{6}$$

Then (5) can be written as:

$$\chi(\lambda, T) = \mathrm{Tr}(\rho_0 e^{iq\mathbb{U}^\dagger\Gamma(\lambda)\mathbb{U}} e^{-iq\Gamma(\lambda)}) \tag{7}$$

Here λ is the matrix $\mathrm{diag}(\lambda_1, \lambda_2, ...)$. To handle this kind of expressions (and to obtain Levitov's formula) we prove in the following section a trace formula.

2. A trace formula

In this section we prove the following:

$$\mathrm{Tr}(e^{\Gamma(A)} e^{\Gamma(B)}) = \det(1 - \xi e^A e^B)^{-\xi} \tag{8}$$

Where $\xi = 1$ for bosons and $\xi = -1$ for fermions (i.e. the creation and annihilation operators satisfy $a_j a_i^\dagger - \xi a_i^\dagger a_j = \delta_{ij}$).

We prove this result for the finite dimensional Hilbert space case, and avoid at this point questions regarding the limit of infinite number of states, to be addressed elsewhere [7].

Proof:
For an N dimensional single particle Hilbert space Γ is a representation of the usual Lie algebra of matrices $gl(N)$. Indeed, substituting the definition (6), together with the relations obeyed by the creation and annihilation operators it is straightforward to check that

$$[\Gamma(A), \Gamma(B)] = \Gamma([A, B]) \tag{9}$$

is true for bosons and for fermions. By Baker Campbell Hausdorf there exists a matrix C such that $e^A e^B = e^C$. C is an element of $gl(N)$ and is given by a series of commutators, since Γ is a representation, it holds that

$$e^A e^B = e^C \rightarrow e^{\Gamma(A)} e^{\Gamma(B)} = e^{\Gamma(C)}. \tag{10}$$

Now let us evaluate $\mathrm{Tr}(e^{\Gamma(C)})$. Any matrix C can be written in a basis in which it is of the form $\mathrm{diag}(\mu_1, ..\mu_n) + \mathrm{K}$ where K is an upper triangular, thus we have

$$\mathrm{Tr}(e^{\Gamma(C)}) = \mathrm{Tr}(e^{\Gamma(\mathrm{diag}(\mu_1,..,\mu_n)) + \Gamma(K)}) = \mathrm{Tr}(e^{\Gamma(\mathrm{diag}(\mu_1,..,\mu_n))}) = \tag{11}$$

$$\mathrm{Tr}(\textstyle\prod_i e^{\mu_i a_i^\dagger a_i}) = \prod_i (1 - \xi e^{\mu_i})^{-\xi} = \det(1 - \xi e^C)^{-\xi}$$

(One may also think of $\mathrm{Tr}(e^{\Gamma(C)})$ as the partition function of a system with Hamiltonian $-C$ at temperature $k_B T = 1$). From this equation (8) follows:

$$\mathrm{Tr}(e^{\Gamma(A)}e^{\Gamma(B)}) = \mathrm{Tr}(e^{\Gamma(C)}) = \det(1 - \xi e^{C})^{-\xi} = \det(1 - \xi e^{A}e^{B})^{-\xi} \quad (12)$$

We remark at this point that this relation can immediately be generalized in the same way to products of more then two operators.

• Let us illuminate our identity with a trivial example: Let $\mathcal{H}$ be an N - dimensional Hilbert space, and choose $A = B = 0$. Then the dimension of the appropriate Fock space is given by

$$\mathrm{Tr}(\mathbb{I}) = \mathrm{Tr}(e^{\Gamma(0)}e^{\Gamma(0)}) = \det(1-\xi)^{-\xi} = \begin{cases} 2^N & \text{Femions} \\ \infty & \text{Bosons} \end{cases}$$

as it should be.

3. Levitov's formula

We now turn to give a novel derivation of Levitov's result for the full counting statistics. In the framework of non interacting fermions the evolution $\mathbb{U}$ in the expression (5) for χ is just the Fock space implementation of the single particle evolution U. That means that $\mathbb{U}^{\dagger}\Gamma(\lambda)\mathbb{U} = \Gamma(U^{\dagger}\lambda U)$ so that by the trace formula (8) for 3 operators, we immediately have

$$\begin{aligned} \chi(\lambda, T) &= \mathrm{Tr}(\frac{e^{-\beta\Gamma(H_0)}}{Z} e^{iq\Gamma(U^{\dagger}\lambda U)}e^{-iq\Gamma(\lambda)}) \\ &\frac{1}{Z}\det(1 + e^{-\beta H_0}(U^{\dagger}e^{iq\lambda}Ue^{-iq\lambda})) = \\ &\det(1 + n(U^{\dagger}e^{iq\lambda}Ue^{-iq\lambda} - 1)) \end{aligned} \quad (13)$$

Where $Z = \det(1 + e^{-\beta H_0})$ and n is the occupation number operator $\frac{e^{-\beta H_0}}{1+e^{-\beta H_0}}$ at the initial time. We note that the result (13) should be viewed as the general expression for the counting statistics of noninteracting fermions, at any given time, and without any approximation. And may be a good start for studying different limits of the problem, as well as regularization difficulties.

Finally, if the scattering time is very small compared with the entire evolution, then one may describe the problem in terms of dynamical scattering operators, $S = \lim_{t\to\infty} e^{iH_0t}U(t,-t)e^{iH_0t}$ where H_0 is the initial free evolution. Using the fact that λ commutes with H_0, one obtains in the limit of $T \to \infty$:

$$\chi(\lambda) = \det(1 + n(S^{\dagger}e^{iq\lambda}Se^{-iq\lambda} - 1)) \quad (14)$$

Which is Levitov's result (4), as promised.

We now add a few remarks:

1. *Convergence and regularization:* First we note that as long as we assume that ρ exists and has trace 1, then trace of ρ times a bounded operator is also

finite, so that χ is well defined. However, problems might arise when taking the thermodynamic limit. Taking the infinite volume limit may cause the Fock space density matrix to be ill defined (i.e it cannot be normalized to trace 1), however, expectation values obtained using it may still have meaning.

If one uses the scattering matrix as in (4), regularization of the determinant is needed, since in this case one uses the static scattering matrix as an approximation for the true evolution. Indeed, in the limit $T \to \infty$ arbitrarily large charges can pass from side to side [1], so that the information about the length of the time interval has to be put in by hand [2]. The equation (13), however, can be shown to be well defined even in the thermodynamic limit, for a finite time interval [7].

2. *Bosons:* It is now straightforward to derive an analogous formula for the full counting statistics of bosons. The result is simply:

$$\chi_B(\lambda, T) = \frac{1}{\det(1 - n_B(U^\dagger e^{iq\lambda} U e^{-iq\lambda} - 1))}. \tag{15}$$

Where n_B is the occupation number operator for bosons.

3. *Rate of charge accumulation:* Here we give an example of how one may compute the rate of charge accumulation in a box A. We choose a box in space, which can be described by a projection P_A in the single particle Hilbert space (with matrix elements $< x|P_A|x' >= \delta(x - x')$ if x is in the box, and zero otherwise). By setting $\lambda = 0$ for all states that are outside the box A, one finds

$$\begin{aligned} \dot{Q_A} &= -i\partial_t\partial_\lambda \log \chi(\lambda, t)|_{\lambda=0} = \\ & q\partial_t \mathrm{Tr}(n(U^\dagger P_A U - P_A)) = q\mathrm{Tr}(n(\dot{U}^\dagger P_A U + U^\dagger P_A \dot{U})) = \\ & q \,\mathrm{Re} < U\dot{U}^\dagger P_A >_t \end{aligned} \tag{16}$$

The angular brackets describe averaging over the distribution at the time of measurement t. Charge accumulation is equivalent to current if the box A is connected via just one contact to the other reservoirs. This equation should be compared to the formula for the current in terms of scattering matrices [8–11] which is of fundamental interest in the field of quantum pumps.

4. Summary

To conclude, we presented a novel derivation of Levitov's determinant formula for the full counting statistics of charge transfer. This was done by introducing a trace formula (8) which is suitable for translating problems of non-interacting particles from Fock space to the single particle Hilbert space. The derivation is general enough to allow consideration of new problems of counting statistics, in

[1] To see this we note that the first moment of (4), which is the transported charge, diverges as time goes to infinity if there is a bias between the reservoirs

particular, further problems involving bosons, or measurement of other operators then charge. We hope that some properties of the determinant (13) under various limits, such as adiabatic and thermodynamic limits will now be easier to address.

I am grateful to J. E. Avron for many discussions and remarks, and especially to O. Kenneth for his help in the proof of (8). I would also like to thank J. Feinberg, L. S. Levitov and M. Reznikov for remarks.

References

1. Ya. M. Blanter and M. Büttiker, Phys. Rep. **336**, 1 (2000).
2. L. S. Levitov and G. B. Lesovik, JETP Lett. **58**, 230 (1993); L. S. Levitov, H.-W. Lee, and G. B. Lesovik, Journal of Mathematical Physics, **37** (1996) 10.
3. D. A. Ivanov, H. W. Lee and L. S. Levitov, Phys. Rev. B **56**,6839 (1997).
4. A. V. Andreev and A. Kamenev, Phys. Rev. Lett. **85**, 1294 (2000).
5. Yu. V. Nazarov and M. Kindermann, cond-mat/0107133.
6. L. S. Levitov and M. Reznikov, cond-mat/0111057.
7. J. E. Avron and I. Klich, unpublished.
8. M. Büttiker, H. Thomas, A. Prêtre, Z. Phys. B **94**, 133 (1994).
9. P.W. Brouwer, Phys. Rev. B **58**, 10135 (1998).
10. J.E. Avron, A. Elgart, G.M. Graf and L. Sadun, Phys. Rev. Lett. **87**, 236601 (2001)
11. Y. Makhlin and A. D. Mirlin Phys. Rev. Lett. **87**, 276803 (2001).

FULL COUNTING STATISTICS IN ELECTRIC CIRCUITS

A short review

MARKUS KINDERMANN
Instituut-Lorentz, Universiteit Leiden, P.O. Box 9506, 2300 RA Leiden, The Netherlands

YULI V. NAZAROV
Department of Nanoscience, Delft University of Technology, Lorentzweg 1, 2628 CJ Delft, The Netherlands

1. Introduction

Full counting statistics (FCS) is one of the most attracting and intellectually invloved concepts in quantum transport. It provides much information about charge transfer (all possible moments, including average current, noise, etc.) in a compact and elegant form. Albeit the study of full counting statistics has begun with some confusion. The first attempt to derive FCS [1] exploited the text-book definion of quantum measurement: the probability of the outcome q of a measurement is given by $\langle n|\delta(q - \hat{Q})|n\rangle$, $\hat{Q}$ being an operator associated with the value measured and $|n\rangle$ the quantum state. However, the choice Q made in [1] resulted in severe interpretation problems. In [2] the authors revised their approach. The new method, that is commonly accepted now, invokes an extra degree of freedom, a detector. The quantum measurement paradigm is applied to the detector degree of freedom.

There is a similarity between the method of [2] and two core approaches to quantum dynamics: the Keldysh technique [3] and the Feynman-Vernom formalism.[4] This similarity was not stressed in [2]. Later it has been noticed that the FCS can be evaluated with a straightforward modification of Keldysh Green function technique [5]. This allows one to extend the studies of FCS to many systems, some results being reviewed in other contributions to this book. This is a good news. A seemingly bad news is that the method of [2] does not always give results that can be interpreted as probabilities of an outcome of a measurement.

Y.V. Nazarov (ed.), Quantum Noise in Mesoscopic Physics, 403–427.

In the statistics of charge transfer, the bad news becomes relevant for systems where gauge invariance is broken. [6, 7] If one tries to generalize FCS to arbitrary variables, these problems arise from the very beginning.

The present contribution consists of two parts. First we address the bad news. For this, we consider a rather general and abstract problem of counting statistics of a general quantum mechanical variable being measured by a linear detector. We find that in this case the FCS is physically meaningful and usefull since it determines the quantum evolution of the detector. This is despite the fact that it can not be interpreted in the form of probabilities.[8]

Thus encouraged, we investigate how far one can go with this approach. An abstract exrecise with a linear detector proves to be very useful to describe a mesoscopic conductor embedded in a linear electric circuit. We reveal the relations between the FCS and the non-equilibrium Keldysh action that describes current and voltage fluctuations in electric circuits. We show how to evaluate the current and voltage counting statistics for any two points of an arbirtary circuit. This appears to be relevant for future experimental activities in the field of quantum noise and statistics.[9]

The article is organized as follows. In section 2 we will pursue a route similar to that proposed in [1], relating the current satistics to the statistics of a charge operator. We follow an alternative route in section 3 and consider a system coupled to a static linear detector. [8]. We show that the result of the detection depends in general on the initial state of the detector. To predict the result, one needs a function of two variables. This is this function that we adopt as the *definition* of the full counting statistics. We will identify situations where the two approaches give the same easy-to-interpret result.

Then we turn to electric circuits. We explain in section 4 why a linear circuit can be viewed as a set of dynamical linear detectors and why the FCS expression provides a building block of the Keldysh action that descibes the quantum dynamics of the circuit. In section 5 we discuss the *low-frequency* limit of the action. In this limit, the action is local in time. This facilitates its evaluation. The general scheme is illustrated in section 6 with a simple model of a Quantum Point Contact in series with an external linear resistor. In this case, we are able to determine the full statistics of current and voltage fluctuations. These results were obtained in collaboration with C.W.J. Beenakker, that we gladly acknoledge.

2. Charge statistics without detector

In this section we define a statistics of time averages of electric current in an elementary manner, without specifying a measurement procedure. The key object in this will be the statistics of charge Q on one side of the cross-section through which the current passes. If Q is fixed at time 0, the statistics of $Q(\tau)$ at time τ corresponds to the statistics of the current averaged over the time interval $[0, \tau]$.

To specify Q we introduce, following [2], a smooth function f that divides the conductor into two parts: left side, where $f(\mathbf{x}) < 0$, and right side, where $f(\mathbf{x}) > 0$. We consider the measurement of the electric current flowing through the boundary $f(\mathbf{x}) = 0$ between these two sides. The operator of electric charge to the right of the cross-section is then $Q = e\theta[f(\mathbf{x})]$ $[\theta(x) = 0$ for $x \leq 0$ and $\theta(x) = 1$ for $x > 0]$, e being the elementary charge. It is convenient to express the statistics of $Q(\tau)$ via the corresponding moment generating function

$$\chi_c(\lambda) = \sum_k \frac{(i\lambda)^k}{k!}\left\langle Q(\tau)^k\right\rangle = \left\langle e^{i\lambda Q(\tau)}\right\rangle = \left\langle e^{iH\tau}e^{i\lambda Q}e^{-iH\tau}\right\rangle. \tag{1}$$

Here H presents the Hamiltonian of the current conductor and the average is taken over the initial density matrix of the conductor.

One advantage of defining χ_c by means Eq. (1) is that it is associated with a (positive) probability distribution. This is because it is a generating function of moments of a Hermitian operator $Q(\tau)$.It has been found [6] that this is not generally the case for other definitions of the generating function for moments of transferred charge. Besides, χ_c evidently predicts for systems of well-localized non-interacting electrons the transfer of charge in integer multiples of the electron charge e, the result to be expected. Also this was found to be violated by other definitions [7]. One buys these advantages by defining a statistics that is only indirectly linked to the statistics of transmitted charge since requires the knowledge of the initial charge state of the conductor.

Using the relation

$$e^{i\lambda Q/2}e^{A}e^{-i\lambda Q/2} = \exp[e^{i\lambda Q/2}Ae^{-i\lambda Q/2}] \tag{2}$$

we rewrite Eq. (1) identically as

$$\begin{aligned}\chi_c(\lambda) = \Big\langle e^{i\lambda Q/2}\exp\left[ie^{-i\lambda Q/2}He^{i\lambda Q/2}\tau\right] \\ \exp\left[-ie^{i\lambda Q/2}He^{-i\lambda Q/2}\tau\right]e^{i\lambda Q/2}\Big\rangle.\end{aligned} \tag{3}$$

The charge operator Q commutes with all position operators $\mathbf{x}$ contained in the Hamiltonian $H = H(\mathbf{p}, \mathbf{x})$. Albeit it does not commute with the momentum operators $\mathbf{p}$. Therefore the momentum operator is transformed as

$$e^{-i\lambda Q/2}\mathbf{p}e^{i\lambda Q/2} = \mathbf{p} - \tfrac{e}{2}\lambda\nabla\theta[f(\mathbf{x})] \equiv \tilde{\mathbf{p}}_\lambda. \tag{4}$$

We define a new Hamiltonian $H_\lambda = H(\tilde{\mathbf{p}}_\lambda, \mathbf{x})$ in the same way as it has been done in [2]. The generating function

$$\chi_c(\lambda) = \left\langle e^{i\lambda Q/2}e^{iH_{-\lambda}\tau}e^{-iH_\lambda\tau}e^{i\lambda Q/2}\right\rangle \tag{5}$$

takes then a form that is very similar to the $\chi(\lambda)$ found in[2] with the spin-$\frac{1}{2}$ current detection model . Tracing back the difference we write χ in terms of charge operators as

$$\chi(\lambda) = \left\langle e^{-i\lambda Q/2} e^{i\lambda Q(\tau)} e^{-i\lambda Q/2} \right\rangle . \tag{6}$$

If the initial state of the conductor is an eigenstate of charge with eigenvalue Q_0, the two generating functions are identical up to the offset charge Q_0, $\chi(\lambda) = e^{-i\lambda Q_0}\chi_c(\lambda)$. In this case both characteristic functions are associated with a positive probability

$$P(q) = \int d\lambda\, e^{-iq\lambda}\chi(\lambda), \tag{7}$$

to transfer q charges during the measurement time. For generic systems of non-interacting electrons this probability has been found to be non-zero only at integer multiples of the electron charge, corresponding to the transfer of an integral number of electrons [2].

If the initial state is a superposition of eigenstates of charge Q, the two generating functions χ and χ_c differ. For example, χ may seem to predict the transfer of half the elementary charge when χ_c does not. This becomes evident if χ is presented in the form (6) since it contains summands of the form $\chi_{mn} = \exp[-i\lambda(m+n)/2]\,\langle m,\alpha|\exp[i\lambda Q(\tau)]|n,\beta\rangle$, $|m,\alpha\rangle$ and $|n,\beta\rangle$ being eigenstates of the charge with additional quantum numbers α, β, $Q|m,\alpha\rangle = m|m,\alpha\rangle$. While the matrix element in the expression for χ_{mn} corresponds to integer charge transfer, the prefactor suggests half-integer charge transfer when $m+n$ is an odd number. This has been pointed out in [7].

For a Josephson junction at the cross-section $f(\mathbf{x}) = 0$ between two superconductors the situation is even more involved. One would like to choose an initial state of well defined phase difference between the two sides of the junction in order to have a well defined current flowing. However, the phase and charge transferred are conjugated variables obeying Heisenberg uncertainty principle. Thus the dispersion of the initial charge Q in such state is infinite. The distribution corresponding to χ_c with such initial state is one of undetermined $Q(\tau)$. Therefore it contains no information about the charge transfer. This is clearly undesirable. Because all the uncertainty of Q is already present in the initial state and has nothing to do with the transfer process, one might hope to be able to eliminate it in a meaningful way. Indeed, we have seen that this is accomplished for the initial charge offset in a state of well defined charge by the factors $e^{-i\lambda Q/2}$ in the definition of χ (6). One could hope that this works also for the initial charge spread in a phase eigenstate. Instead, this gives rise to negative "probabilities" [6].

A way to remedy this problem would be to calculate χ_c in an initial state with a finite charge and phase dispersion. One could also try to find general relations between the charge statistics and the initial state of the conductor that would characterize the charge transfer process more generally.

The alternative is to couple the (super-)conductor to a detector and interpret the detector read-off in terms of charge transfer. We will follow this route in the forthcoming sections. It turns out that in a idealized detector model a function very similar to the generating function χ eventually determines the final state of the detector provided the initial state is known. The charge transfer through the conductor is thus characterized by its effect on the detector.

3. Counting statistics with a static detector

We turn now to the *measurement* process of time averaged quantities. In this section we focus on an idealized detector without its own dynamics, a static one. Within this model we study the measurement of time averages $\int_0^\tau dt\, A(t)$ of an arbitrary operator A. For electric currents, $A = I$, this will allow us to characterize the charge transfer and its statistics by a function similar to the characteristic function introduced in the previous section. The analysis of this function will shed light on the origin of the negative probabilities.

3.1. THE DETECTOR MODEL

The detector in our model consists of one degree of freedom x (with the conjugated variable q) with the Hamiltonian $q^2/2m$. The system measured shall be coupled to the position x of the detector during the time interval $[0, \tau]$ and be decoupled adiabatically for earlier and later times. For this we introduce a smooth coupling function $\alpha_\tau(t)$ that takes the value 1 in the time interval $[0, \tau]$ and is zero beyond the interval $[t_1, t_2]$ ($t_1 < 0$ and $t_2 > \tau$). The values for $t_1 < t < 0$ and $\tau < t < t_2$ are chosen in a way that provides an adiabatic switching. The entire Hamiltonian reads then

$$H(t) = H_{\text{sys}} - \alpha_\tau(t) x A + \frac{q^2}{2m}. \tag{8}$$

The Heisenberg equation of motion for the detector momentum q

$$\dot{q}(t) = \alpha_\tau(t) A(t) \tag{9}$$

suggests, that the statistics of outcomes of measurements of the detector's momentum after having it uncoupled from the system corresponds to the statistics of the time average $\int_0^\tau dt\, A(t)$ that we are interested in.

The coupling term can be viewed as a disturbance of the system measured by the detector. To minimize this disturbance, we would clearly like to concentrate the detector wave function around $x = 0$. The uncertainty principle forbids us, however, to localize it completely. Thereby we would loose all information about the detector momentum, which we intend to measure. This is a fundamental limitation imposed by quantum mechanics, and we are going to explore its consequences step by step. To discern it form a classical back action of the detector we take the limit of a static detector with $m \to \infty$, such that $\dot{x} = 0$ and any classical no back action is ruled out.

3.2. THE APPROACH

Our goal is to relate the density matrices of the detector before and after the measurement. If there were no system to measure we could readily express this relation in the form of a path integral in the (double) variable $x(t)$ over the exponential of the detector action. This is still possible in the presence of a system coupled to the detector [4]. The information about the system to be measured can be compressed into an extra factor in this path integral, the so-called influence functional. This makes the separation between the detector and the measured system explicit. To make contact with [2], we write the influence functional as an operator expression that involves the system degrees of freedom only. We denote the initial detector density matrix (at $t < t_1$) by $\rho^{in}(x^+, x^-)$ and the final one (at $t > t_2$, after having traced out the system's degrees of freedom) by $\rho^f(x^+, x^-)$. R will denote the initial density matrix of the system. The entire initial density matrix is assumed to factorize, $D = R\rho^{in}$.

We start out by inserting complete sets of states into the expression for the time development of the density matrix

$$\begin{aligned} \rho^f(x^+, x^-) &= \underset{\text{System}}{Tr} \langle x^+ | \overrightarrow{T} e^{-i\int_{t_1}^{t_2} dt\, \left[H_{sys} - \alpha_\tau(t) x A + \frac{q^2}{2m}\right]} D \\ &\qquad \overleftarrow{T} e^{i\int_{t_1}^{t_2} dt\, \left[H_{sys} - \alpha_\tau(t) x A + \frac{q^2}{2m}\right]} | x^- \rangle. \end{aligned} \tag{10}$$

Here, $\overrightarrow{T}$ ($\overleftarrow{T}$) denotes (inverse) time ordering. As the complete sets of states we choose product states of any complete set of states of the system and alternatingly complete sets of eigenstates of the position or the momentum operator of the detector. Those intermediate states allow us to replace the position and momentum operators in the time development exponentials by their eigenvalues. We can then do the integrals over the system states as well as the momentum integrals and arrive at the expression

$$\begin{aligned} \rho^f(x^+, x^-) &= \int \underset{x^+(t_2)=x^+}{\mathcal{D}[x^+]} \; \underset{x^-(t_2)=x^-}{\mathcal{D}[x^-]} \; \rho^{in}[x^+(t_1), x^-(t_1)]\, e^{-iS_{\text{Det}}([x^+],[x^-])} \\ & \underset{\text{System}}{Tr} \; \overrightarrow{T} e^{-i\int_{t_1}^{t_2} dt\left[H_{sys} - \alpha_\tau(t) x^+(t) A\right]} R\, \overleftarrow{T} e^{i\int_{t_1}^{t_2} dt\left[H_{sys} - \alpha_\tau(t) x^-(t) A\right]} \end{aligned} \tag{11}$$

with the detector action

$$S_{\text{Det}}([x^+],[x^-]) = -\int_{t_1}^{t_2} dt\, \frac{m}{2}[(\dot{x}^+)^2 - (\dot{x}^-)^2]. \tag{12}$$

We rewrite the expression as

$$\rho^f(x^+,x^-) = \int dx_1^+ dx_1^- \ K(x^+,x^-;x_1^+,x_1^-,\tau)\rho^{in}(x_1^+,x_1^-) \tag{13}$$

with the kernel

$$K(x^+,x^-;x_1^+,x_1^-,\tau) = \int_{x^+(t_2)=x^+,x^+(t_1)=x_1^+} \mathcal{D}[x^+] \int_{x^-(t_2)=x^-,x^-(t_1)=x_1^-} \mathcal{D}[x^-]$$
$$\mathcal{Z}_{\rm Sys}([\alpha_\tau x^+],[\alpha_\tau x^-]) \ \exp\left\{-i\mathcal{S}_{\rm Det}([x^+],[x^-])\right\} \tag{14}$$

that contains the influence functional

$$\mathcal{Z}_{\rm Sys}([\chi^+],[\chi^-]) = \underset{\rm System}{Tr} \ \overrightarrow{T} e^{-i\int_{t_1}^{t_2} dt\left[H_{\rm sys}-\chi^+(t)A\right]} \ R \ \overleftarrow{T} \ e^{i\int_{t_1}^{t_2} dt\left[H_{\rm sys}-\chi^-(t)A\right]}. \tag{15}$$

Taking the limit of an infinite detector mass now, we find that $\mathcal{S}_{\rm Det}$ in Eq. (14) suppresses all fluctuations in the path integral. Therefore, the kernel $K(x^+,x^-,x_1^+,x_1^-,\tau)$ becomes local in position space,

$$K(x^+,x^-,x_1^+,x_1^-,\tau) = \delta(x^+ - x_1^+) \ \delta(x^- - x_1^-) \ P(x^+,x^-,\tau) \tag{16}$$

with

$$P(x^+,x^-,\tau) = \underset{\rm System}{Tr} \ \overrightarrow{T} e^{-i\int_{t_1}^{t_2} dt\left[H_{\rm sys}-\alpha_\tau(t)x^+A\right]} R \overleftarrow{T} e^{i\int_{t_1}^{t_2} dt\left[H_{\rm sys}-\alpha_\tau(t)x^-A\right]}. \tag{17}$$

It is constructive to rewrite now the density matrices in the Wigner representation

$$\rho(x,q) = \int \frac{dz}{2\pi} \ e^{iqz} \ \rho(x+\frac{z}{2},x-\frac{z}{2}) \tag{18}$$

and define

$$P(x,q,\tau) = \int \frac{dz}{2\pi} e^{iqz} \ P(x+\frac{z}{2},x-\frac{z}{2},\tau). \tag{19}$$

This gives the compact relation that summarizes results of the subsection:

$$\rho^f(x,q) = \int dq_1 \ P(x,q-q_1,\tau) \ \rho^{in}(x,q_1). \tag{20}$$

3.3. INTERPRETATION OF THE FCS

We adopt the relations (17), (19) and (20) as the definition of the FCS of the variable A. Let us see why. First let us suppose that we can treat the detector classically. Then the density matrix of the detector in the Wigner representation can be interpreted as a classical probability distribution $\Pi(x,q)$ to be at a certain position

x with momentum q. This allows for a classical interpretation of $P(x, q, \tau)$ as the probability to have measured $q = \int_0^\tau A(t)$. Indeed, one sees from (19) that the final $\Pi(x, q)$ is obtained from the initial one by shifts in q, $P(x, q, \tau)$ being the probability distribution of those shifts.

In general, the density matrix in the Wigner representation cannot be interpreted as a probability to have a certain position and momentum since it is not positive. Concrete calculations given below illustrate that $P(x, q, \tau)$ does not have to be positive either. Consequently, it cannot be interpreted as a probability distribution. Still it predicts the results of measurements according to Eq. (20).

There is an important case where the FCS can indeed be interpreted as a probability distribution. It is the case that $P(x, q, \tau)$ does not depend on x, $P(x, q, \tau) \equiv P(q, \tau)$. Then, integrating Eq. (20) over x, we find

$$\Pi^f(q) = \int dq' \, P(q - q', \tau) \, \Pi^{in}(q') \tag{21}$$

with $\Pi(q) \equiv \int dx \, \rho(x, q)$. Therefore, the FCS is in this special case the kernel that relates the probability distributions of the detector momentum before and after the measurement, $\Pi^{in}(q)$ and $\Pi^f(q)$, to each other. Those distributions are positive and so is $P(q, \tau)$.

When studying the FCS of a stationary system and the measurement time τ exceeds time scales associated with the system, the operator expression in Eq. (17) can be seen as a product of terms corresponding to time intervals. Therefore in this limit of $\tau \to \infty$ the dependence on the measuring time can be reconciled into

$$P(x^+, x^-, \tau) = e^{-\mathcal{E}(x^+, x^-)\tau} \tag{22}$$

where the expression in the exponent is supposed to be big. Then the integral (19) that defines the FCS can be done in the saddle point approximation. Defining the time average $\bar{A} = q/\tau$, that is, $\bar{A}$ is the result of a measurement of $\int_0^\tau A(t)dt/\tau$, the FCS can be recast into the similar from

$$P(x, \bar{A}, \tau) = e^{-\tilde{\mathcal{E}}(x, \bar{A})\tau} \tag{23}$$

where $\tilde{\mathcal{E}}$ is defined as the (complex) extremum with respect to (complex) z:

$$\tilde{\mathcal{E}} = \operatorname*{extr}_z \{\mathcal{E}(x + \frac{z}{2}, x + \frac{z}{2}) + i\bar{A}z\}. \tag{24}$$

The average value of $\bar{A}$ and its variance (noise) can be expressed in terms of derivatives of $\mathcal{E}$:

$$\begin{aligned} \langle \bar{A} \rangle &= -\lim_{z \to 0} \frac{\partial \mathcal{E}(x + z/2, x - z/2)}{i \partial z} \\ \tau \langle\langle \bar{A}^2 \rangle\rangle &= \lim_{z \to 0} \frac{\partial^2 \mathcal{E}(x + z/2, x - z/2)}{\partial z^2}. \end{aligned} \tag{25}$$

More generally, the quantity $P(x^+, x^-, \tau)$ is the generating function of moments of q. It is interesting to note that in general this function may generate a variety of moments that differ in the time order of operators involved, for instance,

$$Q_M^N = i^{2M-N} \lim_{x^\pm \to 0} \frac{\partial^M}{\partial(x^-)^M} \frac{\partial^{N-M}}{\partial(x^+)^{N-M}} P(x^+, x^-, \tau)$$

$$= \int_0^\tau dt_1 ... dt_N \langle \overleftarrow{T}\{A(t_1)...A(t_M)\} \overrightarrow{T}\{A(t_{M+1})...A(t_N)\}\rangle. \quad (26)$$

The moments of (the not necessarily positive) $P(0, q, \tau)$ are expressed through these moments and binomial coefficients,

$$Q^{(N)} \equiv \int dq\, q^N P(0, q, \tau) = 2^{-N} \sum_M \binom{N}{M} Q_M^N. \quad (27)$$

For $A = I$ these moments correspond to the moments generated by the $\chi(\lambda)$ of the previous section, that is contained in $P(x^+, x^-, \tau)$ as $\chi(\lambda) = P(\lambda/2, -\lambda/2, \tau)$ [compare Eq. (5) in the semiclassical approximation, where $H_\lambda = H_0 - \lambda I$, to Eq. (17)]. Interpreting $\chi(\lambda)$ as the characteristic function corresponding to a probability distribution is therefore equivalent to the classical limit discussed above and applicable for systems with an x- independent $P(x, q, \tau)$.

3.4. FCS OF AN EQUILIBRIUM SYSTEM IN THE GROUND STATE

To acquire a better understanding of the general relations obtained we consider now an important special case. We will assume that the system considered is in its ground state $|g\rangle$, so that its initial density matrix is $R = |g\rangle\langle g|$. In this case the FCS is easily calculated. We have assumed that the coupling between the system and the detector is switched on adiabatically. Then the time development operators in (17) during the time interval $t_1 < t < 0$ adiabatically transfer the system from $|g\rangle$ into the ground state $|g(x^\pm)\rangle$ of the new Hamiltonian $H_{sys} - x^\pm A$. In the time interval $0 < t < \tau$ the time evolution of the resulting state has thus the simple form

$$e^{-i\,t\,(H_{sys} - x^\pm A)}\; |g(x^\pm)\rangle \;=\; e^{-itE(x^\pm)}\; |g(x^\pm)\rangle. \quad (28)$$

Here, $E(x^\pm)$ are the energies corresponding to $|g(x^\pm)\rangle$. This gives the main contribution to the FCS if the measurement time is large and the phase acquired during the switching of the interaction can be neglected in comparison with this contribution,

$$P(x^+, x^-, \tau) = e^{-i\tau[E(x^+) - E(x^-)]}. \quad (29)$$

We now assume the function $E(x)$ to be analytic and expand it in its Taylor series. We also rescale q as above, $\bar{A} = q/\tau$. We have then for the FCS

$$P(x, \bar{A}, \tau) = \tau \int dz\; e^{-iz\bar{A}\tau} \cdot e^{-i\tau[E'(x)z + E'''(x)z^3/24 + ...]}. \quad (30)$$

First we observe that $P(x, \bar{A}, \tau)$ is a real function in this case, since the exponent in (30) is anti-symmetric in z. A first requirement for being able to interpret it as a probability distribution is therefore fulfilled. However, the same asymmetry assures that all *even* cumulative moments of $\bar{A}$ are identically zero, whereas the odd ones need not. On one hand, since the second moment corresponds to the noise and the ground state cannot provide any, this makes sense. On the other hand, this situation would be impossible if $P(0, \bar{A}, \tau)$ were a positive probability distribution.

Belzig and Nazarov [6] encountered this situation analyzing the FCS of a general super-conducting junction. In a certain limit the junction becomes a Josephson junction in its ground state. In this limit the interpretation of the FCS as a probability distribution does not work any longer. Fortunately enough, the relation (20) allows us to interpret the results obtained.

Taking the limit $\tau \to \infty$ of Eq. (30) now, we note that all the terms involving higher derivatives of $E(x)$ are negligible and we arrive at

$$\lim_{T\to\infty} P(x, \bar{A}, \tau) = \delta[\bar{A} + E'(x)]. \tag{31}$$

According to the Hellman-Feynman theorem $E'(x) = -\langle g(x)|A|g(x)\rangle$. As one would expect, in this limit the measurement gives the expectation value of the operator A in a ground state of the system that is somewhat altered by its interaction with the detector at position x. Therefore the resulting dispersion of A will be determined by the *initial* quantum mechanical spread of the detector wave function. The error of the measurement stems from the interaction with the detector rather than from the intrinsic noise of the system measured.

3.5. FCS OF THE ELECTRIC CURRENT IN A NORMAL CONDUCTOR

A complementary example is a normal conductor biased at finite voltage. This is a stationary *non-equilibrium* system far from being in its ground state. Here we do not go to microscopic details of the derivation. Our immediate aim is to make contact with the approaches of Refs. [2, 5]. We keep the original notations of the references wherever it is possible.

The FCS of the current in a multi-mode QPC which is characterized by a set of transmission coefficients T_n [c. f. Eq. (37) of [2]] at zero temperature and bias voltage V_{QPC} reads

$$P(x^+, x^-, \tau) = \exp\left\{\frac{e\tau}{2\pi}|V_{\mathrm{QPC}}|S[ie(x^+ - x^-)\mathrm{sign}V_{\mathrm{QPC}}]\right\} \tag{32}$$

with the function

$$S(\xi) = \sum_n \ln[1 + (e^\xi - 1)T_n]. \tag{33}$$

This expression depends on $x^+ - x^-$ only. This is a direct consequence of gauge invariance. Indeed, in each of the Hamiltonians the coupling term is the coupling to a vector potential localized in a certain cross-section of the constriction. The gauge transform that shifts the phase of the wave functions by $x^\pm/e$ on the right side of that cross-section eliminates this coupling term. This transformation was explicitly implemented in [5]. Since there are *two* Hamiltonians in the expression, the coupling terms cannot be eliminated simultaneously when $x^+ \neq x^-$. However, the gauge transform with the phase shift $(x^+ + x^-)/2$ makes the coupling terms depending on $x^+ - x^-$ only.

Since $P(x^+, x^-, \tau)$ depends on $x^+ - x^-$ only, the FCS $P(x, q, \tau)$ is independent of x. As we have seen in section 3.3, under this condition the FCS can be interpreted as a probability distribution. As in section 2 we conclude that for the statistics of the current of non-interacting electrons the characteristic function $\chi(\lambda) = P(\lambda/2, -\lambda/2, \tau)$ is associated with a positive probability distribution.

Superconductivity breaks gauge invariance, thus making such an interpretation impossible.

4. Electric Circuits as General Linear Detectors

The model used in the previous section may seem rather abstract and unrealistic. To make connection to "real world", we notice that i. the time derivative of the detector momentum q is related to the current via the constriction ii. the velocity $\dot{x}$ enters the Hamiltonian in the same way as the voltage applied to the constriction. Next we assume the following definition of the "real world": the only quantities measured are electric voltages and currents between nodes of an electric circuit. The most adequate description of quantum mechanics of the system would thus contain these variables only. This description is hardly possible to achieve within a Hamiltonian formalism, since the latter can contain neither dissipation nor retardation. It is the Feynman-Vernon formalism [4] that allows to formulate quantum mechanics in the form of an action that contains only necessary variables. This action may be derived from Hamiltonian formalism by tracing out irrelevant degrees of freedom. But this is precisely what we have done in the previous sections! The conclusion is that the above results can be used to provide an adequate formulation of quantum dynamics of electric circuits that contain mesoscopic conductors. This formulation clarifies the notion of a detector.

A practical problem can be formulated as follows. Suppose we know the FCS of a mesoscopic conductor. When measured, it is embedded in an electric circuit. Generally speaking, one does not measure a voltage or current directly at the mesocopic conductor but rather somewhere else in the circuit. We present this in Fig. 1 in circuit theory terms. Either voltage or current measurement is performed with using output contacts 1 and $1'$. The mesoscopic conductor (black box) is connected to input contacts 2 and $2'$. The shaded box presents the electromagnetic environment of the mesoscopic conductor. It supposed to be a piece of a linear circuit and thus can be characterized separately from the "black box" by three (frequency-dependent) response functions.

The question is what is the FCS of the measurement.

We answer this question by extending our simple detection model to a set of dynamical detectors. That involves many degrees of freedom, as many as it is requred to present the "environment". The detector dynamics is essentially the dynamics of the environment. This will allow us to study classical back action effects of the detectors. These effects will be more important then the subtle "quantum" influence of the detector on the measured system discussed in the previous section. Detectors representing the environment shall obey linear equations of motion. This restriction to linear circuits is not important for most of the experimental situations. Detector non-linearities are avoided in many experiments because of undesired effects like the mixing of different frequencies.

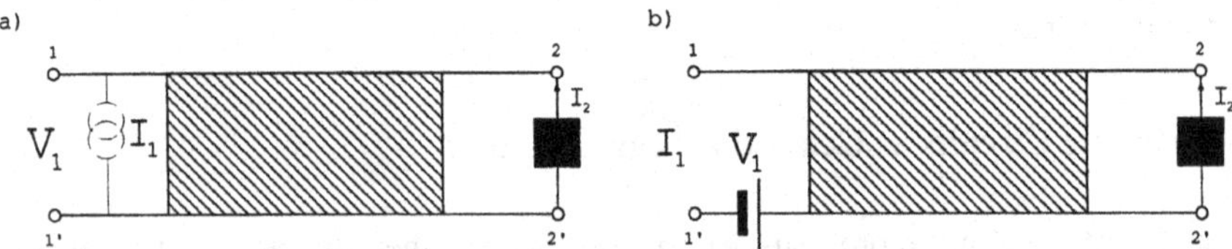

Figure 1. Electric circuit as a general linear detector. The black box denotes the electronic system to be measured. It is embedded into a linear environment (shaded box). Either voltage (a) or current (b) is measured between the output contacts 1 and $1'$ of the circuit.

4.1. THE MODEL OF A LINEAR ELECTROMAGNETIC ENVIRONMENT

It is accustom to model linear systems with a set of non-interacting bosons. [10] The Hamiltonian reads

$$H_{linear} = \sum_m \Omega_m a_m^\dagger a_m - \sum_i h_i(t) H_i \tag{34}$$

where operators H_i present the physical quantities of interest, those are conjugated to generalized forces h_i. These operators are linear combinations of $a_m^\dagger$, a_m

$$H_i = \sum_m c_m^{(i)} a_m + h.c \tag{35}$$

A proper choice of Ω_m, $c_m^{(i)}$ models arbitrary responce functions and enables one to formulate quantum dynamics in terms of quanitities of interest.[10] We are dealing with four variables, $I_{1,2}$ and $V_{1,2}$. Two of this four can be choosen as generalized forces and the remaining two as operators. Although different choices lead to different Hamiltonians, the concrete choice is just a matter of convenience.

Our primary choice is to treat I_2 and $\chi_1 = \int^t d\tau V_1(\tau)$ as generalized forces. The responce functions express I_1 and V_2 in terms of these two:

$$\begin{aligned} I_1(\omega) &= \tilde{Y}(\omega)V_1(\omega) + K(\omega)I_2(\omega) \\ V_2(\omega) &= \tilde{Z}(\omega)I_2(\omega) - K(\omega)V_1(\omega) \end{aligned} \tag{36}$$

More symmetric choices express either currents in terms of voltages via admittances Y_1, Y_2, Y_{12}

$$\begin{aligned} I_1(\omega) &= Y_1 V_1(\omega) + Y_{12}V_2(\omega) \\ I_2(\omega) &= Y_2(\omega)V_2(\omega) + Y_{12}V_1(\omega) \end{aligned} \tag{37}$$

or voltages in terms of currents via impedances Z_1, Z_2, Z_{12}.

$$\begin{aligned} V_1(\omega) &= Z_1 I_1(\omega) + Z_{12}I_2(\omega) \\ V_2(\omega) &= Z_2(\omega)I_2(\omega) + Z_{12}I_1(\omega) \end{aligned} \tag{38}$$

There are obvious relations between these response coefficients: $\tilde{Z} = 1/Y_2, \tilde{Y} = (Y_1Y_2 - Y_{12}^2)/Y_2, K = Y_{12}/Y_2, (Z_1, Z_2, Z_{12}) = (Y_2, Y_1, -Y_{12})/(Y_1Y_2 - Y_{12}^2)$.

Here we assume a passive circuit in thermal equilibrium. The response functions thus satisfy Onsager symmetry relations and the fluctuation-dissipation theorem relates the response functions and fluctuations of the corresponding variables.

4.2. GENERAL RELATION

We derive now the general relation that determines the full counting statistics of the outputs $I_1(t)$ of the linear detector described above *provided* it is coupled to the "black box" to be measured. We will follow the lines of section 3.2. However, now we assume the detector to be in the state of thermal equilibrium at temperature T. In addition, instead of evaluating the final density matrix of the detector, we analyze the expressions for the partition functional $\mathcal{Z}_I$ that generate moments of the read-off environment variable I_1. This functional is defined as

$$\mathcal{Z}_I[\chi_1] = \left\langle \overleftarrow{\mathrm{T}} e^{i\int dt\, [H_I - \chi_1^-(t)I_1]} \overrightarrow{\mathrm{T}} e^{-i\int dt\, [H_I - \chi_1^+(t)I_1]} \right\rangle \tag{39}$$

and generates moments of outcomes of measurements of $I_1(t)$ at any instant of time.

It is advantageous to write the functional in its dependence on the combinations and $V_1^{cl} = (\partial/\partial t)(\chi_1^+ + \chi_1^-)/2$, $\chi_1^q = (\chi_1^+ - \chi_1^-)/2$, that we collect into a "vector" $\chi_1 = (V_1^{cl}, \chi_1^q)$. The 'classical' field V_1^{cl} account for (time-dependent) voltage source between the contacts 1 and 1'. Moments of $I_1(t)$ are generated by differentiation of $\mathcal{Z}_I$ with respect to the anti-symmetric, sometimes called 'quantum', fields:

$$\langle [I_1(t_1)...I_1(t_m)] \rangle = \frac{\delta}{-2i\delta\chi_1^q(t_1)} \cdots \frac{\delta}{-2i\delta\chi_1^q(t_m)} \mathcal{Z}_I[\chi_1]\Big|_{\chi^q{}_1=0} \tag{40}$$

In general, these moments do depend on V_1^{cl}.

H_I presents the full Hamiltonian of the circuit for current measurement. It reads

$$H_I = H_{\text{env}} + H_{\text{Sys}} - \chi_2 I, \tag{41}$$

the first term presenting the boson detector degrees of freedom and the second one incorporating electron degrees of freedom of the "black box". The third term presents the coupling of eletric current and the detector degree of freedom $\chi_2(t) = \int^t dt'\, V_2(t')$. The latter is thus an analogue of x from the section 3.

We note that both χ_2 and I_1 are linear combinations of boson creation/annihilation operators,

$$\chi_2 = \sum_m c_m^{(x)} a_m + c_m^{(x)*} a_m^\dagger, \;\; I_1 = \sum_m c_m^{(I)} a_m + c_m^{(I)*} a_m^\dagger. \tag{42}$$

We rewrite now $\mathcal{Z}_V$ as a path integral in detector variables, like we have done to derive Eq. (11). The integration variables now are $a_m^{(\pm)}(t)$, $\pm$ corresponding to two parts of the Keldysh contour. Operators χ_2, I_1 are replaced by

$$\chi_2^{(\pm)} = \sum_m c_m^{(x)} a_m^{(\pm)} + h.c \;\; I_1^{(\pm)} = \sum_m c_m^{(I)} a_m^{(\pm)} + h.c.^\dagger. \tag{43}$$

the sign depending on the part of the countour they are at. To proceed, we introduce two extra vector variables to the path integral: $\chi_2 = (V_2^{cl}, \chi_2^q)$ and $\mathbf{q}_2 = (I_2^{cl}, q_2^q)$, by inserting an identity of the form

$$\begin{aligned} 1 \simeq \int \prod_t \delta(2V_2^{cl} - \dot{\chi}_2^+ - \dot{\chi}_2^-)\delta(2\chi_2^q - \dot{\chi}_2^+ + \dot{\chi}_2^-)\mathcal{D}[V_2^{cl}]\mathcal{D}[\chi_2^q] \\ \simeq \int \mathcal{D}[V_2^{cl}]\mathcal{D}[\chi_2^q]\mathcal{D}[I_2^{cl}]\mathcal{D}[q_2^q] \\ \exp(i\int dt (I_2^{cl}(2\chi_2^q - \dot{\chi}_2^+ + \dot{\chi}_2^-) - iq_2^q(2V_2^{cl} - \dot{\chi}_2^+ - \dot{\chi}_2^-))) \end{aligned} \tag{44}$$

to the path integral. This allows us to split the integrand onto two factors. One factor is a trace over "black box" variables. It depends on $\boldsymbol{\chi}_2(t)$ only and is in fact the influence functional $\mathcal{Z}_{\text{Sys}}$ introduced in the section 3.2, Eq. (15). It conforms our definition of FCS for "black box" biased by voltage $V^{cl}(t)$. Another factor is a Gaussian form in $a_m(t)$. This enables us to perform integration over a_m, this yeilds

$$\mathcal{Z}_I[\chi_1] = \int \mathcal{D}[\chi_2] e^{-i(\mathcal{S}_{\text{env}}([\chi_1],[\mathbf{q}_2]) + \mathcal{S}_{\text{coup}}([\mathbf{q}_2],[\boldsymbol{\chi}_2]))} \mathcal{Z}_{\text{Sys}}[\chi_2] \tag{45}$$

$$\mathcal{S}_{\text{env}} = \int dt \left[\chi_1 \check{Y} \chi_1 + 2\chi_1 \check{K} \mathbf{q}_2 + \mathbf{q}_2 \check{Z} \mathbf{q}_2\right] \tag{46}$$

$$\mathcal{S}_{\text{coup}} = \int dt \left[I_2^{cl}(t)\chi_2^q(t) - q_2^q(t) V_2^{cl}(t)\right] \tag{47}$$

The environmental part of the action is expressed in terms of responce functions of our primary choice. The 2×2 matrices $\check{\check{Y}}, \check{K}$ are integral kernels depending on time difference. They are defined in frequency representation in terms of corresponding response functions

$$\check{\check{Y}}(\omega) = \begin{pmatrix} 0 & \check{Y}^*(\omega) \\ \check{Y}(\omega) & -2i\omega \coth(\frac{\omega}{2T})\mathrm{Re}\check{Y} \end{pmatrix}; \tag{48}$$

$$\check{K}(\omega) = \begin{pmatrix} 0 & -K^*(\omega) \\ K(\omega) & 2\omega \coth(\frac{\omega}{2T})\mathrm{Im}\check{Y} \end{pmatrix}. \tag{49}$$

Here T stands for temperature of the environment. The matrix $\check{\check{Z}}$ is of the same form as $\check{\check{Y}}$. We can further simplify this relation by integrating over $\mathbf{q_2}$. The result acquires more symmetric form

$$\mathcal{Z}_I[\chi_1] = \int \mathcal{D}[\chi_2] e^{-i\mathcal{S}_{\mathrm{env}}([\chi_1],[\chi_2])} \mathcal{Z}_{\mathrm{Sys}}[\chi_2] \tag{50}$$

$$\mathcal{S}_{\mathrm{env}} = \int dt \left[\chi_1 \check{Y}_1 \chi_1 + 2\chi_1 \check{Y}_{12} \chi_2 + \chi_2 \check{Y}_2 \chi_2\right] \tag{51}$$

where matrices $\hat{Y}_1, \hat{Y}_2, \hat{Y}_{12}$ are defined as in Eq. 48 This is the general relation desired: it expresses FCS of electric current in course of a general electric measurement where the mesoscopic conductor is embedded into a linear electric circuit. It is important that all informanion about the mesoscopic conductor is included into its FCS at voltage bias, $\mathcal{Z}_{\mathrm{Sys}}$.

Let us discuss the relation in some detail. First, let us replace the general four-pole electric circuit by a single two-pole resistor $Z(\omega)$ (see Fig. 2). We do this by setting $Y_1 = Y_2 = -Y_{12} = 1/Z$. The resistor and the "black box" enter the expression on equal foot

$$\mathcal{Z}_I[\chi_1] = \int \mathcal{D}[\chi_2] \mathcal{Z}_{\mathrm{Resistor}}[\chi_2 - \chi_1] \mathcal{Z}_{\mathrm{Sys}}[\chi_2] \tag{52}$$

$$\mathcal{Z}_{\mathrm{Resistor}}[\chi] = \exp\left\{-i \int dt [\chi(\check{Z}^{-1})\chi]\right\} \tag{53}$$

since here $\mathcal{Z}_{\mathrm{Resistor}}$ has the same meaning as $\mathcal{Z}_{\mathrm{Sys}}$ defining the (gaussian) FCS of the linear voltage-biased two-pole resistor. This gives a simple concatenation rule for FCS of compound circuits: the FCS is the convolution of the FCS's of individual elements. The resulting FCS is presented as an integral over fluctuating phase χ_2 defined in the node of the circuit. This is in agreement with known results concerning Keldysh action of electric circuits with Josephson junctions. [11] Moreover, the rule can be easily generalized for more comlicated circuits, for example, the one with two mesoscopic conductors in series.

It is interesting to note that the integration over the fluctuating phase in fact represents a constrain on quantum motion of the compound system. The constrain

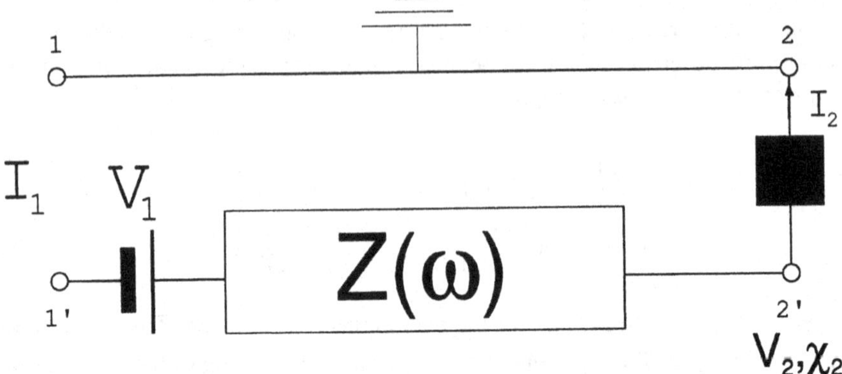

Figure 2. A simple example of linear environment: resistor $Z(\omega)$ in series with the "black box". The circuit is fed with voltage source V. Quantum dynamics of the system is formulated in terms of fluctuating fields V_2, χ_2.

reflects the simple fact that the current through the "black box" equals the input current of our detector.

$$\delta(I_{\rm Sys} - I_1) \to \prod_t \delta(I_{\rm Sys} - I_1^+(t)) \prod_t \delta(I_{\rm Sys} - I_1^-(t)) \simeq \tag{54}$$

$$\int \mathcal{D}[\chi] \exp\left\{ i \int dt(\chi^+(t)(I_{\rm Sys} - I_1^+(t)) - \chi^-(t)(I_{\rm Sys} - I_1^-(t))\right\} \tag{55}$$

the last equality holding for operators. Since these two currents consist of very different operators (electrons for I and bosons for I_2), the constrain would be difficult to handle within a Hamiltonian formalism. This illustrates the use of Keldysh action in this context.

One may have noticed that we disregard the constant factors that arise on various stages of the derivation from the variable change in the path integrals and/or subsequent Gaussian integration. This is not a matter of our carelessness but rather indicates a certain problem with the approach in use. The "correct" factor in the final relation (50) is in fact known from the fact that $Z_I(\chi) = 1$ at $\chi^q = 0$ per definition of generating function. The eventual calculation does not reproduce this, the result being dependent on the details of response functions at high frequencies. The problem is not related to the embedding of the mesoscopic conductor and persists also for compound linear circuits. It stems from the difficulties in defining a proper measure for path integration. One may also draw analogies with known, general and long-standing problems related to quantum mechanics of constrained motion.[12] To our knowledge, this drawback never leads to practical problems. One can always correct for a wrong factor that might arise in a calculation by "normalizing" the result to $Z_I(\chi) = 1$ at $\chi^q = 0$. We will further assume that the correct factor is included into the definition of the metrics in the function space, that is, into definitions of $\mathcal{D}\chi_2$.

4.3. VOLTAGE MEASUREMENT AND PSEUDO-PROBABILITIES

The evaluation of the FCS of elctric current a mesoscopic conductor under condition of voltage bias is a rather straightforward task. One proceeds within Hamiltonian formulation in terms of electronic operators subjected to external voltage. The conjugated problem is the FCS of voltage fluctuations under condition of current bias. The absence of an obvious Hamiltonian formulation for current bias makes this problem less straightforward.

Albeit the problem can be solved in a general way within the approach of the previous subsection. Let us consider the generating function of the voltage fluctuations between 1 and 1' (Fig. 1 b) defined by the relation

$$\mathcal{Z}_V[\mathbf{q}_1] = \left\langle \overleftarrow{\mathrm{T}} e^{i\int dt\,[H_V + q_1^-(t)V_1]} \overrightarrow{\mathrm{T}} e^{-i\int dt\,[H_V + q_1^+(t)V_1]} \right\rangle \tag{56}$$

Similar to Eq. 40, the derivatives of this functional with respect to $q^q(t)$ give the moments of voltage fluctuations. Let us now repeat all derivations of the previous subsection for this particular functional. The answer reads:

$$\mathcal{Z}_V[\mathbf{q}_1] = \int \mathcal{D}[\mathbf{q}_2] e^{-i(\mathcal{S}_{\text{env}}([\mathbf{q}_1],[\mathbf{q}_2]) + \mathcal{S}_{\text{coup}}([\mathbf{q}_2],[\chi_2]))} \mathcal{Z}_{\text{Sys}}[\chi_2] \tag{57}$$

$$\mathcal{S}_{\text{env}} = \int dt \left[\mathbf{q}_1 \check{Z}_1 \chi_1 + \mathbf{q}_1 \check{Z}_{12} \mathbf{q}_2 + \mathbf{q}_2 \check{Z}_2 \mathbf{q}_2\right] \tag{58}$$

$$\mathcal{S}_{\text{coup}} = 2\int dt \left[I_2^{cl}(t)\chi_2^q(t) - q_2^q(t) V_2^{cl}(t)\right]. \tag{59}$$

The matrices $\check{Z}_{1,2,12}(\omega)$ conform to the definitions (49), (38).

Let us now devise the linear detector in such a way that the voltage fluctuations between the 1 and 1' are the same as between 2 and 2' and the conditions of current biase are achieved. Elementary circuit theory tells us that this is achieved in the limit $Z_{12} \to -Z_1 = -Z_2 \to -\infty$. In the expression (57) this correspond to setting $\mathbf{q}_1 = \mathbf{q}_2$ and $\mathcal{S}_{\text{env}} = 0$.

This brings us to the remarkable conclusion: for any conductor, the generating functionals for voltage noise at current bias $I^{cl}(t)$ and for current noise at voltage bias $V^{cl}(t)$ are related by a functional Fourier transform,

$$\mathcal{Z}_V[\mathbf{q}] = \int \mathcal{D}[V^{cl}]\mathcal{D}[\chi^q]\, e^{2i\int dt (q^q V^{cl} - I^{cl}\chi^q)} \mathcal{Z}_I[\chi]. \tag{60}$$

Therefore, the functionals $\mathcal{Z}_V$ and $\mathcal{Z}_I$ are in fact just different forms of the same object. There is some uncertainty in this definition steming from the problem of the previous section. The $\mathcal{Z}_V$ has to be "normalized" so that $\mathcal{Z}_V = 1$ at $q^q(t) = 0$. We assume the normalization factor to be included into the definiton of $\mathcal{D}$.

This simple relation between $\mathcal{Z}_V$ and $\mathcal{Z}_I$ suggests to define a singe functional of the following form:

$$\begin{aligned}\tilde{P}([V],[I]) &= \int \mathcal{D}[q^q]\, e^{2i\int dt\, q^q V}\, \mathcal{Z}_V\left[\begin{pmatrix} I \\ q^q \end{pmatrix}\right] \\ &= \int \mathcal{D}[\chi^q]\, e^{-2i\int dt\, \chi^q I}\, \mathcal{Z}_I\left[\begin{pmatrix} V \\ \chi^q \end{pmatrix}\right] \end{aligned} \tag{61}$$

that depends on "classical" variables only. We show with the help of Eq. (40) that $\tilde{P}$ has the following properties:

$$\langle V(t_1)...V(t_m)\rangle\Big|_{I(t)} = \frac{\int \mathcal{D}[V]\, V(t_1)...V(t_m)\tilde{P}([V],[I])}{\int \mathcal{D}[V]\, \tilde{P}([V],[I])}, \tag{62}$$

$$\langle I(t_1)...I(t_m)\rangle\Big|_{V(t)} = \frac{\int \mathcal{D}[I]\, I(t_1)...I(t_m)\tilde{P}([V],[I])}{\int \mathcal{D}[I]\, \tilde{P}([V],[I])}. \tag{63}$$

Those closely resemble the properties of the probability density. Why $\tilde{P}$ is not a probability? First, $\tilde{P}$ does not have to be positive, just as the $P(x,q,\tau)$ discussed in the section 3. Second, it is dimentionless in contrast to the true probability density of either voltage or current fluctuations. For these reasons we refer to $\tilde{P}$ as to 'pseudo-probability'.

$\tilde{P}([V],[I])$ is a characteristics of a two-pole conductor and takes a simple form for a linear one. In this case $\tilde{P}$ is sure positive and depends only on the impedance of the conductor $Z(\omega)$ and its temperature T,

$$\tilde{P}_Z([V],[I]) = \exp\left\{-\int \frac{d\omega}{4\pi} \frac{|V(\omega) - Z(\omega)I(\omega)|^2}{\omega \mathrm{Re} Z} \tanh\frac{\omega}{2T}\right\}. \tag{64}$$

From Eq. (62) one derives the voltage fluctuation

$$\left\langle V(t)^2 \right\rangle = \int \frac{d\omega}{2\pi}\, \omega\, \mathrm{Re}\, Z(\omega) \coth\frac{\omega}{2T}, \tag{65}$$

that conforms to the fluctuation-dissipation theorem.

The general relation (50) can be rewritten in terms of the pseudoprobabilities. In this case, it will express the pseudoprobability of current/voltage fluctuations between 1 and 1' $\tilde{P}_1([V],[I])$ in terms of the $\tilde{P}_{\mathrm{Sys}}$ of the "black box" and four-pole pseudoprobability $\tilde{P}_{12}$ that depends on two currents and voltages characterizing the linear part of the circuit:

$$\tilde{P}_1([V_1],[I_1]) = \int \mathcal{D}[V_2]\mathcal{D}[I_2]\tilde{P}_{12}([V_1],[I_1];[V_2],[I_2])\tilde{P}_{\mathrm{Sys}}([V_2],[I_2]) \tag{66}$$

We also give here a simpler relation for the "black box" in series with a linear resistor $Z(\omega)$, this corresponds to Eq. 52. The answer is expressed in terms of the pseudoprobability of the resistor defined by Eq. 64:

$$\tilde{P}_1([V],[I]) = \int \mathcal{D}[V_2]\tilde{P}_Z([V-V_2],[I])\tilde{P}_{\mathrm{Sys}}([V_2],[I]) \tag{67}$$

The above relations are transparent if one comprehends them in terms of classical probabilities. They just show that the probability of a certain current/voltage fluctuation is composed of the probabilities of the fluctuations in the "black box" and linear detector, voltage and current satisfying the circuit theory rules. Yet the relations are quantum mechanical ones and are written for pseudoprobabilities. They contain all necessary information about quantum properties of the system under consideration.

To summarize, the difference between the FCS of the "black box" at voltage bias and that of the "black box" embedded into the linear environment is contributed by three factors. First, the voltages and currents are divided according to circuit theory rules. For instance, for the circuit of Fig. 2 the current fluctuation δI in the "black box" results in the current fluctuation $\delta I/(1+ZG)$ in the whole circuit, G being linear conductance of the "black box". This factor is rather trivial from the classical point of view. However, if we express this in terms of quantum mechanical detection, it constitues the main part to the detector back-action. This "classical" back-action could be disregarded for static detectors considered in the first part of this article. Second, the linear part of the circuit produces its own quantum and thermal Gaussin fluctuations. These fluctuations are also detected in course of the FCS of the compound system. Third, the fluctuations in the linear environment affect the fluctuations produced by the "black box". This may result in non-Gaussian response. This factor is least trivial and will be addressed in some detail in section 6.

5. Low-frequency limit

We have shown in the previous section that the FCS of the mesoscopic conductor in general linear environment can be expressed by the relation (50). This relation involves path integration thus defining a quantum field theory with the field $\chi_2(t)$. This quantum field theory is a non-linear one, the non-linearities coming form the mesoscopic conductor. The latter is sometimes underestimated, since $I-V$ curve of a mesoscopic conductor may be perfectly linear. However, even in this case the noise and higher correlators do depend on voltages, this provides non-linearity. Such non-linearity already makes the general problem hardly tractable. To complicate the situation further, the most important part of the action, log $\mathcal{Z}_{Sys}$, is only known for *stationary* χ. In this case it is given by the relation (32). Microscopic Keldysh Green function techniques can be used to evaluate this functional at non-stationary χ in the frame work of perturbation theory. Albeit it would not provide a general answer valid for arbitrary $\chi(t)$.

Fortunately enough, there is a simple limit where the quantum field theory becomes tractable. Moreover, this limit is most likely to occur in course of experimental observation. In this low-frequency limit we consider only quasi-stationary realizations of the fields. The action for these realizations is *local* in time, this enables easy path interation.

Before getting to concrete formulation, let us discuss time scales that occur in our problem and may determine the non-locality of the action. There may be a time scale coming from frequency dependence of the detector response functions $Y_{1,2,12}$; to put it plainly, RC-time. This scale depends on the system layout and can be easily changed. From the experimental point of view it is convenient to choose response functions to be constants in a wide frequency interval. This allows us to disregard this time scale. Another time scale τ_Q is related to quantum mechanics. The environment action 51 becomes non-local at frequencies $\omega \simeq T$, where crossover from classical to quantum noise does occur. For the mesoscopic conductor, similar scale may arise from the energy related to single electron transfer through the conductor, eV. Therefore $\tau_Q \simeq \hbar/\min\{eV, T\}$. Third time scale is a typical time interval τ_I between the electron transfers through the mesoscopic conductor, e/I. Comparing these two scales, we find that $t_I \simeq (\hbar G/e^2)\tau_Q$, G being the conductance of the "black box". Generally, we expect the low-frequency limit to be valid at time scales exceeding both τ_I and τ_Q. These time scales are typically short in comparison with time resolution of measurement electronics. So that a realistic electric measurement is always performed in the low-frequency limit.

In the limit discussed, the general relation (50) takes the following form (all response functions are taken at zero frequency) :

$$Z_I([V_1],[\chi_1]) = \int \mathcal{D}[V]\mathcal{D}[\chi]\exp\{-\int dt(\mathcal{E}_{\text{env}}(V_1(t),\chi_1(t);V_2(t),\chi_2(t)) + \\ +\mathcal{E}_{\text{Sys}}(V_2(t),\chi_2(t)))\}; \quad (68)$$

where

$$\begin{aligned}\mathcal{E}_{\text{env}}(V_1,\chi_1;V_2,\chi_2) &= -2i((V_1Y_1+V_2Y_{12})\chi_1+(V_2Y_2+V_1Y_{12})\chi_2))+ \\ +4T(Y_1\chi_1^2+Y_2\chi_2^2 &+ 2Y_{12}\chi_1\chi_2), \\ \mathcal{E}_{\text{Sys}} &= -\ln Z_{Sys}(V,\chi)/\tau.\end{aligned}$$

In the last equation, the Z_{Sys} is evaluated for stationary V,χ. Thus its log is proportional to the measurement time τ (Eq. 32). For the resistor in series (Fig. 2) $\mathcal{E}_{\text{env}}$ reduces to:

$$\mathcal{E}_{\text{env}} = \mathcal{E}_Z(V_1-V_2,\chi_1-\chi_2);\ \mathcal{E}_Z(V,\chi) = (-2iV+4T\chi)\chi/Z. \quad (69)$$

As expected in the low-frequency limit, the noise of the detector is just Johnson-Nyqist white noise.

A simple estimation shows that the action is not only local in time. It is also big for time intervals in question. This allows us to implement a semiclassical approximation, that is, to evaluate the path integrals in saddle-point approximation. Indeed, the typical values of χ manifesting discretness of charge are of the order of $1/e$. The action can be just estimated as $S \simeq I\chi\tau \simeq I/e\tau \simeq (\tau/\tau_I)$. By definition of low-frequency limit, $\tau \gg \tau_I$.

We concentrate now on the starting problem of this article: distribution of the current averaged over the time interval τ. Taking the path integrals in the saddle-point approximation, we arrive on a simple relation for FCS of the "black box" emdedded into the enviroment:

$$Z_I(V_1,\chi_1) = \exp(-\mathcal{E}_I(V_1,\chi_1)\tau);$$
$$\mathcal{E}_I(V_1,\chi_1) = \underset{V_2,\chi_2}{\text{extr}} \{\mathcal{E}_{\text{env}}(V_1,\chi_1;V_2,\chi_2) + \mathcal{E}_{\text{Sys}}(V_2,\chi_2)\} \tag{70}$$

The corresponding FCS of voltage fluctuations at given current I can be obtained directly from Eq. 70 with using Eq. 60.

$$Z_V(I,q) = \exp(-\mathcal{E}_V(I,q)\tau);$$
$$\mathcal{E}_V(I,q) = \underset{V,\chi}{\text{extr}} \{\mathcal{E}_I(V,\chi) - 2i(I\chi - qV)\}\,. \tag{71}$$

When integrating Eq. 60 we take care of normalization to assure that $Z_V(I,0) = 1$. A similar integration determines pseudoprobability:

$$\tilde{P}(I,V) = \exp(-\mathcal{E}(V,I)\tau);$$
$$\mathcal{E}(I,V) = \underset{\chi}{\text{extr}} \{\mathcal{E}_I(V,\chi) - 2iI\chi\} \tag{72}$$

Finally, we give here expressions for real probabilities to measure voltage V at a given current I, $P_V(I)$, and to measure current I at a given voltage V, $P_I(V)$. By virtue of relations (62),(63) we obtain

$$\left\{ \begin{array}{c} P_V(I) \\ P_I(V) \end{array} \right\} = \sqrt{\frac{\pi}{2\tau}} \left\{ \begin{array}{c} \sqrt{\frac{\partial I^2}{\partial^2 \mathcal{E}}} \\ \sqrt{\frac{\partial V^2}{\partial^2 \mathcal{E}}} \end{array} \right\} \tilde{P}(I,V) \tag{73}$$

Remarkably, these probabilities are in fact *the same* with exponential accuracy. They differ in normalization factors only,

$$\frac{P_V(I)}{P_I(V)} = \sqrt{\frac{\partial I^2}{\partial^2 \mathcal{E}} \frac{\partial^2 \mathcal{E}}{\partial V^2}}. \tag{74}$$

The relations (71-74) are valid for the statistics of the outputs of the compound cirquit as well as for the "black box" biased by either voltage or current source.

The FCS in the low-frequency limit can be thus formulated in terms of stationary functions $\mathcal{E}_I, \mathcal{E}_V, \mathcal{E}(I,V)$. These functions resemble potentials used in thermodynamics. As these potentials, they are related by Legendre transforms. However, the relations presented concern non-equlibrium systems. They provide an effective minimimun principle for fluctuations in such systems.

5.1. WHERE ARE THE CHARGING EFFECTS?

The electron transport in mesoscopic circuits is known to be affected by charging or Coulomb blockade effects, those manifest electron-electron interaction at mesoscopic scale. The simplest case of a tunnel junction embedded into an electromagnetic environment is described in [13] in some detail. Although these charging effects is not the topic of this article, we find it important to note that they are there in our formalism. Indeed, the Keldysh action in use is in fact a generalization of the action introduced in [11] to descibe Coulomb blockade effects in Josephson junctions.

How the charging effects manifest themselfs in the low-frequency limit considered?

The way to understand this is to envoke the field-theoretical concept of renormalization. We have obtained the low-frequency action just by substituting the slow realizations of the fluctuating field χ_2 into the original action. The renormalization procedure gives more accurate way to obtain the low-frequency action. In this procedure, the field is seperated onto slow and fast part. The path integration over fast variables is performed. The resulting functional of slow variables presents a new renormalized action. For the problem under consideration, this mere renormalization presents the charging effects. Since the environment action is a quadratic one, it is not renormalized. The renormalization can be thus ascribed to $\mathcal{E}_{\mathrm{Sys}}$. The typical time scale of relevant fast modes is τ_Q. The importance of the charging effects is governed by the value of environment impedance at the corresponding frequencies, $Z(1/\tau_Q)$.[13] If $Z(1/\tau_Q) \ll \hbar/e^2$, the renormalization is small and charging effects can be disregarged. In the opposite case, the charging effects are important and the renormalized action $\mathcal{E}$ differs significantly from the bare one.

We thus conclude that the charging effects do not change the low-frequency relations given in this subsection. Their only effect is the renormalization of $\mathcal{E}$ in comparison with its bare value at voltage bias. This renormalization does depend on the environment response functions at frequency scale $1/\tau_Q$.

6. Quantum Point Contact in a circuit

In this section, based on Ref. [9], we illustrate the general relations obtained above. We do this by considering a specific example: a quantum point contact in series with an impedance Z (Fig. 2). Further we assume the shot noise limit: $eV_2 \gg T$, and disregard the Jhonson-Nyquist noise produced by the linear resistor. First we address the FCS of current fluctuations in the low-frequency limit discussed above. We will thus make use of the relations given in the section 5. The assumtion $Z(\omega \simeq eV) \ll \hbar/e^2$ allows us to disregard charging effects without putting any restrictions on $Z(0)$.

To employ our scheme, we first need the concrete expression for the FCS of the voltage-biased QPC, $\mathcal{Z}_{\text{Sys}}$. In stationary case, this has been derived in [2]. The QPC is completely characterized by its transmission eigenvalues T_n [14] that are assumed not to depend on energy. In our notations, this expression reads ($|eV| \ll T$)

$$\mathcal{Z}_{\text{Sys}}(V,\chi) = \exp\{-\mathcal{E}_{\text{Sys}}\tau\}\,;\mathcal{E}_{\text{Sys}} = -\frac{e\tau}{2\pi}|V|S[2ie\chi\,\text{sign}(V)] \qquad (75)$$

where dimensionless S is given by Eq. 33.

Since we disregard the noise of the resistor, the corresponding expression for environment action is just $\mathcal{E}_Z(V,\chi) = 2iVZ\chi$. By virtue of Eq.70, the FCS of the circuit is defined by

$$\mathcal{E}_I(V,\chi) = \underset{V_2,\chi_2}{\text{extr}}\left\{2iZ(V-V_2)(\chi-\chi_2) - \frac{e|V_2|}{2\pi}S[2ie\chi\,\text{sign}(V)]\right\} \qquad (76)$$

The simple minimization with respect to V_2 eventually fixes χ_2

$$2iZ(\chi-\chi_2) = -\frac{e}{2\pi}S(2ie\chi_2) \qquad (77)$$

and makes reduntant further minimization with respect to this variable. So that we immediately come to

$$\mathcal{E}_I(V,\chi) = -\frac{e|V|}{2\pi}S(2ie\chi_2(\chi)) \qquad (78)$$

where χ_2 is implicitly given by Eq. 77. In the limit of vanishing external imdedance $Z \to 0$ $\chi \approx \chi_2$ so that $\mathcal{E}_I(V,\chi) \approx \mathcal{E}_{\text{Sys}}$, as it should.

Taylor series of $\mathcal{E}_I$ in χ give the cumulants of the transmitted charge $\langle\langle Q^p\rangle\rangle$ in the circuit given, whereas Taylor series of $\mathcal{E}_I$ give the same cumulants $\langle\langle Q^p\rangle\rangle_0$ in the limit of $Z = 0$. Comparing the series, we obtain the relations between these cumulants. Linear terms give $\langle Q\rangle = (1+ZG)^{-1}\langle Q\rangle_0$, G being the conductance of the QPC. This is a trivial circuit theory relation: the averaged current $\bar{I}$ is reduced by a factor $1 + ZG$ by adding the resistor Z in series. The Langevin approach of Ref. [15] predicted that the same rescaling applies to the fluctuations of the current. Indeed, in second order we find that $\langle\langle Q^2\rangle\rangle = (1+Z_2G)^{-3}\langle\langle Q^2\rangle\rangle_0$, this is in agreement with this prediction.

However, for higher order cumulants we find the terms not consistent with this rescaling hypothesis. For instance, the third cumulant is expressed as

$$\langle\langle Q^3 \rangle\rangle = \frac{\langle\langle Q^3 \rangle\rangle_0}{(1+ZG)^4} - \frac{3Z_2G}{(1+ZG)^5} \frac{(\langle\langle Q^2 \rangle\rangle_0)^2}{\langle Q \rangle_0}. \tag{79}$$

Although the first term on the the right-hand-side does have the scaling form conjectured, the second term does not. This is generic for all $p \geq 3$: $\langle\langle Q^p \rangle\rangle = (1+ZG)^{-p-1} \langle\langle Q^p \rangle\rangle_0$ plus a non-linear (multinomial) function of lower cumulants. To give another example, the forth cumulant reads

$$\langle\langle Q^4 \rangle\rangle = \frac{\langle\langle Q^4 \rangle\rangle_0}{(1+Z_2G)^5} -$$
$$- \frac{10ZG}{(1+Z_2G)^6} \frac{\langle\langle Q^2 \rangle\rangle_0 \langle\langle Q^3 \rangle\rangle_0}{\langle Q \rangle_0} + \frac{15Z^2G^2}{(1+Z_2G)^7} \frac{\langle\langle Q^2 \rangle\rangle_0^3}{\langle Q \rangle_0^2}. \tag{80}$$

Let us now turn to one-channel conductor with transmission probability T_0. In this case, the distribution of the integer charge transmitted $q \equiv I\tau e$ for voltage bias V is known to take a simple binomial form [2] ($\phi \equiv eV\tau/2\pi$)

$$P_\phi(q) = \binom{\phi}{q} T_0^q (1-T_0)^{\phi-q}. \tag{81}$$

Let us find the dual distribution of flux transferred $\phi \equiv eV\tau/2\pi$ for the QPC in the condition of current bias I. We turn to the relation 72 that determines these probabilities. For a QPC in the shot noise limit, the corresponding $\mathcal{E}(I, V)$ reads

$$\mathcal{E}(I, V) = \underset{\chi}{\text{extr}} \left\{ \frac{eV}{2\pi} S(2ie\chi) - 2iI\chi \right\} \tag{82}$$

Expanding this near the point of extremum, we prove that

$$\frac{\partial I^2}{\partial^2 \mathcal{E}} \frac{\partial^2 \mathcal{E}}{\partial V^2} = \frac{I^2}{V^2} \tag{83}$$

for any functional form of S. Recalling Eq. 74 we readily obtain that the distribution function of flux transferred is given by (here we set $q \equiv I\tau/e$)

$$P_q(\phi) = \frac{q}{\phi} P_\phi(q) = \binom{\phi-1}{q-1} T_0^q (1-T_0)^{\phi-q}. \tag{84}$$

This form is known as Pascal distribution.

We complement this derivation by a simple interpretation of Pascal distribution (84) of voltage fluctuations. The binomial distribution (81) of charge transferred can be interpreted [2] in gambling terms: this is the probability to win (transfer an electron) q times in ϕ_0 game slots given a winning chance T_0. The current bias changes the rules of the game. The gambling now goes till q_0 sucsessful attempts. This requries a fluctuating number ϕ of game slots. Indeed, the Pascal distribution derived to originally to quantify gambling gives the probability of the number of independent trials that one has to make to acheive a given number of successes.

References

1. L. S. Levitov and G. B. Lesovik, JETP Lett. **55**, 555 (1992).
2. L. S. Levitov and G. B. Lesovik, JETP Lett. **58**, 230 (1993); L. S. Levitov, H. W. Lee, and G. B. Lesovik, Journ. Math. Phys. **37**, 4845 (1996).
3. J. Rammer and H. Smith, Rev. Mod. Phys. **58**, 323 (1986).
4. R.P. Feynman and F.L. Vernon, Ann. Phys.(N.Y.) **24**, 118 (1963).
5. Yu. V. Nazarov, Ann. Phys. (Leipzig) **8** Spec. Issue SI-193, 507 (1999).
6. W. Belzig and Yu. V. Nazarov, Phys. Rev. Lett. **87**,197006 (2001).
7. A. Shelankov, J. Rammer, cond-mat/0207343 (2002).
8. Yu. V. Nazarov and M. Kindermann, cond-mat/0107133 (2001).
9. M. Kindermann, Yu.V. Nazarov, and C.W.J. Beenakker, cond-mat/0210617 (2002).
10. A.O. Caldeira and A.J. Leggett, Phys. Rev. Lett. **46**, 211 (1981).
11. G. Schön and A.D. Zaikin, Phys. Rep. **198**, 237 (1990).
12. L. Kaplan, N. T. Maitra and E. J. Heller, Phys. Rev. A **56**, 2592 (1997).
13. G.-L. Ingold and Yu. V. Nazarov, in *Single Charge Tunneling*, edited by H. Grabert and M. H. Devoret, NATO ASI Series B294 (Plenum, New York, 1992).
14. C. W. J. Beenakker, Rev. Mod. Phys. **69**, 731 (1997).
15. Ya. M. Blanter and M. Büttiker, Phys. Rep. **336**, 1 (2000). The effect of a series resistance on the noise power is discussed in S 2.5.

MULTITERMINAL COUNTING STATISTICS

A short review

DMITRI A. BAGRETS, YULI V. NAZAROV
Department of Applied Physics and
Delft Institute of Microelectronics and Submicrontechnology,
Delft University of Technology, Lorentzweg 1, 2628 CJ Delft, The Netherlands

1. Introduction

The field of quantum noise in mesoscopic systems has been exploded during the last decade, some achievements being summarized in a recent review article. [1] While in classical systems shot noise is just a straightforward manifestation of discreteness of the electron charge, it can be used in quantum system as unique tool to reveal the information about the electron correlations and entanglement of various kinds. Measurement of fractional charge in Quantum Hall regime [2], noise measurements in atomic-size junctions [3], chaotic quantum dots[4] and superconductors [5, 6] are milestones of the field of quantum noise. Starting from the pioneering work of Büttiker[7], a special attention has been also paid to shot noise and noise correlations in the "multi-terminal" circuits. The correlations of currents flowing to/from different terminals can, for instance, reveal Fermi statistics of electrons. Such cross-correlations have been recently seen experimentally in Ref. [8–10].

A very important step in the field of quantum noise has been made by Levitov *et. al.* in [11, 12] where an elegant theory of *full counting statistics* (FCS) has been presented. This theory provides an efficient way to investigate the current correlations in the mesoscopic systems. The FCS method concentrates on the evaluation of the probability distribution function of the numbers of electrons transferred to the given terminals during the given period of time. It yields not only the noise, but all higher momenta of the charge transfer. The probability distribution also contains the fundamental information about big current fluctuations in the system.

Y. V. Nazarov (ed.), Quantum Noise in Mesoscopic Physics, 429–462.

Original FCS method [11, 12] was formulated for the scattering approach to mesoscopic transport. This made possible the study the statistics of the transport through the disordered metallic conductor [13] and the two-terminal chaotic cavity [14]. Muzykantskii and Khmelnitskii have generalized the original approach by Levitov *et. al.* to the case of the normal metal/superconducting contacts. The very recent application of the scattering approach is the counting statistics of the charge pumping in the open quantum dots [15–17].

However, scattering approach becomes unpractical in case of realistic layouts where the scattering matrix is multi-channel, random and cumbersome. For practical calculations one evaluates the FCS with the semiclassical Keldysh Green function method [18] or with its reduction called the circuit theory of mesoscopic transport [19]. The Keldysh method to FCS was first proposed by one of the authors. It has been applied to treat the effects of the weak localization corrections onto the FCS in the disordered metallic wires and later on to study the FCS in superconducting heterostructures [20].

The above researches address the FCS of the non-interacting electrons in case of common two-terminal geometry. Although the cross-correlations for several multi-terminal layouts have been understood [1], the evaluation of FCS encountered difficulties. For instance, an attempt to build up FCS with "minimal correlation approach" [21] has lead to contradictions [14]. Very few is known about FCS of interacting electrons [22, 23]. Since the interaction may bring correlations and entanglement of electron states the study of FCS of interacting electrons, particularly for the case of multi-terminal geometry, is both challenging and interesting.

In the present work we review the calculational scheme that allows for easy evaluation of FCS in multi-terminal mesoscopic systems. The paper is organized as follows. In the section II we outline the scattering approach by Levitov *et.al* to FCS. Section III is devoted to the multi-terminal FCS of the non-interacting electrons. We show, that this theory appears to be a circuit theory of 2×2 matrices associated with Keldysh Green functions [24]. In the section IV we concentrate on FCS in the opposite situation of mesoscopic systems placed in a strong Coulomb blockade limit. We prove that the FCS in this case turns out to be an elegant extension of the master equation approach. We illustrate both methods by applying them to the various three-terminal circuits. We study the FCS of electron transfer in the three-terminal chaotic quantum dot and compare it with the statistics of charge transfer in the Coulomb blockade island with three leads attached. We demonstrate that Coulomb interaction suppresses the relative probabilities of big current fluctuations. In section V we establish the equivalence of scattering and master equation approach to FCS, by considering the statistics of charge transfer through the single resonance level. Finally, we summarize the results in the section VI.

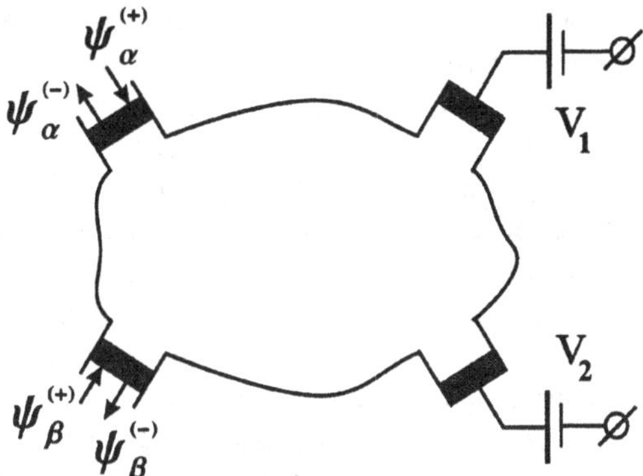

Figure 1. The "multi-terminal" mesoscopic conductor.

2. Scattering approach to FCS

In this section we review the current statistics of non-interacting electrons, using the ideas of general scattering approach to mesoscopic transport. The problem of FCS in this framework has been solved about a decade ago by Levitov *et. al* [11, 12]. Their work, concentrated on the two-terminal geometry [12], became widely cited. But, unfortunately, their preceding paper[11], devoted to multi-terminal systems as well, is only partially known. We will also demonstrate its relation to the current-current correlations in multi-terminal conductors, first investigated by M.Büttiker [7].

The most general mesoscopic system, eligible to scattering approach, is shown in Fig. 1. It is a phase coherent mesoscopic conductor, connected to macroscopic reservoirs (leads, terminals) by means of n external contacts ($n \geq 2$). We concern the situation when the system is placed in the non-equilibrium condition, thus each terminal α is characterized by a distribution function $f_\alpha(\varepsilon)$ of electrons over the energy. When the terminal is in local equilibrium at electro-chemical potential V_α and a temperature T_α, then $f_\alpha(\varepsilon)$ reduces to Fermi distribution $f_\alpha(\varepsilon) = \{1 + \exp[(\varepsilon - eV_\alpha)/kT_\alpha]\}^{-1}$.

The conductor is a region of disordered or chaotic scattering. The scattering within conductor is assumed to be elastic. This is true at sufficiently low energies, such that the energy dependent length of inelastic scattering exceeds the sample size. As to the terminals, it is conveniently to assume, that far from the contacts the longitudinal (along the lead) and transverse (across the lead) motion of electrons are separable. Then at fixed energy ε the transverse motion in the lead α is quantized and described by the integer index $n_\alpha = 1 \dots M_\alpha(\varepsilon)$, with $M_\alpha(\varepsilon)$ being the total number of propagating modes at given energy. Each n_α corresponds to one incoming to the contact and one outgoing from the contact α solution of the

Shrödinger equation, which is usually referred as a scattering channel.

The idea behind the scattering approach is that the electron transport can be described as the one-particle scattering from the incoming channels to the outgoing channel in the leads. Very generally it can be described by a unitary scattering matrix $\hat{S}(\varepsilon)$. At given energy its dimension is equal to M , $M = \sum_\alpha M_\alpha$ being the total number of transport channels. The energy dependence of the scattering matrix is set by the "Thouless energy" $E_{\rm Th} = \hbar/\tau_d$, τ_d being the traversal time through the system. At sufficiently low temperatures, voltages and frequencies, such that $\max\{eV, kT, \hbar\omega\} \ll E_{\rm Th}$, one can conveniently disregard the energy dependence and concentrate on the scattering matrix $\hat{S}(E_F)$, taken at Fermi surface. Then the scattering approach gives the adequate description of the transport, provided the effects of Coulomb blockade can be neglected. This is justified, when the total conductance G of the sample is much larger that the conductance quantum $G_Q = e^2/2\pi\hbar$.

One starts by introducing the scattering matrix $\check{S}$ of electrons at a narrow energy strip $\sim \max(eV, kT) \ll E_{\rm Th}$ in the vicinity of Fermi energy E_F. The S-matrix relates the amplitudes $\psi^{(+)}_{m,\beta}$ of incoming electrons to the amplitudes $\psi^{(-)}_{n,\alpha}$ of the outgoing ones (See Fig. 1):

$$\psi^{(-)}_{n,\alpha} = \sum_{m,\beta} S_{nm,\alpha\beta}\psi^{(+)}_{m,\beta} \tag{1}$$

Here Latin indexes (n, m) refer to the particular quantum scattering channel in given leads (α, β). (See e.g. [1, 7, 30] for a more detailed introduction.) It is worth to mention that at this stage $\psi^{(\pm)}_{n,\alpha}$ correspond to the amplitude of the wave function in the *exterior* area of the contacts. Thus defined S-matrix incorporates the information about scattering in the whole system. This includes the scattering from the disordered/chaotic regions of the conductor as well as multiple reflections from external contacts.

Remarkably, the FCS can be expressed in terms of the scattering matrix in quite a general way. It has been shown by Levitov *et. al.* [11]. The FCS approach deals with the probability distribution $P(\{N_i\})$ for N_i electrons to be transferred through the terminal i during time interval t_0. The detection time t_0 is assumed to be much greater than e/I, I being a typical current through the conductor. This ensures that in average $\bar{N}_i \gg 1$. The probability distribution $P(\{N_i\})$ can be conveniently expressed via generating function ("action") $S(\{\chi_i\})$ by means of Fourier transform

$$P(\{N_i\}) = \int_{-\pi}^{\pi} \prod_i \frac{d\chi_i}{2\pi} e^{-S(\{\chi_i\}) - i\sum_i N_i \chi_i}. \tag{2}$$

Here χ_i are the parameters of this generating function ("counting fields"), each of them being associated with a given terminal i. The generating function $\mathcal{S}(\{\chi_i\})$

contains the whole information about the irreducible moments of charge transfer through the system, as well as the information about big current fluctuations. Indeed, it follows from Exp. (2) that (higher-order) derivatives of $\mathcal{S}$ with respect to χ_i, evaluated at $\chi_i = 0$, give (higher-order) irreducible moments of $P(\{N_i\})$. First derivatives yield average currents to terminals, second derivatives correspond to the noises and noise correlations.

The result of Levitov *et. al.* for the generating function $\mathcal{S}(\{\chi_i\})$ reads as follows. One considers the energy-dependent determinant

$$\mathcal{S}_E(\{\chi_i\}) = \ln \mathrm{Det}\left(1 - \hat{f}_E + \hat{f}_E \hat{S}^\dagger \tilde{S}\right) \tag{3}$$

where the matrix $(\hat{f}_E)_{mn,\alpha\beta} = f_\alpha(E)$ is diagonal in the channel indexes (m, n) and the matrix $\tilde{S}$ is defined as $\tilde{S}_{mn,\alpha\beta} = e^{i(\chi_\alpha - \chi_\beta)} S_{mn,\alpha\beta}$. It has been proved that this expression represents a characteristic function of the probability distribution of transmitted charge for electrons with energies in an infinitesimally narrow energy strip in the vicinity of E. After that the complete generating function $\mathcal{S}(\{\chi_i\})$ is just a sum over energies

$$\mathcal{S}(\{\chi_i\}) = t_0 \int \mathcal{S}_E(\{\chi_i\}) \frac{dE}{2\pi\hbar} \tag{4}$$

This expression reflects the fact that electrons at different energies are transferred independently, without interference. Therefore they yield additive independent contributions to the generating function.

The general results for the FCS, contained in Eqs. (3) and (4), provides the elegant way to derive the pioneering results of M. Büttiker [7], concerning the current-current correlations in the multiterminal conductors. One considers the correlation function

$$P_{\alpha\beta}(t - t') = \frac{1}{2}\left\langle \Delta\hat{I}_\alpha(t)\Delta\hat{I}_\beta(t') + \Delta\hat{I}_\beta(t')\Delta\hat{I}_\alpha(t)\right\rangle \tag{5}$$

of currents fluctuations in contacts α and β. The fluctuation $\Delta\hat{I}_\alpha(t)$ is defined as $\Delta\hat{I}_\alpha(t) = \hat{I}_\alpha(t) - \langle I_\alpha \rangle$, with $\hat{I}_\alpha(t)$ being the current operator in the lead α and $\langle I_\alpha \rangle$ being its mean steady value under given non-equilibrium conditions. The Fourier transform $P_{\alpha\beta}(\omega)$ of current-current correlations functions is sometimes referred to as noise power. Throughout this article we concentrate on zero-frequency (shot-noise) limit of current correlations $P_{\alpha\beta} \equiv P_{\alpha\beta}(\omega = 0)$. At low frequencies both $\langle I_\alpha \rangle$ and $P_{\alpha\beta}$ are readily expressed via the first (second) moments of number of transferred electrons through the corresponding leads

$$\langle I_\alpha \rangle = \frac{e}{t_0}\langle N_\alpha \rangle, \qquad P_{\alpha\beta} = \frac{e^2}{t_0}\langle \Delta N_\alpha \Delta N_\beta \rangle \tag{6}$$

Therefore the second derivative $\frac{\partial^2}{\partial\chi_\alpha\partial\chi_\beta}S(\{\chi_i\})$ of the action yields the result for the shot-noise correlation function $P_{\alpha\beta}$. It can be reduced to the following form

$$P_{\alpha\beta} = \frac{e^2}{\pi\hbar}\int dE \sum_{\gamma\mu} \mathrm{Tr}\left\{A^{\alpha}_{\gamma\mu}A^{\beta}_{\mu\gamma}\right\} f_\gamma(E)(1 - f_\mu(E)) \tag{7}$$

Here the matrix $A^{\alpha}_{\gamma\mu}$ is defined via S-matrix as

$$A^{\alpha}_{\gamma\mu}(E) = \delta_{\alpha\gamma}\delta_{\alpha\mu} - S^{\dagger}_{\alpha\gamma}(E)S_{\alpha\mu}(E) \tag{8}$$

This is precisely the result obtained by M. Büttiker prior the development of FCS approach.

Up to the moment the scattering matrix $\hat{S}$ was not specified. In order to evaluate the FCS, one needs to construct such a matrix, so that to take into account the scattering properties of the concrete mesoscopic system at hand. To accomplish this task, one may proceed as follows. Very generally, one can separate a mesoscopic layout into primitive elements: nodes and connectors. The nodes are similar to the terminals, in the sense, that each of them can be characterized by some non-equilibrium isotropic distribution function of electrons over the energies. The illustrative example of the node is chaotic quantum dot. The scattering within the node is assumed to be random (chaotic). The connectors represent themselves either internal or external contacts in the systems, scattering properties of which are supposed to be known. The actual implementation of connectors can be quite different: quantum point contacts, tunnel junctions, diffusive wires, etc. The purpose of the above separation is to ascribe to each node and to connector the particular unitary scattering matrix with known properties. Then the scattering matrix $\hat{S}$ of the entire system can be unambiguously expressed via these primitive scattering matrices.

We outline this scheme for the simple system, consisting of the single node only. A good example of such system is a multi-terminal chaotic quantum dot, shown in Fig. 2. It is coupled to n external reservoirs via n contacts. Each contact α is described by a unitary scattering matrix $\hat{S}_\alpha$. (Its choice depends on the experimental realization of this given contact.) Let $\phi^{(\pm)}_{m,\alpha}$ be the coefficients of electron wave function, taken at the *interior* boundary of the contact. The amplitudes $\phi^{(+)}_{m,\alpha}$ and $\phi^{(-)}_{m,\alpha}$ correspond to the outgoing (incoming) waves from (to) the the contact α respectively, as shown in Fig. 2. The scattering matrix $\hat{S}_\alpha$ relates the vector $c^{\mathrm{in}} = (\psi^{(+)}_\alpha, \phi^{(-)}_\alpha)^T$ of the incident electron waves to the vector $c^{\mathrm{out}} = (\psi^{(-)}_\alpha, \phi^{(+)}_\alpha)^T$ of the outgoing waves:

$$c^{\mathrm{out}} = \hat{S}_\alpha\, c^{\mathrm{in}} = \begin{pmatrix} r_\alpha & t'_\alpha \\ t_\alpha & r'_\alpha \end{pmatrix} c^{\mathrm{in}} \tag{9}$$

Here the square blocks r_α (r'_α) of the size $(M_\alpha \times M_\alpha)$ describe the electron reflection back to the reservoirs or back to the dot, respectively. The off-diagonal

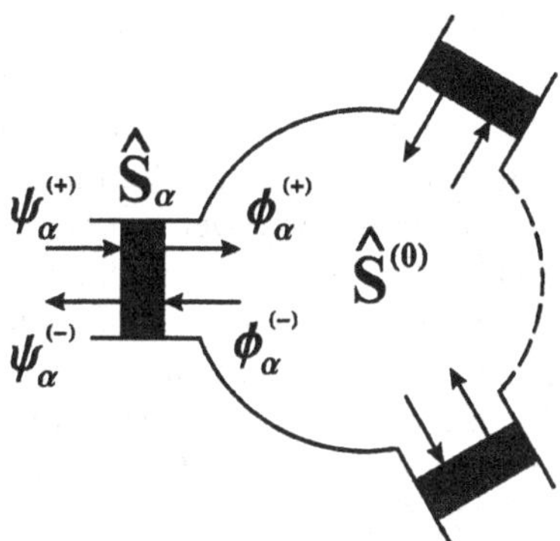

Figure 2. The building blocks of a scattering matrix of the multi-terminal chaotic quantum dot. The matrix $\hat{S}_\alpha$ describes the scattering due to the contact α. The matrix $\hat{S}^{(0)}$ describes the chaotic scattering within the dot.

blocks t_α (t'_α) of the same dimension are responsible for the transmission through the contact α.

Similarly, the chaotic scattering of electrons inside the dots is described by the scattering matrix $\hat{S}^{(0)}$. By analogy to Eq. (1), it relates within the dot amplitudes $\phi^{(+)}_{m,\beta}$ to $\phi^{(-)}_{n,\alpha}$, i.e.

$$\phi^{(-)}_{n,\alpha} = \sum_{m,\beta} S^{(0)}_{nm,\alpha\beta}\phi^{(+)}_{m,\beta} \tag{10}$$

The dimension $(M \times M)$ of $\hat{S}^{(0)}$ is equal to that of the matrix $\hat{S}$.

It is possible now to express the scattering matrix $\hat{S}$ of the entire sample in terms of $\hat{S}_\alpha$ ($\alpha = 1 \ldots n$) and $\hat{S}^{(0)}$. One can use the set of equation (9) and (10) in order to find the linear relations between amplitudes $\psi^{(+)}$ and $\psi^{(-)}$. According to Eq. (1) it will uniquely determine the matrix elements of $\hat{S}_\alpha$. The result reads:

$$\hat{S} = \hat{r} + \hat{t}'\hat{S}^{(0)}\frac{1}{1 - \hat{r}'\hat{S}^{(0)}}\hat{t} \tag{11}$$

Here $\hat{r}$ ($\hat{r}'$) and $\hat{t}$ ($\hat{t}'$) are block diagonal matrices with elements $\hat{r}_{mn,\alpha\beta} = \delta_{\alpha\beta}(\hat{r}_\alpha)_{mn}$ and $\hat{t}_{mn,\alpha\beta} = \delta_{\alpha\beta}(\hat{t}_\alpha)_{mn}$. The denominator of Eq. (11) describes the multiple reflection due to contacts within the dot. It can be verified, that $\hat{S}$ is unitary, $\hat{S}^\dagger\hat{S} = 1$, provided $\hat{S}^{(0)}$ and $\hat{S}_\alpha$ are also unitary. The analogous construction of S-matrix can be in principle implemented for the more complicated layouts.

To evaluate FCS one substitutes $\hat{S}$ into Eqs. (3) and (4) in the form (11). The matrices $\hat{S}_\alpha$ are fixed by the choice of the properties of the contacts. However, the matrix $\hat{S}^{(0)}$ is random describing chaotic scattering inside the dot. This means that the answer should be *averaged* over all possible realization of $\hat{S}^{(0)}$. A reasonable assumption is that $\hat{S}^{(0)}$ is uniformly distributed in the space of all unitary matrices [30].

Thereby one solves the problem for a single node. For less trivial system, comprising two and more nodes, the expression for the $\hat{S}$-matrix, corresponding to Eq. (11), becomes more involved. Moreover one has to average over unitary matrices separately for each node. Besides, to describe a simple system of diffusive wire one has eventually to go to the limit of infinitely many nodes. All this makes the scattering approach extremely inconvenient. Fortunately, one can treat these problems with a more flexible semiclassical Green function method, which we outline in the next section. It is applicable, since we assume that the conductance of the system $G \gg G_Q$.

3. Circuit theory of FCS in multi-terminal circuits

In this section we formulate the semiclassical theory of FCS of non-interacting electrons in case of multi-terminal geometry. Our approach is based on the Keldysh Green function method. We develop the scheme to evaluate the FCS in the arbitrary multi-terminal mesoscopic system. For that we will use the circuit theory of mesoscopic transport [19]. Next, we illustrate this scheme, considering the big fluctuations of current in the three-terminal chaotic quantum dot. In the end of the section we discuss the shot-noise correlations and give the convenient expressions that depend on the scattering properties of connectors only, and do not involve the scattering inside the cavity.

3.1. CIRCUIT THEORY APPROACH TO THE FCS.

We start by introducing current operators $\hat{I}_i$, each being associated with the current to/from a certain terminal i. Extending the method of [18] we introduce a Keldysh-type Green function defined by

$$\left(i\frac{\partial}{\partial t} - \hat{H} - \frac{1}{2}\bar{\tau}_3 \sum_i \chi_i(t)\hat{I}_i\right) \otimes \check{G}(t,t') = \delta(1-1') \tag{12}$$

Here we follow notations of a comprehensive review [25]: χ_i are time-dependent parameters, $\bar{\tau}_3$ is a 2×2 matrix in Keldysh space, $\hat{H}$ is the one-particle Hamiltonian that incorporates all information about the system layout, including boundaries, defects and all kinds of elastic scattering. We use "hat", "bar" and "check" to denote operators in coordinate space, matrices in Keldysh space and operators in direct product of these spaces respectively. The current operator in Eq. (12), associated with the terminal i, is defined as $\hat{I}_i = \int d^3x \Psi^\dagger(\mathbf{p}/m)\Psi\, \nabla F_i(x)$. Here Ψ is the usual Fermi field operator and $\nabla F_i(x)$ is chosen such a way that the spatial integration is restricted to the cross-section and yields the total current through the given lead. The Eq. (12) defines the Green function unambiguously provided boundary conditions are satisfied: $\check{G}(t,t') \equiv \bar{G}(x,x';t,t')$ approaches the common equilibrium Keldysh Green functions $\check{G}_i^{(0)}(t-t')$ provided x, x' are sufficiently far in the terminal i. These $\check{G}_i^{(0)}(t-t')$ incorporate information about the state of the terminals: their voltages V_i and temperatures T_i.

In the following we will operate with the cumulant generating function ("action") $S(\{\chi_i\})$ defined as a sum of all closed diagrams

$$e^{-S(\{\chi_i\})} = \left\langle T_\tau e^{i\sum_{i=1}^{N} \int_{-\infty}^{+\infty} d\tau \chi_i(\tau) \check{I}^{(i)}(\tau)} \tilde{T}_\tau e^{i\sum_{i=1}^{N} \int_{-\infty}^{+\infty} d\tau \chi_i(\tau) \check{I}^{(i)}(\tau)} \right\rangle \tag{13}$$

Here T_τ ($\tilde{T}_\tau$) denotes the (anti) time ordering operator. One can see by traditional diagrammatic methods [25] that the expansion of $S(\{\chi_i\})$ in powers of $\chi_i(t)$ generates all possible irreducible diagrams for higher order correlators of $\hat{I}_i(t)$ and thereby incorporates all the information about statistics of charge transfer. If we limit our attention to low-frequency limit of current correlations, we can keep time-independent χ_i. In this case, the Green functions are functions of time difference only and the Eq.(12) separates in energy representation. Then the action $S(\{\chi_i\})$ can be conveniently expressed via the average χ-dependent currents $I_i(\{\chi_i\})$ in the following way

$$\frac{i}{t_0}\frac{\partial S}{\partial \chi_i} = I_i(\{\chi_i\}) \equiv \int \frac{d\varepsilon}{2\pi} \mathrm{Tr}\,(\bar{\tau}_3 \hat{I}_i \check{G}(\varepsilon)) \tag{14}$$

where t_0 denotes the time of measurement. Thus defined cumulant generating functions allows to evaluate the probability for N_i electrons to be transferred to the terminal i during time interval t_0 in accordance with the general relation (2).

Up to the moment the above Eqs. (12) and (13) can be used to define the statistics of any quantum mechanical variable. However we are interested in the statistics of the *charge* transfer. Since the charge is the conserved quantity, the proper construction of current operators $\hat{I}_i$ requires the gauge invariance of the Hamiltonian. Therefore fully quantum mechanical scheme includes the "counting" fields $\chi_i(t)$ as *gauge* fields. (See the contribution of L.Levitov in this book.) It means, that the initial physical Hamiltonian $H(\mathbf{q},\mathbf{p})$ of the system should be replaced by the χ-dependent Hamiltonian $H_\chi = H(\mathbf{q}, \mathbf{p} - \frac{1}{2}\bar{\tau}_3 \sum_i \chi_i \nabla F_i)$. Thus the appearance of the "counting" fields in the problem is similar to the inclusion of a vector potential $\mathbf{A}(\mathbf{r})$ of an ordinary electromagnetic gauge field. The crucial difference is the change of sign of interaction at the forward and backward branches of the Keldysh contour, that is reflected by the presence of $\bar{\tau}_3$ matrix. Then the χ-dependent Green function obeys the equation of motion, written with the use of χ-dependent Hamiltonian H_χ. Accordingly, the "action" $S(\{\chi_i\})$ is defined as a sum of all closed diagrams with respect to the interaction $H_{\text{int}} = H_\chi - H$.

Thus defined constructions are locally gauge invariant with respect to rotation in the Keldysh space. In particular, the average current (14) is a conserved quantity, provided the current operator is defined as $\hat{I}_i = \partial H_\chi(t)/\partial\chi_i$. Therefore it is possible to perform the local gauge transformation $\psi' = \exp\{-(i/2)\bar{\tau}_3 \sum_i \chi_i F_i\}\,\psi$ in order to eliminate the χ-dependent terms from the equation of motion(12) [18, 20]. The χ-dependence of $\check{G}$ is thereby transferred to the boundary conditions: the gauged Green function far in each terminal shall approach $\check{G}_i(\epsilon)$ defined as

$$\check{G}_i(\epsilon) = \exp(i\chi_i\bar{\tau}_3/2)\check{G}_i^{(0)}(\epsilon)\exp(-i\chi_i\bar{\tau}_3/2) \tag{15}$$

Here $\check{G}_i^{(0)}(\epsilon)$ corresponds to the equilibrium Keldysh Green function sufficiently far in the given terminal i. The precise form of the Green functions $\check{G}_i^{(0)}(\epsilon)$ will be explicitly given below in the text. [See Eq. (19)]

Let us also note that the modification of current operators $\hat{I}_i$ by the "counting" fields can be safely omitted at energies near the Fermi surface, where one can linearize the electron spectrum. This is possible in the semiclassical approximation when the variation of ∇F_i is small at the scale of the Fermi wave length. Expanding the H_χ on the Fermi shell, one arrives to Eqs. (12) and (13). Afterwards it is feasible to verify the result (15), with making use of the semiclassical Eilenberger equations. (For the details we refer the reader to the paper of W.Belzig in this book.)

In the present form, the Eq. (12) with relations (15,14) solves the problem of determination of the FCS for any arbitrary system layout: one just has to find exact quantum-mechanical solution of a Green function problem. This is hardly constructive, and we proceed further by deriving a simplified semiclassical approach. First, we note that even in its exact quantum-mechanical form the Eq.(12) possesses an important property. We consider the quantity defined similar to standard definition of current density, $\bar{j}^\alpha(x,\epsilon) \equiv \lim_{x\to x'}(\nabla'^\alpha - \nabla^\alpha)\bar{G}(x,x';\epsilon)/m$. By virtue of Eq.(12) this quantity conserves so that

$$\partial\bar{j}^\alpha(x,\epsilon)/\partial x^\alpha = 0 \tag{16}$$

This relation looks like the conservation law of particle current at a given energy. However, this relation contains more information since it is a conservation law for a 2×2 *matrix* current.

Next, we construct a theory which makes use of this conservation law. We concentrate on the semiclassical Green function in coinciding points, $\bar{G}(x,\epsilon) \equiv i\bar{G}(x,x';\epsilon)/\pi\nu$, where ν is a density of states at Fermi energy. So defined Green function has been introduced in several semiclassical theories. [25–27] It satisfies the normalization condition $\bar{G}^2 = \bar{1}$. We relate the "current density" $\bar{j}$ to gradients and/or changes of $\bar{G}(x)$, very much like the electric current density is related to the voltage in circuit theory of electric conductance. Following the approach of the circuit theory [19], we separate a mesoscopic layout into elements: nodes and connectors, so that the $\bar{G}(x)$ is constant across the nodes and drops across the connectors. One may associate a graph with each circuit, so that its lines (i,j)

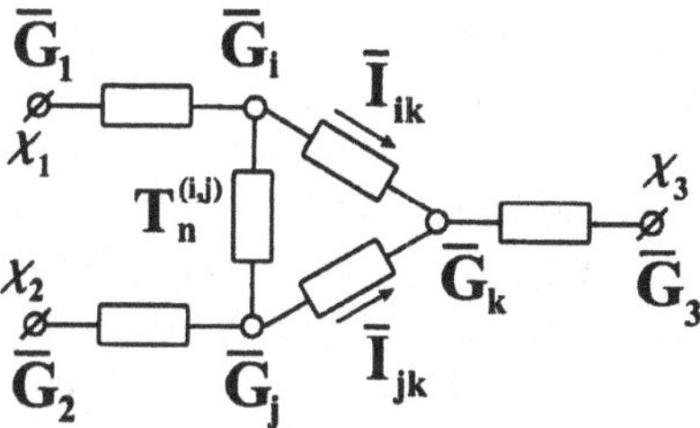

Figure 3. The graph of the circuit theory, associated with a 3 terminal mesoscopic system. $\bar{G}_1$, $\bar{G}_2$, $\bar{G}_3$ in the terminal are fixed by the boundary condition (15). T_n^{ij} define the transport properties of a connector. $\bar{I}_{ik}$ and $\bar{I}_{jk}$ denote the currents, flowing from the nodes i and j into the node k. χ_1, χ_2, χ_3 are counting fields in the terminals.

would denote the connectors, and internal and external vertices correspond to the nodes and terminals, respectively. (See Fig. 3.1) This separation of actual layout is rather heuristic, similar to separation of an electric conductor of a complicated geometry onto nodes and circuit theory elements. The bigger the number and the finer the mesh of the nodes and connectors, the better the circuit theory approximates the actual layout. The idea for this separation is completely analogous to the one considered in the section 2. The nodes are similar to the terminals, the difference is that $\bar{G}$ is fixed in the terminals and yet to be determined in the nodes. The $\bar{G}$ in nodes are determined from Kirchoff rules reflecting the conservation law (16): sum of the matrix currents from the node over all connectors should equal zero at each energy. For this, we should be able to express the matrix current via each connector as a function of two matrices $\bar{G}_{i,j}$ at its ends.

To accomplish this task, consider the connector (i, j), linking to nodes i and j, It can be quite generally characterized by a set of transmission eigenvalues $T_n^{(ij)}$[19, 27]. The problem to solve is to express matrix current $\bar{I}_{ij}$ via the connector in terms of $\bar{G}_{i(j)}$. This problem shall be addressed by a more microscopic approach and was solved in [27] for Keldysh-Nambu matrix structure of $\check{G}$. It is a good news that the derivation made in [27] does not depend on concrete matrix structure and can be used for the present problem without any modification yielding

$$\bar{I}_{ij} = \frac{1}{2\pi} \sum_n \int dE \frac{T_n^{(ij)} \left[\bar{G}_i, \bar{G}_j\right]}{4 + T_n^{(ij)} \left(\{\bar{G}_i, \bar{G}_j\} - 2\right)} . \tag{17}$$

Each connector (i, j) in the layout contributes to the total χ_i-dependent action (14). The corresponding S_{ij} contribution to the action should be found from the relation (14) and reads:[20]

$$S_{ij}(\chi) = \frac{-t_0}{2\pi} \sum_n \int dE \operatorname{Tr} \ln \left[1 + \frac{1}{4} T_n^{(ij)} \left(\{\bar{G}_i, \bar{G}_j\} - 2\right)\right] . \tag{18}$$

Now we are ready to present a set of circuit theory rules that enables us to evaluate the FCS for an arbitrary mesoscopic layout.

(i) The layout is separated onto terminals, nodes, and connectors.

(ii) The $\bar{G}_j$ in each terminal j is fixed by relation (15) thus incorporating information about voltage, temperature and counting field χ in each node.

(iii) For each node k, the matrix current conservation yields a Kirchoff equation $\sum_i \bar{I}_{ik} = 0$, where the summation is going over all connectors (i, k) attached to node k, and $\bar{I}_{ik}$ are expressed with (17) in terms of $\bar{G}_{i(k)}$.

(iv) The solution of resulting equations with condition $\bar{G}_k^2 = 1$ fixes $\bar{G}_k$ in each node.

(v) The total action $S(\chi)$ is obtained by summing up the contributions $S_{ij}(\{\chi_i\})$ of individual connectors, those are given by (18): $S(\{\chi_i\}) = \sum_{(i,j)} S_{ij}(\{\chi_i\})$

(vi) The statistics of electron transfer is obtained from the relation (2).

In the end of this subsection we discuss the limits of applicability of the whole scheme. By virtue of semiclassical approach, the mesoscopic fluctuations coming from interference of electrons penetrating different connectors are disregarded. So that, we assume that conductivities of all connectors are much bigger than conductance quantum $e^2/\pi\hbar$. The same condition provides the absence of Coulomb blockade effects in the system. Besides, we have disregarded the possible processes of *inelastic relaxation* in the system. The latter can be eventually taken into account, since the use of Keldysh Green functions technique allows for perturbation treatment of interaction and relaxation. However, it would considerably complicate the scheme. The point is that the inelastic scattering would mix up the $\bar{G}(\varepsilon)$ at different energies, so that one can not solve the circuit theory equations separately at each energy.

3.2. THE STATISTICS OF CHARGE TRANSFER IN CHAOTIC QUANTUM DOTS

As an illustration of the above scheme, we will consider in the second part of this section the FCS of the 3-terminal chaotic quantum dot. The system is sketched in the inset of Fig. 2. The heuristic circuit, associated with this mesoscopic system is shown by dashed lines. It includes only 3 terminals, 3 arbitrary connectors, associated with the contacts, and the node $\{4\}$, representing the quantum dot itself. This separation is valid provided the cavity is in the *quantum* chaotic regime. (See [28] for definition). This regime corresponds to full isotropization of the Green function $\check{G}(x, x', \epsilon)$ within the dot, so that $\bar{G}_4(\varepsilon)$ can be regarded as a constant at a given energy.

Since the normalization $\bar{G}_k^2 = 1$ holds for each vertex, we use the parametrization $\bar{G}_k = \mathbf{g}_k \cdot \boldsymbol{\tau}$, $\mathbf{g}_k \cdot \mathbf{g}_k = 1$. Here $\mathbf{g}_k$ is a 3-D vector, and $\boldsymbol{\tau} = (\bar{\tau}_1, \bar{\tau}_2, \bar{\tau}_3)$. With the use of this parametrization the anticommutator $\{\bar{G}_i, \bar{G}_k\}$ is proportional to the unity matrix and takes the form of scalar product $\frac{1}{2}\{\bar{G}_i, \bar{G}_k\} = \mathbf{g}_i \cdot \mathbf{g}_k$. In the absence of counting fields the Green functions in the terminals corresponds to the equilibrium Keldysh Green function [25]

$$\bar{G}_k^{(0)} = \begin{pmatrix} 1-2f_k & -2f_k \\ -2(1-f_k) & 2f_k-1 \end{pmatrix}, \tag{19}$$

where Fermi distribution function $f_k(E) = \{\exp[(E - eV_k)/T_k] + 1\}^{-1}$ accounts for the bias voltages V_k and the temperatures T_k in the terminals. The χ_i-dependence of $\bar{G}_k(\chi)$ is then given by Eq. (15).

We see that Green function $\bar{G}_4(\chi) = \mathbf{g}_4 \cdot \boldsymbol{\tau}$ in the dot is in fact the only function to find. For that, we proceed by applying the current conservation law, $\sum_{k=1}^{3} \bar{I}_{k,4} = 0$, inside the dot. We present the currents $\bar{I}_{k,4}$ given by (17) in the form

$$\bar{I}_{k,4} = \frac{1}{2} Z_k(\mathbf{g}_k \cdot \mathbf{g}_4)[\bar{G}_k, \bar{G}_4], \tag{20}$$

where the scalar function $Z_k(x)$ incorporates the information about transmission eigenvalues in each connector k:

$$Z_k(x) \equiv \sum_n T_n^{(k,4)}/[2 + T_n^{(k,4)}(x-1)]. \tag{21}$$

It can be evaluated for any particular distribution $\rho(T)$ of transmission eigenvalues in the given connector and completely defines its scattering properties. For a example, if we denote by $R_{\mathrm{Q}} = \pi\hbar/e^2$ the resistance quantum, then $R_k^{-1} = 2R_{\mathrm{Q}}^{-1} Z_k(1)$ is an inverse resistance of the connector. One can also express the Fano factor $F_k = \langle T(1-T)\rangle/\langle T\rangle$, associated with the given connector as $F = 1 - 2(d/dx)\log Z_k(x)|_{x=1}$.

With the use of $Z_k(x)$ the conservation law can be efficiently rewritten as $[\sum\limits_{k=1}^{3} p^k \bar{G}_k, \bar{G}_4] = 0$, where $p^k = Z_k\,(\mathbf{g}_k \cdot \mathbf{g}_4)$. This relation suggests to look for the vector $\mathbf{g}_4$ in the form $\mathbf{g}_4 = M^{-1} \sum\limits_{k=1}^{3} p^k\, \mathbf{g}_k$, with $M(\chi)$ being an unknown normalization constant. Using the normalization condition $\mathbf{g}_4 \cdot \mathbf{g}_4 = 1$ we obtain the following set of equations

$$p^i = Z_i\Big(M^{-1} \sum_{j=1}^{3} g_{ij}(\chi) p^j\Big), \quad M^2 = \sum_{i,j=1}^{3} g_{ij}(\chi) p^i p^j \tag{22}$$

where the scalar product $g_{ij}(\chi) = \mathbf{g}_i(\chi) \cdot \mathbf{g}_j(\chi)$ between terminal Green function is expressed in terms of Fermi distributions as follows

$$g_{ij}(\chi) = (1-2f_i)(1-2f_j) + 2\,e^{i(\chi_i-\chi_j)} f_i(1-f_j) + 2\,e^{-i(\chi_i-\chi_j)} f_j(1-f_i)$$

The Green function $\bar{G}_4$ then is found from the solution $\{p^i(\chi), M(\chi)\}$ of this set of equations. In the general situation the function $Z_k(x)$ takes the form (21) and therefore Eq. (22) represents the set of non-linear equations. However, their solution can be relatively easy found numerically using the method of subsequent iterations.

The total action can be found by applying the rule (v) of circuit theory and reads

$$S(\chi) = \frac{t_0}{\pi}\sum_{i=1}^{3}\int d\epsilon\, S_i(\mathbf{g}_i \cdot \mathbf{g}_4) \tag{23}$$

where

$$\mathbf{g}_i \cdot \mathbf{g}_4 = M^{-1}(\chi)\sum_{j=1}^{3} g_{ij}(\chi)\, p^i(\chi) \tag{24}$$

Here partial contributions $S_k(x)$ from each connector in Eq. (23) has to be determined from the relation $\frac{\partial}{\partial x}S_k(x) = -Z_k(x)$, provided $S_k(1) = 0$. It follows from the Eqs. (14), (20) and normalization condition $\mathbf{g}_4 \cdot \mathbf{g}_4 = 1$.

We consider three particular types of connectors: tunnel(T), ballistic(B) and diffusive(D). Their corresponding contributions to action read as [18]:

$$S_T(x) = -\frac{1}{2}(R_0/R)(x-1), \tag{25}$$

$$S_B(x) = -(R_0/R)\log[(1+x)/2], \tag{26}$$

$$S_D(x) = -\frac{1}{4}(R_0/R)\log^2(x+\sqrt{x^2-1}) \tag{27}$$

with R being a resistance of the connector. The tunnel connector represents the tunnel junction, so that $T_n \ll 1$ for all n. Ballistic connector corresponds to the quantum point contact, with N open channels. The last expression comes from universal transmission distribution $\rho(T) = R_0/2RT\sqrt{1-T}$ for any diffusive contact.

Analytical results for FCS (23) can be readily obtained only for the system with tunnel connectors. To assess general situation we found $\mathbf{g}_4$ for given χ_i numerically. To find the probability distribution, we evaluated the integral (2) in the saddle point approximation, assuming χ_i has to be complex numbers. Saddle point approximation is always valid in the low frequency limit we consider, since in this case both action S and number of transmitted particles $N_i = I_i t_0/e \gg 1$. Due to the current conservation law only two of three counting fields χ_i are independent, and one can set $\chi_3 = 0$. The relevant saddle point of the function $\Omega(\chi) = S(\chi) + i\chi_1 I_1 t_0/e + i\chi_2 I_2 t_0/e$ always corresponds to purely imaginary numbers $\{\chi_1^*, \chi_2^*\}$. The probability reads $P(I_1, I_2) \approx \exp[-\Omega(\chi^*)]$. Evidently, $\Omega(\chi^*)$ is the Legendre transform of the action, and it can be regarded as implicit function on $I(\chi^*)$.

In the following we assume the shot noise regime $eV \gg kT$ when the thermal fluctuations can be disregarded. The energy integration in (23) becomes trivial, since $f_i(\epsilon) = 0$ or 1, and it is sufficient to consider only the case $V_1 = V_2 = 0$,

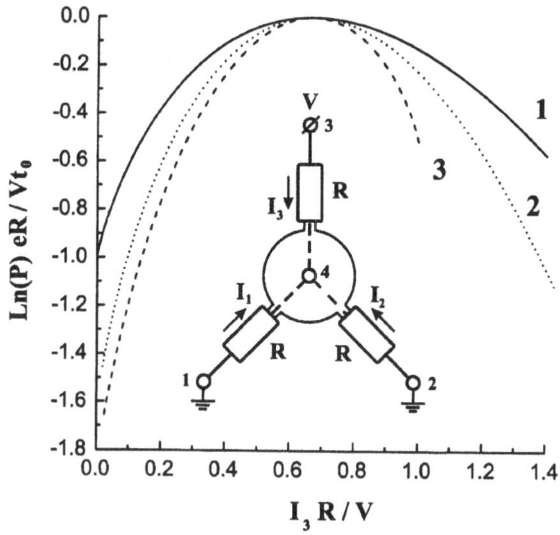

Figure 4. The logarithm of the current probabilities in the 3-terminal chaotic quantum dot as a function of I_3, under condition $I_1 = I_2$. The insert presents the system configuration. The resistances R of all connectors are assumed to be equal. 1 - tunnel connectors, 2 - diffusive connectors, 3 - ballistic connectors.

$V_3 = V$. Any other possible setup can be reduced to the number of previous ones by appropriate subdividing a relevant energy strip. The results of these calculations are shown in Fig. 3.2 and 3.2. We see that the maximum of probability occurs at $I_1 = I_2 = -V/3R$, $I_3 = 2V/3R$. This simply reflects the usual Kirchoff rules. The current distribution $P(I_1, I_2)$ for a ballistic system is bounded. This stems from the fact, that each ballistic contact has a limited number N of open channels. Therefore any plausible current fluctuation in the given terminal may not exceed the threshold value of a current $I_s = NV/R_Q$ via the ballistic connector. Contrary to that, for the tunnel and diffusive type configurations any connector has an infinite number of partially open transmission channels. Therefore the range of current fluctuations is not bounded in this case.

From Fig. 3.2 and 3.2 it follows that the relative probabilities of big current fluctuations increase in the sequence ballistic→diffusive→tunnel. To reveal the origin of this behavior we proceed by considering the shot-noise cross-correlations P_{ij}, defined by Exp. (5). At zero frequency they can be found analytically via the relation $P_{ij} = -\frac{e^2}{t_0}\frac{\partial^2}{\partial\chi_i\partial\chi_j}S\Big|_{\chi=0}$ Taking into account, that the action $S(\{\chi_i\})$, regarded as function of χ_i, is implicitly determined by Eq. (23) via the set of equations (22), we arrive to the following result

$$P_{ij} = -G_{ij}(T + \Theta_4) + R^{-1}\alpha_i\alpha_j(\bar{\Theta} - \Theta_i - \Theta_j) + \delta_{ij}R^{-1}\alpha_i\Theta_i \qquad (28)$$

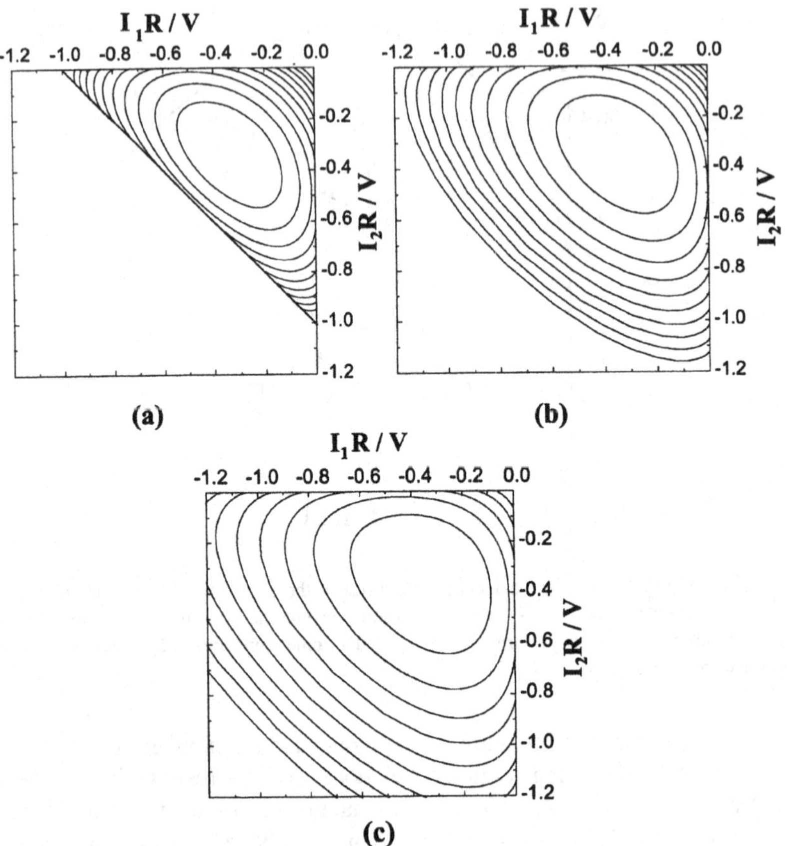

Figure 5. The contour maps of the current distribution $\log[P(I_1, I_2)]$ in the 3-terminal chaotic quantum dot for different configurations of connectors. (a) - ballistic connectors, (b) - diffusive connectors, (c) - tunnel connectors.

Here $\alpha_i = R/R_i$, $R^{-1} = \sum_i R_i^{-1}$,

$$G_{ij} = R^{-1}(\delta_{ij}\alpha_i - \alpha_i\alpha_j) \tag{29}$$

is a conductance matrix of the dot,

$$\Theta_4 = \int d\varepsilon \bar{f}(1-\bar{f}), \quad \Theta_k = F_k \int d\varepsilon (f_k - \bar{f})^2, \quad \bar{\Theta} = \sum_{i=1}^{3} \alpha_i \Theta_i \tag{30}$$

F_k is a Fano factor of the k-th connector and $\bar{f}(\varepsilon) = \sum_{i=1}^{3} \alpha_i f_i(\varepsilon)$ is the non-equlibrium distribution function within the dot.

In case of ballistic contacts $F_k = 0$ and the first two terms of the result (28) reproduce the expression for the noise power obtained with the use of "minimal correlation" approach[21]. At zero temperature it also coincides with the result of random matrix theory [29], as we expected from the our consideration of scattering approach. At equilibrium $P_{ij} = -2TG_{ij}$ in accordance with the fluctuation-dissipation theorem. In the general situation each non-ideal connector with $F_k \neq 0$ gives the additive contribution to the cross-correlation function, which is linear in F_k. Thus we conclude that the current-current correlations in the chaotic quantum dot result from two contributions. The first one, proportional to $(T + \Theta_4)$, corresponds to the cross-correlation function P_{ij}^{b} of the chaotic cavity with ideal ballistic point contacts, which stems from the fluctuation of the distribution function $\bar{f}$ within the dot. The second contribution, proportional to Θ_i, reflects the noise due to connectors.

To present the result for the cross-correlations it is also useful to introduce the (3×3) matrix F with elements $F_{ij} = P_{ij}/eI_\Sigma$, where $I_\Sigma = \sum_{i=1}^{3} |I_i|$. The matrix F is a generalization of the Fano factor for the multi-terminal system. It is symmetric and obeys the relation $\sum_{i=1}^{3} F_{ik} = 0$, which follows from the current conservation law. At zero temperature for the symmetric setup, shown in Fig. 3.2, it reads as

$$F = \frac{1}{81}\begin{pmatrix} 4+3F & -2 & -(2+3F) \\ -2 & 4+3F & -(2+3F) \\ -(2+3F) & -(2+3F) & 4+6F \end{pmatrix} \qquad (31)$$

For a diffusive wire $F_D = 1/3$ and for a tunnel junction $F_T = 1$. Therefore at fixed average currents through connectors the Gaussian's currents fluctuations will increase in the sequence ballistic→diffusive→tunnel. As it was mention before similar behavior is also traced in the regime of the big current fluctuations. The essential point here is that the cross-correlations always persist regardless the concrete construction of the connectors.

4. FCS of Charge Transfer in Coulomb Blockade Systems

In the preceding sections we have considered the FCS of non-interacting electrons. It was assumed that the conductance G of the system is much greater than the quantum conductance $G_Q = e^2/\hbar$. It is known, that under opposite condition, $G \leq G_Q$, the effects of Coulomb blockade become important. This motivates us to study the statistics of charge transfer in the mesoscopic systems, placed in the limit of strong Coulomb interaction, $G \ll G_Q$. The electrons dynamics in this Coulomb blockade limit is fortunately relatively simple, since the evolution of the system is governed by a master equation. The charge transfer is thus a classical stochastic process rather than the quantum mechanical one. Nevertheless the FCS is by no means trivial. In this section we elaborate the general approach to FCS in the systems, governed by master equation.

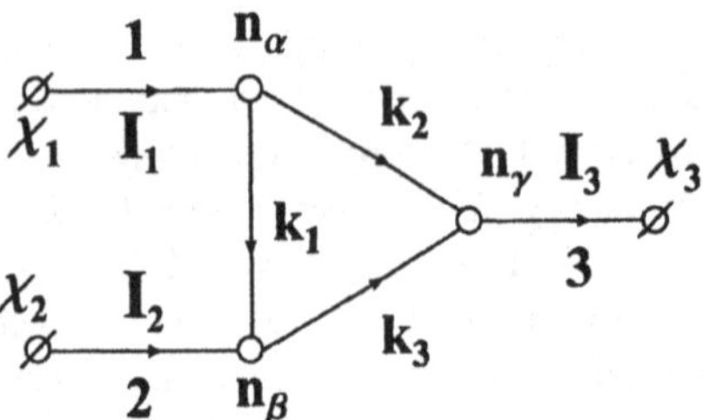

Figure 6. The graph of the general model (See the main text). The terminals are connected with the system via external junctions 1,2 and 3. The nodes α, β and γ are either resonant levels or dots, linked with each other by internal junctions k's. The arrows denote the conventional direction of a current through each junction.

We begin this section by presenting the general model, which dynamics obeys the master equation. Further on we proceed with the proof of the central result of this section for the FCS in the master equation approach. We illustrate the general scheme by considering the statistics of charge transfer through the Coulomb blockade island with 3 leads attached and compare the FCS in this case with the results of the preceding section, i.e. for non-interacting electrons.

4.1. THE GENERAL MODEL

The dynamics of various systems can be described by master equation. For our purposes it is convenient to write it in the matrix form:

$$\frac{\partial}{\partial t}|p(t)\rangle = -\hat{L}|p(t)\rangle \qquad (32)$$

where each element $p_n(t)$ of the vector $|p(t)\rangle$ is the probability to find the system in the state n. The matrix elements of operator $\hat{L}$ are given by

$$L_{mn} = \delta_{nm}\gamma_n - \Gamma_{m\leftarrow n}, \quad \gamma_n = \sum_{m\neq n}\Gamma_{m\leftarrow n} \qquad (33)$$

Here $\Gamma_{n\leftarrow m}$ stands for the transition rate from the state m to the state n, γ_n presents the total transition rate from the state n. The $\hat{L}$ operator thus defined always has a zero eigenvalue, the corresponding eigenvector being the stationary solution of the master equation.

Coulomb blockade mesoscopic systems always obey Eq. (32). The main advantage of the master equation approach is a possibility of non-perturbative treatment of the interaction effects. Below we first consider the master equation description of the general model system. This will prepare us to the next section where we derive the FCS method.

The possible physical realization of the general model includes an array of Coulomb blockade quantum dots and a mesoscopic system with a number of resonant levels. Like in the preceding sections, it is convenient schematically to present the system as a graph (see Fig. 4.1) with each node α corresponding either to a single dot, a single resonant level or an external terminal. The line $k = (\alpha, \beta)$, connecting the nodes α and β, is associated with a possible electron transfer. Let M be the total number of nodes in this graph and L is a total number of lines. For a many-dot systems each line k corresponds to the tunnel junction. For systems with many resonant levels it corresponds to the possible transition between different levels, so that it does not necessary correspond to electron transfer in space. There are N external junction $k = 1 \dots N$, $(N \leq L)$, connecting the terminals with the system. The currents through these junctions are directly measurable and hence are of our interest.

The macro- or microscopic state of the general model is given by a set of occupation numbers $|n\rangle = |n_1, \dots, n_M\rangle$; n_α is equal to any integer for the array of quantum dots and refers to the excess charge on the island α; in case of many resonant level n_α denotes the occupation number of a given level. Owing to the fact that $\sum_n L_{nm} = 0$, the $\hat{L}$ operator has a zero eigenvalue. There are the right, $|p_0\rangle$, and the left, $\langle q_0|$, eigenvectors corresponding to this zero eigenvalue

$$\hat{L}|p_0\rangle = 0, \qquad \langle q_0|\hat{L} = 0 \tag{34}$$

We assume that they are unique. This means that the system does not get stuck in any metastable state. The vector $|p_0\rangle$ gives the steady probability distribution and $\langle q_0| = (1, 1, \dots, 1)$.

It is also useful to present $\hat{L}$ operator in the form

$$\hat{L} = \hat{\gamma} - \hat{\Gamma}, \quad \hat{\Gamma} = \sum_{k=1}^{L} (\hat{\Gamma}_k^{(+)} + \hat{\Gamma}_k^{(-)}) \tag{35}$$

where $\hat{\gamma}$ is the diagonal operator in the basis $|n\rangle$ of the system configuration and $\hat{\Gamma}_k^{(\pm)}$ is associated with the electron transfers through the line $k = (\alpha, \beta)$:

$$\hat{\gamma} = \sum_{\{n\}} |n\rangle \gamma(n) \langle n|, \quad \hat{\Gamma}_k^{(\pm)} = \sum_{\{n\}} |n'\rangle \Gamma_k^{(\pm)}(n) \langle n| \tag{36}$$

The state $|n'\rangle = |n_1, \dots, n'_\alpha, \dots, n'_\beta, \dots, n_M\rangle$ results from the state $|n\rangle$ by appropriate changing the corresponding occupation numbers: $n'_\alpha = n_\alpha - \sigma_k$, $n'_\beta = n_\beta + \sigma_k$, where $\sigma_k = \pm 1$ denotes the direction of the transition.

4.2. THE FCS IN THE MASTER EQUATION

In this section we derive the central result for the FCS of the charge transfer in the system, which dynamics obeys the master equation. We will solve this problem by making use of the property of the system, that its random evolution in time is the Markov stochastic process.

In what follows we will partially use notations of the book [31]. Let us consider the time interval $[-T/2, T/2]$. Suppose the system undergoes s transitions at random time moments τ_i, so that

$$+T/2 > \tau_1 > \tau_2 > \ldots > \tau_{s-1} > \tau_s > -T/2 \tag{37}$$

This gives an elementary random sample $\zeta_s = (\tau_1, k_1, \sigma_1; \ldots; \tau_s, k_s, \sigma_s)$. It corresponds to the set of subsequent events, when at time τ_i the tunneling happens via the junction k_i, $\sigma_i = \pm 1$ being the direction of the transition. The samples ζ_s constitute the set Ω of all possible random samples.

Then one defines the measure (or the probability) $d\mu(\zeta)$ at the set Ω. For this purpose we may very generally introduce the sequence of non-negative probabilities $Q_s(\{\tau_i, k_i, \sigma_i\}) \equiv Q(\tau_1, k_1, \sigma_1; \ldots; \tau_s, k_s, \sigma_s) \geq 0$ defined in Ω so that

$$d\mu(\zeta) = Q_0 + \sum_{s=1}^{+\infty} \sum_{\{k_i, \sigma_i\}} Q_s(\{\tau_i, k_i, \sigma_i\}) d\tau_1 \ldots d\tau_s \tag{38}$$

The functions Q are normalized according to the condition

$$\int_\Omega d\mu(\zeta) \equiv Q_0 + \sum_{s=1}^{+\infty} \sum_{\{k_i, \sigma_i\}} \int \cdots \int_{T/2 > \tau_1 > \ldots > \tau_s > -T/2} Q_s(\{\tau_i, k_i, \sigma_i\}) \prod_{i=1}^{s} d\tau_i = 1 \tag{39}$$

Each term in Exp. (38) corresponds to the probability of an elementary sample ζ_s.

To accomplish the preliminaries, we remind the concept of a stochastic process. Mathematically speaking, it can be any integrable function $\breve{A}(t) \equiv A(t, \zeta)$ defined at the set Ω and parametrically depending on time. It is sometimes convenient to omit the explicit ζ dependence. We will use a "check" in this case to stress that the quantity in question is a random variable. Each stochastic process $A(t, \zeta)$ generates the sequence of time dependent functions $\Big\{A_0(t),\ A_1(t, \tau_1, k_1, \sigma_1), \ldots, A_s(t, \{\tau_i, k_i, \sigma_i\})\Big\}$. Its average $\langle \breve{A}(t) \rangle_\Omega$ over the space Ω is defined as

$$\langle \breve{A}(t) \rangle_\Omega = \int_\Omega A(t, \zeta) d\mu(\zeta) \equiv A_0(t) Q_0 + \sum_{s=1}^{+\infty} \sum_{\{k_i, \sigma_i\}} \int \cdots \int_{T/2 > \tau_1 > \ldots > \tau_s > -T/2} A_s(t, \{\tau_i, k_i, \sigma_i\}) Q_s(\{\tau_i, k_i, \sigma_i\}) \prod_{i=1}^{s} d\tau_i \tag{40}$$

The analogous expression should be used, for instance, to define the correlations $\langle \check{A}(t_1)\check{B}(t_2)\rangle_\Omega$ between any two stochastic processes.

For the subsequent analysis we define the random process $\check{I}^{(k)}(t)$, corresponding to the classical current through the external junction $k \leq N$:

$$I^{(k)}(t,\zeta_s) = \sum_{i=1}^{s} e\sigma_i\, \delta(t-\tau_i)\delta(k-k_i) \tag{41}$$

Here σ_m is included to take into account the direction of the jump and $\delta(k-k_i) \equiv \delta_{k,k_i}$ is the Kronecker δ symbol. Given this definition at hand, we introduce the generating functional $S[\{\chi_i(t)\}]$ depending on N counting fields $\chi_i(\tau)$, each of them associated with a given terminal i:

$$\exp(-S[\{\chi_i(t)\}]) = \left\langle \exp\Big\{ i\sum_{n=1}^{N}\int_{-\infty}^{+\infty} d\tau \chi_n(\tau)\check{I}^{(n)}(\tau)/e\Big\}\right\rangle_\Omega \tag{42}$$

with the average defined by Eq.(40). Let us note the remarkable similarity of this classical expression with the quantum mechanical action (13). As before, in the low-frequency limit of current correlations one may use the time-independent counting fields χ_i. In this case the action $S[\{\chi_i\}]$ can be used to find the probability (2) of N_i electrons to be transferred through the terminal i during the time interval T.

The above definitions were rather general than constructive, since the probabilities Q have not been specified so far. To proceed, one has to relate them to transition rates of the master equation. We assume that at initial time $t = -T/2$ the system was in the state $\{n^{(s)}\}$. Then random sample ζ_s determines the evolution of charge configuration $\{n^{(s)}\} \to \{n^{(s-1)}\}\ldots\{n^{(1)}\} \to \{n^{(0)}\}$ for subsequent moments of time. The choice of ζ_s specifies that the transition between neighboring charge states $\{n^{(i)}\}$ and $\{n^{(i-1)}\}$ occurs at time τ_i via the junction $k_i = (\alpha_i, \beta_i)$. Therefore the sequence $\{n^{(i)}\}$ is given by the relation $n_{\alpha_i}^{(i-1)} = n_{\alpha_i}^{(i)} - \sigma_{k_i}$, $n_{\beta_i}^{(i-1)} = n_{\beta_i}^{(i)} + \sigma_{k_i}$, and $n_\gamma^{(i-1)} = n_\gamma^{(i)}$ for all $\gamma \neq \alpha_i$ and β_i. To determine the probability $Q_s(\{\tau_i, k_i, \sigma_i\})$ we note that (i) the sample ζ_s constitutes the Markov chain (ii) the conditional probability of the system to remain at state $n^{(i)}$ between the times τ_{i+1} and τ_i is proportional to $\exp[-\gamma(n^{(i)})(\tau_i - \tau_{i+1})]$; (iii) the probability that the transition occurs via the junction k_i during the time interval $d\tau_i$ at the moment τ_i is given by $\Gamma_{k_i}^{(\sigma_i)}(n^{(i)})d\tau_i$. These arguments suggest that Q's have the form

$$Q_0 = Z_0^{-1}\exp[-\gamma(n^{(s)})T] \tag{43}$$

$$\begin{aligned} Q_s(\{\tau_i,k_i,\sigma_i\}) = Z_0^{-1}\exp[-\gamma(n^{(0)})(T/2-\tau_1)]\Gamma_{k_1}^{(\sigma_1)}(n^{(1)}) \\ \exp[-\gamma(n^{(1)})(\tau_1-\tau_2)]\Gamma_{k_2}^{(\sigma_2)}(n^{(2)})\ldots\exp[-\gamma(n^{(s-1)}) \\ (\tau_{s-1}-\tau_s)]\Gamma_{k_s}^{(\sigma_s)}(n^{(s)})\exp[-\gamma(n^{(s)})(\tau_s+T/2)] \end{aligned}$$

where the constant Z_0 should be found from the normalization condition (39). As we will see below, $Z_0 = 1$.

The above correspondence between the random Markov chain ζ_s and the probabilities Q's (43) allows one to evaluate the generating function (42). By definition (41) for any given ζ_s we have

$$\exp\left\{i\sum_{n=1}^{N}\int_{-\infty}^{+\infty} d\tau\chi_n(\tau)I^{(n)}(\tau,\zeta_s)/e\right\} = \prod_{i=1}^{s}\exp\{i\sigma_i\chi_{k_i}(\tau_i)\}$$

It is assumed here that $\chi_{k_i} = 0$ if the transition occurs via internal junction, $k_i > N$, thus no physically measurable current is generated in this case. The averaging of the latter expression over all possible configurations Ω with the weight $d\mu(\zeta)$ yields

$$Z[\{\chi_i(\tau)\}] \equiv \exp(-S[\{\chi_i(\tau)\}]) = Q_0 + \tag{44}$$

$$\sum_{s=1}^{+\infty}\sum_{\{k_i,\sigma_i\}}\int\cdots\int_{T/2>\tau_1>\ldots>\tau_s>-T/2} Q_s^{\chi}(\{\tau_i,k_i,\sigma_i\})\prod_{i=1}^{s}d\tau_i$$

The resulting expression resembles the normalization condition (39). Here the χ-dependent functions $Q_s^{\chi}(\{\tau_i,k_i,\sigma_i\})$ are defined similar to probabilities (43) with the only crucial difference that the rates $\Gamma_k^{(\sigma)}(n)$ should be replaced by $\Gamma_k^{(\sigma)}(n)\exp\{i\sigma_k\chi_k(\tau_k)\}$ if $k \leq N$.

The expression (44) can be written in the more compact and elegant way. For that, we introduce the χ-dependent linear operator $\hat{L}_\chi$ defined as

$$\begin{aligned}\hat{L}_\chi(\tau) &= \hat{\gamma} - \hat{\Gamma}_\chi(\tau), \\ \hat{\Gamma}_\chi(\tau) &= \sum_{k=1}^{N}(\hat{\Gamma}_k^{(+)}e^{i\chi_k(\tau)} + \hat{\Gamma}_k^{(-)}e^{-i\chi_k(\tau)}) + \sum_{k=N+1}^{L}(\hat{\Gamma}_k^{(+)} + \hat{\Gamma}_k^{(-)})\end{aligned} \tag{45}$$

In line with consideration above we multiplied each operator $\hat{\Gamma}_k^{(\pm)}$ ($k = 1\ldots N$), that corresponds to the transition through the external junction, by an extra χ-dependent factor $e^{i\chi_k(\tau)}$. The diagonal part and internal transition operators $\hat{\Gamma}_k^{(\pm)}$ with $k > N$ remained unchanged. Then we consider the evolution operator $\hat{U}_\chi(t_1,t_2)$ associated with (45). Since $\hat{L}_\chi(\tau)$ is in general time-dependent, $\hat{U}_\chi(t_1,t_2)$ is given by the time-ordered exponent

$$\hat{U}_\chi(t_1,t_2) = T_\tau\exp\left\{-\int_{t_2}^{t_1}(\hat{\gamma}(\tau) - \hat{\Gamma}_\chi(\tau))d\tau\right\} \tag{46}$$

The similar construction is widely used in quantum statistics. The difference in the present case is that the operator $\hat{U}_\chi(t_1,t_2)$ at $\chi = 0$ gives the evolution of probability rather than the amplitude of probability.

With the use of evolution operator (46) the generating function (44) can be cast into the form

$$Z[\{\chi_i(\tau)\}] = \langle q_0|\hat{U}_\chi(T/2,-T/2)|n_s\rangle \tag{47}$$

To prove it we argue as follows. We exploite the fact that $\hat{\gamma}(\tau)$ and $\hat{\Gamma}(\tau)$ commute under the sign of time-ordering in Eq. (46) and regard $\hat{\Gamma}(\tau)$ as a perturbation. This gives the matrix element $\langle q_0|\hat{U}_\chi(T/2,-T/2)|n_s\rangle$ in the form of series

$$\langle q_0|\hat{U}_\chi(T/2,-T/2)|n_s\rangle = \langle q_0|e^{-\hat{\gamma}T}|p_0\rangle + \tag{48}$$
$$\sum_{s=1}^{+\infty}\langle q_0|T_\tau \exp\Big\{-\int_{-T/2}^{T/2}\hat{\gamma}(\tau)d\tau\Big\}\sum_{\mathbf{k}_s\sigma_s}\int\cdots\int_{T/2>\tau_1>\ldots>\tau_s>-T/2}$$
$$\hat{\Gamma}_{k_1}^{(\sigma_1)}(\tau_1)e^{i\sigma_1\chi_{k_1}(\tau_{k_1})}\ldots\hat{\Gamma}_{k_s}^{(\sigma_s)}(\tau_s)e^{i\sigma_s\chi_{k_s}(\tau_{k_s})}|p_0\rangle\prod_{i=1}^{s}d\tau_i$$

It follows from the definition (43) that each term in this series corresponds to the function $Q_s^\chi(\{\tau_i,k_i,\sigma_i\})$, namely

$$Q_0 = \langle q_0|e^{-\gamma T}|n_s\rangle$$
$$Q_s^\chi(\{\tau_i,k_i,\sigma_i\}) = \langle q_0|T_\tau \exp\Big\{-\int_{-T/2}^{T/2}\hat{\gamma}(\tau)d\tau\Big\} \tag{49}$$
$$\hat{\Gamma}_{k_1}^{(\sigma_1)}(\tau_1)e^{i\sigma_1\chi_{k_1}(\tau_{k_1})}\ldots\hat{\Gamma}_{k_s}^{(\sigma_s)}(\tau_s)e^{i\sigma_s\chi_{k_s}(\tau_{k_s})}|n_s\rangle$$

Therefore Exp. (48) and (47) are reduced to the previous result (44). This completes the proof. Note, that owing to the property (34), $Z_0 = \langle q_0|\exp(-T\hat{L})|n_s\rangle = 1$ identically at $\chi = 0$. Therefore the probabilities (43) are correctly normalized.

The Exp. (47) for the generating function $Z[\{\chi_i(t)\}]$ depends on the initial state $|n_s\rangle$ of the system. It can be shown that the choice of $|n_s\rangle$ does not affect the final results. We assume that $\chi_k(t) \to 0$ when $t \to -T/2$. Physically, it means that the measurement is limited in time. To be specific one may assume that $\chi_k(t) = 0$ when $-T/2 < t < -T/2+\Delta t$ and $\chi_k(t) \neq 0$ if $t > -T/2+\Delta t$. If the time interval Δt is sufficiently large as compared with the typical transition time Γ^{-1}, then the system will reach the steady state during this period of time. The latter follows from the fact that $\exp(-\hat{L}\,\Delta t)|n_s\rangle \to |p_0\rangle$ when $\Delta t \gg \Gamma^{-1}$. Thus one can substitute $|n_s\rangle$ to $|p_0\rangle$ in Exp. (47). Assuming also the limit $T \to \infty$, we arrive to the main result of this section

$$\exp(-S[\{\chi_i(t)\}]) = \langle q_0|T_\tau \exp\Big\{-\int_{-\infty}^{+\infty}\hat{L}_\chi(\tau)d\tau\Big\}|p_0\rangle \tag{50}$$

We see that the generating function can be written in the form of the averaged evolution operator. This operator corresponds to master equation with the rates modified by the counting fields $\chi_i(\tau)$.

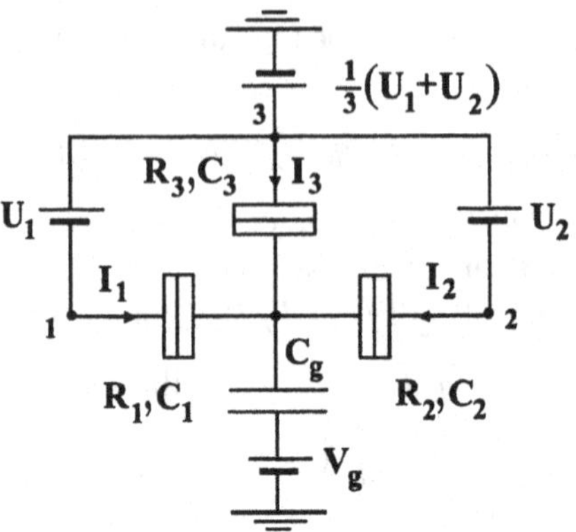

Figure 7. The equivalent circuit of the three-terminal Coulomb blockade island. The voltages $U_{1(2)}$ are used to control the bias between the 3d and the 1st (the 2nd) terminals: $U_{1(2)} = V_3 - V_{1(2)}$. The third terminal is biased at voltage $V_3 = (U_1 + U_2)/3$ with respect to the ground. This setup assures the condition $V_1 + V_2 + V_3 = 0$. Then gate voltage, V_g, can be used to control the offset charge $q_0 = C_g V_g$ on the island.

Further simplification is valid in the low frequency limit of the current correlations, $\omega \ll \Gamma$. (Here Γ is a typical transition rate in the system.) This allows to set $\chi_k(t) = \chi_k$ when $0 \leq t \leq t_0$ and $\chi_k(t) = 0$ otherwise, where t_0 is a time of measurement. The action (50) then reduces to the

$$S(\{\chi_i\}) = t_0 \Lambda_{\min}(\{\chi_i\}) \tag{51}$$

where $\Lambda_{\min}(\{\chi_i\})$ is the minimum eigenvalue of the operator $\hat{L}_\chi$. Thus the problem of statistics in the Coulomb blockade regime, provided the transition rates in the system are known, is merely a problem of the linear algebra.

4.3. THE STATISTICS OF CHARGE TRANSFER IN COULOMB BLOCKADE ISLAND.

In this subsection we use the developed method to study the current statistics in the three-terminal Coulomb blockade island. Its equivalent circuit is shown in Fig. 4.3. This circuit is an extension of the usual set-up of a conventional single electron transistor [32].

The island is in the Coulomb blockade regime, $R_k \gg R_Q = 2\pi\hbar/e^2$. In order to observe the Coulomb blockade effect the condition $k_B T \ll E_c = e^2/2C_\Sigma$ should be also satisfied. Here E_c is the charging energy of the island and $C_\Sigma = \sum_{i=1}^{3} C_k + C_g$. We assume the temperature to be rather high, $k_B T \gg \Delta E$, with ΔE being the mean level spacing in the island, so that the discreteness of the energy spectrum in the island is not important. We also disregard the possible effects of co-tunneling.

Under the above conditions the 3-terminal island is described by the "orthodox" Coulomb blockade theory. In this theory the macroscopic state of Coulomb blockade island is uniquely determined by the excess charge $Q = ne$, which is quantized in terms of electron charge $(-e)$. The charge Q can be changed only by $\pm e$ in course of one tunneling event. Therefore the master equation connects the given macroscopic state n with the neighboring states $n \pm 1$ only. The corresponding rates $\Gamma_{n\pm1\leftarrow n}$ of these transitions are equal to the sum of tunneling rates across all junctions: $\Gamma_{n\pm1\leftarrow n} = \sum_{k=1}^{N} \Gamma^{(k)}_{n\pm1\leftarrow n}$. The tunneling rate $\Gamma^{(k)}_{n\pm1\leftarrow n}$ across the junction k can be expressed via the electrostatic energy difference $\Delta E^{(k)}_{n\pm1\leftarrow n}$ between the initial (n) and final $(n \pm 1)$ configurations

$$\Gamma^{(k)}_{n\pm1\leftarrow n} = \frac{1}{e^2 R_k} \frac{\Delta E^{(k)}_{n\pm1\leftarrow n}}{1 - \exp[-\Delta E^{(k)}_{n\pm1\leftarrow n}/k_B T]} \tag{52}$$

The evaluation of $\Delta E^{(k)}_{n\pm1\leftarrow n}$ can be done along the same lines as in the case of single electron transistor [32].

According to the general result (51) of the preceding subsection, one can find the FCS of the charge transfer through the island, by evaluating the minimum eigenvalue $\Lambda_{\min}$ of the matrix $\hat{L}_\chi$. In case under consideration this problem is reduced to the eigenvalue problem of the three-diagonal matrix:

$$(\Lambda - \gamma_n)p_n + \Gamma^\chi_{n\leftarrow n+1} p_{n+1} + \Gamma^\chi_{n\leftarrow n-1} p_{n-1} = 0 \tag{53}$$

where $\gamma_n = \Gamma_{n\leftarrow n-1} + \Gamma_{n\leftarrow n+1}$, and $\Gamma^\chi_{n\leftarrow n\pm1} = \sum_{k=1}^{N} \Gamma^{(k)}_{n\pm1\leftarrow n} e^{\pm i\chi_k}$. The index $+(-)$ corresponds to electron transition from (to) the island.

To assess the FCS we have treated the related linear problem (53) numerically. We restrict the consideration to sufficiently low temperatures $k_B T \ll E_c$, so that the temperature dependence in rates (52) is non-essential. In this case $\Gamma^{(\pm)}_k(n) = \Delta E^{(k)}_{n\pm1\leftarrow n}/(e^2 R_k)$ when $\Delta E^{(k)}_{n\pm1\leftarrow n} \geq 0$ and $\Gamma^{(\pm)}_k(n) = 0$ otherwise. We have also assumed that $U_2 > U_1$ (See Fig. 4.3). The corresponding χ-dependent rates can be found from Exp. (52) and (45) and read as follows

$$\begin{aligned} \Gamma^\chi_{n+1\leftarrow n} &= \Gamma^{(+)}_3(n) e^{i\chi_3} + \Gamma^{(+)}_1(n)\,\theta(q - 1/2 - n) e^{i\chi_1} \\ \Gamma^\chi_{n-1\leftarrow n} &= \Gamma^{(-)}_2(n) e^{-i\chi_2} + \Gamma^{(-)}_1(n)\,\theta(n - q - 1/2) e^{-i\chi_1} \end{aligned} \tag{54}$$

where

$$\Gamma^{(\pm)}_k(n) = a^{(\pm)}_k \mp \left(n + \frac{C_g V_g}{e} \pm \frac{1}{2}\right) \frac{1}{R_k C_\Sigma} \tag{55}$$

and

$$a^{(+)}_3 = \frac{\tilde{C}_1 U_1 + \tilde{C}_2 U_2}{e R_3 C_\Sigma}, \quad a^{(-)}_2 = \frac{(\tilde{C}_1 + \tilde{C}_3) U_2 - \tilde{C}_1 U_1}{e R_2 C_\Sigma}$$

$$a^{(\pm)}_1 = \pm \frac{\tilde{C}_2 U_2 - (\tilde{C}_3 + \tilde{C}_2) U_1}{e R_1 C_\Sigma}$$

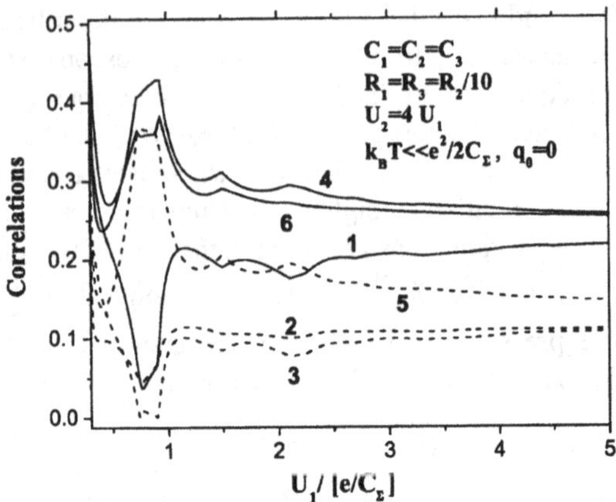

Figure 8. The matrix F of auto- and cross- shot noise correlation versus voltage U_1 for the 3-terminal quantum dot setup. Parameters are shown on the plot. (1) - F_{11}, (2) - $|F_{12}|$, (3) - $|F_{13}|$, (4) - F_{22}, (5) - $|F_{23}|$, (6) - F_{33}.

The effective capacitances C_k are defined as $\tilde{C}_k = C_k + C_g/3$ and the point q is given by the relation

$$eq(U_1, U_2, V_g) = \tilde{C}_2 U_2 - (\tilde{C}_3 + \tilde{C}_2)U_1 - C_g V_g \tag{56}$$

The value q is non-integer in general. It satisfies the condition $\Gamma_1^{(-)}(q+1/2) = \Gamma_1^{(+)}(q-1/2) = 0$. The dimension of the $\hat{L}_\chi$-matrix is equal to $n_{\max} - n_{\min}$, where $n_{\max}(n_{\min})$ can be found from the conditions $\Gamma_3^{(-)}(n) \geq 0$ ($\Gamma_1^{(+)}(n) \geq 0$). The value $e\,n_{\max}$, ($e\,n_{\min}$) gives the maximum (minimum) charge that can be in the island for a given voltages U_1, U_2 and V_g.

We can see from the Exp. (54) that there are four elementary processes of charge transfer in the system at low temperatures, each of them being associated with the pre-factor $e^{\pm i\chi_k}$. The factors $e^{i\chi_3}$ and $e^{-i\chi_2}$ correspond to the charge transfer from the third terminal into the island and from the island into the second terminal, respectively. Hence, the random current through the 3d (2nd) junctions always has the positive (negative) sign. Two factors $e^{\pm\chi_1}$ stem from the charge transfer through the first junction in the direction either from the island into the first contact or vice versa. Therefore the current I_1 fluctuates in both directions.

Let us consider the shot noise correlations in the system. In Fig. 8 we give the illustrative example of the voltage dependence of the shot noise correlations F_{km} for a certain choice of parameters. The definition of matrix F_{km} is the same as we have used in the end of section 3. The Coulomb blockade features are strongly pronounced for an asymmetric setup only. The results shown in Fig. 8 correspond to $R_1 = R_3 = R_2/10$, $\tilde{C}_1 = \tilde{C}_2 = \tilde{C}_3$ and $U_2/U_1 = 4$. In Fig. 4.3 we show

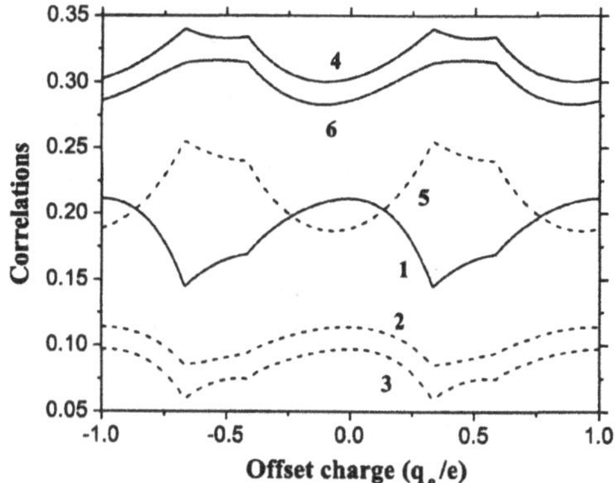

Figure 9. The matrix F of auto- and cross- shot noise correlation versus the offset charge for the 3-terminal quantum dot setup. Parameters are the same as on the Fig. 8. The voltage $U_1 = U_2/4 = 1.25e/C_\Sigma$. (1) - F_{11}, (2) - $|F_{12}|$, (3) - $|F_{13}|$, (4) - F_{22}, (5) - $|F_{23}|$, (6) - F_{33}.

the dependence of the shot noise correlations on the offset charge for the same set of parameters and the value of $U_1 = 1.25\, e/C_\Sigma$. The special points of both these dependences occur when either $n_{\min}$, $n_{\max}$ or the integer part of q are changed by ± 1. As the result one observes multi-periodic Coulomb blockade oscillations in the offset charge dependences.

We now proceed with the evaluation of the FCS. The action $S(\{\chi_i\})$ has been calculated with the use of (51). To find the probability distribution P we have evaluated the Fourier transform (2) in the saddle point approximation. It is applicable here, since we consider the low frequency limit only, $\omega \ll \Gamma$. In this limit both action S and number of transmitted particles $N_i = I_i t_0/e \gg 1$. Due to the current conservation $\sum_k I_k = 0$, only two currents are independent and the action $S(\{\chi_i\})$ depends on the differences $\chi_{ij} = \chi_i - \chi_j$ only. In what follows we have chosen I_1 and I_2 as the independent variables to plot the logarithm of probability $\ln P(I_1, I_2)$. In the saddle point approximation, with the exponential accuracy, it is given by $P(I_1, I_2) \sim e^{-\Omega(\chi^*)}$. Here χ^* is a saddle point of the function $\Omega(\chi) = S(\chi) + i\chi_1 I_1 t_0/e + i\chi_2 I_2 t_0/e$. The results for $\ln P(I_1, I_2)$ are shown in Figs. 4.3 and 4.3(a). From the contour map on Fig. 4.3(a) we see that $P(I_1, I_2)$ is non-zero in the region $I_1 < 0$, $I_2 < 0$ and in the region $I_1 \leq |I_2|$ provided $I_1 > 0$ and $I_2 < 0$. This range of plausible current fluctuations stems from the χ-dependence of rates (54). Any current fluctuation automatically satisfies the constrain $\sum_k I_k = 0$ and conditions $I_2 < 0$ and $I_3 > 0$.

Before discussing the results, let us set the reference point for such discussion. This reference will be the results of the previous section. We consider the FCS in the three-terminal chaotic quantum dot when its contacts are tunnel junctions with

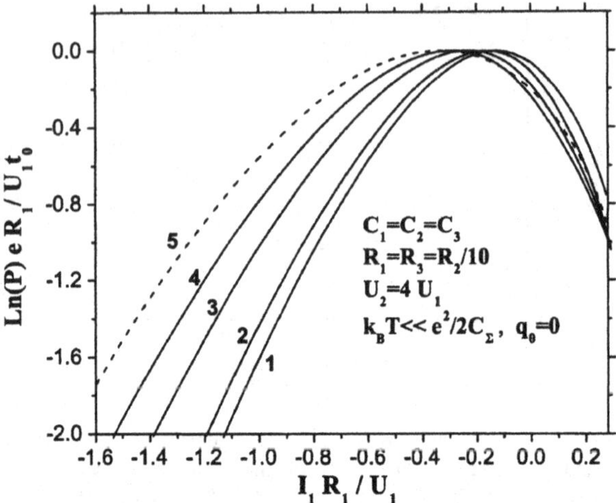

Figure 10. The logarithm of current distribution $\ln P(I_1, I_2)$ in the 3-terminal quantum dot as a function of current I_1, under condition $I_2 = \langle I_2 \rangle$. Parameters are shown on the plot. (1) - $U_1 = 1.25\, e/C_\Sigma$, (2) - $U_1 = 2.0\, e/C_\Sigma$, (3) - $U_1 = 4.0\, e/C_\Sigma$, (4) - $U_1 = 10.0\, e/C_\Sigma$; curve (5) corresponds to the non-interacting regime.

resistances $R_k^{-1} \gg e^2/\pi\hbar$. In this limit the effects of interaction are negligible and electrons are scattered independently at different energies. Provided $U_2 > U_1$, the generating function $S(\{\chi_i\})$ in the given case is a sum of the two independent processes (23)

$$S(\chi_1, \chi_2, \chi_3) = S_1(\chi_1, \chi_2, \chi_3) + S_2(\chi_1, \chi_2, \chi_3) \tag{57}$$

Here

$$S_1(\chi_1, \chi_2, \chi_3) = \frac{U_1 t_0}{2e}\Big\{G_1 + G_2 + G_3 - \sqrt{(G_1 + G_2 - G_3)^2 + 4G_3 e^{i\chi_3}(G_1 e^{-i\chi_1} + G_2 e^{-i\chi_2})}\Big\}$$

$$S_2(\chi_1, \chi_2, \chi_3) = \frac{(U_2 - U_1) t_0}{2e}\Big\{G_1 + G_2 + G_3 - \sqrt{(G_1 + G_3 - G_2)^2 + 4G_2 e^{-i\chi_2}(G_1 e^{i\chi_1} + G_3 e^{i\chi_3})}\Big\}$$

and $G_k = R_k^{-1}$ are the conductances of the junctions.

The logarithm of probability $\ln P_0(I_1, I_2)$, evaluated with the use of statistics (57), is shown by the dashed line in Fig. 4.3. Its contour map for the same

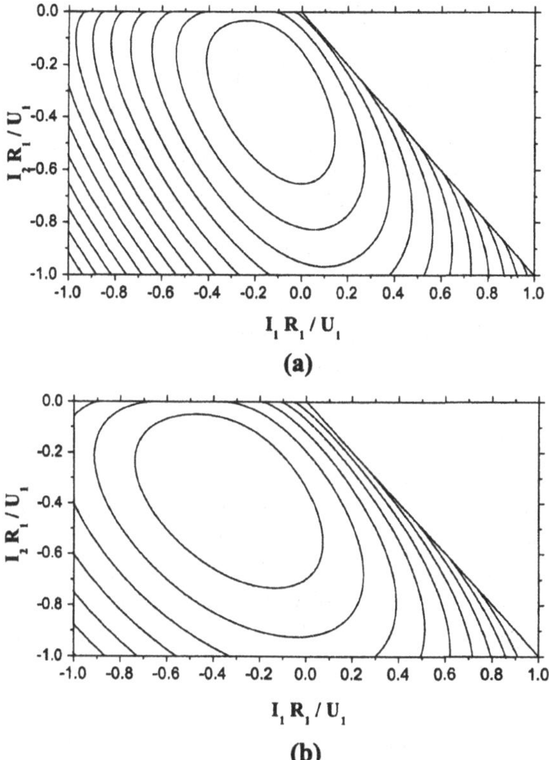

Figure 11. The contour maps of the current distribution $\log[P(I_1, I_2)]$ in the 3-terminal quantum dot. $U_1 = U_2/4.0 = 1.25\, e/C_\Sigma$. (a) - Coulomb blockade island. Parameters are the same as in Fig. 4.3. (b) - Chaotic quantum dot.

values of parameters is separately presented in Fig. 4.3(b). The maximum of $\ln P_0(I_1, I_2)$, as expected, occur at $\bar{I}_1 = \bar{I}_2 = U_1/3R_1$.

Comparing the FCS in the Coulomb blockade and non-interacting limits we can draw the following conclusions. In spite of the different regimes, we see that the qualitative dependence of probabilities versus the currents is similar for both cases. The probability distribution in both cases has a single maximum, corresponding to the average values of currents. The tails of distribution are essentially non-Gaussian both in the weak and strong interacting limit. The statistics approaches to the Gaussian-type one in the strong Coulomb blockade limit only, when the applied voltage to the system is only few above the Coulomb blockade threshold. (See curves 1 and 2 in Fig. 4.3) At higher applied voltages the probability distribution has a tendency to approach to the current distribution of the non-interacting system. However, they never become identical, even in the limit

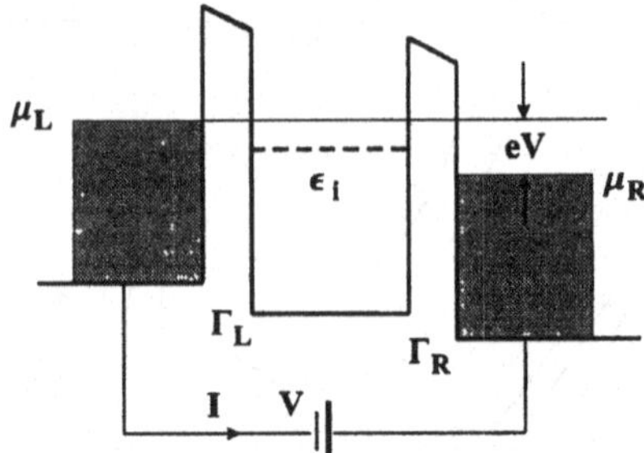

Figure 12. The single resonant level system, formed by the two tunnel barriers. The resonant level in the quantum well is shown by the dashed line.

$U_{1,2} \gg e/C_\Sigma$. (Curves 4 and 5). The same is true for the shot-noise correlations. Generally, we conclude that the Coulomb interaction always suppresses the relative probabilities of big current fluctuations. This stems from the fact that any big current fluctuation in Coulomb blockade dot is related with the large accumulation (or depletion) of the charge on the island. The latter process results in the excess of electrostatic energy. Therefore the relative probability of such fluctuation is decreased, as compared to the probability of the similar current fluctuation in the non-interacting regime.

5. The equivalence of scattering and master equation approaches to the FCS

In this section we evaluate the FCS in the generic case of a single resonant level, shown in Fig. 5. We consider only the non-interacting spinless electrons and demonstrate the equivalence of scattering and master equation approaches to the FCS in the framework of this model.

We start by considering the FCS in the master equation framework. It is applicable, provided the applied voltage or the temperature are not too low, i.e. $\max\{eV, k_BT\} \gg \hbar\Gamma_{L(R)}$. Here $\Gamma_{L(R)}$ denote the quantum-mechanical tunneling rates from the left (right) electrode onto the resonant level. The system can be found in the two microscopic states only: one with no electron on the level, and another with a single electron. Then the transition rates involved reads as

$$\begin{aligned} \Gamma_{1\leftarrow 0} &= \Gamma_L f_L(\epsilon_i) + \Gamma_R f_R(\epsilon_i) \\ \Gamma_{0\leftarrow 1} &= \Gamma_L[1 - f_L(\epsilon_i)] + \Gamma_R[1 - f_R(\epsilon_i)] \end{aligned} \qquad (58)$$

Here the indices $\{0\}$ and $\{1\}$ denote the microscopic state with no and one electron on the level. Fermi function $f_{L(R)}(\epsilon) = (1 + \exp[(\epsilon - \mu_{L(R)})/kT])^{-1}$ accounts for the filling factor in the left (right) lead and ϵ_i is the position of the resonant level.

Following the definition (45) and the expression for the rates (58), the $\hat{L}_\chi$-matrix of the single resonant level model reads as

$$\hat{L}_\chi = \begin{pmatrix} \Gamma_{1\leftarrow 0} & -\Gamma_{0\leftarrow 1}(\chi) \\ -\Gamma_{1\leftarrow 0}(\chi) & \Gamma_{0\leftarrow 1} \end{pmatrix} \tag{59}$$

where

$$\begin{aligned} \Gamma_{1\leftarrow 0}(\chi) &= \Gamma_L f_L e^{-i\chi_1} + \Gamma_R f_R e^{-i\chi_2} \\ \Gamma_{0\leftarrow 1}(\chi) &= \Gamma_L (1-f_L) e^{i\chi_1} + \Gamma_R (1-f_R) e^{i\chi_2} \end{aligned} \tag{60}$$

Evaluating the minimum eigenvalue of this matrix we obtain the current statistics in the following form

$$S(\chi) = \frac{t_0}{2}\left\{\Gamma_L + \Gamma_R - \sqrt{\mathcal{D}(\chi)}\right\} \tag{61}$$

$$\mathcal{D}(\chi) = (\Gamma_L + \Gamma_R)^2 + 4\Gamma_L\Gamma_R\left[f_{(-)}(\epsilon_i)(e^{-i\chi}-1) + f_{(+)}(\epsilon_i)(e^{i\chi}-1)\right]$$

Here $f_{(-)}(\epsilon_i) = f_L(\epsilon_i)[1-f_R(\epsilon_i)]$, $f_{(+)}(\epsilon_i) = f_R(\epsilon_i)[1-f_L(\epsilon_i)]$ and $\chi = \chi_1 - \chi_2$.

Since the electrons above are assumed to be non-interacting, one might have come to the same result in the framework of the pioneering approach by Levitov *et.al.*[12]. We will show now that it is indeed the case.

According to Ref. [12] the general expression for the current statistics through a single contact reads as

$$S(\chi) = -\frac{t_0}{2\pi}\sum_n \int d\epsilon \ln\Big\{1 + T_n(\epsilon) \times \tag{62}$$

$$\Big(f_L(\epsilon)[1-f_R(\epsilon)](e^{-i\chi}-1) + f_R(\epsilon)[1-f_L(\epsilon)](e^{i\chi}-1)\Big)\Big\}$$

It is valid for any two-terminal geometry provided the region between two electrodes can be described by the one-particle scattering approach. $T_n(\epsilon)$ is a set of transmission eigenvalues which are in general energy-dependent. For a resonant level there is a single resonant transmission eigen-value $T_r(\epsilon)$, its energy dependence being given by the Breit-Wigner formula

$$T_r(\epsilon) = \frac{\Gamma_L\Gamma_R}{(\epsilon-\epsilon_i)^2 + (\Gamma_L+\Gamma_R)^2/4} \tag{63}$$

The result (62, 63) is more general, than Exp. (61), obtained by means of master equation. If electrons do not interact, the Exp. (62) is valid for any temperature. It can be simplified in the regime $k_B T \gg \hbar\Gamma$. As we will show, in this limit the general result (62) coincides with the result (61) of master equation. It is easier to perform the calculation if one first evaluates the χ-dependent current $I(\chi) = (ie/t_0)\partial S/\partial\chi$. It reads

$$I(\chi) = \frac{1}{2\pi}\int d\epsilon \left[f_{(+)}(\epsilon)e^{i\chi} - f_{(-)}(\epsilon)e^{-i\chi}\right] \times \tag{64}$$
$$\left\{T_r^{-1}(\epsilon) + \left[f_{(-)}(\epsilon)(e^{-i\chi}-1) + f_{(+)}(\epsilon)(e^{i\chi}-1)\right]\right\}$$

Let us consider the situation when the resonant level is placed between the chemical potentials $\mu_{L\{R\}}$ in the leads. Since we assumed that $k_BT \gg \Gamma_{L(R)}$, the main contribution comes from the Lorentz peak and one can put $\epsilon = \epsilon_i$ in the Fermi functions. Therefore we left only with the two poles $\epsilon_{1(2)} = \epsilon_i \pm i\sqrt{\mathcal{D}(\chi)}/2$ under the integrand (64). Closing the integration contour in the upper or lower half-plane we arrive at

$$I(\chi) = e\Gamma_L\Gamma_R\left[f_{(+)}(\epsilon_i)e^{i\chi} - f_{(-)}(\epsilon_i)e^{-i\chi}\right]/\sqrt{\mathcal{D}(\chi)} \tag{65}$$

Integrating it over χ one finds for the $S(\chi) = (t_0/ie)\int_0^\chi I(\chi')d\chi'$ the result (61) obtained by means of master equation.

Thus, we have verified the correspondence between two approaches to statistics in the non-interacting regime. We have shown that one can reproduce the statistics (61) on substituting $T_r(\epsilon)$ into the Exp. (62) and assuming the regime $k_BT \gg \hbar\Gamma$. As it was discussed previously, this is the condition, when the master equation approach, and hence its consequence (61), are valid.

6. Summary

We have reviewed here a constructive theory of counting statistics for electron transfer in multi-terminal mesoscopic systems. We have covered two opposite limit of weak and strong interaction. For the case of weakly interacting electrons, when the conductance of the system $G \gg G_Q$, the theory of FCS reduces to a circuit theory of 2×2 matrices associated with Keldysh Green functions. In the Coulomb blockade limit, $G \ll G_Q$, the FCS methods turns out to be an extension of the usual master equation approach. We have applied these methods to study the FCS of charge transfer through the three-terminal quantum dot. Surprisingly, the FCS has a similar qualitative features both in weakly and strongly interacting regimes. We found that Coulomb interaction suppresses the relative probabilities of big current fluctuations in the dot. We have also reviewed the scattering approach to FCS in multi-terminal circuits. Then by considering the generic model of a single resonance level, we have established the equivalence of scattering and master equation approaches to FCS.

The theories presented enables one for easy theoretical prediction of the FCS for a given practical layout. Thereby they facilitate experimental activities in this direction. Up to now, only the noise has been measured. In our opinion, the measurements of FCS can be easily performed with *threshold detectors* that produce a signal provided the current in a certain terminal exceeds the threshold value. If the threshold value exceeds the average current, the detector will be switched by this relatively improbable fluctuation of the current. The signal rate will be thus proportional to the probability of these fluctuations $P(I)$, the value given by the theory of FCS.

References

1. Ya. M. Blanter and M. Büttiker, Phys. Rep. **336**, 1 (2000).
2. L. Saminadayar, D. C. Glattli, Y. Jin, and B. Etienne, Phys. Rev. Lett. **79**, 2526 (1997). ; R. de-Picciotto, M. Reznikov, M. Heiblum, V. Umansky, G. Bunin, and D. Mahalu, Nature **389**, 162 (1997).
3. R. Cron, M. F. Goffman, D. Esteve, and C. Urbina, Phys. Rev. Lett. **86**, 4104 (2001).
4. S. Oberholzer, E. V. Sukhorukov, C. Strunk, C. Schönenberger. T. Heinzel, and M. Holland, Phys. Rev. Lett. **86**, 2114 (2001)
5. A. A. Kozhevnikov, R. J. Schoelkopf, and D. E. Prober, Phys. Rev. Lett **84**, 3398 (2000).
6. X. Jehl *et. al.*, Nature (London) **405**, 50 (2000).
7. M. Büttiker, Phys. Rev. B **46**, 12485 (1992).
8. R. C. Liu, B. Odom, Y. Yamamoto, and S. Tarucha, Nature **391**, 263 (1998)
9. W. D. Oliver, J. Kim, R. C. Liu, Y. Yamamoto, Science, **284**, 299 (1999)
10. M. Henny, S. Oberholzer, C. Strunk, T. Heizel, K. Ensslin, M. Holland, C. Schönenberger, Science **284**, 296 (1999).
11. L. S. Levitov and G. B. Lesovik, JETP Lett. **58**, 230 (1993).
12. L. S. Levitov, H.-W. Lee, and G. B. Lesovik, Journal of Mathematical Physics, **37** (1996) 10.
13. H. Lee, L. S. Levitov, A. Yu. Yakovets, Phys. Rev. B, **51**, 4079 (1995)
14. Ya. M. Blanter, H. Schomerus, and C.W.J. Beenakker, Physica E **11**, 1 (2001).
15. A. Andreev and A. Kamenev, Phys. Rev. Lett., **85**, 1294 (2000)
16. L. S. Levitov, arXiv: cond-mat/0103617, see also the contribution to the present book
17. Y. Makhlin and A. D. Mirlin, Phys. Rev. Lett., **87**, 276803 (2001)
18. Yu. V. Nazarov, Ann. Phys. (Leipzig) **8** Spec. Issue, SI-193 (1999), cond-mat/9908143.
19. Yu. V. Nazarov, *Generalized Ohm's Law*, in: Quantum Dynamics of Submicron Structures, eds. H. Cerdeira, B. Kramer, G. Schoen, Kluwer, 1995, p. 687.
20. W. Belzig and Yu. V. Nazarov, Phys. Rev. Lett., **87**, 067006 (2001); W. Belzig and Yu. V. Nazarov, Phys. Rev. Lett., **87**, 197006 (2001).
21. Ya. M. Blanter, E. V. Sukhorukov, Phys. Rev. Lett. **84**, 1280 (2000).
22. A. V. Andreev and E. G. Mishchenko Phys. Rev. B 64, 233316 (2001)
23. M.-S. Choi, F. Plastina, and R. Fazio Phys. Rev. Lett. **87**, 116601 (2001)
24. Yu. V. Nazarob, D. A. Bagrets, Phys. Rev. Lett. 88, 196801 (2002)
25. J. Rammer and H. Smith, Rev. Mod. Phys. **58**, 323 (1986).
26. A. I. Larkin and Yu. V. Ovchinninkov, Sov. Phys. JETP **41**, 960 (1975); Sov. Phys. JETP **46**, 155 (1977).
27. Yu. V. Nazarov, Superlattices Microst. **25**, 1221 (1999).
28. O. Agam, I. Aleiner and A. Larkin, Phys. Rev. Lett., **85**, 3153 (2000).
29. S. A. van Langen, M. Büttiker, Phys. Rev. B., **56**, R1680 (1997)
30. C.W.J. Beenakker, Rev. Mod. Phys., **69**, 731, (1997)

31. N.G. van Kampen, *Stochastic processes in physcics and chemistry*, Rev. and enl. eddition, Elsevier Scinece Publishes B.V., North-Holland, 1992
32. G.-L. Ingold, Yu. V. Nazarov, in *Single Charge Tunneling*, NATO ASI Series B: **294**, ed. H. Grabert, M. H. Devoret (NewYork, 1992)

FULL COUNTING STATISTICS OF SUPERCONDUCTOR-NORMAL-METAL HETEROSTRUCTURES

A short review

W. BELZIG
Department of Physics, University of Basel, Klingelbergstr. 82, 4056 Basel, Switzerland

1. Introduction

In 1918 Schottky discovered that the fluctuations in vacuum diodes can be related to the discrete nature of the charge carriers [1]. His observation was that the power spectrum of the current fluctuations gave direct access to the charge e of the discrete carriers responsible for the current. From his theoretical considerations he found a relation between the noise power of the current fluctuations S_I and the average current I,

$$S_I = 2eI\,, \tag{1}$$

a result nowadays known as the Schottky formula. Its consequence is, that the current noise provides information on the transport process, which is not accessible through conductance measurements only.

Studies of the noise properties of tunnel-junctions renewed the interest in noise later [2]. Since about ten years the investigation of transport in quantum coherent structures has boosted the interest in the theory of current noise in mesoscopic structures [3, 4]. Correlations in the transport of fermions have lead to a number of interesting predictions. For example, the noise of a single channel quantum contact of transparency T at zero temperature has the form $S_I = 2eI(1-T)$ [5, 6]. The noise is thus suppressed below the Schottky value, Eq. (1). The suppression is a direct consequence of the Pauli principle, and is therefore specific to electrons. Particles with bosonic statistics or classical particles show a different behaviour, e. g. for Bosons the noise is enhanced by a factor $1+T$. A convenient measure of the deviation from the Schottky result is the so-called Fano factor $F = S_I/2eI$.

Y.V. Nazarov (ed.), Quantum Noise in Mesoscopic Physics, 463–496.

For a number of generic conductors, it turns out that the suppression of the Fano factor is universal, i. e. it does not depend on details of the conductor like geometry or impurity concentration. A diffusive metal with purely elastic scattering leads to $F_{diff} = 1/3$ [7, 8], which is independent on the concrete shape of the conductor [9]. In a symmetric double tunnel junction, on the other hand, $F_{dbltun} = 1/2$ [10]. A chaotic cavity (a small region with classical chaotic dynamics, connected to two leads by open quantum point contacts) shows a suppression of $F_{cavity} = 1/4$ [11]. Thus, we conclude that from a noise measurement two kinds of information can be obtained. First, we can get information on the statistics of the carriers, e. g. if they are fermions. Second, provided the statistics is known, the comparison of the magnitude of the noise power with the average current gives information on the internal structure of the system. However, the picture described here is a little bit too simplified. In many real experiments the structure is much less well defined, the temperature is finite, or other complications make the interpretation of experimental data less trivial.

Having motivated the interest in the noise, we may ask, what we can learn further from current fluctuations. *Higher correlators* of the current will provide even more information on the transport process. However, theoretical calculations of higher correlators become increasingly cumbersome and one should find a different concept to obtain this information. This step was performed by a transfer of concepts from the field of quantum optics. Here it is possible to count experimentally the number of photons occupying a certain quantum state. This number is subject to quantum and thermal fluctuations and requires a statistical description: the *full counting statistics* (FCS). Lesovik and Levitov adopted this terminology to mesoscopic electron transport [12, 13], in which the electrons passing a certain conductor are counted. Since then the FCS has been studied in the field of mesoscopic electron transport. It the aim of this article to review some of the progress, which has been recently made.

The general problem of counting statistics has been considered before on a heuristic level. If one makes the *ad hoc* assumption that individual transfers of charges are *uncorrelated* and unidirectional, simple calculation show that the probability distribution of the number of transfered charges is Poissonian

$$P_{t_0}(N) = e^{-\bar{N}(t_0)} \frac{\bar{N}(t_0)^N}{N!} . \tag{2}$$

Here t_0 is the time period during which the charges are counted and $\bar{N}(t_0) = It_0/e$ is the average number of transfered charges. Schottky's result (1) for the current noise power can be easily derived from Eq. (2). This kind of counting statistics is usually found in tunnel junctions, where the charge transfers are rare events, or at high temperatures, when the mean occupation of the states is small and the statistics of the particles plays no role. However, in a degenerate electron gas one encounters a completely different situation: all states are filled due to

Fermi correlations. If we now consider a quantum transport channel of transmission probability T and applied bias voltage V, the rigidity of the Fermi sea leads to a *fixed* number of electrons, which are sent into each quantum channel. A charge is transfered to the other side with probability T and the statistics is therefore binomial

$$P_{t_0}(N) = \binom{M(t_0)}{N} T^N (1-T)^{M(t_0)-N} . \tag{3}$$

This is the result of the quantum calculation of Levitov and coworkers [12]. Note, that arguments based on the FCS have been used already to interpret current noise calculations or measurements. However, in general these interpretations are not unique, and a calculation of the FCS is required to interpret the results for the noise unambiguously. Thus, obtaining the FCS is a theoretical task, which leads to a better understanding of quantum transport processes.

The structure of this article is as follows. In the next section we introduce the theoretical basis, necessary to obtain the FCS. Our approach is based on an extension of the well-known Keldysh-Green's function technique [14] and described in Section 2.2. A convenient simplification is the *circuit theory of mesoscopic transport*, developed by Nazarov [15], which allows to obtain the FCS for a large variety of multi-terminal structures with minimal calculational overhead. In the two following sections, we will discuss several concrete examples. First, we show that for phase coherent two-terminal conductors the counting statistics can be obtained in quite a general form [16]. This is illustrated for normal and superconducting constrictions as well as two-barrier structures. A somewhat special case is the diffusive conductor coupled to a superconductor (i. e. in the presence of proximity effect), where intrinsic decoherence between electrons and holes influences the transport properties [17]. Second, we turn to more complex mesoscopic structures with more than two terminals. An analytic solution is obtained for the case of an arbitrary number of terminal connected by tunnel junctions to one central node [18]. As specific example we calculate the counting statistics of a beam splitter, in which a normal current or a supercurrent is divided into two normal currents. Afterwards we present some conclusions. In the appendices we give some theoretical background information to the methods used in this article.

In this article we concentrate on counting statistics in two- or multi-terminal devices with either normal- or superconducting contacts. More aspects of FCS are in other parts of this book. We would like to mention works related to FCS not covered here. Normal-superconductor transport at finite energies and magnetic fields was recently addressed in [19, 20]. Time-dependent transport phenomena in normal contacts have been studied in [13]. Fluctuations of the current in adiabatic quantum pumps have attracted some attention (see e.g. [21]). A connection to photon counting or photon transport can be found in Refs. [22, 23]. Counting statistics has also been addressed in the context of the readout process of a qubit [24] or to study spin coherence effects [25]. Results for the FCS of entangled electron pairs [26] and of resonant Cooper pair tunneling [27] have also been published.

2. Counting Statistics and Green's function

2.1. BASICS OF CURRENT STATISTICS

We introduce some basic formulas, relevant for the theory of FCS. The quantity we are after is the probability $P_{t_0}(N)$, that N charges are transfered in the time interval t_0. Equivalently, we can find the *cumulant generating function* (CGF) $S(\chi)$, defined by

$$\exp(-S(\chi)) = \sum_N P_{t_0}(N) \exp(iN\chi) \,. \tag{4}$$

To keep notations simple in this section, we will limit the discussion to the two-terminal case, in which only the number N of transfered charges in one terminal matters. In the other terminal it is given by $-N$, since the total number of charges is conserved. However, most relations are straightforwardly generalized to many terminals. Note that normalization of the FCS requires that $S(0) = 0$. Also, we will suppress the explicit dependence of $S(\chi)$ on the measuring time t_0. In the static case considered mostly in this article, we have $S(\chi) \sim t_0$.

From the full counting statistics one obtains the cumulants

$$C_1 = \overline{N} \equiv \sum_N N P_{t_0}(N) \;,\; C_2 = \overline{(N-\overline{N})^2} \,, \tag{5}$$

$$C_3 = \overline{(N-\overline{N})^3} \;,\; C_4 = \overline{(N-\overline{N})^4} - 3\overline{(N-\overline{N})^2}^2 \,, \tag{6}$$

and so on. The meanings of the various cumulants are depicted in Fig. 1a. Alternatively, the cumulants can be obtained from the CGF

$$C_n = -\left(-i\right)^n \frac{\partial^n}{\partial \chi^n} S(\chi)\bigg|_{\chi=0} \,. \tag{7}$$

The relation to the average current and the noise power of current fluctuations is obtained as follows. Almost trivially one has

$$C_1 = \langle N \rangle = \frac{1}{-e}\int_0^{t_0} dt \langle I(t)\rangle = -\frac{t_0}{e}\overline{I} \,, \tag{8}$$

where we have denoted the charge of the electrons with $-e$. The relation between the second cumulant and the current noise power

$$S_I \equiv \int_{-\infty}^{\infty} d\tau \, \langle\{\delta I(\tau), \delta I(0)\}\rangle \tag{9}$$

is less obvious. We write for the second cumulant

$$C_2 = \overline{(N - \overline{N})^2} = \frac{1}{2e^2} \int_0^{t_0} \int_0^{t_0} dt dt' \langle \{ \delta I(t), \delta I(t') \} \rangle \,, \tag{10}$$

where $\delta I(t) = I(t) - \langle I \rangle$ is the current fluctuation operator and $\langle ... \rangle$ denote the quantum statistical average. We transform the time coordinates to average $\bar{t} = (t + t')/2$ and relative time $\tau = t - t'$. Assuming the observation t_0 is much larger than the correlation time of the currents, the correlator in Eq. (10) does not depend on T, and we find the desired relation between current noise power and the second cumulant

$$S_I = \frac{2e^2}{t_0} \left. \frac{\partial^2 S(\chi)}{\partial \chi^2} \right|_{\chi=0} . \tag{11}$$

For higher correlators similar formulas can be derived.

2.2. EXTENDED KELDYSH GREEN'S-FUNCTION TECHNIQUE

The task to measure the number of charges transfered in a quantum transport process has, in general, to be formulated as a quantum measurement problem. A thorough derivation goes beyond the scope of this article and we refer to Ref. [28]. The quantum-mechanical form of the cumulant generating function is given by [14, 17, 28]

$$e^{-S(\chi)} = \langle \mathcal{T} e^{-i \frac{\chi}{2e} \int_0^{t_0} dt I(t)} \tilde{\mathcal{T}} e^{-i \frac{\chi}{2e} \int_0^{t_0} dt I(t)} \rangle \,. \tag{12}$$

Here $\mathcal{T}(\tilde{\mathcal{T}})$ denotes (anti-)time ordering and $\hat{I}(t)$ the operator of the current through a certain cross section. As preliminary justification of Eq. (12) we note, that it is easily shown that expansion of Eq. (12) in χ yields the various current-correlators. The expectation value in Eq. (12) can be implemented on the Keldysh contour, see Appendix 6.1, which makes it possible to use diagrammatic methods [29]. Equation (12) has a form similar to the thermodynamic potential in an external field. From the linked cluster theorem (see Appendix 6.2) it follows that the CGF is the sum of all connected diagrams.

To connect the CGF to accessible field-theoretical quantities we consider the nonlinear response of our electronic circuit to the time-dependent perturbation

$$H_c(t) = \frac{\chi}{2e} I_c(t) = \mp \frac{\chi}{2e} \int d^3x \Psi^\dagger(\mathbf{x}, t) \hat{j}_c(\mathbf{x}) \Psi(\mathbf{x}, t) \,, \tag{13}$$

where the $+(-)$ sign is taken on the lower(upper) part of the Keldysh contour. The operator $\hat{j}_c(\mathbf{x})$ is the operator of the current through a cross section c, depicted in Fig. 1. We allow here for multi-component field operators, such as spin or Nambu for example. Matrices in this subspace are denoted with a $\hat{}$. Since we are aiming at the total charge counting statistics, we will assume that χ is nonzero and constant in a finite time interval $[0, t_0]$.

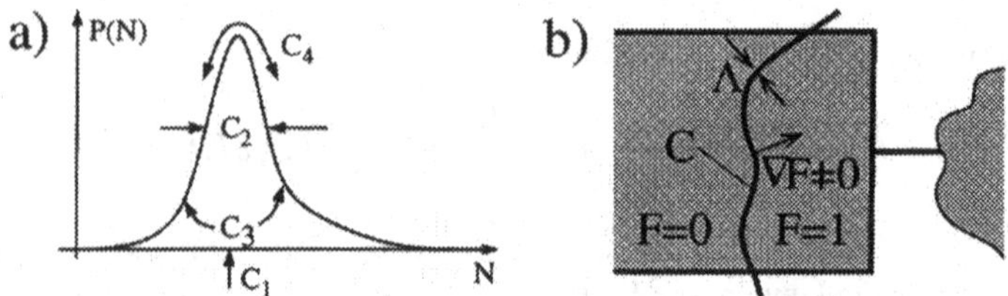

Figure 1. Left panel: An example of a probability distribution illustrating the meaning of the different cumulants. The average is given by C_1, the width by C_2, the 'skewness' by C_3 and the 'sharpness' is related to the C_4. Right panel: schematics of an ideal charge counter in a terminal. The number of charges passing the cross section C are counted.

The unperturbed system evolves according to a Hamiltonian

$$H_0 = \int d^3x \Psi^\dagger(\mathbf{x})\hat{h}_0(\mathbf{x})\Psi(\mathbf{x}), \tag{14}$$

where $\hat{h}_0$ is the usual single-particle Hamiltonian of the system. The equation of motion for the matrix Green's function subject to $H = H_0 + H_c(t)$ reads (in the Keldysh matrix representation)

$$\left[i\frac{\partial}{\partial t} - \hat{h}_0(\mathbf{x}) + \frac{\chi}{2e}\bar{\tau}_3\hat{j}_c(\mathbf{x})\right] \check{G}(\mathbf{x}, t; \mathbf{x}', t'; \chi) = \delta(t - t')\delta(\mathbf{x} - \mathbf{x}'), \tag{15}$$

Here $\bar{\tau}_3$ denotes the third Pauli matrix in the Keldysh space. The relation of the Green's function (15) to the CGF (12) is obtained from a diagrammatic expansion in χ. One finds the simple relation (see App. 6.2)

$$\frac{\partial S(\chi)}{\partial \chi} = \frac{it_0}{e} I(\chi). \tag{16}$$

The *counting current* $I(\chi)$ is obtained from the χ-dependent Green's function via

$$I(\chi) = \int d^3x \operatorname{Tr}\left[\bar{\tau}_3\hat{j}_c(\mathbf{x})\check{G}(\mathbf{x}, t; \mathbf{x}', t; \chi)\right]\Big|_{\mathbf{x}\to\mathbf{x}'}. \tag{17}$$

Since we are assuming a static situation, the r.h.s. of Eq. (17) is time-independent. The relations (15)-(17) offer a very general way to obtain the full counting statistics of any system. It is, however, difficult to find the Green's function in the general case.

For a mesoscopic transport problem there is a particularly simple way to access the full counting statistics, based on the separation into terminals (or reservoirs) and an active part, the first providing boundary conditions, and the second being responsible for the resistance. Let us consider the equation of motion for the Green's function *inside* a terminal for the following parameterization of the current operator in Eq. (13)

$$\hat{j}_c(\mathbf{x}) = (\nabla F(\mathbf{x})) \lim_{\mathbf{x}\to\mathbf{x}'} \frac{ie}{2m} (\nabla_{\mathbf{x}} - \nabla_{\mathbf{x}'})\hat{\sigma}_3. \tag{18}$$

$F(\mathbf{x})$ is chosen such that it changes from 0 to 1 across a cross section C, which intersects the terminal, but is of arbitrary shape, see Fig. 1. Here we have added a matrix $\hat{\sigma}_3$ to the current operator, which accounts for possible multicomponent field operators like in the case of superconductivity. The change from 0 to 1 should occur on a length scale Λ, for which we assume $\lambda_F \ll \Lambda \ll l_{imp}, \xi_0$ (Fermi wave length λ_F, impurity mean free path l_{imp}, and coherence length $\xi_0 = v_F/2\Delta$). Under this assumption we can reduce Eq. (15) to its quasiclassical version (see Ref. [30] and App. 7.1). This is usually a very good approximation, since all currents in a real experiment are measured in normal Fermi-liquid leads.

The Eilenberger equation in the vicinity of the cross section reads

$$\mathbf{v}_F \nabla \check{g}(\mathbf{x}, \mathbf{v}_F, t, t', \chi) = \left[-i\frac{\chi}{2}(\nabla F(\mathbf{x}))\mathbf{v}_F \check{\tau}_K \, , \, \check{g}(\mathbf{x}, \mathbf{v}_F, t, t', \chi) \right] . \qquad (19)$$

Here $\check{\tau}_K = \bar{\tau}_3 \hat{\sigma}_3$ is the matrix of the current operator. Other terms can be neglected due to the assumptions we have made for Λ. The counting field can then be eliminated by the gauge-like transformation

$$\check{g}(\mathbf{x}, \mathbf{v}_F, t, t', \chi) = e^{-i\chi F(\mathbf{x})\check{\tau}_K/2} \check{g}(\mathbf{x}, \mathbf{v}_F, t, t', 0) e^{i\chi F(\mathbf{x})\check{\tau}_K/2} . \qquad (20)$$

We assume now that the terminal is a diffusive metal of negligible resistance. Then the Green's functions are constant in space (except in the vicinity of the cross section C) and isotropic in momentum space. Applying the diffusive approximation [31] in the terminal leads to a transformed terminal Green's function

$$\check{G}(\chi) = e^{-i\chi \check{\tau}_K/2} \check{G}(0) e^{i\chi \check{\tau}_K/2} , \qquad (21)$$

on the right of the cross section C (where $F(\mathbf{x} = 1)$ with respect to the case without counting field. Consequently the counting field is entirely incorporated into a *modified boundary condition* imposed by the terminal onto the mesoscopic system. Note, that it follows from (18) and (19), that the counting field for a particular terminal vanishes from the equations of motion in the mesoscopic system and the other terminals.

The generalization of this method to the counting statistics for multiterminal structures was performed in [32]. The surprisingly simple result is, that one has to add a separate counting field for each terminal, in which charges are counted. In Sec. 4.1 we demonstrate this for an example, in which an analytic solution can be found. This concludes the derivation of our theoretical method to obtain the FCS.

What are the achievements of this method? We should emphasize that it does not simplify the solution of a specific transport problem, i.e. we still have to know the solution corresponding to the Hamiltonian H_0. If this solution is not known, the counting field makes this situation not easier. Rather, the method paves a very general way to obtain the FCS, if a method to find the average currents, i.e. for $\chi = 0$, is known. In the next section we will introduce such a method, the circuit theory of mesoscopic transport. Initially it was invented to calculate average currents, however the method to obtain the FCS introduced in this section is straightforwardly included.

What is the price to pay? Loosely speaking, the method to obtain the average currents has to be sufficiently general. Usually the absence of a field, which has different signs on the upper and lower part of the Keldysh contour, allows some simplification. For example, in the Keldysh-matrix representation all Green's functions can be brought into a tri-diagonal form, which is obviously simpler to handle than the full matrix. The method above does not allow this simplification anymore. Or, in other words, the counting rotation (21) destroys the triangular form. Thus, the price we have to pay for an easy determination of the FCS is that we need a method, which respects the *full Keldysh-matrix structure* in all steps. The circuit theory, which we describe in the next section fulfills this requirement.

2.3. CIRCUIT THEORY

A concise formulation of mesoscopic transport is the so-called circuit theory [15, 33]. Its main idea, borrowed from Kirchhoff's classical circuit theory, is to represent a mesoscopic device by discrete elements. These approximate the layout of an experimental device with arbitrary accuracy, provided one chooses enough elements. In practice one has to find the balance between a small grid size and the computational effort.

We briefly repeat the essentials of the circuit theory. Topologically, one distinguishes three elements: terminals, nodes and connectors. Terminals are the connections to the external measurement circuit and provide boundary conditions, specifying externally applied voltages, currents or phase differences. Besides, they also determine the type of the terminal, i.e. if it is a normal metal or a superconductor. The actual circuit, which is to be studied consists of nodes and connectors, the first determining the approximate layout, and the second describing the connections between different nodes.

The central element of the circuit theory is the arbitrary connector, characterized by a set of transmission coefficients $\{T_n\}$. Its transport properties are described by a *matrix current* found in [33]

$$\check{I}_{12} = -\frac{e^2}{\pi}\sum_n \frac{2T_n\left[\check{G}_1, \check{G}_2\right]}{4+T_n\left(\{\check{G}_1, \check{G}_2\}-2\right)}\,. \tag{22}$$

Here $\check{G}_{1(2)}$ denote the matrix Green's functions on the left and the right of the contact. We should emphasize that the matrix form of (22) is crucial to obtain the FCS, since it is valid for any matrix structure of the Green's functions. The *electrical current* is obtained from the matrix current by

$$I_{12} = \frac{1}{4e}\int dE \mathrm{Tr}\check{\tau}_K \check{I}_{12}\,. \tag{23}$$

A special case is a diffusive wire of length L in the presence of proximity effect. If L is longer than other characteristic lengths like ξ_0, decoherence between electrons and holes becomes important and the transmission eigenvalue ensemble is not known. Instead, one solves the diffusion-like equation [31]

$$\nabla D(\mathbf{x})\check{G}(\mathbf{x})\nabla\check{G}(\mathbf{x}) = \left[-iE\hat{\tau}_3, \check{G}(\mathbf{x})\right] . \tag{24}$$

The matrix current is now given by

$$\check{I}(\mathbf{x}) = -\sigma(\mathbf{x})\check{G}(\mathbf{x})\nabla\check{G}(\mathbf{x}) . \tag{25}$$

In these equations, $D(\mathbf{x})$ the diffusion coefficient and $\sigma(\mathbf{x}) = 2e^2N_0D(\mathbf{x})$ is the conductivity. In general this equation can only be solved numerically, but in some special cases (e.g. for $E = 0$) an analytic solution is possible.

If the circuit consists of more than one connector, the transport properties can be found from the circuit theory by means of the following circuit rules. We take the Green's functions of the terminals as given and introduce for each internal node an (unknown) Green's function. The two rules determining the transport properties of the circuit completely are

1. $\check{G}_j^2 = \check{1}$ for the Green's functions of all internal nodes j.
2. The total matrix current in a node is conserved: $\sum_i \check{I}_{ij} = 0$, where the sum goes over all nodes or terminals connected to node j and each matrix current is given by (22) or (25), depending on the type of the connector.

An important feature of the circuit theory in the form presented above is that it accounts for any matrix structure (i.e. Keldysh, Nambu, Spin, etc.). Thus, we can straightforwardly obtain the FCS along the lines of Sec. 2.2. If the charges in a terminal are counted, we have to apply a counting rotation (21) to the terminal Green's function. The counting-rotation matrix has the form $\check{\tau}_K = \hat{\sigma}_3\bar{\tau}_3$, where $\hat{\sigma}_i(\bar{\tau}_i)$ denote Pauli-matrices in Nambu(Keldysh)-space. Then we can employ the circuit rules to find the χ-dependent Green's function and finally obtain the total CGF by integrating all currents into the terminals over their respective counting fields (see [32] and [18] for more details).

3. Two-Terminal Contacts

In this chapter we demonstrate several examples of the FCS of contacts between two terminals, see Fig. 2. One can easily derive a number of rather general results, such as Poisson statistics in the case of tunnel junction, or binomial statistics for single channel contacts of transparency T. All these results can be found from a general CGF [16], which depends on the ensemble of transmission eigenvalues $\{T_n\}$. To illustrate specific examples, we will compare the cases of transport between two normal terminals or between one superconducting and one normal terminal. In the end, the statistics of an equilibrium supercurrent is discussed.

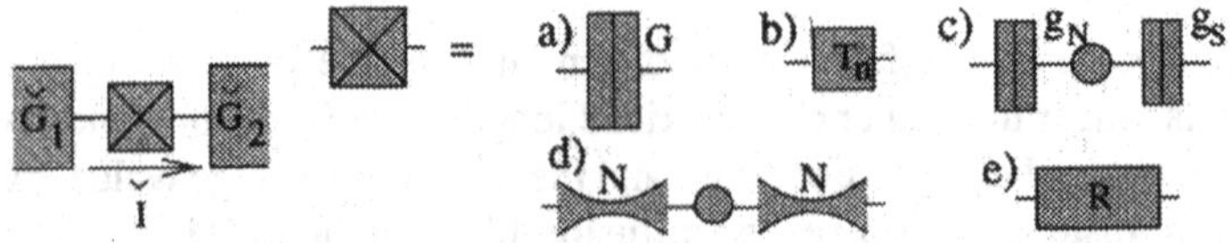

Figure 2. Two terminal contacts. The leftmost picture shows a general two-terminal contact. A matrix current $\check{I}$ is flowing between terminals, characterized by Green's functions $\check{G}_{1,2}$. Several connectors (a-e) are described in the text and depicted as: a) tunnel junction, b) arbitrary connector with transmission eigenvalues $\{T_n\}$, c) two tunnel junctions in series with conductances $g_N(g_S)$, respectively, d) a chaotic cavity connected by N_{ch} channels to the leads, and e) a diffusive wire, characterized by a resistance R_N.

3.1. TUNNEL JUNCTION

The counting statistics of a tunnel junction contact can be obtained from a direct expansion of the matrix current (22), if $T_n \ll 1$ for all n. It coincides with the result obtained by means of the tunneling Hamiltonian [34]. The matrix current takes the form

$$\check{I}_{\text{tun}}(\chi) = \frac{G_T}{2}\left[\check{G}_1(\chi), \check{G}_2\right]. \tag{26}$$

Here the matrix current depends only on the tunneling conductance $G_T = (e^2/\pi)\sum_n T_n$ of the contact, and we have arbitrarily chosen to count the charges in terminal 1. Now, the counting current is

$$I(\chi) = \frac{G_T}{8e}\int dE \mathrm{Tr}\left(\check{\tau}_K\left[\check{G}_1(\chi), \check{G}_2\right]\right). \tag{27}$$

We use the pseudo-unitarity $\check{\tau}_K^2 = \check{1}$ to express the counting rotation as $\exp(i\chi\check{\tau}_K/2) = \check{\tau}_+\exp(i\chi/2) + \check{\tau}_-\exp(-i\chi/2)$, where $\check{\tau}_\pm = (1 \pm \check{\tau}_K)/2$. Then the current has the form

$$I(\chi) = \frac{e}{t_0}\left[N_{12}e^{-i\chi} - N_{21}e^{i\chi}\right], \tag{28}$$

$$N_{ij} = \frac{t_0 G_T}{4e^2}\int dE \mathrm{Tr}\left[\check{\tau}_+\check{G}_i\check{\tau}_-\check{G}_j\right]. \tag{29}$$

The CGF follows from integrating (28) with respect to χ and we obtain

$$S(\chi) = -N_{12}(e^{i\chi} - 1) - N_{21}(e^{-i\chi} - 1). \tag{30}$$

It is easy to see that the even and odd cumulants obey

$$C_{2n+1} = N_{12} - N_{21} \quad , \quad C_{2n} = N_{12} + N_{21}. \tag{31}$$

If only tunneling processes in one direction occur (say from 1 to 2), $N_{21} = 0$ and the average $\bar{N} = N_{12}$. The statistics is Poissonian

$$P_{t_0}(N) = e^{-\bar{N}}\frac{\bar{N}^N}{N!} \tag{32}$$

In particular, for the current noise we find the Schottky result $S_I = 2eI$.

We conclude that the charge counting statistics of a tunnel junction (or more precisely, if the transfer events are rare) is of a generalized Poisson type [34]. If only tunneling events in one directions are possible, the statistics is Poissonian.

3.2. QUANTUM CONTACT – GENERAL CONNECTOR

We consider now a quantum contact characterized by a set of transmission eigenvalues $\{T_n\}$. It turns out that the CGF can be obtained in a quite general form, valid for arbitrary junctions between superconductors and/or normal metals.

The matrix current through a quantum contact is described by Eq. (22) and the CGF can then be found from the relation $\partial S(\chi)/\partial\chi = (-it_0/4e^2)\int dE$ $\mathrm{Tr}[\check{\tau}_K \check{I}(\chi)]$. Using that $[\check{A},\{\check{A},\check{G}_2\}] = 0$ for all matrices with $\check{A}^2 = \check{1}$ the following identity holds

$$\mathrm{Tr}\frac{\partial}{\partial\chi}\left\{\check{G}_1(\chi),\check{G}_2\right\}^n = \frac{i}{2}\mathrm{Tr}\check{\tau}_K\left[\check{G}_1(\chi),\check{G}_2\right]\left\{\check{G}_1(\chi),\check{G}_2\right\}^{n-1}. \tag{33}$$

We can therefore integrate (23) with respect to χ and obtain [16]

$$S(\chi) = -\frac{t_0}{2\pi}\sum_n\int dE\,\mathrm{Tr}\ln\left[1+\frac{T_n}{4}\left(\{\check{G}_1(\chi),\check{G}_2\}-2\right)\right]. \tag{34}$$

This is a very important result. It shows that the counting statistics of a large class of constrictions can be cast in a common form, independent of the contact types. Another important property of Eq. (34) is, that the CGF's of all constrictions are linear statistics of the transmission eigenvalue distribution, and can therefore be averaged over by standard means (see e.g. [35]). Examples are given in the following sections. Note also, that an expansion of Eq. (34) for $T_n \ll 1$ yields the result (30) of the tunnel case.

We will now discuss several illustrative examples. Consider first two normal reservoirs with occupation factors $f_{1(2)} = [\exp((E-\mu_{1(2)})/T_e)+1]^{-1}$ (T_e is the temperature). We obtain the result [12, 13]

$$S(\chi) = -\frac{t_0}{\pi}\sum_n\int dE\ln\left[1+B_{1n}(E)\left(e^{i\chi}-1\right)+B_{-1n}(E)\left(e^{-i\chi}-1\right)\right]. \tag{35}$$

Here we introduced the probabilities $B_{1n} = T_n f_1(E)\,(1-f_2(E))$ for a tunneling event from 1 to 2 and B_{-1n} for the reverse process. The terms with *counting factors* $e^{\pm i\chi}-1$ obviously correspond to charge transfers from 1 to 2 (2 to 1). At zero temperature and $\mu_1-\mu_2 = eV \geq 0$ the integration can easily be evaluated and we obtain

$$S(\chi) = -\frac{et_0|V|}{\pi}\sum_n\ln\left[1+T_n\left(e^{i\chi}-1\right)\right]. \tag{36}$$

The corresponding statistics for a single channel with transparency T is binomial

$$P_{t_0}(N) = \binom{M}{N} T^N (1-T)^{M-N} . \tag{37}$$

Here we have introduced the *number of attempts* $M = et_0V/\pi$, which is the maximal number of electrons that can be sent through one (spin-degenerate) channel in a time interval t_0 due to the exclusion principle for Fermions.

The FCS of an superconductor-normal metal-contact also follows from Eq. (34) (for a definition of the various Green's functions see App. 7.4). Evaluating the trace in Eq. (34) the CGF can be expressed as [36]

$$S(\chi) = -\frac{t_0}{2\pi} \sum_n \int dE \ln \left[1 + \sum_{q=-2}^{2} A_{nq}(E) \left(e^{iq\chi} - 1 \right) \right] . \tag{38}$$

The coefficients $A_{nq}(E)$ are related to a charge transfer of $q \times e$. For example a term $\exp(2i\chi) - 1$ corresponds just to an Andreev reflection process, in which two charges are transfered simultaneously [37]. Explicit expressions for the various coefficients are given in Appendix 8.1. For the BCS case, they reproduce the results of Ref. [36]. In the fully gapped single-channel case at energies $k_B T_e \ll eV \ll \Delta$ only terms corresponding to Andreev reflection (A_2) are nonzero and the CGF becomes

$$S(\chi) = -\frac{et_0|V|}{\pi} \ln \left[1 + R_A \left(e^{i2\chi} - 1 \right) \right] , \tag{39}$$

where $R_A = T^2/(2-T)^2$ is the probability of Andreev reflection. The CGF is now π-periodic, which means that only *even* numbers of charges are transfered, a consequence of Andreev reflection. The corresponding statistics is binomial

$$P_{t_0}(2N) = \binom{M}{N} R_A^N \left(1 - R_A\right)^{M-N} , \; P_{t_0}(2N+1) = 0 . \tag{40}$$

The number of attempts M is, however, the same as in the normal state.

It is interesting to see how the CGF for normal transport, i. e., Eq. (35), emerges from Eq. (38). Putting $f_{R,A} = 0$ and $g_{R,A} = \pm 1$ the coefficients in Eq. (38) can be written as

$$\begin{aligned} A_{\pm 1} &= B^+_{\pm 1} + B^-_{\pm 1} - 2B^+_{\pm 1}B^-_{\pm 1} - B^+_{\pm 1}B^-_{\mp 1} - B^-_{\pm 1}B^+_{\mp 1} , \\ A_{\pm 2} &= B^+_{\pm 1}B^-_{\pm 1} \; ; \; B^{\pm}_1 = TB_1(\pm E) , B^{\pm}_{-1} = TB_{-1}(\pm E) , \end{aligned}$$

The argument of the 'ln' in Eq. (38) factorizes in positive and negative energy contributions

$$\ln \left[1 + \sum_{q=-2}^{2} A_q \left(e^{iq\chi} - 1 \right) \right] = \sum_{s=\pm} \ln \left[1 + \sum_{q=-1}^{1} B^s_q \left(e^{iq\chi} - 1 \right) \right] . \tag{41}$$

Integrating over energy both terms give the same contribution, and the CGF results in Eq. (35). This shows explicitly how the positive and negative energy quasiparticles are correlated in the Andreev reflection process.

3.3. DOUBLE TUNNEL JUNCTION

We now consider a diffusive island (or a chaotic cavity) connected to two terminals by tunnel junctions with respective conductance g_1 and g_2 [18]. We assume for the conductance of the island $g_{island} \gg g_{1,2} \gg e^2/h$, so we can neglect charging effects. This provides a simple application of the circuit theory. The layout is shown in Fig. 2c. The central node is described by an unknown Green's function $\check{G}_c$. We have two matrix currents entering the node, which obey a conservation law:

$$0 = \check{I}_1 + \check{I}_2 = \frac{1}{2}\left[g_1\check{G}_1 + g_2\check{G}_2, \check{G}_c\right] . \tag{42}$$

Using the normalization condition $\check{G}_c^2 = 1$ the solution is

$$\check{G}_c = \frac{g_1\check{G}_1 + g_2\check{G}_2}{\sqrt{g_1^2 + g_2^2 + g_1 g_2\left\{\check{G}_1, \check{G}_2\right\}}} . \tag{43}$$

We can integrate the current $I(\chi) \sim \mathrm{Tr}\check{\tau}_K\check{I}_1$ and obtain the CGF

$$S(\chi) = -\frac{t_0}{4e^2}\int dE\mathrm{Tr}\sqrt{g_1^2 + g_2^2 + g_1 g_2\left\{\check{G}_1, \check{G}_2\right\}} . \tag{44}$$

Again, this result is valid for all types of contacts between normal metals and superconductors.

We first evaluate the trace for two normal leads and find

$$S(\chi) = -\frac{t_0}{2e}\int dE \times \tag{45}$$
$$\sqrt{(g_1 + g_2)^2 + 4g_1 g_2\left(f_1(1 - f_2)(e^{i\chi} - 1) + f_2(1 - f_1)(e^{-i\chi} - 1)\right)} .$$

We observe that the CGF contains again counting factors corresponding to charge transfer from 1 to 2 and vice versa. In contrast to the Poissonian case for a tunnel junction, Eq. (30), the charge transfers are not independent, but correlated by the square-root function. At zero temperature and $\mu_1 - \mu_2 = eV > 0$ we find the result,

$$S(\chi) = -\frac{t_0 V}{2e}\sqrt{(g_1 + g_2)^2 + 4g_1 g_2(e^{i\chi} - 1)} . \tag{46}$$

There are two relatively simple limits. If the two conductances are very different (e.g. $g_1 \ll g_2$), we return to Poissonian statistics:

$$S(\chi) = -\frac{t_0 V g_1}{e}(e^{i\chi} - 1) . \tag{47}$$

On the other hand, in the symmetric case $g_1 = g_2 = g$ we find [38]

$$S(\chi) = -\frac{t_0 V g}{e}(e^{i\chi/2} - 1)\,, \tag{48}$$

and the cumulants are

$$C_n = \frac{\bar{N}}{2^{n-1}} \quad , \quad \bar{N} = \frac{t_0 g |V|}{2e}\,. \tag{49}$$

The suppression factor $1/2$, which occurs already in the Fano factor, carries forward to all cumulants. Note, that the same kind of statistics (45) follows also from a Master equation [38, 39].

The CGF (44) for the transport between a normal metal and a superconductor (at zero temperature, for simplicity) reads [18]

$$S(\chi) = -\frac{t_0 V}{e\sqrt{2}}\sqrt{g_1^2 + g_2^2 + \sqrt{(g_1^2 + g_2^2)^2 + 4g_1^2 g_2^2 (e^{i2\chi} - 1)}}\,. \tag{50}$$

Thus, the influence of the superconductor is two-fold: charges are transfered in units of $2e$ (indicated by the π-periodicity) and another square root is involved in the CGF, resulting from the higher order correlations. In the limit that both conductances are very different (e.g. $g_1 \ll g_2$) we obtain again Poissonian statistics

$$S(\chi) = -\frac{t_0 V}{e}\frac{g_1^2}{g_2}\left(e^{i2\chi} - 1\right)\,. \tag{51}$$

This corresponds to uncorrelated transfers of pairs of charges. Consequently we obtain for the cumulants

$$C_n = 2^{n-1}\bar{N} \quad , \quad \bar{N} = \frac{2 t_0 V g_1^2}{e g_2}\,, \tag{52}$$

and the effective charge $2e$ can indeed be found from the Fano factor. The transport properties at finite energies and magnetic fields of this structure have recently been addressed in [19] and [20].

3.4. SYMMETRIC CHAOTIC CAVITY

Another interesting system is the chaotic cavity, i. e. a small island coupled to terminals by perfectly transmitting contacts (with $N_{ch} \gg 1$ channels). This system is described by the circuit depicted in Fig. 2d. The matrix current between terminal 1 and the cavity is $\check{I}_1 = N_{ch}(e^2/\pi)[\check{G}_1, \check{G}_c]/(2 + \{\check{G}_1, \check{G}_c\})$. Similar as in the previous chapter, the current conservation reads now

$$0 = \frac{\left[\check{G}_1, \check{G}_c\right]}{2 + \left\{\check{G}_1, \check{G}_c\right\}} + \frac{\left[\check{G}_2, \check{G}_c\right]}{2 + \left\{\check{G}_2, \check{G}_c\right\}}\,, \tag{53}$$

which is solved by the solution (43) for $g_1 = g_2$. The integration of (23) leads to

$$S(\chi) = -N_{ch}\frac{t_0}{2\pi}\int dE \mathrm{Tr}\ln\left[2+\sqrt{2+\left\{\check{G}_1(\chi), \check{G}_2\right\}}\right] . \tag{54}$$

The interpretation of this result is straightforward. As we have seen in Sec. 3.2, the ln appeared already in the FCS of a quantum point contact (leading to binomial statistics). The square-root we encountered already in the previous section and we attribute it to inter-mode mixing on the central node ('cavity noise').

For normal leads at zero temperature and applied bias voltage V we obtain (with the number of attempts $M = N_{ch}t_0eV/\pi$)

$$S(\chi) = -2M\ln\left[1+e^{i\chi/2}\right] . \tag{55}$$

On the other hand, in the case of Andreev reflection we find

$$S(\chi) = -2M\ln\left[2+e^{i[\chi \mathrm{mod}\pi]}+2\sqrt{1+e^{i[\chi \mathrm{mod}\pi]}}\right] , \tag{56}$$

where the π-periodicity reflects the fact that charges are transfered in pairs.

3.5. DIFFUSIVE CONNECTOR

A metallic strip of length L with purely elastic scattering, characterized by an elastic mean free path l, is called a diffusive connector if $l \ll L$. Its transport properties are governed by the diffusion-like Usadel equation, Eq. (24). In the case of proximity transport the r.h.s. of Eq. (24) accounts for decoherence of electrons and hole during their diffusive motion along the normal wire. This term has the form of a *leakage current*, if the l.h.s. is considered as a conservation law for the matrix current [33]. Note that the electric current is still conserved, it is only loss of coherence which occurs. In general the solution of the full equation is rather complicated and can only be found numerically for the full parameter range. There is, however, one case, in which an analytic solution is possible, namely if the r.h.s. of Eq. (24) vanishes. This is either the case for purely normal transport, when electrons and holes are transported independently, or for $E = 0$, which means we are restricted to low temperatures and voltages. The scale here is set by the Thouless energy E_{Th} (given by $\hbar D/L^2$ for a wire of uniform cross section). At this scale the famous reentrance effect of the conductance occurs [40]. This regime was studied in Ref. [17] and will be discussed in connection with experimental results for the current noise in the article by Reulet *et al.* in this book. Numerical results for equilibrium counting statistics in the full parameter range are discussed below.

We now concentrate on the analytic solution in a quasi-one-dimensional geometry, i. e. we assume a wire of uniform cross section connects two reservoirs located at $x = 0$ and $x = L$. It is characterized by a conductivity $\sigma(x)$, which in general could depend on x, e. g. due to an inhomogeneous concentration of scattering centers. The diffusion equation is then indeed a conservation law for the matrix current density

$$\frac{\partial}{\partial x}\check{j} = 0 \quad , \quad \check{j} = -\sigma(x)\check{G}(x)\frac{\partial}{\partial x}\check{G}(x) . \tag{57}$$

This equation has to be solved with the boundary condition that $\check{G}(0) = \check{G}_1$ and $\check{G}(L) = \check{G}_2$. It follows from Eq. (57) and the normalization condition $\check{G}^2(x) = \check{1}$ that $\check{G}(x)$ obeys the equation

$$\check{j} = \text{const.} \quad , \quad \frac{\partial}{\partial x}\check{G}(x) = -\frac{1}{\sigma(x)}\check{G}(x)\check{j} . \tag{58}$$

This is an homogeneous first-order differential equation, which is easily solved. Using the boundary conditions we find the solution

$$\check{G}(L) = \check{G}(0)e^{\check{I}/g_d} \ , \ g_d = \frac{A}{\int_0^L dx/\sigma(x)} , \tag{59}$$

where g_d is the conductance of the wire and A its cross section. The current is thus given by [14]

$$\check{I}(\chi) = -g_d \ln\left(\check{G}_1(\chi)\check{G}_2\right) , \tag{60}$$

where we have reinserted the dependence on the counting field χ. To find the CGF we have to find the integral $\int d\chi \mathrm{Tr}\check{\tau}_K \check{I}(\chi)$ with respect to the counting field. Expanding the ln and using repeatedly the normalization condition the results is

$$S(\chi) = -\frac{t_0 g_d}{8e^2}\int dE\mathrm{Tr}\left[\mathrm{acosh}^2\left(\frac{1}{2}\left\{\check{G}_1(\chi), \check{G}_2\right\}\right)\right] . \tag{61}$$

This is the counting statistics for a general diffusive contact (under the restrictions mentioned above). The first thing to note is that from the properties of the wire only the conductance enters and this holds for all cumulants. In this sense Eq. (61) shows that the entire FCS is universal. In our derivation, we have assumed a wire of uniform cross section, but it has been shown [9] that this also holds for an arbitrary shape of the wire (as long as it can be considered as quasi one-dimensional). We should also mention, that one could have obtained the same result, by averaging the CGF (12) over the bimodal distribution of transmission eigenvalues [41, 9].

Now we evaluate the trace in Eq. (61) for normal metals at zero temperature and applied bias voltage eV. We obtain

$$S(\chi) = -\frac{t_0 g_d V}{4e}\mathrm{acosh}^2\left(2e^{i\chi} - 1\right) , \tag{62}$$

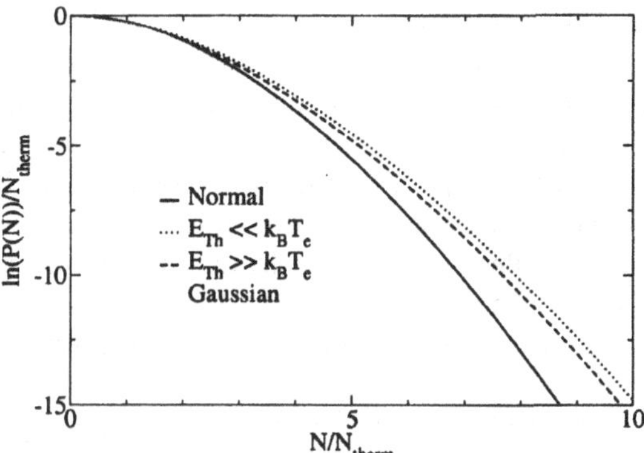

Figure 3. Equilibrium distribution of the current fluctuations in a diffusive SN-contact. We observe a) a strong deviation of all distributions from a Gaussian b) enhanced fluctuations in the superconducting case, and c) differences between proximity effect and coherent transport. Note, that the second cumulant (i.e. the thermal noise) is the same for all displayed curves.

which coincides with the results of Refs. [14, 42].

In the case of Andreev transport the easiest way to obtain the CGF is as follows. We have already previously noted, that Eq. (62) follows from averaging Eq. (36) with the transmission eigenvalue distribution $\rho(T) = (2e^2/g_d\pi)/T\sqrt{1-T}$ for a diffusive metal [9, 41]. Now, the CGF for Andreev transport Eq. (39) has the same form as in the case of normal transport, provided we replace χ with 2χ and the transmission eigenvalue T_n with the Andreev reflection probability $R_A(T_n)$. A simple calculation shows, that the R_A are distributed according to the *same* distribution as the normal transmission eigenvalues (up to a factor of 1/2). Thus, we can immediatly read off the CGF for the diffusive SN-wire in the limit of zero temperature and $eV \ll E_{Th}$ from Eq. (62) and obtain

$$S(\chi) = -\frac{t_0 g_d V}{8e} \text{acosh}^2\left(2e^{2i\chi} - 1\right) . \tag{63}$$

As a consequence the relation between the cumulants in the SN-case, C_n^{SN}, and the normal case, C_n^{NN}, is

$$C_n^{\text{SN}} = 2^{n-1} C_n^{\text{NN}} . \tag{64}$$

We observe that we can read off the effective charge from the ratio $C_n^{\text{SN}}/C_n^{\text{NN}} = (q_{eff}/e)^{n-1}$ and, indeed, find $q_{eff} = 2e$. We should emphasize, however, that this is a special property of the *diffusive connector*. Our prove of Eq. (64) is valid as long as $eV \ll E_{Th}$, and it is not clear, wether Eq. (64) is true also for $eV \gg E_{Th}$

The counting statistics in equilibrium for arbitrary temperature T_e was studied in Ref. [17]. By a numerical solution of Eq. (24), it is possible to evaluate the integral over χ in the inversion of Eq. (4) in the saddle point approximation, i.e. we take χ as complex and expand the exponent around the complex saddle point $\chi = ix_0$. The integral yields then $P(N) \approx \exp(-S(ix_0) - x_0 N)$, which we plot implicitly as a function of $N(x_0) = \partial S(ix_0)/\partial x_0$. Results of this calculation are displayed in Fig. 3. The charge number N is normalized by $N_{\text{therm}} = g_d t_0 k_B T_e/e^2$. Note that the conductance (and due to the fluctuation-dissipation theorem the noise) is the same in all cases. The solid line shows the distribution in the normal state, which does not depend on the Thouless energy. In our units, this curve is consequently independent on temperature. In the superconducting state the Thouless energy does matter, and the distribution depends on the ratio $E_{\text{Th}}/k_B T_e$. We observe that large fluctuations of the current in the superconducting case are enhanced in comparison to the normal case, and in both cases are enhanced in comparison to Gaussian noise $\sim \exp(-N^2/4N_{\text{them}})$. The differences between the normal and the superconducting state occur in the regime of non-Gaussian fluctuations.

3.6. SUPERCURRENT

The CGF of a quantum contact, i. e. Eq. (12), can be used to find the counting statistics for a supercurrent between two superconductors at a fixed phase difference ϕ. This has been done in Ref. [16]. The result can be represented in a form similar to the CGF of the SN-contact (38)

$$S(\chi, \phi) = -\frac{t_0}{2\pi} \sum_n \int dE \ln \left[1 + \sum_{q=-2}^{2} \frac{A^S_{n,q}(E,\phi)}{Q_n(E,\phi)} \left(e^{iq\chi} - 1 \right) \right] . \tag{65}$$

Explicit expressions for the coefficients are given in Appendix 8.2. To find the statistics of the charge transfer, we will treat two separate cases.

Gapped superconductors. If the two leads are gapped like BCS superconductors, the spectral functions are given by Eq. (116). Here we account for a finite lifetime δ of the Andreev bound states, *e. g.* due to phonon scattering. The supercurrent in a one-channel contact of transparency T_1 is solely carried by Andreev bound states with energies $\pm\Delta(1 - T_1 \sin^2 \phi/2)^{1/2} \equiv \pm E_B(\phi)$. The importance of these bound states can be seen from the coefficient $Q(E, \phi)$ (see Eq. (125)), which may become zero and will thus produce singularities in the CGF.[1] The broadening δ shifts the singularities of $Q(E, \phi)$ into the complex plane and allows an expansion of the coefficients $A^S_{1,q}$ close to the bound state energy. Performing the energy integration the CGF results in

$$S(\chi, \phi) = -2t_0\delta \sqrt{1 - \chi^2 \frac{I_1^2(\phi)}{4\delta^2} - i\chi \frac{I_1(\phi)}{\delta} \operatorname{th} \left(\frac{E_B(\phi)}{2k_B T_e} \right)} , \tag{66}$$

[1] It is interesting to note that in the limit $\delta \to 0$ the CGF has poles for energies $E_B^2(\chi) = \Delta^2 \left(1 - T \sin^2 \left(\frac{\phi \pm \chi}{2}\right)\right)$. The counting field therefore couples directly to the phase sensitivity of the Andreev bound states.

where $I_1(\phi) = \Delta^2 T_1 \sin(\phi)/2E_B(\phi)$ is the supercurrent carried by one bound state. In deriving (66) we have also assumed that $\chi \ll 1$. This corresponds to a restriction to "long trains" of electrons transfered, and the discreteness of the electron transfer plays no role here. Fast switching events become less probable at low temperatures and are neglected here. In the saddle point approximation and for $\gamma(\phi) \equiv 1/\operatorname{ch}(E_B(\phi)/2T_e) \ll 1$ the FCS can be found. We express the transfered charge in terms of the current normalized to the zero temperature supercurrent: $j(\phi) = eN/t_0 I_1(\phi)$. We find for the current distribution in the saddle point approximation

$$P(j,\phi) \sim \frac{1}{\gamma} \exp\left[2\delta t_0 \left(\gamma(\phi)\sqrt{1-j^2(\phi)} - j(\phi)\sqrt{1-\gamma^2(\phi)}\right)\right]), \tag{67}$$

for $|j(\phi)| \leq 1$ and zero otherwise. At zero temperature Eq. (67) approaches $P(j,\phi) \to \delta(j-1)$. Thus the charge transfer is noiseless. At finite temperature, on the other hand, the distribution (67) confirms the picture of switching between Andreev states which carry currents in opposite directions, suggested in Ref. [43]. The previous result is valid under the following conditions. In the energy integration it was assumed that the bound states are well defined. For small transmission the distance of the bound state to the gap edge is $\approx T_1\Delta$. Thus, to have well-defined bound states we have to require $\delta < T_1\Delta$. Similarly, for a highly transmissive contact and a small phase difference we require $\phi \sim eN/t_0 I_c > \delta/\Delta$. The statistics beyond these limits is similar to what is discussed in the following.

Tunnel junction/gapless superconductors. Let us now consider the supercurrent statistics between two weak superconductors, where the Green's functions can be expanded in Δ for all energies. One can see that this is equivalent to the tunnel result (30). It has the form

$$S(\chi,\phi) = -N_+(\phi)(e^{i\chi}-1) - N_-(\phi)(e^{-i\chi}-1). \tag{68}$$

where

$$N_\pm(\phi) = \frac{t_0}{2}\left(P_s(\phi) \pm I_s(\phi)\right). \tag{69}$$

This form of the CGF shows that the FCS is expressed in terms of supercurrent $I_s(\phi)$ and noise $P_s(\phi)$ only. Supercurrent and current noise are

$$I_s(\phi) = -\frac{G_T}{4}\mathrm{Re}\int dE \mathrm{Tr}\left\{\hat\sigma_3\left[\hat R_1(E,\phi),\hat R_2(E)\right]\right\}\operatorname{th}\left(\frac{E}{2k_BT_e}\right), \tag{70}$$

$$P_s(\phi) = -\frac{G_T}{4}\mathrm{Re}\int dE \mathrm{Tr}\left\{\hat\sigma_3\hat A_1(E,\phi)\hat\sigma_3\hat R_2(E)\right\}\frac{1}{\operatorname{ch}^2\left(\frac{E}{2k_BT_e}\right)}. \tag{71}$$

Here G_T is the normal-state conductance of the contact. Eq. (71) shows that P_s vanishes at zero temperature, whereas I_s vanishes at T_c. Therefore, there is some crossover temperature below which $P_s < |I_s|$. In this limit one of the coefficients $N_\pm$ becomes negative and the interpretation, that the CGF (68) corresponds to a generalized Poisson distribution, makes no sense anymore. In fact, the CGF leads to 'negative probabilities' and does not correspond to any probability distribution, The origin of this failure is the broken U(1)-symmetry the superconducting state. Nevertheless the FCS can be used to predict the outcome of any charge transfer measurement, as is discussed in detail in Refs. [16, 28, 44].

4. Multi-Terminal Structures

Many mesoscopic transport experiments are performed in multi-terminal configurations. An example is shown in Fig. 4. Due to the quantum nature of the charge carriers, interesting non-local effects can appear, such as sensitivity of measured voltage differences to changes in the setup outside the current path. Obviously, the same is true for current fluctuations and for the full counting statistics. These are sensitive to the quantum correlations between the charge carriers, which can have a nonlocal character. In terms of counting statistics this means, for example, that the joint probability to count N_1 particles in terminal 1 and N_2 particles in terminal 2 can not be factorized into separate probabilities for the two events.

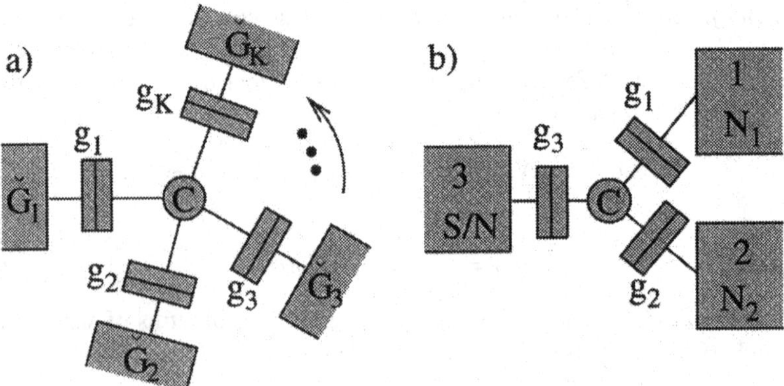

Figure 4. a) An example for a multi-terminal structure. K terminals are connected to a central node by tunnel junctions b) Three-terminal structure. Two different voltage configurations are considered. In the first case, all terminals are normal metals. A bias voltage is applied between terminal 1 and terminal 2, whereas terminal 3 is operated as a voltage probe (no mean current). In the second case we consider a beam splitter configuration. A supercurrent or normal current in terminal 3 is divided in (or merged from) two normal currents from terminals 1 and 2. Here, we assume that the same potential is applied to terminals 1 and 2.

4.1. GENERAL RESULT FOR MULTI-TUNNEL GEOMETRY

The generalization of the method, introduced in Section 2 for two terminals, to many terminals is straightforward [32]. The counting field χ is replaced by a vector χ, with dimension equal to the number of terminals. For brevity we collect the charges passing each terminal into a vector $\mathbf{N}$. The current in each terminal is coupled to the respective component of the counting field by an expression like (13). Following the procedure outlined in Sections 2.2 and 2.3 the result is, that the Green's function of each terminal acquires its own counting field χ_n. The rules, that determine the transport properties, remain essentially unchanged.

The procedure outlined above is best illustrated by an example. We consider a node connected to K terminals via tunnel junctions with conductances g_k. This setup is shown in Fig. 4a. Each terminal is described by a Green's function $\check{G}_k(\chi_k)$ $(k = 1..K)$, which is related to the terminal's usual Green's function by a counting rotation (21). We do not need to specify yet, whether the terminals are superconducting or normal. The goal is to find (for arbitrary applied voltages and temperatures) the joint probability $P_{t_0}(N_1, N_2, ..., N_K)$, that N_1 particles enter through terminal 1, N_2 particles through terminal 2, ..., and N_K particles through terminal K. Correspondingly we define the cumulant generating function

$$e^{-S(\chi)} = \sum_{N_1,N_2,\cdots,N_K} P_{t_0}(\mathbf{N})e^{i\mathbf{N}\chi} . \tag{72}$$

The central node is described by a Green's function $\check{G}_c(\chi)$, which has to be determined from the circuit rules. The matrix currents through terminal k is given by $\check{I}_k = -(g_k/2)[\check{G}_k(\chi_k), \check{G}_c(\chi)]$ and current conservation on the node can be written as

$$0 = \sum_{k=1}^{K} \check{I}_k = \frac{1}{2}\left[\sum_{k=1}^{K} g_k\check{G}_k(\chi_k), \check{G}_c(\chi)\right] . \tag{73}$$

This equation (together with the normalization condition $\check{G}_c^2(\chi) = \check{1}$) is solved by

$$\check{G}_c(\chi) = \frac{\sum_{k=1}^{K} g_k\check{G}_k(\chi_k)}{\sqrt{\sum_{k,l=1}^{K} g_k g_l \left\{\check{G}_k(\chi_k), \check{G}_l(\chi_l)\right\}}} . \tag{74}$$

The CGF is found from integrating the relations $\partial S(\chi)/\partial\chi_k = (-it_0/e)\, I_k(\chi)$, where $I_k(\chi) = (1/4e)\int dE \mathrm{Tr}\check{\tau}_K \check{I}_k(\chi)$. The CGF is then determined up to an additive constant, which is fixed by the normalization $S(\mathbf{0}) = 0$ and neglected in the following. The result is [18]

$$S(\chi) = -\frac{t_0}{4e^2}\int dE\ \mathrm{Tr}\sqrt{\sum_{k,l=1}^{K} g_k g_l \left\{\check{G}_k(\chi_k), \check{G}_l(\chi_l)\right\}} . \tag{75}$$

This provides the counting statistics for an arbitrary multi-terminal structure of the type shown in Fig. 4. Below we discuss several examples.

4.2. NORMAL METAL MULTI-TERMINAL STRUCTURES

In the case that all terminals are normal metals, we can evaluate Eq. (75) further. All terms under the square root are proportional to the unit matrix and the trace can be taken easily with the result [45]

$$\begin{aligned} S(\chi) &= -\frac{t_0 g_\Sigma}{2e^2} \int dE \times \\ & \sqrt{1 + \sum_{k,l=1}^{K} t_{kl} \left[f_k(1-f_l) e^{i(\chi_k - \chi_l)} + f_l(1-f_k) e^{i(\chi_l - \chi_k)} \right]}, \end{aligned} \tag{76}$$

where $g_\Sigma = \sum_{k=1}^{K} g_k$ and $t_{kl} = 2g_k g_l / g_\Sigma^2$. The argument of the square root is the sum over all tunneling events from between the terminals. For two terminals we recover the result (45).

Let us now consider three normal-metal terminals, one of which is operated as voltage probe, i.e. no average current enters the terminal. This layout is depicted in Fig. 4. We assume for the applied potentials that $V_1 < V_3 < V_2$. Terminal 3 is operated as a voltage probe and it follows, that $V_3 = (g_1 V_1 + g_2 V_2)/(g_1 + g_2)$. At zero temperature the energy integration yields

$$\begin{aligned} S(\chi) &= -\frac{t_0 g_\Sigma V}{2e} \left[\frac{g_2}{g_1 + g_2} \sqrt{1 + t_{13}(e^{i(\chi_3 - \chi_1)} - 1) + t_{12}(e^{i(\chi_2 - \chi_1)} - 1)} \right. \\ & \left. + \frac{g_1}{g_1 + g_2} \sqrt{1 + t_{23}(e^{i(\chi_2 - \chi_3)} - 1) + t_{12}(e^{i(\chi_2 - \chi_1)} - 1)} \right]. \end{aligned} \tag{77}$$

The CGF separates into two terms. The first term corresponds to the energy window $eV_3 > E > eV_1$, in which transport is only possible between terminals 1 and 3 or between 1 and 2. The second term results from the energy window $eV_2 > E > eV_3$, in which no electrons can enter into terminal 2.

We now consider a different configuration: an instreaming current is divided into two outgoing currents. This corresponds to the voltage configuration $V = V_1 = V_2 > V_3 = 0$. The corresponding CGF is

$$S(\chi) = -\frac{t_0 g_\Sigma V}{2e} \sqrt{1 + t_{13}(e^{i(\chi_1 - \chi_3)} - 1) + t_{23}(e^{i(\chi_2 - \chi_3)} - 1)}. \tag{78}$$

In the limit $g_2, g_1 \gg g_3$ or vice versa the CGF takes the form

$$S(\chi) = -\frac{t_0 g_\Sigma V}{4e} \left(t_{13}(e^{i(\chi_1 - \chi_3)} - 1) + t_{23}(e^{i(\chi_2 - \chi_3)} - 1) \right), \tag{79}$$

and the corresponding counting statistics is

$$P(N_1, N_2) = e^{-(\bar{N}_1 + \bar{N}_2)} \frac{\bar{N}_1^{N_1}}{N_1!} \frac{\bar{N}_2^{N_2}}{N_2!}. \tag{80}$$

Here $\bar{N}_i = (t_0 V/e) g_i g_3 / 4 g_\Sigma^2$. The statistics is a product of two Poisson distributions, i. e. the two transport processes are uncorrelated.

4.3. SUPERCONDUCTING MULTI-TERMINAL STRUCTURES

Let us now consider the beam splitter configuration, if the incoming current originates from a superconductor. Here, we have to use the Nambu×Keldysh matrix structure. The layout is as shown in Fig. 4. We choose terminal 3 as superconducting terminal with $V_3 = 0$ and the potential in the two normal terminal is assumed to be the same, $V_1 = V_2 = V$. We consider the limit $T \ll eV \ll E_{Th}, \Delta$. Transport occurs then only via Andreev reflection, since no quasiparticles in the superconductor are present. The Green's functions for the terminals can be found in Appendix 7.4. We note, that the pair breaking effect due to a magnetic field in a chaotic dot with the same terminal configuration was studied by Samuelsson and Büttiker [19].

We now evaluate Eq. (75) for the three-terminal setup. The CGF depends only on the differences $\chi_1 - \chi_3$ and $\chi_2 - \chi_3$, which is a consequence of charge conservation and allows to drop the explicit dependence on χ_3 below. Introducing $p_i = 2g_3 g_i / (g_3^2 + (g_1 + g_2)^2)$ we find

$$S(\chi_1, \chi_2) = -\frac{V t_0 \sqrt{g_3^2 + (g_1 + g_2)^2}}{\sqrt{2} e} \times \tag{81}$$
$$\sqrt{1 + \sqrt{1 + p_1^2 (e^{i2\chi_1} - 1) + p_2^2 (e^{i2\chi_2} - 1) + 2p_1 p_2 (e^{i(\chi_1 + \chi_2)} - 1)}}\,.$$

The inner argument contains counting factors for the different possible processes. A term $\exp(i(\chi_k + \chi_l) - 1)$ corresponds to an event in which two charges leave the superconducting terminal and one charge is counted in terminal k and one charge in terminal l. The prefactors are related to the corresponding probabilities. For instance, p_1 is proportional to the probability of a coherent tunneling event of an electron from the superconductor into terminal 1. A coherent pair-tunneling process is therefore weighted with p_1^2. This is accompanied by counting factors which describe either the tunneling of two electrons into terminal 1(2) (counting factor $\exp(i(2\chi_{1(2)})) - 1)$) or tunneling into different terminals (counting factor $\exp(i(\chi_1 + \chi_2) - 1)$). The nested square-root functions show that these different processes are non-separable.

It is interesting to consider the limiting case if $g_3/(g_1 + g_2)$ is not close to 1. Then, $p_{1,2} \ll 1$ and we can expand Eq. (81) in $p_{1,2}$. The CGF can be written as

$$S(\chi_1, \chi_2) = -\frac{t_0 V}{e} \frac{g_3^2}{(g_3^2 + (g_1 + g_2)^2)^{3/2}} \times \tag{82}$$
$$\left[g_1^2 (e^{i2\chi_1} - 1) + g_2^2 (e^{i2\chi_2} - 1) + 2g_1 g_2 (e^{i(\chi_1 + \chi_2)} - 1) \right]\,.$$

The CGF is composed of three different terms, corresponding to a charge transfer event of $2e$ either into terminal 1 or terminal 2 (the first two terms in the bracket) or separate charge transfer events into terminals 1 and 2. The same form of the CGF appears if the proximity effect is destroyed by other means, e. g. a magnetic field [19]. According to the general principles of statistics, sums of CGFs of independent statistical processes are additive. Therefore, the CGF (82) is a sum of CGFs of independent Poisson processes. The total probability distribution $P(N_1, N_2)$ corresponding to Eq. (82) can be found. It vanishes for odd values of $(N_1 + N_2)$ and for even values it is given by

$$P(N_1, N_2) = \frac{e^{-\frac{\bar{N}}{2}} \left(\frac{\bar{N}}{2}\right)^{\frac{N_1+N_2}{2}}}{\left(\frac{N_1+N_2}{2}\right)!} \binom{N_1 + N_2}{N_1} T_1^{N_1} T_2^{N_2} . \tag{83}$$

Here we have defined the average number of transfered electrons $\bar{N} = (t_0 V/e)(g_1 + g_2)^2 g_3^2/((g_1 + g_2)^2 + g_3^2)^{3/2}$ and the probabilities $T_{1(2)} = g_{1(2)}/\ (g_1 + g_2)$ that one electron leaves the island into terminal 1(2). If one would not distinguish electrons in terminals 1 and 2, the charge counting distribution can be obtained from Eq. (82) by setting $\chi_1 = \chi_2 = \chi$ and performing the integration. This leads to $P_{tot}^S(N) = \exp(-\bar{N}/2)(\bar{N}/2)^{N/2}/(N/2)!$, which corresponds to a Poisson distribution of an uncorrelated transfer of electron pairs. The full distribution Eq. (83) is given by $P_{tot}^S(N_1 + N_2)$, multiplied with a *partitioning factor*, which corresponds to the number of ways to distribute $N_1 + N_2$ identical electrons among the terminals 1 and 2, with respective probabilities T_1 and T_2. Note, that $T_1 + T_2 = 1$, since the electrons have no other possibility to leave the island.

5. Concluding Remarks

Full Counting Statistics is a new fundamental concept in mesoscopic electron transport. The knowledge of the full probability distribution of transfered charges completes the information on the transport mechanisms. In fact, the FCS represents *all* information, which can be gained from charge counting in a transport process - a clear progress in our understanding of quantum transport.

In this article we have reviewed the state of the field with particular emphasis on superconductor-normal metal heterostructures. We have introduced a theoretical method, which combines full counting statistics with the powerful Keldysh Green's function technique. This method allows to obtain the FCS for a large variety of mesoscopic systems. For a two-terminal structure a general relation can be derived which contains the FCS of all kinds of constrictions between normal metals and superconductors. Our method is readily applicable to multi-terminal structures and we have discussed one example. An important advantage of the method is that it allows a direct numerical implementation, which means that we are able to find the FCS of arbitrary mesoscopic SN-structures.

I would like to thank the Alexander von Humboldt-Stiftung, the Dutch FOM, the Swiss SNF, and the NCCR Nanoscience for financial support in different stages of this work. Special thanks go to Yuli V. Nazarov, who introduced many of the concepts discussed here. Valuable insights emerged from discussions with D. Bagrets, J. Börlin, C. Bruder, M. Büttiker, M. Kindermann, P. Samuelsson, and F. Taddei.

Appendix

6. Field Theoretical Methods

We summarize some methods and definitions of quantum field theory. In the first part we review briefly the standard Keldysh-Green's function technique. We follow essentially the review [29]. In the second part we explicitly perform the linked cluster expansion for the cumulant generating function, which establishes the relation between our definition of FCS and the Green's function method.

6.1. KELDYSH GREEN'S FUNCTIONS

One commonly used method to study nonequilibrium phenomena is the so-called Keldysh technique. Quite generally time-dependent problems are cast into the form of calculating expectation values of some operator A of the form $\langle A(t)\rangle$, where $A(t)$ is the time-dependent operator in the Heisenberg picture with respect to some Hamiltonian: $A(t) = U^\dagger(t,t')A(t')U(t,t')$. In this expression both the time-ordered evolution operator $U(t,t') = \mathcal{T}\ \exp(-i\int_{t'}^{t} d\tau H(\tau))$ and the anti-time-ordered evolution operator $U^\dagger(t,t') = \tilde{\mathcal{T}}\exp(-i\int_{t}^{t'} d\tau H(\tau))$ appear. A diagrammatic theory requires to account for the various time orderings, which is rather complicated. A considerable simplification arises from Keldysh's trick. We introduce two time coordinates (t_1,t_2), which live on the upper and lower part of the contour (C_1,C_2), and an ordering prescription along the closed time path C_K, depicted in Fig. 5.

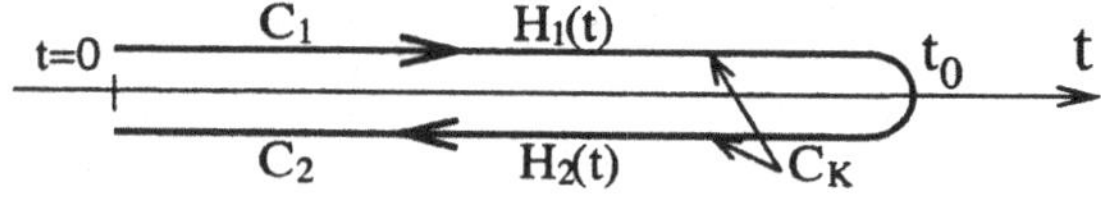

Figure 5. The Keldysh time ordering contour C_K

In the context of full counting statistics we introduce *different* Hamiltonians $H_{1(2)}$ for the two parts of the contour $C_{1(2)}$. The actual different parts are given by Eq. (13). The rest of the Hamiltonian $H_{\rm sys}$ coincides on both parts, just as in the usual formulation [29].

We define a *contour-ordered* Green's function ($\mathcal{T}_K$ denotes ordering along the Keldysh-contour)

$$\hat{G}_{C_K}(t,t') = -i\langle \mathcal{T}_K \Psi(t)\Psi^\dagger(t')\rangle . \tag{84}$$

Here the field operators can have multiple components, such as spin or Nambu for example. This Green's function can be mapped onto a matrix space, the so-called Keldysh space, by considering the time coordinates on the upper and lower part of the Keldysh contour as formally independent variables

$$\check{G}(t,t') \equiv \begin{pmatrix} \hat{G}_{11}(t,t') & \hat{G}_{12}(t,t') \\ -\hat{G}_{21}(t,t') & -\hat{G}_{22}(t,t') \end{pmatrix}, \tag{85}$$

which reads in terms of the field operators in the Heisenberg picture

$$-i\begin{pmatrix} \langle \mathcal{T}\Psi_{H_1}(t)\Psi^\dagger_{H_1}(t')\rangle & \langle \Psi^\dagger_{H_1}(t')\Psi_{H_2}(t)\rangle \\ -\langle \Psi_{H_2}(t)\Psi^\dagger_{H_1}(t')\rangle & \langle \tilde{\mathcal{T}}\Psi_{H_2}(t)\Psi^\dagger_{H_2}(t')\rangle \end{pmatrix}. \tag{86}$$

The current is obtained from the Green's functions (with spatial coordinates reinserted) as

$$\begin{aligned} \hat{\mathbf{j}}(\mathbf{x},t) &= -\frac{e}{2m}\lim_{\mathbf{x}\to\mathbf{x}'}(\nabla_{\mathbf{x}}-\nabla_{\mathbf{x}'})\,\hat{G}_{12}(\mathbf{x},t;\mathbf{x}',t) \qquad (87)\\ &= -\frac{e}{2m}\lim_{\mathbf{x}\to\mathbf{x}'}(\nabla_{\mathbf{x}}-\nabla_{\mathbf{x}'})\left(\hat{G}_{11}(\mathbf{x},t;\mathbf{x}',t)+\hat{G}_{22}(\mathbf{x},t;\mathbf{x}',t)\right), \end{aligned}$$

where the second form is the one we have used in the context of counting statistics. The current has still a matrix structure in the subspace of the components of the field operators. How the electric current is obtained, depends on the definition of the subspace.

In the usual Keldysh-technique $H_1 = H_2$ and we have the general property $G_{11} - G_{22} = G_{12} - G_{21}$. One element of the Green's function (85) can be eliminated by the transformation $\underline{G} = \mathcal{L}\check{G}\mathcal{L}^\dagger$, where $\mathcal{L} = (1 - i\bar{\tau}_2)/\sqrt{2}$ and $\bar{\tau}_i$ denote Pauli matrices in Keldysh space. Then the matrix Green's function takes the form

$$\underline{G}(t,t') = \begin{pmatrix} \hat{G}^R(t,t') & \hat{G}^K(t,t') \\ 0 & \hat{G}^A(t,t') \end{pmatrix}, \tag{88}$$

where

$$\hat{G}^{R(A)}(t,t') = \mp i\theta(\pm(t-t'))\langle[\Psi(t),\Psi^\dagger(t')]\rangle , \tag{89}$$

$$\hat{G}^K(t,t') = -i\langle\{\Psi(t),\Psi^\dagger(t')\}\rangle . \tag{90}$$

The bulk solutions for normal metals and superconductors in App. 7.4 are given in this form.

6.2. THE RELATION BETWEEN CUMULANT GENERATING FUNCTION AND GREEN'S FUNCTION

We establish the connection between the diagrammatic expansions of the Green's function, defined in (15), and the cumulant generating function in (12). We first show how the diagrammatic expansion of the CGF is obtained. In fact, the same expansion occurs in the expression for the thermodynamic potential [46]. According to the definition of the CGF (12) we write

$$e^{-S(\chi)} = \langle T_{C_K} e^{-i\frac{\chi}{2e}\int_{C_K} dt I_c(t)} \rangle . \tag{91}$$

We consider a term of the order n in the expansion of the exponent. Such a term has a form

$$\frac{1}{n!}\left(-i\frac{\chi}{2e}\right)^n \int_{C_K} \cdots \int_{C_K} dt_1 \cdots dt_n \langle T_{C_K} I_c(t_1) I_c(t_2) \cdots I_c(t_n) \rangle . \tag{92}$$

We abbreviate for the sum of connected diagrams of order n

$$Q_n = \frac{1}{n!}\int_{C_K} \cdots \int_{C_K} dt_1 \cdots dt_n \langle T_{C_K} I_c(t_1) I_c(t_2) \cdots I_c(t_n) \rangle_{con} . \tag{93}$$

To count the possible diagrams, which contain p_1 connected diagrams of order m_1, p_2 of order m_2, and so on, where $p_1 m_1 + p_2 m_2 + \cdots + p_k m_k = n$, we note that their number is equivalent to the number of possibilities to assign n operators to p_1 cells containing m_1 places, p_2 cells containing m_2 places, ... This number is given by

$$\frac{n!}{p_1!(m_1!)^{p_1} p_2!(m_2!)^{p_2} \cdots p_k!(m_k!)^{p_k}} \tag{94}$$

and it follows that the CGF can be written as

$$e^{-S(\chi)} = \sum_{p_1, p_2, \cdots} \frac{1}{p_1!}\left[-i\frac{\chi}{2e} Q_1\right]^{p_1} \frac{1}{p_2!}\left[\left(-i\frac{\chi}{2e}\right)^2 Q_2\right]^{p_2} \cdots \tag{95}$$

$$= \exp\left[-i\frac{\chi}{2e} Q_1 + \left(-i\frac{\chi}{2e}\right)^2 Q_2 + \cdots\right] . \tag{96}$$

Thus, we find that the CGF $S(\chi)$ is directly given by the sum

$$-S(\chi) = \sum_{n=1}^{\infty} \left(i\frac{\chi}{2}\right)^n Q_n \tag{97}$$

over the connected diagrams only.

The Green's function introduced in Eq. (15) has a perturbation expansion

$$G(\mathbf{x}, t; \mathbf{x}', t', \chi) = -i \sum_{n=0}^{\infty} \frac{1}{n!}\left(-i\frac{\chi}{2e}\right)^n \int_{C_K} \cdots \int_{C_K} dt_1 \cdots dt_n \times \tag{98}$$
$$\langle T_{C_K} \Psi(\mathbf{x}, t) \Psi^{\dagger}(\mathbf{x}', t') I_c(t_1) I_c(t_2) \cdots I_c(t_n) \rangle_{con} ,$$

recalling that all disconnected diagrams are canceled from this expression. Now we calculate the current from this Green's function with the same current operator used in I_c. We find

$$I(\chi, t) = -\sum_{n=1}^{\infty} \frac{1}{(n-1)!} \left(-i\frac{\chi}{2e}\right)^{n-1} \int_{C_K} \cdots \int_{C_K} dt_1 \cdots dt_{n-1} \times \quad (99)$$
$$\langle T_{C_K} I_c(t_1) I_c(t_2) \cdots I_c(t_{n-1}) I_c(t) \rangle_{con} \,.$$

In a static situation (as we consider) this current does not depend on t and we integrate along the Keldysh contour between 0 and t_0. Using Eq. (93) it follows that

$$2t_0 I(\chi) = -\sum_{n=1}^{\infty} n \left(-i\frac{\chi}{2e}\right)^{n-1} Q_n \quad (100)$$

By comparing the right-hand sides of (100) and (97) we finally obtain the relation between the χ-dependent current and the CGF:

$$I(\chi) = i\frac{e}{t_0}\frac{\partial S(\chi)}{\partial \chi}\,. \quad (101)$$

The constant contribution to $S(\chi)$ follows from the normalization $S(0) = 0$.

7. Quasiclassical Approximation

In practice an exact calculation of Green's functions is impossible in virtually all mesoscopic transport problems. An important simplification is the quasiclassical approximation [30], which makes use of the smallness of the most energy scales involved in transport with respect to the Fermi energy E_F. We briefly summarize the derivation of the basic equations [29]. Here we concentrate on the derivation in the context of superconductivity. The inclusion of spin-dependent phenomena is straightforward.

7.1. EILENBERGER EQUATION

The starting point is the equation of motion for the real-time single-particle Green's function. We consider here the static case, in which the equations can be considered in the energy representation. The equation of motion reads

$$\left[E\hat{\sigma}_3 - \frac{\mathbf{p}^2}{2m} + E_F - \check{\Sigma}\right] \check{G}(\mathbf{x}, \mathbf{x}'; E) = \delta(\mathbf{x} - \mathbf{x}')\,. \quad (102)$$

Here the $\check{}$ denotes matrices in the combined Nambu$\times$Keldysh-space and we use $\hat{\sigma}_i(\bar{\tau}_i)$ to denote Pauli matrices in Nambu(Keldysh)-space. $E_F = p_F^2/2m$ is the

Fermi energy and the selfenergy $\hat{\Sigma}$ includes scattering processes. The complicated dependence on two spatial coordinates in Eq. (102) can be eliminated by the following procedure. We introduce the Wigner transform

$$\check{G}(\mathbf{r},\mathbf{p};E) = \int d^3s \exp(i\mathbf{ps})\, \check{G}(\mathbf{r}+\mathbf{s}/2, \mathbf{r}-\mathbf{s}/2; E)\,, \tag{103}$$

for which the equation of motion reads

$$\left[E\hat{\sigma}_3 - i\mathbf{v}\nabla_{\mathbf{r}} - \check{\Sigma}(\mathbf{r},\mathbf{p})\right] \check{G}(\mathbf{r},\mathbf{p},E) = \check{1}\,. \tag{104}$$

In this equation, we can neglect the dependence on the absolute value of the momentum in the expression in the brackets, since $\check{G}(\mathbf{r},\mathbf{p},E)$ is strongly peaked at $\mathbf{p} = \mathbf{p}_F$. We now subtract the inverse equation and integrate the resulting equation over $\xi = p^2/2m$. We are lead to the definition of the quasiclassical Green's function

$$\check{g}(\mathbf{r},\mathbf{p_F},E) = \frac{i}{\pi}\int \frac{dpp}{m} \check{G}(\mathbf{r},\mathbf{p};E)\,. \tag{105}$$

The quasiclassical Green's function obeys the Eilenberger equation [30]

$$\frac{1}{e^2 N_0}\nabla \check{\mathbf{j}}(\mathbf{r},\mathbf{v_F},E) = \left[-iE\hat{\sigma}_3 + i\check{\sigma}(\mathbf{r},\mathbf{v}_F)\,,\, \check{g}(\mathbf{r},\mathbf{v}_F;E)\right]\,. \tag{106}$$

The current density is obtained from the generalized matrix-current

$$\check{\mathbf{j}}(\mathbf{r},\mathbf{v_F},E) = e^2 N_0 \mathbf{v_F} \check{g}(\mathbf{r},\mathbf{v_F},E) \tag{107}$$

$$\mathbf{j}(\mathbf{r}) = \frac{1}{4e}\int dE \mathrm{Tr}\check{\tau}_K \langle \check{\mathbf{j}}(\mathbf{r},\mathbf{v_F},E)\rangle_{\mathbf{p_F}}\,, \tag{108}$$

where $\langle\rangle_{\mathbf{p}_F}$ denotes angular averaging of the momentum direction and $\check{\tau}_K = \hat{\sigma}_3\bar{\tau}_3$.

Physically, we have in this way integrated out the *fast* spatial oscillation on a scale of the Fermi wave length λ_F, and the remaining function $\check{g}$ is slowly varying on this scale. We have also assumed that the selfenergy $\check{\Sigma}$ does not depend on the momentum p. This has two important consequences. First, the equation is now much simpler and has the intuitive form of a transport equation along classical trajectories. It is in fact easy to show, that Eq. (106) reproduces the well known Boltzmann equation for normal transport. Second, we have a price to pay. Obviously the above derivation fails in the vicinity of interfaces or singular points. This means, that the boundary conditions have to be derived using the underlying non-quasiclassical theory. Fortunately, this can be done quite generally and one simply has to use effective boundary conditions. Related to the problem of boundary conditions is the homogeneity of Eq. (106). In fact, it was shown by Shelankov [47] that the condition ensuring the regularity for $x \to x'$ in the general Green's function leads to a normalization condition

$$\check{g}^2(\mathbf{r},\mathbf{v_F},E) = \check{1}\,. \tag{109}$$

This condition is very important, since almost any further manipulation of Eq. (106) relies on it.

In the context of superconductivity the most important contribution to the self energy is the pairpotential

$$\hat{\sigma}^{R(A)} = -i\hat{\Delta}\ ,\ \hat{\Delta} = -i\frac{\lambda}{4}\int dE\langle \hat{g}^K_{\text{offdiag}}\rangle_{\mathbf{p}_F}\ , \tag{110}$$

where 'offdiag' denotes that only the off-diagonal components in Nambu space should be considered. λ is the attractive BCS interaction constant.

7.2. THE DIRTY LIMIT – USADEL EQUATION

An important simplification arises if the system is almost homogeneous in the momentum direction. This is e.g. the case for diffusive systems, in which the dominant term in the self-energy arises from impurity scattering. In (self-consistent) Born approximation the impurity self-energy has the form

$$\check{\sigma}(\mathbf{r}, E) = -\frac{i}{2\tau_{imp}}\langle \check{g}(\mathbf{r}, \mathbf{p}_F, E)\rangle_{\mathbf{p}_F}\ . \tag{111}$$

In the limit $1/\tau_{imp} \gg E, \Delta$, etc., the Green's functions will be nearly isotropic and we make the Ansatz $\check{g}(\mathbf{r}, \mathbf{p}_F, E) = \check{G}(\mathbf{r}, E) + \mathbf{p}_F\check{\mathbf{G}}(\mathbf{r}, E)$. Using Eq. (106) together with the normalization condition we obtain the so-called Usadel equation [31]

$$\frac{1}{e^2 N_0}\nabla\check{\mathbf{j}}(\mathbf{r}, E) = \left[-iE + \check{\Delta}, \check{G}(\mathbf{r}, E)\right]\ , \tag{112}$$

where the generalized matrix current is now

$$\check{\mathbf{j}}(\mathbf{r}, E) = \sigma\check{G}(\mathbf{r}, E)\nabla\check{G}(\mathbf{r}, E) \tag{113}$$

The conductivity is given by the Einstein relation $\sigma = e^2 N_0 D$, with the diffusion coefficient $D = v_F^2\tau_{imp}/3$. The current density is

$$\mathbf{j}(\mathbf{r}) = \frac{1}{4e}\int dE\mathrm{Tr}\left[\check{\tau}_K\check{\mathbf{j}}(\mathbf{r}, E)\right]\ . \tag{114}$$

7.3. BOUNDARY CONDITIONS

Close to boundaries the quasiclassical equations are invalid and have to be supplemented by boundary conditions. In the general case these boundary conditions have been derived by Zaitsev [48]. They are rather complicated and we will not treat them here. In diffusive systems a very concise form of the boundary conditions was obtained by Nazarov[33]. Under the assumption that two diffusive pieces of metals are connected by a quantum scatterer, the matrix current depends only on the ensemble of transmission eigenvalues and has the form (22). The boundary condition is equivalent to the conservation of matrix currents in the adjacent metals.

7.4. BULK SOLUTIONS

We summarize the necessary ingredients for the various circuits treated in this review. As terminals we consider only normal metals or superconductors. They are determined by external parameters like applied potentials or temperature. The matrix structures are obtained from the bulk solutions of the Eilenberger- or Usadel-equations. We give below their form in the triangular Keldysh-matrix representation (88).

A normal metal at chemical potential μ and temperature T_e is described by a Green's function

$$\underline{G}_N(E) = \hat{\sigma}_3\bar{\tau}_3 + (\bar{\tau}_1 + i\bar{\tau}_2)\left(1 - f(E) - f(-E) + \hat{\sigma}_3(f(-E) - f(E))\right) \quad (115)$$

with the Fermi distribution $f(E) = (\exp((E-\mu)/k_BT_e) + 1)^{-1}$.

A superconducting terminal at chemical potential $\mu_S = 0$ is described by

$$\underline{G}_S(E) = \frac{1}{2}\left(\hat{R} + \hat{A}\right) + \frac{1}{2}\left(\hat{R} - \hat{A}\right)\left(\bar{\tau}_3 + (\bar{\tau}_1 + i\bar{\tau}_2)\,\mathrm{th}\left(\frac{E}{2k_BT_e}\right)\right), \quad (116)$$

where the retarded and advanced functions are

$$\hat{R}(\hat{A}) = \begin{pmatrix} g_{R(A)} & f_{R(A)} \\ f^{\dagger}_{R(A)} & -g_{R(A)} \end{pmatrix} = \frac{(E \pm i\delta)\hat{\sigma}_3 + i\hat{\Delta}}{\sqrt{(E \pm i\delta)^2 - |\Delta|^2}}, \quad (117)$$

where δ is a broadening parameter. The gap matrix contains the dependence on the superconducting phase ϕ

$$\hat{\Delta} = \begin{pmatrix} 0 & |\Delta|e^{i\phi} \\ |\Delta|e^{-i\phi} & 0 \end{pmatrix}. \quad (118)$$

In the limit of zero temperature and $|E| \ll \Delta$ the bulk solutions simplify to

$$\underline{G}_N(E) = \begin{cases} \hat{\sigma}_3\bar{\tau}_3 - (\bar{\tau}_1 + i\bar{\tau}_2)\mathrm{sgn}(eV) & , \ |E| < |eV| \\ \hat{\sigma}_3\bar{\tau}_3 - (\bar{\tau}_1 + i\bar{\tau}_2)\hat{\sigma}_3\mathrm{sgn}(E) & , \ |E| > |eV| \end{cases} \quad (119)$$

$$\underline{G}_S(\phi) = \hat{\sigma}_1\cos(\phi) - \hat{\sigma}_2\sin(\phi). \quad (120)$$

8. Appendix: CGF for the Single Channel Contact

8.1. SUPER-NORMAL CONTACT: COEFFICIENTS A_N

We present the coefficients in the CGF (38) for a channel of transparency T_n. We assume the superconductor to be in equilibrium and the normal metal at a chemical potential μ_N. The occupation factors are $f_\pm^N = f_0(\pm E - \mu_N)$ and $f_\pm^S = f_0(\pm E)$, where f_0 is the Fermi-Dirac distribution. The coefficients take the form

$$\begin{aligned} A_{n,1} &= T_n(1-T_n/2)\frac{2(g_R-g_A)}{(2-T_n(g_R-1))(2-T_n(g_A+1))} \\ &\quad\times\left[f_+^N(1-f_+^S)+f_-^N(1-f_-^S)\right] \\ &\quad+2T_n^2\frac{1-f_Rf_A-g_Rg_A}{(2-T_n(g_R-1))(2-T(g_A+1))} \\ &\quad\times\Big[(f_+^N-f_-^N)(f_+^S-f_-^S)(1-(f_+^N-f_-^N)(f_+^S-f_-^S)) \\ &\qquad+2(f_+^S-f_-^S)^2(1-f_+^N)(1-f_-^N)\Big], \end{aligned} \tag{121}$$

$$\begin{aligned} A_{n,2} &= \frac{T_n^2}{2}f_+^Nf_-^N\times \\ &\quad\frac{1+f_Rf_A-g_Rg_A-(f_+^S-f_-^S)^2(1-f_Rf_A-g_Rg_A)}{(2-T_n(g_R-1))(2-T_n(g_A+1))}. \end{aligned} \tag{122}$$

The coefficients $A_{n,-q}$ can be obtained from Eq. (121) and Eq. (122) by the substitution $f_+^{S(N)} \leftrightarrow (1-f_-^{S(N)})$, i.e. interchanging electron-like and hole-like quasiparticles.

8.2. SUPER-SUPER CONTACT: COEFFICIENTS A_N^S

Introducing $q=(1-g_\mathrm{R}g_\mathrm{A})(1-h^2)+f_\mathrm{R}f_\mathrm{A}(1+h^2)$ the coefficients may be written as

$$A_{n,\pm 2}^S = \frac{T_n^2}{64}q^2, \tag{123}$$

$$\begin{aligned} A_{n,\pm 1}^S &= \frac{T_n}{4}q-\frac{T_n^2}{16}q\left[q-4f_\mathrm{R}f_\mathrm{A}\sin^2\frac{\phi}{2}\right] \\ &\quad+\frac{T_n}{8}\left[(f_\mathrm{R}+f_\mathrm{A})h\cos\frac{\phi}{2}\mp i(f_R-f_A)\sin\frac{\phi}{2}\right]^2, \end{aligned} \tag{124}$$

$$Q_n = \left[1-T_nf_\mathrm{R}^2\sin^2\left(\frac{\phi}{2}\right)\right]\left[1-T_nf_\mathrm{A}^2\sin^2\left(\frac{\phi}{2}\right)\right]. \tag{125}$$

References

1. W. Schottky, Ann. Phys. (Leipzig), **57**, 541 (1918).
2. A. I. Larkin and Yu. V. Ovchinikov, Sov. Phys. JETP **26**, 1219 (1968); A. J. Dahm, A. Denenstein, D. N. Langenberg, W. H. Parker, D. Rogovin and D. J. Scalapino, Phys. Rev. Lett. **22**, 1416 (1969); G. Schön, Phys. Rev. B **32**, 4469 (1985).
3. Ya. M. Blanter and M. Büttiker, Phys. Rep. **336**, 1 (2000).
4. M. J. M. de Jong and C. W. J. Beenakker, in: L. L. Sohn, L. P. Kouwenhoven, and G. Schön (Eds.), *Mesoscopic Electron Transport*, NATO ASI Series E, **345** (Kluwer, Dordrecht, 1997).
5. V. A. Khlus, Sov. Phys. JETP **66**, 1243 (1987).
6. G. B. Lesovik, JETP Lett. **49**, 592 (1989).
7. C. W. J. Beenakker and M. Büttiker, Phys. Rev. B **46**, 1889 (1992).
8. K. E. Nagaev, Phys. Lett. A **169**, 103 (1992)
9. Yu. V. Nazarov, Phys. Rev. Lett. **73**, 1420 (1994).
10. L. Y. Chen and C. S. Tin, Phys. Rev. B **43**, 4534 (1991).
11. R. A. Jalabert, J.-L. Pichard, and C. W. J. Beenakker, Europhys. Lett. **27**, 255 (1994).
12. L. S. Levitov and G. B. Lesovik, JETP Lett. **58**, 230 (1993).
13. L. S. Levitov, H. W. Lee, and G. B. Lesovik, J. Math. Phys. **37**, 4845 (1996).
14. Yu. V. Nazarov, Ann. Phys. (Leipzig) **8**, SI-193 (1999).
15. Yu. V. Nazarov, Phys. Rev. Lett. **73**, 134 (1994).
16. W. Belzig and Yu. V. Nazarov, Phys. Rev. Lett. **87**, 197006 (2001).
17. W. Belzig and Yu. V. Nazarov, Phys. Rev. Lett. **87**, 067006 (2001).
18. J. Börlin, W. Belzig, and C. Bruder, Phys. Rev. Lett. **88**, 197001 (2002).
19. P. Samuelsson and M. Büttiker, cond-mat/0207585 (unpublished).
20. P. Samuelsson, cond-mat/0210409 (unpublished).
21. A. Andreev and A. Kamenev, Phys. Rev. Lett. **85**, 1294 (2000); Yu. Makhlin and A. Mirlin, *ibid.* **87**, 276803 (2001); L. S. Levitov, cond-mat/0103617 (unpublished).
22. C. W. J. Beenakker and H. Schomerus, Phys. Rev. Lett. **86**, 700 (2001).
23. M. Kindermann, Yu. V. Nazarov, and C. W. J. Beenakker, Phys. Rev. Lett. **88**, 063601 (2002).
24. Yu. Makhlin, G. Schön, and A. Shnirman, Phys. Rev. Lett. **85**, 4578 (2000).
25. H.-A. Engel and D. Loss, Phys. Rev. B **65**, 195321 (2002).
26. F. Taddei and R. Fazio, Phys. Rev. B **65**, 075317 (2002).
27. M.-S. Choi, F. Plastina, and R. Fazio, cond-mat/0208318 (unpublished).
28. Yu. V. Nazarov and M. Kindermann, cond-mat/0107133 (unpublished).
29. J. Rammer and H. Smith, Rev. Mod. Phys. **58**, 323 (1986).
30. G. Eilenberger, Z. Phys. **214**, 195 (1968); A. I. Larkin and Yu. N. Ovchinnikov, Sov. Phys. JETP **26**, 1200 (1968).
31. K. D. Usadel, Phys. Rev. Lett. **25**, 507 (1970).
32. Yu. V. Nazarov and D. Bagrets, Phys. Rev. Lett. **88**, 196801 (2002).
33. Yu. V. Nazarov, Superlattices Microst. **25**, 1221 (1999).
34. L. S. Levitov and M. Reznikov, cond-mat/0111057 (unpublished).
35. C. W. J. Beenakker, Rev. Mod. Phys. **69**, 731 (1997).
36. B. A. Muzykantskii and D. E. Khmelnitzkii, Phys. Rev. B **50**, 3982 (1994).
37. M. J. M. de Jong and C. W. J. Beenakker, Phys. Rev. B **49**, 16070 (1994); K. E. Nagaev and M. Büttiker, *ibid.* **63**, 081301(R) (2001).
38. M. J. M. de Jong, Phys. Rev. B **54**, 8144 (1996).
39. D. Bagrets and Yu. V. Nazarov, cond-mat/0207624 (unpublished).
40. Yu. V. Nazarov and T. H. Stoof, Phys. Rev. Lett. **76**, 823 (1996).
41. O. N. Dorokhov, Solid State Comm. **51**, 381 (1984).
42. H. Lee, L. S. Levitov, and A. Yu. Yakovets, Phys. Rev. B **51**, 4079 (1996).

43. D. Averin and H. T. Imam, Phys. Rev. Lett. **76**, 3814 (1996).
44. A. Shelankov and J. Rammer, cond-mat/0207343.
45. J. Börlin, Diploma Thesis (Basel, 2002).
46. A. A. Abrikosov, L. P. Gorkov, and I. E. Dzyaloshinski, *Methods of Quantum Field Theory in Statistical Physics* (Dover, New York, 1963).
47. A. L. Shelankov, J. Low Temp. Phys. **60**, 29 (1985).
48. A. V. Zaitsev, Sov. Phys. JETP **59**, 1015 (1984).

HIGH CUMULANTS OF CURRENT FLUCTUATIONS OUT OF EQUILIBRIUM

The classical kinetic equation vs. the microscopic approach

D.B. GUTMAN
Department of Condensed Matter Physics, The Weizmann Institute of Science
Rehovot 76100, Israel

YUVAL GEFEN
Department of Condensed Matter Physics, The Weizmann Institute of Science
Rehovot 76100, Israel

A.D. MIRLIN
Institut für Nanotechnologie,
Forschungszentrum Karlsruhe,
76021 Karlsruhe,
Germany

Abstract. We consider high order current cumulants in disordered systems out of equilibrium. They are interesting and reveal information which is not easily exposed by the traditional shot noise. Despite the fact that the dynamics of the electrons is classical, the standard kinetic theory of fluctuations needs to be modified to account for those cumulants. We perform a quantum-mechanical calculation using the Keldysh technique and analyze its relation to the quasi classical Boltzmann-Langevin scheme. We also consider the effect of inelastic scattering. Strong electron-phonon scattering renders the current fluctuations Gaussian, completely suppressing the $n > 2$ cumulants. Under strong electron-electron scattering the current fluctuations remain non-Gaussian.

1. INTRODUCTION

The fact that electric current exhibits time dependent fluctuations has been known since the early of 20th century. Still it remains an active field of experimental and theoretical research [1]. Among other features, non-equilibrium shot noise can teach us about the rich many-body physics of the electrons, and may serve

Y.V. Nazarov (ed.), Quantum Noise in Mesoscopic Physics, 497–524.

as a tool to determine the effective charge of the elementary carriers. While the problem of non-interacting electrons is practically resolved by now, the issue of noise in systems of interacting electrons remains largely an open problem. In this paper we focus on two main issues: (1) The role of both inelastic electron-electron and electron-phonon scattering for current correlations. (2) In the absence of interference effects one is tempted to employ the semi-classical kinetic theory of fluctuations [2, 3], also known as the Boltzmann-Langevin scheme. This approach has been originally developed to study pair correlation functions, hence is not naturally devised for higher order cumulants. A naive application of this approach turns out problematic. To study high order cumulants, we first need a reliable scheme, which is why we resort to a microscopic quantum mechanical approach. The comparison between microscopic calculations and the application of the Boltzmann-Langevin scheme for the study of higher order cumulants is the second objective of this paper.

The outline of this work is the following: Section 1 addresses a few introductory issues concerning current fluctuations in non-interacting systems. These brief reminders are necessary for the following analysis. In Section 1.1 we consider the two main sources of noise in non-interacting systems: thermal fluctuations in the contact reservoirs and the stochasticity of the elastic scattering process involved. For this purpose a simple stochastic model is studied. In Section 1.2 we explain why the study of high order correlation functions is of interest. In Section 1.3 we recall some elements of the kinetic theory of fluctuations, explaining the difficulty in applying it to high cumulants. Section 2 is devoted to the analysis of the third order current cumulant. We consider two limiting cases: that of independent electrons (2.1), and that of high electron-electron collision rate (2.2). In Section 3 we compare the results of the microscopic quantum mechanical analysis with those of the Boltzmann-Langevin scheme. In Section 4 we discuss higher moments of current cumulants the full counting statistics, and comment briefly on the effect of electron-phonon scattering.

1.1. NOISE IN NON-INTERACTING SYSTEMS: PROBABILISTIC SCATTERING AND THERMAL FLUCTUATIONS

Before discussing the problem of interacting electrons we would like to recall some important features of fluctuations in a non-interacting electron gas (for a review see [1]). One of the issues addressed in the present work is whether what has become to be known as "*quantum noise* " can be properly discussed within the (semi)-classical framework of the kinetic theory of fluctuations. Surely, non-equilibrium shot noise depends on the discreteness of the elementary charge carriers (and this charge is quantized). Also, at low temperatures one is required to employ the Fermi-Dirac statistics governing the occupation of the reservoir states. In addition, the channel transmission and reflection probabilities are governed by quantum mechanics. But other than that, interference effects appear to play a very minor role in the formation of current fluctuations. Hence the appeal of a semi-classical approach. To understand the origin of the current fluctuations we consider a simple stochastic model, following earlier works [4–6]. This model is a caricature of the physics underlying low frequency transport through a quantum point contact (QPC) with a single channel having a transmission probability $\mathcal{T}$. The QPC is connected to two reservoirs with respective chemical potentials μ_L and μ_R (this model is readily generalized to the multi-channel case). For our

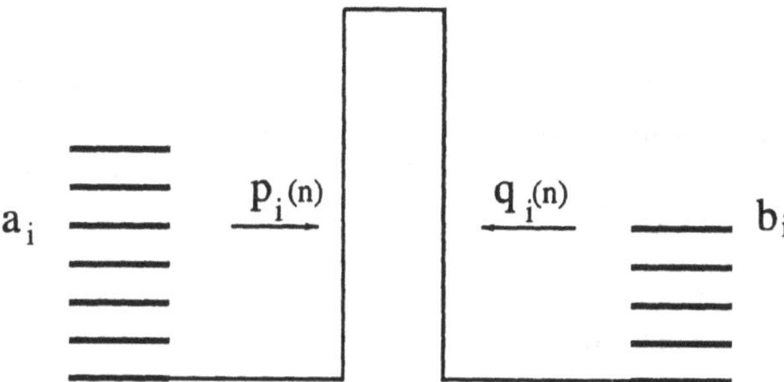

Figure 1. A simplified view of transport through a single channel quantum point contact. For definitions consult the text.

needs electrons occupying (in principle broadened) single particle levels (depicted in a Fig.1) can be perceived as classical particles attempting to pass through the barrier (with transmission probability $\mathcal{T}$). In principle $\mathcal{T}$ may be a function of the level's energy ($\mathcal{T} \to \mathcal{T}_i$). For the mean level spacing $\Delta\epsilon$ the time $\Delta t = h/\Delta\epsilon$ can be interpreted as a time interval between two consecutive collisions. Since the electron charge is discrete (and is well-localized on either the left or the right side of the barrier) the results of a transmission attempts have a binomial statistics.

Consider the fluctuation in the net charge transmitted through the QPC over the time interval $\bar{t} \gg \Delta t$. The number of attempts $N = \bar{t}/\Delta t$ occurring within this interval is large. Since the attempts are discrete events, they can be enumerated. To describe the result of the nth attempt of an electron originating from level i on the left (right) we define the quantity $p_i(n)$ and $q_i(n)$ (cf. Fig.1). For an attempt ending up in the transmission of an electron from left to right (from right to left) the quantity $p_i(n)$ $(q_i(n))$ is assigned the value 1. Otherwise $p_i(n)$ $(q_i(n))$ are zero. This implies that $\{p_i(n)\}$'s and $\{q_i(n)\}$'s have each binomial statistics with the expectation value $\langle p_i(n)\rangle = \langle q_i(n)\rangle = \mathcal{T}_i$. We also assume that neither different attempts from the same level, nor attempts of the electrons coming from different levels are correlated. In addition to fluctuations in the transmission process, there are also fluctuations of the occupation numbers a_i (b_i) of a single particle level i. These represent thermal fluctuations in the reservoirs. For fermions they possess a binomial distribution [7]. We end up with

$$P(x) = (1 - \langle x\rangle)\delta(x) + \langle x\rangle\delta(1 - x) \quad \text{for } x = p_i(n), q_i(n), a_i, b_i, \qquad (1)$$

and the correlations

$$\langle p_i(n)p_j(n')\rangle = \delta(i,j)\delta(n,n')T_i;$$
$$\langle q_i(n)q_j(n')\rangle = \delta(i,j)\delta(n,n')T_i. \tag{2}$$

Also

$$\langle a_i a_j\rangle = \begin{cases} n_i^L, & i=j \\ 0, & i\neq j. \end{cases} \quad \langle b_i b_j\rangle = \begin{cases} n_i^R, & i=j \\ 0, & i\neq j. \end{cases} \tag{3}$$

Here $n_i^L \equiv \langle a_i\rangle$ and $n_i^R \equiv \langle b_i\rangle$. The typical time scale over which the occupation number of a level fluctuates is usually dictated by the interaction among the electrons or with external agents. We assume that at the leads the time fluctuations of the occupation numbers (a_i and b_i) over the interval $\bar{t}$ are negligible.

Motivated by this picture, we consider fluctuations in the net number $Q_{\bar{t}}$ of electrons transmitted through the constriction within the time interval $\bar{t}$. Employing the Pauli exclusion principle one can write

$$Q_{\bar{t}} = \sum_{n=1}^{N}\left[\sum_i [p_i(n)a_i(1-b_i) - q_i(n)b_i(1-a_i)]\right]. \tag{4}$$

According to eqs.(2,3 and 4) the expectation value of the transmitted charge is

$$\langle Q_{\bar{t}}\rangle = TN\sum_i [n_i^L(1-n_i^R) - n_i^R(1-n_i^L)]. \tag{5}$$

Evidently it is related to the value of the d.c. current through $\langle Q_{\bar{t}}\rangle = \bar{t}I/e$. We next discuss the higher order moments of current fluctuations. As was shown in Refs.[8] (see also [9]) the experimentally measured high order current cumulants should be defined in quite a subtle manner. To make contact between the measured observables and theoretically calculated quantities, we recall that within the Keldysh formalism [10] the time axes is folded. For any given moment of time there are two current operators, one for the upper and another for the lower branch of the contour. The product of the symmetric combination of these two operators, I_2, time ordered along the Keldysh contour (T_c) and averaged with respect to the density matrix yields the proper current correlation function. In a stationary situation the pair current correlation function [11]

$$\mathrm{S}_2(t-t') = \langle T_c I_2(t) I_2(t')\rangle \tag{6}$$

depends only on the difference between the time indices. Similarly we define the third order current correlation function

$$\mathrm{S}_3(t-t';t'-t'') = \langle T_c I_2(t) I_2(t') I(t'')\rangle\,. \tag{7}$$

In Fourier space it can be represented as

$$\mathrm{S}_2(\omega_1) = \int_c d(t-t')e^{-i\omega_1(t-t')}\mathrm{S}_2(t-t') \; .$$

$$\mathrm{S}_3(\omega_1, \omega_2) = \int_c d(t-t')d(t'-t'')e^{-i\omega_1(t-t')-i\omega_2(t'-t'')}\mathrm{S}_3(t-t', t'-t''). \quad (8)$$

Usually it is more interesting to consider only the cumulants, i.e. irreducible parts of the correlation functions. For the pair current correlation function we have

$$\mathcal{S}_2(\omega = 0) = \mathrm{S}_2 - I^2, \quad (9)$$

while the third order current cumulant is given by

$$\mathcal{S}_3(\omega_1 = 0, \omega_2 = 0) = \mathrm{S}_3 - 3I\mathrm{S}_2 + 2I^3, \quad (10)$$

where $\mathrm{S}_2, \mathrm{S}_3$ are taken at zero frequencies. Microscopic calculations of the current cumulants will be performed in the Sections 2 and 4 below. Here we evaluate the cumulants within the stochastic model described above. The results agree with the low frequency current cumulants of the non-interacting electrons in the QPC [4]. The temperature dependence of those cumulants is qualitatively similar to the dependence of the current cumulants in a disordered junction, analyzed below.

To find the variance of the stochastic variable Q we employ eqs.(2, 3 and 4)

$$\langle Q_{\bar{t}}^2 \rangle - \langle Q_{\bar{t}} \rangle^2 = \mathcal{T}(1-\mathcal{T})\sum_i [n_L^2(i) - n_R^2(i)]^2 +$$
$$\mathcal{T}\sum_i [n_L(i) + n_R(i) - n_L^2(i) - n_R^2(i)]. \quad (11)$$

The first term on the r.h.s. of eq.(11) vanishes at thermal equilibrium. We associate this part with the shot noise of the electrons. The second term vanishes at zero temperature, and we associate it with the thermal noise. As expected the "shot noise" part vanishes in the limit of perfect conductor ($\mathcal{T} \to 1$). Going along the same procedure for the third order cumulant one obtains [5] in the limit of low temperature ($eV \gg T$)

$$\langle (Q_{\bar{t}} - \langle Q_{\bar{t}} \rangle)^3 \rangle = \mathcal{T}(1-\mathcal{T})(1-2\mathcal{T})\langle Q_{\bar{t}} \rangle \quad (12)$$

and in the limit of high temperature ($eV \ll T$)

$$\langle (Q_{\bar{t}} - \langle Q_{\bar{t}} \rangle)^3 \rangle = \mathcal{T}(1-\mathcal{T})\langle Q_{\bar{t}} \rangle. \quad (13)$$

Note that with increasing the temperature the third order cumulant of the transmitted charge approaches a constant. This is a robust feature of all odd order cumulants, which can be understood from quite general arguments. Indeed, since the current operator changes sign under time reversal transformation, any even-order correlation function of the current fluctuations (e.g. S_2) taken at zero frequency is invariant under this operation. Assuming that current correlators are functions of the average current, I, it follows that even-order correlation functions depend only on the absolute value of the electric current (and are independent of the direction of the current). In the Ohmic regime this means that even-order current correlation functions (at zero frequency) are even functions of the applied voltage. Evidently, this general observation agrees with the result eq. (11).

By contrast, odd-order current correlation functions change their sign under time reversal transformation. In other words, such correlation functions depend on the direction of the current, and not only on its absolute value. Therefore in the Ohmic regime, odd-order correlation functions of current are odd-order functions of the applied voltage. This condition automatically guarantees that odd-order correlation functions *vanish* at thermal equilibrium.

One can show that by considering high order moments of the stochastic model one reproduces the correct results for any cumulants of a current noise in a QPC. Of course, the solution of a toy model can not replace the real microscopic calculations. But the fact that the results of the latter and the stochastic model agree suggests that the underlying physics is rather simple (and it is basically captured by such a simple model). It is the combination of thermal fluctuations (in the occupation of the single electron states) and random transmission of particles through the barrier that gives rise to current fluctuations. These two sources of stochasticity remain there when the electron can no longer be considered non-interacting. In that case, though, one cannot consider fluctuations at different energy levels to be independent.

1.2. WHY ARE HIGH ORDER CUMULANTS INTERESTING?

Low frequency current fluctuations give rise to a large (in general infinite) number of irreducible correlation functions (cumulants). The pair current correlation function provides us with only partial information about current fluctuations. To obtain the complete picture one should consider high order cumulants as well. As we have explained in Section 1.1 the symmetry-dictated properties of odd and even correlation function are very different from each other; in particular odd order current cumulants are not masked by thermal fluctuations. For this reason they can be used for probing non-equilibrium properties at relatively high temperatures. Shot noise has been used to detect an effective quasi-particle charges in the FQHE regime [12, 13]. Potentially, odd moments can be used to measure the effective quasiparticle charge in other strongly-correlated systems [6]. This may be needed for systems undergoing a transition controlled by temperature (for example normal-super-conductor) and having different fundamental excitations at different temperature regimes. One may hope therefore, that better understanding of current correlations will teach us more about the many-body electron physics. At this moment this remains a challenge. Before trying to reach this goal, we need to understand the genuine properties of the high order cumulant in the relatively simple physical models. This is done next.

1.3. BACKGROUND AND ISSUES TO BE DISCUSSED

As was mentioned above it is quite appealing to try to discuss current noise in terms of the semi-classical kinetic equation. To be more specific, let us consider a disordered metallic constriction. Its length L is much greater than the elastic mean free path l. The disorder inside the constriction is short-ranged, weak and uncorrelated. Under these conditions the electrons kinetics can be described by the Boltzmann equation:

$$\hat{\mathcal{L}}\bar{f}(\mathbf{p},\mathbf{r},t) = \mathrm{Col}\{\bar{\mathrm{f}}\}. \tag{14}$$

Here $\mathrm{Col}\{\bar{\mathrm{f}}\}$ is the collision integral and

$$\hat{\mathcal{L}} = \left(\frac{\partial}{\partial t} + \mathbf{v}\cdot\nabla_{\mathbf{r}} + e\mathbf{E}\cdot\nabla_{\mathbf{p}}\right) \tag{15}$$

is the Liouville operator of a particle moving in an external electric field $\mathbf{E}$. The pair correlation function of any macroscopic quantity may be found from the kinetic theory of fluctuations. Within this theory the distribution function (f) consists of the coarse-grained $(\bar{f})$ and fluctuating (δf) parts

$$f(\mathbf{p},\mathbf{r},t) = \bar{f}(\mathbf{p},\mathbf{r},t) + \delta f(\mathbf{p},\mathbf{r},t). \tag{16}$$

Since f is a macroscopic quantity (a quantity associated with a large number of particles) it must satisfy the Onsager's regression hypothesis (see Ref. [14], Section 19).

To sketch this hypothesis for the case of interest we consider a perturbation of the equilibrium distribution function $\delta\bar{f}(\mathbf{p},\mathbf{r},t)$, small but substantially larger than the typical fluctuations of the distribution function. It follows then that the distribution function (with a high probability) will evolve toward the equilibrium state. Its relaxation dynamics is governed by the Boltzmann equation and because the perturbation is small, the collision integral can be linearized. According to Onsager the correlation function of *any* macroscopic quantity, and in particular $\langle\delta f(\mathbf{p},\mathbf{r},t)f(\mathbf{p}',\mathbf{r}',\mathbf{t}')\rangle$, is governed by the *same* equation as the one governing its relaxation (i.e. as equation governing the quantity $\delta\bar{f}(\mathbf{p},\mathbf{r},t)$)

$$\left(\hat{\mathcal{L}}(\mathbf{p},\mathbf{r},t) + \mathcal{I}(\mathbf{p},\mathbf{r},t)\right)\langle\delta f(\mathbf{p},\mathbf{r},t)\delta f(\mathbf{p}',\mathbf{r}',t')\rangle = 0,\ t > t' \tag{17}$$

Here $\mathcal{I}$ is a linearized collision integral. It was later suggested by Lax [15] (and can be proven within Keldysh formalism) that eq.(17) does hold for any stationary, not necessarily equilibrium, state. However, we need to recall that the Onsager hypothesis was formulated only for a *pair* correlation function. There is no obvious way to apply this logic for higher cumulants.

An alternative route of describing fluctuations within kinetic theory which is seemingly free of this difficulty had been proposed by Kogan and Shul'man[2]. Their picture is the following. The real space is divided into small volumes (as explained in a Ref.[14]). The function $\bar{f}(\mathbf{p}, \mathbf{r}, t)$ represents the average number of particles in the state $\mathbf{p}$ of a unit volume element (cell) labeled by index ($\mathbf{r}$). The total number of the electron in every cell must be large. It was suggested in Ref. [2] that fluctuations of this number can be taken into account by adding a random (Langevin) source term to the Boltzmann equation. The resulting stochastic equation (including this *additive* noise) is called the Boltzmann-Langevin equation.

$$\left(\hat{\mathcal{L}}(\mathbf{p}, \mathbf{r}, t) + \mathcal{I}(\mathbf{p}, \mathbf{r}, t)\right) \delta f(\mathbf{p}, \mathbf{r}, t) = \delta J(\mathbf{p}, \mathbf{r}, t) \,. \tag{18}$$

The Langevin source $\delta J(\mathbf{p}, \mathbf{r}, t)$ denotes a random number of particles incoming into the given state in some interval (around time t). Since eq.(18) is a linear one, the statistics of the distribution function is determined by the random source term δJ. To establish its properties Kogan and Shul'man had used a rather simple physical picture. To be consistent with the Boltzmann equation they have assumed that interference effects are weak. The collision events are local in space and time. Since the typical number of electrons inside the cell is large one can ignore the correlation between the scattering of different electrons. The electron scattering is a Poissonian process, with the number of scattered particles within any given cell (over a microscopic time interval) being large. While this picture yields a correct result for the pair correlation function, it substantially underestimates all high order correlators (starting from $\mathcal{S}_3$).

2. THIRD ORDER CURRENT CUMULANT

In this section we use microscopical calculations to evaluate the third order current cumulant for a quasi one-dimensional system of a length L with diffusive disorder[16]. We start with the coordinate-dependent correlation function

$$\mathrm{S}_3(x, t; x', t'; x'', t'') = \langle T_c \hat{I}_2(x, t) \hat{I}_2(x', t') \hat{I}_2(x'', t'') \rangle \,. \tag{19}$$

Here x is a coordinate measured along a quasi one-dimensional wire ($0 \le x \le L$) of cross-section $\mathcal{A}$; T_c is the time ordering operator along the Keldysh contour. Next we perform the Fourier transform with respect to the time difference as in eq.(8). For small values of the frequencies ω_1, ω_2, (small compared with the inverse diffusion time along the wire), the current fluctuations are independent of the spatial coordinate. We next evaluate the expression, eq. (19), for a disordered junction, in the hope that the qualitative properties we are after are not strongly system dependent. In the present section we consider non-interacting electrons in the presence of a short-range, delta correlated and weak disorder potential ($\epsilon_f \tau \gg \hbar$, where τ is the elastic mean free time and ϵ_f is the Fermi energy). To calculate S_3 we employ the σ-model formalism, recently put forward for dealing with non-equilibrium diffusive systems (for details see Ref.[17]).

The disorder potential is δ-correlated:

$$\langle U_{\rm dis}(\mathbf{r})\, U_{\rm dis}(\mathbf{r}')\rangle = \frac{1}{2\pi\nu\tau}\,\delta(\mathbf{r}-\mathbf{r}'), \tag{20}$$

where ν is the density of states at the Fermi energy.

The Hamiltonian we are concerned with is:

$$H = H_0 + H_{\rm int}. \tag{21}$$

The motion of electrons in the disorder potential is described by:

$$H_0 = \int_{\rm Volume} d\mathbf{r}\bar{\Psi}(\mathbf{r})\Big[-\frac{\hbar^2}{2m}(\nabla - i\mathbf{a})^2 + U_{\rm dis}\Big]\Psi(\mathbf{r}). \tag{22}$$

Here $c\mathbf{a}/e$ is a vector potential. The Coulomb interaction among the electrons is described by

$$H_{\rm int} = \frac{1}{2}\int d\mathbf{r}d\mathbf{r}'\bar{\Psi}(\mathbf{r})\bar{\Psi}(\mathbf{r}')V_0(\mathbf{r}-\mathbf{r}')\Psi(\mathbf{r})\Psi(\mathbf{r}')\ , \tag{23}$$

where

$$V_0(\mathbf{r}-\mathbf{r}') = \frac{e^2}{|\mathbf{r}-\mathbf{r}'|}\ . \tag{24}$$

2.1. WEAK INELASTIC COLLISIONS

Following the procedure outlined in Ref. [17], we introduce a generating functional and average it over disorder. Next we perform a Hubbard-Stratonovich transformation, integrating out fermionic degrees of freedom. Employing the diffusive approximation one obtains an effective generating functional expressed as a path integral over a bosonic matrix field Q

$$Z[\mathbf{a}] = \int \mathcal{D}Q\exp(iS[Q,\mathbf{a}])\ . \tag{25}$$

Here the integration is performed over the manifold

$$\int Q(x,t,t_1)Q(x,t_1,t')dt_1 = \delta(t-t'), \tag{26}$$

the effective action is given by

$$iS[Q,\mathbf{a}] = -\frac{\pi\hbar\nu}{4}\mathrm{Tr}\{D\,(\nabla Q + i[\mathbf{a}_\alpha\gamma^\alpha, Q])^2 + 4i\hat{\epsilon}Q\}\ , \tag{27}$$

and

$$\gamma_1 = \begin{pmatrix} 1 & 0 \\ 0 & 1 \end{pmatrix},\ \gamma_2 = \begin{pmatrix} 0 & 1 \\ 1 & 0 \end{pmatrix}. \tag{28}$$

Tr represents summation over all spatio-temporal and Keldysh components. Here $\mathbf{a}_1$ and $\mathbf{a}_2$ are the Keldysh rotated classical and quantum components of $\mathbf{a}$. Hereafter we focus our attention on $\mathbf{a}_1$, $\mathbf{a}_2$, the components in the direction along the wire. The third order current correlator may now be expressed as functional differentiation of the generating functional $Z[a]$ with respect to a_2

$$\mathrm{S}_3(t_1-t_2, t_2-t_3) = \frac{ie^3}{8}\frac{\delta^3 Z[a]}{\delta a_2(x_1,t_1)\delta a_2(x_2,t_2)\delta a_2(x_3,t_3)}. \tag{29}$$

Performing this functional differentiation one obtains the following result

$$\mathrm{S}_3(t_1-t_2, t_2-t_3) = \frac{e^3\mathcal{A}(\pi\hbar\nu D)^2}{16}\Big\langle \hat{\mathrm{M}}(x_1,t_1)\hat{\mathrm{I}}^D(x_2,x_3,t_2,t_3) +$$
$$(x_1,t_1 \leftrightarrow x_3,t_3) + (x_1,t_1 \leftrightarrow x_2,t_2) + \frac{\pi\hbar\nu D}{4}\hat{\mathrm{M}}(x_1,t_1)\hat{\mathrm{M}}(x_2,t_2)\hat{\mathrm{M}}(x_3,t_3)\Big\rangle. \tag{30}$$

Here we have defined

$$\hat{I}^D(x,x',t,t') = \mathrm{Tr}^K\Big\{Q_{x,t,t'}\gamma_2 Q_{x',t',t}\gamma_2 - \delta_{t,t'}\gamma_1\Big\}\delta_{x,x'}, \tag{31}$$

$$\hat{\mathrm{M}}(x,t) = \mathrm{Tr}^K\Big\{\int dt_1\Big([Q_{x,t,t_1};\nabla]\,Q_{x,t_1,t}\Big)\gamma_2\Big\}. \tag{32}$$

We employ the notations $Q(x,t,t') \equiv Q_{x,t,t'}$; Tr^K is the trace taken with respect to the Keldysh indices; $\langle\rangle$ denotes a quantum-mechanical expectation value. The matrix Q can be parameterized as

$$Q = \Lambda \exp(W)\ \text{, where}\ \ \Lambda W + W\Lambda = 0 \tag{33}$$

and Λ is the saddle point of the action (27)

$$\Lambda(x,\epsilon) = \begin{pmatrix} 1 & 2F(x,\epsilon) \\ 0 & -1 \end{pmatrix}. \tag{34}$$

The function F is related to the single particle distribution function f through

$$F(x,\epsilon) = 1 - 2f(x,\epsilon)\ . \tag{35}$$

The matrix $W_{x,\epsilon,\epsilon'}$, in turn, is parameterized as follows:

$$W_{x,\epsilon,\epsilon'} = \begin{pmatrix} F_{x,\epsilon}\bar{w}_{x,\epsilon,\epsilon'} & -w_{x,\epsilon,\epsilon'} + F_{x,\epsilon}\bar{w}_{x,\epsilon,\epsilon'}F_{x,\epsilon'} \\ -\bar{w}_{x,\epsilon,\epsilon'} & -\bar{w}_{x,\epsilon,\epsilon'}F_{x,\epsilon'} \end{pmatrix} . \tag{36}$$

It is convenient to introduce the diffusion propagator

$$(-i\omega + D\nabla^2)\mathrm{D}(x, x'\omega) = \frac{1}{\pi\hbar\nu}\delta(x - x') . \tag{37}$$

The absence of diffusive motion in clean metallic leads implies that the diffusion propagator must vanish at the end points of the constriction. In addition, there is no current flowing in the transversal direction (hard wall boundary conditions). It follows that the component of the gradient of the diffusion propagator in that direction (calculated at the hard wall edges) must vanish as well. The correlation functions of the fields $w, \bar{w}$ are then given by:

$$\begin{aligned} &\langle w(x,\epsilon_1,\epsilon_2)\bar{w}(x',\epsilon_3,\epsilon_4)\rangle = 2(2\pi)^2\delta(\epsilon_1-\epsilon_4)\delta(\epsilon_2-\epsilon_3)\mathrm{D}(x,x',\epsilon_1-\epsilon_2)\,, \\ &\langle w(x,\epsilon_1,\epsilon_2)w(x',\epsilon_3,\epsilon_4)\rangle = -g(2\pi)^3\delta(\epsilon_1-\epsilon_4)\delta(\epsilon_2-\epsilon_3) \\ &\int dx_1 \mathrm{D}_{\epsilon_1-\epsilon_2,x,x_1}\nabla F_{\epsilon_2,x_1}\nabla F_{\epsilon_1,x_1}\mathrm{D}_{\epsilon_2-\epsilon_1,x_1,x'}\,, \\ &\langle \bar{w}(x,\epsilon_1,\epsilon_2)\bar{w}(x',\epsilon_3,\epsilon_4)\rangle = 0\,. \end{aligned} \tag{38}$$

To evaluate S_3 one follows steps similar to those that led to the derivation of S_2, see Ref. [17]. If all relevant energy scales in the problem are smaller than the transversal Thouless energy ($E_{Th} = D/L_T^2$, where L_T is a width of a wire), the wire is effectively quasi-one dimensional. In that case only the lowest transversal mode of the diffusive propagator can be taken into account, which yields

$$\mathrm{D}(x_1, x_2) = \frac{1}{2\pi g}\left[|x_1 - x_2| - x_1 - x_2 + \frac{2x_1x_2}{L}\right] . \tag{39}$$

Here $g = \hbar\nu D$. The electron distribution function in this system is equal to

$$F(x,\epsilon) = \frac{x}{L}F_{eq}\left(\epsilon - \frac{eV}{2}\right) + \left(1 - \frac{x}{L}\right)F_{eq}\left(\epsilon + \frac{eV}{2}\right) . \tag{40}$$

The quantities F and D determine the correlation functions, eq. (38). We can now begin to evaluate S_3, (c.f. eq. (30)), performing a perturbative expansion in the fluctuations around the saddle point solution, eq. (34). After some algebra we find that in the zero frequency limit the third order *cumulant* is given by

$$\begin{aligned} &S_3(\omega_1 = 0, \omega_2 = 0) = \frac{3e^3 A\pi g^2}{\hbar L^3}\int_0^L dx_1 dx_2 \\ &\int_{-\infty}^{\infty} d\epsilon F(\epsilon, x_1)\mathrm{D}[0, x_1, x_2]\nabla\left(F^2(\epsilon, x_2)\right) . \end{aligned} \tag{41}$$

Integrating over energies and coordinates we obtain

$$\mathcal{S}_3(\omega_1 = 0, \omega_2 = 0) = e^2 I y(p) \ ,$$
$$y(p) = \frac{6(-1+e^{4p}) + (1 - 26e^{2p} + e^{4p})p}{15p(-1+e^{2p})^2} \ , \tag{42}$$

where $p = eV/2T$. The function y is depicted in Fig.1, where it is plotted on a logarithmic scale.

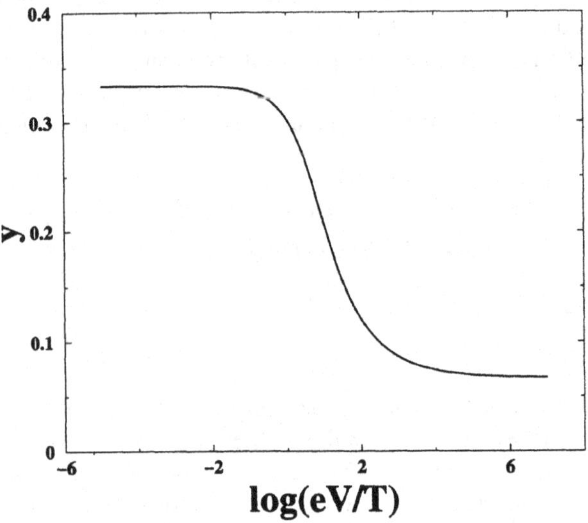

Figure 2. The scaling function y plotted on a logarithmic scale, cf. eq. (42)

Let us now discuss the main features of the function $\mathcal{S}_3$. In agreement with symmetry requirements $\mathcal{S}_3$ is an odd function of the voltage (the even correlator $\mathcal{S}_2$ is proportional to the absolute value of voltage), and vanishes at equilibrium. The zero temperature result (high voltage limit) has already been obtained by means of the scattering states approach for single-channel systems [4], and later generalized by means of Random Matrix Theory (RMT) to multi-channel systems (chaotic and diffusive) [18]. In our derivation we do not assume the applicability of RMT [19]. Our result covers the whole temperature range. We obtain that at low temperatures the third order cumulant is linear in the voltage

$$\mathcal{S}_3 = \frac{e^2}{15} I \ . \tag{43}$$

At high temperatures the electrons in the reservoirs are not anymore in the ground state, so the correlations are partially washed out by thermal fluctuations. One may then expand $y(p)$, eq. (42), in a series of $eV/2T$. The leading term in this high temperature expansion is linear in the voltage

$$\mathcal{S}_3 = \frac{e^2}{3} I \ . \tag{44}$$

Note that although thermal fluctuations enhance the noise (compared with the zero temperature limit), eqs.(43) and (44) differ only by a numerical factor. The experimental study of $\mathcal{S}_3$ (and higher odd cumulants) provides one with a direct probe of non-equilibrium behavior, not masked by equilibrium thermal fluctuations.

2.2. STRONG INELASTIC COLLISIONS

In our analysis so far we have completely ignored inelastic collisions among the electrons. This procedure is well justified provided that the inelastic length greatly exceeds the system's size. However, if this is not the case, different analysis is called for. To understand why inelastic collisions do matter for current fluctuations, we would like to recall the analysis of $\mathcal{S}_2$ for a similar problem. The latter function is fully determined by the effective electron temperature. Collisions among electrons, which are subject to an external bias, increase the temperature of those electrons. This, in turn, leads to the enhancement of $\mathcal{S}_2$, cf. Refs. [20, 21]. In the limit of short inelastic length

$$l_{\mathrm{e-e}} \ll L \ , \tag{45}$$

the zero frequency and zero temperature noise is

$$\mathcal{S}_2(0) = \frac{\sqrt{3}}{4} eI \ . \tag{46}$$

In the present section we consider the effect of inelastic electron collisions on $\mathcal{S}_3$. We assume that the electron-phonon collision length is large, $l_{\mathrm{e-ph}} \gg L$, hence electron-phonon scattering may be neglected. The Hamiltonian we are concerned with is

$$H = H_0 + H_{\mathrm{int}}. \tag{47}$$

The Coulomb interaction among the electrons is described by

$$H_{\mathrm{int}} = \frac{1}{2} \int d\mathbf{r} d\mathbf{r}' \bar{\Psi}(\mathbf{r}) \bar{\Psi}(\mathbf{r}') V_0(\mathbf{r}-\mathbf{r}') \Psi(\mathbf{r}) \Psi(\mathbf{r}') \ , \tag{48}$$

where

$$V_0(\mathbf{r}-\mathbf{r}') = \frac{e^2}{|\mathbf{r}-\mathbf{r}'|} \ . \tag{49}$$

We need to deal with the effect of electron-electron interactions in the presence of disorder and away from equilibrium. Following Ref. [23] one may introduce an auxiliary bosonic field

$$\Phi = \begin{pmatrix} \phi_1 \\ \phi_2 \end{pmatrix}, \tag{50}$$

which decouples the interaction in the particle-hole channel. Now the partition function (eq. (25)) is a functional integral over both the bosonic fields Q and ϕ ,

$$\langle Z \rangle = \int_{Q^2=1} \mathcal{D}Q\mathcal{D}\phi \exp(iS_{total}) \ . \tag{51}$$

The action is

$$iS_{total} = iS[\Phi] + iS[\Phi, Q] \,, \tag{52}$$

$$iS[\Phi] = i\mathrm{Tr}\{\Phi^T V_0^{-1} \gamma^2 \Phi\} \,, \tag{53}$$

$$iS[\Phi, Q] = -\frac{\pi\nu}{4\tau}\mathrm{Tr}\{Q^2\} + \mathrm{Tr}\ln\left[\hat{G}_0^{-1} + \frac{iQ}{2\tau} + \phi_\alpha\gamma^\alpha\right]. \tag{54}$$

It is convenient to perform a "gauge transformation" [23] to a new field $\tilde{Q}$

$$Q_{t,t'}(x) = \exp\left(ik_\alpha(x,t)\gamma^\alpha\right)\tilde{Q}_{t,t'}(x)\exp\left(-ik_\alpha(x,t')\gamma^\alpha\right) \ . \tag{55}$$

Introducing the long derivative

$$\partial_x \tilde{Q} \equiv \nabla\tilde{Q} + i[\nabla k_\alpha \gamma^\alpha, \tilde{Q}] \,, \tag{56}$$

one may write the gradient expansion of eq. (54) as

$$\begin{aligned} iS[\tilde{Q}, \Phi] &= i\nu\mathrm{Tr}\{(\Phi - i\omega K)^T \gamma_2 (\Phi + i\omega K)\} - \\ &\frac{\pi\nu}{4}\left[D\mathrm{Tr}\{\partial_x\tilde{Q}\}^2 + 4i\mathrm{Tr}\{(\epsilon + (\phi_\alpha + i\omega k_\alpha)\gamma^\alpha)\tilde{Q}\}\right] \ . \end{aligned} \tag{57}$$

At this point the vector $K^T = (k_1, k_2)$ that determines the transformation (55) is arbitrary. The saddle point equation for Q of the action (57) is given by the following equation

$$D\partial_x(\tilde{Q}\partial_x\tilde{Q}) + i[(\epsilon + (\phi_\alpha + i\omega k_\alpha)\gamma^\alpha), \tilde{Q}] = 0 \,. \tag{58}$$

Let us now choose the parameterization

$$\tilde{Q} = \tilde{\Lambda}\exp(\tilde{W}), \tag{59}$$

where $\tilde{W}$ represents fluctuation around the saddle-point

$$\tilde{\Lambda}(x, \epsilon) = \begin{pmatrix} 1 & 2\tilde{F}[\phi](x, \epsilon) \\ 0 & -1 \end{pmatrix} \ . \tag{60}$$

Eq. (60) implies that the solution of the saddle point equation (58), determines $\tilde{F}$ as a functional of ϕ. We do not know, though, how to solve it. Instead we average over ϕ the eq.(58). The solution of this averaged equation, denoted by $\bar{F}$, is determined by:

$$-D\nabla^2\bar{F}(\epsilon) = I^{ee}\{F\}, \tag{61}$$

where the r.h.s. is given by

$$I^{ee}\{\bar{F}\} = D\int\frac{d\omega}{\pi}[\langle\nabla k^1(\omega)\nabla k^1(-\omega)\rangle(\bar{F}(\epsilon)-\bar{F}(\epsilon-\omega)) + (\langle\nabla k^1(\omega)\nabla k^2(-\omega)\rangle - \langle\nabla k^2(\omega)\nabla k^1(-\omega)\rangle)(\bar{F}(\epsilon)\bar{F}(\epsilon-\omega)-1)]. \tag{62}$$

Since $(\bar{\Lambda})$ is not a genuine saddle point of the action there is coupling between the fields $\tilde{W}$ and ϕ, (∇k) in the quadratic part of the action. However the coupling constant between those fields is proportional to the gradient of the distribution (cf. eq.(84)). Therefore this term can be treated as a small perturbation.

Taking variation of the action with respect to w, $\bar{w}$, we obtain the following gauge, determining $k[\phi]$:

$$\begin{aligned} D\nabla^2 k_2 - \phi_2 - i\omega k_2 &= 0 \\ D\nabla^2 k_1 + \phi_1 + i\omega k_1 &= 2\mathrm{B}[\omega,x]\nabla^2 k_2, \end{aligned} \tag{63}$$

where

$$\mathrm{B}[\omega,x] = \frac{1}{2\omega}\int d\epsilon[1-\bar{F}(\epsilon,x)\bar{F}(\epsilon-\omega,x)]. \tag{64}$$

Though we have failed to find the true saddle point the linear part of the action expanded around $(\bar{\Lambda})$ is zero. It is remarkable to notice that under conditions (63) eq.(61) becomes a quantum kinetic equation [24] with the collision integral being $I^{ee}\{F\}$. Coming back to our calculations we note that the correlation function of current fluctuations is a gauged invariant quantity (does not depend on the position of the Fermi level). This means that momenta $q \le \sqrt{\omega/D}$ do not contribute to such a quantity [22]. In this case the Coulomb propagator is universal, i.e. does not depend on the electron charge. The fact that we address gauge invariant quantities allows us to represent the generating functional Z in terms of the fields Q and ∇k (rather than Q and ϕ), as in Ref [23].

$$\langle Z\rangle = \int\mathcal{D}\nabla K\,\exp\left(-i\nu D\mathrm{Tr}\{\nabla K^T\mathcal{D}^{-1}\nabla K\}\right)\int\mathcal{D}\tilde{Q}\,\exp\left(\sum_{l=0}^{2} iS_l[\tilde{Q},\nabla K]\right). \tag{65}$$

Here we define

$$\mathcal{D}^{-1} = \begin{pmatrix} 0 & -D\nabla_x^2 + i\omega\delta_{x,x'} \\ -D\nabla_x^2 - i\omega\delta_{x,x'} & -2i\omega\delta_{x,x'}B_\omega(x) \end{pmatrix}, \tag{66}$$

where the expansion $S = S^0 + S^1 + S^2$, is in powers of ∇K; the $l-th$ power ($l = 0, 1, 2$) is given by

$$iS^0[\tilde{Q}] = -\frac{\pi\nu}{4}\left[D\mathrm{Tr}\{\nabla\tilde{Q}\}^2 + 4i\,\mathrm{Tr}\{\epsilon\tilde{Q}\}\right], \tag{67}$$

$$iS^1[\tilde{Q},\nabla K] = -i\pi\nu\left[D\mathrm{Tr}\{\nabla k_\alpha\gamma^\alpha\tilde{Q}\nabla\tilde{Q}\}+\mathrm{Tr}\{(\phi_\alpha + i\omega k_\alpha)\gamma^\alpha\tilde{Q}\}\right], \tag{68}$$

$$iS^2[\tilde{Q},\nabla K]=\frac{\pi\nu D}{2}\left[\mathrm{Tr}\{\nabla k_\alpha\gamma^\alpha\tilde{Q}\nabla k_\beta\gamma^\beta\tilde{Q}\}-\mathrm{Tr}\{\nabla k_\alpha\gamma^\alpha\tilde{\Lambda}\nabla k_\beta\gamma^\beta\tilde{\Lambda}\}\right]. \tag{69}$$

From eq.(66) we obtain the gauge field correlation function

$$\langle\nabla k_\alpha(x,\omega)\nabla k_\beta(x',-\omega)\rangle = \frac{i}{D}Y_{\alpha,\beta}(\omega,x,x'), \tag{70}$$

where

$$Y(\omega,x,x')=\begin{bmatrix} -2i\pi\nu\omega\int dx_1 \mathrm{D}[-\omega,x,x_1]\mathrm{B}[\omega,x_1]\mathrm{D}[\omega,x_1,x'] & \mathrm{D}[-\omega,x,x'] \\ \mathrm{D}[\omega,x,x'] & 0 \end{bmatrix}, \tag{71}$$

Using eqs.(70,71) we rewrite eq.(62) for the quasi-one-dimensional wire as:

$$D\nabla^2\bar{F}(\epsilon) = I^{ee}\{\epsilon,x\}\,, \text{where} \tag{72}$$

$$\begin{aligned} I^{ee}(\epsilon,x) = \frac{i\pi}{2}\int d\omega[-2i\omega\pi\nu D[x,x_1,-\omega]B[\omega,x_1] \\ \mathrm{D}[x_1,x,\omega](\bar{F}(\epsilon)-\bar{F}(\epsilon-\omega)) + \\ (\mathrm{D}[x,x,\omega]-\mathrm{D}[x,x,-\omega])(1-\bar{F}(\epsilon)\bar{F}(\epsilon+\omega))]. \end{aligned} \tag{73}$$

The total number of particles and the total energy of the electrons are both preserved during electron-electron and elastic electron-impurity scattering. The collision integral, eq.(73), satisfies then

$$\int_{-\infty}^{\infty} I^{ee}(\epsilon,x)d\epsilon = 0, \tag{74}$$

$$\int_{-\infty}^{\infty} \epsilon I^{ee}(\epsilon,x)d\epsilon = 0. \tag{75}$$

We now consider the limit $l_{ee} \ll L$. The solution of eq.(61) assumes then the form of a quasi-equilibrium single-particle distribution function

$$\bar{F}(\epsilon,x) = \mathrm{th}\left(\frac{\epsilon - e\phi(x)}{2T(x)}\right). \tag{76}$$

Here ϵ is the total energy of the electron, and $e\phi$ is the electrostatic potential and $T(x)$ is the effective local temperature of the electron gas. To find the electrostatic potential ϕ we employ eq. (74). To facilitate our calculations we further assume that conductance band is symmetric about the Fermi energy and that the spectral density of single-electron energy levels is constant. Integration over the energy, eq.(74) yields

$$\partial_x{}^2\phi(x) = 0. \tag{77}$$

Solving eq.(77) under the condition that the voltage difference at the edges of a constriction is V, we find

$$e\phi(x) = eV\left(\frac{x}{L} - \frac{1}{2}\right) + \bar{\mu}. \tag{78}$$

Multiplying eq. (72) by energy and and employing eq.(75) we obtain an equation

$$\partial_x^2\left(\frac{\pi^2}{6}(kT(x))^2 + \frac{1}{2}(e\phi(x))^2\right) = 0. \tag{79}$$

The boundary condition of eq.(79) is determined by the temperature of the electrons in the reservoirs. Combining eqs.((78) and (79)) we find the electron temperature in two opposite limits:

$$T(x) = \begin{cases} \frac{\sqrt{3}eV}{\pi L}\sqrt{x(L-x)} & eV \gg T\,, \\ T, & eV \ll T\,. \end{cases} \tag{80}$$

Eqs.((76), (78) and (80) determine the function $\bar{F}$ uniquely. We now replace the right-corner element of the matrix $\tilde{\Lambda}$ (i.e. $\tilde{F}[\phi]$, cf. eq.(60)) by its average value $\bar{F}$.

To calculate S_3 under conditions of strong electron-electron scattering (eq.(45)) one needs to replace the operators $\hat{\mathbf{I}}^D$ and $\hat{\mathrm{M}}$ in eq. (30) by their gauged values

$$\mathrm{S}_3(t_1-t_2, t_2-t_3) = \frac{e^3(\pi\hbar\nu D)^2}{8}\Big\langle\frac{1}{2}\hat{\tilde{\mathrm{M}}}(x_1,t_1)\hat{\tilde{\mathbf{I}}}^D(x_2,x_3,t_2,t_3) + (x_1,t_1 \leftrightarrow x_3,t_3) +$$
$$(x_1,t_1 \leftrightarrow x_2,t_2) + \frac{\pi\hbar\nu D}{8}\hat{\tilde{\mathrm{M}}}(x_1,t_1)\hat{\tilde{\mathrm{M}}}(x_2,t_2)\hat{\tilde{\mathrm{M}}}(x_3,t_3)\Big\rangle_{\nabla k,\tilde{Q}}\,, \tag{81}$$

where the averaging is taken over the entire action S and the Gaussian weight function for ∇K, as in eq. (51). Here we define (cf. eqs.(31), (32) with eqs. (82),(83))

$$\hat{\tilde{\mathbf{I}}}^D(x,x',t,t') = \mathrm{Tr}\Big\{\tilde{Q}_{x,t,t'}\gamma_2\tilde{Q}_{x',t',t}\gamma_2 - \delta_{t,t'}\gamma_1\Big\}\delta_{x,x'}\,, \tag{82}$$

$$\hat{\tilde{\mathrm{M}}}(x,t) = \mathrm{Tr}\Big\{\int dt_1\Big([\tilde{Q}_{x,t,t_1};\partial_x]\,\tilde{Q}_{x,t_1,t}\Big)\gamma_2\Big\}\,, \tag{83}$$

where the "long derivative", ∂_x, is presented in eq.(56). In order to actually perform the functional integration over the matrix field $\tilde{Q}$ we use the parameterization of eq. (59). We need to find the Gaussian fluctuations around the saddle point of the action (67,68,69). Though we did not find the exact saddle point, the expansion of Q around $\bar{\Lambda}$ works satisfactorily. The coupling between the fields ∇k and W which appears already in the Gaussian (quadratic) part is small, since it is proportional to the gradient of the distribution function:

$$iS_1^1 = -2i\pi g \mathrm{Tr}\Big\{\bar{w}_{x,\epsilon,\epsilon'}[\nabla k_{1x,\epsilon'-\epsilon}\nabla \bar{F}_{x,\epsilon} - \nabla \bar{F}_{x,\epsilon'}\nabla k_{1x,\epsilon'-\epsilon} + \nabla \bar{F}_{x,\epsilon'}\nabla k_{2x,\epsilon'\epsilon}\bar{F}_{x,\epsilon} + \bar{F}_{x,\epsilon'}\nabla k_{2x,\epsilon'-\epsilon}\nabla \bar{F}_{x,\epsilon}]\Big\}. \tag{84}$$

Here the upper index refers to the power of the ∇k fields; the lower refers to the power of $w, \bar{w}$ fields in the expansion. Considered as a small perturbation, iS_1^1 does not affect the results.

The more dramatic effect on the correlation function arises from the non-Gaussian part of the action, eqs. (68,69) (by this we mean non-Gaussian terms in either $w, \bar{w}$ or ∇K). After integrating over the interaction an additional contribution to the Gaussian part (proportional to $w\bar{w}$) of the action arises. To find the effective action $iS^{\text{eff}}[W]$ we average over the interaction [16]. One notes that

$$iS_0^1 = iS_0^2 = 0\,, \tag{85}$$

(where, again, S_0^1 refers to the component of the action, eq.(68), that has zero power of the $w, \bar{w}$ fields and one power of the ∇k field). In addition, due to the choice of the gauge, eq.(63), and the condition ($l_{\text{ee}} \ll L$), the averaging over ∇k does not generate terms linear in $w, \bar{w}$ in the effective action:

$$\langle iS_1^2 \rangle_{\nabla k} = -2i\pi\nu \int \frac{d\epsilon}{2\pi} \bar{w}_{\epsilon,\epsilon} I_{\text{ee}}[F] = 0\,. \tag{86}$$

Combining eqs. (85 and 86) we find that the effective action acquires an additional contribution:

$$\Big\langle \exp\left(iS^1 + iS^2\right) \Big\rangle_{\nabla k} \simeq \exp\left(\langle iS_2^2 \rangle + \frac{1}{2}\langle iS_1^2 iS_1^2 \rangle\right) \tag{87}$$

The general form of the effective action is rather complicated, however for the low frequency noise only diagonal part of the action matters:

$$iS_2^{\text{eff}}[w, \bar{w}] = \frac{\pi\nu}{2} \mathrm{Tr}\Big\{\bar{w}_{x,\epsilon,\epsilon}\Big[-D\nabla^2 + \hat{\mathcal{I}}^{ee}\Big] w_{x,\epsilon,\epsilon} - \bar{w}_{x,\epsilon,\epsilon} D\nabla \bar{F}_{x,\epsilon}\nabla \bar{F}_{x,\epsilon}\bar{w}_{x,\epsilon,\epsilon}\Big\}. \tag{88}$$

Here the operator

$$\hat{\mathcal{I}}^{ee} w_{x,\epsilon,\epsilon} \equiv \int d\omega [Y_{11}(\omega)[w_{\epsilon,\epsilon} - w_{\epsilon-\omega,\epsilon-\omega}] +$$

$$(Y_{12}(\omega) - Y_{21}(\omega))\left[F_\epsilon w_{\epsilon-\omega,\epsilon-\omega} + F_{\epsilon-\omega} w_{\epsilon,\epsilon}\right] +$$
$$\int d\bar{\epsilon}\frac{1}{2\omega}\left(F_\epsilon - F_{\epsilon-\omega}\right)\left(Y_{12}(\omega) - Y_{21}(\omega)\right)\left(F_{\bar{\epsilon}+\omega} + F_{\bar{\epsilon}-\omega}\right) w_{\bar{\epsilon},\bar{\epsilon}} \qquad (89)$$

is a *linearized* collision integral, i.e. a variation of the collision integral (73) with respect to the distribution function. Substituting eqs.(82 and 83) into eq.(81) and calculating the Gaussian integrals with the action (88), we find

$$\mathcal{S}_3(\omega_1 = 0, \omega_2 = 0) = \frac{3e^3 \mathcal{A}\pi g^2}{\hbar L^3} \int_0^L dx_1 dx_2$$
$$\int_{-\infty}^{\infty} d\epsilon_1 d\epsilon_2 \bar{F}(\epsilon_1, x_1) \mathcal{D}[x_1, \epsilon_1; x_2, \epsilon_2] \nabla\left(\bar{F}^2(\epsilon_2, x_2)\right), \qquad (90)$$

where the "inelastic diffusion" propagator, $\mathcal{D}$, is the kernel the equation

$$[-D\nabla^2 - \hat{\mathcal{I}}^{ee}]\mathcal{D}[x_1, \epsilon_1; x_2, \epsilon_2] = \frac{1}{\pi\nu}\delta(\epsilon_1 - \epsilon_2)\delta(x_1 - x_2). \qquad (91)$$

For weak electron-electron scattering the collision integral is small, yielding the standard propagator of the diffusion equation, D (cf. eq. 39). In the presence of strong electron-electron scattering the collision integral dominates eq.(91). In this limit we evaluate the leading asymptotic behavior for ($l_{\text{in}}/L \ll 1$). On scales longer than the inelastic mean free path the distribution function has a quasi-equilibrium form

$$f(\epsilon, x) = f_0\left(\frac{\epsilon - \mu(x) - \delta\mu(x)}{T(x) + \delta T(x)}\right), \qquad (92)$$

where

$$f_0(x) = \frac{1}{1 + \exp(x)}. \qquad (93)$$

The values of the local temperature and electro-chemical potential can fluctuate. To find the correlations of these fluctuations we consider the equation (with the same Kernel as in eq.(91))

$$[-D\nabla^2 - \hat{\mathcal{I}}^{ee}]\delta f(\epsilon, x) = \delta J(\epsilon, x). \qquad (94)$$

Integrating equation (94) with respect to energy and using the particle-conservation property of the collision integral (eqs.(74)) we find:

$$\mu(x) = \int dx' \mathrm{D}[x, x'] \int_{-\infty}^{\infty} d\epsilon J(\epsilon, x'). \qquad (95)$$

Using energy conservation (eq. 75) we find

$$\delta T(x) = \frac{3}{\pi^2}\frac{1}{T(x)} \int dx' \mathrm{D}[x, x'] \int d\epsilon(\epsilon - \mu(x))\delta J(\epsilon, x'). \qquad (96)$$

Combining eqs. (95) and (96) we find:

$$\mathcal{D}[x_1,\epsilon_1;x_2,\epsilon_2]=\left(\frac{\partial}{\partial\epsilon}f_0\left(\frac{\epsilon_1-\mu(x_1)}{T(x_1)}\right)\right)\bullet$$

$$\mathrm{D}[x_1,x_2]\left(-1-\frac{3}{\pi^2}\frac{\epsilon_1-\mu(x_1)\epsilon_2-\mu(x_1)}{T(x_1)\quad T(x_1)}\right). \tag{97}$$

Evaluating $\mathcal{S}_3$ explicitly we find that the third order current cumulant is

$$\mathcal{S}_3(\omega_1=0,\omega_2=0)=\frac{36e^3Ag^2eV}{L^4\pi}\int_0^L dx_1dx_2$$

$$\mathrm{D}[x_1,x_2]\left[\frac{T(x_1)}{T(x_2)}+(x_1-x_2)\frac{1}{T(x_2)}\frac{\partial}{\partial x_1}T(x_1)\right]. \tag{98}$$

At high temperatures (cf. eq.80) one obtains

$$\mathcal{S}_3(\omega_1=0,\omega_2=0)=\frac{3}{\pi^2}e^2I\,, \tag{99}$$

while at low temperatures

$$\mathcal{S}_3(\omega_1=0,\omega_2=0)=\left(\frac{8}{\pi^2}-\frac{9}{16}\right)e^2I. \tag{100}$$

Our analysis was performed for a simple rectangular constriction. However, our results hold for any shape of the constriction, provided it is quasi-one dimensional (we have considered a single transversal mode only).

3. COMPARISON WITH THE KINETIC THEORY OF FLUCTUATIONS

As was mentioned in Section 1.3 the applications of the kinetic equation in the study of high order cumulants is not straight forward. To compare our microscopic quantum mechanical calculation with the semi-classical Boltzmann-Langevin scheme we consider the diagrams for the pair and third order correlation function, depicted in Fig.(3).

The diagram in Fig. (3-b) corresponds exactly to the results obtained above in the framework of the σ-model formalism. Indeed, evaluating this diagrams, we recover the results (41) and (90). Comparing Fig. (3-b) with diagram (3-a) (the latter determines the correlator $\langle\delta J\delta J\rangle$ of Langevin sources), we conclude that the third cumulant corresponding to Fig. (3-b) can be expressed in the form

$$\langle\delta J(1)\delta J(2)\delta J(3)\rangle=\hat{p}\int d4d4'\frac{\delta\langle\delta J(1)\delta J(2)\rangle}{\delta f(4)}D(4,4')\langle\delta J(4')\delta J(3)\rangle\,. \tag{101}$$

Indeed, the block on the right hand side of the diagram corresponds to the pair correlation function of random fluxes, while the left part can be obtained by functional differentiation of the diagram (3-a) with respect to the function $F=1-2f$.

This justifies the regression scheme proposed recently by Nagaev [26]. Diagram (3-a) for the pair correlation function is local in space; the diagram for the third order correlator is not. Despite this non-locality the use of the Langevin equation (with non-local random flux) remains convenient; in that case the correlation function of the random fluxes needs to be calculated from first principles. This is somewhat analogous to the description of mesoscopic fluctuations through the Langevin equation, proposed by Spivak and Zyuzin [25].

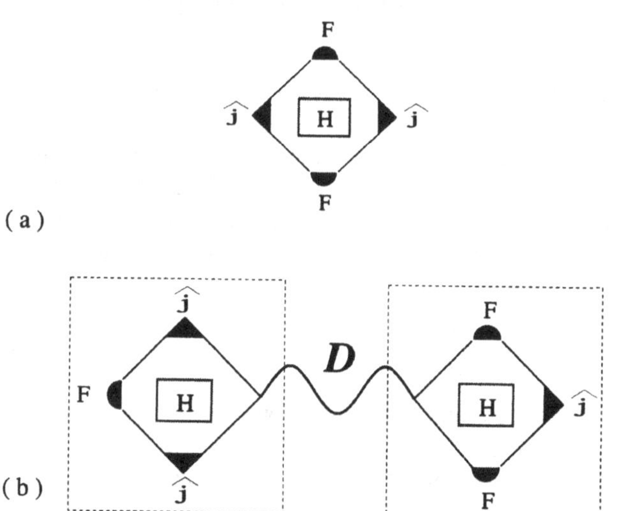

Figure 3. The second (a) and the third (b) current correlation functions.
The vector vertices $\hat{j}$ represent the current operators, while collisions with static disorder correspond to the Hikami box (H). $F = 1 - 2f$, where f is the single-electron distribution function. D is the diffuson.

Following Ref[17] the revisited version of the kinetic theory of fluctuations has been proposed by Nagaev [26]. He has noticed that for a diffusive system there exists a regression scheme of high order cumulants, expressed in terms of pair correlators [27]. However, a truly classical theory addressing high moments of noise needs to be taken on the same footing as the Boltzmann-Langevin theory, i.e. without appealing to quantum mechanical diagrammatics. Close inspection of the diagrams depicted in Fig. 3 and the dimensionless parameter by which corrections are small implies that the regression recipe (eq. 101) has the same range of applicability as the Boltzmann-Langevin scheme itself.

As a simple example of a classical problem for which the regression procedure can be applied, we consider the following simple model[6]. A heavy molecule of mass M and cross-section $\mathcal{A}$ is embedded in a gas of light classical particles. The gas consists of N particles of mass $m \ll M$, in thermal equilibrium with temperature T. It is enclosed in a narrow tube of volume V. The fluctuating velocity u of the molecule here is the analogue of the fluctuating electron distribution function δf. The collisions of the molecule with the light particles is the counterpart of the electron's scattering on the random disorder. The motion of the molecule is governed by Newton law

$$M\frac{du}{dt} = F(t). \tag{102}$$

Here u is the velocity and F is the force acting on the molecule. The latter can be calculated from

$$F(t)=2mA\left[\int_0^{\infty} dvv^2 P_1(v+u(t),t) - \int_{-\infty}^{0} dvv^2 P_2(v+u(t),t)\right], \tag{103}$$

where P_1 and P_2 are the fluctuating distribution functions of particles on the left and on the right sides of the system respectively. Since the time between two consecutive collisions is much shorter than the relaxation time of the molecule (τ_m) one can average over fast collisions while considering the relaxation dynamics of the heavy molecule:

$$\frac{d\bar{u}}{dt} = -\frac{1}{\tau_m}\bar{u}. \tag{104}$$

To find the various velocity correlation functions, one needs to know the corresponding correlation functions of the random forces. Using the fact that the equilibrium fluctuations of the distribution function in a Boltzmann gas are Poissonian ([7], Section 114) and neglecting the small velocity u relatively to v, one finds the pair correlation function

$$\langle \delta F(t)\delta F(t')\rangle = \frac{16}{\sqrt{2\pi}}\frac{AN}{V} m^2 \langle v^2\rangle^{3/2}\delta(t-t'). \tag{105}$$

Eq. (105) yields the correct value for the thermal fluctuations $\langle u^2\rangle = T/M$. To find the higher order correlations of the random force one *has* to take into account the dependence of the force on the velocity of the molecule. We expand eq.(103) up to second order in u. After averaging over all possible pair correlations (such as $\langle u\delta P\rangle$ and $\langle \delta P\delta P\rangle$) we obtain

$$\langle \delta F(t_1)\delta F(t_2)\delta F(t_3)\rangle = \frac{512}{\sqrt{2\pi}}\left(\frac{AN}{V}\right)^2 m^3 \langle u^2\rangle\langle v^2\rangle \hat{p}\Big\{\delta(t_1-t_2)\theta(t_2-t_3)\Big\}. \tag{106}$$

Here $\hat{p}$ denotes all permutation over (t_1, t_2, t_3). The triple force correlator was reduced to pair correlators only. Bearing in mind the analogy between u and δf on one hand, and the scattering of the electrons on the disorder and the scattering of the molecule by light particles on the other hand, we note that the reduction of the third order cumulant of the random forces to the pair correlation functions is similar in both cases.

We finally come back to the question of whether it is possible to calculate high order cumulants employing the classical kinetic equation (Boltzmann-Langevin) rather than resorting to the diagrammatic reduction scheme depicted above. To be able to answer this question we first note that the applicability of the kinetic theory requires that both $1/g, \hbar/\tau_{\text{corr}} k_B T \ll 1$. Here τ_{cor} is the correlation time for the current signal and g is the dimensionless conductance. For non-interacting electrons Onsager relation and Drude formula yield $\tau_{\text{cor}} = \tau$, where τ is the transport time. It follows that the condition

$$z \equiv \frac{\hbar}{g\tau_{\text{cor}} k_B T} \ll 1 \tag{107}$$

must be satisfied. The reason for theses inequalities are first, that g needs to be large in order for quantum interference effects to be negligible. Secondly, the value $\hbar/k_B T\tau$ needs also to be small; the transport equation is applicable for times longer than the duration time of an individual collision $\delta\tau$, ($\tau \gg \delta\tau$). For a degenerate electron gas the uncertainty relation requires that $\delta\tau \geq \hbar/k_B T$ [28, 29].

We now evaluate the relative magnitude of, e.g., the fourth and the second cumulants. We consider the ratio

$$\tilde{z} \equiv \frac{<< I^4(0) >>}{<< I^2(0) >>^2}, \tag{108}$$

where $<< \ >>$ denotes the irreducible part of the correlator. At equilibrium

$$\tilde{z} \simeq z. \tag{109}$$

As we see, for the kinetic theory to be valid, the value of the parameter z needs to be small. But this is exactly the parameter ($\tilde{z}$) by which the high order correlation functions are smaller than the lower ones (cf. eq.107). In other words, the evaluation of high order cumulants goes *beyond* the validity of the standard Boltzmann-Langevin equation. This is why we have to resort to the diagrammatic approach: either to justify the reduction of high order cumulants to pair correlators, or, alternatively, to introduce non-local noise correlators in the Boltzmann-Langevin equation.

4. COUNTING STATISTICS

So far we have studied the second and third order cumulants. In the present section we discuss the whole distribution function of the low frequency electron current (so-called counting statistics). The zero temperature limit had been studied by Levitov et. al. [18]. The full temperature regime was addressed by Nazarov[31]. Here we present a different derivation based on the σ-model approach.

Being a stochastic process, the charge transmission can be characterized by the probability distribution function (PDF) $P_{\bar{t}}(n)$ of the probability for n electrons to pass through the constriction within the time window $\bar{t}$. In practice it is more convenient to work with the Fourier transform of the PDF, the characteristic function

$$\kappa_{\bar{t}}[\lambda] = \sum_n e^{i\lambda n} P_{\bar{t}}(n). \tag{110}$$

Below we evaluate $\kappa_{\bar{t}}[\lambda]$ for the case of elastic scattering. By expanding the logarithm of characteristic function over its argument we can find the cumulants (irreducible correlation functions) of a transmitted charge

$$\ln(\kappa_{\bar{t}}[\lambda]) = \sum_k \frac{(i\lambda)^k}{k!} S_k. \tag{111}$$

For the problem of diffusive junction the disorder average characteristic function can be represented as:

$$\bar{\kappa}_{\bar{t}}[\lambda] = \int \mathcal{D}Q \exp(iS[Q, \mathbf{a}]), \tag{112}$$

where the external source is given by

$$\mathbf{a}_2(t) = \begin{cases} \frac{\lambda}{2} \ , & 0 < t < \bar{t} \\ 0 \ , & \text{otherwise} \end{cases} . \tag{113}$$

The applied bias enters the problem through boundary conditions on Q at the edges of the constriction:

$$\Lambda(0,\epsilon) = \begin{pmatrix} 1 & 2F\left(\epsilon + \frac{eV}{2}\right) \\ 0 & -1 \end{pmatrix}, \Lambda(L,\epsilon) = \begin{pmatrix} 1 & 2F\left(\epsilon - \frac{eV}{2}\right) \\ 0 & -1 \end{pmatrix}, \tag{114}$$

and F is defined by eq.(35).

Inasmuch as we are not interested in spatial correlations, the external source term is a function of time only. Therefore, by performing the transformation

$$Q(x,t,t') = e^{-ix a_2(t)\gamma_2} \tilde{Q}(x,t,t') e^{ix a_2(t')\gamma_2}. \tag{115}$$

one gets:

$$iS[\tilde{Q}, \mathbf{a}] = -\frac{\pi\nu}{4\tau} \mathrm{Tr}\Big\{ D\left(\nabla\tilde{Q}\right)^2 - \left(\frac{\partial}{\partial t} + \frac{\partial}{\partial t'}\right)\tilde{Q}(t,t') +$$
$$i\left(\frac{\partial}{\partial t} a_2(t)\right)\gamma_2 \tilde{Q}(t,t') - \tilde{Q}(t,t') i \frac{\partial}{\partial t'} a_2(t')\gamma_2 \Big\}, \tag{116}$$

and the boundary conditions change correspondingly:

$$\tilde{Q}(0,t,t') = \Lambda(0,t-t'),$$
$$\tilde{Q}(L,t,t') = e^{-ia_2(t)\gamma_2}\Lambda(L,t-t')e^{ia_2(t')\gamma_2}, \tag{117}$$

with

$$\Lambda(t-t') = \int \frac{d\epsilon}{2\pi} e^{i\epsilon(t-t')}\Lambda(\epsilon). \tag{118}$$

We use the saddle point approximation to calculate the characteristic function, Eq. (112). In the presence of an external potential the minimum of the action, eq. (116), satisfies

$$D\nabla(\tilde{Q}\nabla\tilde{Q}) - \left(\frac{\partial}{\partial t} + \frac{\partial}{\partial t'}\right)\tilde{Q} +$$
$$i\frac{\partial}{\partial t}a_2(t)\gamma_2\tilde{Q}(t,t') - \tilde{Q}(t,t')i\frac{\partial}{\partial t'}a_2(t')\gamma_2 = 0. \tag{119}$$

Let us define a parameter that can roughly be regarded as "the number of attempts per channel", $M = \max\{eV, T\}\bar{t}/\hbar$. We will focus on the case where the time window is much larger than the Thouless time $\bar{t} \gg t_{Thouless}$ and the conductance $g \gg M \gg 1$. Inside the region $0 < t, t' < \bar{t}$ eq. (119) becomes (up to the corrections $O(t_{\text{Thouless}}/\bar{t})$)

$$D\nabla(\tilde{Q}\nabla\tilde{Q}) = 0. \tag{120}$$

This can be represented in the form resembling a current conservation law

$$D\nabla J = 0, \tag{121}$$

where the current is defined as

$$J \equiv \tilde{Q}\nabla\tilde{Q}. \tag{122}$$

The solution of Eq. (121), $\tilde{Q}_{sp}(x)$ can be written as

$$\tilde{Q}_{sp} = \tilde{Q}(0)\exp(Jx), \tag{123}$$

and from the boundary conditions (eq. 117) it follows that

$$J = \ln(\tilde{Q}(0)\tilde{Q}(L)). \tag{124}$$

One may show that the anticomutator of the matrix Q with J vanishes

$$\{\tilde{Q}(0), J\}_+ = 0. \tag{125}$$

Based on this property, we can show that the following statements do hold:
(1) The square of the matrix $\tilde{Q}$ is still equal to unity

$$\tilde{Q}^2_{sp}(x) = 1, \tag{126}$$

(2) The determinant of the matrix $\tilde{Q}$ satisfies

$$\mathrm{Det}[\tilde{Q}_{sp}(x)] = -1, \tag{127}$$

(3) The trace of the matrix $\tilde{Q}$ is equal to zero

$$\mathrm{Tr}\{\tilde{Q}_{sp}(x)\} = 0, \tag{128}$$

(4) The trace of the matrix J is zero

$$\mathrm{Tr}\{J\} = 0, \tag{129}$$

(5) The square of the matrix J is proportional to the unit matrix

$$J^2 = \delta[\epsilon, \lambda] I, \tag{130}$$

where $\delta[\epsilon, \lambda]$ is given by (there is small discrepancy with the result obtained in Ref.[31])

$$\delta[\epsilon, \lambda] = \ln\Big[(2f_1 - 1)(2f_2 - 1) + 2f_1(1 - f_2)e^{-i\lambda} + 2f_2(1 - f_1)e^{i\lambda} +$$
$$2\sqrt{(e^{i\lambda/2} - e^{-i\lambda/2})(1 - f_1 + f_1 e^{-i\lambda})(1 - f_2 + f_2 e^{i\lambda})(f_2(1 - f_1)e^{i\lambda/2} - f_1(1 - f_2)e^{-i\lambda/2})}\Big]. \tag{131}$$

Using these properties we find the disorder averaged counting statistics

$$\bar{\kappa}[\lambda] = \exp\left(\frac{G\bar{t}}{\hbar}\int d\epsilon \delta^2[\epsilon, \lambda]\right). \tag{132}$$

The zero temperature limit coincides with the results derived previously by Lee et. al.[18].

At finite temperatures we find:

$$S_2 = 2GT$$
$$S_3 = \frac{e^2}{3} I$$
$$S_4 = \frac{2}{3} e^2 GT. \tag{133}$$

It is worthwhile to note that the value of S_4 agrees with the one obtained in Ref.[26]. As we see the counting statistics of the current in a disordered wire is not Gaussian. Remarkably, *all the information* contained in the counting statistics can be extracted from the pair correlation function (of the distribution function), eq.(17).

Finally we would like to discuss the role of *inelastic electron-phonon scattering*. As has been already realized [1], such an interaction suppresses shot noise. Moreover, based on our approach, one can show that in the limit of $l_{\mathrm{e-ph}} \ll L$ (macroscopic conductor) the current fluctuations are Gaussian (to leading order in $L/l_{\mathrm{e-ph}}$). To show this, we repeat our analysis concerning $\mathcal{S}_3$. We find that eq. (90) still holds, but the distribution function and the inelastic diffusion propagator need to be calculated taking electron-phonon collisions into account. In the presence of both electron-electron and electron-phonon interaction one obtains

$$[D\nabla^2 \bar{F} + I^{ee} + I^{e-ph}]\bar{F} = 0\,, \tag{134}$$

(cf. eq.61) where $\hat{\mathcal{I}}^{e-ph}$ is the linearized electron-phonon collision integral. The inelastic diffuson (cf. eq.91) is now determined by

$$[-D\nabla^2 - \hat{\mathcal{I}}^{ee} - \hat{\mathcal{I}}^{e-ph}]\mathcal{D}[x_1, \epsilon_1; x_2, \epsilon_2] = \frac{1}{\pi\nu}\delta(\epsilon_1 - \epsilon_2)\delta(x_1 - x_2). \tag{135}$$

In the limit $L/l_{\mathrm{e-ph}} \gg 1$ fluctuations of the chemical potential is the only long-range propagating mode in the problem (no fluctuations of the $k_B T$ along the system). Solving eqs.(134, 135 and 90) we find that $\mathcal{S}_3$ vanishes as $(l_{e-ph}/L)^2$. A conductor longer than the electron-phon... length ($l_{\mathrm{e-ph}}$) can be viewed as a number ($L/l_{\mathrm{e-ph}}$) of resistors connected in series. The current fluctuations of a single resistor are given by eq.(132) and would render the corresponding voltage fluctuations. Since the fluctuations at different resistors are *uncorrelated* (local fluctuations in the chemical potential do not change the resistance significantly) the large number of the resistors results in *Gaussian* current fluctuations. By contrast, temperature fluctuations (in the case of solely electron-electron interactions) have long distance correlations, and give rise to non-Gaussian fluctuations.

Acknowledgements

The authors wish to thank M. Reznikov and Y. Levinson for inspiring discussion and A. Kamenev for his critical comments in the earlier stages of this work. This work was supported by GIF foundation, the Minerva Foundation, the European RTN grant on spintronics, by the SFB195 der Deutschen Forschungsgemeinschaft, and by RFBR gr. 02-02-17688.

References

1. Ya. M. Blanter, M. Büttiker Physics Reports **336**, 2, (2000).
2. A.Ya. Shulman Sh.M. Kogan Sov. Phys. JETP **29** , 3 (1969).

3. S.V. Gantsevich, V.L. Gurevich and R. Katilius Sov. Phys. JETP **30**, 276 (1970).
4. L.S. Levitov and G.B. Lesovik Sov. Phys. JETP Lett. **58**, 230, (1993).
5. L.S. Levitov and M. Reznikov "Electron shot noise beyond the second moment" cond-mat/0111057.
6. M. Reznikov, private comm.
7. L.D. Landau and E.M. Lifshitz *Statistical Mechanics*, (Pergamonn Press, Oxford, 1980), Pt. 1.
8. L.S. Levitov, H. Lee and G.B. Lesovik Journal of Math. Phys. **37**, 4845 (1996).
9. I. Klich "Full Counting Statistics: An elementary derivation of Levitov's formula" cond-mat/0209642, to appear in these proceedings.
10. L.V. Keldysh JETP **20**, 1018 (1965).
11. We note that even physical observables corresponding to the second current correlator may be associated with different current correlation functions, such as symmetrised, antisymmetrzied or combinations thereof. For discussion of this point see G.B. Lesovik and R. Loosen, Pism'ma Zh. E'ksp. Teor. Fiz. **65**, 280 (1997) [JETP Lett. **65**, 295, (1997)]; U. Gavish, Y. Levinson and Y. Imry Phys. Rev. B. **62**, R10637 (2000).
12. R de-Picciotto, M. Reznikov, M. Heiblum, V. Umansky, G. Bunin and D. Mahalu Nature **389**, 6647 (1997).
13. L. Saminadayar, D. C. Glattli, Y. Jin and B. Etienne Phys.Rev. Lett. **79**, 2526 (1997).
14. E.M. Lifshitz and L.P. Pitaevskii *Physical Kinetics* (Pergamonn Press, Oxford, 1981).
15. M. Lax Rev. Mod. Phys. **32**, 25 (1960).
16. D.B. Gutman and Yuval Gefen "Shot noise at high temperatures", cond-mat/0201007
17. D.B. Gutman and Y. Gefen Phys. Rev. B. **64**, 205317 (2001).
18. H. Lee, L.S. Levitov and A.Yu. Yakovets Phys. Rev. B **51**, 4079, (1995).
19. for the discussion of application of random-matrix theory to shot noise see M. J. M. de Jong, C. W. J. Beenakker in "Mesoscopic Electron Transport," edited by L.L. Sohn, L.P. Kouwenhoven, and G. Schoen, NATO ASI Series Vol. 345 (Kluwer Academic Publishers, Dordrecht, 1997).
20. K.E. Nagaev Phys. Rev. B **52**, 4740, (1995).
21. V.I. Kozub and A.M. Rudin Phys. Rev. B, **52**, 7853, (1995).
22. A.M. Finkel'stein Physica B **197**, 636 (1994).
23. A. Kamenev, A. Andreev Phys. Rev. B **60**, 2218 (1999).
24. A. Schmid Z. Physik **271** 251 (1974); B.L. Altshuler and A.G. Aronov JETP Lett. **30** 483 (1979).
25. A.Ju. Zyuzin, B.Z. Spivak, Zh. Eksp. Theor. Fiz. **93**, 994 (1987) (Sov. Phys. JETP **66**, 560 (1987)).
26. K. E. Nagaev "Cascade Boltzmann - Langevin approach to higher-order current correlations in diffusive metal contacts", cond-mat/0203503.
27. Recently the regression procedure has been applied for a closed disordered cavity: K.E. Nagaev, P. Samuelsson, S. Pilgram "Cascade approach to current fluctuations in a chaotic cavity", cond-mat/0208147
28. R. Peierls Helv. Phys. Acta **7**, Suppl. 2, 24 (1934).
29. As a side remark, we note that if the condition $\hbar/k_BT\tau \ll 1$ is not satisfied the deviation from the kinetic theory may be substantial. In particular the renormalization of conductance by electron-electron interaction (Altshuler-Aronov effect) is pronounced when $\tau k_BT/\hbar$ is small[30].
30. B.L. Altshuler and A.G. Aronov in *Electron Electron interaction in the disordered metals*, (Elsevier Science Publishers B.V., New-York, 1985).
31. Yu. V. Nazarov Ann. Phys. (Leipzig) **8**, Spec. Issue, 193 (1999). We believe that the result in this paper contains a small error.

9 781402 012402